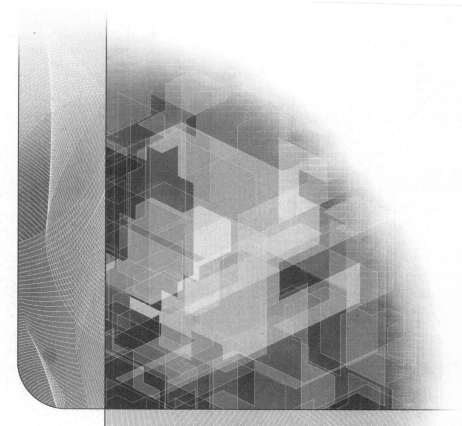

The Aubin Academy Master Series: Revit® Architecture 2012

autodesk Press

PAUL F. AUBIN

DELMAR
CENGAGE Learning

Australia • Brazil • Japan • Korea • Mexico • Singapore • Spain • United Kingdom • United States

**The Aubin Academy Master Series: Revit®
Architecture 2012**
Paul F. Aubin

Vice President, Editorial: Dave Garza

Director of Learning Solutions: Sandy Clark

Acquisitions Editor: Stacy Masucci

Managing Editor: Larry Main

Senior Product Manager: John Fisher

Editorial Assistant: Andrea Timpano

Vice President, Marketing: Jennifer Baker

Marketing Director: Deborah Yarnell

Associate Marketing Manager: Jillian Borden

Senior Production Director: Wendy Troeger

Art Director: David Arsenault

Senior Content Project Manager:
Angela Sheehan

Technology Project Manager: Joe Pliss

Cover image: stock-photo-beautiful-modern-
elevator-in-the-office-building. © 2012 Buchan.
Used under license from shutterstock.com.

For product information and technology assistance, contact us at
Cengage Learning Customer & Sales Support, 1-800-354-9706

For permission to use material from this text or product,
submit all requests online at **www.cengage.com/permissions.**
Further permissions questions can be e-mailed to
permissionrequest@cengage.com

Library of Congress Control Number: 2011930743

ISBN-13: 978-1-111-64848-0
ISBN-10: 1-111-64848-4

Delmar
5 Maxwell Drive
Clifton Park, NY 12065-2919
USA

Cengage Learning is a leading provider of customized learning solutions with
office locations around the globe, including Singapore, the United Kingdom,
Australia, Mexico, Brazil, and Japan. Locate your local office at:
www.cengage.com/region

Cengage Learning products are represented in Canada by Nelson Education, Ltd.

To learn more about Cengage Learning, visit **www.cengage.com**

Purchase any of our products at your local college store or at our preferred
online store **www.cengagebrain.com**

Printed in the United States of America
1 2 3 4 5 6 7 12 11

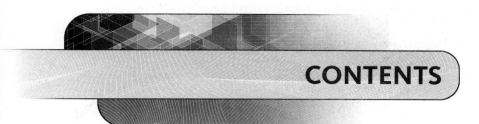

CONTENTS

CHAPTER 4 SETTING UP PROJECT LEVELS AND VIEWS 173

CHAPTER 5 COLUMN GRIDS AND STRUCTURAL LAYOUT 236

CHAPTER 6 GROUPS AND LINKS 281

CHAPTER 7 VERTICAL CIRCULATION 334

CHAPTER 8 FLOORS AND ROOFS 387

CHAPTER 9 DEVELOPING THE EXTERIOR SKIN 429

CHAPTER 10 WORKING WITH THE FAMILY EDITOR 485

III CONSTRUCTION DOCUMENTS

CHAPTER 11 DETAILING AND ANNOTATION 579

CHAPTER 12 WORKING WITH SCHEDULES AND TAGS 637

CHAPTER 13 CEILING PLANS AND INTERIOR ELEVATIONS 688

CHAPTER 14 PRINTING, PUBLISHING AND EXPORTING 712

CHAPTER 15 WORKSHARING 726

IV CONCEPTUAL MASSING AND RENDERING

V APPENDICES

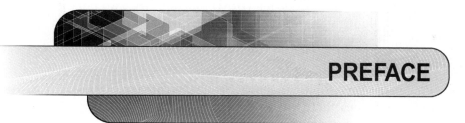

PREFACE

WELCOME

Within the pages of this book you will find a comprehensive introduction to the methods, philosophy, and procedures of the Revit Architecture software. Revit is an advanced and powerful architectural design and documentation software package. By following the detailed tutorials contained in this book, you will become immersed in its workings and functionality.

WHO SHOULD READ THIS BOOK?

The primary audience for this book is users new to Revit Architecture. However, it is also appropriate for existing Revit users who wish to expand their knowledge. You need not be an experienced computer operator to use this book. Only basic knowledge of the Windows operating system and basic use of a mouse and keyboard are assumed. No prior computer-aided design software knowledge is required. If part of your job requires that you design buildings and produce architectural construction documentation or design drawings, facilities layouts, or interior design studies and documentation, then this book is intended for you. Architects, interior designers, design build professionals, facilities planners, and building industry CAD professionals will benefit from the information contained within. Prior knowledge and familiarity with architectural practice, procedures, and terminology are assumed.

FEATURES IN THIS EDITION

Aubin Academy Master Series: Revit Architecture 2012 is a concise manual focused squarely on the rationale and practicality of the Revit process. The book emphasizes the process of creating projects in Revit rather than a series of independent commands and tools. The goal of each lesson is to help readers complete building design projects successfully. Tools are introduced together in a focused process with a strong emphasis on "why" as well as on "how." The text and exercises seek to give the reader a clear sense of the value of the tools, and a clear indication of each tool's potential. The Aubin Academy Master Series provides resources designed to shorten your learning curve, raise your comfort level, and, most importantly, give you real-life, tested, and practical advice on the usage of the software to create architectural Building Information Models.

What You Will Find Inside

Section I of this book focuses on the underlying theory and user interface of Revit Architecture. This section is intended to get you acquainted with the software and put you in the proper mindset. Section II relies heavily on tutorial-based exercises to present the process of creating a building model in Revit, relying on the software's powerful Building Information Modeling (BIM) functionality. Two projects are developed concurrently throughout the tutorial section: one residential and one commercial. Detailed explanations are included throughout the tutorials to identify clearly why each step is employed. Annotation and other features specific to construction documentation are covered in Section III. Section IV includes coverage of the conceptual modeling features and rendering. Section V contains appendices with many additional resources useful to the book's content.

What You Won't Find Inside

This book is not a command reference. This book approaches the subject of learning Revit by both exposing conceptual aspects of the software and extensive tutorial coverage. No attempt is made to give a comprehensive explanation of every command or every method available to execute commands. Instead, explanations cover broad topics of how to perform various tasks in Revit, with specific examples coming from architectural practice. References are made within the text wherever appropriate to the extensive online help and reference materials available on the Web. The focus of this book is the design development and construction documentation phases of architectural design. Chapter 16 briefly covers conceptual design tools, and rendering is explored in Chapter 17.

STYLE CONVENTIONS

Style Conventions used in this text are as follows:

Text	Revit Architecture
Step-by-step tutorials	I. Perform these steps.
Menu picks	**SaveAs > Project**
On screen input	For the length type **10'-0"** [**3000**].
Ribbon Tabs	On the Home Tab, on the Build panel, click the **Wall** tool
File and Directory names	*C:\MasterRAC 2012\Chapter10\Sample File.rvt*

UNITS

This book references both imperial and metric units. Symbol names, scales, references, and measurements are given first in imperial units, and are then followed by the metric equivalent in square brackets[ ]. For example, when there are two versions of the same file, they will appear like this within the text:

Curtain Wall Dbl Glass.rfa [*M_Curtain Wall Dbl Glass.rfa*].

When the scale varies, a note like this will appear: **1/8" = 1'-0"** [**1:100**].

If a measurement must be input, the values will appear like this: **10'-0"** [**3000**].

Please note that in many cases, the closest logical corresponding metric value has been chosen, rather than a "direct" mathematical translation. For instance, 10'-0" in imperial drawings translates to 3048 millimeters; however, a value of 3000 will be used in most cases as a more logical value.

NOTE

> Every attempt has been made to make these decisions in an informed manner. However, it is hoped that readers in countries where metric units are the standard will forgive the American author for any poor choices or translations made in this regard.

All project files are included in both imperial and metric units on the book's online companion unless noted otherwise. See the "Files Included with the Student Companion" topic below for information on how to install the dataset in your preferred choice of units.

HOW TO USE THIS BOOK

The order of chapters has been carefully thought out with the intention of following a logical flow and architectural process. If you are relatively new to Revit, it is recommended that you complete the entire book in order. However, if there are certain chapters that do not pertain to the type of work performed by you or your firm, feel free to skip those topics. But bear in mind that not every procedure will be repeated in every chapter. For the best experience, it is recommended that you read the entire book, cover to cover. For example, the early chapters cover the detailed procedures for drawing Walls, step-by-step with each click. Later chapters may simply say, "Draw a Wall from this point to this," without detailing exactly how to draw a Wall. Most importantly, even after you have completed your initial pass of the tutorials in this book, keep your copy of *Aubin Academy Master Series: Revit Architecture 2012* handy, as it will remain a valuable resource in the weeks and months to come.

A NOTE ABOUT COMPUTER HARDWARE AND OPERATING SYSTEMS

If Revit is your primary production application, you may want to consider maximizing your hardware and operating system to boost performance. Two important considerations are your processor and the amount of random access memory (RAM). Most systems today have multi-core processors. This essentially means the system has two, four (or more) processors working in tandem on the same chip. In order to take advantage of such a configuration, an application must be "multi-threaded," which means that it can actually make use of all processor cores. An application that is not multi-threaded will only make use of a single core. With each release, more functions in Revit become multi-threaded. This currently includes the Mental Ray rendering engine, loading lements into memory, silhouette edge graphics, and other graphical view display items. Despite the fact that this list does not include all Revit functions, multi-core machines are often still the best choice because most people run several applications simultaneously and many other programs can utilize all cores simultaneously, or the load of several applications can be spread among the various cores.

The amount of memory your system has will have a more direct impact on Revit performance. 32bit hardware and operating systems (OS) can only address a limited amount of RAM. Depending on your current configuration, this will be between 3 and 4 gigabytes maximum. However, 64bit hardware and operating systems are becoming much more popular in recent years, and this hardware is available from many manufacturers. Microsoft Windows XP, Vista, and Windows 7 all come in 64bit editions. The 64bit version of Revit is functionally the same as the 32bit version. Users will notice no difference in the interface or function of the product. The primary benefit of the 64bit version is its ability to access significantly more memory than 32bit versions. Many users of 64bit OS have machines with 8 or 16 gigs of RAM, but it can go much higher (128 GB in Vista and 192 GB in Windows 7). If you frequently work on large projects, it will not be unusual for your Revit models to exceed 200 MB in file size. At some firms, models of 500 MB and larger are not unheard of. (The author has even seen models of nearly 1 gig in size!) If you work on projects with file sizes in this range, the 64bit version with as much RAM as you can justify economically is a must.

Using 64bit and having more RAM in your system will give you the following benefits:

- You are able to open and work in larger models
- You can print more Sheets at a Time (this is useful even in small Projects)
- Speed increases of approximately 20% have been reported with very big Models (some reported higher gains)
- Even if you don't realize these speed gains you will not crash large projects due to limited RAM.
- Large files and renderings will not fail on save.
- More physical RAM reduces the amount of hard disk swapping required.
- You can export more views to AutoCAD at a time.
- Intense operations such as updating Groups will process faster.

Please note that in some cases, not all of your existing software will run properly on 64bit OS. This may be the case for older programs or custom database applications. Check with the program's manufacturer to see if 64bit is supported. While most firms report good compatibility with 64bit systems, be sure that all of your hardware supports it and that printer drivers are available. Also, it is not recommended that you mix Revit 32bit and Revit 64bit on the same project. This can cause problems, as the 32bit system runs out of memory and is unable to save the project.

FILES INCLUDED WITH THE STUDENT COMPANION

Files used in the tutorials throughout this book are available for download from the accompanying Student Companion site online at CengageBrain.com. Most chapters include files required to begin the lesson and in many cases a completed version is provided as well that you can use to check your work. This means that you will be able to load the files for a given chapter and begin working. When you install the files from the student companion, the files for *all* chapters are installed automatically. The files will install into a folder on your *C:* drive named *MasterRAC 2012* by default, but you can install the files to a different location (such as *My Documents*) if you prefer. Inside this folder will be a folder for each chapter. Please note that in some cases a particular chapter or subfolder will not have any Revit files. This is usually indicated by a text file (TXT) within this folder. For example, the *Chapter01* folder contains no Revit files, but instead contains a text document named, *There is no Complete version of Chapter 1.txt.* This text file simply explains that this folder was left empty intentionally.

NOTE Please note that the Student Companion contains only Revit (RVT, RTE, RFA) and other related resource files necessary to complete the tutorial lessons in this book. The Student Companion does *not* contain the Revit Architecture software. Please contact your local reseller if you need to purchase a copy of Revit.

Accessing the Student Companion site from CengageBrain

You must have your own copy of Revit Architecture to follow along with the lessons in this book. However, several dataset files (mostly RVT and RFA files) are required if you wish to follow along. Dataset files are available for download from CengageBrain free of charge. To download the files, do the following:

1. In your web browser, visit: **http://www.cengagebrain.com**
2. Type author, title, or ISBN in the Search window. (see the back cover)
3. Locate the desired product (i.e. *Aubin Academy Master Series: Revit Architecture 2012*) and click on the title.
4. When you arrive at the Product Page, in the access to free study tools area, click the **Access Now** button.
5. Use the "Click Here" link to access the Companion site.

| You will only see the Click Here link if there is a companion product available. |

NOTE

6. Click on the "Student Resources" link in the left navigation pane to access the resources.
7. Download and unzip the files to your C Drive.

The default unzip folder is named *C:\MasterRAC 2012* on your hard drive. You can move this folder to another location if you wish.

If you wish to install both the imperial and metric datasets, return to the student companion and repeat the steps above for the other units. Installation requires approximately 450 MB of disk space per unit type (approximately 850 MB if you install both). If you install both datasets, some files will be the same. Click OK if WinZip asks to overwrite any files.

KEEP YOUR SOFTWARE CURRENT

It is important to keep your software current. Be sure to check online at **www. autodesk.com** on a regular basis for the latest updates and service packs to the Revit Architecture software. Having the latest version installed will ensure that you benefit from the latest features and enhancements. If you are on the Autodesk Subscription program, you will be entitled to new releases as they become available. You will also have access to extensions as they are released. Extensions add powerful functionality to the Revit software. Visit the Autodesk web site or talk to your local reseller for more information.

WE WANT TO HEAR FROM YOU

We welcome your comments and suggestions regarding *Aubin Academy Master Series: Revit Architecture 2012*. Please forward your comments and questions to:

The CADD Team
Cengage Learning
Executive Woods
5 Maxwell Drive
Clifton Park, NY 12065-8007
Web site: **www.autodeskpress.com**

ABOUT THE AUTHOR

Paul F. Aubin is the author of many CAD and BIM book titles including the widely acclaimed: *The Aubin Academy Master Series*: *Revit Architecture*, *AutoCAD Architecture*, *AutoCAD MEP*, and *Revit MEP*. Paul has also authored video training in Revit for lynda.com (www.lynda.com/paulaubin). Paul is an independent architectural consultant who travels worldwide providing Revit® Architecture and AutoCAD® Architecture implementation, training, and support services. Paul's involvement in the architectural profession spans over 20 years, with experience that includes design, production, CAD management, mentoring, coaching, and training. He currently serves as Moderator for Cadalyst magazine's online CAD Questions forum, is an active member of the Autodesk user community, and has been a top-rated speaker at Autodesk University (Autodesk's annual user convention) for many years. This year Paul speaks at the Revit Technology Conference in both the United States and Australia. His diverse experience in architectural firms, as a CAD manager and as an educator, gives his writing and his classroom instruction a fresh and credible focus. Paul is an associate member of the American Institute of Architects. He lives in Chicago with his wife and three children.

Contact Paul directly at: **www.paulaubin.com** (click the Contact link).

Visit Paul's Blog: **www.paulaubin.com/blog/**.

DEDICATION

This book is dedicated to my son Marcus. I am so proud of you. Good luck in your first year in High School!

ACKNOWLEDGMENTS

The author would like to thank several people for their assistance and support throughout the writing of this book.

Thanks to Stacy Masucci, John Fisher, and all of the Delmar/Cengage team. It continues to be a pleasure to work with so dedicated a group of professionals.

Thanks to Robert Guarcello Mencarini, Architect, who was the technical editor for this and all previous editions.

Technical contributors to portions of this text include Matt Dillon, Zach Kron, Heather Lech, AIA, David Baldacchino, James Smell, Velina Mirincheva, Robert Guarcello Mencarini, Architect, AIA, and Mark Schmieding. A special thanks to each of you. Additional contributions and quotations have been noted within the text. Thank you to those contributors as well.

A very special thank you for significant contributions to the following chapters in the previous edition. These contributions continue to prove invaluable in this updated edition. I could not have completed them without your invaluable assistance:

Thank you to contribtors for content to the following chapters: For Chapter 15, Revit Server information: David Baldacchino. Chapter 16, Conceptual Massing: Heather Lech, AIA, LEED AP and Zach Kron. For Chapter 17, Rendering: James Smell and Jeff Hanson, Subject Matter Expert, Autodesk.

Thanks to the following members of the Autodesk User Group International forums for their input on the 64bit version of Revit and 64bit machines: need4mospd, whgeiger1, Michael Ruehr, clog boy, Scott Womack, josh.made4worship, Steve

Stafford, swalton.161301, brenehan, Jamie Spartz, Cliff B. Collins, Scott Davis, Eric Vieleand, and Jaroslaw Janiszewski.

A special acknowledgment is due the following instructors who reviewed the chapters in detail:

Matt Dillon–DC CADD Company

Mel Persin, Coordinator—Chicago Autodesk Revit Users Group

Stephen K. Stafford II—Stafford Consulting Services (thanks for the Workset/Library analogy)

For taking the time to discuss this project personally and offer suggestions and feedback, thanks to Jeff Millett, AIA—Vice President and Director of Information Technology; Eddie Barnett—LEED, Interior Designer; Sarah Vekasy—LEED, Architect; Marc Gabriel—LEED, Architect; John Jackson—LEED, Architect; and Marwan Bakri—Stubbins Associates, Boston, MA, and also to Mark Dietrick—CIO and Senior Associate and Michael DeOrsey—Graduate Architect of Burt Hill Kosar Rittlemann Associates, Boston, MA.

There are far too many folks at Autodesk to mention (my apologies in advance for any omissions). Thanks to all of them, but in particular, Christie Landry, David Mills, Kelcy Lemon, Lillian Smith, Jason Winstanley, Tatjana Dzambazova, David Conant, Matthew Jezyk, Steve Crotty, Erik Egbertson, Greg Demchak, Chico Membreno, Trey Klein, Tobias Hathorn, Cindy Xingchen Wang, Scott Blouin, Andy Parrella, Joe Charpentier, Ken Marsh, Zach Kron, Michael Juros, Brian Fitzpatrick, and all of the folks at Autodesk Tech Support.

I am ever grateful for blessings I have received from my many friends and family. Finally, I am most grateful for the constant love and support of my wife, Martha, and our three wonderful children.

Introduction and Methodology

Most common words in Section I

This section introduces the methodology of Revit Architecture. The concept of "Building Information Modeling" (BIM) is introduced and defined as are many other important topics and concepts. Within this section you will gain valuable experience using Revit by exploring its interface and overall conceptual underpinnings.

Section I is organized as follows:

Quick Start   General Revit Architecture Overview
Chapter 1   Conceptual Underpinnings of Revit Architecture
Chapter 2   Revit Architecture User Interface

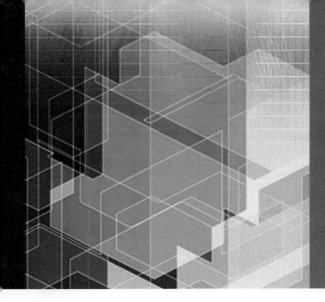

General Revit Architecture Overview

INTRODUCTION

This Quick Start provides a simple tutorial designed to give you a quick tour of some of the most common elements and features of Revit Architecture. You should be able to complete the entire exercise in one to two hours. At the completion of this tutorial, you will have experienced a first-hand look at what Revit Architecture has to offer.

OBJECTIVES

In this chapter you will:

- Experience an overview of the software.
- Create your first Revit Architecture model.
- Receive a first-hand glimpse at many Revit Architecture tools and methods.
- Gain some basic experience with the Revit interface.

CREATE A SMALL BUILDING

Let's get started using Revit Architecture right away. For this tutorial, we will take a "whirlwind" tour of the Revit Architecture tool set. All of the tools covered in the following steps use simple, and often default, settings. The chapters that follow cover each of these tools and settings in detail. So don't worry if a particular topic is not covered in depth at this point. Think of this Quick Start chapter as akin to warm-up exercises before a full workout. This book was authored using Microsoft Windows 7 Professional 64bit, but the exercises and tutorials will perform equally well in other versions of Windows. Please refer to the Preface for complete details on prerequisites and assumptions.

Install the Dataset Files and Open a Project

The lessons that follow require the dataset included on the Aubin Academy Master Series student companion. If you have already installed all of the files from this site, skip to step 3 to begin. If you need to install the files, start at step 1.

1. If you have not already done so, download the dataset files located on the CengageBrain website.

   Refer to "Accessing the Student Companion site from CengageBrain" in the Preface for information on installing the dataset files included in the Student Companion.

2. Launch Autodesk Revit Architecture from the icon on your desktop or from the *Autodesk* ≥ *Revit Architecture 2012* group in *All Programs* on the Windows Start menu.

TIP	In Windows 7, you can click the Start button, and then begin typing Revit in the Search field. After a couple letters, Revit Architecture should appear near the top of the list. Click it to launch to program.

3. On the Recent Files screen, click the Open link beneath Projects.

TIP	The keyboard shortcut for Open is CTRL+O. You can also click the Open icon on the Quick Access Toolbar (QAT) at the top left corner of the screen.

   • In the "Open" dialog box, browse to the location where you installed the *MasterRAC 2012* folder, and then double-click the *Quick Start* folder.

4. Double-click *Pavilion.rvt*.

You can also select it and then click the Open button (see Figure Q.1).

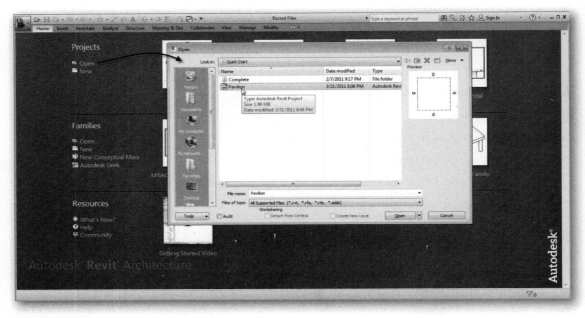

FIGURE Q.1 *Open the Pavilion project to get started*

NOTE	In this Quick Start tutorial, only one project has been provided and it uses Imperial units. The remainder of the book provides a Metric dataset as well.

The project will open in Revit Architecture with the *Level 1* floor plan view visible on screen. This project has been started already and contains a Property Line element (dashed square) in the middle of the screen. There is also a Toposurface terrain model element in this file that represents the site for the building (you will learn about Toposurface elements in Chapters 4 and 6). Let's start by displaying this item so we know where to place the Walls of our building.

Begin a New Model

To get started, we need to begin with the basics: Walls, Doors, and Windows. These elements are the basic building blocks of any architectural model. Adding these elements in Revit Architecture is simple and straight forward.

Create an Underlay

On the left side of the screen running vertically is a panel named: **Pavilion– Project Browser**. In it are listed several representations of our project including views (drawings), schedules, and sheets. Four floor plan views are provided here: *Level 1*, *Level 2*, *Roof*, and *Site*. The *Level 1* first floor plan view is bold indicating that it is the currently active view and open onscreen. (An annotated overview of the Revit interface is shown in Figure 2.1 in Chapter 2. Chapter 2 covers the interface in detail.)

5. On the Project Browser, double-click to open the *Site* plan view.

Notice that the Site Plan includes site contours and a shape in the middle of the plan. This shape represents the building footprint and its entrance patio.

-   On the Project Browser, double-click *Level 1* to return to the first floor plan view.

We can display any one of the other levels (such as the *Site* that we just viewed) as an underlay to this view to help us coordinate elements at different levels. Also on the left side of the screen, above the Project Browser is the Properties palette. Near the top of the palette, it should read "Floor Plan: Level 1" indicating that we are seeing the properties of the floor plan view (see Figure Q.2).

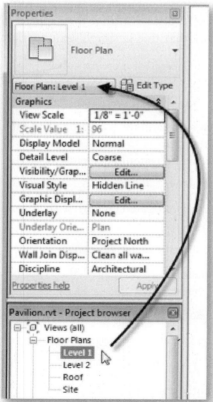

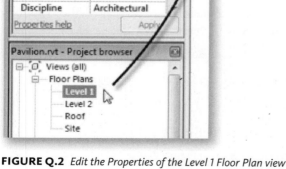

FIGURE Q.2 *Edit the Properties of the Level 1 Floor Plan view*

6. On the Properties palette, within the "Graphics" grouping locate the "Underlay" Item.

- Click on the Underlay value (currently "None") and from the pop-up menu that appears, choose **Site** (see Figure Q.3).

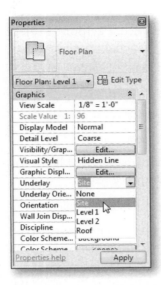

FIGURE Q.3 *Assign the Site plan view as an underlay to the Level 1 plan*

- Click Apply to see the results.

Notice that only the patio and building footprint outline appeared. This is because Topography is turned off in the *Level 1* plan view. Notice also that they appear in 50% halftone gray as well. This reinforces visually that this is simply an underlay; much like underlays in traditional hand drafting.

Create Walls

We begin our building model with some simple Walls.

Along the top of the screen appears a collection of icons and tools organized on tabs. This is called the "ribbon." This style of interface is common in many newer software packages such as Microsoft Office. The "Home" tab should currently be active.

7. On the ribbon, click the Home tab (if not already active) and then click the **Wall** tool.

A "Modify | Place Wall" tab appears on the ribbon (tinted green) with several options.

- On the Draw panel (right side of the ribbon), click the rectangle icon.

Just beneath the ribbon, additional settings appear on a bar (called the "Options Bar") running horizontally across the screen.

- From the "Location Line" list on the Options Bar, choose **Finish Face:Exterior**.

In addition to the ribbon and the Options Bar, you can also find pertinent settings on the Properties palette allowing you to interact with and change the settings of the Walls as you draw them. For example, at the top of the Properties palette, a drop-down list appears (known as the Type Selector) which currently reads Basic Wall Generic – 5".

- From the Type Selector, choose: ***Generic - 8*"** (see Figure Q.4).

FIGURE Q.4 *Pick the Wall tool and set it to draw Basic 8" Walls in a rectangular shape*

> you can also change the Location Line setting on the Properties palette as well.   **NOTE**

8. With the mouse pointer (now shaped like a cross hair) click the lower right corner of the gray shape on screen (see the left side of Figure Q.5).

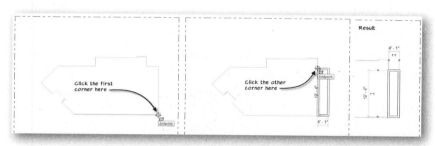

FIGURE Q.5 *Click the opposite corners to draw the Walls in a rectangle shape*

- Pick the Endpoint of the short horizontal edge indicated on the right side of Figure Q.5.

You will now have four Walls on screen. However, the "room" they define is very narrow (see the "Result" in the figure). We can easily adjust this. Before we can manipulate the Walls, however, we must cancel the current Wall creation command.

9. On the ribbon, on the left, click the **Modify** tool (or press the ESC key twice).

> Either method can be used anytime to cancel the current command and return to the **Modify** (selection pointer) tool. In Revit Architecture there is always one active tool. The default tool is the "Modify" tool, which is really just the standard mouse pointer.   **NOTE**

10. Click on the vertical Wall on the left to select it.
    Objects turn light blue on screen and shade in when they are selected.
11. On the dimension that appears, click directly on the blue text (see Figure Q.6).

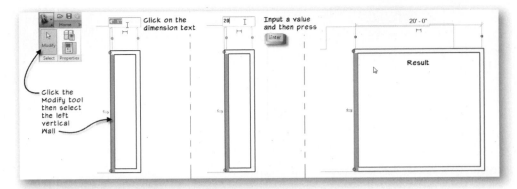

FIGURE Q.6 *Click the text of a temporary dimension to edit it*

12. In the text field that appears, type **20** and then press ENTER.

Notice that the Wall moved to a new location, an amount equal to the value we input and that the two horizontal Walls stretched with it to remain attached. Please note that when you edit this way, the *selected* Wall moves. The dimensions that we used for this edit are referred to as "temporary dimensions."

13. On the Home tab of the ribbon, click the **Wall** tool again.

   • From the "Location Line" list, choose **Wall Centerline**.
   • From the Type Selector choose: **Generic - 5"**.
   • Draw a vertical wall from top to bottom of the room, dividing it approximately into thirds (see Figure Q.7).

NOTE   The exact dimensions are unimportant at this point, we will move the Wall next.

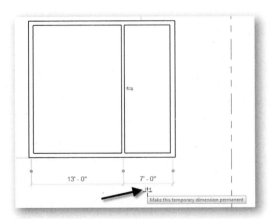

FIGURE Q.7 *Draw a Wall in a random location in the space*

   • On the ribbon, click the **Modify** tool or press the ESC key twice.
   • Select the Wall you just drew and then beneath the temporary dimension that appears, click the small icon (indicated in Figure Q.7) to make the dimension permanent.

Now that the dimension is permanent, notice that it remains onscreen when the Wall is no longer selected.

14. Click to select the dimension.

- Click the small "EQ" (Toggle Dimension Equality) icon beneath the dimension (see Figure Q.8).

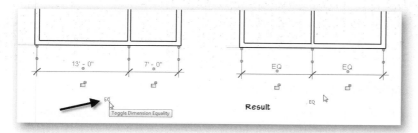

FIGURE Q.8 *Toggle the Dimension Equality and watch the middle Wall move to equal distances*

> This icon is part of the Revit constraint system. The constraint system is used to "lock-in" design intent. This notion is an integral part of the underlying concepts inherent to Revit.   **NOTE**

15. From the Application menu in the upper left hand corner of the Revit screen (the button with the Revit "R" icon) choose **Save**.

It is important to remember to save every so often to preserve your work. Revit Architecture is configured by default to remind you to save at regular intervals. You can edit the interval, but if the message asking you to save appears, you should always heed the suggestion and perform the save.

> You can find more information and tutorials on working with Walls in Chapter 3.   **NOTE**

Insert Doors and Windows

Next we'll add some openings (inserts) in our Walls.

16. On the Home tab of the ribbon, click the **Door** tool.

- Accept all of the defaults. Move the pointer near the top horizontal Wall to begin placing the Door.

Move the mouse around without clicking it yet. Notice how the Door follows the cursor and also stays attached to the Wall as it does. If you move the mouse away from the Wall, the Door will disappear. This is because elements such as Doors are "hosted" elements. The Door will be "hosted" by the wall. In other words, it is not possible to place a Door freestanding without a Wall host. Also notice that moving the mouse from one side of the Wall to the other will flip the Door in or out relative to the Wall. (You can also press the SPACEBAR to flip the Door before placement.) The gray underlay we added previously indicates a patio shape to the left of the plan and wrapping around the top. Our first door will be out to that passageway along the top.

17. Position the mouse on the top Wall near the right side of the passageway so it swings out and then click (see Figure Q.9).

FIGURE Q.9 *Place a Door to the outside near the passageway*

18. Place another near the top of the interior vertical Wall.

Notice that Door tags have automatically appeared and the numbers have filled in sequentially. We will learn more about tags later.

Sometimes after placing a Door, it is not positioned or oriented correctly. Just like the Walls above, we can select a Door, and then edit its temporary dimensions to move it to the desired location. There are also small flip control icons on the Door to control its orientation.

19. Click the Flip control to change the interior Door's orientation (see Figure Q.10).

Repeat if desired on the other Door.

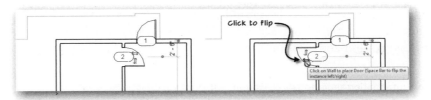

FIGURE Q.10 *Use the flip controls to change Door orientation*

Adding Windows works the same way as adding Doors.

20. On the ribbon, click the Home tab and then click the **Window** tool.

   • Accept all of the defaults. Move the pointer near the top horizontal Wall and move up then down.

Again notice how this controls the placement orientation of the Window. As with the Door, you can always flip it later if necessary.

21. Place Windows in the two horizontal Walls only (see Figure Q.11).

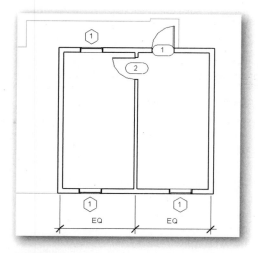

FIGURE Q.11 *Place Windows in the horizontal Walls*

Let's place one more Door in the Wall at the left. For this door we will use a different style of door (called a Family in Revit) and to do so, we will load it from an external library.

22. On the Home tab of the ribbon, click the ***Door*** tool.

A tab labeled Modify | Place Door will appear to the right of the other ribbon tabs. This tab is tinted in a green color. This is referred to as a context tab. Such tabs appear whenever you create a new element or select an existing element for editing.

- On the Modify | Place Door tab, to the right on the Mode panel, click the Load Family button.
- Navigate to the folder where you installed the book dataset files.
- From there, double-click the *Library* folder, then the *Quick Start* sub-folder.
- Select *Double-Glass 2.rfa* and then click Open.

This action loads the Door component into the currently active project making it available to place in the model. (This Door Family file is a copy of one provided with the out-of-the-box content in the Imperial install of the product.)

- Place the Door in the center of the left Wall swinging out. Use the temporary dimensions to assist you in placement (see Figure Q.12).

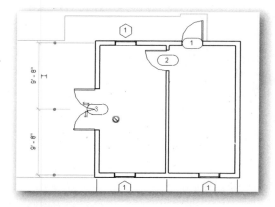

FIGURE Q.12 *Place a double entry Door*

- On the ribbon, click the ***Modify*** tool or press the ESC key twice.

23. Save the project.

NOTE   You can find more information and tutorials on working with Doors and Windows in Chapter 3.

WORKING IN OTHER VIEWS

We can work in many types of views in Revit; not just floor plans. Our project includes elevation views and ceiling plan views already. We can also add section views and 3D views. Many other view types are also available and discussed in future chapters. One thing to keep in mind is that all of these views representing the project are taken from the same Revit model; they are *not* separate drawings.

View the Model in 3D

Opening a three-dimensional (3D) view will give us a good overall look at the model and reveal that our Wall height needs adjustment.

1. On the Project Browser, double-click to open the *{3D}* 3D view.

This is the default three-dimensional view in Revit Architecture. You can modify it as you like or create others from it. We will use this one for our tutorial, but make some simple adjustments to its vantage point. Notice that the *{3D}* view is an isometric view of our building model. We can see the Walls, Doors, and Windows we added from a bird's eye vantage point. We also see the Toposurface terrain model that was included in this project at the start. You can change the vantage point of a 3D view interactively on screen.

On the right side of the screen, locate the small Navigation Bar.

2. On the Navigation Bar, click the Steering Wheels icon. (You can also press F8).

A "steering wheel" control will appear on screen and follow the position of your cursor. Several types of 3D navigation are possible.

- Position the steering wheel over the middle of the model.
- Move your mouse to the left side of the steering wheel over the Orbit option.
- Click and drag the mouse in the view window to spin the model around interactively.

Drag side-to-side to move around the building.

Drag up or down to change height of the vantage point.

- Spin the model around so that the front double Door is visible (see Figure Q.13).

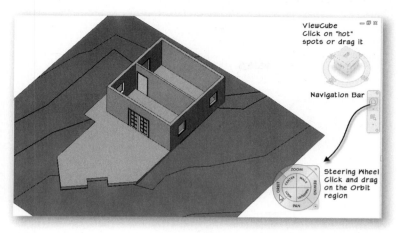

FIGURE Q.13 *Dynamically Modify the {3D} view window*

Several other options are possible. For now just try Orbit, Zoom, and Pan. To use each option, click and drag. Another option is the view cube in the upper right corner of the screen. You can click several hot spots on the cube to orient the model to that position. You can drag the ViewCube as another way to orbit.

You can also spin the model by holding down the shift key and dragging with the middle wheel button on your mouse or simply click and drag the ViewCube.	**TIP**

- On the ribbon, click the **Modify** tool or press the ESC key.

Create a Section View

Let's create a section view to help us understand the relationships built into our Revit model very clearly as we make some simple edits.

3. On the Project Browser, double-click to open the *Level 1* floor plan view.
4. On the ribbon, click the View tab and then click the **Section** tool.

- Click to the left of the double Door.

Move the pointer through the model to the right keeping the section line horizontal.

- Click outside the model to the right to complete the section line (see Figure Q.14).

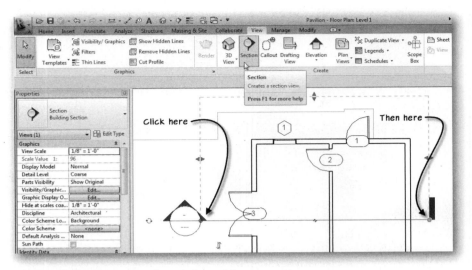

FIGURE Q.14 *Cut a section through the model*

A section line, with a section head and tail will appear. Three blue dashed lines with drag handles will also appear. The Section Head will currently be light blue indicating that it is selected.

5. Click next to the section line in the white area of the view background being careful not to click on any geometry.

TIP	This is a quick way to deselect the selected element(s). You can also just press the ESC key.

Notice that the Section Head now turns dark blue. It is no longer selected, but the dark blue color indicates that it is an interactive element. In this case, it has a linked view associated with it and acts as a "hyperlink" to the associated section view. In the Project Browser, notice that there is now a *Sections* category included in the list. Revit automatically creates such branches in the Project Browser as needed.

• Double-click the dark blue Section Head to open the associated view.

You should now see the *Section 1* view on screen.

• On the Project Browser, double-click to open the *West* Elevation view.

We now have five views open. Your ribbon should still be on the View tab. If you click on the Switch Windows drop-down button, a drop-down menu will appear listing all the open views. (In the default installation, this tool is also on the QAT.) You can choose them off this list to bring them to the front of the pile, or you can simply double-click the view name in Project Browser again to display them. You can also tile them all on screen at once. Let's do that now.

6. From the Windows panel, click the **Tile** tool. Close the *Floor Plan:Site* view and then click the **Tile** tool again (see Figure Q.15).

• On the keyboard, type the letters ZA.

This is the keyboard shortcut for zoom all views to fit. If you prefer, you can use the Zoom controls on the Navigation Bar to adjust the window's display.

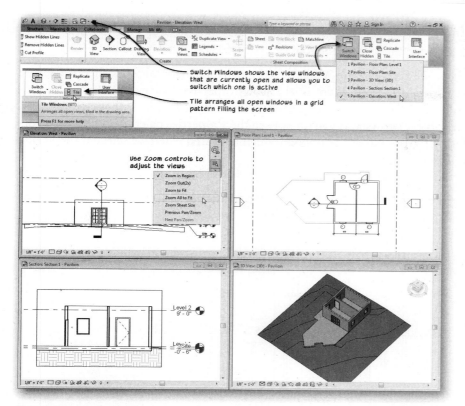

FIGURE Q.15 *Tile the views on screen to view them all at once*

You can find more information and tutorials on working with views in Chapter 4.

NOTE

EDIT IN ANY VIEW

When you edit your model, you may perform the edits in *any* view. Changes will automatically be applied to *all* views. This is the power of the Revit platform! You are describing a single virtual building model or *Building Information Model* (BIM) (see Chapter 1 for complete details.) You can "view" it in an unlimited number of ways. Regardless of where you make the edit—plan, section, elevation, 3D, or even schedules, all views are completely coordinated. These graphical and tabular views are just different ways of portraying the data within the single BIM.

Editing Levels

With our current screen configuration, take a look at the elevation and section views in particular. This project has been set up to have two stories plus a roof. Currently our Walls only go up one story and do not interact with the second floor level at all. Let's fix both problems.

1. Click in the *Level 1* floor plan view window, to make it active.
2. Place your mouse pointer (the **Modify** tool) over one of the exterior Walls.

Notice the way that it highlights under the cursor. (If you move the mouse away without clicking, the Wall will no longer highlight.) This is called "pre-highlight" and is a useful aid in proper selection.

- Pre-highlight one exterior Wall—do not click yet.
- Press the TAB key.

Notice how all of the exterior Walls now pre-highlight. This is called a chain selection.

- Click the left mouse button to select the pre-highlighted elements.

Notice how all four exterior Walls are now shaded light blue in all open views. They are also a little transparent in the views other than plan. The Properties palette should be visible onscreen. By default, it is on the left side of the screen docked above the Project Browser. If you do not see the Properties palette, on the green colored Modify | Walls tab of the ribbon (which appeared when we selected the Walls) click the **Properties** tool. Make sure you only click the tool if the Properties palette is *not* currently displayed. Clicking it when it is displayed will hide it.

- On the Properties palette, from the "Top Constraint" list, choose **Up to level: Roof** and then click Apply (see Figure Q.16).

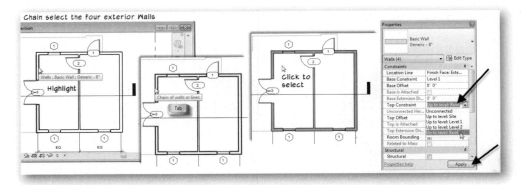

FIGURE Q.16 *Use the Properties palette to edit the selected Walls*

Notice how the Walls project up to the Roof Level line in all views. The Walls are now set relative to this Level in the project. If we were to change the height of the Roof Level, the Walls would also adjust accordingly. Let's try that now.

> **NOTE** The section view will likely not show the top of the Walls as it is currently cropped to the first floor. You can adjust this with the round light blue drag control at the top of the Crop Boundary. Click the rectangular box surrounding the section. Click the control at the top edge of the box and drag it upward. Zoom or Pan the view as required. The Walls should now show.

3. Click anywhere in the *West* elevation view.

Zoom In Region to get a better look if you need to (right-click to access **Zoom In Region**, or use the Navigation Bar to access it). Drag a rectangular region on screen to indicate which portion of the view to zoom in on.

- Click to select the Roof Level line.

Notice the temporary dimensions that appear. Like the Walls and other elements drawn so far, we can edit the blue dimension value to move the Level lines to a new location. We will move both Level 2 and the Roof level; starting with the Roof.

- Click the blue text of the temporary dimension between Level 2 and Roof, type **11** and then press ENTER (see Figure Q.17).

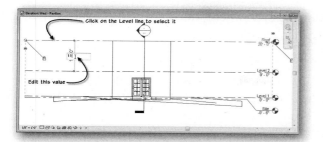

FIGURE Q.17 *Move the Level line with the temporary dimension*

> Be sure to use the temporary dimension between Level 2 and Roof. If you edit the height of the Roof level directly (the height shown on the level head symbol), you will need to type 20 instead. The height on the Level shows its height from zero.

Notice that not only does the Roof Level line move, but since we constrained the Walls to the Roof Level, the top edge of the Walls adjusts as well!

Repeat for Level 2.

- Click the blue text of the temporary dimension between Level 1 and Level 2, type **10** and press ENTER.

Notice that this moves Level 2 relative to Level 1, and also the distance between Level 2 and the Roof has changed as well. This is because only the item that is selected will move, and in this case, we only had Level 2 selected. This change also only affected the Level. It had no impact on the exterior Walls. This is because none of the exterior Walls have their Top Constraint set to Level 2.

> You can find more information and tutorials on working with Walls in Chapter 3 and with Levels in Chapter 4.

Modify a Window

Let's edit a Window next. Again, we can edit in whatever view is convenient with confidence that the edit will appear in all appropriate views automatically.

4. Spin the 3D model view to show the north Wall.

You can hold down the SHIFT key and drag with the middle wheel button on your mouse, or click the small corner hot spot on the top surface on the ViewCube.

5. Select the Window on the north Wall. (You can select it in any view—try plan).

- Click on the titlebar of the plan view, then the section view.

Not only is the Window highlighted light blue (indicating that it is selected) in all views, but in each of these views where it is visible, the temporary dimensions appear when the view is made active.

- Edit the temporary dimension values to move the Window. Click on the dimension text, type any number (a little bigger or smaller than the current number is fine), and then press ENTER.

Notice how the Window moves instantly in *all* views. When using Revit, you will never have to worry about chasing down a change in several different drawings to be certain that it has been coordinated everywhere. This will boost productivity and help reduce the number of costly change orders.

Move additional Windows in the same way if you wish.

At this point you may wish to line up the Window on the North Wall across the plan with the one on the South. You can do this and have Revit maintain the relationship with the Align tool.

Work in the plan view.

6. On the ribbon, on the Modify | Windows tab, on the Modify panel, click the **Align** tool.

You first indicate the point of reference (what you want to align to). We'll use the Window we just moved.

- Click near the center of the Window you just moved to set the point of alignment.
- Click near the center of the opposite Window to align it to the reference point (see Figure Q.18).

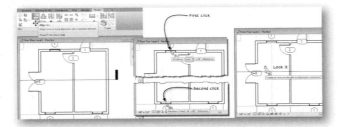

FIGURE Q.18 *Use the Align Tool to align Windows to one another*

- Click the lock icon to constrain the alignment of the two Windows together.
- On the ribbon, click the **Modify** tool or press the ESC key twice.

If you now move either Window, they will move together. Try it!

Add Openings on the Second Floor

Now that our Walls span the height of both floors, we should add some fenestration on the second floor.

7. On the Project Browser, double-click to open the *Level 2* floor plan view.

- Following the above procedures, add Windows and a Door to *Level 2* as shown in Figure Q.19.

| TIP | You can use the temporary dimensions to help you in placing the openings in logical locations. |

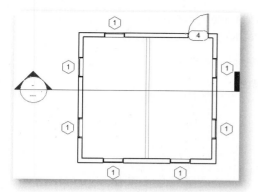

FIGURE Q.19 *Add Doors and Windows to Level 2*

Fine-tune the placement of any of these openings and feel free to repeat the alignment procedure on any of the pairs of Windows as well.

ROUND OUT THE PROJECT

Our project is coming along. Let's keep going and enclose it with a Floor and Roof and refine the Walls a little.

Add a Floor

The second floor of our building will have an interior balcony on the right overlooking the space to the left.

Work in the *Level 2* floor plan view.

1. On the ribbon, click the Home tab and then click the **Floor** tool.

Floor is a "split button" tool. This means that the top half of the button performs the most common command or action. If you click the lower half of the button, a pop-up menu appears with several alternate choices for the tool. In this case, the default top half of the button gives us the **Floor** tool we need. Split buttons and other interface conventions are discussed in more detail in Chapter 2.

When you click the Floor tool, the floor plan will turn gray placing the drawing editor into "Sketch mode." The ribbon will also change to show a Modify | Create Floor Boundary tab. Sketch mode is a special two-dimensional drawing mode used when the element that you are creating has a shape that Revit Architecture cannot easily "guess." In this case, it would not be possible for Revit to assume the size and shape of the Floor that we want, so instead, we will sketch it. This is easy to do, given that we already have several Walls and can use them for reference.

On the Draw panel of the ribbon (on the right), the "Boundary Line" and "Pick Walls" icons will already be enabled (selected).

- Click one of the horizontal exterior Walls, then the other.
- Click the vertical exterior Wall on the right.

Notice that with each Wall you click a magenta sketch line will appear on the Wall. Also notice that the sketch will appear on either the inside face or the outside face depending on which side of the Wall you clicked. Notice also that the right side corners automatically formed clean corners to one another. If the sketch lines appear on the outside face of a Wall click the double arrow flip control (appearing on one of the sketch lines) to reposition the sketch line to the other face. Repeat as required to locate all three sketch lines on the inside face as indicated on the left of Figure Q.20.

The Floor will only cover the right half of the plan, so for the last Wall, we will use the vertical one in the center on the plan rather than the exterior one on the left.

- Click on the vertical Wall in the center (see the middle of Figure Q.20).

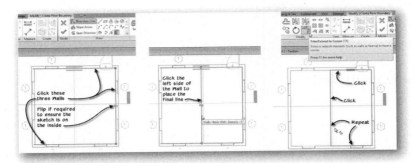

FIGURE Q.20 *Create Floor sketch lines from the existing Walls and use Trim/Extend to clean it up (Sketch lines in the figure enhanced for clarity)*

2. On the Modify panel, click the ***Trim/Extend to Corner*** tool.

- Click the vertical sketch line in the center of the plan, and then click the right side of the horizontal line at the top.

NOTE   When using the ***Trim/Extend*** tool you always select the portion of the lines that you want to keep.

- Repeat by clicking the vertical again, then the right side of the horizontal one on the bottom (see the right side of Figure Q.20).

You cannot finish the sketch if it does not form a closed shape. In this case, we created a closed rectangle. It doesn't matter what the shape is, but all corners must be closed (joined) to one another.

- On the ribbon, click the Finish Edit Mode (Green Check Mark) button.
- In the dialog that appears, click Yes (see the right side of Figure Q.21).

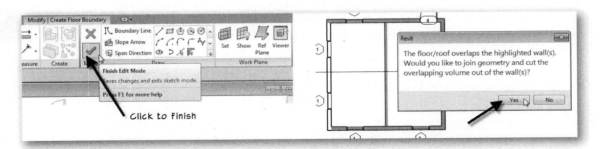

FIGURE Q.21 *Finish the Sketch and then answer Yes to join the Floor to the Walls*

3. On the Project Browser, double-click to open the *{3D}* view.

NOTE   If you are still working with four tiled view windows, simply click the titlebar of the *{3D}* view to make it active. Double click the view's titlebar to maximize the view if desired.

- Spin the model around and angle down to see the new Floor.

You can also study it in the section view (see Figure Q.22).

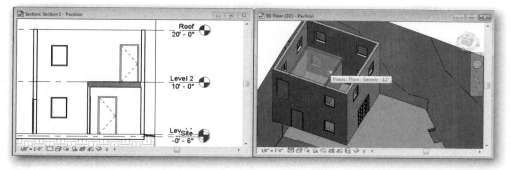

FIGURE Q.22 *Study the new Floor in the {3D} and Section 1 views*

> You can find more information and tutorials on working with Floors in Chapter 5.

NOTE

Add a Roof

We can sketch a roof in much the way as we sketched the Floor. There are a few Roof techniques available. If you want a shed, gable or hip Roof, use the Roof by Footprint option. For this exercise, we will make a curved sweeping shape using the Roof by Extrusion option.

4. On the Project Browser, double-click to open the *West* elevation view.

> If you are still working with four tiled view windows, simply click the titlebar of the *West* elevation view to make it active.

NOTE

> In the default Revit project template, North is toward the top of the screen in plan view. Therefore, the top elevation mark symbol is the North elevation, the bottom one is South, the right one is East, and the left one is West. From a plan view, you can double-click the triangle portion of the marker to open the associated elevation in a similar fashion to the way we opened the section above.

NOTE

5. On the Home tab of the ribbon, click the drop-down arrow on the **Roof** tool.
   - From the drop-down menu that appears, choose **Roof by Extrusion**.

With Roof by Extrusion, we sketch a simple 2D shape that will be used to form the Roof's profile. The Roof will extrude this shape along the building. Since we are working in an elevation, Revit needs us to establish the plane in which we wish to sketch. This is known as the "Work Plane."

   - In the "Work Plane" dialog, accept the defaults (Pick a plane) and click OK.

If you look at the bottom of the screen on the Status Bar, the prompt reads: "Pick a vertical plane."

   - Click the Wall facing us (see Figure Q.23).

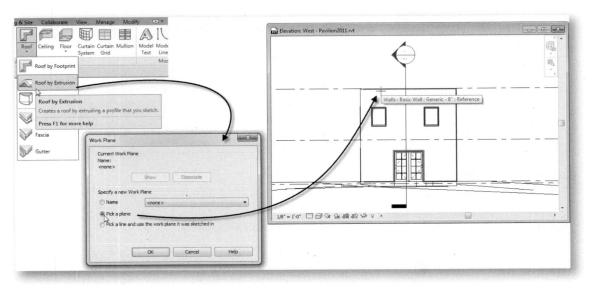

FIGURE Q.23 *Click the Wall to set a Reference Plane*

- In the "Roof Reference Level and Offset" dialog that appears, verify that the Level is set to Roof with a 0 offset (this is the default) and then click OK.

We are placed into sketch mode as with the Floor element above. We will now create a curved shape for the Roof. On the ribbon, the Modify | Create Extrusion Roof Profile tab appears, and on the Draw panel, the *Line* tool should be active.

- On the Draw Panel, directly beneath the *Line* tool, click the *Start-End-Radius Arc* tool.
- Click the first point of the arc at the top left corner of the Wall.
- Set the next point about 6° below the right top corner (see Figure Q.24).

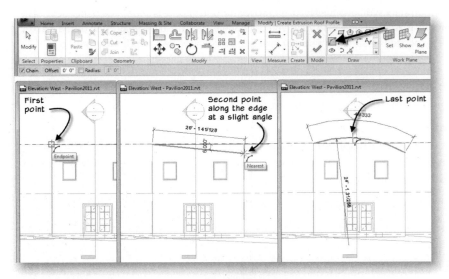

FIGURE Q.24 *Create a three-point arc shape for the Roof Extrusion*

- Set the final point approximately where indicated in the figure. (It does not need to be precise).
- On the ribbon, click the Finish Edit Mode (Green Check Mark) button.
- Switch to the *{3D}* view.

> If you are still working with four tiled view windows, simply click the titlebar of the *{3D}* view to make it active. Otherwise, double-click it on Project Browser.

The Roof automatically spanned over the entire building model. However, its eaves are flush with the Walls and the Walls pass through the Roof. Let's add an overhang to the Roof and then we will fix the Walls.

The Roof should still be selected; if it is not, click on it now.

Notice the small arrow handles pointing away from the Roof on two ends of the extrusion.

6. Click and drag each of these handles slightly away from the Walls to create an overhang (see Figure Q.25).

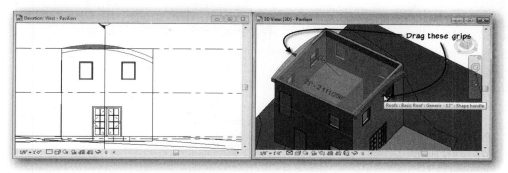

FIGURE Q.25 *Stretch Roof Extrusion handles to make overhangs*

The previous edit was simple because we were actually changing the overall length of the extrusion. To add overhangs in the other direction, we can edit the sketch.

- With the Roof still selected, on the Modify | Roofs tab, on the Mode panel, click the **Edit Profile** button.

The sketch line will reappear and the Roof will temporarily disappear. You can switch back to the elevation view or edit the sketch directly in 3D. Remember, you can edit in any view and the change will occur in all views.

- Drag the ends of the line away from the Walls slightly as shown in Figure Q.26.

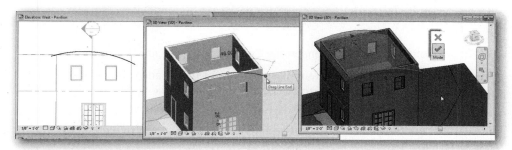

FIGURE Q.26 *Edit the sketch line to re-shape the Roof to include overhangs*

- On the ribbon, click the Finish Roof button.

7. Use the TAB select method above to chain-select all the exterior Walls. You can do this in any view including the *{3D}* view.

   - On the Modify | Walls tab of the ribbon, on the Modify Wall panel, click the Attach Top/Base button.

   - Click the Roof (see Figure Q.27).

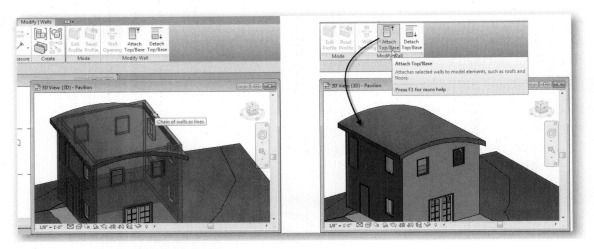

FIGURE Q.27 *Attach the Walls to the Roof*

> **NOTE**
>
> If you edit the shape of the Roof, the Walls will remain attached. Try it out if you like. Select the Roof. On the ribbon click the Edit Profile button and change the shape of the arc. Finish the sketch to see the change to the roof shape and how the Walls remain attached. Undo any change before continuing.

- Save the model.

> **NOTE**
>
> You can find more information and tutorials on working with Roofs in Chapter 8.

Add a Stair

We have no way to reach our second floor balcony. Let's add a Stair.

8. On the Project Browser, double-click to open the *Level 1* floor plan view.

> **NOTE**
>
> Whether the view is open already or not, double-clicking its name on the Project Browser will make the view active.

9. On the Home tab of the ribbon, on the Circulation panel, click the **Stairs** tool.

The Modify | Create Stairs Sketch tab will appear on the ribbon and will be placed in sketch mode again. The Stair will go to the north (top) side of the building. There is a little bump out on the existing patio for this purpose.

- Click near the middle of the patio bump out and drag up (see Figure Q.28).

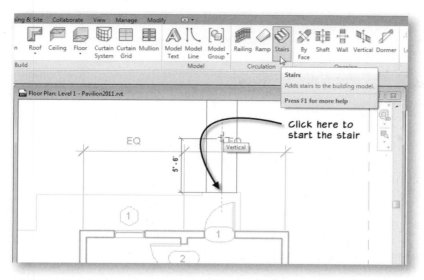

FIGURE Q.28 *Click the Stairs tool and then create half the risers*

A small label in gray text will appear on screen indicating how many risers have been created and how many remain. Revit calculates this based on the settings built into the Stair element and its type. For more information on Stairs, please refer to Chapter 7.

- Drag straight up until the gray label reads "9 Risers created, 9 remaining" and then click to create the first 9 risers.
- Click a point next to the first run of stairs at the location indicated in Figure Q.29.

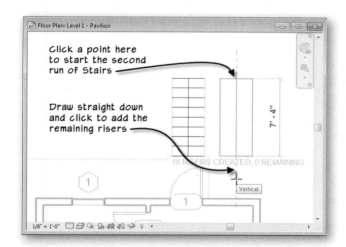

FIGURE Q.29 *Create a second run using the remaining Risers*

- Drag straight down until the message indicates that zero risers remain and then click to create the remaining risers.

This will give us the basic Stair but it will not "hook up" with the second floor. We need to extend the top riser to make it a landing at the top.

10. Select the riser line at the bottom right (the last riser of the Stair).

- Drag it down until it snaps to the building.
- On the Edit panel, click the **Split Element** tool.
- Click on each of the green lines to split them where indicated in Figure Q.30.

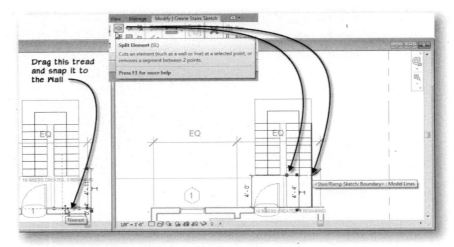

FIGURE Q.30 *Split the Stringer Lines to make a landing*

The green lines represent the stringers of the Stair. It is necessary to split them (break them into two segments) so that one part can slope with the Stair treads and the other part can be flat and follow the landing. Now that we have split the lines, we need to change the slope to flat for the landing portion.

- Click the **Modify** tool.
- Select one of the stringer lines (the ones we just split closest to the building) and on the Options Bar; choose **Flat** from the Slope list.
- Repeat for the other side (see Figure Q.31).

NOTE   You have to do these one at a time.

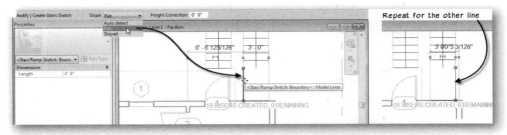

FIGURE Q.31 *Set the slope of the sketch lines to Flat*

- On the ribbon, click the Finish Edit Mode button.

When you are finished, repeat the process for the two Railings. Railings are separate from Stairs, but by default Revit automatically adds them when you create a Stair. Deselect the Stair. Select one of the Railings and click the Edit Path button on the ribbon. You do not need to split the sketch line, it is already split, but you will need to set the slope to Flat. Finish the sketch and then do the other Railing.

11. On the Project Browser, double-click to open the *{3D}* view.

- Spin the model around (if necessary) to see the Stair (see Figure Q.32).
- Move the Door on the second floor if necessary to line up with the landing.

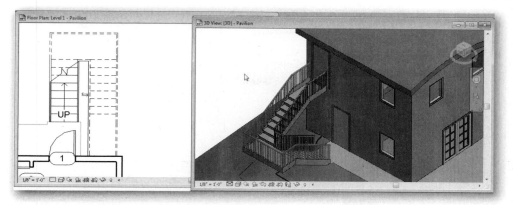

FIGURE Q.32 *Study the results in the {3D} view*

You can find more information and tutorials on working with Stairs and Railings in Chapter 7.   **NOTE**

Create a Custom Wall Type

There is plenty more that we could do to enhance our simple model. Many ideas and techniques will be explored in the chapters and topics that follow. However, let's make one quick modification to the exterior Walls. Let's see what our building would look like if we added a brick veneer.

12. In the {3D} view, click to select one of the exterior Walls.
    • On the Properties palette, near the top, click the Edit Type button.
    • In the "Type Properties" dialog, click the Duplicate button. In the dilaog that appears, type **Wall w Brick Veneer** and then click OK.

We now have a new Wall type. Next we will edit its structure to add the Brick veneer.

13. In the "Type Properties" dialog, click the Edit button next to Structure.
    • In the Thickness column next to Structure [1], change the value to 4" (type **0 4**).
    • Click the Insert button and then click the Up button.

This will add a new Layer and then move it to the top of the list.

    • Change the Function of item 1 to Finish [4].
    • Click in the Material column and then click the small browse button to open the "Materials" dialog.
    • Choose Masonry - Brick and then click OK.
    • Change the Thickness to 4".
    • Click OK twice to complete the Wall Type (see Figure Q.33).

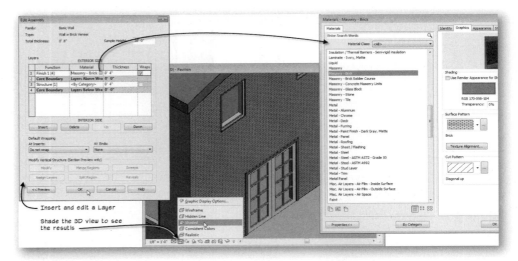

FIGURE Q.33 *Insert new Wall layers and configure the materials*

14. In the {3D} view, from the View Control Bar, choose **Shaded** for the Visual Style and zoom in on the brick.

You can also try the Realistic Visual Style if you wish. If your video card is capable, the Realistic Visual Style will display the render surface textures in the the view window. To see the brick in plan views, you have to change the level of detail to Medium. There is a control for this also on the View Control Bar between the Visual Style pop-up that we edited here and the Scale list. If you like, you can select the other three exterior Walls and on the Properties palette, choose the new Type from the Type Selector to apply brick around the entire building. The choice is up to you.

This completes the basic geometry of the model. We could add many more embellishments like railings on the patio and second floor balcony, adjustments to the windows and roof eaves, and additional materials to name a few. For now we'll shift our attention to some construction documentation items like Door and Window Schedules and some sheets for printing.

15. Save the project.

Create a Schedule

We can create automated schedules of most element categories in Revit. All we need to do is generate a Schedule view which, while not graphical like the plan, section, and elevation views, is just like the other views in Revit and dynamically reports the items in the building model. Plans, sections, elevations, and 3D views, etc. are graphical views. Schedules are tabular views. You can view information related to the model and even edit it directly from a schedule view.

16. On the ribbon, click the View tab and then click the *Schedules* tool.

    • From the drop-down menu that appears, choose **Schedule/Quantities**.

The "New Schedule" dialog will appear.

- From the "Category" list, choose Doors and then click OK.

The "Schedule Properties" dialog will appear.

- In the "Schedule Properties" dialog, on the "Fields" tab, click "Mark" in the "Available Fields" list and then click the Add -≥ button in the middle of the dialog.

"Mark" is the Door's number field.	**NOTE**

- Repeat for the following Fields: Level, Width, Height, Frame Type, Frame Material, Family and Type, and Comments (see Figure Q.34).

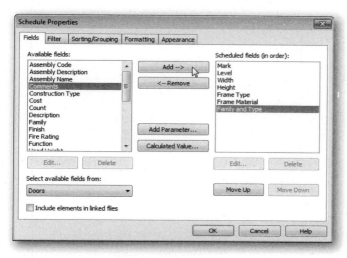

FIGURE Q.34 *Add Fields to the Door Schedule*

- Click OK to create the Schedule.

A Schedule view will appear on screen. The Schedule view appears much like a spreadsheet. Let's look at the Schedule tiled next to one of the floor plans. However, before we tile, let's close some of the other views.

If you still have the windows tiled on screen, maximize the current window by clicking the Maximize icon in the upper right corner of the window, or simply double-click the titlebar.

17. Click the View tab of the ribbon and then click the **Close Hidden** tool.

- On the Project Browser, double-click to open the *Level 1* floor plan view.
- From the View tab, click the **Tile Windows button** (or type WT).
- Zoom in on the plan view.

You should now have just the *Level 1* floor plan view and the *Door Schedule* view open on screen side by side.

18. In the Door Schedule view, click on Door number 3.

The Door number will highlight in the Schedule and the Door itself will highlight in the plan (see Figure Q.35).

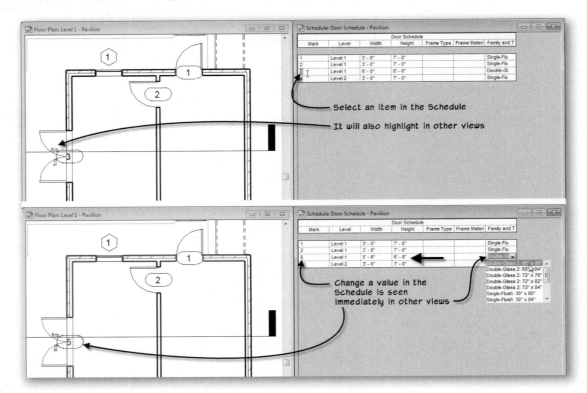

FIGURE Q.35 *Selected elements highlight in graphical views and Schedules*

- Highlight the value in the Mark field, type **5**, and then press ENTER.

Notice that the value changes in the Door Tag on the floor plan as well.

- For the same Door, click in the Family and Type column and choose ***Double-Glass 2: 68" × 80"*** from the pop-up list.

Notice that the size of the Door changes in both the Schedule and the floor plan. This is considered the correct way to change the size of a Door since most Doors use "Type-based" parameters to control their width. To illustrate this point, try an experiment on the other Doors in this project. Click in the Width column for Door 1. Note that it will again highlight in the plan as well. In the Width field, change the value to **2**. A message will appear stating: This change will be applied to all elements of type Single-Flush: 36" × 84". This means that *all* Doors of this Type will be affected when you click OK. Go ahead and click OK to see this. Notice that all single Doors in this project (including the one on the second floor) are now 2'-0". This is likely not the desired result, particularly since the Family and Type value still indicates that the name of this Door is Single-Flush: 36" × 84", which is likely to foster confusion. This topic will be discussed in more detail in later chapters. For now, it suffices to say that if you wish to change the width or height of the Door, you should choose a different Door Family and/or Type instead rather than simply edit the dimensions directly in the schedule (unless you really want to edit all Doors at once). Undo the change to the single Doors before continuing.

There is a Schedules/Quantities branch on the Project Browser. If you expand it, you will see the Door Schedule listed there. New in this release, you can add a new Schedule directly from here.

19. On Project Browser, right-click on Schedules/Quantities and choose **New Schedules/ Quantities**.

- In the "New Schedule" dialog, from the "Category" list, choose Windows and then click OK.
- In the "Schedule Properties" dialog, on the "Fields" tab, add the Type Mark, Level, Family and Type, Width, Height, and Count fields.
- Click the Sorting/Grouping tab.
- For "Sort by," choose **Level** and then place a checkmark in the "Header" checkbox.
- For "Then by," choose **Family and Type**.
- At the bottom of the dialog, place a checkbox in the "Grand totals" checkbox and clear the "Itemize every instance" checkbox.
- Click OK to create the Schedule (see the left side of Figure Q.36).

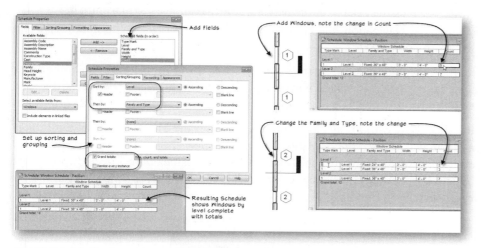

FIGURE Q.36 *Configure the Window Schedule to sort and total the Windows in the project by Level*

20. In the *Level 1* floor plan view, add Windows to the right Wall.

    Notice that the new Windows appear immediately in the Schedule as reflected by the new increased total on Level 1.
    - Change any Window in the model to a different Type.

    Notice that the Schedule will adjust to show an additional line for this Type and adjust the quantities accordingly (see the right side of Figure Q.37).
21. Save the model.

You can find more information and tutorials on working with Schedules in Chapter 12.	**NOTE**

PREPARING OUTPUT

At some point in your project, you will need to output your designs and produce some form of deliverable. This will often be a collection of printed drawings. In Revit Architecture you use special "sheet" views for this purpose. These views emulate the final paper output and allow us to compose the completed sheets formatted the way we wish, complete with titleblocks.

Add a Sheet

While it is possible to directly print the views we already have, we will get more polished and professional results by creating sheet views. Sheet views are basically pieces of paper upon which we drag and drop the floor plans, elevations and sections within a titleblock border. Both graphical (drawings) and tabular (schedule) views can be combined in sheet views. A sheet can contain one or several views. Sheets are of full size, that is, the scale is 1:1. Graphical views added to sheets can be at different scales. The exact composition is up to you and your office standards.

1. On the ribbon, click the View tab and then on the Sheet Composition panel, click the **Sheet** tool.

   • In the "New Sheet" dialog accept the defaults and then click OK.

There is currently only one titleblock choice available, but you can load other title-block borders in your own projects. You can even create your own custom titleblock including your company logo, information, and other standard graphical elements. Please refer to the "Create a Custom Titleblock Family" topic in Chapter 4 for more information.

   • Double-click the titlebar of the new window that appears to maximize it.
   • From the Navigation Bar, click the small pop-up arrow on the Zoom icon and choose **Zoom to Fit**.

If the window does not zoom after doing this, click the Zoom icon again. Sometimes you have to set the Zoom mode with the pop-up menu first, and then click the icon to execute the command. You can also type the keyboard shortcut ZF to Zoom to Fit.

A blank sheet view with a Titleblock appears. (Beneath the Sheets branch of the Project Browser, you will also see the new sheet listed.) The new sheet is ready to receive views. There are two ways to do this. Right-click the sheet view name on the Project Browser and choose **Add View**. A dialog will appear listing all available views, or simply drag and drop the desired view from Project Browser to the sheet on screen. We'll use drag and drop here.

2. From the Project Browser, drag the *Level 1* floor plan view over the Sheet area and drop it on the sheet.

After you drop, an outline of the view will appear attached to the cursor. This is called the viewport boundary and it automatically sizes itself to the extents of the dragged view's contents. You can use this to place it on the sheet where desired.

   • Position the view in the upper left corner of the sheet and then click to place it.
   • Repeat the drag and drop process for each of the remaining floor plans.

It will be a close fit, but the first, second and roof plans should fit across the top of the sheet.

   • Drag each of the Schedule views to the sheet as well (see Figure Q.37).

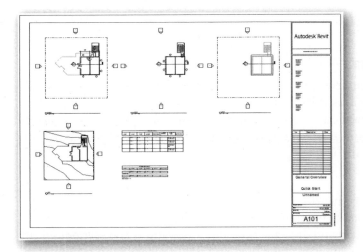

FIGURE Q.37 *Drag all the plans and schedules to the sheet view and position them*

If you need to re-position a view after you drag it, you can click on it directly on the sheet, and then drag it again to move it or use the arrow keys on your keyboard to nudge it slightly. When you add the Site plan, notice that it is a little smaller than the others. Each view has its own scale setting that is used when it is added to a sheet. A title bar appears beneath each view as you place it. To make these more legible, you can click to select the viewport on screen, and then use the round handles to adjust the length of the title bars.

3. Repeat the entire process to create another sheet and add the elevations and section views.

Suppose that you wish to change the scale of a view after it is added. For instance after adding all of the elevations and the section view to this new sheet, you may wish to enlarge the Section view.

4. Select the Section view on the sheet, right-click and choose **Activate View**.

This makes the view editable as if you had opened it from the Project Browser and edited the original section view.

Look for the Properties palette on your screen (it is docked above the Project Browser by default). If you closed it, right-click in the Section view and choose **Properties**.

5. On the Properties palette, beneath the "graphics" grouping, change the "View Scale" to **1/4" = 1'-0"** and then click the Apply button.

   • Right-click in the view again and choose **Deactivate View**.
   • Reposition the view as required.

Notice that the change in scale only affected the graphics of the view. The graphics of the model elements such as walls, doors, and windows enlarged but the text and annotations remained the same size, relative to the sheet & titleblock. You may also notice that the titlebar may no longer match the viewport. You can click on the viewport to access round grips at either end of the titlebar to change its length. If you want to move the entire titlebar, do not click the viewport first. Rather click on the titlebar directly and then drag it.

6. Perform any other edits and explorations you wish.

If you look at the right corner of the titleblocks or at the sheets listed on the Project Browser, you will notice that logical numbers have been automatically assigned, but the names of the two sheets are "Unnamed." You can rename (or renumber) the sheets by clicking directly on the titleblock. Select the titleblock, and then click on the blue text of the title or number to edit it. An alternative method is to right-click the sheet on the Project Browser and choose **Rename**.

7. Zoom in on the titlebars beneath the elevations.

Notice that each drawing has been automatically numbered sequentially.

   • Switch back to the other sheet (A101).

Notice that the section marks and elevation marks in the floor plan views correctly reflect the numbers of elevations and sections on sheet A102 (see Figure Q.38).

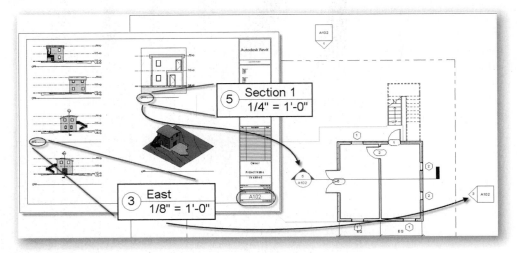

FIGURE Q.38 *View tag references and titlebars are automatically updated*

ADD DIMENSIONS

Typically when you issue sheets, you also need to include some dimensions on the plans and other views. We have a few basic dimensions on the plan already, but let's add a few more to calrify the design a bit.

1. On the Project Browser, double-click on the *Level 1* floor plan view to open it.
2. On the Annotate tab, on the Dimension panel, click the ***Aligned*** tool.

   - On the Options Bar, choose Wall Faces from the first drop-down list.
   - In the plan window, click on the outside face of the top horizontal Wall.
   - Move the mouse down and then click at the center of the double Door.
   - Continue down and then click the outside face of the bottom horizontal Wall.
   - Move the pointer outside the building to the left and click in an empty space to place the dimension (see Figure Q.39).

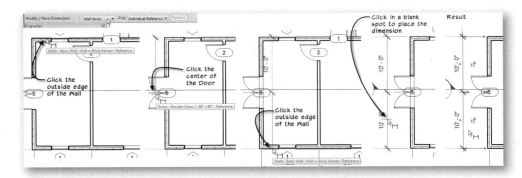

FIGURE Q.39 *Add dimensions by picking individual elements*

Compare this dimension to the equal one at the bottom of the plan. Using the Wall Faces option allows us to dimension to exactly the points we desire. Let's add another dimension across the top Wall using a slightly different technique.

3. If you canceled the command, on the Annotate tab, on the Dimension panel, click the ***Aligned*** tool again.
   - On the Options Bar, from the Pick drop-down list, choose Entire Walls from the first drop-down list.

- Next to the drop-down, click the small Options button.
- Check the Openings checkbox, choose the Widths option, and then click OK.
- Select the top horizontal Wall, move the mouse up, and click in an empty space to place the dimension (see Figure Q.40).

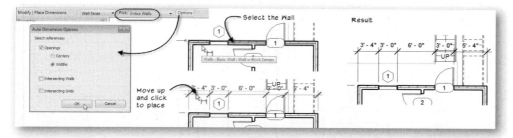

FIGURE Q.40 *Add dimensions by selecting entire Walls*

Notice how this adds a complete string of dimensions including the openings in the Wall. Feel free to experiment further before continuing. Also note that you can select elements in the model and use these newly placed dimensions to adjust their locations just like you can with temporary dimensions. For example, you can select the Window and edit the dimension value to move it onto a proper brick dimension. Give it a try. Also notice that when you return to the Sheet, all of these dimensions appear on the corresponding view. Remember, edits in one view appear in all appropriate locations in Revit.

OUTPUT

Feel free to print your two sheets out to your printer or plotter. You can also "go green" and generate a digital plot instead using the **Export ≥ DWF/DWFx** command on the Application menu. You can choose the **≤In session view/sheet set≥** option and then select the **Sheets in the Model** option to print both sheets to a single DWF file. This will create a compact portable file that can be opened and redlined on any computer using the freely available Design Review software from Autodesk. You can further edit and refine this model if you wish to add additional elements and annotations. Everything will remain coordinated and the sheets will automatically receive all the updates as we have seen.

SOLAR STUDIES

Giving you tools for producing traditional document sets and plotting are key strengths of the Revit software package. However, there is so much more that Revit can do. Schedules can be used to quantify elements in the model and assist professionals in cost estimating, reducing waste, and coordinating data with other applications outside of Revit. If you work with consultants who use Revit MEP or Revit Structure, the models you produce in Revit Architecture can be used to facilitate clash-detection studies and help catch conflicts before they become costly change orders in the field. Revit can even help us study how our building will interact with its environment by enabling us to create very accurate shadow studies. In our final exercise in this quick start tour of Revit's capabilities, let's take a brief look at the Sun Path feature.

1. On the Project Browser, double-click to open the {3D} view.

   - Orbit the model around so that you can see the Stairs and front door. (You can click the North-West corner of the ViewCube).

- On the View Control Bar at the bottom of the view window, click the Sun Path icon and choose Sun Path On.

The "Sun Path - Sun Not Displayed" dialog will appear. This message tells us that we need to adjust some settings in order for the Sun to appear.

- In the "Sun Path - Sun Not Displayed" dialog, choose the "Use the specified project location, date and time instead" option.
- Next to the Sun Path icon, click the Shadows icon (see Figure Q.41).

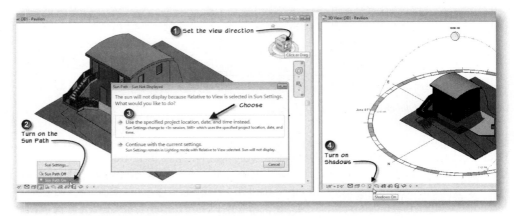

FIGURE Q.41 *Turn on the Sun Path and Shadows in the {3D} view*

A circular compass will appear surrounding the model and a shadow will appear in front of the building. We can change a few settings and have Revit display shadows using the project's location in the world and the actual position of the Sun.

2. Click-the Sun Path icon again and choose Sun Settings.

- Position the "Sun Settings" dialog next to the 3D view window.
  Several presets are available.
- Click on Summer Solstice and then click Apply (see Figure Q.42).

FIGURE Q.42 *Enable the Sun Path by choosing an option other than "Lighting"*

Notice that the shadow shifts on the stair side of the building. Attached to the compass we have a yellow icon representing the Sun and its path of travel through the sky on the date of the Summer Solstice. In this dialog, you also have the "Ground

Plane" control. If you check this box, you can cast the shadows on any level in your project. If you uncheck this, the shadows will cast on the surrounding geometry such as the terrain model. Feel free to experiment.

- In the "Sun Settings" dialog try a few more presets. Settle on Spring Equinox and then click OK.

This tool is completely interactive. You can click on the Sun icon in the view window and drag it to alternate positions. When you drag, a yellow-shaded area (called the Total Sun Area) will appear representing the Sun's total movements throughout the year for your project's geographic location. You can drag the Sun along the yellow arc to change the time of day, or along the analemma (figure-8 shaped path) to change the month. There are also text labels indicating both the date and time of day that you can click on and manipulate as an alternate way to change the Sun's position and alter the shadows accordingly. You can find more detail on features of the Sun path in the online Help. Let's try one more variation.

3. Return to the "Sun Settings" dialog.

- Beneath Solar Study, choose Single Day.
- Click the small drop-down icon in the Date field on the right and choose the "Today" option.
- Set the times from 6:00 AM to 6:00 PM.
- From the Time Interval list, choose 15 Minutes.
- Uncheck the "Ground Plane at Level" checkbox and then click OK.

These settings tell Revit that we want to study how the Sun moves across the sky from morning to evening on today's date. Now let's create an AVI file of the results.

4. From the Application Menu, choose Export ≥ Images and Animations ≥ Solar Study.
- Change the Frames/Sec to 5 and then click OK.
- In the "Export Animated Solar Study" dialog, choose a folder where you wish to save the file, accept the suggested file name and then click Save.
- Click OK in the next dialog to complete the export.

5. In Windows Explorer, browse to the location where you saved the AVI and double-click it.

The animation file will open in Windows Media player (or other media player on your system). You will see the shadows move across the ground as the day goes by! A sample of the file is provided in the *Quick Start\Complete* folder. Feel free to experiment further with any of the settings and export another file if you like.

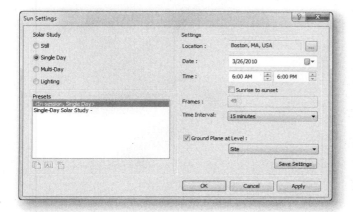

FIGURE Q.43 *Configure a single day solar study from early morning to early evening*

6. Save and close the project.

Congratulations! You have completed your first Revit Architecture project! Your journey into Revit Architecture and Building Information Modeling awaits you in the coming chapters.

SUMMARY

- Getting started with Revit Architecture is easy—click a tool on the ribbon and place the item in the view window.
- Walls, Doors, and other elements interact with each other as you place them in the model.
- Relationships and constraints are maintained automatically as you work. You can even "lock in" design intent if you choose.
- Build or edit your model from any view and changes are fully coordinated in all views. Edits *cannot* get out of sync!
- Views include graphical representations like plans and sections and non-graphical tabular representations like schedules.
- Sheets are special views designed for composing a titleblock layout and printing.
- Drag and drop to add views to sheets.
- All view references are fully coordinated on the views and sheets.

INTRODUCTION

Revit Architecture is an advanced software package for architects and other professionals involved in the design, documentation and construction of buildings. It facilitates the creation of a "Building Information Model" (BIM) in which plans, sections, elevations, 3D models, quantities, and other data are fully coordinated and can be readily manipulated, accessed and shared in a variety of meaningful ways. From the BIM database, one can perform design tasks, query quantities and takeoffs and generate drawing sheets for construction documentation needs. The advantages of this approach are many. From a production point of view, it means less time drafting and coordinating building data, because all drawings and reports come from the same source model. If the model changes, all "views," whether floor plans, sections, elevations, or schedules reflect the change immediately. To work effectively in such an environment, it is important to understand a bit about what it means to create and work within a Building Information Model. This is the primary goal of this chapter.

OBJECTIVES

In this chapter, we will explore the meaning of Building Information Modeling and take a high level look at the Revit Architecture software package. Working in a provided dataset, you will learn how to view a single model in many different ways that serve a variety of architectural drawing and documentation needs. Topics in this chapter include:

- Building Information Modeling
- The fully coordinated nature of a Revit Architecture Project File
- The basics of Revit Architecture Elements
- An introduction to Revit Families and Types
- Core concepts within the context of a Project

BUILDING INFORMATION MODELING

In the Autodesk Revit Curriculum & Student Workbook, author Simon Greenwold concisely and elegantly presents the concept of Building Information Modeling. A portion of that text is reproduced here to help explain BIM and how it compares to more traditional computer aided design (or drafting) (CAD) technology. The following topic is excerpted from the above-mentioned publication which is © 2005 Autodesk, Inc. It is used here with permission.

Editor's Note: While this essay is now several years old, it deals with the concept of BIM from the so-called "thousand foot level." As such, the points discussed remain relevant to BIM today even as the specifics of BIM continue to evolve each year.

Building Information Modeling [BIM] is a process that fundamentally changes the role of computation [and delineation] in architectural design. It means that rather than using a computer to help produce a series of drawings and schedules that together describe a building, you use the computer to produce a single, unified representation of the building so complete that it can generate all necessary documentation. The primitives from which you compose these models are not the same ones used in CAD (points, lines, curves). Instead you model with building components such as walls, doors, windows, ceilings, and roofs. The software you use to do this recognizes the form and behavior of these components, so it can ease much of the tedium of their manipulation. Walls, for instance, join and miter automatically, connecting structure layers to structure layers, and finish layers to finish layers.

Many of the advantages are obvious—for instance, changes made in elevation propagate automatically to every plan, section, callout, and rendering of the project. Other advantages are subtler and take some investigation to discover. The manipulation of parametric relationships to model coarsely and then refine is a technique that has more than one career's worth of depth to plumb.

BIM design marks a fundamental advance in computer-aided design. As the tools improve, ideas spread, and [practitioners] become versed in the principles, it is inevitable that just as traditional CAD has secured a deserved place in every office, so will BIM design.

CAD versus Building Information Modeling

Modeling Is Not CAD—BIM is entirely unlike the CAD tools that emerged over the last 50 years and are still in wide use in today's architectural profession. BIM methodologies, motivations, and principles represent a shift away from the kind of assisted drafting systems that CAD offers.

To arrive at a working definition of Building Information Modeling first requires an examination of the basic principles and assumptions behind this type of tool.

Why Draw Anything Twice?— You draw things multiple times for a variety of reasons. In the process of design refinement, you may want to use an old design as a template for a new one. You are always required to draw things multiple times to see them in different representations. Drawing a door in plan does not automatically place a door in your section. So the traditional CAD program requires that you draw the same door several times.

Why Not Draw Anything Twice?— There is more to the idea of not drawing anything than just saving time in your initial work of design representation. Suppose you have drawn a door in plan and have added that same door in two sections and one elevation. Now should you decide to move that door, you suddenly need to find every other representation of that door and change its location too. In a complicated set of drawings, the likelihood that you can find all the instances of that door the first time is slim unless you are using a good reference tracking system.

Reference and Representation—But doesn't reference tracking sound like something a computer ought to be good at? In fact, that's one of about three fundamental things a computer does at all. And it is exceedingly good at it. The basic principle of BIM design is that you are designing not by describing the forms of objects in a *specific* representation, but by placing a *reference* to an object into a holistic model. When you place a door into your plan, it automatically appears in any section or elevation or other plan in which it ought to be visible. When you move it, all these views update because there is only *one* instance of that door in the model. But there may be *many* representations of it.

You Are Not Making Drawings—That means that as you create a model in modeling software, you are not making a drawing. You are asked to specify only as much geometry as is necessary to locate and describe the building components you place into the model. For instance, once a wall exists, to place a door into it, you need to specify only the type of door it is (which automatically determines all its internal geometry) and how far along the wall it is to be placed. That same information places a door in as many drawings as there are. No further specification work is required.

That means that the act of placing a door into a model is not at all like drawing a door in plan or elevation, or even modeling it in 3D. You make no solids, draw no lines. You simply choose the type of door from a list and select the location in a wall. You don't draw it. A drawing is an artifact that can be automatically generated from the superior description you are making.

Even More Than a 3D Model—You are making a model—a *full* description of a building. This should *not* be confused with making a full three-dimensional (3D) model of a building. A 3D model is just another representation of a building model with the same incompleteness as a plan or section. A full 3D model can be cut to reveal the basic outlines for sections and plans, but there are drawing conventions in these representations that cannot be captured this way. How will a door swing be encoded into a 3D model? For a system to intelligently place a door swing into a plan but not into a 3D model, you need a high-level description of the building model separate from a 3D description of its form. This is the model in BIM design.

Encoding of Design Intent—This model encodes more than form; it encodes high-level design intent. A staircase is modeled not as a rising series of 3D solids, but as a staircase. That way if a level changes height, the stair automatically adjusts to the new criterion.

Specification of Relationships and Behavior—When a design changes, BIM software attempts to maintain design intent. The model implicitly encodes the behavior necessary to keep all relationships relative as the design evolves. Therefore the modeler is required to specify enough information that the system can apply the best changes to maintain design intent. When you move an object, it is placed at a location relative to specific data (often a floor level). When this [datum] moves, the object moves with it. This kind of relativity information is not necessary to add to CAD models, which are brittle to change.

Objects and Parameters—You may be troubled by the idea that the only doors you are allowed to place into a wall are the ones that appear on a predefined list. Doesn't this limit the range of possible doors? To allow variability in objects, they are created with a set of parameters that can take on arbitrary values. If you want to create a door that is nine feet high, it is only necessary to modify the height parameter of an existing door. Every object has parameters—doors, windows, walls, ceilings, roofs, floors, even drawings themselves. Some have fixed values, and some are modifiable. In advanced modeling you will also learn how to create custom object types with parameters of your choosing.

How Do BIM Tools Differ from CAD Tools?

Clearly, because modeling is different from CAD, you are obliged to learn and use different tools.

Modeling tools don't offer such low-level geometry options—As a general rule modeling deals with higher-level operations than CAD does. You are placing and modifying entire objects rather than drawing and modifying sets of lines and points. Occasionally you must do this in BIM, but not frequently. Consequently, the geometry is generated from the model and is therefore not open to direct manipulation.

Modeling tools are frustrating to people who really need CAD tools—For users who are not skilled modelers, modeling can feel like a loss of control. This is much the same argument stick-shift car drivers make about control and feel for the road. But automatic transmission lets you eat a sandwich and drive, so the choice is yours. There are also ways to layer on low-level geometric control as a post-modeling operation, so you can regain control without destroying all the benefits of a full building model.

Modeling entails a great deal of domain-specific knowledge—Many of the operations in the creation of a Building Information Model have semantic content that comes directly from the architectural domain. The list of default door types is taken from a survey of the field. Whereas CAD gets its power from being entirely syntactical and agnostic to design intent, BIM design is the opposite. When you place a component in a model, you must tell the model what it is, *not* what it looks like.

Or else requires you to build it in yourself—Adding custom features and components to a BIM design is possible but requires more effort to specify than it does in CAD. Not only must geometry be specified, but also the meanings and relationships inherent in the geometry.

Is Modeling Always Better Than CAD?

As in anything, there are trade-offs.

A model requires much more information—A model comprises a great deal more information than CAD drafting. This information must come from somewhere. As you work in a modeling tool, it makes a huge number of simplifying assumptions to add all the necessary model information. For instance, as you lay down walls, they are all initially the same height. You can change these default values before object creation or later, but there are a near infinitude of parameters and possible values, so the program makes a great many assumptions as you work. The same thing happens whenever you read a sketch, in fact. That sketch does not contain enough information to fully determine a building. The viewer fills in the rest according to tacit assumptions.

Flouting of convention makes for tough modeling—This method works well when the building being modeled accords reasonably well with the assumptions the modeler is making. For instance, if the modeler makes an assumption that walls do not cant in or out but instead go straight up and down, that means that vertical angle does not need to be specified at the time of modeling. But if the designer wants tilted walls, it's going to require more work—potentially more work than it would to create these forms in a CAD program. It is even possible that the internal model that the software maintains does not have the flexibility to represent what the designer has in mind. Tilted walls may not even be representable in this piece of software. Then a workaround is required that is a compromise at best. Therefore unique designs are difficult to model.

Editor's Note: The author of this passage used the example of tilted walls simply to make the accompanying point. Tilted Walls are possible in Revit Architecture, but admittedly with a little more effort than common vertical ones.

Whereas CAD doesn't care—In CAD, geometry is geometry. CAD doesn't care what is or isn't a wall. You are still bound by the geometric limitations of the software (some CAD software support nonuniform rational b-splines [NURBS] curves and surfaces, and others do not, for instance), but for the most part there is always a way to construct arbitrary forms if you want.

Editor's Note: In the last few releases of Revit, the Massing Environment has added many ways to easily model complex forms. These forms in the Massing Environment can be constructed parametrically and given rules and behaviors, but unless designated specifically as such, will not have any knowledge of what sort of element they are. Chapter 16 explores the Massing Environment.

Modeling can help project coordination—Having a single unified description of a building can help coordinate a project. Because drawings *cannot* ever get out of sync, there is no need for concern that updates have not reached a certain party.

A single model could drive the whole building lifecycle—Increasingly, there is interest in the architectural community for designing the entire lifecycle of a building, which lasts much longer than the design and construction phases. A full building description is invaluable for such design. Energy use can be calculated from the building model, or security practices can be prototyped, for instance. Building systems can be made aware of their context in the whole structure.

BIM potentially expands the role of the designer—Clearly this has implications for the role of architects. They may become the designers of more than the building form, but also specifiers of use patterns and building services.

Modeling may not save time while it's being learned—It is likely that while designers are learning to model rather than to draft, the technique will not save time in an office. That is to be expected. The same is true of CAD. Switching offices from hand drafting to CAD occurred only as students became trained in CAD and did not therefore have to learn the techniques on the job. The same is likely to be true of BIM design. But students are beginning to learn it, and it is rapidly becoming an integral part of practice in most offices.

Potential hazards exist in BIM that do not exist in CAD—Because design *intelligence* is embedded into a model, it is equally possible to embed design *stupidity*.

Editor's Note: For example, if the computer operator inputs a floor to floor height of only 4 feet [1200 millimeters], BIM software will not automatically correct or even flag this as an error. The knowledge of the architect is still the driving force behind the design intent that creates the model. Building Information Modeling software by itself does not replace the knowledge and experience of an Architect.

Improperly structured models that look fine can be unusable—It is possible to make a model that looks fine but is created in such a way that it is essentially unusable. For instance, it may be possible to create something that looks like a window out of a collection of extremely tiny walls. But then the program's rules for the behavior of windows would be wasted. Further, its rules for the behavior of walls would cause it to do the wrong things [with the "windows" modeled this way] when the design changed.

What is an engineering technique doing in architecture?—BIM design comes from engineering techniques that have been refined for many years. Many forces are acting together to bring engineering methodologies like BIM into architecture. First, the computing power and the basic ability to use computers have become commonly available. Second, efficiencies of time and money are increasingly part of an architect's concern. BIM offers a possible edge in efficiency of design and construction.

Editor's Note: You can find the original version of the preceding text as part of the Instructor Lecture Notes document on Autodesk's Web site in PDF format. The quoted portion here is only a small part of the entire document. At the time of this writing, the URL to this document is: http://usa.autodesk.com/adsk/servlet/index?siteID=123112&id=8029689& linkID=9243097

DEFINING BIM

This quoted passage does a wonderful job at outlining the high-level concepts involved in BIM and comparing and contrasting its tenets and techniques to traditional CAD and drafting methods. Using the points raised in this article, let's try to synthesize them into a working definition of Building Information Modeling.

Building Information Model(ing) is frequently misunderstood. "BIM" is often assumed to be synonymous with simply generating a three-dimensional (3D) model of a building—whether that model has any useful non-graphical information or not, and regardless of the level to which the 3D model is detailed. There is much more to BIM than that. Building Information Modeling *is* an evolving concept; one that will continue to change as the capabilities of technology and our own ability to manipulate technology improve. These issues make it difficult to formulate a simple definition for BIM. However, as the popularity of BIM is growing, reaching a consensus on its meaning and intent is increasingly important.

3D models are certainly important in BIM; in many ways critically so. However, summarizing all of the points made so far, the emphasis really belongs on the "I"— Information in BIM. That information can be either graphical (3D or even 2D) or non-graphical; either contained directly in the building model or accessible from the building model through linked data that is stored elsewhere. As has already been mentioned in the quoted passage above, when you exercise BIM, you are making a model which is a *full* description of a building—*not* just a 3D model. A data model is just as valid a model as a geometric model. (This is not a new concept in Architecture: consider the existing requirement in most jurisdictions of both a set of drawings *and* a written specification to complete a construction documents package.) Despite the importance of these distinctions, when we think of a Building Information Model, a three-dimensional geometric model of the building is often what comes to mind. So the first step to fully understanding BIM is to realize that "BIM" and "3D Model" are *not* the same thing.

In simple terms, a Building Information Model is a complete representation or depiction of a building that aids in its design, construction and potentially ongoing management. Such representation will often employ any combination of 3D graphics, 2D abstractions and/or non-graphical data as required for conveying full intent. The coordination and delivery of information and intent is the most important goal in BIM. To be considered BIM, all deliverables should be fully coordinated and provide a platform for meaningful computation.

Some of the content of the previous topic is paraphrased from Matt Dillon's Web Log (Blog). You can find the complete article at the following URL: http://modocrmadt.blogspot.com/2005/01/bim-what-is-it-why-do-i-care-and-how.html. Portions used here were used with permission.

REVIT KEY CONCEPTS

So now that you've got a good idea of the BIM concept you may be wondering how it specifically relates to Revit. Even more importantly, you may also be wondering how BIM will improve the way you work. Throughout the course of this book, and even more as you begin working with Revit on your own projects, you will gain comfort and familiarity with key Revit concepts. In this topic, we will identify and describe some of the most important Revit concepts, including Revit Elements, Families & Types, and Editing Modes.

ONE PROJECT FILE—EVERYTHING RELATES

In a Revit project (which is often contained within a single computer file), you will notice that no matter the location within the project where you perform your work, no matter what kind of view you are working in, all changes occur immediately in all views, and all elements retain their relationships with each other. This complete *bidirectional* coordination is perhaps the most significant benefit to using Revit. You can make a change in any view (plan, section or schedule) with complete confidence that the change is instantly reflected throughout the entire project file in all other views—no updates or synchronization required.

> Changes to "Drafting" views occur only in the edited view. Drafting views by definition are not linked to the model directly. These concepts are discussed in Chapter 11.

NOTE

The most talked about examples of the bidirectional coordination usually have to do with the "physical" aspects of your building project. If you move a door in a plan view, for example, the same door will also move wherever it appears in an interior elevation or perspective view.

Another example is in the "informational" aspects of that door. If you go into a schedule view where that door is listed and change it from wood door to a glass door, not only will the calculations like quantities or costs for that door change in the schedule, but also the change will be reflected in all graphical views—for example, in shaded views, the door will now appear transparent. Likewise, if this data is linked to cost estimation or green building calculations, the change to a glass door type will have other important impacts as well. This is a good example of how the "I" in BIM often deserves equal or even more emphasis than the "M."

Another important aspect of Revit is the Project Browser. Every view, Family and Group is clearly listed and organized in the Project Browser. The advantage of this is that all pieces of the project are always accessible and neatly organized. It is not possible for a team member to accidentally save project data in the wrong folder or location. Views and other elements in a Revit project know automatically in which location they should appear on the Project Browser.

REVIT ELEMENTS

In Revit you create a virtual representation (or model) of your project, using various types of elements. An element in Revit is simply a discrete building block or piece of data like an object or drawing sheet. The four basic types of elements are Model Elements (Walls, Doors, Roofs), Datum Elements (Levels, Grids, Reference Planes), View Elements (Plans, Sections, Schedules), and Annotation Elements (Tags, Text, Dimensions). Your graphical model primarily contains Model Elements.

Model Elements graphically represent "real" items in your building. View Elements enable you to display, study, and edit the model in depictions that represent traditional architectural drawing types. All views automatically appear in logical categories in the Project Browser tree. Specific architectural scale, level of detail, and other display characteristics are the province of views. Views can be placed on sheets and plotted to produce presentation drawings or drawing sets. To prepare views of your Revit model to appear on and print from sheets, Annotation Elements such as text, dimensions, and tags are used to notate and clarify the information shown. Annotation Elements are fundamentally different than Model Elements in that they belong to the view in which they are added and appear only in that view. Model elements appear in all views in which they logically should appear. (A door placed in plan will appear also in elevation, but a door tag placed in plan will not appear automatically in elevation). Datum Elements establish planes of reference in three-dimensional space. They are useful for establishing benchmarks and working planes for your project's geometry.

An illustration of the major element types with examples and their relationship to Revit and each other appears in Figure 1.1.

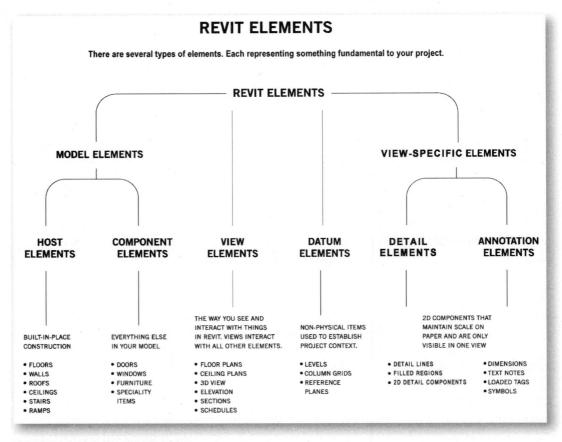

FIGURE 1.1 *Revit Elements Flow Chart*
Image courtesy of Autodesk

MODEL ELEMENTS

You can think of Model Elements as the items you use to describe the physical aspects of your building project. These elements in Revit can be separated into two sub-categories (as shown in Figure 1.1)—Host Elements and Component Elements. To use an analogy, hosts represent items that are typically constructed in place at the job site in the actual building like Walls and Roofs. Component Elements are delivered and installed in the building like Doors and Furniture. The figure shows further examples of each of these element types. For the purpose of our discussion here, let's consider the most common element of each category: a Wall (Host) and a Door (Component).

TABLE 1.1 *Comparing Host and Component Elements*

Host Elements	Component Elements
Here are the similarities:	
Wall	**Door**
Both Categories include Families.	
Basic Wall	Single-Flush
Their Families can both have multiple Types.	
Generic – 4" Brick	30" × 84"
Exterior – Brick on CMU	36" × 80"
Here are the differences:	
Their Basic Role in Projects	
Define Spaces and Enclosures	Modify/Detail Spaces and Enclosures
Their Relationship in Hosting	
Can *Host* Components (A Wall can *Host* a Door)	Must be *Hosted* (A Door is *Hosted* by a Wall)
They are saved in different places	
Saved in Project Files (*transferred* to new projects)	Saved in Family Files (*loaded* into new projects)

The primary concept to understand about Model Elements is that they *are* the elements that you use to create a (virtual) physical model that depicts your building, and these same elements are the ones that appear within *any* of the views of your project.

When adding a wall to your project, you start by deciding the type of Wall you are creating and in what location the Wall should appear. Determining how the lines and patterns that represent that wall will look on a particular drawing is handled automatically by the software. In this way, you are actually constructing a model of the required Walls rather than drafting a specific representation of them as you might in manual or CAD drafting.

The specific graphical settings can be modified if required in the Object Styles dialog box (on the Manage ribbon tab).

To get a complete listing of all of the Model Elements available in Revit, and how they will appear graphically in your project, open the Object Styles dialog box (click the Manage tab on the ribbon and then click the Object Styles button), and study the list on the "Model Objects" tab (see Figure 1.2).

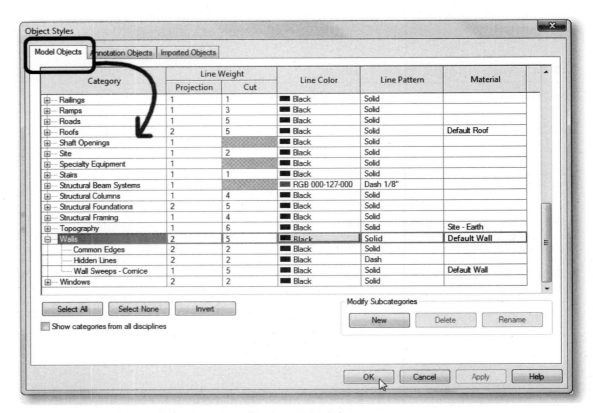

FIGURE 1.2 *A list of Model Elements appears in the Object Styles dialog*

Datum Elements

Datum Elements include three items. Each shares a common purpose of helping you establish a frame of reference for the geometry in your building projects. Levels represent horizontal planes cutting through the building at each floor level. They literally represent building levels. Levels are added from section or elevation views. Grids are vertical planes used to represent column grid lines (see Figure 1.3). The intersection of two grids typically indicates the location of a column. Grids are seen in plan, elevation, and section views. Reference Planes are used for more generic purposes. When you wish to establish a reference benchmark that is neither a level nor a grid, use a reference plane. Reference planes can be added to any orthogonal view. None of the Datum Elements appears in 3D views (unless you are in the Massing Environment, explored in Chapter 16).

You will learn more about Levels in Chapter 4, Grids in Chapter 5, and Reference Planes in Chapter 10.

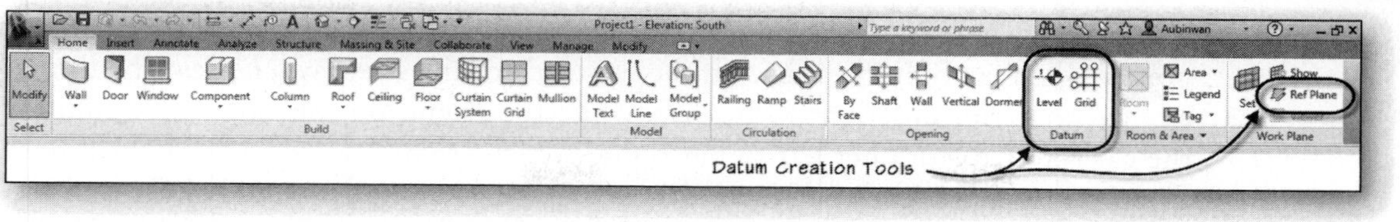

FIGURE 1.3 *Datum Elements include Levels, Grids, and Reference Planes available on the Home tab*

View-Specific Elements (Annotation and Detail Elements)

If you return to Figure 1.1, you will see that the View-Specific Elements category contains two sub-categories like the Model Elements category does. The first sub-category is Annotation Elements and the other is Details. Annotation Elements appear *only* in the views in which they are added. This makes their behavior unique in relation to the other types of Revit elements. Annotation Elements include all of the text and other descriptive architectural symbology that is required on drawings to explain and clarify the intent of the graphics, such as tags, dimensions, and view-specific detail components.

Annotation Elements are the closest thing to what you might consider "drafting" in Revit. They rarely cause changes in other views, or to the project model as a whole. In some cases, Annotation Elements can be used to manipulate the objects from within the view where they are placed. Dimensions are one such example. But the dimension itself, while capable of "driving" the geometry, appears only in the view to which it is added. We will see examples of dimensions used to manipulate model geometry in the lessons that follow.

Annotation Elements include objects such as Dimensions, Text, Tags, and more. To get a complete listing of all of the Annotation Elements available in Revit, and how they will appear graphically in your project, open the Object Styles dialog box again (shown above in Figure 1.2) and this time study the list on the "Annotation Objects" tab.

View Elements

As you develop your project in Revit, you will work with and create several views. Every Revit project begins with at least some views already in the Project Browser. The specific views that are available are a function of the template project from which the project is created. (A project template provides the framework and settings for a new project. More information on project templates is available in Chapter 4). To open and work in a view, you simply double-click its name in the Project Browser. Views are available for every architectural drawing type traditionally included in Architectural documentation sets. Like the drawings they represent, views allow you to interface with your model and edit its contents and composition within a particular context like plan, elevation, or section. Unlike traditional drawings, (as mentioned above) an edit in one view is instantly reflected in all appropriate views throughout the project.

Multiple Views-One Model—Imagine two friends living on opposite sides of the same street. Let's assume that the street runs north-south and that one friend lives on the west side of the street while the other lives on the east side. One day, both friends were looking out their window at the same time, as a car was passing by on the street below. The car was traveling from the south to the north. Which way would the friends say that the car was driving relative to their respective vantage points? The friend in the house on the west side of the street would describe the car as traveling from his right to his left, while the friend on the east side would say the car traveled from her left to her right. Which friend was correct? What if both friends snapped a photo at the same time? The two photos would show a different "view" of the same car and its travel pattern. Upon comparing photos with one another, would the two friends describe the respective scenes as two different cars? Or would they rather describe them as two different ways of seeing the same car?

This hypothetical scenario illustrates how project views work in Revit. When working in Revit, we frequently switch from view to view to edit and create elements; and although the specific graphics displayed on screen may vary (like showing the driver or passenger side of the car in the scenario above), they convey aspects of the *same* model. Therefore, the specific view in which you make an edit is irrelevant. A change to the model occurs in only one place—the model. You can study the change from any number of vantage points as represented in the various views available to you in the project browser.

To see the various view types available in Revit, you can look at the Create panel on the View tab (see Figure 1.4).

FIGURE 1.4 *Use the View tab to see the view types available*

> **NOTE**
>
> Revit includes both graphical views like plans, sections, and elevations and tabular views like Schedules and Material Takeoffs. Both types provide the means to study and manipulate models. Examples occur throughout this book.

Components, Model Lines, and Detail Lines

Most elements in Revit Architecture are purpose-built elements that have obvious functions based upon their namesakes like Wall, Door, Room Tag, and Section. The function of each of these elements is easily inferred from their respective names. However, there are also more generic elements whose function is not as specific that serve important functions in Revit Architecture as well. These elements include items like components, text, and lines. A Component is a model element that does not have a pre-defined function like a Door or a Window. Components are employed to create Furniture, Fixtures, and other items that are placed in models. Components are based on predefined templates so that while there is no specific "Furniture" button, there is a Furniture category and Family Template. (The term Family will be discussed in more detail below).

Text and Lines on the other hand can actually be either Model or Annotation elements. The same distinction (discussed above) between model and annotation applies to these elements as well—Model Text and Model Lines are actually part of the model and used to represent real items in the model. For instance, if you want to create signage you create it with Model Text. Model Lines might be employed to create inlaid patterns on Walls or Floors or to represent control joints on a masonry wall or other items that you don't find necessary to model three-dimensionally, but still want to see in more than one view.

While the creation process and graphical appearance on screen of Model and Detail lines may appear very similar and therefore make it difficult to distinguish them from one another, they are in fact very different. For example, if you draw a floor pattern using Model Lines; it *will* appear in all appropriate views like the floor plan view and a 3D view of the space. Model Lines added to the surface of a Wall would appear in both elevation and section views as if you actually painted these lines on the surfaces of the model.

On the other hand, a Detail Line, like other annotation, will appear in *only* the view in which it is added. Adding a Detail Line is like drawing on a piece of paper covering the view of your project, and will not be added *physically* to your building model. It is treated like other annotation as a simple embellishment to that particular view only. The most common use of such embellishment would be on enlarged details created from the model. Rather than meticulously model components that would only be practical to show in large scale drawings of a design, Detail Lines can be employed to represent those elements that would otherwise take too much time and effort to model throughout and would also add unnecessary overhead to the model without a commensurate amount of benefit. Such items might include building paper in a wall or roof section, nails, screws or other fasteners, reinforcing, flashing or even moldings, and trim in some cases. It would certainly be possible to model any of these elements, but in most cases, the additional overhead and effort required to model them would not be justified. Understanding what *not* to model is perhaps even more important than knowing what or how to model. This is a very important issue and understanding it is critical to understanding BIM and using Revit Architecture in the most efficient and practical manner.

If necessary, you can select any Model Line and convert it to a Detail Line and vice versa. To do so, simply select the line and then, on the content ribbon tab that appears, click the Convert Lines button. Revit will automatically convert the line to the other type (see Figure 1.5).

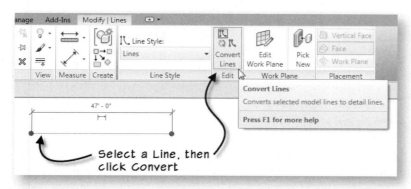

FIGURE 1.5

FAMILIES & TYPES

One of the most common terms in Revit is the term "Family." A Family in Revit is an object designated for a particular purpose that has a specific collection of parameters and behaviors. Within the limits established by the Family and its parameters, a potentially endless number of "Types" can be spawned. Types are basically saved and named variations in the Family and its parameters. Where a Family establishes a set of available variable parameters, a Type is a specific version of the Family with actual values for each parameter. The term "Family" was selected to characterize objects in Revit that have an inherited-property relationship. The main idea comes from the notion of a Parent-Child relationship—which is a central idea in object-based computer programming.

In order to illustrate how this concept works in Revit, we will look at some common categories of Families and describe how they are created and used in a project. You will notice these follow along similar lines as the Elements hierarchy covered above, with an additional Element classification called "System Families."

Model Element Families

Everything in the Revit software belongs to a Category and a Family. Each of the element types shown in Figure 1.1 above has a corresponding Family classification. Recall that the model element branch includes both Host and Component elements. This means that we have both Host Families and Component Families in Revit. The most common example of a Host Family is the Wall Family. The organization of Revit elements follows a hierarchical progression from global (Category) to specific (Instance) parameters and can be referred to as a "Family Tree." Here is an example of a Wall Family tree:

TABLE 1.2 *Family Tree Examples*

Family Tree Hierarchy	Sample Wall Family Tree
Element Category	Walls
Family/System Family	Basic Wall
Type	Exterior – Brick on CMU
Instance	Specific Instance of a Wall Selectable in the Project

You can also see the hierarchy interactively in a Revit project. To do so, expand the Families branch of the Project Browser. Here is the Wall Family Tree illustrated in a Revit Architecture project (see Figure 1.6):

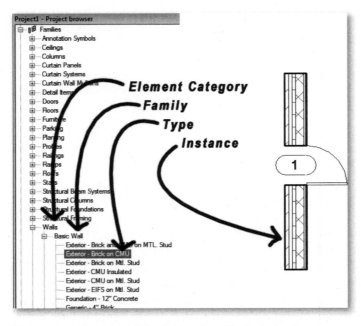

FIGURE 1.6 *A typical Wall Family Tree viewed in the Project Browser*

Recalling our discussion of element categories above, a Wall Family is a host Family. Host elements represent "built in place" construction. (This includes items that are assembled from raw materials on the job site). Host Families are hard-coded into the software and cannot be created or edited by the user. Users can however add, edit, or delete Types associated with host Families. If the Wall Type that you wish to use is not present in the current project, you must either duplicate and modify a similar Type that is resident in the file or transfer one from another project. Wall Types and other host element Types always live in project files. Component Families, as defined above, represent items that are purchased and installed in a project (not assembled in place). Perhaps the most common component element Family is a Door Family. Here is an example of a Door Family tree:

TABLE 1.3 *Family Tree Examples*

Category	Item
Element Category	Doors
Family/System Family	Single-Flush
Type	36" × 84"
Instance	Specific Instance of a Door selectable in the project

Here is this Door Family tree illustrated in a Revit Architecture project (see Figure 1.7):

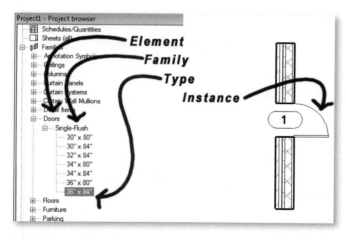

FIGURE 1.7 *A typical Door Family tree viewed in the Project Browser*

Unlike host Families, component Families do not have to be resident in the project file before you can use them. You can load component Families from external Family (RFA) files (as such, they are also referred to as "Loadable Families"). Loading a Component Familiy will typically load all Types associated with that Family into the current project (unless the Family file has a "Type Catalog" associated with it like Structural shapes—see the online Help and Chapter 5 for more information). You can also create new Types within the project just as you can with host Families or by editing the Family file. Barring these subtle differences, host and component Families are similar to one another. There are other less subtle differences as well, but for now those remain out of the scope of the current discussion. Families will be covered in more detail throughout the coming chapters, and Chapter 10 is devoted entirely to the topic of Families and the Family Editor.

Annotation Families

Every element in Revit belongs to a Family. Annotation Families are similar to model Families in that they are also objects that have a specific collection of parameters and behaviors. However, those parameters and behaviors are specific to annotation rather than the model. For example, there are Families for Dimensions, Text, Tags, and Datums. You can see most Annotation families loaded in your project by expanding the Annotation Symbols branch in the Project Browser (see Figure 1.8).

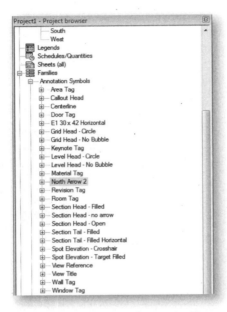

FIGURE 1.8 *Annotation Element Families appear beneath Families > Annotation Symbols on the Project Browser*

Many Annotation Families serve special purposes in a project. A Title block for instance is a special kind of Annotation Family that is used when you create a sheet view. It can include text fields, called Labels that automatically report project data and can also contain company logos and other graphics. Some Annotation Families, like Level Heads, Section Heads, and Elevation Heads are typically associated directly to some other view in the project and provide a means to navigate from one view to another. For example, add a Section line in a plan view to indicate where the section is cut. Double-click the Section head to open the associated section view. Tags usually contain a symbol created from simple geometry and a label that reports a specific parameter or parameters from an associated object. Examples include the door number, room number, or area of a room. Some are simple symbols that you can add to any view. One such example is the Centerline symbol included in the default template.

System Families

In general terms, a System Family is any Family that is built into the Revit software by the programmers and cannot be edited by users. Many of the elements on which we have already commented are System Families such as Walls, Roofs, Text, and Dimensions. System Families include more than just Annotation and Model elements. All aspects of the software are organized hierarchically into a category and Family structure. Families even include items designed to help organize the way we work with the software. Such is the case with the Browser Organization System

Family. Take a look at the Project Browser. At the top is a heading labeled "Views (all)" (see Figure 1.9).

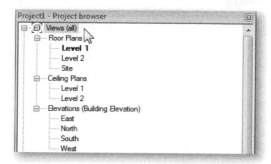

FIGURE 1.9 *Even the organization of the Project Browser is governed by a Family*

In the parenthesis, (all) is actually a Type associated with the Browser Organization System Family. Like most Families, the Browser Family includes other Types. You can see and edit those Types by using the Browser organization command. You can find it on the View ribbon tab on the User Interface drop-down button or more directly on the Properties palette (see the left side of Figure 1.10). Many of the other Types available are quite useful depending on the kind of project on which you are working and/or the phase or composition of the design team.

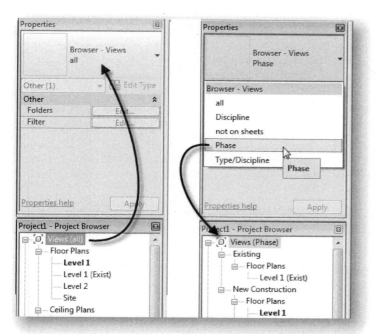

FIGURE 1.10 *The Browser Organization Family contains other very useful Types.*

To activate a different organization, select the Views (all) item in the Project Browser and then on the Properties palette, choose the desired organization Type from the Type Selector. If you wish to modify an existing one or create a new one, click the View tab of the ribbon, and then on the Windows panel, click User Interface and then select Browser Organization. In the dialog that appears, select one of the items in the list, and then click the Edit button. If the Type of Browser Organization that you wish to use is not included in the list, you can create your own using the New button.

Changing the Browser Organization can help you find a specific view you need on large projects with many views. For example, if you are working in a project that will have many phases (existing, new, etc.) the Phase organization Type can be helpful. Discipline organization will sort the views by various disciplines like Architectural, Structural, or Mechanical. Not on sheets will hide all the views that have already been added to a sheet in the project, thus showing only those views not currently on a sheet. This is a handy way to be sure you don't forget to include an important view on a sheet in your document set.

EXPLORE AN EXISTING PROJECT

We have covered a lot of important concepts in this chapter so far. However, we have only discussed these topics in the abstract. If you did the tutorial in the Quick Start chapter, you have had some hands-on exposure to the basics of the software. To help reinforce and solidify the concepts presented let's open and explore a dataset provided with the files from the book's student companion site. We will still be discussing topics at a high level in this exercise, but having a project open as a backdrop will make the concepts easier to understand.

Install the Dataset Files and Open a Project

The lessons that follow require the dataset included on the Aubin Academy Master Series student companion. If you have already installed all of the files from this site, skip to step 3 to begin. If you need to install the files, start at step 1.

1. If you have not already done so, download the dataset files located on the CengageBrain website.

    Refer to "Accessing the Student Companion site from CengageBrain" in the Preface for information on installing the dataset files included in the Student Companion.

2. Launch Autodesk Revit Architecture from the icon on your desktop or from the *Autodesk > Revit Architecture 2012* group in *All Programs* on the Windows Start menu.

TIP	You can click the Start button, and then begin typing Revit in the Search field. After a couple letters, Revit Architecture should appear near the top of the list. Click it to launch to program.

3. On the Recent Files screen, click the Open link in the Projects section. (Be sure you don't click the Open link beneath Families for this exercise.)

TIP	The keyboard shortcut for Open is CTRL + O. **Open is also located on the Application menu and the Quick Access toolbar.**

*   In the "Open" dialog box, browse to the location where you installed the *MasterRAC 2012* folder, and then the *Chapter01* folder.

4. Double-click *MasterRAC Chapter01.rvt* to open the project.

    You can also select it and then click the Open button.

For this brief tutorial, only an Imperial units dataset has been provided. For the remainder of the book, Metric datasets are provided as well.

NOTE

The project will open in Revit Architecture with the last opened view visible on screen.

The dataset for this chapter provided courtesy of Mark Schmieding.

GETTING ACQUAINTED WITH THE PROJECT

For this tutorial, we will explore a series of sheet views included in the project. A sheet view is a special kind of view that emulates a sheet of paper from which drawing sets can be printed to output devices. Sheet views typically include a title block which contains project and drawing information.

Revit remembers the last view that was open when the project was saved. In this case, it is a three-dimensional aerial view of the entire project composed on a simple titleblock border. This is a small one-floor project for a youth center. It includes offices, exam and counseling rooms, a multipurpose room, and media rooms. Let's take a closer look (see Figure 1.11).

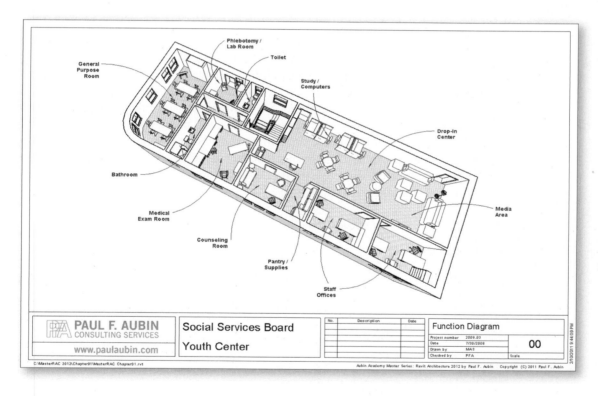

FIGURE 1.11 *The Youth Center dataset shown from the "Function Diagram" sheet*

View Navigation

You can use the wheel on your mouse to zoom in and out in any view. You can hold the wheel in and drag to pan the screen. You can also use the scroll bars at the right and bottom for this purpose. If you do not have a wheel mouse, you should consider purchasing one. However, with or without a wheel mouse, you can use the commands on the Navigation Bar (located by default in the upper right corner of the view window) to navigate in any view. Depending on the kind of view active on screen, you will have access to differing tools on the Navigation Bar (see Figure 1.12). Among these are the Steering Wheels, the Zoom pop-up, and the ViewCube (not shown in the figure).

FIGURE 1.12 *Zoom the sheet to Sheet Size and pan around to see it as it will print*

> Revit 2012 introduces 3D connexion device support. If you have one of these devices connected to your computer, you can now use it to navigate in 2D and 3D views. Additional icons will appear on the Navigation Bar indicating that the device is detected and available for use in Revit.

The zoom pop-up offers many ways to zoom the current window. Most of these commands will be available in all kinds of views, like Zoom To Fit (which fits the screen to the extent of the model) and Zoom In Region (which allows you to drag a rectangular region on screen to zoom). We also have the handy **Zoom Sheet Size** available. This command zooms a view to a size roughly comparable on screen to the actual size it will appear when printed. Since Revit displays line weights and other graphics accurately on screen, this can give you a good preview of how the sheet will look when printed. Each of the zoom commands has a command shortcut that you can execute via the keyboard. These shortcuts are two characters and you simply type both characters in succession to execute the appropriate command. For example, to issue Zoom to Fit, you can simply type ZF.

TABLE 1.4 *Zoom Command Keyboard Shortcuts*

Zoom Command	Keyboard Shortcut
Zoom in Region	ZR
Zoom Out (2x)	ZO
Zoom to Fit	ZF
Zoom All to Fit	ZA
Zoom Sheet Size	ZS
Previous Pan/Zoom	ZP

5. From the Zoom pop-up on the Navigation Bar, choose **Zoom in Region**.

You can also type ZR to issue this command. If Zoom in Region is already selected (a checkmark appears next to it) then simply click the zoom icon to execute the command.

- Drag a rectangular region around the upper left corner of the drawing.
- Hold in the wheel on the mouse and drag around to pan the model.

If you prefer, or if you don't have a wheel, use the scroll bars instead.

The image you see on screen is actually the view named *Large Overview* in the *3D Views* branch of the project. It has been added to the current sheet and displays in a "Viewport" on the sheet.

6. Zoom back out. The easiest way is to choose **Zoom to Fit** (shortcut ZF) from the Zoom pop-up menu.

The Steering Wheel offers an alternative to wheel mouse navigation with such commands as dynamic zoom and pan. Click the Steering Wheel icon to make it appear. In this case, since we have a sheet active, only the two-dimensional commands will appear. (This is true even though a 3D view is placed on the sheet; the sheet itself is still two-dimensional.)

7. Click on the Steering Wheel icon (the tool tip will read "2D Wheel").

Each function works the same way. Place your mouse on the area of the wheel for the function you want. It will highlight as your cursor passes over it. You are also simultaneously moving the wheel around the screen. Click and drag with the mouse to begin the function. For example, if you wish to zoom, move the wheel to the area of the screen that you wish to center your zoom on, move the pointer over the Zoom part of the wheel, click and hold down the mouse and begin to drag. Dragging up zooms in, dragging down zooms out. Varying the speed of your dragging varies the speed of the zooming. Release the mouse button to stop zooming and make the wheel reappear to change functions. Pan works the same way except that panning occurs in the direction that you drag (see Figure 1.13).

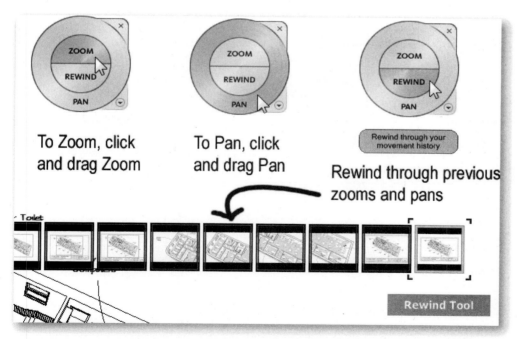

FIGURE 1.13 *Steering Wheels offer many view navigation functions. Click and drag on the part of the wheel labeled for the function you want*

As you perform several zooms and pans, they are stored in memory. You can use the Rewind function to back up through previous zooms and pans in a visual way. Move your mouse pointer over the Rewind function, click and hold down. A ribbon of thumbnail previews will appear, each representing a previous zoom or pan. Drag to the left to highlight previous zooms and pans, drag back to the right to move forward. Release the mouse to stop rewinding or forwarding. When you are done with the wheel, click the small close box ("X") in the upper right corner of the wheel or press ESC.

When finished experimenting with Steering Wheel, close it to continue.

Understanding Screen Tool Tips

8. Using the zoom icon, Zoom to Fit.
9. Move your mouse pointer into the middle of the screen and pause it there—pause over the drawing, not a text note.

   Do not click the mouse.

Notice how a rectangular border highlights around the image. As you pause the mouse, an onscreen tool tip should appear as well. In this case, this tip will read: **Viewports : Viewport : No Title** (see Figure 1.14).

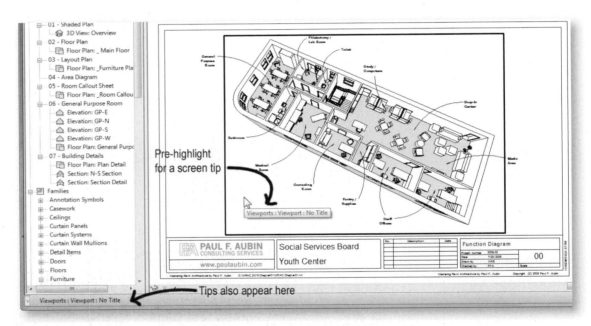

FIGURE 1.14 *Tool Tips will indicate the Element category, Family, and Type*

The tool tip conveys three bits of information about the element highlighted— **Element category : Family Name: Type Name**. So in this case, the Element category is "Viewports," the Family is "Viewport" and the Type is "No Title."

NOTE	The same information appears in the status line at the bottom left corner of the Revit interface.

Now hover the pointer over a piece of text but do not click. This is called "Pre-highlighting." The tool tip for a piece of text will read—Text Notes : Text : 3D Notes. Here, **Text Notes** it the Element category, **Text** is the Family and **3D Notes** is the Type.

Since the 3D view is actually a viewport containing one of our project views, you do not see the elements within the model pre-highlighting. However, you can choose to "Activate" a Viewport that will give you access to the building model elements shown within the view. Editing them from a Viewport is no different than opening the view from the Project Browser and editing them there; the results are the same view either way. Let's take a look.

10. Pre-highlight the Viewport, and then click to select it this time.

   • Right-click and choose **Activate View**.

Notice that the sheet title block and the text labels have grayed out. While they are still visible, this graying effect indicates that they are currently inactive.

   • Move the mouse around the model.

Notice that the elements within the model now pre-highlight (see Figure 1.15).

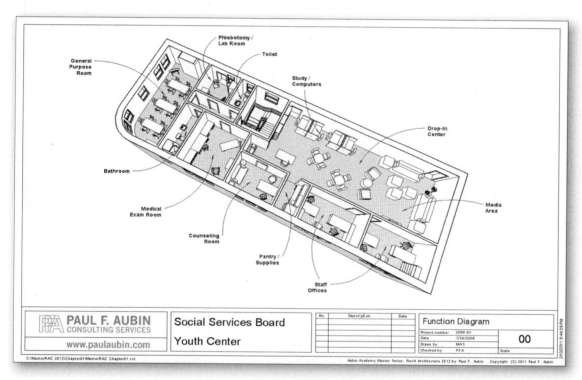

FIGURE 1.15 *Once the Viewport is activated, you can pre-highlight the elements in the model*

We will not actually edit any model objects in this view, but do take notice of the tool tips at this level. The interior partitions, for example, display Walls : Basic Wall : Generic – 6". The Element category is **Walls**, the Family is **Basic Wall** and the Type is **Generic – 6"**(see Figure 1.15).

Feel free to select objects if you like, but don't edit anything. If you accidentally move or change an element, click the undo icon on the Quick Access Toolbar at the top left corner of the Revit interface.

Notice that pre-highlighted elements highlight in purple by default and selected elements highlight in light blue. Both of these defaults can be changed in the Options dialog. Access Options via the Application menu (the big "R" at the left of the Quick Access Toolbar—see Chapter 2 for more details.)

You may also notice that with the three-dimensional view now active, in addition to the Navigation Bar, the ViewCube is also displayed. The ViewCube is a 3D navigational tool available in all Autodesk products. Clicking on any of the labeled sides of the cube will orient the view to that direction such as top, front, or right. There are also several active regions between faces that will orient the view at an angle between the two adjacent faces. For example, click the edge between front and right to orient the view to the southeast. Click the corner between three faces to orient the view to an axonometric orientation. You can also click and drag any edge of the cube to orbit the model in real-time. Feel free to experiment with the ViewCube to get the hang of it (see Figure 1.16).

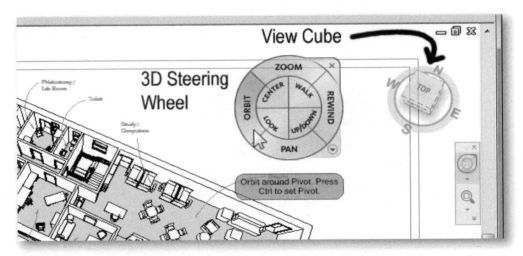

FIGURE 1.16 *Three-dimensional views show the ViewCube and 3D Steering Wheel*

In addition to the ViewCube, the Steering Wheel has more options in a three-dimensional view. You can orbit the view, change the center of rotation, move the vantage point up and down, and walk and look around the model. Consult the online help for more information on these options and the many ways you can customize the Steering Wheels to suit your preferences. Feel free to experiment with the 3D Steering Wheel as well. Remember to get the view back to the original, you can use the Rewind tool on the Steering Wheel and/or right-click on the Steering Wheel and choose **Undo View Orientation Changes**.

11. When you are done exploring in the model, right-click in the Viewport again and choose **Deactivate View**.

This returns you to the sheet and the elements in the view are no longer selectable.

Views and Detailing

Earlier we discussed how Model and Annotation elements were handled in distinct ways. Using this dataset, let's explore the difference between Model and Annotation elements a bit further.

12. On the Project Browser, beneath the *Views (all)* branch, double-click to open the *_Main Floor* plan view.

This is the basic floor plan view for this project.

13. On the Project Browser, double-click to open the *_Room Callouts* plan view.

This plan is very similar to the *_Main Floor* view except that it also includes call-outs around the General Purpose Room on the left and some elevation and section markers. A sheet has been provided showing each of these views.

14. On the Project Browser, beneath the *Sheets (all)* branch, double-click to open the *05 – Room Callout* sheet view.

Notice how the only visual difference here is that the plan appears on a title block sheet in this view.

15. On the Project Browser, double-click to open the *02 – Floor Plan* sheet view.

This is the sheet presentation of the *_Main Floor* plan view. In other words, this sheet composes the *_Main Floor* plan view on a title block for printing. You can easily see which views appear on a sheet in the Project Browser.

16. On the Project Browser, beneath the *Sheets* branch, expand the tree (click the small plus (+) sign) beneath the *01 – Shaded Plan* sheet (see Figure 1.17).

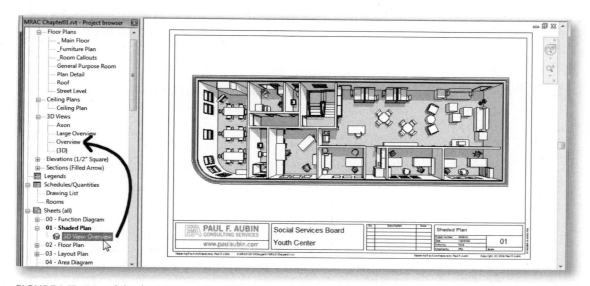

FIGURE 1.17 *Expand the sheet entries in the Project Browser to see the views they contain*

This provides an easy way to see which views are inserted on particular sheets. Another useful tool (noted above) gives us a way to see which views have not yet been placed on sheets.

17. On the Project Browser, scroll to the top and click the Views (all) branch.

- On the Properties palette, from the Type Selector (drop-down list at the top), choose **not on sheets** (similar to Figure 1.10 above).

Notice that the list of views on the Project Browser filters to show only those views that are not yet assigned to a sheet. In this particular project, there are only a couple views not placed on sheets. Expand each sub-group to see.

- Make sure that "Views (not on sheets)" is selected, and then on the Properties palette, change back to **all**.

This sets the default browser organization back to showing all views regardless of their placement on sheets.

18. On the Project Browser, double-click to return to the _Main Floor_ plan view.

Suppose that we needed to create another floor plan that was similar to this one, but that was to convey a different type of information on the printed sheet or that we were planning to use simply as a convenient place in which to edit the model with no intention of adding it to a sheet. To achieve either goal, we simply duplicate an existing view.

- On the Project Browser, right-click the _Main Floor_ plan view and choose **Duplicate View > Duplicate**.

A new floor plan view named *Copy of _Main Floor* will appear and become active. Notice that none of the room labels or dimensions were copied in this operation. This might be useful if you were creating a "working" view. A "working" view is intended as a view in which you manipulate the model only and do not plan to add to a sheet for printing. Bear in mind that nothing prevents the working view from being used on a sheet; rather it is simply not intended for that purpose by our project team. If we want to duplicate the view, including the tags and dimensions, we choose a different command.

- On the Project Browser, right-click the _Main Floor_ plan view and choose **Duplicate View > Duplicate with Detailing**.

NOTE "Duplicate with Detailing" is short for "Duplicate with view-specific detailing elements and annotation elements." Remember that the "Detailing" is being copied, while the Model Elements are simply being viewed.

Be sure to right-click on _Main Floor_ and not *Copy of _Main Floor* in this step.

A new floor plan view named *Copy (2) of _Main Floor* will appear and become active. Notice that this copy includes the room tags and dimensions.

- Right-click *Copy (2) of _Main Floor* and choose **Rename**.
- In the "Rename View" dialog, type **Area Diagram** and then click OK.

19. With the CTRL key held down, select each of the Dimensions in the view.

- Press the DELETE key.

We do not need Dimensions for the new view we are creating. However, there is no way to duplicate only the room tags and not the dimensions, so simply deleting them achieves the desired result.

20. On the Home tab of the ribbon, in the Room & Area panel, click the Legend tool (see Figure 1.18).

FIGURE 1.18 *Create a Legend for the new Area Plan*

A small square with a tag will appear attached to the cursor.

- Click a point above the plan to place the Color Scheme Legend.
- In the dialog that appears, Scheme 1 will be selected. Accept this by clicking OK.

As you can see, the Scheme 1 color scheme color codes each room based on its name. The legend itself is currently overlapping the plan. To make it fit better, we can resize and/or move it.

- Click on the Color Fill Legend and then drag the small round Control at the bottom up to make the legend two columns (see Figure 1.19).

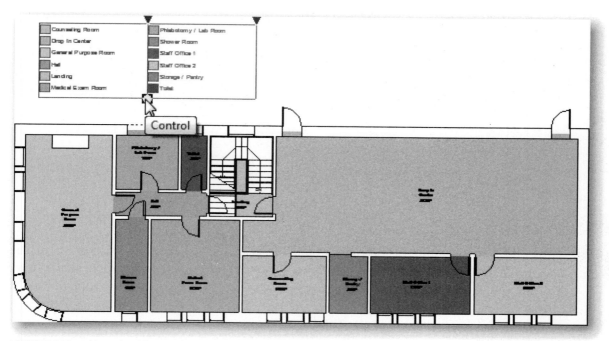

FIGURE 1.19 *Add a Color Fill Legend and then drag it to two columns*

21. On the Project Browser, double-click to open the *04 – Area Diagram* sheet view.

A sheet appears on screen, which does not yet have a drawing on it. Let's add our new shaded plan to this sheet.

- On the Project Browser, right-click the *04 – Area Diagram* sheet and choose **Add View**.
- From the "Views" dialog, choose *Floor Plan: Area Diagram* view and then click the Add View to Sheet button.
- Click to place the view on the sheet.

Notice that the view is a little too big for the sheet. We can adjust the scale of the view and it will update on the sheet.

22. On the Project Browser, reopen the *Area Diagram* view.

- At the bottom of the screen, on the View Control bar choose **1/8" = 1"-0"** from the scale pop-up menu (see Figure 1.20).

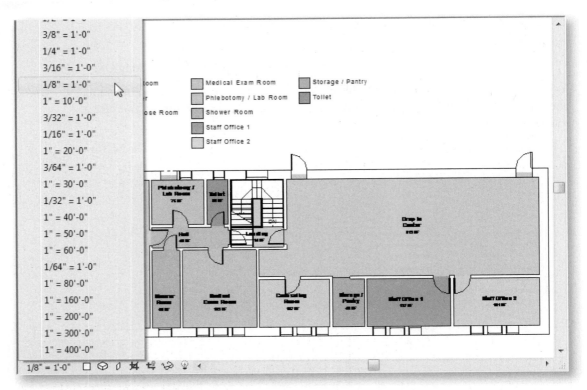

FIGURE 1.20 *Change the scale of the view*

23. Return to the *04 – Area Diagram* sheet to see the change.

You can move the viewport around as needed to make a nicer composition on the sheet. You may also want to remove the titlebar under the viewport. Select the viewport on the sheet.

- On the Properties palette, click on the Type Selector (drop-down list at the top) and then choose **No Title**.

You should also take a look at the *_Furniture Plan* floor plan view and the *03 – Layout Plan* sheet next. In this view and sheet, you will notice that the plan is displayed with furniture. Therefore creating plans with and without detailing (text and other annotation) is *not* the only way to vary the specifics of what we see. We can also control the visibility of each type of element in *any* Revit view. The visibility settings are a parameter of the view itself. This is how we can choose to display the furniture in the *_Furniture Plan* floor plan view and not display it in the *_Main Floor* view. On the View tab of the ribbon, you can choose the Visibility/Graphics tool on the Graphics panel. This will display a dialog listing all element categories and enabling you to turn on and off these categories within the current view. While we will discuss the specifics of this process in later chapters, the important point for this exercise is that this sort of control *is* possible and extremely useful. If you wish to explore the Visibility/Graphics Overrides dialog, please feel free to do so. Simply undo your changes before continuing with the lesson.

Edit in any View

Perhaps the most powerful feature of Revit is the ability to edit in any view and see the results instantly in all views.

24. On the Project Browser, double-click to open the *06–General Purpose Room* sheet view.

This sheet shows the views that are associated to the callouts we saw on the *05 – Room Callout Sheet* above.

- Select the plan view on the left, right-click and choose **Activate View**.
- On the Home tab of the ribbon click the Window tool.

The Modify | Place Window tab will appear on the ribbon. On its right, the Tag on Placement tool is highlighted in blue.

- On the right side of the Modify | Place Window tab, click the Tag on Placement tool.

This clears the highlight from the tool.

- Click a point on the exterior Wall on the left to add a new Window (see Figure 1.21).

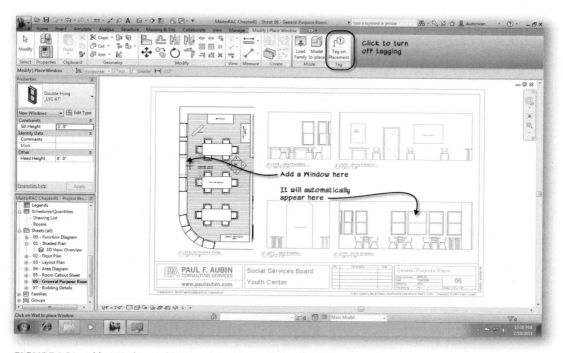

FIGURE 1.21 *Add a Window and it appears in all appropriate Views automatically*

- Right-click in the plan view again and choose **Deactivate View**.

Explore a Detail View

As we have noted above, a Detail view is a little different than the other Views. Typically it will include a live view of the model—usually a callout of some part of a section or plan—and various types of annotation and other graphical embellishments drawn on top. One such detail view has been included in this sample dataset.

25. On the Project Browser, expand (click the plus [+] sign) the *07 – Building Details* sheet view.

Beneath this sheet in the Project Browser a listing of three views will appear that are already placed on the sheet.

- Beneath the *07 – Building Details* sheet view entry, double-click to open the *Section : Section Detail* view.
- Pre-highlight some of the elements in this view (see Figure 1.22).

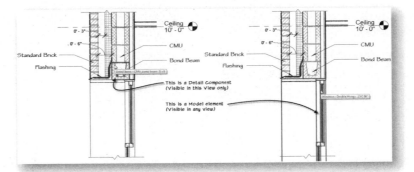

FIGURE 1.22 *Explore a Detail view—Note the combination of Detail and Model elements*

Notice that the Detail view contains both Model Elements (which would appear in all Views) and Detail elements (which appear only in this view). Even though the Detail elements represent items like concrete blocks, brick, flashing, and bond beams, the level of detail required in a construction detail is much higher than that required in nearly any other view. Therefore, these types of items are typically drawn as Detail elements on top of the Model view geometry in this way to keep overhead low and reduce the amount of time and effort required to build your overall model. Complete coverage of this technique can be found in Chapter 11.

Continue to explore in this dataset as much as you wish to get a better feel of how the various elements and Views in a Revit Architecture project interact. Close Revit when you are finished exploring. You do not need to save the file.

SUMMARY

- Building Information Modeling is the process of creating an accurate representation of the building including its form, functions, and systems from which detailed and accurate information can be extracted.
- Using BIM successfully requires a firm understanding of the concepts and techniques enabling its use.
- All items in Revit Architecture belong to Families.
- There are System Families, Model Families, and Annotation Families.
- Model Families can be either Host or Component (Loadable) Families.
- In Revit Architecture, all views show the same source data stored in a single source model. If changes are made in one view, the change is immediately available in all views. With few exceptions, nearly all elements in Revit can be classified as either Model Elements or Annotation Elements.
- Model Elements are those items that represent real building components and are visible in all views.
- Annotation or Drafting elements appear only in the view to which they are added and include notes, dimensions, and detail embellishments.

CHAPTER 2

Revit Architecture User Interface

INTRODUCTION

This chapter is designed to acquaint you with the user interface and work environment of Revit Architecture. The Revit user interface is logically organized and easy to learn. In this chapter, we will explore its major features. Many features of the interface follow standard Microsoft Windows™ conventions. Some aspects of the user interface are unique to Revit Architecture. In this overview, our goal is to make you comfortable with all aspects of interacting with and receiving feedback from your Revit software. Many of the lessons that follow are descriptive in nature and some are tutorial-based. Feel free to follow along in Revit Architecture as you read the descriptions.

OBJECTIVES

To get you quickly acquainted with the Revit user interface, topics we will explore include:

- An overview of the Revit Architecture user interface
- Interface terminology
- Working with the Ribbon, Quick Access Toolbar, and the Options Bar
- Moving around a Revit Architecture model

UNIT CONVENTIONS

Throughout this book, Imperial units and files will be listed first, followed by metric in brackets, for example, **Imperial [Metric]**. See the Preface for complete details on style conventions used throughout this book.

Imperial dimensions throughout this text appear in the "Feet and Inch" format for clarity. However, when typing imperial values into Revit, neither the foot symbol (') (when typing whole feet) nor the hyphen separating the feet from inches (when typing both) is required. Therefore, to type values of whole feet, simply type the number. To type values of both feet and inches, type the number of feet with the foot symbol (') followed immediately by the number of inches; the inch symbol is not required. You can also separate feet from inches with a space (press the SPACEBAR) and omit the unit symbol. When typing only inches, the inch (") symbol *is* required unless you preface the value with a leading 0'. For example, 4'-0" can be typed in Revit Architecture as simply: **4** (or **48"**). To input four feet six inches, type: **4.5**, **4'6** or **54"**. You can also type **4 6** (that is **4** SPACE **6**). To type 10 inches, type: **10"** or **0'10** or **0 10** (that is **0** SPACE **10**). Hyphens are not required. When separating inches from fractions, use a SPACE. Consult Table 2.1 for additional examples.

| NOTE | In all cases, avoid using a dash (or hyphen) to separate values as Revit will try to interpret such input as a mathematical expression. |

TABLE 2.1 *Acceptable Imperial Unit Input Formats*

Value Required	Type This:
Four feet	**4 or 48"**
Six inches	.5 or **6" or 0'6** or **0 6**
Five feet six inches	**5'6** or **66"** or **5.5** or **5 6**
Four feet six and one half inches	**4'6 ½** or **54.5"** or **4 6.5** or **4 6 1/2**

| NOTE | Typing the foot (') mark is acceptable when typing whole feet as well; however, it is not required. |

| NOTE | Dimensions throughout this text are given in the "Feet and Inch" format for clarity. However, feel free to enter dimension values in whatever of the above acceptable formats you prefer. Eliminating the inch or foot marks where possible reduces keystrokes and is recommended despite their inclusion in this text. |

If using metric units, all values in this text are in millimeters and can be typed in directly with no unit designation required. More information on style conventions used in this book can be found in the Preface.

UNDERSTANDING THE USER INTERFACE

Revit Architecture offers a clean and streamlined work environment designed to put the tools and features that you need to use most often within easy reach. In addition to the many onscreen tools and controls, many of the most common tools also have keyboard shortcuts. The topic of shortcuts will be explored below. Figure 2.1 shows the Revit Architecture screen with each of the major interface elements labeled for your reference.

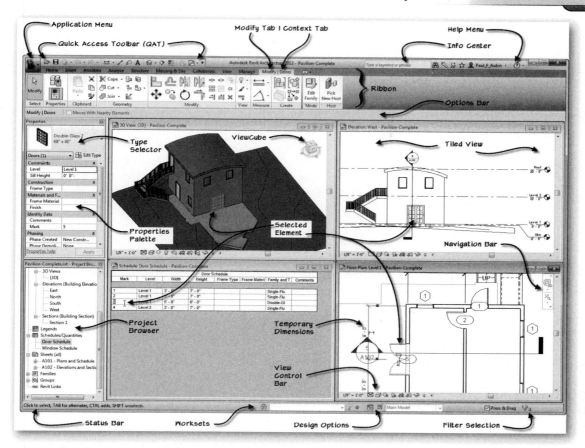

FIGURE 2.1 *The Revit Architecture User Interface*

The Revit interface includes many key features such as the Properties palette and Project Browser docked to the left side of the appliction by default, the ribbon, and the main drawing area or "canvas". The ribbon is the tabbed panel across the top of the screen that includes all of the software's buttons and controls. This ribbon interface is consistent with many of Microsoft's latest offerings and is becoming widely adopted by other vendors as well. The concept of the ribbon-style interface is to make commands and tools easily accessible and to present tools to you in context with the task you are performing.

Above the ribbon, in the top left corner of the Revit screen is where the Application menu (adorned by the large Revit "R" icon) and the Quick Access Toolbar are located. At the top right are the standard Windows minimize, maximize, and close icons, accompanied immediately to the left by the Info center and Help icons. New in this release, you can use the WebServices Button to sign in to your Autodesk account and access services that integrate with Revit. Consistent with most Windows applications, a status bar frames the bottom edge of the application frame. On the left side of the screen (usually docked/attached) are the Project Browser and the Properties palette stacked atop one another. The Properties palette allows immediate and ongoing access to the properties of any view or selected element in your Revit environment. The Project Browser can be thought of as the "table of contents" for your Revit project. It reveals all of the various representations of your project data—referred to in Revit as "Views." Views in Revit can be graphical (drawings, sketches, diagrams) or non-graphical (schedules, legends, takeoffs) and offer the means to both query your project (output) and to manipulate and edit it (input). The Project Browser also lists the Families, Groups, and linked files that reside in the project. When you open one or more of these views, they occupy view windows that can be tiled, cascaded, or maximized (full screen). Finally, stretched across the top of the screen just beneath the ribbon is the Options Bar. Options appear in this space as you work depending on the item or tool you have selected. (All of these items are labeled in Figure 2.1). Take some time to acquaint yourself with the various user interface elements. You can also right-click many items to see additional context-sensitive menus. Do not choose any right-click menu commands at this time. You can simply click away from the menu to close it or press the ESC key.

Recent Files

When you first launch Revit, the "Recent Files" screen will appear (see Figure 2.2). This is like a welcome page that offers you options to open or create Revit files. In the "Projects" area, you can use the Open link to open an existing Revit project. Use the New link to create a new Revit project from the default template. The same two links appear for the "Families" area. In addition, there is also a "New Conceptual Mass" item in Families. Open existing Family files to edit content in your library and create new ones to add to your library. Families will be discussed in detail in Chapter 10. Up to four of your recently opened files will appear in each area with preview icons. You can simply click the preview image to open that file.

FIGURE 2.2 *The Recent Files screen greets you when you launch Revit*

The third area is Resources. Here you will find links to "What's New," the Help system, and a getting started video showcasing various features of the Revit Architecture software. All of these resources are located on the Autodesk Web site, so Internet access is required.

Application Menu

File access and management tools are grouped under the Application menu (adorned by the large Revit "R" icon). Click on the big "R" to open the Application menu. If this is your first time launching Revit, the right side of the Application menu will be empty. But as you open and close files, the list of recent files will begin to populate. Revit remembers the last several files you had opened and shows them here. You can even click the pushpin icon to permanently "pin" a particular file to the menu, making it easier to load next time (see the left side of Figure 2.3). Right at the top of the Application menu are two icons to switch the list from Recent Documents to currently Open Documents. These icons are pointed out in the figure. If you switch to Open Documents and you have several project files and/or view windows open, you can use the Application menu to switch between open windows (see the middle of Figure 2.3). There is also an icon on the Quick Access Toolbar (see below) for this purpose.

On the left side of the menu, you will find commands like New, Open, Save, and Save As. Sub-menus on each item list the various file formats that can be opened or created. For example, a Revit project file contains all the work you do on your building projects. It is the primary file type in Revit and has an RVT file extension. A Revit Family file has an RFA extension and each Family file contains a single piece of Revit content that can be loaded and used in one or more projects. Revit can also open files with the ADSK and IFC extensions. Search the online help for more information on these file formats. When creating new Revit files, separate options are offered for Conceptual Mass, Annotations, and Titleblocks. While these are all Family files and will use the RFA extension, each is typically created from different template files and settings. Having the separate commands on the Application menu makes it easier to load these templates when creating such files (see the right side of Figure 2.3).

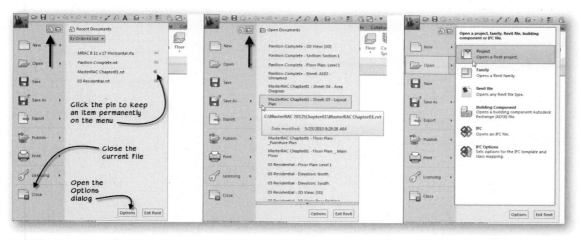

FIGURE 2.3 *The Application Menu*

On the Save As sub-menu, options will be available based on the current file you have loaded. For example, you cannot save a project file (RVT) as a Family file (RFA). Therefore, the Save As > Family option is grayed out when you are in a project. However, you can save a project as a new project template. Project Templates have a RTE extension and are discussed in detail in the next chapter. In addition to the options for saving, Revit files can be exported to a number of industry standard formats, such as DWG (AutoCAD drawing file), DGN (MicroStation drawing file), and IFC (Industry Foundation Classification).

At the bottom of the Application menu two buttons appear: Options and Exit Revit. Exit Revit is self-explanatory. Revit will prompt you to save any unsaved work. Use the Options button to open the Options dialog. This dialog has many program preferences that you can configure. Most of the out-of-the-box settings are suitable for the beginner. There may be some items that you or your CAD/BIM Manager will want to adjust. Refer to the "Settings" topic below and the online help for more information.

Quick Access Toolbar

The Quick Access Toolbar (QAT) as its name implies is a location for commonly used tools to which you wish to have easy and "quick access." The default QAT includes many common tools such as: Open, Save, Synchronize with Central, Undo, Redo, Close Hidden Windows, and Switch Windows. You can add buttons to the QAT with the menu on the right end of the QAT itself. For example, the New command is not part of the default QAT. Simply choose it from the pop-up menu to add it. At the bottom of this menu, you can also choose the Customize Quick Access Toolbar command to open a dialog with more options. In this dialog, you can rearrange tools on the QAT, add separators, and remove commands. For other commands not included on the list, locate them on the ribbon (see the next topic), right-click the tool, and choose **Add to Quick Access Toolbar** (see Figure 2.4).

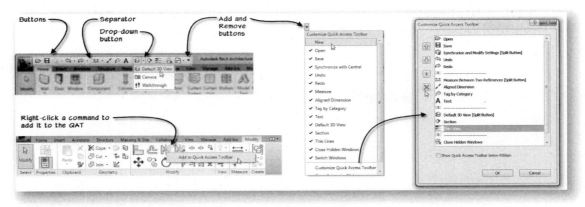

FIGURE 2.4 *The Quick Access Toolbar*

Ribbons

You issue commands in Revit by clicking their tools on the ribbon. The ribbon includes a series of tabs that appear just beneath the QAT. Each tab is separated into one or more Panels. Each Panel contains one or more Tools (see Figure 2.5).

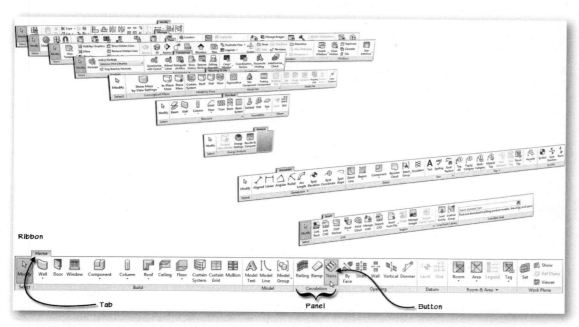

FIGURE 2.5 *A look at the Revit ribbon tabs*

To navigate the ribbon, click a tab, locate the panel and tool you need, and then just click the tool to execute a command. When tutorial instructions are given in this text, you will be directed first to the tab, then the panel, and finally the tool. For example, instructions to execute the Wall tool might look something like this:

On the Home tab of the ribbon, on the Build panel, click the ***Wall*** tool.

In the context of the exercise, when it is obvious which tab or panel, the description might be shortened to something like:

On the Build panel, click the ***Wall*** tool.

Or just:

Click the ***Wall*** tool.

Look to the online help for a description of each of the default ribbon tabs.

Contextual Ribbon Tabs

In addition to the default ribbon tabs, certain actions you perform in the software will cause other ribbon tabs to appear. These "contextual" ribbon tabs contain tools and commands specific to the item you are creating or editing. Contextual tabs will often be attached to the standard Modify tab. For example, if you select a Wall element in the model, a Modify | Walls contextual ribbon tab will appear. If you execute the Floor tool and begin creating a Floor element, a Modify | Create Floor Boundary tab will appear with the tools and options required to enable you to build a Floor. In both cases, the standard Modify tab and its tools will remain on the left side of the ribbon (see Figure 2.6).

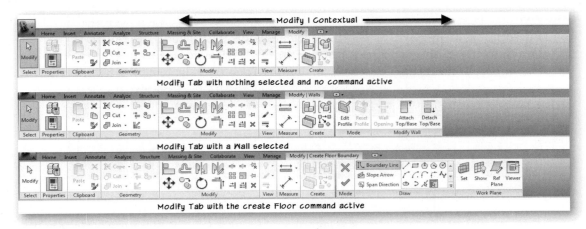

FIGURE 2.6 *Tear off ribbon panels and drag them anywhere you like on screen*

Please note that if you install any third-party add-on applications, you may also get an Add-Ins tab on your ribbon.

Panels

Ribbons are segregated into panels to help further classify and group the various tools. Panels simply group common tools and make locating the tool you need easier to accomplish. If you use a certain tool frequently, you can right-click on it and add it to the QAT as noted above in the "Quick Access Toolbar" topic. If you use all of the tools on a particular panel frequently, you can "tear off" the entire panel. This makes the panel into a floating toolbar on your screen. You can drag such a floating panel anywhere you like, even to a secondary monitor if you have one attached to your system. If you "tear off" any panels, Revit will remember the custom locations of the panels the next time you launch the application.

You can reset the QAT and all the custom positions of ribbon panels by deleting the *UIState. dat* file on your system. You can locate this file in the local profile folders on your system. Search the online help for the exact paths for Windows XP, Vista, and Windows 7.

If you tear off a panel and later wish to restore it, simply move your mouse over the floating panel. This will make gray bars appear on each side. On the left side, there is a drag bar that you can use to drag the panel around your screen to a new location. On the right side, there is a small icon that if clicked will restore the panel to its original ribbon tab and location.

Feel free to customize your interface by tearing off panels if you wish; however, all instructions in the tutorials that follow assume that panels are in their default locations on the ribbon tabs and refer to them as such.

You can only tear off panels on the permanent default ribbon tabs. Panels on contextual ribbon tabs cannot be torn off and left floating on screen. You can tell if a panel can be torn off by placing your mouse over the panel's titlebar. If the titlebar pre-highlights, you can tear it off.

On the panel titlebar (bottom edge of the panel), most panels simply show the name of the panel. In some cases, however, a small icon will appear on the right side of the title. This can be one of two icons. The left side of Figure 2.7 shows a "Dialog Launcher" icon. Clicking an icon such as this will open a dialog. Usually these are settings dialogs that you use to configure several options for a particular type of element.

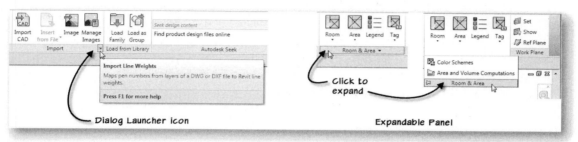

FIGURE 2.7 *Panel with a Dialog Launcher icon on the left and an expanded panel on the right*

On the right side of the figure an expanded Panel is shown. In this case, clicking panel title expands the panel temporarily to reveal additional related tools. Such tools are typically used less frequently than the ones always visible on the main panel. Expanded panels are not ideal, but provide a compromise to what would otherwise be overcrowded ribbon panels in those instances where utilized.

Ribbon View State

The ribbon has four viewing states. The default state shows the complete ribbon and panels. A portion of the top of the screen is reserved for the ribbon. Click the tabs to switch which tools display, but the same amount of screen space is used regardless of the current tab. This mode makes it easiest to see the tools but uses more precious screen space (see the top of Figure 2.8).

Three alternative states are available that use less screen space. The small icon to the right of the Manage tab is used to toggle to the next state. Click it once to switch to the "Minimize to Panel Buttons" state. In this state, the panel titles of the current ribbon are displayed as buttons; pass your mouse over a panel button to reveal a pop-up with that panel's tools. Move your mouse (shift focus) away from the panel and it will disappear. A similar mode is "Minimize to Panel Titles." This mode uses even less screen space by showing only the title of the panels on the current ribbon tab. Hover over them in the same way as Panel Buttons to reveal the hidden tools (see the middle of Figure 2.8).

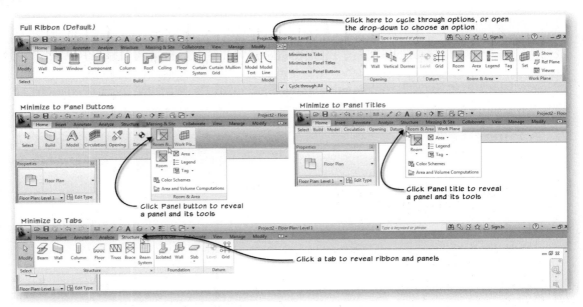

FIGURE 2.8 *Cycle through several minimized ribbon options*

The final display state shows only the ribbon tabs (see the bottom of Figure 2.8). Click on a ribbon tab to make the tab pop up. Like the panel titles state, if you shift focus away from a tab, it will disappear. It is easy to experiment with each mode and discover the one that you prefer. Simply click the toggle icon once to switch to panel titles, and click it again to switch to tabs. If you wish to return to the full ribbon, click it again. Each time you click, it toggles to the next state. You can also use the drop-down on the minimize icon to choose the mode you want directly.

NOTE   Instructions throughout this text use the full ribbon display mode. Adjust accordingly if you choose to use one of the minimized modes.

Tools

Ribbon panels contain tools. These tools will appear using one of three types of buttons. These are: Buttons, Drop-down buttons, and Split buttons. Examples of each type of button can be seen on most ribbon tabs. On the Home tab, examples of a button are the ***Door*** and ***Window*** tools (see the top-left of Figure 2.9). Clicking a button simply invokes that tool. In the case of either ***Door*** or ***Window***, the Modify | Place Door or Modify | Place Window tab will appear on the ribbon and you can configure and place the element in your model.

On the Model panel, the ***Model Group*** tool is an example of a drop-down button. In this case, if you click the tool, a drop-down list will appear showing the various options for the tool. In the case of the Model Group tool, we can choose from the ***Place Model Group***, ***Create Group***, or ***Load as Group into Open Projects*** tools (see the bottom left of Figure 2.9).

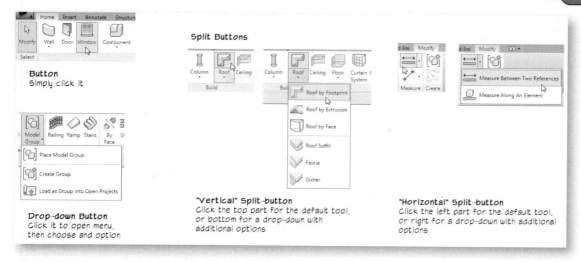

FIGURE 2.9 *Examples of each kind of button on the Home tab*

Split buttons can be either vertical or horizontal. They appear like the other buttons until you pass your mouse over them, at which point it will be clear that that only part of the button highlights under the mouse. The portion of the button with the small pop-up indicator (small triangle) behaves like a drop-down button. The other side behaves like a normal button. On the Home tab, the **Wall**, **Roof**, and **Floor** tools are examples of split buttons (see the right side of Figure 2.9).

Some tools will appear grayed out if the particular command is not available in the current context. For instance, if you have a sheet view active on screen, most tools such as **Wall**, **Door**, or **Roof** on the Home tab will not be available. There are plenty of other examples; Walls cannot be drawn in an elevation view and Levels cannot be drawn in a plan view. If a tool you want to use is grayed out, try opening a different view and try again.

User Interface Options

Several aspects of the user interface have customizable options. From the Application menu, click the Options button (shown in Figure 2.3 above) to open the "Options" dialog. Click the User Interface tab. The first item you can change is the theme of the interface. Revit offers a choice of Dark or Light. You can also use the controls on this tab to decide if the Recent Files screen should appear when Revit launches. If you prefer not to see Recent Files, uncheck this box. In the "Tab Display Behavior" area, you can decide what happens when you clear your selection or exit a tool. Revit will either stay on the Modify tab, or return to the tab you previously had open. Furthermore, you can even prevent Revit from switching to the Modify tab when you select objects onscreen. The author recommends that you select "Return to previous tab" and check the "Display the contextual tab on selection" checkbox (see the left side of Figure 2.10). You can of course choose alternate settings if you prefer.

When you pause your mouse over tools, a tooltip usually appears. Tooltips give you the name of a tool, its keyboard shortcut (in parenthesis), a short description, a long description, and often a descriptive image. Tooltip assistance can be configured to different levels of information. You can find settings to control how much tooltip assistance you want on the User Interface tab in the Tooltips area. If you choose **None**, no tooltip assistance will appear. Figure 2.10 on the top right shows an example of the tooltip you will receive with the tooltip assistance set to **Minimal**. If you choose **High**, you will get a tip like the one pictured on the bottom right of the figure. The **Normal** options will display a tip like the one on the top first, and then, after a few moments, the more detailed tip will appear.

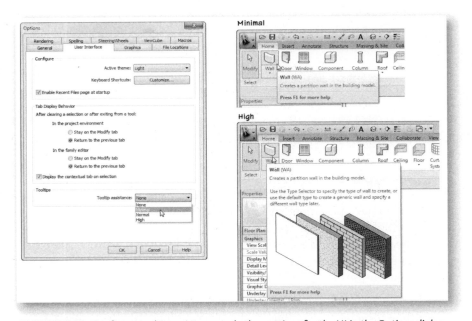

FIGURE 2.10 *Configure tooltip assistance and other settings for the UI in the Options dialog*

As we have seen, all Revit commands and functions are included on the ribbons. Keyboard shortcuts provide a way to execute commands without first locating them on the ribbon. They are simple keystroke combinations that can be typed as an alternative way to issue a command. To use a keyboard shortcut, simply type the two letters on the keyboard in succession. There is no need to press ENTER following the keystrokes. Figure 2.10 shows that the keyboard shortcut for the *Wall* tool is WA. This means you can invoke the Wall tool by typing WA instead of clicking the tool on the ribbon. Tooltip assistance can be a useful way to learn the keyboard shortcuts for your most frequently used tools and commands. In general, once you learn the shortcuts, they will usually be the fastest way to issue commands.

You can easily modify the existing keyboard shortcuts and add shortcuts to commands that do not already have them. To do this, click the Customize button on the User Interface tab of the "Options" dialog (you can also find it on the User Interface drop-down button on the View tab or keyboard shortcuts even has a shortcut of its own: just type KS). In the "Keyboard Shortcuts" dialog, you can scroll through all of the currently assigned shortcuts. In some cases like the Properties command, there will be more than one shortcut. All work the same, so choose the one you prefer. If you wish to add a new shortcut, locate the desired command and select it in the list. Next, type the desired shortcut in the "Press new keys" field and then click the Assign button (see Figure 2.11). If you wish to edit an existing shortcut, first select and then remove it, then assign a new one. Be careful not to assign a shortcut to a command that is already in use. Revit will use the first instance of a shortcut that it finds ignoring it for other commands.

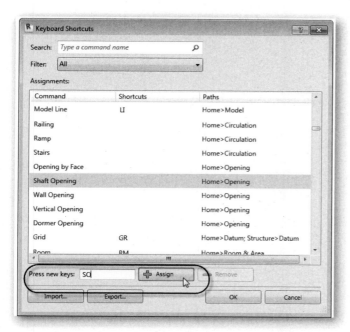

FIGURE 2.11 *Use the "Keyboard Shortcuts" dialog to add and edit shortcuts*

Another Windows convention supported by Revit is the ability to issue menu commands with the keyboard using the ALT key and a key letter combination from the desired command. To try this, press the ALT key. Doing so will place a small label on each tool and ribbon tab. Numbers appear on each of the tools on the QAT. Simply press this number to execute that command. Letters appear on each of the ribbon tabs. To invoke a tool on a tab, first press the letter for the tab. This will make a new set of letters appear on all the tools. Next press the key or keys shown on the tool. For example, to access the ***Align*** tool via the ALT key, press the ALT key, then the letter M, and then the letters AL. If a drop-down button is involved, use the arrows on the keyboard to choose the desired command and then press ENTER to complete the selection.

> Please note, even if the tab you want is current, when using the ALT key, you must still press the keystroke for that tab first.

NOTE

Project Browser

When you open a Revit project file, its contents will be displayed in the Project Browser. Think of the Project Browser as the table of contents for your project. It is the primary organizational tool for a Revit project. It is typically docked on the left side of the screen but you can also tear it off and place it anywhere you like (see Figure 2.12).

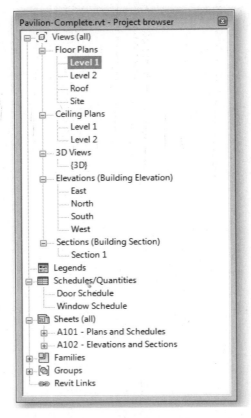

FIGURE 2.12 *The Project Browser (shown undocked)*

You can re-size and move the Project Browser depending on your screen resolution and space needs. Changes you make to the size and location of the Project Browser *will* be retained when you close Revit and launch a new session—even moving it to a secondary monitor! While it is possible to close the Project Browser, it is *not* recommended. Since the Project Browser is the primary means of interacting with and navigating between your project's views, closing it makes these functions difficult and inefficient. Should you inadvertently close the Project Browser, you can restore it by clicking View tab on the ribbon and then clicking on the User Interface drop-down button. A pop-up menu will appear from which you can select the Project Browser option to redisplay it (see Figure 2.13).

FIGURE 2.13 *Restore the Project Browser (or other items like Properties) if it is hidden*

Longer view names in the Project Browser can get truncated. It is therefore recommended that you widen the Project Browser window by stretching the right edge as much as your screen size will permit and to the extent of your own personal preferences. This will make it easier to read the full names of your views. However, if you are not able to widen the Project Browser or you prefer not to, you can simply hover your mouse over a view to see a tooltip of its full name (see Figure 2.14).

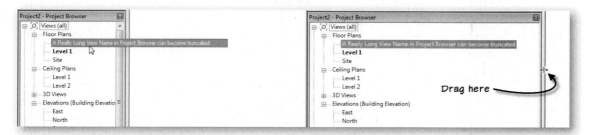

FIGURE 2.14 *Partially hidden names can be seen with tooltips or you can widen the Browser window*

As already noted, you can also tear the Project Browser away from the side of the screen, undocking it, and place it anywhere on screen you like. This has the effect of making your drawing window larger while allowing you to make the browser a more convenient size as well. The Project Browser enables you to view and work with two distinct parts of your project: views of your model and elements within your model. Views are listed at the top in the *Views*, *Legends*, *Schedule/Quantities*, and *Sheets* branches. In the previous chapter, we explored this part of the Project Browser and its organization options. Please refer to that discussion for more details. To interact directly with elements within your model from the Project Browser, you will work within the *Families*, *Groups*, and the *Revit Links* branches of the Project Browser tree. Every Family, Group, and Revit file link in a project is listed among these items. From these branches, you can edit existing items, or even create new ones (see Figure 2.15).

FIGURE 2.15 *Understanding the various branches of the Project Browser*

Right-click options are sometimes available on each node and sub-branch of the Project Browser. Please take a moment to right-click on each item and study the menus and commands that are available. If you right-click an item and no menu appears, this simply indicates that there are no commands associated with that node or branch of the Project Browser. New in this release, *Legends, Schedules/Quantities, and Families* all sport right-click options. You can learn more about Groups and Links in Chapter 6. Families are explored in detail in Chapter 10.

In the previous chapter, we explored the *Views* branch of the Project Browser within the context of a provided dataset. Here we will reiterate certain behaviors that have more to do with understanding the user interface associated with Project Browser. Beneath the *Views* branch of the tree, certain categories will appear automatically as various views are added to the project. For instance, a *Floor Plans* branch is automatically created to group the various floor plan views of your project. Likewise, "*Elevation*" and "*Section*" branches will appear as their respective view types are added. If you delete all views in a particular category, the branch for that category will also disappear. This behavior is maintained automatically by the software.

The name of the currently active view will appear **bold** in the Project Browser. You can open any view by simply double-clicking on its name in the Project Browser. This is the most common function of the Project Browser. In addition, you can right-click views in the Browser to access commands specific to that item (see below for more on right-clicking). It is also important to note that more than one item may be selected on the Project Browser at the same time. To select multiple views, select the first view, then hold down the CTRL key and select additional views. You can also use the SHIFT key to select everything between any two items. Once you have more than one view selected, you may right-click on any of the highlighted view names to access a menu that will apply to the entire selection of views. This might be useful if you wish to make a global change such as applying a View Template to several views at once (see Figure 2.16).

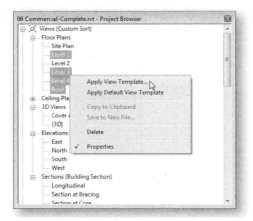

FIGURE 2.16 *Select multiple items in the Project Browser with the* CTRL *or* SHIFT *keys*

Another important aspect of the Project Browser is its presentation of the project's sheets. Although sheets are technically classified as views in Revit they have some unique properties not shared by other views. They are also presented in their own branch of the Project Browser. Expand the *Sheets* branch to see the sheet views in your project. Since sheets can actually have other views placed on them, a plus (+) will appear next to the sheet name. Expanding this will reveal the names of the views that have been placed upon the sheet. A sheet appearing in the Project Browser without a plus sign indicates that the sheet does not yet contain references to any other views. (The dataset from Chapter 1 provides a good example of each condition.) Like other views, you can open a sheet view by double-clicking on its name in the Project Browser. In addition, you can open referenced views placed on sheets by double-clicking their names beneath the expanded sheet name (see Figure 2.17). New in this release, Schedule views and Legends placed on sheets now appear indented beneath their sheet names as well.

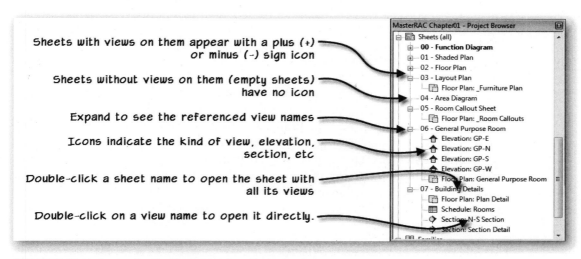

FIGURE 2.17 *Understanding the Sheets branch on the Project Browser*

When using these techniques to navigate your project, recall from the previous chapter our discussion of changing the way that the Project Browser is organized. In that exercise, we noted how we could organize our views by discipline, phase, or even by those not yet placed on sheets. Look back to Figure 1-9 in Chapter 1 and the accompanying discussion for more details.

When you choose to organize the Browser using the "Not on Sheets" type, you do not loose anything. You simply change the way things are displayed. If you wish to work on a view that is already on a sheet, you first expand the sheet on which it is placed, and then double-click the view name beneath it. If the view is not yet on a sheet, it will be listed beneath the *Views* branch at the top of the Browser instead. Once you add a view to a sheet, it will disappear from the *Views* branch and appear instead beneath its host sheet. If you wish to experiment with this, try reopening the dataset from Chapter 1 and explore further.

PROPERTIES palette and the TYPE SELECTOR

The Properties dialog is a modeless palette. "Modeless" is a computer term that refers to a persistent interface element that does not rely on a "mode". In other words, the Properties palette will stay onscreen as you work and does not need to be closed in order to continue working. You simply interact with the properties you wish to edit and either click the Apply button or simply shift focus away from the palette to accept the changes and continue working. This makes work more fluid with fewer disruptions from having to go in and out of modal (non-persistent) dialogs.

You can edit element properties on the palette both when you are creating new elements and when you select existing elements. Like the Project Browser, the Properties palette can be docked or floating. By default, it appears docked on the left side of the screen above the Project Browser. Every Element in Revit, from Model Elements like Walls and Floors, to Annotation Elements like text, tags, and Levels, and even views themselves can be edited via this palette. Most elements also have type properties. Editing the Type Properties affects all elements in the model of that type. The Properties palette will allow access to instance properties of any selected element, new elements as they are being created, and the current view properties. To switch between the properties of the selected elements on screen (including the view properties), use the drop-down on the Properties palette. If you wish to edit the Type properties of a selected element, click the Edit Type button on the Properties palette (see Figure 2.18).

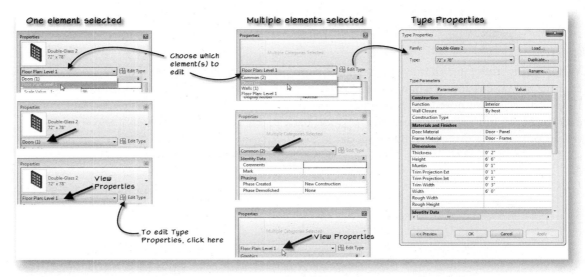

FIGURE 2.18 *The Properties palette can edit instance properties of elements or views. Click Edit Type to access Type Properties*

At the top of the Properties palette, we have the Type Selector. The Type Selector is a drop-down list of Types available in your project for the kind of item you are creating or editing. Families have at least one, but can have several Types. Think of a Type as a "saved variation" of a particular Family. For example, a double flush door might have several standard size configurations. Each such variation would be saved as a separate Type that we could select from the Type Selector list. Whenever you create elements, you will choose an appropriate Type from this list. You can also use the Type Selector to change the Type of elements already in the model. Simply select an element and then choose a different Type from the list (see the left side of Figure 2.19).

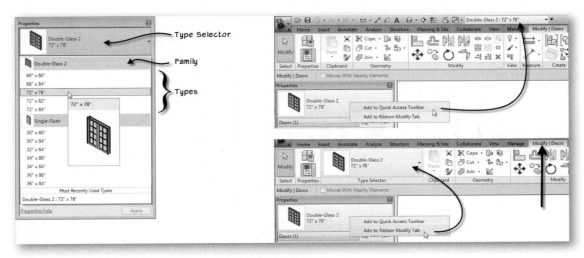

FIGURE 2.19 *The Type Selector appears at the top of the Properties palette*

New in this release, you can right-click the Type Selector and clone it to the Quick Access Toolbar (QAT) or the Modify tab of the ribbon. Both of these options are shown on the right side of Figure 2.19. Feel free to use these options if you like, but instructions throughout this text will assume that you are accessing the Type Selector from its default location atop the Properties palette.

You will interact with both the Type Selector and Properties palette when creating or modifying elements. When you are creating an element, you first choose the tool for the item you wish to create on the ribbon. Next choose a Type from the Type Selector and if necessary edit other properties on the Properties palette. If you are modifying an existing element, simply select the element (or elements) and then choose the desired Type from the Type Selector and/or modify the desired parameters on the Properties palette and then click Apply. If you wish to modify all elements of a particular type globally across the entire project, click the Edit Type button.

When you select multiple elements, if they all share the same Type, the Type Selector will display the name. If the selection of items is the same kind of element (all Doors, for example) but are not currently the same Type, then the Type Selector will remain active, but will display a message reading "Multiple Types Selected" or "Multiple Families Selected." If you select multiple elements and the Type Selector reports "Multiple Categories Selected," this indicates that you have made a selection of dissimilar elements (like a Wall and a Door), and you cannot make changes to them using the Type Selector. You may be able to edit other Properties, but the options available on the Properties palette will be limited only to those properties that the selection of elements shares (examples of this can be seen in Figure 2.13).

Options Bar

The Options Bar runs horizontally across the screen and is located directly below the ribbon. The Options Bar displays options available for elements you are creating. On occasion, some options will also display for existing elements you have selected as well (see Figure 2.20).

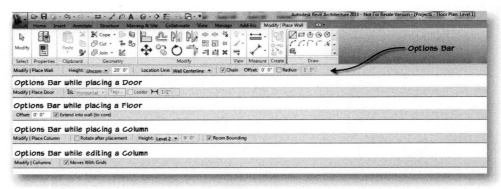

FIGURE 2.20 *Examples of the Options Bar when creating or editing various model elements*

You will interact with the Options Bar regularly as you work in Revit. Keep this in mind and always remember to look there for context-specific options relating to the task at hand. Remember to also look at the contextual ribbon tabs as well. In many cases, you will need to configure options on both the Options Bar and the contextual ribbon.

The Canvas (View Window Display Area)

The most interactive portion of the interface is the canvas area where you work with the views of your building model. When you open a view, it appears in a window in the canvas area. At least one view window must be open to work in a project. You can however open several views for a project at one time. If the view windows are maximized, you can see which are opened on the Application menu (click the Open Documents icon shown in Figure 2.3 above) or using the *Switch Windows* tool on the View tab or on the QAT. You can also tile (type WT) or cascade (type WC) the windows onscreen. These commands are also on the View tab. If you have many views open at once, tiling can make the actual windows very small and hard to work with. To prevent this, you can cascade them or maximize them instead. Be sure to close windows when you no longer need them. An easy way to do this is to maximize the current view window and then choose *Close Hidden* tool on the View tab or QAT.

Status Bar

The Status Bar is the gray bar along the bottom edge of the Revit application frame. If you glance down at the Status Bar, you will notice a constant readout of feedback appears there. In some cases, the information provides prompts and clues as to what actions are required within a particular command; in other cases, the feedback may simply describe an action taking place or describe an element beneath the cursor (see Figure 2.21 for examples).

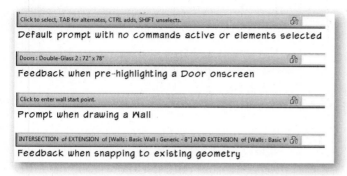

FIGURE 2.21 *The Status Bar provides ongoing feedback and guidance as you work*

In many cases, the same or similar feedback is available on screen in the form of tool tips. The extent to which tooltips appear is controlled by a setting that you can modify. You can opt for a high level of tooltip prompting, a moderate level, or none at all. To edit the degree of Tooltip Assistance, choose **Options** from the Application menu and edit the "Tooltip Assistance" item on the User Interface tab. You cannot edit the Status Bar messages in any way.

In addition to the Status messages, the Status Bar includes two other optional controls: Status Bar – Worksets and the Status Bar – Design Options. These two controls provide shortcuts to these two Revit features. Worksets is the collection of tools that enable multiple users to access the same Revit model simultaneously. You can learn more about them in Chapter 15. Design Options provides a mechanism to consider alternative design schemes in the same Revit model. To learn more, please visit the online help.

View Control Bar

Every graphical view window has a View Control Bar located at the bottom edge of the window. When windows are maximized, this will appear directly above the Status Bar adjacent to the horizontal scroll bar. If the view windows are tiled, it will appear in the lower left corner of the view window (see Figure 2.22). The View Control bar serves two purposes—it displays at a glance the most common view settings of the window and provides a simple and convenient way to change them if required.

Like so many other aspects of Revit, the View Control Bar is context sensitive. Notice the two bottom windows in Figure 2.22 have different View Control Bars

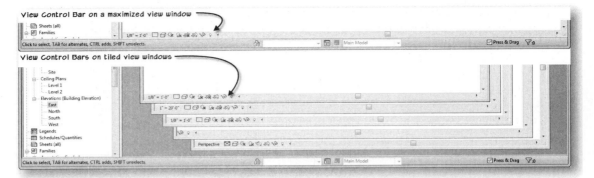

FIGURE 2.22 *The View Control Bar maximized (top) and tiled (bottom)*

than the others (the lower one says Perspective and does not show scale, the other only has the Hide/Isolate settings). The one with only the hide/isolate settings is a sheet view and sheet views do not have as many settings available as other views. Perspective views naturally do not have scale. Click on any of the icons on the View Control Bar to access a pop-up menu of available choices. For example, to change the scale of a view, click on the scale item, or pick the Detail Level icon to change the level of detail in which the view is being displayed (see Figure 2.23).

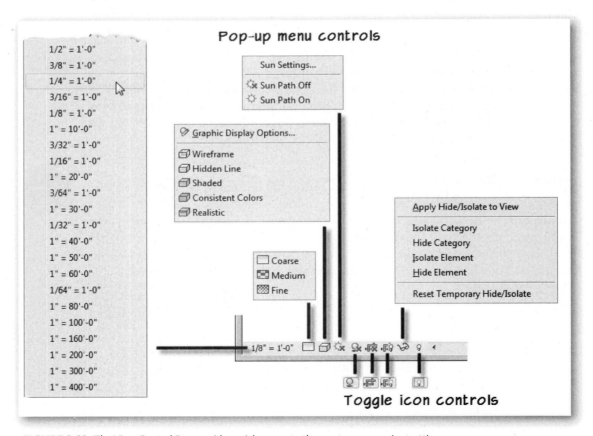

FIGURE 2.23 *The View Control Bar provides quick access to the most common view settings*

All of the settings, except Hide/Isolate, on the View Control Bar are also accessible from the Properties palette. To access these settings in Properties, choose the name of the current view from the pop-up on the Properties palette (examples shown in Figure 2.18 above).

The last two icons on the right are used to hide and show elements in your model to make it easier to see things as you work. There are two ways to hide objects. You can use the icon on the Temporary Hide/Isolate View Control Bar (looks like sunglasses) to hide elements temporarily. You can also hide elements permanently. To use the temporary Hide/Isolate tools, select an element or elements in the model and then click the Hide/Isolate icon and choose an option. The **Hide Element** option temporarily makes the selected element(s) invisible. The **Isolate Element** option leaves the selected elements visible and hides everything else. There are also options to hide or isolate the entire category of objects based upon the category of objects you have selected. For example, if you select a Door in the model and then choose **Hide Category**, all Doors in the view will temporarily hide. When you use the temporary Hide/Isolate command, a cyan border will appear around the current view window with a label in the upper left corner until the mode is disabled.

The important thing to know about using the Temporary Hide/Isolate function in a view is that the changes it makes to the view window will not be saved outside the current work session. Furthermore, if you were to print or export the current view, Hide/Isolate settings will be ignored. If you want to hide an object permanently on screen or when printing, use the "Hide in View" command on the right-click menu (or the Modify ribbon) instead. If you wish to make the Hide/Isolate settings permanent, select the Temporary Hide/Isolate icon and choose Apply Hide/Isolate to View from the pop-up.

To reveal hidden elements, both temporary and permanent, you click the Reveal Hidden Elements toggle icon (light bulb on the far right). If there are temporarily hidden elements in the current view, this mode will reveal them in a cyan color. Permanently hidden elements will be revealed in maroon. A maroon border will appear around the current view with a label in the left corner of the window as long as this mode is active.

Remember also that while the View Control Bar shows the most common View Properties, using the Properties palette will give access to all view properties—many more than on the View Control Bar.

RIGHT-CLICKING

Like most Windows software, in Revit, you can right-click on almost anything and receive a context-sensitive menu. In fact, we have already seen examples of this in the previous chapters and in this chapter. Let's take a few minutes here to explore some of the more common right-click menus.

Right-click on the Quick Access Toolbar—As mentioned above, you can customize the QAT. To do so, use the small menu icon on the right side of the QAT or simply right-click on any tool (see Figure 2.24).

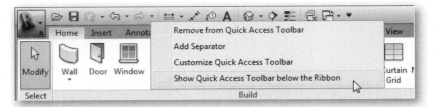

FIGURE 2.24 *Right-click on the QAT to customize it*

Right-click on a particular button on the QAT if you want to remove it. You can also add a separator which is a small line between buttons to help visually group them. The separator appears after the button you right-click. You can also move the QAT to below the ribbon if you prefer. Do this if you wish to add many tools to the QAT as you will have more space to do so if it is below the ribbon. However, the trade-off is that it will reduce the drawing area slightly.

Right-click to add to the QAT—If you wish to add a tool to the Quick Access Toolbar, locate it on the ribbon and right-click. You will see an option to add it to the Quick Access Toolbar, AT if available for that command (see Figure 2.25).

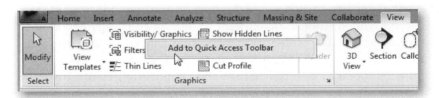

FIGURE 2.25 *Right-click on the Ribbon to change display states*

Right-click on the Options Bar—One choice appears when you right-click on the Options Bar. You can have it docked at either the bottom or the top. The choice on the right-click toggles between the two as appropriate. This is a personal preference and otherwise makes no change in the Options Bar behavior or size on screen. Throughout this text it will reamain in the default position at the top.

Right-Click Items in Project Browser—Items on Project Browser often have right-click options as well. Context menus are not available for every branch of the tree. Examples of where you can right-click can be seen in Figures 2.15 and 2.16 above. You can always right-click directly on a view name to receive a context menu of options for that view (or views). Several such examples have already been discussed.

Right-Click in the View Window—When you right-click directly in the view window, you get a different menu depending on whether there are any objects selected or not. If there are no objects selected, the menu contains basic commands (see the left side of Figure 2.26). This menu includes items like: zooming and scrolling commands, repeat recent commands, and the Find Referring Views command. The "Find Referring Views" command is particularly useful. When choosing this command, a dialog will appear like the one shown on the right side of Figure 2.26.

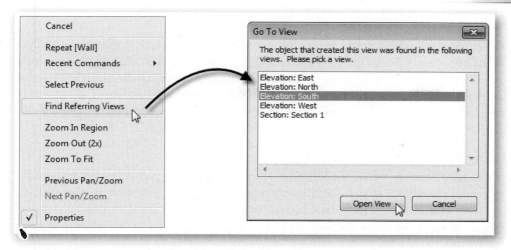

FIGURE 2.26 *Find Referring Views locates all views that refer to the current view*

You can choose any view listed in the dialog and then click the Open View button. The figure illustrates an example of running the command from a plan view. If you run the command from a section or elevation view, a similar dialog will appear listing different views including plans and reflected ceiling plans.

When you right-click with one or more objects selected on screen, you get a different menu than if there is no selection active. However, some commands appear in either case. One such command is **Properties**. This is simply another way to hide or show the Properties palette. If Properties appears with a checkmark next to it, then the palette is displayed. If it appears with no checkmark, then it is hidden (see Figure 2.27).

FIGURE 2.27 *Properties is visible if the checkmark appears*

NAVIGATING IN VIEWS

When working with models on a computer screen, you need more than the standard Windows scroll bars to navigate your model. You will need to change the magnification of the model and frequently scroll other parts of the model into view on the limited screen space available. The act of changing the magnification of the screen is referred to as "zooming," and moving the image on screen within the borders of the view window to see parts off screen is referred to as "panning."

Using a Wheel Mouse to Navigate

If you have a mouse with a middle wheel button, Revit provides zooming and scrolling using the wheel! If you don't have a wheel mouse, this might be a good time to get one. This modest investment in hardware will pay for itself in time saved and increased productivity by the end of the first day of usage. Using the wheel you have the following benefits:

- To **Zoom**—Roll the wheel. Roll up to zoom in (closer), roll down to zoom out (farther away). Move the pointer before you roll to control the center of the zooming.
- To **Pan**—Push and hold the wheel down and then drag the mouse.
- To **Dynamic Zoom**—Hold down the CTRL key and then drag the wheel (this is similar to rolling the wheel, except the zoom is centered).
- To **Dynamic Orbit**—Hold down the SHIFT key and then drag the wheel (3D Views only).

Navigation Bar

The Navigation Bar includes a small toolbar with navigation commands and the ViewCube. The Navigation Bar and ViewCube were both introduced in the previous chapter. Rather than reiterate their function in detail here, you are directed to the "View Navigation" topic in Chapter 1 for more information. You can also find a useful video in the online help that showcases the common functions of the ViewCube and Navigation Bar. New in this release, Revit now supports 3D Connexion devices. If you have one of these "3D Mice" you can install the driver and use it to navigate your 3D views in Revit. An icon will appear with options on the Navigation Bar if such a device is installed.

Zoom

In addition to the wheel mouse, Revit provides a few other zooming methods. Access these commands from the right-click menu or from the zoom icon on the Navigation Bar (pictured in Figure 1-12 in Chapter 1).

Zoom In Region—This command will enlarge an area of the model that you designate by dragging a box around the area on screen.

Zoom Out (2x)—This is also a zoom (or reduction in magnification) command that simply reduces the size of the image on screen by half.

Zoom To Fit—The function of this command is to fit the entire extents of model and any annotation objects into the available view window space on screen. A similar command is available on the Navigation Bar: **Zoom All to Fit**. This command performs the "Zoom To Fit" command in all of the open view windows (when tiled or cascaded) rather than just the active one.

Previous Pan/Zoom—You can backtrack through your navigation changes on screen using this command. Each time you choose it, the screen will go back one step to a previous zoom level or pan location.

Next Pan/Zoom—Works the same as Previous Pan/Zoom except that it will move forward in the progression of view changes. This command is only available following the use of Previous Pan/Zoom.

Previous and Next Pan/Zoom are similar to the Rewind function on the Steering Wheel except that it is not as interactive. The Steering Wheel and Rewind functionality was discussed in the "View Navigation" topic in Chapter 1. In addition to the right-click menu and Navigation Bar, don't forget that each zoom command also has keyboard shortcuts (as seen in Table 1-4 in Chapter 1).

SELECTION METHODS

Before elements can be modified in Revit, they must be selected. There are various methods used to select elements, some of which are similar to other software. When modifying elements in a model, the basic workflow is as follows:

1. Select the Element(s) to be manipulated with the Modify tool (mouse pointer).
2. Issue a command to perform on the selection, usually by clicking a tool on the ribbon or typing a Keyboard Shortcut like MV for Move.
3. Indicate on screen where and how to perform the command.

To summarize this workflow, choose (select) *what* you want to modify and then indicate *how* you want to modify it—"to this selection of elements I wish to perform this action."

Creating a Selection Set

A selection set is one or more selected elements. As you move the Modify tool around a view window, you will notice that any elements available for selection will temporarily highlight while under the cursor—this is known as "pre-highlighting." The purpose of pre-highlighting is to preview what will be selected if you click the mouse. This is particularly helpful when working in a complex model with many elements close together. Use the pre-highlighting (and often the TAB key—see below) as a tool to assist you in accurate selection.

The simplest way to select an element is to click on it. When you select an element it will display in light blue on screen—this indicates that the element(s) is selected. Once selected, an element will remain selected until you deselect it. You can deselect elements in three ways: selecting another element will automatically deselect the current selection set (unless you hold down the CTRL key) and become the new selection set. You can deselect all elements without creating a new selection set by clicking on a blank portion of the screen (where there are no elements) or by pressing the ESC key. (You can also click the Modify tool on the ribbon.)

To create a selection set containing more than one element, use the following techniques:

- Hold the CTRL key down while clicking on another element. The new element will be added to the current selection set (and highlight light blue as well).

- Hold the SHIFT key down while clicking on a selected element (highlighted light blue) to remove the element from the current selection set (it will no longer be highlighted in light blue).

> **TIP**
>
> If you accidentally pick an object without holding down the CTRL key (or if you accidentally click in the white space), you will lose your selection set (which will be replaced with only the one element just picked or nothing if you click in white space). You can restore your previous selection set by right-clicking in the view window and then choosing **Select Previous**. You can also select previous by holding down the CTRL key and then pressing the LEFT ARROW key.

Either action will restore your previous selection without having to start all over again. New in this release, selected elements also appear shaded and semi-transparent. Selection and pre-selection colors and other settings can be customized on the Graphics tab of the Options dilaog.

Selection Boxes

Even with the CTRL and SHIFT keys, making selections of multiple elements can be time consuming. The easiest way to create a large selection set is by using a selection box. To create a selection box with the Modify tool, click down with the left mouse button next to an element, hold the button down and drag a rectangular box around the elements you wish to select. Elements will pre-highlight as you make the selection box.

The direction in which you drag the selection box determines which specific elements are selected. If you create your selection box by dragging the Modify tool from left to right on the screen, the edge of the selection box will appear solid as you drag and only elements *completely* within the box will be selected. If you create your selection box by dragging from right to left on the screen, the edges of the selection box will appear dashed and any element completely or partially included in the box will be selected (see Figure 2.28).

You can always combine methods—first create a selection box, and then use the CTRL and SHIFT keys to add and remove from the basic selection.

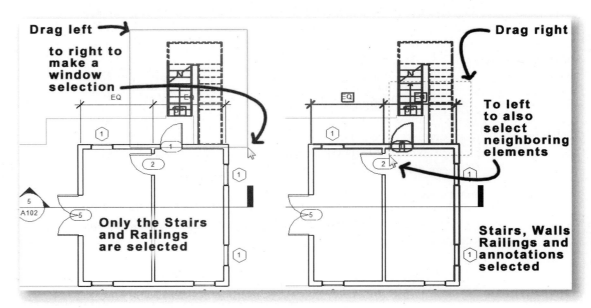

FIGURE 2.28 *Selection boxes vary with the direction you drag*

Filter Selection

Another approach to building a large selection set is to deliberately select too many elements and then use the *Filter* tool to remove items of categories that you designate from the selection. The *Filter* selection tool is located on the Status Bar and on the ribbon. Remember, when using the *Filter* selection tool, you always start with a selection that includes *more* than the items you want, and you then filter *out* the undesired elements by category (see Figure 2.29).

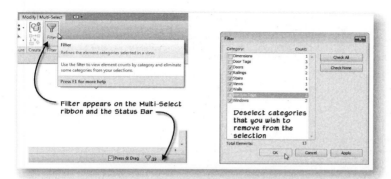

FIGURE 2.29 *Use the Filter Selection dialog to remove categories of elements from the selection set*

Each of the selection methods will require a little practice to become second nature to you. Take the time to practice each one since selection is such an important and frequent part of nearly all tasks in Revit.

THE ALMIGHTY TAB KEY

You will find that the TAB key is probably the most used and most useful of all the modifier keys on the keyboard when working in Revit. As mentioned in the previous topic, being able to quickly add or remove from a selection set can greatly increase your efficiency during a work session. In general, you can think of the TAB key as a "toggle switch," allowing you to cycle through potential selections. When you attempt to pre-highlight a particular element on screen, sometimes a neighboring element will pre-highlight instead. No matter how subtly you move your mouse in an attempt to capture the desired element, it often proves difficult or impossible to highlight the right one. The TAB key provides the solution to this situation. Use it to cycle through adjacent or stacked items to highlight the particular element you desire.

"When in doubt, press TAB."	**TIP**

While there are dozens of examples of using the TAB key in a Revit Architecture session, a few examples are presented here that will give you an idea of where using the TAB key proves most handy.

TAB to Pre-Highlight Elements for Selection

If the tool, whether it is the Modify, Dimension, Split, or another tool, is near two or more Elements in a view window, you can either move your mouse around the view to pre-highlight the different objects for selection, or use the TAB key to pre-highlight each element in succession. Each time you press TAB, a different element will pre-highlight until all items have been cycled through. Once you have tabbed through all elements, the cycling will repeat (see Figure 2.30).

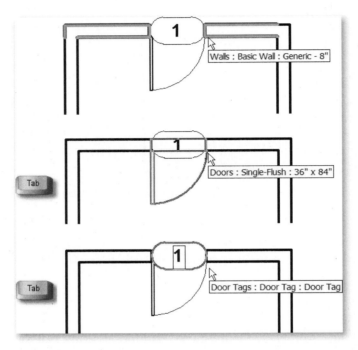

FIGURE 2.30 *Use the* TAB *to cycle through and pre-highlight nearby elements*

When the element that you wish to select is pre-highlighted, click the mouse as normal to select it. This is important. Do not forget to click once the items you want to select are highlighted.

TAB During Dimensioning

When you are using existing geometry for reference points such as when adding dimensions, using the align command or tracing a background, you can use the TAB key to cycle through possible reference points (see Figure 2.31).

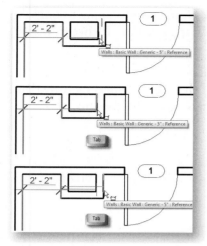

FIGURE 2.31 *Use the* TAB *to cycle through dimension reference points*

TAB for Chain Selection

Chain selection is a very powerful feature of Revit. Chain selection works with Walls or Lines. Like an actual chain, chain selection highlights all of the Walls or Lines that touch one another end to end. To make a chain selection, you first pre-highlight a single Wall or Line. Next you press the TAB key and in so doing, any walls or lines that form a chain (i.e., are located end-to-end) with that wall or line are pre-highlighted together. Finally, you click the mouse to make the selection with a single mouse-click (see Figure 2.32). Don't forget to click when the chain pre-highlights. If you don't click, when you move your mouse away nothing will be selected.

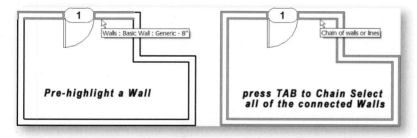

FIGURE 2.32 *Use the* TAB *key to chain select a collection of Walls*

There are dozens of additional ways to use the TAB key in Revit. One of the easiest things to do is simply try tabbing in various situations while you work in the software. Other examples will be presented throughout this book in the tutorials that follow. For more discussion on the topic, consult the Revit Architecture online help.

SETTINGS

While not strictly related to interface, this section covers some of the settings you can use to manipulate your overall Revit preferences. Most of the following settings apply to your Revit Architecture application (global) in general (in other words, you can set it once and then forget about it regardless of the project you open) and in other cases the setting is saved with the project file and can vary from one project to the next.

Examples of global settings include the Username, the Tooltip Assistance, or the Path settings. Any of these settings applies to your installation of Revit Architecture not to a particular project file. On the other hand, Units settings apply to only the current project file. So if you change the Project Units from Feet-Inches to Inches, this would apply only to the current project file.

There are many Options and Project Settings available. What follows here is not a comprehensive look. We will only look at a few of the more common settings that will directly impact your experience working through this book's lessons.

The Options dialog General Tab

From the Application menu, click the **Options** button to open the "Options" dialog and view or edit the settings therein. The dialog is divided into several tabs, each containing various options (see Figure 2.33).

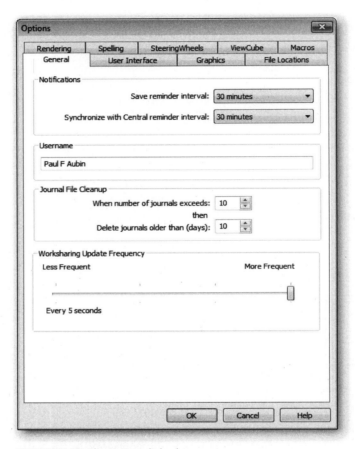

FIGURE 2.33 *The Options dialog box*

Save Reminder Interval—The default setting is 30 minutes. This means that every 30 minutes during your work session, Revit will prompt you to save your file. You can change the increment here to suit your preference.

TIP	Regardless of your Save Reminder frequency, you should make it a habit to save your work often.

Synchronize with Central Reminder Interval—The default setting is also 30 minutes. Central files are used in team environments when multiple users need to work in the same project model simultaneously. Synchronizing with central frequently is as important as saving your local file frequently. Refer to Chapter 15 for more information on Central files.

You should set both reminders at an amount of time that will be helpful in reminding you to save, but not so frequent as to become disruptive to your work.

Username—By default, this is set to your Windows login name. The only time this is important is if you are working with Worksharing and Worksets. Worksharing and Worksets enable you to work with teams on the same project. Refer to Chapter 15 for more information on using Worksharing. The important thing to remember about Username is that Revit will only allow you to open and work in a Workset-enabled local file if your username matches the username that was active when that file was saved.

Journal File Cleanup—Journal files are created by Revit as a troubleshooting tool. Should you experience problems with your project file, Autodesk tech support may ask you to send them your journal files. A new journal file is created each time you launch Revit. The settings in this section control how many previous journal files are kept.

Worksharing Update Frequency—New in this release, this feature provides visual cues onscreen when items have been modified in a Worksharing project. Refer to Chapter 15 for more information.

The User Interface tab was discussed earlier in this chapter.

Graphics

If you have a video card capable of hardware acceleration, you can configure Revit to use hardware acceleration on the Graphics tab. You can also change the colors used for selection, pre-highlighting, and alerts. If you are unsure about your computer's hardware specifications, you should consult your IT support person before making changes on the Graphics tab.

When you have temporary dimensions appearing on screen, the size of the text is fixed and does not adjust as you zoom in or out. This is unlike the behavior of permanent dimensions which have a size and scale set in final plotting units and therefore do adjust as you zoom in or out. On the Graphics tab, at the bottom, you can change the overall text size for all temporary dimensions in your project. If you are finding them difficult to read, increase the value here. You can also make the background opaque if you wish so that the text covers model elements. Leave it transparent if you do not wish to have the dimensions obscure the model. A Size value of about 12 or 14 is a good everyday setting.

File Locations

The File Locations tab of the "Options" dialog is where you configure the hard drive and/or network locations of the template and library files used by Revit Architecture.

Default template file—Project Template files are discussed in detail in Chapter 4. Use the setting here to point to your firm's preferred project template file. This might be an out-of-the-box template provided by Autodesk or a custom one developed by your CAD or BIM Manager.

Default path for user files—If you have a specific location on your computer that you typically keep your project files, you can browse to that location here. The folder written here will be used by Revit when the Open dialog is used. If you are working in a team environment, using the Worksharing functionality, Revit will automatically save local files to this location. You can learn more about Worksharing and local files in Chapter 15.

Default path for family template files—Family Template files are discussed in detail in Chapter 10. Use the setting here to point to your firm's preferred location for Family template files.

Places—This is a very handy feature. Click the Places button to see a list of Places. Each entry in this list will show up as an icon on the left side of the Open and Save dialog boxes. In this way, you can jump from one library location to another with a single click.

TIP	Add a Library pointer for the folder in which you installed your dataset files from the student companion. In this way, you can navigate to this location quickly at the start of each tutorial.

To add a path, click the Places button to open the "Places" dialog. Click the Add Value icon (plus) on the left side. In the Library Name field, type a description for the place such as "MRAC 2012." Click in the Library Path field and use the browse icon to point to the location where you installed the dataset files from the student companion files (see Figure 2.34).

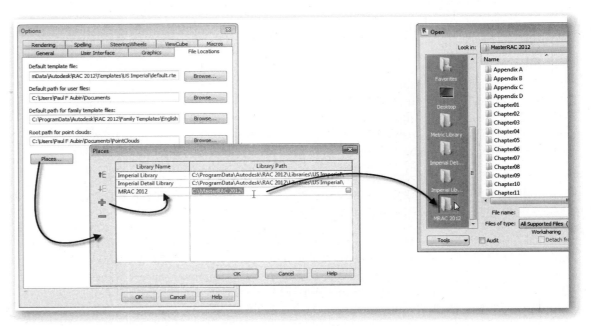

FIGURE 2.34 *Add a path in the "Places" dialog for the Book dataset files*

Click OK twice to close the dialog. To test it out, click the Open tool on the QAT. You should see your new icon on the left side among the other places.

There are several other tabs in the Options dialog. There are some user-configurable options on the Steering Wheels and ViewCube tabs that you may wish to explore. However, the default settings are a good place to start and offer the best experience for most users. It is recommended that you leave these settings at the default values for now. The same is true for Spelling. Your CAD Manager or IT support person may have configured some options under Rendering and Macros. For most beginning users, you do not need to configure anything here. Check with your CAD or BIM Manager or IT support person for more information.

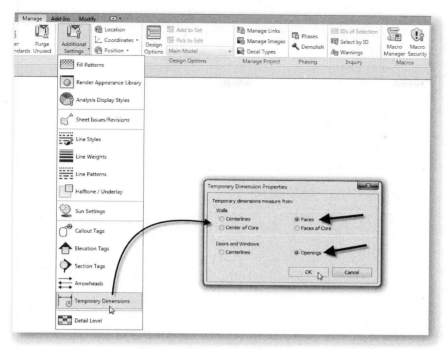

FIGURE 2.35 *Configure temporary dimension behavior*

Temporary Dimensions

Although you can toggle the reference point used by temporary dimensions using the TAB key, it is often more efficient to change the default behavior instead. Defaults for Temporary Dimensions can be configured for Walls, Doors, and Windows. Depending on your preferences, you can have the Temporary Dimensions default to either the centers (default) or edges of objects. Click the Manage tab of the ribbon. On the Settings panel, click the ***Additional Settings*** tool. From the drop-down menu that appears, choose **Temporary Dimensions** (see Figure 2.35).

Choose your preference for both the "Walls" and "Doors and Windows" settings. Revit Architecture defaults to centerlines for both. For Walls, you can dimension faces, centerlines, and/or Wall Cores. For Doors and Windows, choose between centerlines and openings. The author recommends Faces for Walls and Openings for Doors and Windows. This setting is saved with the project file. So you will need to check it in each new project.

Snaps

Many project settings are accessible from both the Settings drop-down button and directly on the Manage ribbon. Feel free to look at any of the others. We will end our explorations in this chapter with a look at the "Snaps" dialog. The Snaps button is on the Manage tab on the Settings panel. Revit includes dimension snaps and object snaps. Dimension snaps adjust with your zoom level on screen. As you zoom out, the snap increment becomes larger. As you zoom in, it becomes smaller. You can customize the increment upon which it adjusts at the top of the dialog for both length and angles. In the middle of the dialog, all of the object snap modes are listed. You can turn them on and off to suit your preferences. They are all on by default. You do not need to change anything in this dialog, but do become familiar with all of the items it contains. In particular, make note of the keyboard shortcuts next to each object snap. These are two character codes in parenthesis (see Figure 2.36).

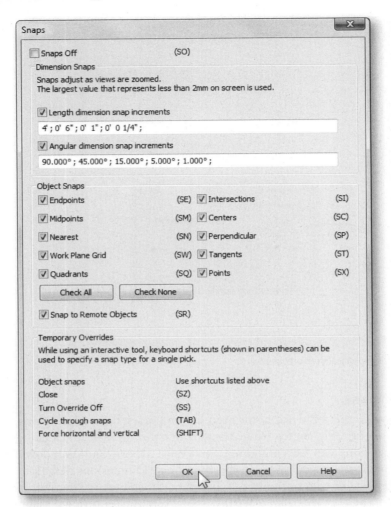

FIGURE 2.36 *Study the settings in "Snaps" and make note of the keyboard shortcuts for object snaps*

As you work on your models, you can type these shortcuts as needed to force Revit to use a particular object snap. This amounts to an override that lasts for one click of the mouse.

SUMMARY

Now that you have completed the Quick Start Tutorial, read the high-level overview of Building Information Modeling in Chapter 1, and have been guided through an overview of the user interface in this chapter, you are now ready to begin creating your first Revit projects. In the chapters that follow, we will build two projects from scratch: one residential and one commercial. In this chapter, you have learned:

- The User Interface of Revit Architecture uses many common Windows conventions and some unique ones as well.
- The ribbon along the top of the screen contains several Tools organized into Tabs.
- Each ribbon tab is divided into panels.

- Tools come in three varieties: buttons, drop-down buttons, and split buttons.
- The Options Bar changes to reflect the current tool or command.
- One or more views of the project can be opened at one time in the Workspace area. They can be tiled, cascaded, or maximized.
- The Project Browser, along the left side by default, contains all of the views of the project as well as the Families, Groups, and Linked files. (Project Browser is like a "table of contents" for your project).
- The View Control Bar can be used to manipulate the display settings of the active view.
- The Status Bar along the bottom of the screen shows prompts for the current command.
- Many common commands are available from the context-sensitive right-click menus.
- The easiest way to navigate a view is using the wheel of your mouse and the appropriate modifier keys.
- Roll the wheel to zoom, drag the wheel to pan, hold the SHIFT key, and drag to orbit the 3D model.
- Most editing commands first require a selection set.
- Select elements with the modify tool (mouse pointer), use the CTRL key to add to the selection set, and the SHIFT key to remove from the selection set.
- Drag left to right to select all elements within the selection box, right to left to select all those in contact with the box.
- The TAB key is used to cycle through potential selections when more than one is possible.
- Use the TAB key to "Chain" select Walls or Lines. (Select Walls or Lines touching end to end.)
- Configure settings in the "Options" and "Temporary Dimensions" dialogs to suit your personal preferences.
- Settings in "Options" persist from session to session regardless of the open project; settings in "Temporary Dimensions" apply only to the current project.

Create the Building Model

Section II

The first step toward reaping the benefits of Building Information Modeling is to construct a virtual building model in the software. The tools provided for this purpose in Revit are many and varied. In this section we will explore the many building modeling tools available such as Walls, Doors, Windows, Columns, Beams, Stairs, Railings, Roofs, Floors, and Curtain Walls. In Chapter 10, we will take a detailed look at Revit Family Components and the Family Editor.

Section II is organized as follows:

Creating a Building Layout

INTRODUCTION

Revit Architecture can be used successfully in all types of architectural projects and within all project phases. Most of our attention in this book is given to the design development and construction documentation phases. The tutorial exercises in this book will explore two building types concurrently, starting at different points in the project cycle. This will give a sense of the multiple ways you can approach the design process while also keeping some variety in the lessons. Don't feel limited to the techniques covered here. The aim of the tutorials is to get you thinking in the right direction. Exploration is highly encouraged and in so doing, you may discover equally valid ways to achieve the same end results.

Architectural projects start in many different ways. You may be commissioned to design a new building from its early conception all the way through construction and beyond, or you may be asked to join a project already in progress as a member of a larger team. Likewise, design data comes in many forms. If the project is new construction, you may have only the site plan and some other contextual data. If the project is a renovation, or if you join a project already in progress, you may use or be given existing design data in formats that vary from traditional hand-drawn sketches to digitally created images and models saved in all manner of software formats.

The first two chapters were intended to get you comfortable with the theoretical underpinnings of Revit Architecture. Now that you have the correct mind-set and a level of comfort with the user interface, get ready to roll up your sleeves—it is time to begin creating our first model.

It should be noted that the major focus of this book is on the design development and construction documentation project phases of an architectural project. Therefore, in this book we start our models directly in the Revit project environment beginning with Wall layout. However, Revit also includes a complete conceptual modeling environment where you can design your building form as a series of masses and forms. Massing studies can then become the basis for a building design as your project moves from conceptual design to design development. The conceptual design environment offers an excellent way to perform conceptual design studies during the early schematic design phases of the project. While we will not begin our projects in this chapter with conceptual massing, if you would like to learn more, you can find an introduction to the conceptual environment in Chapter 16.

OBJECTIVES

As noted above, we will explore two building types in this book. One is a residential renovation project and the other is a new small-scale commercial office building. Throughout the course of the hands-on tutorials in this chapter, we will lay out the existing conditions for our residential project. We will explore the various techniques for adding and modifying Walls, Doors, and Windows. In addition, we will add plumbing fixtures and other elements to make the layout more complete. After completing this chapter, you will know how to:

- Add and modify Walls and work with Temporary Dimensions
- Explore Wall properties
- Add and modify Doors and Windows
- Assign Phase parameters to model components
- Add plumbing fixtures
- Add an In-Place Family

WORKING WITH WALLS

Most building objects are found on the Home tab of the ribbon. Beneath the ribbon is a small space called the Options Bar (see the "Options Bar" topic and Figure 2.20 in Chapter 2) utilized by certain commands and functions. We will begin by working with Walls. You can add Walls point by point or enable the "Chain" option to create them in series (each segment beginning where the previous one ended). Walls will "join" automatically with intersecting Walls at corners and intersections. Walls have many parameters such as length, height, and type; can have custom shapes and profiles; and can receive (i.e., host) Doors and Windows and automatically create openings for them.

Basic Wall Options

Since Walls are the basic building blocks of any building, we will start with them. We will create a new temporary project and sketch some Walls to get comfortable with the various options. Later we will create one of the actual projects that will be used throughout the rest of the book.

Create a New Project

We will begin our work in a new project created from the Revit Architecture default template file.

1. Launch Revit Architecture from the icon on your desktop or from the *Autodesk > Revit Architecture* group in *All Programs* on the Windows Start menu.

TIP	In Windows 7 or Vista, you can click the Start button, and then begin typing **Revit** in the Search field. After a couple letters, Revit Architecture should appear near the top of the list. Click it to launch to program.

The Recent Files screen should appear. Let's be sure that we are starting from the out-of-the-box default template.

2. From the Application menu (big "R"), choose **New > Project**.

    - In the New Project dialog, in the "Template File" area, be sure that the lower radio button is selected (the one that lists a path and a template file).
    - Verify that the *default.rte* [*DefaultMetric.rte*] template file is listed in the text field.

The default template file name and location for the Imperial template file is:
C:\ProgramData\Autodesk\RAC 2012\Templates\US Imperial\default.rte
The default template file name and location for the Metric template file is:
[C:\ProgramData\Autodesk\RAC 2012\Templates\US Metric\DefaultMetric.rte]

The path indicated is for Windows 7 and Vista; in Windows XP, the folder varies slightly. Also the path may vary in countries outside the United States.

If the template shown on your screen does not match the location and name listed above, click the Browse button to locate it before continuing.

NOTE

If you are in a country for which your version of Revit Architecture does not include these template files, they have both been provided with the dataset files from the student companion for *The Aubin Academy Master Series: Revit Architecture 2012*. Please browse to the *\MasterRAC 2012\ Templates* folder in the location where you installed the dataset files to locate them.

- In the "Create New" area, verify that the Project radio button is selected and then click OK (see Figure 3.1).

FIGURE 3.1 *Create a new project based upon the default template file*

Depending on your system's settings, the same result can normally be achieved by simply clicking the NEW link beneath the Projects heading on the recent files screen. This "New" link automatically creates a new project using the default template file. To configure the default template file, click the Options button on the Application menu. Click the File Locations tab and then edit the location in the Default template file field. For more information on the Options dialog, refer to the "Settings" heading in Chapter 2.

Getting Started with Walls

As discussed in the previous chapter, the ribbon (which is the primary interface for nearly all Revit commands) can be shown in four formats: The full ribbon, minimized to buttons, minimized to panel titles and minimized to tabs. While you may prefer to use one of the minimized options, instructions and images throughout this text will assume you are using the full ribbon. (Please refer to the "Ribbon View State" heading in Chapter 2 for more details.) Nearly every building component in Revit Architecture is created following the same basic procedure. Locate the tool for the object you wish to create on the ribbon. Use the Options bar, Properties palette and contextual ribbon tabs to configure the settings of the object as you create it. Use your mouse to create the object in the current view window.

3. On the Home tab of the ribbon, on the Build panel, click the **Wall** tool (see Figure 3.2).

FIGURE 3.2 *The Wall tool on the Home tab of the ribbon*

A new "Modify | Place Wall" tab will appear on the ribbon, several settings will appear on the Options Bar, the Properties palette will change to reflect the Wall you are drawing, the pointer will change to a crosshair cursor and the Status Bar prompt will read: "Click to enter wall start point."

- At the top of the Properties palette, from the Type Selector, choose ***Generic - 8"*** [***Generic – 200mm***].

  If the Properties palette is not open on screen, you can open it in a variety of ways. You can right-click and choose Properties or click the Properties tool on the ribbon or type the shortcut P P.

The following list explains the major fields and controls that you can manipulate while adding a Wall. You can find these items on the Properties palette, on the Options Bar or both (see Figure 3.3).

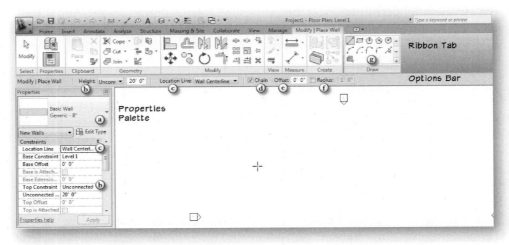

FIGURE 3.3 *Options available while placing Walls*

a. **Type Selector**—Use this drop-down to choose from a list of Wall types in the current file. The specific list is populated by the template we used to create the project. Always choose a type from this list before configuring other options.

Walls are a System Family. System Families have built-in parameters predefined within the software that are not editable by the end user. (They typically represent building components that are constructed on-site like walls, floors or roofs as opposed to those created in a factory or shop and delivered to the site and installed like doors and light fixtures.) System Family instances cannot be edited outside of the Project Editor. Furthermore, the System Family itself cannot be renamed, deleted, or edited. System Families do have types like component Families. User edits are permitted to the System Family's types. The list of choices you see on the Type Selector is determined by the template used to create the project file. In order to have customized System Family types available automatically to new projects, you must add and edit them in your project template files. You can create your own project templates by choosing the "Project Template" radio button when creating a new project. You can also transfer Wall types, other System Family types and general system settings from existing projects to your current project by opening both projects and then choosing the Transfer Project Standards command on the Project Settings panel of the Manage ribbon tab. For more information, search for "Transfer Project Standards" in the online help.

b. **Height**—There are two options for Height; a drop-down list of parametric height options and a text field (available when "Unconnected" height is chosen). When you choose "Unconnected" you are able to simply type in a fixed height for the Wall. The heights of Walls can also be connected to the project's levels instead. (Our project created from the default template has a Level 1 and Level 2.)

c. **Location Line**—The point within the width of the Wall that is used as a reference. Choices include: Wall Centerline, Interior and Exterior Finish Faces and several "Core" options. The Core of the Wall will be discussed in detail in later chapters. Location Line is useful when changing a Wall to a type whose width is larger or smaller.

d. **Chain**—This option creates Walls in a sequence automatically joined end to end. If you deselect the Chain option, you will need to indicate the start and end points of each Wall segment you draw. With Chain enabled, the end of your first Wall will automatically become the start point of the next Wall and so on. This option is selected by default.

e. **Offset**—Use this field to input a dimension value. The Walls you create will be placed parallel (offset) to the points you click at a distance equal to this value.

f. **Radius**—This option can be used to create rounded corner joins as walls are created. Place a checkmark in the box before the Radius field becomes available. Input the desired radius in units.

g. **Draw tools**—Walls can be drawn on-screen in a variety of shapes or created from existing model components. Choose to draw Walls line by line, as a rectangle, circle, arc, etc. The Pick Lines option will allow you to pick existing linework on screen and create Walls from those lines. The Pick Faces option is similar except that it creates Walls from the faces of 3D geometry.

Now that we have an overview of the available options, let's see some of them in practice.

Verify that you selected *Generic - 8"* [*Generic – 200mm*] from the Type Selector above and that the "Line" icon is selected in the Draw panel. Accept the defaults for all remaining options.

4. Click anywhere on screen to place the first point of the Wall.

Move your mouse to the right, keep it horizontal, but don't click yet (see the top of Figure 3.4).

Notice the dimension on screen as you move the mouse. As you move your pointer, this dimension automatically snaps to whole unit increments. Depending on the size and resolution of your screen, the exact increment may vary. For example, if you are using Imperial units, the increment is likely 4'-0" and the increment for Metric is likely 100mm.

As you zoom in, the increment of the dimension will reduce. Depending on the unit type you are using, you may need to zoom in or out to see this. This is easy to do with the wheel. Simply roll the wheel up or down to zoom in or out. If you zoom off screen, keep the command active, press and hold the wheel button in and then drag. This will pan the screen. You can also access zoom commands on the Steering Wheel (F8) or the right-click menu.

5. Roll the wheel of your mouse up a few clicks to zoom in a bit and continue to move the pointer left or right (see the bottom of Figure 3.4).

NOTE   If you do not have a wheel mouse, you might want to consider purchasing one. You can use the Steering Wheel instead of the wheel, but the wheel makes quick zooming much easier. Refer to Chapters 1 and 2 for more information.

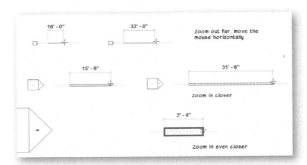

FIGURE 3.4 *The dimension increment varies with the level of zoom (The exact values on your screen may not match the figure)*

- Click to set the other point of the Wall. (Keep it horizontal for now, but the length is unimportant.)
- A single Wall segment will be created.

Often you will want to create more than one Wall segment, each beginning where the previous one ended. You can achieve this with the "Chain" option on the Options Bar.

6. On the Options Bar, verify that there is a checkmark in the "Chain" checkbox to enable this option.

- Click any point on screen to begin placing the next Wall.
- Click two more points at any locations on screen (see Figure 3.5).

Notice that the corner where the two Walls meet has formed a clean intersection (called a "Wall Join").

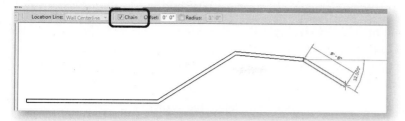

FIGURE 3.5 *Use the Chain option to create continuous Wall segments*

7. Click the Start-End-Radius Arc icon on the Draw panel.

The Start-End-Radius Arc tool first places each of the endpoints and then adds the intermediate point last to define the degree of curvature. Other Arc options are also available.

- Move the pointer in any direction and click to place the end point of the Arc.
- Following the cue at the Status Bar, click to place the intermediate point of the Arc.

Continue to experiment by adding additional Wall segments as desired.

8. To finish adding Walls, click the **Modify** tool on the Selection panel of the ribbon (or press the ESC key twice).

Think of these few Walls as a simple "warm up" exercise. Drawing them helped us explore some of the basic Wall options. However, these Walls have been placed a bit too randomly to be useful. Let's delete them now and create some new ones.

Using a Crossing selection box

Before we can delete the existing Walls, we need to select them. There are many ways to select objects in Revit. Several methods were explored in the "Selection Methods" topic in Chapter 2. As we saw in that topic, a convenient way to select multiple objects at one time is with the Window and Crossing selection methods. In either technique, you create a box by dragging two opposite corners with your mouse. To create a Window selection, click and drag from left to right. To make a Crossing selection, click and drag the opposite direction—from right to left. A Window selection selects only those items completely surrounded by the box, while a Crossing selects anything touched by or within the box. This was discussed in brief in Chapter 2. Let's review it now.

9. Click a point below and to the right of the Walls you have on screen.

- Hold down the mouse button and drag up and to the left far enough to touch all objects with the dashed (Crossing) selection box (see Figure 3.6).

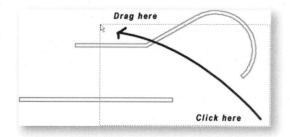

FIGURE 3.6 *Make a Crossing selection by dragging from right to left*

- When all of the Walls highlight, release the mouse button.

All of the Walls will turn blue to indicate that they are now selected. Once you have a selection of objects, you can manipulate their properties, move, rotate, or mirror them (notice that these and other tools appear on a Modify | Walls tab of the ribbon) or you can delete them.

10. Press the DELETE key on your keyboard to delete the Walls.

Adding Walls with the Rectangle Option

When you draw Walls, tools are available to create Walls that form closed geometric shapes like rectangles, circles, and polygons. Often using these is the quickest way to lay out such regular shapes. Let's try the rectangle option now.

11. On the Home tab of the ribbon, click the **Wall** tool.
    - Verify that **Generic - 8"** [**Generic – 200mm**] is chosen for the Element Type.
    - Accept the defaults for Height and Location Line and then click the Rectangle shape icon on the Draw panel (see Figure 3.7).

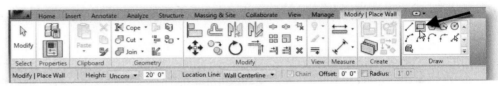

FIGURE 3.7 *Using the rectangle draw option for Walls*

Our project contains four elevation markers. Let's zoom the screen to fit these to the screen so they are visible.

12. On the keyboard, type **ZF**.

NOTE   If you prefer, you can also choose **Zoom To Fit** from the *Zoom* tool on the Navigation Bar. However, where available, the keyboard shortcuts like the one suggested in the previous step are usually quicker once you learn them. Shortcuts for zoom commands are listed in Table 1.4 in Chapter 1.

The mouse pointer will show a small rectangle next to it indicating that we are in rectangle drawing mode.

13. Click a point within the upper left region of the space surrounded by the elevation markers.
    - Move the mouse down and to the right and watch the values of the dimensions.

14. Click a point in the lower right region of the space surrounded by the elevation markers.

    Notice that the Temporary Dimensions continue to display on the Walls just drawn. Furthermore, these dimensions appear in blue. The blue color indicates that their values may be edited dynamically; or put another way, blue means "interactive."

15. Click on the blue numerical value of the horizontal dimension.

    The value will become an editable text field.

- Type a new value into this field and then press ENTER. (The exact value is unimportant—see Figure 3.8).

The value on your screen will vary from that shown in the figure.	NOTE

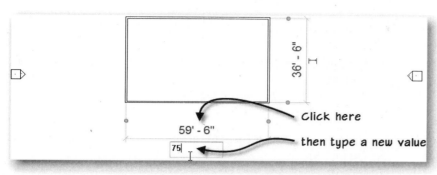

FIGURE 3.8 *Use the Temporary Dimensions to edit the locations of the Walls*

16. The Wall command is still active. Using the same technique, draw another rectangle overlapping the first.

Notice that the Temporary Dimensions now reference points on the first rectangle to points on the new one. You can edit these values in the same way that we edited the ones above. You can also move the witness lines of the dimensions to gain more control over their exact locations. We will explore this technique below. If you wish, try some of the other sketch shapes like circle or polygon. In the case of the polygon, additional controls will appear on the Options Bar to control the quantity of sides.

17. When you are finished exploring Walls, choose **Close** from the Application menu to close the current project. When prompted to save, choose No.

Create an Existing Conditions Layout

Now that we have practiced adding a few Walls and seen some of the options available while doing so, let's begin creating an actual model. In this book we will follow two projects from the early schematic phase through to the construction document phase. We will start with the first floor existing conditions for our residential project. The residential project is an 800 SF [75 m²] residential addition. This project will require a little bit of demolition and new construction and will require plans, sections, elevations, details, and schedules. In this project we will explore the Phasing tools in Revit—Demolition, Existing and New Construction will be articulated later in the tutorial. This will give us the required separation between construction phases of the project. The completed files for the residential project are available in the *Chapter03\Complete* folder with the dataset files installed from the student companion. You can open the completed version at any time to compare it to your progress.

Create a New Residential Project

We will use the same template file that we used above to begin our residential model.

1. Create a new project file using the *default.rte* [*DefaultMetric.rte*] template file as we did in the "Create a New Project" heading above.

Be sure that the *Level 1* floor plan view is open onscreen. You can see this indicated in bold (under *Views (all) > Floor Plans*) on the Project Browser and in the title bar of the Revit Architecture window.

2. On the Home tab of the ribbon, click the **Wall** tool (shown in Figure 3.2 above).

   - From the Type Selector (at the top of the Properties palette), choose **Generic - 12"** [**Generic – 300mm**].
   - For Height choose **Unconnected** and set the value to **18'-0"** [**5500**].

TIP	If you are using Imperial units, simply type **18** and then press ENTER. No unit symbol or zero inches is necessary. Also note that to enter a value like 18'-2 1/2" you can type it in several ways. For example, you can type 18' 2 1/2" or 18' 2.5" or 18 2 1/2 or 18 2.5 or 18.20833. Refer to the "Unit Conventions" topic in Chapter 2 for more information.

   - For the Location Line, choose **Finish Face: Exterior** and then click the Rectangle sketch icon (see Figure 3.9).

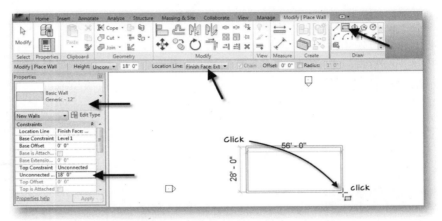

FIGURE 3.9 *Set the Options for the exterior Walls of the residential project*

3. Click two opposite corners on screen within the space bounded by the elevation markers (the exact size is not important for initial placement).

Notice that even though we have chosen a Location Line of Finish Face: Exterior, the Temporary Dimensions still have witness lines at the centerlines of the Walls. This is simply a default behavior independent of the Wall's individual location line setting. If you want to input a value for the dimension based upon the face of the Walls, you can simply move the witness lines.

4. Click on the small blue circle handle on one of the horizontal dimension's witness lines. Zoom into the small blue circle to see this better.

   Notice that the witness line moves to one of the Wall faces. If you click it again, it will move again, this time to the opposite face. One more click returns it to the centerline (see Figure 3.10).

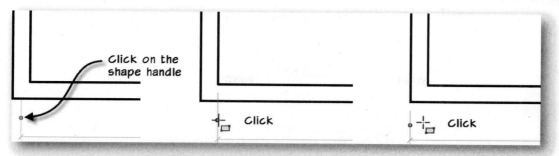

FIGURE 3.10 *Move the witness lines of the Temporary Dimension*

- Click the witness line shape handle (of the horizontal dimension) until both sides reference the outside edges of the rectangle.
- Click the blue numeric value of the horizontal dimension, type: **33'-0"** [**10000**] and then press ENTER.

5. Repeat this process on the vertical dimension making the outside face to face dimension equal to: **24'-0"** [**7300**] (see Figure 3.11).

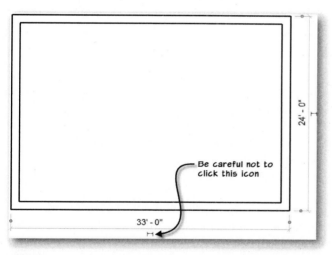

FIGURE 3.11 *Edit the size of the rectangle to match the desired outside dimensions*

| If your witness lines are in the wrong locations when you edit the Temporary Dimension, simply repeat the process to move the witness lines and then repeat the dimension edit process. | **TIP** |

| Be careful not to click the small "permanent" dimension icon when editing the dimension values. Clicking this icon will make the Temporary Dimension a "permanent dimension" (it will remain in the current view even after the associated Wall is deselected). If this happens, simply select the dimension and delete it. | **CAUTION** |

6. On the Select panel, click the ***Modify*** tool, or press ESC twice to complete the operation.

You should now have four walls in a rectangular configuration measuring 33'-0" × 24'-0" [10000 × 7300] outside dimensions.

NOTE	If you need to modify the dimensions after deselecting them, you can reselect any Wall and the Temporary Dimensions will reappear. However, you will need to select one Wall at a time to make edits. Select one Wall, edit witness lines if required, and then type in your desired distance in the Temporary Dimension. Repeat in the other direction. Also, keep in mind that the element you select is the one that will move when editing the dimensions.

• From the Application menu, choose **Save As > Project.** In the Save As dialog, navigate to your *Documents* folder (*My Documents* in Windows XP.)
• For the File name, type: **03 Residential** and then click the Save button.

Offset Walls

Let's begin adding the interior partitions. There are several techniques that we could employ to do this. In this sequence, we will use the ***Offset*** tool. The ***Offset*** tool moves or copies the selected objects parallel to the original by an amount that you input. This tool is located on the Modify tab of the ribbon.

7. On the Modify tab of the ribbon, click the ***Offset*** tool.

TIP	The shortcut for offset is **OF**.

The Options Bar for Offset has two modes: Graphical and Numerical. When you choose Graphical, the numeric input field is disabled and you use a Temporary Dimension on screen to indicate the distance of the offset. When you choose the Numerical option, you input the offset distance first, and then use the pointer on screen to indicate the side of the offset. In both cases, you can enable the "Copy" checkbox, which will create a copy of the object as it offsets. If you disable this option, the object you offset will be moved parallel to itself by the amount you indicate.

• On the Options Bar, verify that Numerical is chosen (if it is not, choose it now).
• In the Offset field on the Options Bar, type **12'-5 1/2" [3794]** and verify that "Copy" is selected.

The mouse pointer will change to an Offset cursor and the Status Bar will prompt for a selection.

• In the model canvas move the Offset cursor over the upper edge of the bottom horizontal Wall.

A dashed line will appear indicating the location of the Offset. If you move your mouse up and down this line will shift up and down as well (see Figure 3.12).

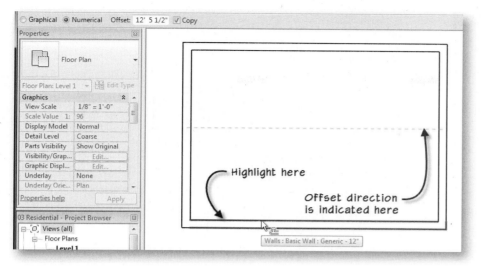

FIGURE 3.12 *Offsetting a new wall*

As in most places in Revit, if you press the TAB key, the pre-highlighted selection will cycle to the next available option. In this case, all four Walls will pre-highlight and the offset result would be a concentric ring inside or outside the building. Go ahead and try it if you like. Press the TAB key once, and then move the pointer to indicate outside or inside. The green dashed line will indicate a rectangle now instead of a single edge. If you actually click to create the Walls, be sure to undo (Quick Access Toolbar) after this experimentation.

- When the dashed line appears above the Wall (inside the house) click the mouse.

A new Wall will appear inside the house.

- Click the **Modify** tool, or press the ESC key twice to complete the operation.

Copying Walls

We can also copy Walls. You select the Wall to copy first and then click the **Copy** tool.

8. Select the vertical Wall on the left.

    A Modify | Walls tab will appear on the ribbon.

    - On the Modify panel, click the **Copy** tool.

| The keyboard shortcut for Copy is **CO**. | **TIP** |

It should be noted that many editing tools in Revit allow you to first click the tool and then select the element(s) that you wish to edit. In this case, you would click the **Copy** tool on the Modify tab first. Then following the prompt on the Status Bar, you would select the Wall (or Walls) you wish to copy and press ENTER to complete the selection process. The rest of the command sequence would be the same.

With the **Copy** tool you input the distance of the offset on the screen. At the Status Bar, a prompt reads "Click to enter move start point."

- Click a point anywhere on screen. (The exact start point is not important, but for convenience, you can click directly on the Wall). Begin moving the Wall to the right, type **12'-8 1/2"** [**3818**] and then press ENTER (see Figure 3.13).

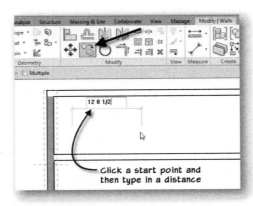
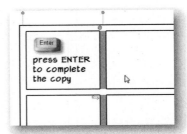

FIGURE 3.13 *Copy a vertical Wall inside*

9. On the Select panel, click the ***Modify*** tool or press the ESC key twice.
10. Save your project.

Changing a Wall's Parameters

One of the features that makes Revit such a powerful tool is the ability to easily change an object's parameters at any time, as design needs change. Let's take a look at modifying some of the walls as we continue with the layout of the first floor existing conditions for the residential project.

11. Select the two internal Walls created in the previous steps.

To select the Walls, you can either click just inside the house near the lower right corner and then drag up and to the left until both Walls highlight or you can hold down the CTRL key and click each of the Walls one at a time (see Figure 3.14).

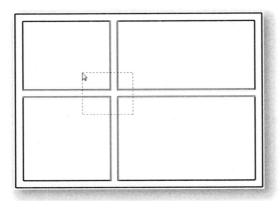

FIGURE 3.14 *Select the interior Walls*

NOTE The Properties palette should be open onscreen already. (The default location is docked to the left side of the Revit application frame.) If it is not, there are several methods to open it: click the Properties tool on the Modify tab of the ribbon, type the shortcut: PP, press CTRL + 1 or right-click and choose **Properties**.

12. On the Properties palette, click on the Type Selector at the top.

- Choose **Generic – 5″** [**Interior - 135mm Partition (2-hr)**] from the Type list.
- Beneath the Constraints grouping, choose **Wall Centerline** for the Location Line (see Figure 3.15).

> **TIP**
>
> To make this change, click in the Location Line field, and then click the small down arrow that appears to choose the Wall Centerline option from the popup menu.

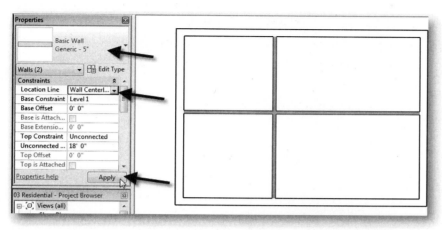

FIGURE 3.15 *Change the parameters of the interior Walls on the Properties palette*

To apply such a change, you either click the Apply button on the Properties palette, or simply shift focus away from the palette by moving your mouse away from the palette. This will apply the change automatically.

13. Shift focus away from the Properties palette or click Apply to accept the changes.

- Notice the change to the Walls.

Let's verify that everything is in the correct place. As noted above, we are laying out the existing conditions of our residential project. Therefore, the locations of the Walls that we just placed are based on field dimensions. The room in the lower left corner of the plan is supposed to be 11'-6″[3500] × 11'-9″[3658]. We can verify these dimensions and easily make adjustments as necessary.

Adjust Wall Locations

In the "Create a New Residential Project" heading above, we learned to use the Temporary Dimensions and the handles on the witness lines to edit the location of Walls. We can use the same technique to verify our dimensions now.

14. Click on the interior horizontal Wall.

The Temporary Dimensions will appear indicating its current location relative to the other Walls. However, depending on the settings of your system, the dimensions may be from the centerlines. The desired size of the room that we are verifying are to the inside faces of the Walls (as you might expect from field dimensions). Click the "Move Witness Line" grip controls as shown above in Figure 3.10 to move the witness lines to the inside faces of one of the rooms. You can repeat to verify other rooms. This is effective, but we can also change the default behavior of the dimensions from now on so that they show to the inside faces instead of centers. We do this with the Temporary Dimensions command.

NOTE

Revit will remember the modified location of witness lines during your work session. To try it out, select a Wall and edit the witness line of a Temporary Dimension as indicated above in the "Create a New Residential Project" topic. Deselect the Wall. Reselect the same Wall and notice that the witness lines remember the modified location. When you close the project and reopen, they will reset to their defaults.

15. On the Manage tab of the ribbon, on the Settings panel, click the Additional Settings drop-down button and choose **Temporary Dimensions**.

   • In the Walls area, click Faces and then click OK (see Figure 3.16).

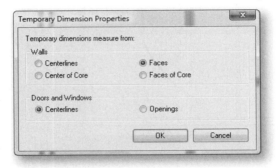

FIGURE 3.16 *Change the Temporary Dimensions to measure from Wall faces*

16. Click the horizontal interior Wall again.

The dimension from the inside face of the bottom horizontal exterior Wall to the inside face of the selected Wall should be 11'-5 1/2" [3658].

17. Click on the vertical interior Wall.

This time, the dimension from the inside face of the left exterior Wall to the inside face of the selected Wall should read 11'-8 1/2" [3650]. However, this value is not correct; the distance should be 11'-6" [3500]. Despite best efforts to transfer field measurements to our building model, an error crept into our calculations. Fortunately, errors like this are very easy to correct in Revit.

   • Click the blue text of the left-hand dimension and type the correct value **11'-6"** [**3500**] (see Figure 3.17).

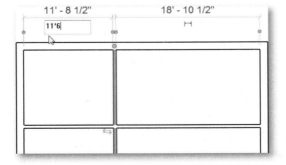

FIGURE 3.17 *Edit the Temporary Dimension to move the Wall to the correct location*

18. Save the project.

Sketching Walls

The offset and copy techniques covered above are effective ways to add new Walls based upon Walls already existing in the model. In many cases, however, it is easier to sketch the new Walls and then use the Temporary Dimensions to position them correctly. Let's try that technique next.

19. On the Home tab of the ribbon, click the **Wall** tool.

   - From the Type Selector (on the Properties palette), choose **Generic – 5" [Interior - 135mm Partition (2-hr)]**.
   - Verify that the Line icon is chosen on the Draw panel.
   - For Location Line, choose Wall Centerline.

20. Move the mouse over the exterior vertical Wall on the left.

   Notice that the centerline of the wall highlights and the Temporary Dimension that appears.

   - When the value of the dimension above the interior horizontal Wall is about **4'-0" [1200]** click to set the first point.
   - Move the pointer horizontally to the right and click to set the second point just past the middle of the plan (see Figure 3.18).

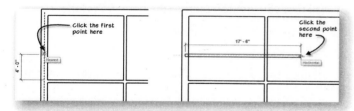

FIGURE 3.18 *Sketch a Wall segment above the horizontal interior Wall*

   - Click the Modify tool or press ESC twice to complete the command and then click to select the Wall just created.
     Several Temporary Dimensions will appear on screen.

21. Click the blue text of the vertical Temporary Dimension between the new Wall that you just created and the other horizontal interior Wall (the **4'-0" [1200]** space from above).

   - Type **3'-10" [1170]** and then press ENTER (see the left side of Figure 3.19).

NOTE	This is a face to face dimension. Once you edited the Temporary Dimension setting above, it should remain persistent.

22. Using the same process, create a vertical Wall approximately **3'-0"** [**900**] to the left of the existing interior vertical Wall.
    - Start at the lower exterior Wall and draw up until it intersects with the upper horizontal interior Wall (the one drawn in the last step).
    - Using the Temporary Dimensions, edit the face to face distance between the two Walls to be **2'-6"** [**750**] (see the right side of Figure 3.19).

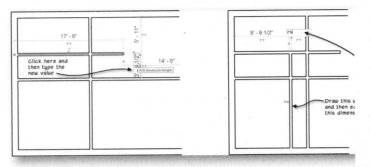

FIGURE 3.19 *Sketch Walls and use Temporary Dimensions to fine-tune their placement*

- Click the **Modify** tool or press the ESC key twice to complete the edit.

23. Select the vertical interior Wall that you just completed (the one on the left).

    At each end of the Wall is a small round blue grip handle.

    - Click and drag the handle at the bottom upward.
    - Using the temporary guidelines that appear, drag up and snap to the horizontal Wall (see Figure 3.20).

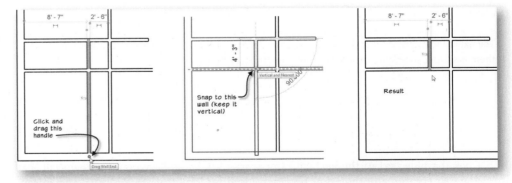

FIGURE 3.20 *Drag the Wall End Handle up to the intersection with the horizontal Wall*

Trim Walls

While the control handles are quick and easy, another common way to edit a plan layout is using the Trim and Extend tools.

24. On the Modify tab of the ribbon, click the **_Trim/Extend to Corner_** tool.

The keyboard shortcut for Trim is **TR**.	**TIP**

The **_Trim/Extend to Corner_** tool will create a clean corner from the two segments by either lengthening or shortening the selected segments. Click the vertical Wall that was just edited with the handle.

Pay attention to the prompt at the Status Bar, in this case instructing you to select the first Wall that you want to Trim or Extend. Please note that you are further prompted to select the portion that you wish to keep.	**NOTE**

25. Next click the horizontal Wall to the right of the intersection (see Figure 3.21).

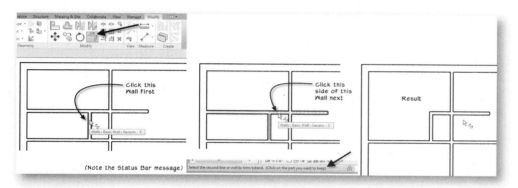

FIGURE 3.21 *Using Trim to create the closet corner*

26. Click the **_Modify_** tool or press the ESC key twice.

Adding Walls Using the Pick Lines Mode

Let's continue adding to our wall layout by using a technique that combines features of both the previous techniques. We will add Walls using the **_Wall_** tool and its "Pick Lines" mode. We will use this in conjunction with the Offset option on the Options Bar to place the Wall close to where we need it.

27. On the Home tab of the ribbon, click the **_Wall_** tool.

- On the Type Selector, verify that the Type is set to: **Generic – 5" [Interior - 135mm Partition (2-hr)]**.
- On the Draw panel, click the Pick Lines icon to activate this mode.
- Verify that the Location Line is Wall Centerline.
- On the Options Bar, in the Offset field, type **6'-0" [1800]**.

There are two number fields on the Options Bar. Be sure to input in the Offset field, not the Height field.	**NOTE**

28. Move the mouse next to the right side of the interior vertical Wall in the middle of the plan.

    The right face of the Wall will pre-highlight.

    Click the wall to place the new wall.

    - Be sure that the temporary guideline indicates a new Wall to the right and then click.
    - Click the **Modify** tool or press the ESC key twice.

29. Select the new Wall and edit the dimension between this Wall and the one from which it was offset to **6'-6"** [**1983**] (see Figure 3.22).

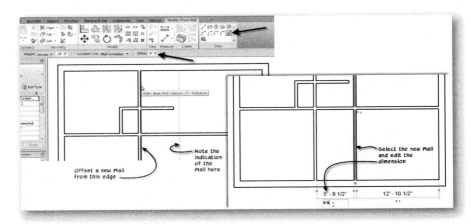

FIGURE 3.22 *Adding a Wall using the Pick Lines mode*

In this case we used two steps: first offsetting to an approximate amount and then using the Temporary Dimensions to fine-tune the result precisely. If you want to calculate the exact offset to use instead of editing the Temporary Dimension, you can do so. The Offset value will be measured from the line you pick to the Location Line of the new Wall—in this case Wall Centerline. Therefore, half of the Wall's width would have to be added to the dimension. Despite the relative ease of this calculation, it is typically easier to place the Wall close and then edit the dimension to the exact value as we have done here. The choice is yours.

Trim/Extend Single Element Mode

Let's modify a few more Wall segments using the *Trim/Extend Single Element* tool.

30. On the Modify tab of the ribbon, click the *Trim/Extend Single Element* tool.

In this mode you first select the element to use as a boundary and then the click the Wall or line you wish to extend or trim to the selected boundary.

Remember to read the prompts on the Status Bar.	**TIP**

- Click the vertical interior Wall on the right as the boundary (the one we just created).
- Click the short horizontal Wall at the top of the closet to extend (see the top row of panels in Figure 3.23).

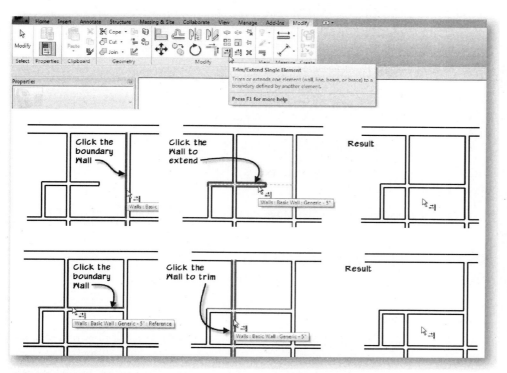

FIGURE 3.23 *Use the Trim/Extend Single Element to create "T" Intersections*

31. Use the same technique to trim away the vertical Wall at the top of the closet (Remember to click the piece you want to keep—read the Status Bar—shown in the bottom row of panels in the figure.)

- Click the **Modify** tool or press the esc key twice.

Using the Split and Move Tools

The layout is coming along. Next let's focus on the stair hallway in the middle of the plan.

32. On the Modify tab, click the **Split Element** tool.

The keyboard shortcut for Split is **SL**.	**TIP**

- Place the cursor (now shaped like a knife) over the long horizontal Wall in the middle of the plan where it intersects the left vertical Wall in the middle.
  The horizontal Wall should highlight. If it does not, press the TAB key a few times until it does.
- When the horizontal Wall is highlighted and small vertical line appears at the cursor, click the mouse (see the left side and middle of Figure 3.24).

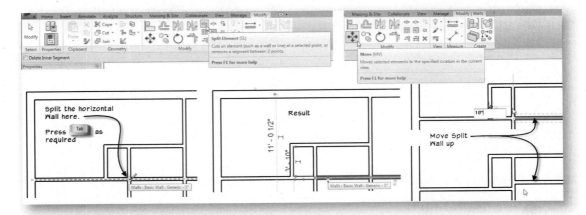

FIGURE 3.24 *Split the horizontal Wall in the middle into two segments*

- Click the **Modify** tool or press the ESC key twice.

33. Click the right side of the Wall we just split.

Notice how the Wall is now two segments. Use Split anytime you need to break a single Wall into two parts.

34. With the Wall selected, click the **Move** tool on the Modify Walls tab of the ribbon.

TIP   The keyboard shortcut for Move is **MV**.

- Pick a start point on screen, such as the Endpoint of the Wall.
- Begin moving the mouse up (do not click) and then type **10"** [**250**] and press ENTER (see the right side of Figure 3.24).

If you are working in Imperial units, be sure to type the inch symbol ("). If you prefer, you can also zoom in and snap to this value as you move the mouse. In this case, simply click the mouse when the desired value appears on screen. Or, you can also move randomly at first, and then use the Temporary Dimension to edit it to the correct value. The exact technique you use is up to you, as long as the Wall moves up 10" [250].

35. Select the long vertical Wall on the left (the one intersecting where we just split) and then click the **Copy** tool on the Modify Walls tab.

- Pick a start point on the selected Wall and then move to the right.
- Move the mouse to the right **3'-6"** [**1067**].

TIP   You can begin moving the mouse to the right, type the desired value, and then press ENTER. Or use any other technique covered so far. The dimension given in the previous step is a center-to-center distance.

- Verify that the face to face dimension between the new Wall and the original is **3'-1"** [**942**]; if it is not, edit the Temporary Dimensions (see the left panel of Figure 3.25).
- Using the middle panel of Figure 3.25 as a guide, Trim the two new Walls to form a corner.
- Using the right panel of Figure 3.25 as a guide, drag the handle up and edit the Temporary Dimension as shown: **8'-3"** [**2500**].

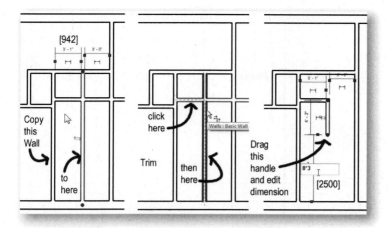

FIGURE 3.25 *Copy, Trim and Edit Walls for the hallway stairs*

36. Save your Project.

Sketch the Remaining Walls

At the top middle of the plan is a small half-bath. Let's sketch these Walls to complete the Wall layout of our first floor existing conditions.

37. Select one of the interior Walls on screen. On the Create panel of the ribbon, click the Create Similar button.

   This will run the Wall command with the settings matched from the selected Wall. Verify that the Wall Type is **Generic – 5" [Interior - 135mm Partition (2-hr)]**, the Draw shape is Line and the Location Line is: Wall Centerline.

   • Working at the top middle of the plan, sketch two Wall segments to create a small half-bath.

   • Edit the Temporary Dimensions to make the inside clear dimensions of the half-bath **4'-4"** [**1300**] × **3'-6"** [**1050**] (see Figure 3.26).

Remember: Select the Wall that you want to move before editing a Temporary Dimension. The highlighted wall will move relative to the non-selected Wall.	**HINT**

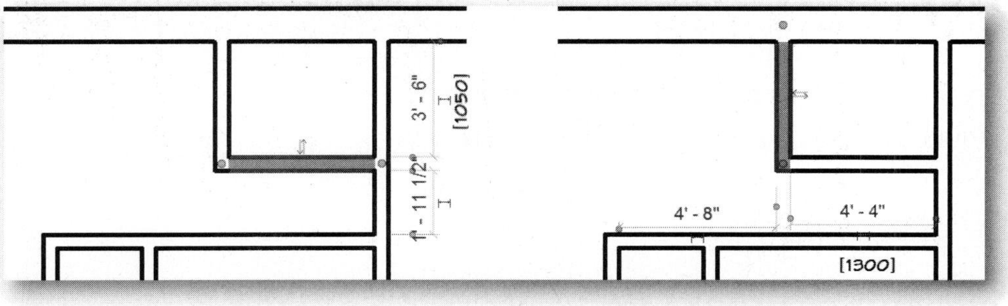

FIGURE 3.26 *Sketch the small half-bath and edit the dimensions to the desired values*

38. Extend the vertical Wall in the middle of the stair hallway up to the half-bath.

   • Use the **Split** tool with the "Delete Inner Segment" option on the Options Bar to remove the small piece of Wall shown in the middle panel of Figure 3.27.

   • Use the **Trim** tool to complete the layout as shown in the right panel of Figure 3.27.

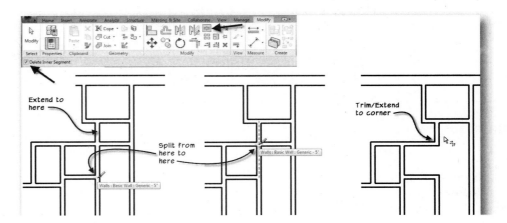

FIGURE 3.27 *Complete the Wall Layout using Trim/Extend and Split*

Figure 3.28 shows the completed Wall layout for the residential project's first floor existing conditions. There is still plenty of work to do, but use the illustration to check your work before you continue.

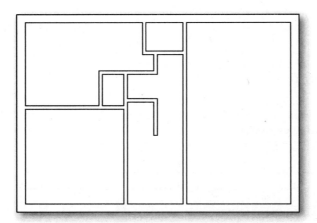

FIGURE 3.28 *Your completed Wall Layout should look like this*

39. Save the Project.

WORKING WITH PHASING

We have completed the layout of existing walls on the first floor of the house. Assigning the Walls to a particular construction phase will help us distinguish them as existing construction as the project progresses. Revit includes robust Phasing tools allowing us to quickly and easily designate these Walls as existing construction. Every element in the model has two phasing parameters: "Phase Created" and

"Phase Demolished." In this way you can track the "life span" of any element in the model. In the case of the Walls that we have drawn here, all of them are existing construction and none will be demolished yet. We will be demolishing some of the Windows and Doors later on in the chapter.

Assign Walls to a Construction Phase

1. Using a window selection (click and drag to surround all Walls), select all of the Walls in the model (but not the elevation markers).

   The Properties palette should still be open. If it is not, press CTRL + 1 to open it now. The drop-down list near the top of the Properties palette should read: Walls (14). If instead it says Common with a different quantity, reselect just the Walls.

   - On the Properties palette, scroll down and locate the Phasing grouping.
   - For "Phase Created," choose **Existing**.
   - For the "Phase Demolished," verify that **None** is chosen and then click the Apply button (see Figure 3.29).

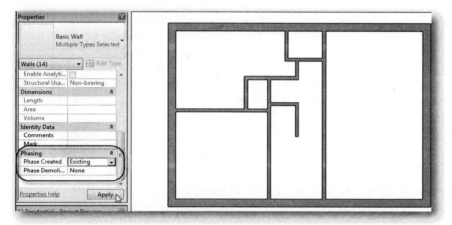

FIGURE 3.29 *Assign the Existing Phase to the selected Walls*

2. Deselect all of the Walls to see the change.

   - You can deselect the Walls by simply clicking in the white space next to the Walls. You can also click the **Modify** tool or press the ESC key twice.

The Walls will display in a lighter lineweight and will also be colored gray. This indicates that they are existing construction. Assigning a phase to an element automatically assigns a graphic override to it. You can control the settings of these overrides as well as edit and add phases in the Phasing dialog. To open this dialog, click the Manage tab and on the Phasing panel, click the Phases button. Feel free to explore this dialog now if you wish. A good example of a typical task that you could perform in this dialog would be to set up phases of construction in a large project. For example, you could create a "Foundations and Caissons" phase, "Phase 1" and "Phase 2 New Construction" phases. For our residential project, the out-of-the-box phases of "Existing" and "New Construction" are sufficient. Therefore, if you make any changes in the Phases dialog, please do not apply them at this time, or undo them after you are finished exploring.

3. Save the project.

Exploring the Properties of the Floor Plan View

At this time, it is appropriate to explore Phasing a bit deeper to get a full understanding of the tools available. Many of the changes we are about to make are not required by our residential project, therefore we will save a copy of the project in which to experiment and then reopen the original when we finish exploring.

Make sure you have saved the project as instructed in the previous step.

1. From the Application menu, choose **Save As > Project**.

   • Name the project *03 Residential – Temp.rvt* and then click Save. On the Home tab, click the **Wall** tool.

   • Accept all the default settings and draw a single Wall segment anywhere in the model, but joining or crossing one or more of the existing Walls.

   • Click the **Modify** tool or press the ESC key twice.

Notice that the new Wall came in as New Construction. Notice also that it did join with the existing Wall. Elements you create automatically use the current phase assigned to the view. In this case, while we changed the phase of the Walls to Existing, the *Level 1* floor plan view is still assigned to New Construction. Therefore all newly drawn elements will default to New Construction as well.

2. Select any Existing (phase) Wall.

   • On the Properties palette locate the Phase Demolished list, choose **New Construction** and then click Apply.

   • Deselect the Wall.

Notice that the Wall is now displayed using a dashed line style.

3. Select a different Existing (phase) Wall.

   • From the Phase Demolished list, choose **Existing** and then click Apply.

Notice that this time, the Wall has disappeared. This is because each view (*Level 1* floor plan in this case) has their own phasing parameters that control display: "Phase" and "Phase Filter." To understand why the Wall disappeared, let's change the current phase of the *Level 1* floor plan view to Existing and see how the display of elements changes.

4. Make sure that there are no elements selected and look carefully at the Properties palette.

Notice that the palette still has many active parameters. Look carefully at the top of the palette; the Type Selector reads "Floor Plan" just beneath it the drop-down list reads: Floor Plan: Level 1. This tells us that instead of seeing the properties of a selected element, the palette is now displaying the properties of the view itself: "Level 1" in this case. To Revit both items have properties we can manipulate. (For more information on this concept, you can review the "Properties Palette and the Type Selector" topic and Figure 2.18 in Chapter 2). As you can see, like our Walls above, views also have two phasing parameters in a "Phasing" grouping at the bottom of the palette.

   • Beneath the Phasing grouping, for "Phase," choose **Existing**.

   • Click Apply to see the change (see Figure 3.30).

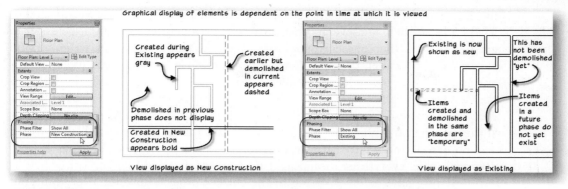

FIGURE 3.30 *Change the current phase of a view and study the results on the elements shown*

This change makes the "Existing" phase the current phase for the *Level 1* view. Therefore, the view will now show only items that existed in that phase. This means that no elements assigned to the New Construction phase will show (because they do not yet exist during the Existing phase time period). The Wall that we specified as demolished in New Construction will now appear solid again because during the Existing phase, it was not yet demolished. Furthermore, all of the previously light gray Walls now appear bold and black. That is because relative to the current phase (Existing), they are (were) new. Phasing essentially gives us a timeline with which to view our projects. When you set a view to a different phase, the relative "definition" of "Existing," "New" and "Demolished" shift accordingly. Finally, the Wall that we set to "demolished" in the Existing Phase now appears again, but it is dashed and cross hatched. This style indicates "temporary" construction, which occurs when an element is both created and demolished within the same phase. Examples of temporary elements are temporary dust walls, construction barrier, or possibly furniture that is relocated to a temporary space.

While we can create as many phases as we need for a particular project, Revit gives us four conditions (Phase Status) that can be used to describe (and graphically convey) an element at any point in time. These are built into the software. You can edit their display characteristics but you cannot add or delete them. The four states are "Existing," "New," "Demolished" and "Temporary."

5. Draw another Wall.

    • Select the new Wall and study its phasing Properties.

Notice that the newly drawn Wall was automatically assigned to the Existing phase. Any elements that we draw will automatically be added to the current phase of the view. We will make use of this technique below when we add Doors and Windows.

In addition to the current Phase setting, we can also assign a "Phase Filter" in the View Properties dialog. A Phase Filter does not change the active phase; rather it changes which elements display and how they appear in the current view based on their phase "lifespan" designations. By default, all views are set to "Show All." Several Phase Filters are included in the default templates from which our project was created. To see some of these, be sure to deselect all elements or select Floor Plan: Level 1 from the selection list at the top of the Properties palette. Scroll down and you will see the Phase Filter drop-down for the current view. For example, if you choose None, all elements will display (from all phases) and their graphics will display the same regardless of phase setting. This is probably not the most useful of Phase Filters. There are several other possibilities available; some might be easier to grasp with more than phases in the project than the two default ones. Keep in mind as you explore further, that the Phase Filter and Phase setting work together. For example, if you choose Show Complete for the Phase Filter and leave the Phase of the view set to Existing, you will see different results than if you set the Phase back to New Construction. Give it a try. Show Complete displays how the project will look at the completion of the current phase. This means that all elements will appear bold and no demolition will be displayed. Another example is "Show Previous Phase." This will make everything disappear if the current phase is Existing since there is currently no phase before Existing in the current project, however, everything but any newly drawn Walls would display if you change the Phase to New Construction. Like Phases, Phase Filters can be edited in the Phasing dialog.

> When you are finished experimenting with Phasing, return to the previously saved residential project file.

6. From the Application menu, choose Close.
   - It is not necessary to save *03 Residential-Temp*.

7. From the Application menu, choose *03 Residential* from the Recent Documents list to re-open it.

It may be possible to click the undo icon several times to restore the drawing to the state before we began demolishing and adding extraneous Walls. It is likely easier to simply reopen the saved version. Regardless, the Wall layout should look like it did in Figure 3.29 with all Walls assigned to the Existing phase before continuing.

If you prefer, a version of the project file in the correct state is provided for your convenience in the *Chapter03* folder. The file is named: *03 Residential-Walls.rvt* [*03 Residential-Walls-Metric.rvt*].

WORKING WITH DOORS AND WINDOWS

Continuing with our layout of the residential existing conditions, let's add some Doors and Windows. Doors and Windows automatically interact with Walls when inserted to create the opening and attach themselves to the Wall in an intelligent way. Doors and Windows cannot be placed free-standing in a Revit model. They *must* be inserted in an appropriate host Wall.

Add a Door

Doors are "Wall-hosted" elements. This means that Doors must be placed within Walls. Like Walls, Doors have both type and instance parameters. Changing an instance property affects the Door (or Doors) selected. Changing a Type property affects all Doors of that Type whether selected or not. A Door Family with several types has been included in the out-of-the-box template file from which our residential project was based.

1. Be sure that the *Level 1* view of the Residential project is open and then zoom out to see the entire plan layout.

   You can do this with the mouse wheel or the commands on the Navigation Bar.

2. On the Home tab, click the **_Door_** tool (see Figure 3.31).

TIP	The keyboard shortcut for Door is **DR**.

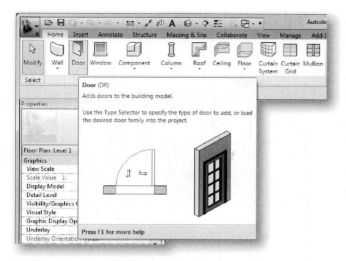

FIGURE 3.31 *The Door tool on the Home tab*

Much like the ***Wall*** tool, when you click the ***Door*** tool, a context tab is appended to the Modify tab, some options appear on the Options Bar and the Properties palette also shows relevant settings. The Status Bar prompt will read: "Click on a Wall to place Door." This project has one Door Family currently loaded named Single Flush [M_Single-Flush]. This Family has several types.

- From the Type Selector (on the Properties palette), choose ***36" × 80"*** [***0915 × 2032mm***].

The following list explains the major fields and controls shown on the ribbon, Options Bar and Properties palette when you are adding a Door (see Figure 3.32).

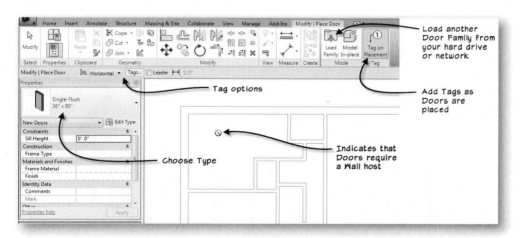

FIGURE 3.32 *Choose your options for adding Doors on the ribbon, the Options Bar and the Properties palette*

- **Load Family**—Use this button to access the external libraries and load Door Families and their types not already available in the current project. "Libraries" are folders containing Family files on your hard drive or network server.

- **Tag on Placement**—This button is a toggle setting. When selected (highlighted blue) Revit will add a Door Tag as Doors are added to the project. Even if you deselect this option, you can still add Tags to Doors later.
- **Tag Options** (on the Options Bar)

  - **Tag Orientation**—When "Tag on Placement" is selected, this option sets the orientation of the Tag as either horizontal or vertical.
  - **Tags**—Opens the Tag dialog where you can see which Tags are currently loaded in your project and also choose to load additional Tag Families from a library location.
  - **Leader**—Check this to add a Leader to your Tags as they are placed. Tags without leaders are placed in the center of the object. With a leader, you can move the Tag away from the element being tagged while the leader points back to the object.
  - **Leader Length**—When you enable the Leader option, you can input the desired length of the leader in this field.

- **Type Selector (Change Element Type)**—Use this drop-down to choose from a list of Door Families and Types in the current file. The specific list is populated by the Door Families that are already loaded into the template we used to create the project. Always choose a type from this list before configuring other options on the Properties palette. If the Family you want is not on the list, you can use the Load Family button (on the ribbon) to load additional Families and Types from an external library.

NOTE

The Model In-place button on the ribbon allows you to create an in-place Door Family. In-place Families are not designed to be moved, copied, rotated, etc. They are meant to be used only once. If you need to use it more than once within this project or in a different project, a regular door family should be created in the Family Editor, saved to a library and then loaded into your project as needed. It would be very rare to create an In-place Door Family. Below, in the "Create an In-Place Family" topic, we will learn how to create an In-Place Family for a fireplace, which is a more acceptable use of in-place Families. Fortunately, many Door Families are included with the software and many more are available online on various Revit-themed web sites. Simply do a web search for "Revit Door Families," and you are bound to come up with plenty of options. The Family Editor (for creating regular component Families) will be explored in Chapter 10.

As you move your mouse pointer around on screen, a Door will only appear when you move the pointer over a Wall. If you are unhappy with the direction of the door swing, press the SPACE BAR to flip it before you click to place the Door. (This is noted on the Status Bar—you remembered to look there, right?)

3. On the ribbon, click the Tag on Placement tool to disable that option. (Remember, this is a toggle; you could click it again to turn it back on).
4. Move the mouse to the horizontal exterior Wall at the top right side of the plan.

   - Position the Door roughly in the center of the room on the right and then click the mouse (see Figure 3.33).
   - As with Walls, Temporary Dimensions will guide your placement.

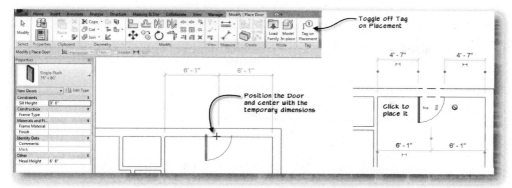

FIGURE 3.33 *Click to place the first Door*

Notice that the Door appears in the drawing and cuts a hole in the Wall. However, notice that the hole in the Wall is filled with dashed lines. This is Phasing at work again. At the start of this segment, we returned to the saved copy of the file, which had the *Level 1* floor plan view's phase set to "New Construction." Therefore, Revit Architecture is showing this Door as being a new Door placed into an existing Wall. This requires the opening for the Door to be shown as demolition while the Door appears as new construction.

5. Click the **Modify** tool or press the ESC key twice.

Change the Phase Filter

Let's try repeating the Phase Filter exercise above to see the different ways that this condition will display in each phase.

6. Make sure that there are no elements selected; on the Properties palette, make sure that Floor Plan: Level 1 is listed at the top.

    -   Beneath the Phasing grouping, for "Phase Filter," choose **Show Previous + Demo**.
    -   Click Apply to see the change (see the top of Figure 3.34).

With this phase filter active, you only see the previous phase (Existing in this case since it is the phase that occurs before New Construction) and any demolition. In this case, the only demolition is the opening for the Door.

7. Edit the view Properties again.

    -   Beneath the Phasing grouping, for "Phase Filter," choose **Show Complete**.
    -   Click Apply to see the change (see the bottom of Figure 3.34).

This filter shows what the view would look like when the final phase is complete. So in this case, all the Walls appear bold and there is no demolition shown.

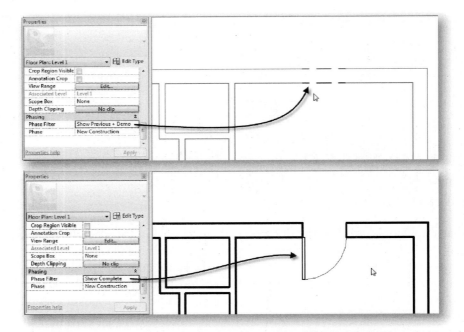

FIGURE 3.34 *Change the Phase Filter to view the model at different points in time relative to the current Phase*

8. On the Properties palette, choose **Show All** for the Phase Filter.

The power and potential of the Phasing parameters was seen when we explored these options with just Walls. Now that we have added a Door, we can truly see the full potential of these tools. If this Door truly were a new Door being added to these existing Walls, all of these graphical behaviors would be managed for us automatically by Revit simply by assigning the Door to the New Construction Phase parameter as we have done here. It turns out that this Door is actually an existing Door. Therefore, we need to change its Phase parameter to make it (and the Wall) display properly.

9. Select the Door in the model.

   Notice that the Properties palette now displays the properties of the Door.

   • Beneath the Phasing grouping, for "Phase Created," choose **Existing**.

   • Click Apply to accept the change to this instance of the door and then deselect the Door by clicking in the white background of the Canvas area.

The Door now displays the same as the Wall in which it is inserted and the dashed demolition lines no longer display. This is because the Door and Wall now belong to the same Phase, therefore there is no demolition required. Since we are going to add several more existing Doors, let's change the view's active Phase to Existing (as we did above) to save us the trouble of having to edit phase property of the Doors (and Windows that we will add below) later on.

10. On the Properties palette, be sure that the properties of the floor plan view are displayed and then choose **Existing** for the Phase.

The Walls will turn bolder to reflect this change.

A typical set of construction documents requires existing conditions, demolition and new construction drawings. In Revit Architecture this is easily achieved by duplicating the views (plans, sections and/or elevations, even schedules) and editing the views' Phase and Phase Filter parameters to display the correct data. Refer to the "Create an Existing Conditions View" heading in Chapter 6 for an example of this.

Place a Door with Temporary Dimensions

Let's add several more existing Doors to our model. For the next several Doors, it will be easier to place them if the Temporary Dimensions are set to the openings rather than the centers. (This is similar to the change we made for Walls above).

11. On the Manage tab of the ribbon, on the Settings panel, click the Additional Settings drop-down button.

    • Choose **Temporary Dimensions**.

    • In the Temporary Dimension Properties dialog, in the Doors and Windows area, choose Openings and then click OK (the dialog is pictured above in Figure 3.16).

12. On the Home tab, click the **Door** tool.

    • From the Type Selector, choose **30" × 80" [0762 × 2032mm]**.

    • Verify that "Tag on Placement" is *not* selected.

    • Move the cursor to the upper left corner of the plan and position it so that the Door is being added to the topmost horizontal exterior Wall.

    • When the Temporary Dimension reads 2'-0[600] from the upper left corner, click the mouse.

      As before, the Temporary Dimensions will remain on screen until you cancel the command or place another Door.

13. Click the blue value of the Temporary Dimension on the left and type **2'-4" [762]** and then press ENTER (see Figure 3.35).

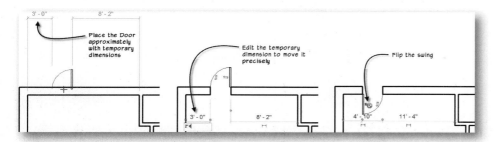

FIGURE 3.35 *Place the Door in the approximate location, and then use the Temporary Dimensions to fine-tune placement*

Notice there are two sets of small arrow handles, one group horizontal and the other vertical. Use these handles to flip the Door swing.

14. Click the flip controls on the Door to swing it into the plan as indicated.

The Door will shift the indicated amount.

Load a Door Family

The next Door that we are going to add is a bi-fold door for the small closet in the middle of the plan. There is no bi-fold Family available in the current project. Therefore, we will use the Load function to access the Revit Architecture library and load a bi-fold Door Family and its Types.

- You should still be in the Door command. If you have canceled it, click the **Door** tool again.

As noted, if you open the Type Selector, there is only the Single Flush Family and its types loaded.

15. On the Modify | Place Door tab, on the Mode panel, click the Load Family button.

    Your default library folder will load.

- Double-click the *Doors* folder, choose *Bifold-2 Panel.rfa* [*M_Bifold-2 Panel.rfa*] and then click Open.

    There will be a pause while Revit loads the Family and its types. If during this process a Save Reminder appears, click Save the Project.

16. Open the Type Selector.

Notice that there are now two Families shown on the list, each with its own types indented beneath (see Figure 3.36).

- Choose **Bifold-2 Panel : 30" × 80" [M_Bifold-2 Panel : 0762 × 2032mm].**

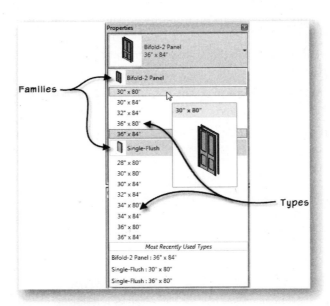

FIGURE 3.36 *Each Family is listed with an icon preview and its types indented beneath.*

- Verify that "Tag on Placement" is *not* selected.

17. Move the pointer over the left vertical Wall of the closet in the middle of the plan.

- Using the Temporary Dimensions, get the Door centered on the vertical Wall.

    Do NOT click the mouse to place the door yet.

- Slowly move the mouse left to the right.

Notice that you can control whether the Door swings into or out of the closet with the mouse, but not which side of the opening (up or down in this case) that it swings. Take note of the Status Bar and in some cases a tip that appears on screen. There it notes that you can use the SPACE BAR to flip the swing.

- Press the SPACE BAR to flip the swing.
- Press it again to flip back (see Figure 3.37).

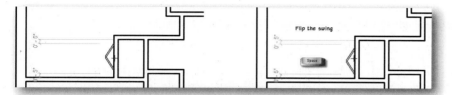

FIGURE 3.37 *Move the mouse to control Door placement, and press the* SPACE BAR *to flip the Door*

- When the Door opens to the lower portion of the Wall, click the mouse.

Don't worry if you added the Door with the wrong swing. We can easily edit this after the Door is placed.

18. Click the **Modify** tool or press the ESC key twice.
19. Save your project.

Flip a Door with Flip Arrows

As we saw above, if you want to flip a Door after it is placed, you can use its handles.

20. Click on the hinged Door at the top right (the first Door placed in the previous steps above).
21. Click either one of the Flip handles.

- It will pre-highlight and a tip will appear to indicate its function (see Figure 3.38).

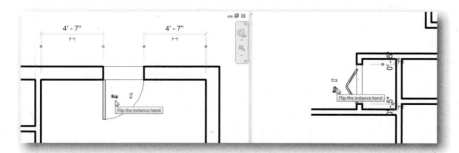

FIGURE 3.38 *Flip the instance hand of a Door*

Repeat this process on other Doors if you wish.

Change the Door Size

Door size is governed by a Door's type. If you wish to change the size of Door in your model, simply choose a different type. If you wish to use a size that is not on the list, you must either edit an existing door type or create a new type with a different size. This process will be covered below.

22. Select the Door in the top right of the plan.

    - On the Properties palette, from the Type Selector, choose a new size, such as **Single-Flush : 34" × 80"** [**M_Single-Flush : 0864 × 2032mm**].

23. Click next to the drawing to deselect the Door (or press ESC).

You can edit other instances or type parameters for the selected element (Door in this case) on the Properties palette as well. For example, we used this technique above to set the Phase Created parameter of our first Door. You could input the Frame Material or Finish. (These are just text fields and will accept any input typed in. To change the actual materials represented in a shaded or rendered view the Family Component needs to be edited in the Family Editor.) You could adjust the Sill or Head height of the Door, which would shift its position vertically in elevation.

Edit Door Placement with Dimensions

We have seen several examples so far of the use of Temporary Dimensions to control the placement of elements in the model. We can also create permanent dimensions, which give us even greater flexibility and control over element placement.

NOTE The following examples would be more appropriate in a New Construction model since it is unlikely that we would need to maintain constraints in an existing layout. These tools are covered here merely for the educational value, not as a recommendation of their usage for existing conditions plans.

24. Select the Door in the top left side of the plan.

    Two Temporary Dimensions will appear; one on either side—take notice of the small blue icon that looks like a dimension itself (see Figure 3.39).

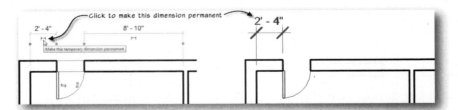

FIGURE 3.39 *You can convert a Temporary Dimension to permanent with the dimension icon*

25. On the left side dimension, click this icon to make the dimension permanent.

When you do this, a permanent dimension will appear. It will be colored black. Various colors are used in Revit to indicate various states of selection and/or if the element is editable. Here is a summary of those colors and what they signify:

- **Black**—The element is not selected, nor active for editing.
- **Outlined in Light Blue**—The element is pre-highlighted (caused by the mouse passing over it, or the TAB key being used to cycle to the element).
- **Light Blue**—The element is selected.
- **Dark Blue**—The element is interactive (for example, Temporary Dimensions, handles and view tags like sections and elevations).
- **Orange**—Alert color. Indicates an element with an error or warning.
- **Magenta**—Used for sketch lines in sketch-based elements such as Floors, Roofs and Railings.

> Selection color can be customized. On the Application menu, click the Options button. On the Graphics tab of the Options dialog, change the colors to suit your preference. Element colors can be customized in the Object Styles dialog (Manage tab, Settings panel, **Object Styles** Button).

NOTE

When you make the Temporary Dimension permanent, the new dimension appears in black. However the text stays dark blue as long as the element it is attached to (the Door in this case) is selected. The new permanent dimension is a new element in the current view (*Level 1*). To access the options of the new permanent dimension, we must select it instead of the Door.

26. Click on the new permanent dimension.

    The Door will deselect (turning black) and the permanent dimension will select (turning light blue) instead.

    - Click the small padlock icon (see the left side of Figure 3.40).

    The padlock icon will "close" to indicate that the dimension is now constrained.

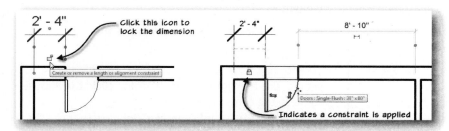

FIGURE 3.40 *Apply a constraint to the Door position using the padlock on the permanent dimension*

27. Click on the Door to select it.

    The permanent dimension will deselect (turning black) and the Door will select (turning light blue) instead.

Notice the padlock icon that displays beneath the dimension (see the right side of Figure 3.40). If you click the blue dimension value and attempt to edit it, it will not work. The only way that you could edit the value would be to first remove the constraint. You can do this by clicking the closed padlock icon. Let's test our new constraint with a quick experiment. We will undo it when we are finished.

28. Select the left vertical exterior Wall.

   • Move the pointer over the selected Wall (the pointer changes to a small move cursor), click and drag the Wall to the left.

   The exact amount is not important.

Notice that the Wall moved, and the Door also moved and maintained its dimension.

   • Undo the change. (Click the Undo icon on the QAT or press CTRL + Z).

29. Try the same experiment with the right vertical exterior Wall. Move it to the right.

Notice that the Door on that side did not move. It has no constraints applied to it.

   • Undo the change.

NOTE Dimensions are annotation elements. Therefore, they are view-specific—meaning that they appear only in the view in which they are added. Walls and Doors on the other hand are model elements and appear in all views in which they would normally be seen. However, although the display of dimensions is view specific, their constraints are not. If the dimension is locked, the constraint is enforced throughout the model, even if the dimension is not displayed in the current view.

Applying an Equality Constraint

Let's apply another kind of constraint to a different Door. The Door that we have on the right should be centered in the room. We can add a permanent dimension with an equality constraint to maintain this relationship automatically for us.

30. On the Annotate tab, on the Dimension panel, click the Aligned dimension tool.

We are going to place the dimension ourselves this time because we want more control over the specific witness lines. When you place a dimension, pre-highlight the elements that you wish to dimension. If the wrong element pre-highlights, use the TAB key to cycle to the one you want.

31. Move the mouse over the right exterior vertical Wall.

   The Wall centerline will pre-highlight.

   • Move the mouse pointer over the inside face of the Wall and then press TAB. Repeat if necessary until the inner face of the Wall pre-highlights and then click.

   • Click the center of the Door next.

   The center of the Door should pre-highlight, when it does, click the mouse. Otherwise, use the TAB to cycle to the center first.

   • Set the final point at the inside face of the vertical Wall on the other side of the room—remember to use the TAB key if necessary (see Figure 3.41).

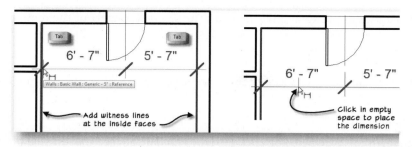

FIGURE 3.41 *Dimension from the faces of the Walls to the center of the Door using* TAB *to cycle to the correct points*

32. To place the dimension, click anywhere in the blank white space of the room.
33. Click Modify or press ESC twice to exit the dimension command.
34. Click on the new dimension to select it.

Notice the small "EQ" icon with a line though it (see the left side of Figure 3.42). This indicates that the permanent dimension is *not* set to maintain an equal distance. If you click this icon, you enable the equality constraint and force the elements attached to the dimension to remain equally spaced.

- Click the EQ icon to enable the equal constraint.

Notice that the Dimension Equality toggle icon no longer has the slash through it (see the right side of Figure 3.41). If your Door was not previously centered, it will move to the center now.

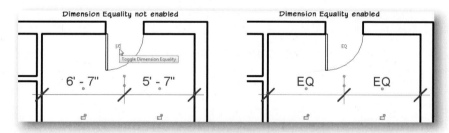

FIGURE 3.42 *Set the dimension to an equal constraint*

35. Repeat the experiment above, and move the exterior Wall.

Notice that the Door remains constrained to the center of the room. If your Door was not previously centered, it will move to the center now.

- Undo the Wall move.

36. Save the Project.

Use CTRL + S to save quickly.	**TIP**

Add a New Door Size

The Door to the small half-bath at the top of the plan is 2'-4" [710]. This size is not available in our current list of Door Types. Since it is the same Door Family, we will simply create a new type with the desired size. To do this, we need to choose one of the Types of this Family, duplicate it, and then edit the size parameters of the duplicate Type.

37. On Home tab, click the **Door** tool.

    • From the Type Selector, choose **Single Flush: 30" × 80" [M_Single-Flush : 0762 × 2032mm]**.

    Don't place one yet.

    • Directly beneath the Type Selector, click the Edit Type button (see the left side of Figure 3.43).

The Type Properties dialog will appear. At the top you can see that the Door Family is *Single Flush* [*M_Single-Flush*] and the Type is *30" × 80"* [*0762 × 2032mm*]. The Family list includes all of the Door Families loaded in the current project. This currently includes only the Single Flush and Bi-fold Families. Next to the Family list is the Load button. You can use this button to load additional Families from external libraries as we did above. The Type list includes all of the Types available for the selected Family. If you scan that list, you will not find the size we need which is 28" × 80" [0710 × 2032mm]. Next to the Type list is the Duplicate button. You use this button to create a copy of the currently selected type. This is how you make a new size. Duplicate an existing one and edit it. Let's do that now.

38. Next to the Type list, click the Duplicate button.

TIP	A shortcut to this is to press ALT + D.

A new Name dialog will appear suggesting the name: Single **Flush: 30" × 80" (2)** [**M_Single-Flush : 0762 × 2032mm (2)**].

It is a good idea to change this name. Let's follow the convention established with the out-of-the-box Families and name this type after its size.

    • Change the name to **28" × 80" [0710 × 2032mm]** and then click OK (see the middle of Figure 3.43).

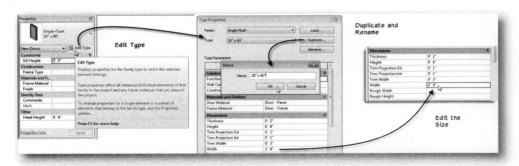

FIGURE 3.43 *Create a new Type by Duplicating an existing one*

Now that we have created and named a new type, let's edit it to the size we need.

39. Beneath the Dimensions grouping, for "Width," type **2'-4" [762]** and then click OK (see the right side of Figure 3.43).

We are now ready to add a Door using our new Door type.

40. On the ribbon, verify that "Tag on Placement" is not selected.

    • Add a Door to the small half-bath at the top of the plan.

    • Use the SPACE BAR to swing the door in and to the right (see Figure 3.44).

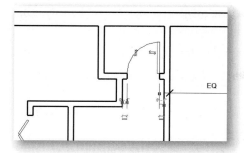

FIGURE 3.44 *Add a Door in the new Type to the half-bath*

Add the Remaining Doors

41. Using Figure 3.45 as a guide and the techniques covered above, place the remaining doors in the plan.

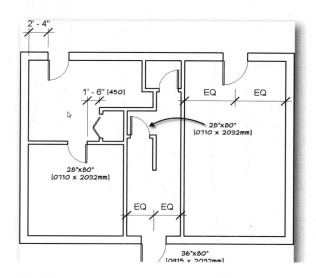

FIGURE 3.45 *Add the remaining Doors as shown*

> You can select a Door, right-click, and choose **Create Similar** (or use the Create Similar button on the ribbon) as a way of quickly adding Doors of the same type (with similar parameters).

HINT

42. Save the project.

Add Windows

Working with Windows is nearly identical to working with Doors. Many of the parameters are the same, and placement and manipulation of Windows works the same as with Doors.

43. On the Home tab, click the ***Window*** tool.

All of the options for placing Windows match those of Doors. For detailed descriptions of each of these options, see Figure 3.32 above and the accompanying descriptions.

The only Window Family loaded in the current project is the *Fixed* Family. This house has double hung windows. Let's see what we have in the library.

44. On the Place Window tab, on the Mode panel, click the Load Family button.

- In the Load Family dialog, double-click the *Windows* folder, choose *Double Hung.rfa* [*M_Double Hung.rfa*] and then click Open.

The size that we need is not included on the list. So as we did above for the Door, we will create a new Double Hung window type.

- From the Type Selector, choose **Double Hung : 36" × 48"** [**M_Double Hung : 0915 × 1220mm**].

45. On the Properties palette, right beneath the Type Selector, click the Edit Type button.

- Next to the Type list, click the Duplicate button (or press ALT + D).
- Change the name to **36" × 56"** [**0915 × 1422mm**] and then click OK.
- Beneath the Dimensions grouping, for "Height," type **4'-8"** [**1422**] and then click OK.
- On the ribbon, click the Tag on Placement button to deselect it.

46. Using Figure 3.46 as a guide and the techniques covered above, place the Windows indicated.

- Use the new type that you just created for all of these Windows.

Note that the Window in the lower left corner is placed at the midpoint of the room. The Temporary Dimensions should automatically snap to a centered configuration as you move the mouse nearby. This can also be done easily by making a permanent dimension as above and then clicking the EQ constraint.

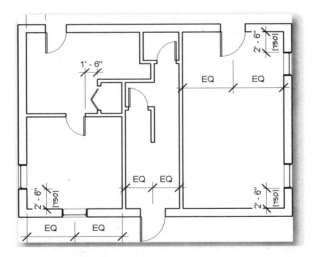

FIGURE 3.46 *Add several Windows to the plan*

47. Using the process covered here, create two new Window types:

- **36" × 32"** [**0915 × 0800mm**] (width x height)
- **24" × 32"** [**0600 × 0800mm**] (width x height)

48. Place the remaining two windows as shown in Figure 3.47.

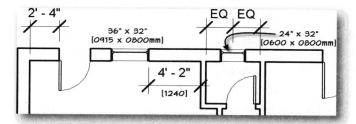

FIGURE 3.47 *Create two new custom types and then place the remaining two Windows*

49. Save the project.

Add Cased Openings

There are several passages between the hallway and the neighboring rooms that are simply cased openings. You place these the same way as Doors—in fact they are stored in the *Doors* folder in the library, but they are categorized as Generic Models so that they will not appear in Door Schedules. The element includes the casing and a hole in the Wall, but no Door panel. Since they are not actually Doors, we use a different tool to add them.

50. On the Home tab, click the **Component** tool.

Components are used to represent many things including furniture, fixtures, parking spaces, and nearly anything else you can imagine. In this case, the Cased Opening components that we need are not currently loaded in the model.

- On the Modify | Place Component tab, on the Mode panel, click the Load Family button.
- In the Load Family dialog, double-click the *Doors* folder, choose *Opening-Cased.rfa* [*M_Opening-Cased.rfa*] and then click Open.

Remember, the Cased Opening Families are stored in the *Doors* folder and are essentially Door elements, but they have been categorized as Components (Generic Models) instead because typically you would not want cased openings to appear on Door Schedules.	**NOTE**

- From the Type Selector, choose **Cased Opening : 30" × 80"** [**M_Opening-Cased : 0762 × 2032mm**].

51. Using Figure 3.48 as a guide, place four Openings in the locations indicated.

Use the Temporary Dimensions to place them **6" [150]** from the intersection with the other Walls. The horizontal opening at the top should be centered.	**TIP**

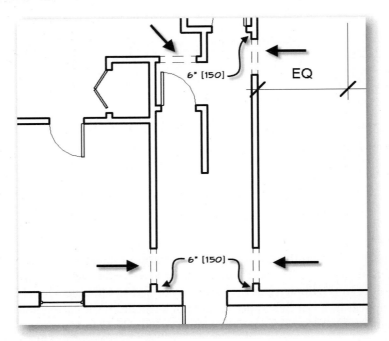

FIGURE 3.48 *Add Openings to the hallway*

After placement, you can click on a Cased Opening and move it or edit it just like the Doors and Windows.

Load a Custom Family

There is one additional Window in the existing house that we need to add. In the front of the house in the living room (right side at the bottom) is a picture window comprised of a fixed Window flanked by two double-hung units. We could add this as three Windows simply enough. However, it is better to use a single Family that contains the three Windows. To save time, this configuration of three Windows has been pre-assembled and saved as a Family file and included with the Chapter 3 dataset files from the student companion. In this sequence, we will load this Family into our current project and add it to the front of the existing house.

52. On the Home tab, click the **Window** tool.

    • On the Modify | Place Window tab, on the Mode panel, click the Load Family button.

53. In the Load Family dialog, browse to the location where you installed the dataset files from the student companion and open the *MasterRAC 2012\Chapter03* folder.

    • Choose *Existing Living Room Front Window.rfa* [*Existing Living Room Front Window-Metric.rfa*] and then click Open.

    • Add the Window to the bottom horizontal Wall in the room at the right (see Figure 3.49).

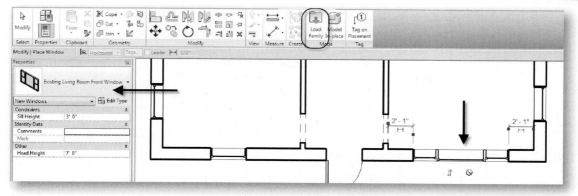

FIGURE 3.49 *Create an Instance of the imported Family at the front of the house*

This Family is a very simple one that contains three Window elements. This can be thought of as a "nested" Family because the Family file contains instances of other Families. We will explore Families in more detail in Chapter 10.

- Click Modify or press ESC twice.

54. Save the project.

Modifying Door Size and Type

Modifying Doors and Windows is easy. We have already covered most of the basic techniques above. It turns out that the Door in the top right of the plan (the first one we added) is supposed to be a double door. Furthermore, it is actually a French door leading out to an existing patio. Let's make this edit to our model.

55. On the Insert tab, on the Load from Library panel, click the Load Family button.

- In the Load Family dialog, double-click the *Doors* folder, choose *Double-Glass 2.rfa* [*M_Double-Glass 2.rfa*] and then click Open.

Accessing the Load Family button from the Insert tab is an alternative to the approach used above. It simply loads the Family into the project and makes it available for use. It is a good option for the task at hand since the Door already exists in the model.

56. Select the Single Hinged door at the top right corner of the plan (the first one we added).

- On the Properties palette, from the Type Selector, choose Double-Glass 2: **68" × 80"** [**M_Double-Glass 2: 1730 × 2032mm**].

You can use this technique for Windows and other elements as well.	**NOTE**

The new Door Type has replaced the previous one and remains nicely centered since we added the equality constraint to this Door earlier. In plan view, it is not obvious what effect choosing a glass Door Family had. To see the glass, we will have to view the model in 3D.

VIEWING THE MODEL IN 3D

We have spent the entire chapter working in a single plan view. However, as we saw in previous chapters, you can add or edit elements in any view you wish and then view them or edit them in elevation, section, and/or three-dimensional views. Revit

Architecture maintains all of the parameters of our building model including those required to view it in 3D, even if we don't explicitly ask it to!

Open the Default Three-Dimensional View

Let's have a look at how the model is shaping up in the third dimension.

1. On the View tab, click the Default 3D View icon. (Choose Default 3D, not Camera or Walkthrough).

TIP	You can also find this icon on the QAT.

A new view will appear on screen named: *{3D}*. This name is assigned automatically to the default 3D view in Revit. The default 3D view is an axonometric of the complete building model looking from the south-east. This view appears in the Project Browser beneath the *3D Views* category. From the default view, we can see the ganged Window that we added above in the front of the house. Even though the graphics in the Floor Plan view indicated that the middle Window was different from the two flanking Windows, from the 3D vantage point it is now very clear that the middle Window is fixed while the two flanking ones are double-hung.

You can dynamically pan and/or zoom any view as we have already seen. In a 3D view, you can also "orbit" the view. When you do this, you can change the angle and height from which we are viewing the model and virtually spin it in all three dimensions.

2. Within the view window on the Navigation Bar, click the Steering Wheel icon (see Figure 3.50).

TIP	The keyboard shortcut for Steering Wheels is the F8 key.

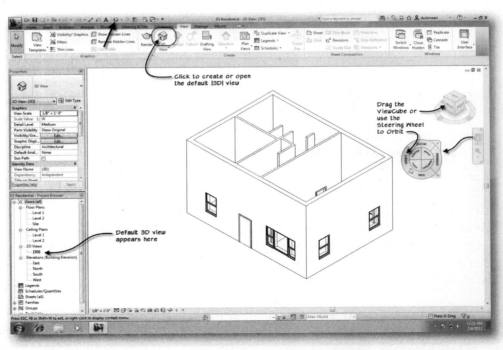

FIGURE 3.50 *Display the Default 3D view and manipulate it dynamically with the Steering Wheel*

Note the Steering Wheel appears on screen and follows the movement of the cursor. The Steering Wheel has several controls. Among them are: Pan, Zoom, and Orbit. We explored several of these back in Chapter 1. By now you should already be comfortable with panning and zooming. Orbit is only available in a 3D view. In other words, you can open the Steering Wheel from any view such as the Floor Plan view *Level 1* that we have worked in throughout the chapter. When you do, only Zoom and Pan will appear. Let's give Orbit a try in our 3D view.

- Move your mouse pointer over the Orbit portion of the Steering Wheel; click and hold down the mouse button and then drag to orbit.

  The pointer will change to an orbit icon to indicate that this mode is active. A Pivot point icon will also appear on screen.

- In the middle of the Steering Wheel, move your mouse pointer over the Center portion; click and drag over the model to relocate the pivot point.

  In the Orbit mode, drag side to side to rotate the model and drag up or down to tilt the vantage point up or down. The Look mode functions similarly. Give it a try.

3. Orbit the model around so that you can see the back of the house.

- Pan and Zoom to fine-tune the view to your liking.

4. On the View Control Bar (bottom left corner of each view window) click the Model Graphic Style icon and choose **Shaded** from the pop-up menu that appears (see Figure 3.51).

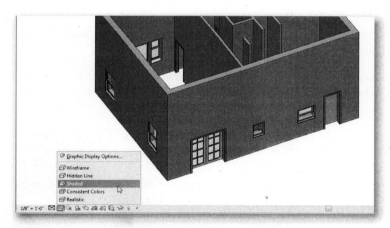

FIGURE 3.51 *Change the view to Shading with Edges*

The dynamic viewing functions can also be performed with a wheel mouse and the CTRL and SHIFT keys. Drag with the mouse wheel held down to pan, Hold down the CTRL key and drag with the wheel, or simply roll the wheel to zoom. Hold down the SHIFT key and drag with the wheel to orbit. Your specific mouse and mouse driver will determine the exact behavior. Mouse drivers by many manufacturers allow customization of specific button functions. New in 2012: 3D Connexion devices are now supported!

TIP

The keyboard shortcut for Shaded is **SD**. The keyboard shortcut for Hidden Line is **HL**.

TIP

Two additional shading modes are: Consistent Colors and Realistic. Consistent Colors shows the edges like Shaded with Edges but uses a consistent tone for each color rather than tinting them as the geometry changes angle to the vantage point.

Realistic takes advantage of the rendering features of Materials in Revit and displays any photorealistic textures in the viewport onscreen. In our case, the Walls we are using do not have any textures applied, so this mode would not enhance our viewing much at this time. However, as the project progresses, we will revisit the shading modes and try our Realistic once we have some Materials assigned to our model.

The default three-dimensional view: *{3D}* is always available. You could certainly use this view exclusively and simply orbit, scroll, and zoom the view as needed each time you displayed it. However, once you get a 3D view displaying just the way you like, you are encouraged to save the view. The new view will be added to the Project Browser beneath the 3D views category along with any other 3D views your project has. To do this, right-click on *{3D}* beneath *3D views* in the Project Browser and choose **Rename**. Give a descriptive name. If you then click the Default 3D view icon again, the default *{3D}* view will be created anew. If you click the 3D view icon and the default *{3D}* view already exists, it will open this existing view instead of creating a new one.

5. Save your modified 3D view.

- Right-click the name *{3D}* beneath *3D Views* on the Project Browser and choose **Rename**.

- Give the new view a unique name such as "**Rear Existing French Door**."

Edit in Any View

From our explorations in 3D, you may have noticed that many of the interior Walls are the wrong height. If you can't tell, try orbiting the model or simply select one of the exterior Walls. New in 2012, elements selected in a 3D view will display transparently while the element is selected. We can correct this easily.

6. Close your "Rear Existing French Door" view and then click the Default 3D view icon to create a new {3D} view.

- On the View tab, on the Windows panel, click the Window Tile icon.

TIP	The keyboard shortcut for Tile is **WT**.

- Orbit the view down slightly so that you can clearly see the heights of the interior partitions.

Remember, in addition to the techniques already covered, you can orbit 3D views by dragging the ViewCube. On the Navigation Bar, choose **Zoom > Zoom All to Fit**.

TIP	The keyboard shortcut for Zoom All to Fit is **ZA**.

7. Activate the View: *Floor Plan: Level 1* by clicking anywhere in the view or the view's title bar and then use the **Modify** tool to create a selection window, click within the exterior walls and drag from right to left to select all interior Walls.

Be sure to select only the interior Walls and not any of the exterior ones. Do not worry about selecting Doors and Openings. We will remove them from the selection next.

All of the selected elements will highlight in light blue in both views (see Figure 3.52).

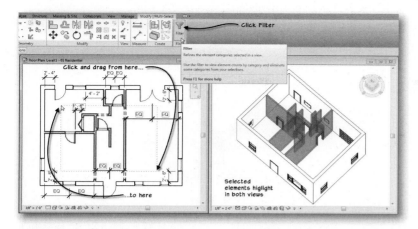

FIGURE 3.52 *Select all interior Walls*

8. On the ribbon click the Filter button.

> The quantity of selected elements will appear next to the Filter icon on the Status Bar at the lower right.

NOTE

- In the Filter dialog, remove the checkmarks from all boxes *except* "Walls" and then click OK (see Figure 3.53).

FIGURE 3.53 *Now only the Walls will be highlighted.*

9. On the Properties palette, beneath the Constraints grouping, for "Top Constraint," click in the empty cell to the right, click the drop-down arrow, choose **Up to level: Level 2** and then click Apply (see Figure 3.54).

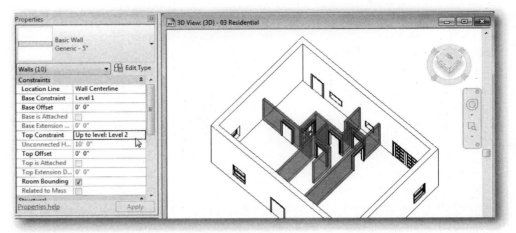

FIGURE 3.54 *Set the Top Constraint of the interior Walls to Level 2*

Note that even though we made the change from a 2D floor plan view, the edit appeared in the 3D view immediately. In Revit, every view is *always* up to date. There is no need to coordinate or refresh anything. Regardless of the view you are working in, all edits apply directly to the model. You simply choose the most convenient and logical view in which to work and Revit takes care of the rest.

In this exercise, we have constrained the top of the interior Walls to the height of the second floor. It turns out that the default template from which we started our project included two levels. However the height of those levels does not match the existing conditions of our residential project. We will adjust the height of the levels, and in so doing, all of these interior partitions will automatically adjust with the level height. This is the primary reason to apply such a constraint.

10. On the Project Browser, double-click the *East* elevation view to open it.

    Your windows should still be tiled on screen.

    • Position the view so that you can clearly see Level heads on the left side and the 3D view in the background.

11. Click on Level 2 to select it.

    • Click again directly on the blue dimension text beneath the Level 2 label.
    The text will highlight—ready to receive input.

    • Change the value to **9'-0"** [**1800**] and then press ENTER (see Figure 3.55).
    Notice the immediate change to all of the interior Walls in the 3D view.

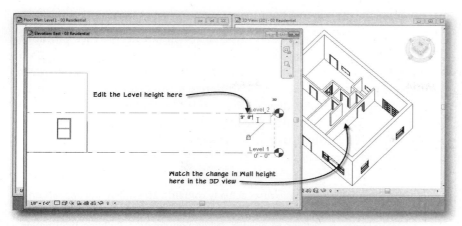

FIGURE 3.55 *Adjust the height of Level 2 and see the immediate effect on the interior Walls*

12. Maximize the *Floor Plan: Level 1* view.

- Zoom the window to fit. On the View tab, in the Windows panel, click the Close Hidden button.

When you have many open windows, they can all be closed except the current one using this command. This helps keep your interface organized and preserves memory and computer resources.

13. Save your project.

ADDING PLUMBING FIXTURES

The small half-bath in the top of our plan could use some fixtures. The library of components provided with Revit includes a variety of items such as furniture, toilets, trees, parking spaces, equipment, electrical fixture symbols, targets, tags, and much more. All of these items are Families and many have special behaviors and parameters appropriate to the object that they represent. In this exercise, we will simply load the Families we need and insert them in the model much like we did for the Doors, Windows, and cased openings above.

Add Components to the Model

You add Components like plumbing fixtures and furniture in nearly the same way as Doors, Windows and Openings. Each Family may vary slightly depending on the parameters built into it. However, the basic process is the same: click the ***Component*** tool on the Home tab, choose an item from the Type Selector or click the Load Family button to access the library. Place the item in your model.

1. On the Home tab, click the ***Component*** tool.

   If you open and scroll through the Type Selector, you will note that no plumbing fixture Families currently appear on the list.

As we did above for Doors and Windows, we will simply load the items we need from the library.

2. On the Modify | Place Component tab, on the Mode panel, click the Load Family button.

- In the Open dialog, double-click the *Plumbing Fixtures* folder.

You will note that two Family files exist for domestic toilets, one for 2D and the other for 3D. The reason for this is that in many cases, you will not need to see plumbing fixtures in 3D. By using a 2D symbol in those cases, you can reduce demand on computer resources in the project file. In reality, in a model the size of the one that we are building here, there would be no noticeable difference in performance from using either Family. However, let's begin developing good BIM habits right away. Since we have no need for a 3D toilet in this particular project, let's load the 2D Family for this exercise. Like all Families, you can swap out the 2D Component later with a 3D one should project needs change.

It should be noted that the "3D" plumbing fixtures actually contain a 2D symbol that appears only in floor plan views and a 3D symbol that appears only in the 3D views. Revit automatically switches between the appropriate representations as needed. When you create a Building Information Model, it is important to recognize techniques such as this and to realize that "Model" does not always mean "3D." Nor does "Model" imply that every facet and screw of an item should be painstakingly represented in graphical form. Always remember to strike a balance between what is conveyed graphically and what is conveyed by other means such as with attached data parameters—this is the "I" in BIM. In many cases, an "Information" model is much more practical and useful to a project team than a "3D" model. There are many types of models. Statisticians and economists refer to their spreadsheets as "models." Meteorologists refer to their predictions and weather simulations as "models." As Architects, we tend to only think "3D" when the word model is mentioned. Learning to embrace BIM means understanding that a model is not always 3D, and that the "I" is just as important as, and sometimes more important than, the "M." A really good BIM will include both a graphical and data model tightly integrated with one another. Model in this context therefore can more accurately be described as a "representation" or a "simulation" of our building project.

- Choose *Toilet-Domestic-2D.rfa* [*M_Toilet-Domestic-2D.rfa*] and then click Open.

  If you get a warning: "No tag Loaded," simply click No to continue. Like Doors and Windows above, the Component tool can Tag on Placement as well.

3. Zoom in on the half-bath at the top of the plan.

- Move the pointer around the four Walls of the half-bath.

Notice how the Toilet automatically orients itself to the Walls (see Figure 3.56). This is because the 2D toilet is a wall-hosted Family. The Doors and Windows we added already were also wall-hosted Families, but they also cut holes in the Walls. Naturally the toilet does not need to cut the Wall, but requiring the attachment is a nice feature. If you move your mouse away from the Wall, the cursor changes to indicate that placement is not allowed. It is also possible to create Families that are hosted and not hosted. Refer to Chapter 10 for more information.

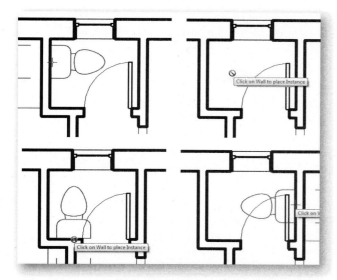

FIGURE 3.56 *The Toilet Family is wall-hosted, so it automatically attaches to the Walls*

- Click a point on the left vertical Wall to place the toilet in the model.

This toilet room is clearly not up to code! Such is the nature of existing conditions. Just place the toilet very close to the intersection with the exterior Wall. We still need room for a small sink.

4. On the Place Component tab, click the Load Family button again.

Return to the same location and Open the *Sink–Single–2D.rfa* [*M_Sink–Single–2D. rfa*] Family this time.

Move the pointer around again and notice that this Component is also hosted. However, it will be too big for the space that we have available. Sometimes what we find in the field does not meet current building codes. The sink in this existing half-bath is *very* small.

5. On the Properties palette, click the Edit Type button.

   - Next to the Type list, click the Duplicate button (or press ALT + D).
   - Change the name to 18" × 15" [450 × 375mm] and then click OK.
   - Beneath the Dimensions grouping, for "Depth," type **1'-3"** [**375**].

Remember, if you are using Imperial units, you can simply type 13 with a space between the numbers, and Revit Architecture will interpret this as 1'-3".

TIP

   - Beneath the Dimensions grouping, for "Width," type **1'-6"** [**450**] and then click OK.

6. Place the sink next to the toilet. Use the Temporary Dimensions to fine-tune the placement of both the sink and the toilet.

It will be a tight fit. You will need to set the values of the Temporary Dimensions to about **1"** [**25**] between the Walls and the Components to get everything to fit (see Figure 3.57).

You can also select the element and "nudge" it with the arrow keys on your keyboard to move it. The element will move approximately 2mm on screen. The more you are zoomed into the view the smaller the distance each press of the arrow key will move the object.

TIP

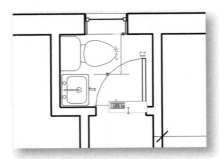

FIGURE 3.57 *Use the Temporary Dimensions to fine-tune placement of the Components*

7. Click the **Modify** tool or press the ESC key twice.
8. Save the project.

CREATE AN IN-PLACE FAMILY

The first floor existing conditions plan is nearly finished. We still need to add the fireplace in the living room and stairs in the middle of the plan. While it would be possible to create a fireplace Family in the Family Editor and save it in our library for use in any project, this would only make sense if we used the same fireplace design often. If you design a lot of homes that use the same fireplace, this is exactly what you should do. Refer to techniques in Chapter 10 to learn how to create a loadable component Family.

In this case, we will create the exact fireplace we need for this project directly in-place. This is called an "In-Place Family."

NOTE In-Place Families are not designed to be moved, copied, rotated, etc. They are meant to be used only once. If you need to use it more than once within this project or in a different project, a regular loadable component Family should be created in the Family Editor, saved to a library and then loaded into your project as needed. The Family Editor will be explored in detail in Chapter 10.

Create an In-Place Family and Choose a Category

To get started, we will create a new In-Place Family and assign it to a pre-defined category. Revit has a long list of predefined categories. Categories are at the top of the hierarchy discussed in the "Families & Types" topic in Chapter 1.

1. Zoom in on the middle of the right vertical exterior Wall.
   This is where our fireplace will go.
2. On the Home tab, click the drop-down button on the **Component** tool and choose Model In-Place.

   • In the "Family Category and Parameters" dialog, choose Generic Models and then click OK (see Figure 3.58).

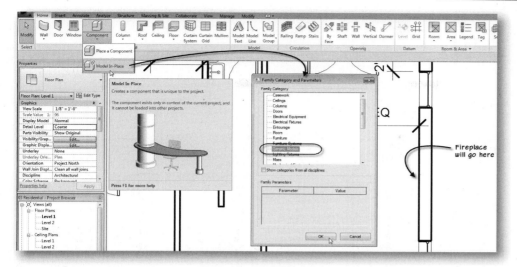

FIGURE 3.58 *Create an In-Place Family and choose its category*

The Family Category list is a fixed list built into the software. When you create a Family, you must assign it one of these categories. The Family you create will gain the characteristics of the category to which it is assigned. In general, when choosing a category, try to select the one that most closely matches the actual object that you are creating. The Construction Specifications Institute (CSI) spec section for fire-places is Division 10—Specialties (10300 Fireplaces and Stoves), which would tempt us to choose "Specialty Equipment." However, your choice of category does impart certain behaviors to your Family. Specialty Equipment is intended more for free-standing equipment items and does not have a "cut" representation. Items like 10340 Manufactured Exterior Specialties, 10500 Lockers or 10670 Storage Shelving are all examples of things that would work well in the Specialty Equipment category. Items in Revit that are "cutable" interact with the cut plane of floor plan and section views and show bold when cut and lighter when viewed in projection. Since we will want our fireplace to interact with the Wall and appear bolder when cut in plan, we need a category that supports the cut plane.

To get the cutting behavior, we will use the "Generic Models" category instead. This is sort of a "catch all" category. You typically choose Generic Models when the item you are modeling does not fit neatly into any of the other categories. Generic Models does not impart any specialized parameters that might be available from other more descriptive categories, but aside from the need for interaction with the cut plane, our existing fireplace has no other specialized needs.

- In the Name dialog, type: **Existing Fireplace** and then click OK.

You are now in "In-Place Family Editing" mode. The model will gray out but remain visible for reference. The ribbon tabs will change showing a collection of In-Place Family editing tools in place of their usual tools. Take a look at the Home tab for example. If you expand the Project Browser, you will find the new Family listed under Generic Model (see Figure 3.59).

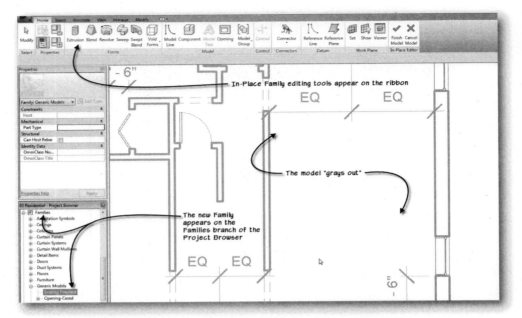

FIGURE 3.59 *The Family Editor mode is enabled when you create a new In-Place Family*

The Home tab includes many Family editing tools. You can create solid and void forms, insert Components, or add connectors. Simply click on the other tabs to access these tools as normal. Note that several tools like Walls, Doors and Floors are not available in Family editing mode. You cannot place (nest) a System Family within another Family. Also notice that the In-Place Editor panel with its Finish and Cancel buttons appears on the right side of the ribbon in all tabs.

Adding Reference Planes

When you construct complex geometry, it is often useful to have guidelines to assist in locating elements. Reference Planes are used for this purpose in Revit. You sketch a Reference Plane similar to the way you sketch Walls or lines. You can snap and constrain other elements to Reference Planes, making them useful tools for design layout. You can add Reference Planes in any orthographic view of the model. (Reference Planes do not show in 3D.) In this example, we will add them within our In-Place Family. When you add them in this way, the Reference Planes will become part of the In-Place Family and will be visible only when editing the In-Place Family.

3. On the Home tab, on the Datum panel, click the ***Reference Plane*** tool. (Do not click Reference Line; make sure you click Reference Plane).

   • Click a point inside the large room on the right just above the lower Window.

   • Move the pointer horizontally to the right past the exterior Wall and then click outside.

   • The exact locations of either click are not critical so long as you draw horizontally and above the Window.

A small Reference Plane (green dashed line with round blue handles at the ends) will appear.

   • Edit the Temporary Dimension from the bottom horizontal Wall to **7'-11"** [**2400**] (see the left side of Figure 3.60).

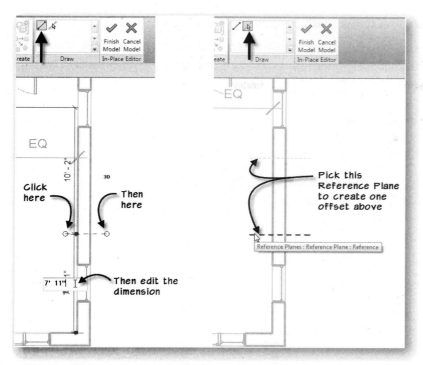

FIGURE 3.60 *Create two horizontal Reference Planes*

- On the Modify | Place Reference Plane tab, on the Draw panel click the Pick Lines tool.
- On the Options Bar, type **6'-2"** [**1900**] in the Offset field.
- Pre-highlight the first Reference Plane and move the mouse so that the offset line appears above.
- Click to create the new Reference Plane (see the right side of Figure 3.60).

Repeat the process to create two more vertical Reference Planes. These will frame out the rectangular footprint of the fireplace.

4. On the Draw panel, switch back to the **Line** tool, and then type **4"** [**100**] in the Offset field.
   - Snap to the endpoint of the lower Window on the inside edge of the Wall.
   - Snap to the endpoint of the upper Window on the inside edge of the Wall (see the left side of Figure 3.61).

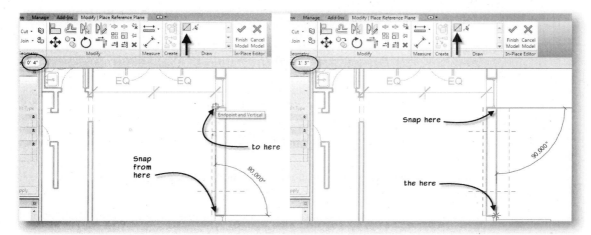

FIGURE 3.61 *Create vertical Reference Planes relative to the horizontal ones*

5. Change the Offset to **1'-3"** [**380**].

- Snap to the endpoint of the upper Window on the outside edge of the Wall.
- Snap to the endpoint of the lower Window on the outside edge of the Wall (see the right side of Figure 3.61).

TIP	The start and end points suggested here will make the first Reference Plane fall to the inside of the house and the second to the outside. If you click the points in the wrong order, do not cancel, simply tap the SPACEBAR to flip the line.

Figure 3.62 shows the completed Reference Plane layout. The dimensions in the figure are for your reference. They do not need to be added to the model.

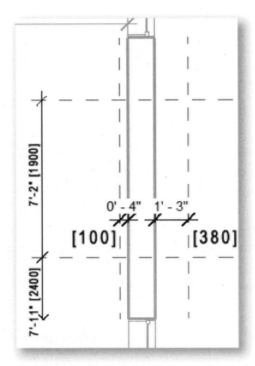

FIGURE 3.62 *The completed Reference Plane layout*

We now have four Reference Planes that we can use to guide the creation of our fireplace's form. This is common best-practice. Complete details of using Reference Planes in Families will be discussed in Chapter 10.

Create a Solid Form

Using our Reference Planes as a guide, let's create the overall mass of the fireplace.

6. On the Home tab, on the Forms panel, click the ***Extrusion*** tool (see Figure 3.63).

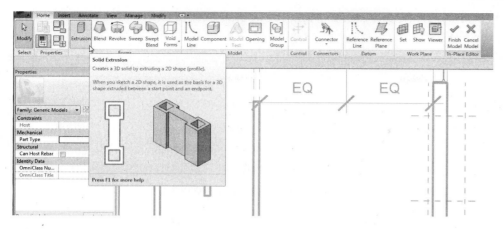

FIGURE 3.63 *Create an Extrusion.*

7. On the Options Bar, in the "Depth" field, type **9'-0"** [**2750**].

   • On the Draw panel, for the sketch shape, click the Rectangle icon.

8. Snap to the intersection of two of the Reference Planes and then snap to an opposite intersection to define the rectangular shape (see Figure 3.64).

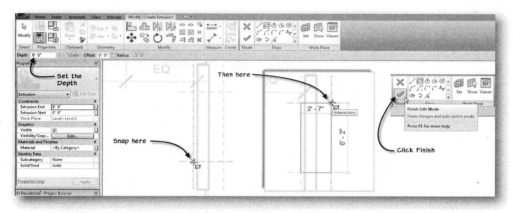

FIGURE 3.64 *Sketch the overall shape of the extrusion*

An open padlock icon will appear on each side of the shape. If we wanted to constrain the sides to the Reference Planes, we could click these locks. There is no need for explicit (deliberate) constraints here so we will not do that for this exercise.

9. On the Modify | Create Extrusion tab, on the Mode panel, click the Finish button (large green checkmark).

This gives us our basic fireplace mass. We now need to carve out the firebox.

Create a Void Form

Using the same basic process, we can create a Void form that will carve away from the solid form in our Family giving us the firebox opening.

10. On the Home tab, click the **Void Forms** drop-down button and then choose **Void Extrusion**.

The Modify | Create Void Extrusion tab will appear with the same Sketch tools as before.

11. On the Options Bar, in the "Depth" field, type **4'-0"** [**1200**].

   - On the Draw panel, click the Pick Lines icon.
   - Click the left vertical edge of the Solid Extrusion.

     A magenta sketch line will appear directly on top of this edge.

12. On the Options Bar, change the Offset value to **1'-1"** [**325**].

   - Click the right edge of the Solid Extrusion to create a magenta sketch line within the fireplace structure.
   - Change the Offset value to **1'-5"** [**430**] and click the top edge to create a sketch line below it.
   - Click the bottom edge of the Solid Extrusion to create a sketch line above it (see Figure 3.65).

The side of the offset will pre-highlight before you click so that you can be sure to offset the sketch line to the correct side. Notice that the lines automatically trim when you offset more lines. If yours do not trim exactly the way they are illustrated in the figure, you can simply use the **Trim/Extend to Corner** tool on the Modify tab to clean up the sketch.

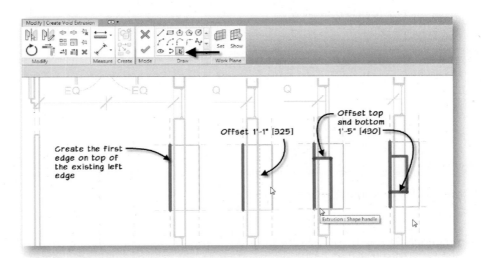

FIGURE 3.65 *Offset sketch lines to form the firebox shape*

Now we will use the **Trim/Extend to Corner** tool (the same one we used for Walls at the start of the chapter) to cleanup the sketch (form a rectangle).

13. On the Modify tab, click the **Trim/Extend to Corner** tool (or Type **TR**).

   - Trim any segments as required to make a rectangular shape.

HINT	Remember: Select the portion of the sketch line that you wish to keep.

- Click the **Modify** tool or press the ESC key to finish trimming (see the left side of Figure 3.66).

14. Click the lower horizontal sketch line and drag the handle on the right up till the Temporary Dimension reads 20° (see the middle of Figure 3.66).

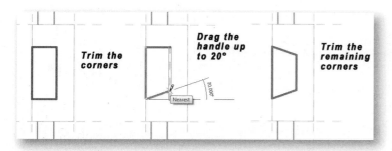

FIGURE 3.66 *Edit the sketch lines to finalize the shape*

- Repeat by stretching the top line down 20°.
- Use Trim once more to cleanup the remaining corners (see the right side of Figure 3.66).

15. On the Mode panel, click the Finish Edit Mode button.

While the void is still selected, it will appear solid. However, when you deselect it, it will cut away from the previously drawn solid to form the fireplace shape.

16. On the In-Place Editor panel, click the Finish Model button (big green checkmark).

Join the Fireplace with the Wall

The Fireplace Family is finished but it is modeled inside of the Wall making both the Wall and the Fireplace hard to read. Let's fix this.

17. On the Modify tab, click the **Split** tool (or type **SL**).

- On the Options Bar, place a checkmark in the "Delete Inner Segment" checkbox.
- Split the exterior vertical Wall on both sides of the fireplace (see Figure 3.67).

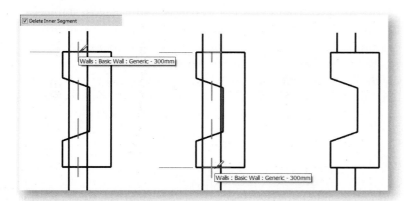

FIGURE 3.67 *Split the exterior Wall*

- Click the **Modify** tool or press the ESC key twice.

This is close to what we want but let's make one more edit.

18. On the Modify tab, on the Geometry panel, click the Join tool (see the top of Figure 3.67).

- Click one of the exterior Walls.
- Then click the Fireplace to join them (see Figure 3.68).

TIP	Remember to watch the Status Bar for detailed prompts.

- Repeat for the other Wall.

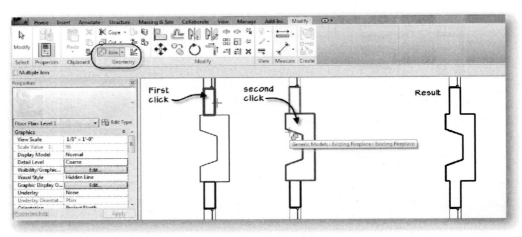

FIGURE 3.68 *Use Join Geometry to join the Walls to the Fireplace*

19. Click the Modify tool or press the ESC key twice to cancel the Join command.
20. On the QAT, click the Default 3D view icon.

- Use the techniques covered above and orbit the model around so that you can see the Fireplace.

We modeled the fireplace a bit too short. However, for now we will leave this alone. In later chapters we will address the height of the fireplace as well as how it changes on the second floor. The fireplace could also use a mantel and a hearth. However, because there will be no new work done in the living room of this project and therefore no sections or elevations are needed of the fireplace, that extra level of detail is unnecessary for this tutorial. What we have created works well for floor plans. If you wish to try it anyway for the practice, feel free. Select the fireplace, and then on the Modify | Generic Models tab, click the Edit In-Place button. This will return you to the In-Place Family editor where you can add these accoutrements using additional solids.

Reset the Current Phase

Congratulations! Our work on residential project first floor existing conditions layout is complete for now (see Figure 3.69). We still need to add the Stairs to this model. However, Stairs will be covered in a dedicated chapter. Therefore, we will save our layout without the Stairs for now.

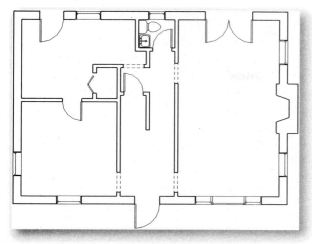

FIGURE 3.69 *The final first floor existing conditions layout*

21. Deselect all objects and then on the Properties palette, beneath the Phasing grouping, for "Phase," choose **New Construction**.

    • Verify that "Phase Filter" is set to **Show All**.

    • Click Apply to see the change.

Later in Chapter 6, we will actually duplicate this view and create a permanent Existing Conditions view. For now, we have simply returned the view to its Phase settings at the start of the chapter.

NOTE

22. Save the project.

SUMMARY

• The basic process for adding elements in Revit Architecture is to choose a tool on the ribbon, choose a Type from the Type Selector, set additional options on the Properties palette and Options Bar and then click to add the item in the view window.

• Walls can be added one segment at a time or chained to draw them end to end.

• Sketch Walls quickly and then use Temporary Dimensions to fine-tune size and placement precisely. You can edit the witness lines of temporary dimensions by clicking or dragging the handles. Permanent dimensions can be used the same as temporary dimensions while elements are selected.

• Assign elements to construction Phases to show Existing, Demolition and New Construction.

• Doors, Windows and Openings automatically "cut" a hole in, and remain attached to, the receiving Wall.

• Add Walls, Doors and Windows quickly, and modify their properties to add detail later. Use temporary and permanent dimensions to fine-tune Door and Window placement.

• Use Trim, Extend, Split, Offset, Move and Copy to quickly layout a series of Walls. Use the component tool to add additional items like plumbing fixtures, furniture, and equipment.

- Door, Window and Component Types can be included with the project template or loaded as needed from library files.
- It is not always necessary to model all elements in 3D. 2D Components can save overhead and computer resources when the 3D is not required.
- View the model interactively in 3D, plan, section, etc. to study and investigate possibilities in the design.
- Use In-Place Families for unique conditions that must be modeled in context and will not be reused in other projects.
- Edit in any view and see the change immediately in all views.
- Use In-Place Families to model custom or project-specific components directly in the model where they are required. Do not use In-Place Families for items that must be reused or placed in several locations or other projects.

Setting Up Project Levels and Views

INTRODUCTION

Setting up a new project can involve many steps and some careful decisions. Among the items for consideration are the floor levels, column grids, project settings and, most importantly, the many views your project will require to facilitate editing, querying and publishing project data. The most common use of views is to display your model graphically on screen. However, views are not just for display; views are also used to create and edit data and graphics in the building model. Each task you perform in your building model will occur in one or more views. This means that having the right views in place early in the project will make the entire team's job easier. Another important aspect of project setup is the floor levels in your building. Each significant horizontal datum in the building will be represented by a level in Revit. Usually you will have a level for each actual building level, but in some cases there will be additional ones for other important horizontal datums like "top of steel," "grade level" or "bottom of footing."

When you create a new project, the software does not presume to know the type of building you wish to create, how many levels or the kinds of views you intend to use to view and edit it. Every project must have at least one level, and some basic views are common to nearly all projects like floor plans and elevations, but mostly you must create the specific levels and views you need in your projects. To save time and ensure consistency from one project to the next, project templates are available when creating new projects. A project template is simply a Revit project that has been pre-configured to contain the most common and useful views, Families, settings (and even levels) required to start a new project. Revit ships with several sample project templates, and you can customize and save your own templates based upon your preferred standards. A project template will help get you started, but since each project is unique, a certain degree of setup is always required. Exploration of common tasks required during project setup is the primary topic of this chapter.

OBJECTIVES

In this chapter, we will explore project templates, levels and views. We will explore many of the out-of-the-box project templates provided with the software. Following this exploration, we will begin our commercial project. We'll add the required floor levels and we'll add some simple geometry to the project for basic context and proceed to explore the many different types of views available in Revit Architecture. After completing this chapter you will know how to:

- Open and explore several project templates
- Create a new project from a template
- Create levels
- Set up preliminary views
- Understand switching between and working with views
- Work with sheet views
- Print a digital cartoon set

NOTE There are many other aspects to successful project setup such as column Grids, Object Style settings, project units, project parameters, shared parameters, location of project Families and libraries, and Worksharing setup. While it is possible to set up all such items before project work begins, in many cases changes to project setup occur as needed in the project timeline. To simulate this reality and to keep this chapter from becoming too long, basic setup tasks will be performed here with many other realted topics being discussed in later chapters and appendices.

UNDERSTANDING PROJECT TEMPLATES

Project template files provide a means to quickly apply project setup information, enforce company standards and project-specific settings and save time. The concept is not unlike other popular Windows software packages such as Microsoft Word. Please note that template files apply only at the time of project creation. Projects do *not* remain linked to the template. However, you can transfer project standards from one project to another later, as required. A project template is simply a project file pre-configured for a particular type of building or task that has been saved in the template format. Template files have an RTE extension and are available when you choose the **New > Project** command from the application menu. In addition to the time saved when creating projects, templates help to ensure file consistency and office standards by giving all projects the same basic starting point. Several pre-made template files ship with the product and are ready to use. However, because office standards and project-specific needs vary, feel free to modify the default templates to better suit your needs. The exact composition of the template used to create a project is not as important as ensuring that templates are used consistently on all new projects. Table 4.1 lists some of the Revit Architecture project templates included with the software.

NOTE The table lists imperial and metric templates typically installed in the United States. When you install Revit, you are given the option to install several other templates and content files appropriate to other parts of the world. The exact items you have available may therefore vary from the table.

TABLE 4.1 *Out-of-the-Box US Project Templates*

Template File Name	Drawing Units	Rounding	Description
Imperial Templates			
Default.rte	Feet and fractional inches	To the nearest 1/32"	Default template for Imperial Units. Intended for any project type.
Commercial-Default.rte	Feet and fractional inches	To the nearest 1/32"	Intended for Commercial Projects
Construction-Default.rte	Feet and fractional inches	To the nearest 1/32"	Intended for Construction Projects
Residential-Default.rte	Feet and fractional inches	To the nearest 1/32"	Intended for Residential Projects
Metric Templates			
DefaultMetric.rte	Millimeters	0 decimal places	Default template for Metric Units. Intended for any project type.
DefaultUS-Canada.rte	Millimeters	0 decimal places	Default template for Metric Units in the United States and Canada.
Construction-DefaultMetric.rte	Millimeters	0 decimal places	Intended for Construction Projects
Construction-DefaultUS-Canada.rte	Millimeters	0 decimal places	Intended for Construction Projects in the United States and Canada

---

CAUTION

Please avoid creating projects without a template, because doing so requires an enormous amount of user configuration before serious work can begin.

MANAGER NOTE **BIM**

Revit Architecture saves virtually all data and configuration within the project file. This includes Families, settings, levels and views. Therefore, template files provide an excellent tool for promoting and maintaining office standards. Several sample template files ship with the product to help you get started. These include project templates tailored to different types of projects such as construction or commercial in both Imperial and Metric units. Refer to the table above for examples. You may find one of these templates suitable for your firm's needs as-is or more likely useable as a good starting point for customization. When you establish and configure office standards, it is highly recommended that you start with one of these default templates. Modify them to suit individual project or office-wide needs. Frequently used Families, title blocks, and other resources can also be stored in the template. Others can be stored in separate library files saved on the office network. Externally saved resources can be accessed using the Load Family button on the ribbon. Loading items on demand when needed can help keep the template file size smaller. Best-practice dictates the "80/20" rule. Include only those items needed in all (or most—80%) of the projects in the template. The other 20% can be saved in an easily accessible central repository containing additional office standard items and loaded into project files as needed by project team members.

This method also provides additional ongoing flexibility because new items can be easily added to the office standard library without requiring frequent changes to the office-standard template. In addition, Revit provides the **Transfer Project Standards** command for copying System Family items such as Wall Families and other System Families from one project to another. To use this tool, first open the source project (the one with the standards that you want to copy), and then open the project to which you wish to transfer. Execute this command from the Manage tab of the ribbon, and check the boxes for the items you wish to transfer in the "Select Items To Copy" dialog. Some firms even do this using a simple copy and paste. In this case, create a "warehouse file" containing instances of all the office standard elements (they can be System or Component Families in this case). For example, there might be one of each Wall Type inserted into the warehouse project. To access one of these Wall Types, you would open the warehouse file, copy the instance of the Wall you want, and then simply paste it into your active project. Such pasting will automatically import the required Family Types and settings.

The best way to demonstrate the importance of using project templates is to create some test projects using some of them.

Create a New Project with the Default Template

Let's create a new project using the default template (the same one we used in the previous chapter) and explore some of its settings.

1. Launch Revit Architecture from the icon on your desktop or from the Windows Start menu choose *All Programs > Autodesk > Revit Architecture 2012*.

TIP You can click the Start button, and then begin typing **Revit** in the Search field. After a couple letters, Revit Architecture 2012 should appear near the top of the list. Click it to launch to program.

Revit Architecture launches to the Recent Files screen.

2. From the application menu, choose **New > Project**. (Access the application menu by clicking on the big "R" at the top left corner of the Revit window).
   • In the New Project dialog, in the "Template file" area, be sure that the lower radio button is selected (the one that lists a template file).
   • Click the Browse button.

   The default template file name and location for the Imperial template file is:

   C:\ProgramData\Autodesk\RAC 2012\Templates\US Imperial\default.rte

   The default template file name and location for the Metric template file is:

   [C:\ProgramData\Autodesk\RAC 2012\Templates\US Metric\DefaultMetric.rte]

NOTE These paths are for the US installation on Windows 7. Paths may vary slightly for other regions and previous versions of Windows..

   • From this location select the *default.rte* [*DefaultMetric.rte*] template file and then click Open (see Figure 4.1).

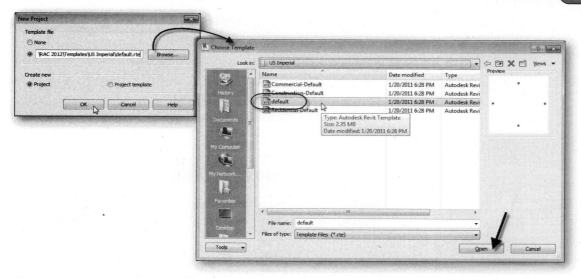

FIGURE 4.1 *Creating a new Project with the default template*

> If your version of Revit Architecture does not include the template files cited here, both have been provided with the dataset files from the student companion. Please browse to the *Templates* folder in the location where you installed the dataset files to locate them.

NOTE

- In the "Create New" area, verify that Project is selected and then click OK.

The imperial and metric templates are very similar in composition. Other than units, the only obvious difference is in the shape of the elevation tags. The imperial file uses a square elevation tag, while the metric template uses a round tag (see Figure 4.2).

default.rte   **DefaultMetric.rte**

FIGURE 4.2 *The shape of the Elevation tag varies in the North American and non-North American templates*

> You can customize the look of the elevation tags if you wish. To do so, select one, edit its Type properties. In the Type Properties dialog, you can change the Elevation Tag type. Click OK to update all Elevation tags in the project. The template includes two choices (types) for the Elevation Tag Family setting. You can create additional types if you wish. To do so, click the Manage tab of the ribbon and then click the Settings drop down menu icon on the Project Settings panel. Choose Elevation Tags and then click the Duplicate button to create a new type. Modify the settings of the new type to fit your preferences. If you want to make any changes a permanent office standard, re-save your default template file. You can even build your own custom elevation tags. To learn more, see the "Loading Custom Elevation Tags" topic below or the "Creating Custom Elevation Tags" topic in Chapter 10.

MANAGER NOTE

Elevation tags are an obvious on-screen difference between the metric and imperial templates. There are some other minor differences as well.

3. On the Manage tab of the ribbon, on the Project Settings panel, click the Materials icon (see Figure 4.3).

Some of the material names vary between the North American and non-North American templates. However the overall list of materials is largely similar.

FIGURE 4.3 *The names of some of the Materials vary between metric and imperial*

- Close the Materials dialog when you are finished exploring.

Despite these minor differences, the "default" imperial and metric templates are virtually the same. They contain the same starter levels and views. They have the same basic Families pre-loaded, and the other settings are similar but with unit-specific or regional vernacular differences (like "Cast-in-Place" vs. "Cast In Situ," shown in the figure as an example).

4. On the Manage tab, on the Settings panel, click the Project Units button.

TIP	The keyboard shortcut for Project Units is **UN**.

- Explore the various settings without making any permanent changes. Refer to the table above for the default settings of each Template.

 **MANAGER NOTE** | Imperial Units in Revit use feet as the primary unit by default. (Refer to the "Unit Conventions" topic in Chapter 2 for more information.) This means that numbers typed without unit designations will be interpreted as feet. If you have experience with AutoCAD, this will take some getting used to as the default unit in AutoCAD is the inch. Using the Project Units command discussed here, you could conceivably change the default of your project template to inches rather than feet. However, on-screen temporary dimensions would display in inches only. Further, you would also need to make adjustments to the settings of your dimension types if you want them to display both feet and inches. Feel free to experiment with such changes if you wish; realize, however, that a complete emulation of the AutoCAD behavior will be somewhat elusive. It might be easier to simply ask users to take some time to get used to the "new" default.

- Click Cancel when finished.

5. On the Project Browser, under Views (all), examine the contents of *Floor Plans*, *Ceiling Plans* and *Elevations (Building Elevation)*.

    There will be three Floor Plan views: *Level 1*, *Level 2* and *Site*.

    There will be two Ceiling Plan views: *Level 1* and *Level 2*.

    And beneath Elevations, there are four elevation views: *East*, *North*, *South* and *West* (see Figure 4.4).

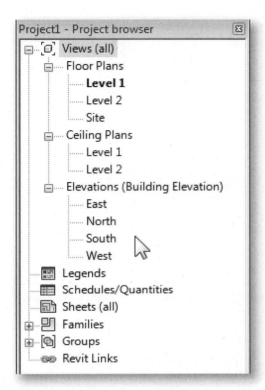

FIGURE 4.4 *There are several pre-made plan and elevation views*

6. On the Home tab of the ribbon, on the Build panel, click the ***Wall*** tool.

    • On the Properties palette, open the Type Selector.

There will be several Wall types available in the project (see Figure 4.5).

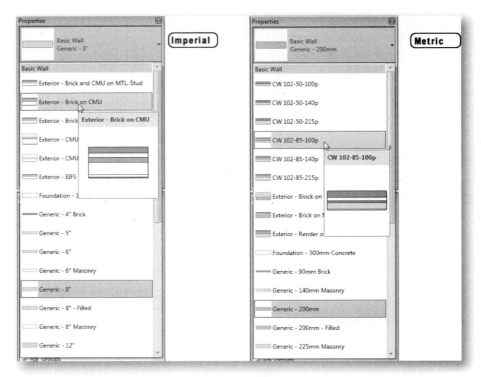

FIGURE 4.5 *Several Types are present in the project*

NOTE The list varies slightly between imperial and metric, but as noted above the two lists are largely comparable.

- Select any Wall Type and draw a short segment of Wall.
- On the ribbon, click the **Modify** tool or press ESC twice.

Section, Elevation and Level Heads are loaded in the project (however, these may also vary slightly in imperial and metric).

7. On the View tab, on the Create panel, click the **Section** tool.

- Click a point anywhere on screen, move the mouse in a straight line and click again (see Figure 4.6).

Notice the Section Head and the Section Tail that appear.

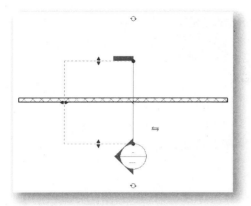

FIGURE 4.6 *Section lines are pre-configured in the template with a section head and tail*

8. On the Project Browser, expand *Elevations (Building Elevation)* and then double click *East*.

   • Zoom in on the Level Heads to the right (see Figure 4.7).

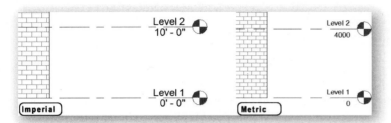

FIGURE 4.7 *Levels have Level Heads*

You can see what other tags are loaded in the project by using the Loaded Tags command. Click the Annotate tab of the ribbon, and then click the Tag panel label. This will expand the Tag panel to reveal hidden icons. Click the Loaded Tags icon. Each category of element can have one or more tags loaded in the project. Office standard tags can be preloaded using this dialog and saved with the office standard template.

Many other settings can be explored if you wish. If you are interested in building a custom template for your office, you may also like to explore the Families branch of the Project Browser and most of the commands on the Settings panel of the Manage tab. Pay particular attention to the Additional Settings drop-down and its many commands and options.

Create a Project from the Construction and Structural Templates

Let's repeat the exploration process in some of the other provided project templates.

9. From the Application menu, choose **Close**.

   • When prompted to save changes, click No.

10. From the Application menu, choose **New > Project**.

   • In the New Project dialog, in the "Template File" area, be sure that the lower radio button is selected (the one that lists a template file).

   • Click the Browse button.

As above, the dialog should open directly to your default template folder. If it does not, browse to the template location listed at the start of the tutorial above.

- Select the Construction-Default.rte [Construction-DefaultMetric.rte] template file and then click Open.
- In the "Create New" area, verify that Project is selected and then click OK.

The first thing you should notice about this template is that there are many more views in the Project Browser; particularly beneath the *Schedules/Quantities* branch (see Figure 4.8).

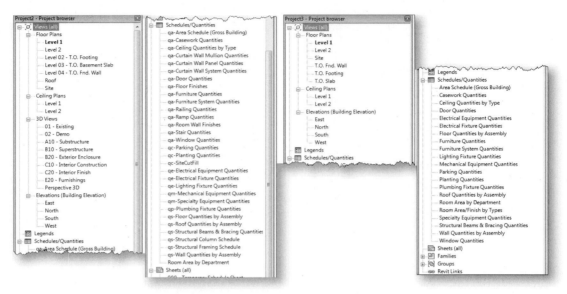

FIGURE 4.8 *The Construction template is pre-loaded with dozens of schedules and views*

Spend some time in this template exploring the floor plan, ceiling plan and elevation views like we did above. The same four elevation views are provided here. There are a few additional floor plan views, however, and if you open one of the elevation views, you will notice that there are Levels defined for the foundation and footing. The most significant addition to this template, which was not included in the Default template, is the inclusion of dozens of schedule views. As you can see from the names of these schedules, you can determine quantities for nearly every element in the project.

11. On the Project Browser, double-click the floor plan View: *Level 1* to make it active.

- Create a Wall anywhere on the screen. Click **Modify** or press ESC twice when done.

12. On the Project Browser, double-click the *qs-Wall Quantities by Assembly* [*Wall Quantities by Assembly*] Schedule view to open it (see Figure 4.9).

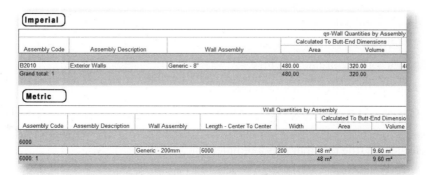

FIGURE 4.9 *Schedules will populate automatically as elements are added to the model*

Take a little time exploring these schedules now if you like. Add a few more Walls or some Doors and then open some of the schedules to see how they update. Schedules will be covered in detail in later chapters. The exact list and names of schedules in the Imperial and Metric templates vary. Therefore, despite your unit preference, you may wish to open both templates and look at each of them.

> The Imperial template provides some sheets as well. In particular, there is a sheet named *000 – Temporary Schedule Sheet*. This sheet has each of the Schedules inserted onto it. There is a piece of text noting "Temporary schedule sheet to allow easy copy and pasting into other projects." This sheet remains part of this template, but is no longer required. In earlier versions of Revit, you could not easily copy and paste a schedule between projects; it could only be done if it was placed on a sheet. This limitation no longer exists. You can now simply right-click a schedule name in Project Browser and then choose Copy to Clipboard. Switch to the project where you want to paste, and on the Modify tab of the ribbon, click the Paste icon. You can also still copy and paste directly from and to a sheet, so the presence of the *000 – Temporary Schedule Sheet* can still be useful.

13. On the Project Browser, expand *Elevations (Building Elevation)* and then double click *East*.

   - Zoom in on the Level Heads to the right (see Figure 4.10).

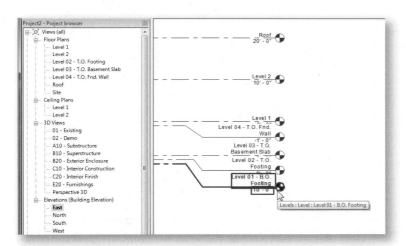

FIGURE 4.10 *Blue level heads indicate an associated floor plan view while black level heads have no plan views*

If you compare the list of floor plan views in the Project Browser with the level heads in the elevation, you will notice that there is no "Footing" floor plan. Further, the Footing level head is colored black rather than blue like the others. The color blue on screen indicates interactivity. We have seen this in previous chapters with the temporary dimensions and section heads. Here, when you see a tag like the level heads colored blue, it is also interactive. Double-click on the level head to open the associated floor plan view. Think of it like a hyperlink in a web browser. The same behavior is true with elevation tags and section tags. If the symbol is black, it means that it is just annotation and there is no associated view.

OTHER TEMPLATES

The imperial content contains a *Commercial-Default.rte* and a *Residential-Default.rte* template as well. You can create a project from each of these and explore them the same way. There are also some Canadian variations in the Metric folder. Even if you do not have imperial templates installed, you can find copies of these templates with the files installed from the dataset files from the student companion.

Starting a Project without a Template (Not Recommended)

It is possible to begin a project without a template; doing so is not recommended, however. After exploring several of the provided templates in the exercise above, you should be getting a sense of the variety of settings and elements that are resident in a typical template. If you start a project without one, you are forced to either configure/create all of these items on your own manually as needed, or import them from other projects or libraries. While it is possible to do this, the amount of extra work that it adds to your project makes it an ill-advised approach to creating a new project. If you have any doubt as to the validity of these claims, try it out for yourself to see.

14. Create a New Project, but in the New Project dialog, choose None for "Template file" and then click OK. When prompted for "Initial Units" make your choice of either Imperial or Metric.

15. Repeat the process that was followed above to explore the template.

- Check the Project Browser, and notice that only a plan and ceiling plan are included. Check the Families and Types. Check the Materials. Both will contain minimal choices.
- Add a Section or Elevation and notice that they do not even include Section or Elevation Tags.

16. Close the Project(s) without saving when you are satisfied that you understand the differences between it and a Project created from a template.

Create Your Own Template

Hopefully, you are beginning to see the benefits to starting new projects with a template. There is really no compelling reason to begin projects any other way. As you work through the exercises in the coming chapters, you will certainly discover areas where the default templates could be enhanced and improved. Make note of these observations as you go. When you are ready, try your hand at creating your own template file. The basic steps are simple:

- From the Application menu, choose **New > Project**.
- Load the existing template that is most similar to the one you wish to create.
- Choose the "Project Template" option in the New Project dialog.
- Edit any settings as you see fit. (Change Settings, add or delete views, load or delete Families, etc.)
- Add, edit, or delete Materials.
- Configure settings on the Manage tab such as units, fill patterns, line styles, objects styles, etc.
- Also configure the settings from the Additional Settings pop-up menu.
- When finished configuring, choose **Save As > Template** from the Application menu.
- From the "Save as type" list, verify that **Template Files (*.rte)** is selected.
- Be sure to browse to your *Template* folder, type a name for your new template and then click the Save button (see Figure 4.11).

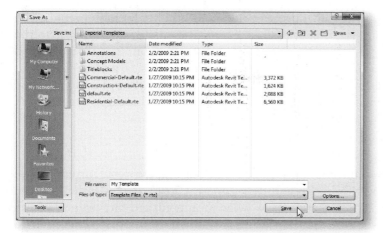

FIGURE 4.11 *Creating a new Template*

The templates shipping with Revit Architecture are ready to use straight out-of-the-box. They are excellent starting points for developing your own office-standard project template file(s). It is highly recommended that you become comfortable with the provided offerings, and then if necessary, customize them to meet your firm's specific needs. Consider including those settings and views that people will use most frequently. Include unique items like your office standard titleblock. Once you have a standard template in place, it is imperative that all users be required to use it. In some cases, a project will have special needs not addressed by this office standard template. In such cases, project-based derivatives can be created.

SETTING UP A COMMERCIAL PROJECT

Now that we have seen some of the features and benefits of creating projects from templates, let's create an actual project that we will follow (as well as the one started in the last chapter) throughout the remainder of the book. This project is a 30,000 SF [2,800 SM] small commercial office building. The project is mostly core and shell with some build-out occurring on one of the tenant floors. The tutorials that follow walk through the startup of the commercial project and the setup of several views including: plans, elevations, sections and schedules. At the end of the chapter, we will generate sheet views and print a cartoon set. The completed files for the project are available with the dataset files you installed from the student companion in the *Chapter04\Complete* folder on your hard drive.

Getting started with the commercial project gives us a nice practical exercise to further the goals of this chapter: namely, the understanding of project templates, levels, and views. Even though we have explored many templates thus far, we'll use the default template for this project. This is because the default template gives a common starting point in both Imperial and Metric units. We will adjust the project levels to suit the needs of our commercial building and then begin creating views that we will need. In some cases, we will create these items ourselves, and in others we will borrow them from additional templates as appropriate. When we are finished with the set up in this chapter, you can optionally save the current project as a new template file.

It is a really good idea to take the opportunity early in the project cycle to develop a digital "cartoon set." Like a traditional paper-based cartoon set, a digital one is simply a collection of preliminary sheets that will allow you to assess the quantity and composition of each of the drawings, schedules and other deliverables required in the final document package. This will assist the team in allocating resources to the project. Remember, like everything else in Revit, the cartoon set will evolve and update automatically as the project progresses. The goal at this early stage is to assist with planning and gain a jump-start on production. This chapter will explore several setup topics, but the final task of the chapter is to produce this digital cartoon set. Keep this in mind as you work through the exercises that follow.

WORKING WITH LEVELS

Before we start our commercial project, close all projects that you currently have open (open the Application menu and then choose **Close**). If it is easier, you can exit Revit Architecture, choose not to save when prompted, and then launch a fresh session. In either case, be sure that all projects are closed before proceeding. (You do not need to save anything from the previous exercises).

Adding and Modifying Levels

1. From the Application menu, choose **New > Project**.

    • In the New Project dialog, in the "Template File" area, be sure that the lower radio button is selected (the one that lists a template file).

    • Verify that the *default.rte* [*DefaultMetric.rte*] template file is listed in the text field (use the Browse button if necessary to choose it).

NOTE If you are in a country for which your version of Revit Architecture does not include these template files, they have both been provided with the dataset files from the student companion. Please browse to the *Template* folder to locate them.

    • In the "Create New" area, verify that Project is selected and then click OK.

By now we have started a few projects this way, and the familiar *Level 1*, *Level 2* and *Site* plan views will show on the Project Browser. Normally, the names of the Levels in the project will also be used as the names of the Floor Plan views. As our first task in our new commercial project, let's modify and add to the project Levels.

A Level is a horizontal datum element used to represent each of the actual floor levels or other significant reference heights in the building. Levels are used to organize projects vertically, in the Z axis. Typically you will add a Level for each story of the building ("First Floor," "Second Floor," "Roof" etc). In addition Levels can also be used to define other meaningful horizontal planes such as the "Top of Structure" or the "Bottom of Footing." We saw examples of this in several of the sample templates. For our commercial project, we need four stories, a roof and the grade level.

Take a look at the Home tab of the ribbon. Notice that the Level tool is grayed out. This is because the currently active view: *Level 1*, is a floor plan view. You cannot add or edit Levels in a floor plan view. To do so, you must switch to an elevation or section view.

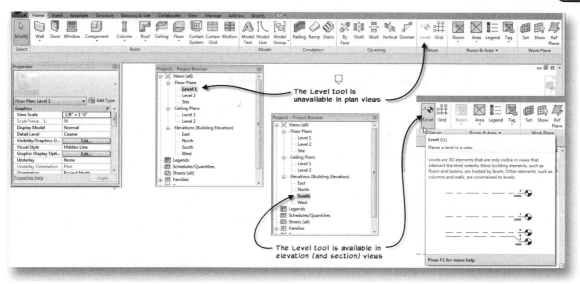

FIGURE 4.12 *Levels cannot be added or edited in Plan views*

2. On the Project Browser, expand *Elevations (Building Elevation)* and then double-click *South* (see Figure 4.12).

In this view, you can see two Levels: Level 1 and Level 2.

> **NOTE**
>
> Notice that even though there is a "*Site*" Floor Plan view in the Project Browser, there is *not* a "Site" Level. The *Site* view is actually associated to Level 1. In other words, there can be multiple plan views associated with the same Level. In addition, as we saw above, you can also have Levels that have no associated plan views. We will explore these concepts further below.

3. Zoom in on the Level Heads at the right side of the screen so that you can see them clearly.

Most elements in Revit will show a Tool Tip when you pass the mouse over them. Please note that you can change the behavior of these tips in the Options dialog from the Application menu.

- Move your pointer over the Level line to see a Tool Tip.
- Move your pointer over the Level Head to see a different tip (see Figure 4.13).

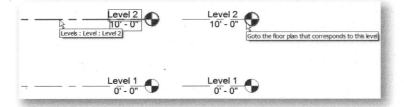

FIGURE 4.13 *Level Heads can be used to navigate to the associated Plan view*

> **TIP**
>
> In general, the dark blue color in Revit indicates that the item is interactive. In the case of Level, Section, and Elevation tags, double-click to open the associated view. In the case of blue text, labels, or dimensions, click to edit the value. We saw examples of this in the previous chapter.

To select the Level, click the dashed level line as shown on the left side of Figure 4.13. To open the associated floor plan view for the level, double-click the blue Level Head (or right-click and choose **Go to Floor Plan**) as shown on the right side. This is analogous to clicking on a link on a web page (except you must double-click in Revit), and it opens the associated view. If the Level Head is black, that means that there is no associated view (seen with the construction template above). The Level by itself is a datum for documentation and modeling purposes only and should not be confused with the actual plan view, which is one of many ways to view and interact with your Revit model. Levels typically have associated floor plan views, but it is not required that they do so. On the other hand, floor plans *must* be associated to a Level.

4. Select Level 2 (click the dashed line, not the Level Head symbol or the text).

When you select the Level, it will highlight in light blue (indicates that it is selected) and several additional drag controls and handles appear. As you move your pointer (referred to as the *Modify* tool) over each one (don't click them yet; simply hover the pointer over each one), the handle will temporarily highlight and the tool tip will appear (see Figure 4.14).

Moving left to right in Figure 4.14, the following briefly describes each control.

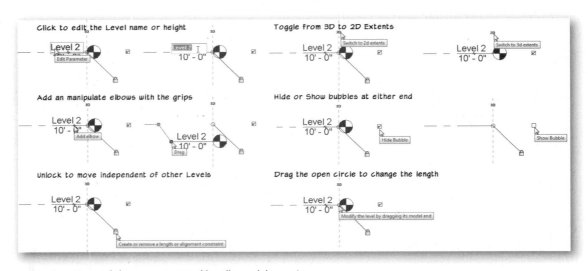

FIGURE 4.14 *Levels have many control handles and drag points*

- **Edit Parameter**—Click the blue text to rename the Level or edit its height. If you rename a Level, you will be prompted to also rename associated plan views. You can accept or reject this suggestion. This means that the Level and its associated views can have different names if you wish.

BIM MANAGER NOTE — The height is measured from an "Elevation Base," which is a Type parameter of the Level System Family. The Elevation Base is the zero point from which the Level heights are measured. The Elevation Base can be either "Project" or "Shared." With Elevation Base set to Project, the values will be relative to the origin of the project (Level 1 is at zero in the default template). If you choose the "Shared" Elevation Base, values will report relative to a shared origin (in the shared coordinate system). This could be the height at Sea Level or some other appropriate datum height. You can find more information on this topic in the on-line help, and shared coordinates are covered later in Chapter 6.

- **Add Elbow**—This small "squiggle-shaped" handle creates an elbow in the Level line. This is useful when the annotation of two Level Heads overlap one another in a particular view. Click this handle to create the elbow, and then drag the resultant drag handles to your liking. This is a graphical effect in the view only. The level datum height is unchanged.

- **Modify the Level Drag Control**—This round handle is used to drag the extent of the Level. If the length or alignment constraint parameter is also active, dragging one Level will affect the extent of the other constrained Levels as well. Levels are constrained to one another by default.

- **Hide/Show Bubble**—Use this checkbox control to hide or show the Level Head bubble at either end of the Level line. This control toggles from hide to show.

- **2D/3D Extents Control**—When 3D Extents are enabled, editing the extent of the Level line literally changes the extent of the datum plane representing the Level and affects all views. If you toggle this to 2D Extents, dragging the extent of the Level line affects only the graphical representation in the current view only.

- **Length and Alignment Constraint**—This padlock icon is used to constrain the length and extents of one Level line to the others nearby. This is useful to keep all of your Level lines lined up with one another.

5. Click anywhere next to the Level (where there are no objects) to deselect it (or press the ESC key).

6. On the ribbon, click the Home tab and then click the Level tool (on the Datum panel).

- Zoom out so that you can see the entire length of the Levels.
- Move the cursor above the left ends of the existing levels.

A temporary dimension will appear with an alignment vector above the top Level.

- When the temporary dimension reads 12'-0" [3600], click the mouse to set the first point (see the left side of Figure 4.15).
- Move the pointer over to the right side; when an alignment vector appears above the existing Level Heads, click to set the other point (see the right side of Figure 4.15).

FIGURE 4.15 *Add a Level aligned with the existing ones*

Remember, you can zoom in to change the temporary dimension snap increment if necessary. Also, you can simply type the value that you wish while the temporary dimension is active and then press ENTER to apply it. Finally, like other tools that we have already seen, you can place the Level at any dimension and then edit the value of the temporary dimension later to the desired value.

TIP

A new Level is added as well as two new plan views: a *Level 3* Floor Plan view and a *Level 3* Ceiling Plan view (see Figure 4.16). To control whether adding a new Level also creates new plan views, use the "Make Plan view" checkbox on the Options Bar. Furthermore, if you click the "Plan View Types" button, you can control which type of plan view(s) are created. By default, a floor plan and a ceiling plan view are created. Notice also that the "Length and Alignment" constraint is also automatically applied as you add Levels.

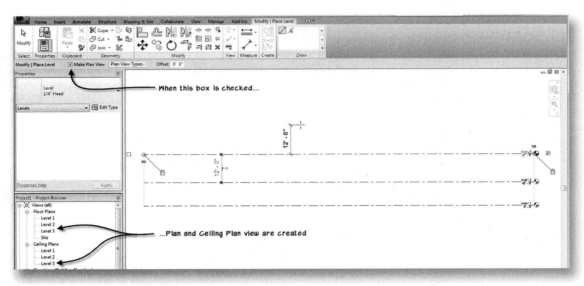

FIGURE 4.16 *As you add new Levels, by default new floor plan and ceiling views are created*

The Level command should still be active. If you canceled it, click the Level tool on the ribbon to restart it.

7. On the Modify | Place Level tab of the ribbon, on the Draw panel, click the Pick Lines icon. In the Offset field on the Options Bar, type: **12'-0" [3600]**.

   • Place your mouse pointer over the Level 3 line, and when the dashed offset line appears above, click to create a new Level 4.

   • Repeat once more to create an additional Level above Level 4.

You should now have five Levels. Our project already has a *Site* plan view. If you edit its properties, you will see that the associated Level is actually Level 1 though. There is nothing wrong with that approach, but in our project, Level 1 is actually raised above the grade level a few feet. Even if this weren't the case, it is sometimes preferable to have dedicated Level for the site plan regardless. To do this, we must first delete the existing site plan view and recreate it with a new Level.

8. On the Project Browser, right-click the *Site* floor plan and choose **Delete**.
9. On the ribbon, on the Home tab, click the *Level* tool.

   • On the Options Bar, click the Plan View Types button.

   • Click on Ceiling Plan to deselect it (you only want Floor Plan highlighted) and click OK.

   • On the Draw panel, click the Pick Lines icon. In the Offset field, type: **3'-0" [900]**.

   • Offset a Level beneath Level 1 (see Figure 4.17).

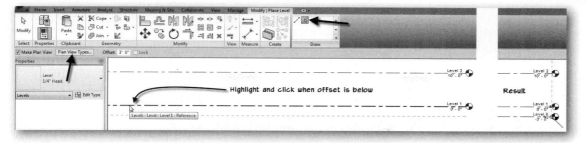

FIGURE 4.17 *Add the final Level below Level 1 with only an associated floor plan view*

- On the ribbon, click the **Modify** tool or press the ESC key twice.

Note that a new Level 6 has appeared in only the *Floor Plans* category.

<table>
<tr><td>If you forget to click on the Plan View Types button, you can always delete the unwanted view (Ceiling plan in this case) later.</td><td>**TIP**
</td></tr>
</table>

Renaming Levels

We can accept the default level names as they appear, or we can change them to something more suitable for our specific project. The names of Levels can be anything that makes sense to the project team. Levels in the model will appear in their correct physical locations; however, in the Project Browser they will sort alphabetically by default. On larger projects with many levels, it might make sense to use a naming scheme with a numerical prefix such as 01 Level, 02 Level, etc. (In projects with more than 10 stories, you should consider placing a leading zero in the level names so that Level 10 does not inadvertently sort before Level 1). We'll just rename the top and bottom levels in this project to something a bit more descriptive.

<table>
<tr><td>Browser Organization controls which views are shown in the Project Browser and how they are organized. It is possible to modify the default organization to make the plans sort by associated level height instead of alphabetical. An example of customizing the Browser Organization is presented below in the "Working with Browser Organization" topic.</td><td>**TIP**
</td></tr>
</table>

10. Select Level 5 (the one at the top).

    - Click on the blue text of the name to edit it. Type: **Roof** for the new name.
    - A dialog will appear asking you if you like to rename the corresponding plan views; click Yes (see Figure 4.18).

FIGURE 4.18 *Rename the top Level to Roof and accept the renaming of corresponding views*

When you choose "Yes" in this dialog, the two associated plan views, *Level 5* floor plan and *Level 5* ceiling plan, become "*Roof.*" Since we don't really need a ceiling plan for the roof, however, let's delete this view.

11. On the Project Browser, right-click the *Roof* ceiling plan view and choose **Delete**.

12. Repeat the rename process on Level 6 (the one at the bottom). Rename it to **Street Level**.

    • In the dialog that appears, answer No this time.

      You do not need to name the Levels and the views the same. In this example, we will find it useful to call out the level on the elevations and sections as "Street Level," but the plan associated to this level will be our site plan. Therefore, in this case, it is better to name the Level and the plan view separately.

13. Right-click Level 6 beneath *Floor Plans* on the Project Browser and choose **Rename**.

    • Name the view: **Site** and then click OK.

    • In the elevation view on screen, double-click the Street Level head.

    • Notice that this opens the *Site* view. So while the names are no longer the same, they are still linked to one another.

Apply a View Template

Look carefully at this view. When you create a floor plan view, certain defaults are automatically applied to it like display settings and scale, for example. Typically, a site plan is drawn at a scale smaller than the default 1/8"=1'-0" [1:100] used here. Further, there are certain object categories that typically do not display on site plans and other less obvious settings as well. While we could manually configure all of the various settings needed to make this view a more suitable site plan, we can do this more efficiently by applying a view template. A view template is simply a collection of saved view settings that can be applied to a view in one step. Many view templates are included in the template we used, and you can make them your own as well.

14. On the View tab of the ribbon, on the Graphics panel, click on the View templates drop-down button.

    • From the pop-up that appears, choose **Apply Template to Current View**.

    • In the "Apply View Template" dialog, select Site Plan from the list and then click OK (see Figure 4.19).

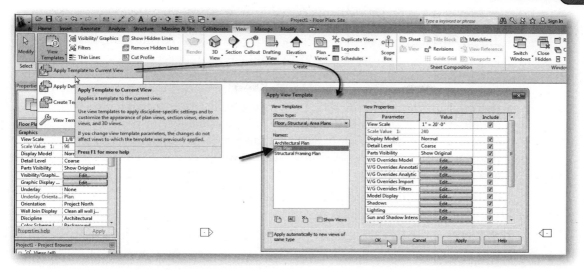

FIGURE 4.19 *Apply a view template to the Site Plan view*

The most obvious change will be the increase in size of the elevation tags. This reflects that the scale of the view has changed. There are other settings that were changed as well. For example, two icons appear in the center of the view. They currently occupy the same location. One is the Project Base Point icon and the other the Survey Point icon. The Project Base Point icon can be used to relocate the project, and the Survey Point can be used to adjust a reference point relative to a linked site file. We will discuss these icons further in Chapter 6.

15. From the Application menu, choose **Save**.

- Browse to the *Chapter04* folder, give the project a name such as **MRAC-Commercial** and then click Save.

You can create your Levels and plan views at the same time as we did here. You can also remove the option to create plan views completely when new Levels are added. You may choose to do this to create datum views to demarcate Top of Steel, Top of Plate, or Bottom of Footing, etc. New plan views can be created and associated to any and all existing Levels at any time. To add a view to an existing Level, click the Plan Views tool on the View tab of the ribbon. By default, Revit selects the "Do not duplicate existing views" option in the "New Plan" dialog. To create another floor or ceiling plan on a level that already has one, clear this checkbox. Select the level where you wish to add a plan and select an appropriate scale before clicking OK. If you are creating a duplicate plan, you can also right-click an existing view at any time in the Project Browser and duplicate it directly.

MANAGER NOTE **BIM**

CREATE A SIMPLE SITE

Now that we have created and named all of our Levels, we can move on to adding some basic building elements to our model. Having some simple geometry in the file will make planning views and setting up preliminary sheets easier. Since the site plan is already open, this is a good place to start.

Create a Toposurface

The first thing we need in a site plan is our site! Revit Architecture has some simple site design tools that enable us to build a topographical surface upon which our building model can sit. We can also add planting, parking, property lines, and other site accoutrements. You can build a Toposurface in two ways: by manually placing points or by importing site data from external files. In this chapter, we will create a simple surface manually as a temporary ground plane for our project setup. Later in Chapter 6 we'll get into more detail on the Site tools, build a site model and link it into our project replacing the simple site that we are creating here. This one will serve as a temporary stand-in.

1. On the ribbon, click the Massing & Site tab.
2. Click the ***Toposurface*** tool.

The Modify | Edit Surface tab will be displayed on the ribbon. The existing items in the view (such as the elevation tags) will gray out. he ***Place Point*** tool on the Tools panel should be selected. (If it is not, click it now.)

- On the Options Bar, verify that the Elevation field is set to **0**.
- Zoom out slightly and click four or five points across the top of the screen slightly above the top elevation marker.
- Change the Elevation to **-1'-0" [-300]**. Click four or five more points running horizontally beneath the elevation marker.
- Change the Elevation to **-2'-0" [-600]**. Click the next set of points horizontally beneath the east and west elevation markers.
- Change the Elevation to **-3'-0" [-900]**. Click a final set of points horizontally beneath the south elevation marker (see Figure 4.20).

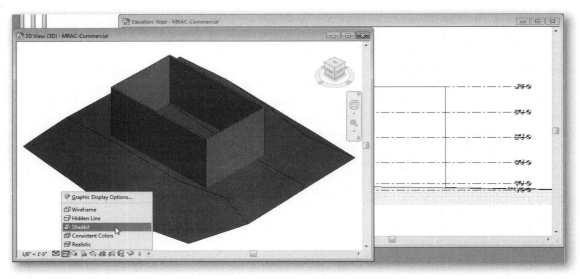

FIGURE 4.20 *Create a Toposurface from sketched points*

Do try to build the surface so that it is larger than the elevation markers all the way around. Do not worry about making it exactly like the figure. If you want to edit any points, click the Modify tool and simply drag them where you like.

3. On the ribbon, click the Finish Surface button (large green checkmark).

   • Open the *East* elevation view.

Zoom in as necessary and study the results. Notice that the surface appears with a gentle slope and an earth pattern applied where the grade is cut (see Figure 4.21).

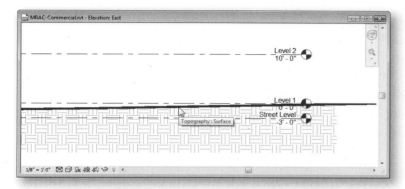

FIGURE 4.21 *Finish the sketch and view the Toposurface in elevation*

If you only see the edge of the Toposurface and not the earth hatching, this means that the points you placed above were inside the elevation zone (to the left of the East elevation marker). The hatching only shows when the elevation line actually cuts through the Toposurface (making it essentially a section at that point).

NOTE

Like any sketch-based object in Revit, you can select the Toposurface at any time and click the Edit Surface button on the ribbon. This will return you to sketch mode where you can select and modify the points from which the topo is derived. You can move the points, change their elevations, add new points, and delete existing points. If you choose to add a new point, you add it at an absolute elevation like the ones we added above, or when editing an existing surface, you can also choose to make them relative to the surface. You will find this on the Options Bar after you click the ***Place Point*** tool. Feel free to experiment with this Toposurface before continuing. We will be deleting it in favor of a new one we will create from an imported CAD file later in Chapter 6. So have some fun exploring the options and don't worry about "messing" this one up. We only need some sort of ground plane for now to give our building model a base upon which to sit. The exact shape of the base is not terribly important.

ROUGH OUT THE BUILDING FORM

Since our goal in this chapter continues to be the overall setup of our commercial project, we will take this opportunity to add some simple geometry to the project to suggest building form. The purpose is not to arrive at complete design solution, but rather to provide enough of a building form to give the various views and sheets that we will establish below some context. The tasks that follow are therefore simple suggestions of project setup workflow. Feel free to vary the steps as appropriate when following the process in setting up your own projects.

Adding Walls and Constraining Walls to Levels

We have a simple Toposurface. Next we'll add some geometry to rough out the overall form of our building. At this early stage of the project, some simple Generic Walls will be sufficient. As the design evolves the wall types, layout and shapes can be changed and edited to suit our needs.

1. On the Project Browser, double-click the *Site* floor plan view.
2. On the ribbon, click the Home tab and then click the Wall tool.

    - From the Type Selector, (on the Properties palette) choose **Basic Wall: Generic - 12"** [**Basic Wall: Generic - 300mm**]
    - On the Modify | Place Wall, on the Draw panel, click the rectangle icon.
    - On the Options Bar, from the Height list, choose **Roof**.
    - For the Location Line, choose **Finish Face: Exterior** (see Figure 4.22).

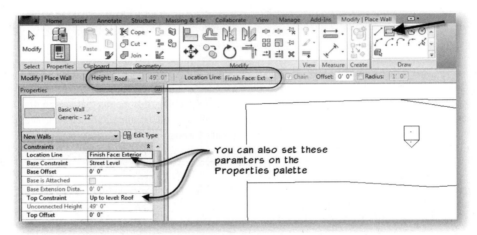

FIGURE 4.22 *Set the options to create new Walls*

3. In the view window, click two opposite corners to create a rectangle in the middle of the elevation markers.

    - Temporary dimensions will remain on screen. Edit the horizontal temporary dimension to a value of: **104'-0"** [**31500**].
    - Edit the vertical temporary dimension to a value of: **64'-0"** [**19200**] (see Figure 4.23).
    - Make sure the rectangle is roughly centered in the space defined by the elevation markers. If you need to move the Walls, select all four Walls and then drag them to reposition.

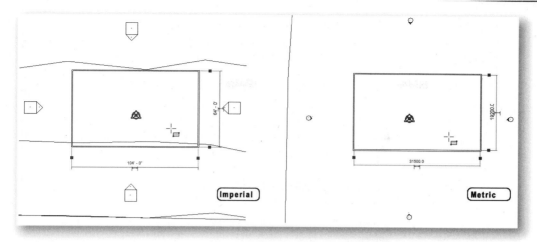

FIGURE 4.23 *Edit the temporary dimensions of the Walls and ensure that they fall within the elevation markers*

4. On the Project Browser, double-click to open the *West* elevation view.

   • Zoom in near the bottom of the building

Since we drew the Walls on the *Site Plan* view, the bottom edges should meet the Toposurface nicely.

5. On the Quick Access Toolbar (QAT), click the Default 3D View icon.

   • On the View Control Bar, click the Model Graphics Style icon and choose **Shaded with Edges** (see Figure 4.24).

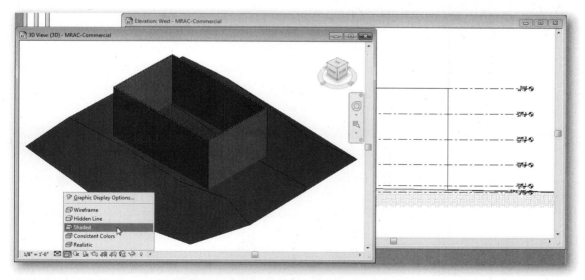

FIGURE 4.24 *Study the model in elevation and 3D to see the results of adding the four Walls*

We now have a big hollow box sitting on our terrain. Let's add a simple Roof.

Add a Simple Roof

6. On the Home tab, click the drop-down arrow on the **Roof** tool and then choose **Roof by Footprint**.

A message will appear requesting that we select the desired Level to build the Roof. This message appears because we are in a 3D view. Naturally we will want to create the Roof at the Roof Level.

- In the "Lowest Level Notice" dialog, choose **Roof** from the drop-down list and then click Yes (see Figure 4.25).

FIGURE 4.25 *When creating a Roof in 3D, clarify the desired creation Level*

- On the Modify I Create Roof Footprint tab, on the Draw panel, verify that the Boundary Line tool is active and that the **Pick Walls** icon is selected.
- On the Options Bar, clear the checkmark from the "Defines slope" checkbox.
- Using the TAB key, chain select all four Walls and then click to create sketch lines (see Figure 4.26).

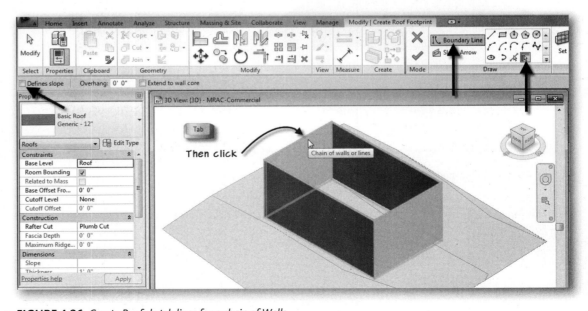

FIGURE 4.26 *Create Roof sketch lines for a chain of Walls*

- Zoom in as required and verify that the sketch lines are on the inside edge of the Walls. If they are not, click the small flip control to shift the sketch lines.
- On the ribbon, click the Finish Edit Mode button.

Since we cleared the "Defines slope" checkbox, you will get a simple flat slab of a Roof placed at the top edge of the Walls. If you got a hip roof, you forgot to clear this checkbox. Select the Roof, and on the ribbon that appears, click the Edit Footprint icon. Select all four sketch lines, and then, on the Options Bar, clear the "Defines Slope" checkbox.

Adjust Wall Height

Our goal at this stage is to create a very rough model that we can use to help set up and understand the views in our project. We don't need to get too concerned with the details of the design at this stage. However, one quick edit to the Walls is immediately obvious. Let's adjust the Wall height to suggest a parapet at the Roof.

7. Chain select all four Walls (pre-highlight one Wall, press TAB to chain highlight and then click to select all four).

   - On the Properties palette, for the Top Offset parameter, type **4'-0" [1200]** and then click Apply (see Figure 4.27).

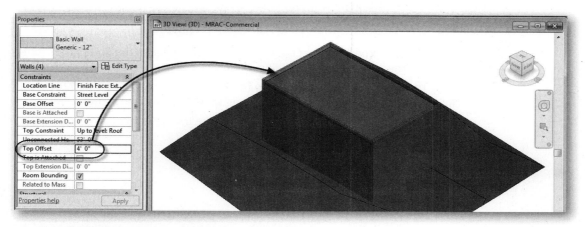

FIGURE 4.27 *Add a Top Offset to the Walls to represent a parapet*

8. Save the Project.

Add an In-Place Mass

In the coming chapters, the design of both our projects will evolve. At this stage of our commercial project we have little information other than the number of levels, a vague notion of the slope of the site and few ideas of the building's overall form. It is desired that some sort of special treatment be given to the front façade. We don't have the specifics about this treatment yet but at this stage can make a few assumptions. To do this, we will sketch out a simple Massing element on the front façade of our building to suggest that some design element will later occur in this location. You can design the entire form of your building using Masses or just a portion of it as we are doing here.

NOTE The In-Place Massing tools use the conceptual modeling toolset. As such, some of the behavior of the specific tools will vary slightly from other sketch-based objects that we have seen so far. For detailed coverage of the Conceptual Modeling environment, refer to chapter 16.

9. On the Project Browser, double-click to open the *Level 1* floor plan view.

   Be sure to open the *Level 1* floor plan and not the *Level 1* ceiling plan.

10. On the ribbon, click the Massing & Site tab and then click the **In-Place Mass** tool.

A message will appear indicating that the "Show Mass" mode will be enabled. Massing tools are meant as design tools and as such are not visible by default. We can enable their visibility in any view that we wish. This message is simply a courtesy indicating that Massing display will be enabled temporarily for us. There is also a tool on the Conceptual Mass panel next to the **In-Place Mass** tool that toggles massing display on and off.

   • In the message dialog, click Close after you have read it.

   • In the Name dialog, type: **Front Façade** and then click OK.

The Home tab will become active, and the tools available will change slightly to reflect the Massing environment. Some of the tabs will disappear, and the In-Place Editor panel will appear at the right side of the ribbon. This panel is visible on any tab while you are in the In-place modeling mode.

11. On the Home tab, on the Draw panel, click the Rectangle icon.

   • Click the first corner of the rectangle on the outside edge of the bottom horizontal Wall near the left side of the building.

   • Using the temporary dimension, keep the height of the rectangle 8'-0" [2400] and click down and to the right to create a long thin rectangle (see Figure 4.28).

   • On the ribbon, click the **Modify** tool or press the ESC key twice.

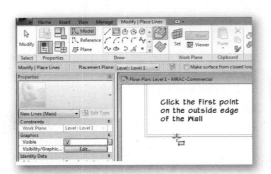
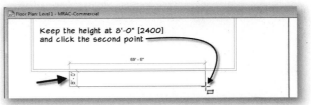

FIGURE 4.28 *Sketch a rectangle using temporary dimensions as a guide*

Working in the conceptual modeling environment, selection is a little different than in the traditional sketch mode. For example, when you pre-highlight a shape, chain selection is the default. Therefore, Revit will select the entire rectangle in a single click. To select just one edge, use the TAB key.

12. Pre-highlight the vertical edge of the rectangle on the left and then press TAB.

   • The vertical edge should pre-highlight. Click to select the vertical edge and then change the temporary dimension from the left edge of the building to 24'-0" [7200] (see Figure 4.29).

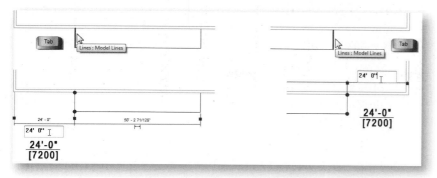

FIGURE 4.29 *Adjust the size of the rectangle using temporary dimensions*

- Repeat the process on the other side as indicated in the figure.
- On the QAT, click the Default 3D view icon.

You should see the rectangle down near the bottom of the front Wall of the building. If you have orbited your 3D view to another vantage point, use the view cube to rotate it back to a southeast isometric view.

13. Click to select the rectangle (remember that in conceptual mass mode, chain select is the default, so simply click any edge of the rectangle).

- On the ribbon, click the Create Form icon.

Notice that the rectangle has been extruded up to form a 3D box. A form control appears at the top of the box. It has three colored arrows that allow you to stretch the shape in any direction. In this case, we can drag the blue arrow to change the height of the box.

- Drag up on the blue arrow on the form control (see the left side of Figure 4.30).

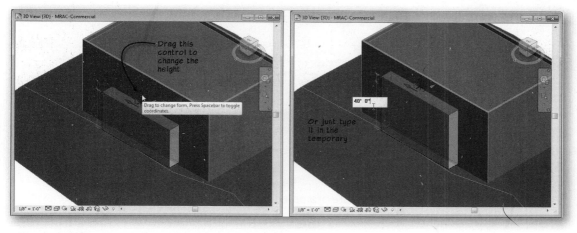

FIGURE 4.30 *The completed Mass at the front of the building*

- You can also use the temporary dimensions. Click on the vertical temporary dimension, and type **40'-0" [12,000]** for the height of the box (see the right side of Figure 4.30).

This short exercise barely scratches the surface of the possibilities in the conceptual modeling environment. We will have opportunities to return to it later and make adjustments and explore other tools. For now, however, we will finish our simple Mass and continue with our project setup.

- On the ribbon, click the Finish Mass icon.

Masses are intended as a quick way to create and study building form. They are primarily a design tool but have some functions that allow them to easily transition into design development. In particular, looking at the Massing & Site tab of the ribbon, you will notice four tools on the Model by Face panel. We can create a Curtain System, Roof, Wall and Floor using these tools. Each of these object types can be created directly from the faces of a Mass. We will try a few of these tools here, and we will refine our Front Façade Mass element later, in Chapter 9. Also remember that Chapter 16 is devoted to the Massing environment.

Continue in the default 3D view.

14. On the Massing & Site tab, on the Model by Face panel, click the **Curtain System** tool (see Figure 4.31).

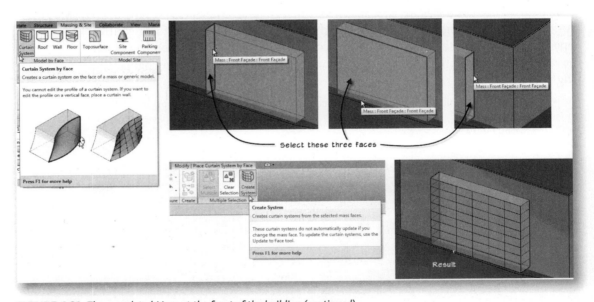

FIGURE 4.31 *The completed Mass at the front of the building (continued)*

At the Status Bar, you are prompted to select a face. Click each of the three exposed sides of the Mass.

- With the three surfaces selected (highlighted light blue) click the **Create System** tool on the ribbon (see the bottom of Figure 4.31).
- On the ribbon, click the **Modify** tool or press the ESC key twice.
- On the Model by Face panel, click the **Roof** tool.

At the Status Bar, you are prompted to select a face.

- Click the top surface of the Mass.
- On the ribbon, click the Create Roof button.
- On the ribbon, click the **Modify** tool or press the ESC key twice.

This quick example shows how you can use the Massing tools in a schematic design stage to suggest design elements that you are considering. Once you have completed working with the Mass, you can toggle their display off again using the tool on the Conceptual Mass panel of the ribbon. The other elements we created on the faces of our Mass will continue to display. If we later edit the Mass, an update to face option appears on the ribbon for the Walls, Curtain Systems, Floor, and Roofs that we create (refer to the "Editing Massing Elements" topic in Chapter 9, for example). This makes it easy to keep your Mass model and building model coordinated.

15. On the Conceptual Mass panel, click the Show Mass drop-down button and choose Show Mass by View Settings.

    This will toggle the display of Masses back off and leave just the Curtain System and Roof displayed.

16. Save the project.

WORKING WITH ELEVATION VIEWS

The default templates that we used to create our project already included four building elevations. As our project progresses, we can work with these elevations, delete them, and/or add additional ones. Elevation views are vertical slices through the building model cut from a certain point and projected orthographically. Level heads will display automatically in elevation views, and elevation view tags will appear by default in all plan views to indicate where the elevations are cut. The four default elevation views are sufficient to our needs at this time, but we will make a series of adjustments to them to fine-tune them.

Adjusting Elevation Tags

We have added enough geometry to our model to begin studying the effectiveness of our elevations. As we study them, we can see that it might be beneficial to adjust a few things.

1. On the Project Browser, double-click each elevation view in succession.

Zoom and pan within each elevation to study them carefully. As you open each elevation, take note of the edges of the Toposurface. Most of the surface appears in section with a bold top edge and an earth fill pattern beneath. This result is due to the locations of the elevation tags in the plan views. Let's take a closer look at the elevation tags to get a better understanding of their behavior.

2. On the Project Browser, double-click to open the *Level 1* plan view.

- Zoom in on the bottom elevation tag.
- Move the mouse pointer over the tag.

Notice that it is actually made of two pieces. The elevation tag is the square [round] shape, and the arrow (triangle) portion indicates the viewing direction and extent of the associated elevation view (see Figure 4.32).

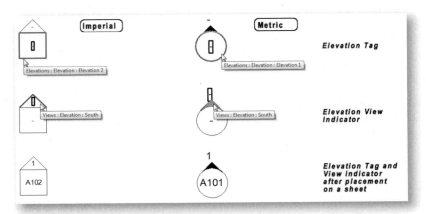

FIGURE 4.32 *Elevation tags and elevation view indicators*

A small dash appears within each tag. These are text labels that fill in automatically when the elevation view is placed on a sheet. (An example is shown at the bottom of the figure.) We will add sheets later in this chapter, so you do not have to create them now.

- Select the Elevation Tag (square in Imperial and round in Metric).

By default one elevation arrow is active per elevation tag. Each elevation tag can have up to four elevation arrows and corresponding elevation views. Three additional elevation view Indicator arrows will appear temporarily as long as the Tag is selected. Small checkboxes appear that you can use to add new elevation views. If you remove a checkmark, the associated elevation view will be deleted (see the left side of Figure 4.33). A prompt will appear to warn you of this. To test this out, check one of the empty boxes, and then note the addition of a new elevation view on the Project Browser. Remove the checkmark from the same box, and a warning will appear indicating that this new view will be deleted. There is also a rotate handle that appears when the tag is selected. This will rotate the entire elevation tag and all of its associated views. As the elevation tag is rotated it will snap to 90° increments and perpendicular to adjacent walls and other geometry. Try it out if you like; just be sure to undo before continuing.

- Deselect the Elevation Tag and then select the Elevation view Indicator arrow.

When this arrow is selected, it turns blue. You can double-click the blue arrow to open the associated elevation view. First select the arrow, and then double-click it if you want to open the view. Selecting the elevation arrow also reveals a line perpendicular to the direction of the arrow (see the right top of Figure 4.33).

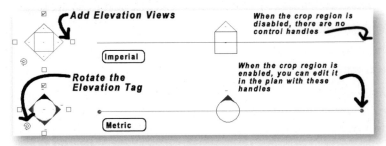

FIGURE 4.33 *Handles and drag controls on the elevation tags and elevation view Indicators*

This line indicates the extent of the cut plane of the elevation view. An elevation view is very similar to a section cut at this line. Objects crossing this line will show as cut in the corresponding view. We have seen this above with the Toposurface. Objects behind this line will not be shown in the elevation view. Objects in front of the line appear in projection. The extents of the elevation line can actually crop the view. To do this, you must enable the crop view parameter in the associated view properties. Let's explore a few of these parameters.

3. Type **ZF** on the keyboard, or choose **Zoom to Fit** from the Zoom pop-up on the Navigation Bar.

4. Select both parts of the south elevation Tag (the Tag and the arrow for the bottom-most one pointing up).

   - Drag it to up into the building just above the front façade (see the left side of Figure 4.34).

If you tile both the plan and elevation views side by side on the screen at once while performing these steps, you can see the elevation view change instantly as you move the tag around.

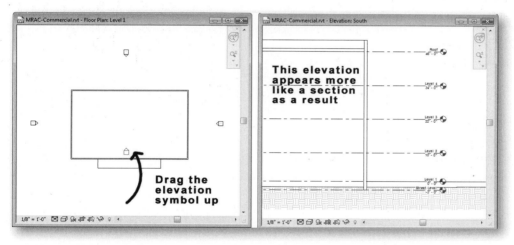

FIGURE 4.34 *Moving the South elevation tag and arrow inside the building results in a section*

- Undo the change or simply move the symbol back to its original location.
- Nudge each elevation view tag closer to the building on all four sides. (Select the elevation tag, and then use the arrow keys on the keyboard to nudge).
- Click the **Modify** tool or press the ESC key.

Loading Custom Elevation Tags

You may find that the elevation tag used in the out-of-the-box template does not match your office standard. We can customize the elevation tag to match virtually anything we like. To do so, we have to build two custom Annotation Families: one for the arrow and the other for the tag itself, called the "body." Once we have both pieces, we combine them to create the final elevation tag. The Family Editor is the environment in Revit where such customizations occur. Since use of the Family Editor can be complicated and sometimes intimidating, in this lesson, we will dispense with the steps to create the custom elevation tag and simply load one that has been provided for you. If you wish to try your hand at creating it yourself, refer to the "Creating Custom Elevation Tags" topic in Chapter 10 for the complete process.

5. On the Insert tab, on the Load from Library panel, click the Load Family button.

- In the "Load Family" dialog, browse to the *Chapter04* folder.
- Select the *NCS Elevation Tag.rfa* file and then click Open.

This process loads an external Revit Family file (RFA) into our current project. This file is an elevation tag drawn to the specifications outlined in the US National CAD Standard (NCS). The next step is to swap out the existing elevation tags in our project with the new one we have just created.

6. Select any one of the existing elevation tags onscreen.

Remember the tag is actually the square [round] part in the middle. The triangular part is the elevation view itself. Be sure to select the tag. On the Type Selector, it will say: Elevation: Building Elevation.

- On the Properties palette, click the Edit Type button (see the left side of Figure 4.35).

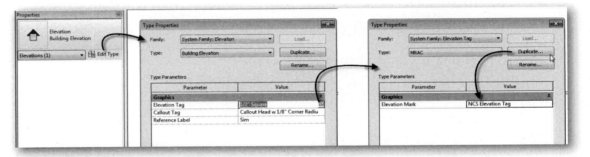

FIGURE 4.35 *Assign the MRAC Elevation Tag Type to the Exterior Elevations*

- In the "Type Parameters" area, click in the field next to Elevation Tag.
- A small browse button will appear. Click it (see the middle of Figure 4.35).

A second "Type Properties" dialog will open. This one is the properties of the Elevation Tag itself. At the top of the dialog the Family and Type are shown as drop-down lists. Next to the Type are some buttons.

7. Click the Duplicate button to copy the Type.

- For the name, type: **MRAC** and then click OK.
- Beneath Type Parameters, change the Elevation Mark to: **NCS Elevation Tag**.
- Click OK to dismiss the dialog.
- Click OK again to see the result.

Adjusting Elevation Cropping

The elevation views by default do not use or show the crop regions. This means that the view continues to dynamically resize itself as the building geometry grows and shrinks. If you wish to limit the size of the elevation, you can display the crop region and then resize it to crop away the unwanted portion of the view.

8. Open the *South* elevation view.

Take note of the odd way the edges of the Toposurface display. They are displaying accurately relative to the position of the elevation, but do not give a nice clean edge that would be more desirable in a finished elevation drawing. Adjusting the Crop Region can help.

9. On the View Control Bar, click the Show Crop Region view icon.

The Crop Region will appear as a long rectangle surrounding the elevation.

- Select the Crop Region.
- Drag the handles to crop away the unwanted portion of the view (see Figure 4.36).

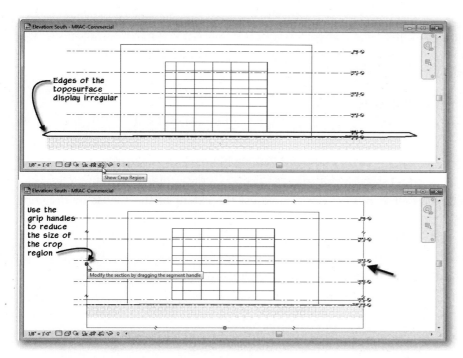

FIGURE 4.36 *Showing and adjusting the Crop Region*

- Click the Hide Crop Region icon. (same icon used to show it)

If you like the results, repeat the process on the North elevation. Otherwise, click the Do Not Crop View icon (to the left of the Show/Hide Crop Region icon) to turn it off and return to the full un-cropped view. Please note that the "Crop View" icon toggles to become the "Do Not Crop View" icon after you click it and likewise the "Show Crop Region" icon toggles to become the "Hide Crop Region" icon. When you have the view uncropped, all new geometry will appear automatically as the view dynamically adjusts to display the entire model. When you crop it, items added later that appear outside the Crop Region will be cropped. For example, later in the book we will add a stair tower to the roof of the building. The current cropping will need to be adjusted at that time to display this feature. Finally, take note of the way the Level lines adjust with the Crop Region. Try dragging the Crop Region in toward the middle of the building. Notice that the Level Heads will shorten but remain just outside the edge of the Crop Region. When the Crop Region is larger than the extents of the Levels (or Grids—see the next chapter) they will not extend to meet the Crop Region.

10. Repeat the cropping process on the *North* Elevation.

Dragging the crop region controls is a quick and easy way to adjust the crop region. You can also select the crop region, and then on the ribbon, click the Size Crop tool. This will open a dialog where you can adjust the width and height numerically. You can use this method to input a value that will fit within the available size of your sheet titleblock.

Adjusting Level Heads

Now that we have moved and resized the elevations, all of the views will display correctly. However, the Level Heads still need adjustment.

11. On the Project Browser, double-click to open the *West* elevation view.

In some views the Levels are a little off-center or a bit too long. Enabling Cropping might take care of it, but we can also adjust the extent of the Levels by dragging the ends.

- Click on one of the Level lines.
- Click and drag the small hollow circle handle at the end of the Level line and stretch the end closer to the elevation (see Figure 4.37).

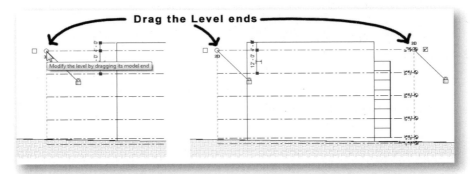

FIGURE 4.37 *Drag the end of the Level Heads to center them on the Elevation*

- Repeat on the other side.
- Adjust view cropping if desired.

The precise amount of the stretch does not matter. You simply want the end result to have the Level heads comfortably centered on the elevation.

12. Perform similar adjustments in the *North* (or *South*) elevation view if necessary.

Notice that you only need to adjust the extents of the Levels in two elevation views. Since elevations (like all Revit Architecture views) are live views of the model, changes you made in the *West* view can be seen automatically in the *East* view and likewise the changes in the *North* view appear automatically in the *South* and vice versa. You will also notice that when you drag one Level line, all of them will adjust together. This is because they are all constrained to one another (see above).

Recall the definition of 2D/3D Extents in the "Adding and Modifying Levels" topic above. Since the Levels currently use 3D extents, modifying their extent can be seen in any parallel view such as *West* and *East* in this case. Were we to change the extents to 2D, the change would apply only to the view in which it was edited. Further, the same Level can have 2D extents enabled in one view and continue to display its 3D extents in other views. Should you have occasion to modify the 2D extents or add or remove elbows from your Levels (or Grids), you can "push" these changes to other parallel views using the Propagate Extents tool. (Select the Level in question and then on the Modify Level tab, click the Propagate Extents tool).

- Review all four elevations and make any final adjustments to Level extents and cropping.

13. Save the Project.

It may be necessary to adjust the extent of the Levels in the *North/South* elevations as well. This edit is likely necessary only if you are working in Metric units as the default Metric template from which we began the project has smaller Level extents than the Imperial template.

Add Elbows

To fine-tune your elevations a little further, you can add an elbow to the Street Level line. Select the Street Level line and then click the Add Elbow handle near the Level Head. This is pictured above in top middle frame of Figure 4.14. Once you add the elbow, two handles appear that you can use to adjust its position. Move them in such a way as to make the labels more legible. Perform this action on each elevation, or you can do it in the *North* elevation and then use the Propagate Extents tool to push the change to the *South*. Repeat for *West* and *East*.

TIP

By now you likely have several windows open. To conserve computer resources and clear up visual clutter on screen, you can close unused windows. A quick way to do this is to maximize the current window and then click the **Close Hidden** tool on the View tab of the ribbon.

Edit Level Heights

Earlier when we set up all of the Levels, we neglected to adjust the default height between Level 1 and Level 2. By default in the Imperial template it is 10'-0" and it is 4000mm in the Metric template. However, for this project, the heights of all four

Levels should be the same at 12'-0" [3600]. The change was postponed till now so that we can get a sense of the power of some of the simple constraints that we have already built into our model.

14. On the Project Browser, double-click to open the *West* elevation view.

- Maximize the window if it is not done already. On the View tab, click the **Close Hidden** tool.
- Select Level 2.

Notice the temporary dimensions that appear. You can certainly edit those values to move the Levels to their correct height. However, the problem with that technique is that you must move the Levels one at a time. A better approach is to select and move them all together.

15. Select Level 2, Level 3, Level 4 and the Roof Level.

> **TIP**   Remember, you can hold down the CTRL key and click each one, or drag a window from right to left touching each Level you wish to select.

At this point, we could click the **Move** tool on the Modify tab. This would require us to click a start point and then an end point to indicate the amount we wish to move. To move this way, you would have to calculate the amount of move required. The math required in this case is simple, but in general, temporary dimensions might be a nicer approach since no math is required. Unfortunately the temporary dimensions do not appear automatically with a multiple selection such as this. However, we can enable them manually.

- On the Options Bar, click the "Activate Dimensions" button.
- In the temporary dimension that appears between Level 1 and Level 2, type **12'-0" [3600]** (see Figure 4.38).
- Watch the top edge of the Wall as you press ENTER.

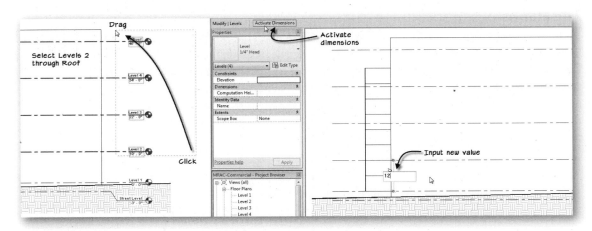

FIGURE 4.38 *Activate Dimensions on a multiple selection to move them together*

You should notice that the top edge of the Wall moves with the Levels. This is because earlier, we constrained the top of the Wall to the Roof Level. This is another example of the power of the parametric relationships between elements in a Revit model. The Roof over the front facade moved too. For now we'll ignore that.

CREATING SECTION VIEWS

Unlike elevations, section views are not included in the default template; instead, we create them where needed. Section views are also vertical cuts through the building model. They typically show the building in a cutaway fashion in an orthographic view. Level Heads also appear in Sections automatically. Section Heads appear in all intersecting views automatically. Adding Sections to our model is a simple task.

Create a Section Line and Associated View

Let's add two building sections, one longitudinal and one transverse.

1. On the Project Browser, double-click to open the *Level 1* floor plan view.
2. On the ribbon, click the View tab and then click the **Section** tool.

   - Click the first point on the outside left of the building near the middle.
   - Move the pointer horizontally to the right and then click the second point outside the building to the right (see Figure 4.39).

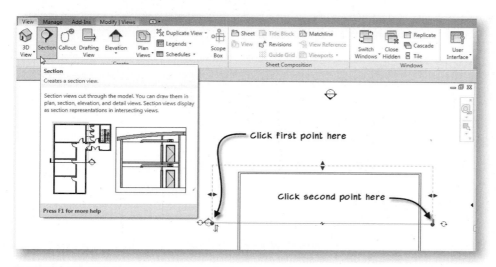

FIGURE 4.39 *Click two points to create a Section Line*

A section line will appear with several Drag controls and Handles on it. Like the other tags and symbols that we have seen so far, simply hover the mouse pointer over each control; the handle will temporarily highlight and the tool tip will appear indicating the function (see Figure 4.40). Brief descriptions of each control follow.

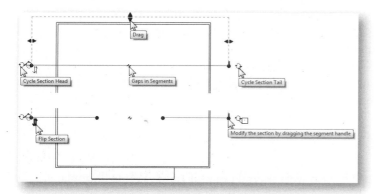

FIGURE 4.40 *Section Line Drag Controls and Handles*

- **Cycle Section Head/Tail**—These controls appear at each end; one for the head and another for the tail. Each time you click one of these handles, the Section head or tail will cycle to a different symbol.
- **Flip Section**—Click this control to flip the Section Line and symbols to look the opposite way.
- **Segment Drag Handle**—This handle (round dot) controls the length of the section line and moves the section heads with it. Dragging this handle does not change the size of the section crop box.
- **Drag**—Drag Handles appear on three sides of the Section Crop Box. Use them to define the precise extent of the associated Section view. If you prefer to set these values numerically, edit the properties of the section.
- **Gaps in Segments**—This control breaks the Section Line and removes the inner segment. You can then drag the exact length of each end segment.

Notice that when you add the Section marker, a new "Sections" branch appears in the Project Browser and that an associated Section view appears on this branch.

3. Using the Drag Handles, make any desired adjustments to the extent of the section box or the position of the section heads.

   - If you wish, click the "Gaps in Segments" handle to remove the middle section of the Section Line.

4. On the Project Browser, expand *Sections (Building Section)*.

   - Right click *Section 1* and choose **Rename**.
   - Type Longitudinal and then click OK.

5. Double-click **Longitudinal** to open the view.

TIP	You can double-click the blue Section Head in any plan view to jump to the associated Section view.

Notice the box that surrounds the section in the view. This is the Crop Region for this section, and it matches the size of the section box that we created in plan. This is functionally identical to the crop region we manipulated for the elevations above. By default crop regions are turned on in section views.

6. Repeat the steps above to create a vertical section looking to the right.

   - Rename the view to **Transverse**.

Make any additional adjustments that you wish. When you open other views (like plans and elevations), the section lines will appear automatically. You can make the same sort of adjustments to them in each view: add gaps to the line, drag the Section Head to a different location. These adjustments are view specific. The same Section marker in the other views can look different. The location of the Section marker line (perpendicular to its length) is in the exact same place. Or you can adjust the size of the Crop Region from any view. This is not view specific. If you adjust the Crop Region or the location of the Section marker in any view, it will move accordingly in all other views.

7. Save the project.

SCHEDULE VIEWS

We have now created and worked with several types of views. While floor plans, ceiling plans, sections and elevations may be the most common views that our projects will contain, Revit Architecture includes over a dozen distinct view types. (You can see them all on the View tab.) One of the most powerful view types at our disposal is the schedule view. A schedule view is not a drawing, but rather a non-graphical view of a collection of related data queried from the model and presented in tabular format. You do not need to "draw" and manually compile your schedules when you work in Revit. Working them up in Excel and importing them in is not required. You simply create a schedule view directly in the software by indicating which data fields you wish to include and how you wish it to be formatted and the data required to populate the schedule comes directly from the model and is tabulated into a list view.

Add a Schedule View

In this sequence, we will add a few simple schedule views to our project. We will not get into a detailed explanation of scheduling features at this time. Detailed information on schedules is found in Chapter 11.

1. On the Project Browser, locate the Schedule/Quantities branch and right-click it.

   • From the pop-up that appears, choose **New Schedule/Quantities**.

   • In the "New Schedule" dialog, choose Walls from the Category list and then click OK (see Figure 4.41).

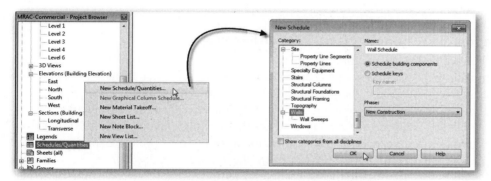

FIGURE 4.41 *Create a New Wall Schedule*

   • In the "Available fields" column, select "Type Mark" and then click the Add button.

   • Repeat this process for the "Length," "Width," "Family and Type" and "Comments" fields (see Figure 4.42).

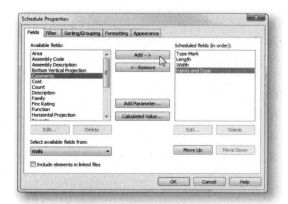

FIGURE 4.42 *Add Fields to the Schedule*

   • Click OK to complete field selection and open the Schedule view.

The *Wall Schedule* view will appear on screen with each of the four Walls that we currently have in our project listed. The Walls do not yet have "Type Marks" or "Comments," so these fields are empty. Later, as we edit the data parameters of these Walls, these fields will update to reflect the latest information. The Schedule view also appears in the "Schedules/Quantities" node of the Project Browser (see Figure 4.43). You can open it from there any time just like the other views of the project that we explored above.

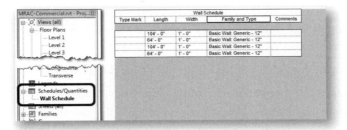

FIGURE 4.43 *The Wall Schedule appears in the Project Browser and opens on screen*

Importing Schedule Views from Other Projects

When we began this chapter, we opened several out-of-the-box Revit Architecture template projects. It was noted above that starting a project from a template was the preferred method of beginning a Revit Architecture project. In some cases, you will begin a project with one template and then later realize that a particular Family, Type or view required already exists in another template file or project. Rather than recreate the item, it is easier to borrow the element from the other project. In this topic, we will create a new project from the *Commercial–Default.rte* template file (installed with the out-of-the-box Imperial templates) and borrow some Schedule views from that project to use in our own commercial project.

NOTE The *Commercial-Default.rte* template file is installed with the out-of-the-box Imperial templates. If you did not install the Imperial templates, or your version of Revit Architecture does not give you access to this file, it has been provided for you in the *Templates* folder with the dataset files installed from the student companion.

2. On the Application menu, choose **New > Project**.

   • In the New Project dialog, in the "Template File" area, be sure that the lower radio button is selected (the one that lists a template file) and then click the Browse button.

   • Browse to the *Imperial Templates* folder.

   • Select the *Commercial-Default.rte* template file and then click Open and then OK.

Take a look around this project. Many views similar to those that we have spent time configuring above also appear in this project template. Some are unique. In particular, you will note that the Level structure of this project includes Levels for bottom and top of footing. Levels such as this are not required, but can be helpful when creating and presenting your model. In addition to the Levels, this project has some useful Schedule views and a large collection of Sheets. We are going to borrow the Schedules to use in our project, and then we'll have a look at some of the Sheets.

3. On the Project Browser, expand *Schedules/Quantities.*

- Double-click on *Door Schedule.*

Take note of the formatting of the schedule view. Several fields appear, and some columns contain headers. As noted above, we will learn how to build such a schedule in Chapter 11; for now, it will be useful to simply copy this view with all its formatting and headers for use in our project.

- On the Project Browser, right-click on *Door Schedule* and choose **Copy to Clipboard**.

- On the QAT, click the Switch Windows tool. Choose **MRAC-Commercial – Floor Plan:Level 1**.

This will return you to your original project so that you can paste the Schedule.

4. On the Modify tab, click the Paste tool (or press CTRL + V). Click OK in the "Suplicate Types" dialog.
5. On the Project Browser, beneath *Schedules/Quantities*, the pasted *Door Schedule* view will appear and it will open automatically.

Notice that this Schedule contains all of the pre-formatted columns, but lists no Doors. This is because our project does not yet contain any Doors.

Add a Door to the Project

When you add elements to the project, they will automatically appear in *all* appropriate views—including the Schedule views. Let's add a Door to our project to see this.

6. On the Project Browser, double-click to open the *Level 1* floor plan view.

- Zoom in on the middle of the top horizontal Wall in the plan.
- On the ribbon, click the Home tab and then click the Door tool.
- On the Type Selector, choose ***Single-Flush: 36' × 84" [M_Single-Flush: 0915 × 2134mm]***.

Make sure Tag on Placement *is* selected.

- Add a Door to the middle of the top horizontal Wall.
- On the ribbon, click the ***Modify*** tool or press the ESC key twice.

The Door number in the tag reads: 1.

7.On the Project Browser, double-click to open the *Door Schedule* view.

Notice that the Door also appears in the Schedule.

8. Click in the Door Number field on the Schedule.

- Change the door number to **101** and then press ENTER.
- Return to the *Level 1* plan view and confirm that the door tag reflects the new number (see Figure 4.44).

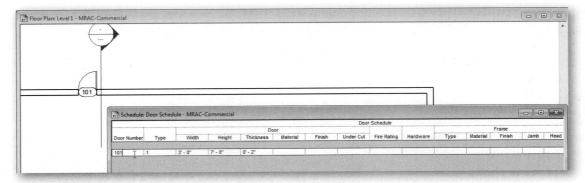

FIGURE 4.44 *A change in any view is reflected automatically in all views*

In Revit, a change made in one view applies in all associated views. You are not changing a graphic representation in a drawing. You are actually editing the parameters of an element in your virtual building model. This change must therefore be reflected in all views that show that component of the model—regardless if they are graphical views or tabular views.

9. Return to the other project and repeat the process to copy and paste the other two schedule views (*Room Schedule* and *Window Schedule*) to your *MRAC-Commercial* project.

Be sure to choose a floor plan view from the Window menu before pasting. The Paste command will not be available if you open a schedule view first.

TIP	You can switch between open views using the CTRL & TAB keys. Hold down CTRL and then press TAB. Each time you press tab, you will cycle to another open view.

BIM MANAGER NOTE	If you use the same Schedule views (or any views) repeatedly from one project to the next, you should create your own project template that already includes the views you require. In this way, new projects will have these views immediately upon starting.

10. Save your *MRAC-Commercial* project.

    • Close the other project without saving.

SHEET VIEWS AND THE CARTOON SET

A sheet is a special type of view that is intended for publishing your document set—they are printing views. When you create a sheet, you can add a title block to it, configure the print setup to match the paper size and other required printing parameters and add viewports containing the various views of your project. Since most architectural projects will require printed output (often at several points) in the life of the project, it is a good idea to set up a set of sheet views to create professional looking results.

Project submission time is usually a high-stress endeavor characterized by panic as project team members scramble to assemble the required sheets and print everything required for a particular submission. This frenzy often occurs at the last minute, which contributes to the stressful environment and often to errors and omissions. One simple way to help alleviate this situation is to create the sheet views early in the project so that they are ready to print at any time. Since all Revit views are live, you can create a sheet with views inserted early in schematics that have very little information on them (like the views we have created here), and over the life of the project, these sheets will reflect all project changes immediately as they occur. This means that you can print a sheet or set of sheets at a moment's notice with complete confidence that the information portrayed on those sheets represents the latest state of the project.

> Consider adding a number of pre-configured sheet views to your office-standard project template. This will ensure consistency from one project to the next and simplify project setup. The *Commercial-Default.rte* template file used above gives a good example of this approach. Take some time to study the sheets included therein to glean ideas for your own templates.

Create a Floor Plan Sheet

1. On the Project Browser, right-click *Sheets* and choose **New Sheet**.

   • In the "New Sheet" dialog, accept the defaults and click OK (see Figure 4.45).

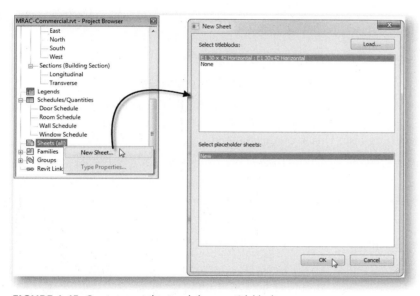

FIGURE 4.45 *Create a new sheet and choose a titleblock*

A plus (+) sign will appear next to the *Sheets* node indicating that a new sheet has been added. The sheet will also open on screen.

2. Expand the *Sheets* branch.

The sheet will have a default name and number that we can change. Zoom in on the lower right corner of the titleblock before performing the next step.

   • Right-click on the sheet (currently named A101 - Unnamed) and choose **Rename**.
   • Change the sheet Name to **Floor Plans**, and then click OK.

> If you prefer, you can rename and re-number Sheets directly in the titleblock. Click on the titleblock to highlight it and then click the blue text to edit the value. The result will be the same.

TIP

Make sure that the *A101 – Floor Plans* sheet is open. It should have opened automatically when it was created.

- From the Project Browser, drag the *Level 1* floor plan view and drop it on the sheet (see Figure 4.46).

- Click to make the final placement.

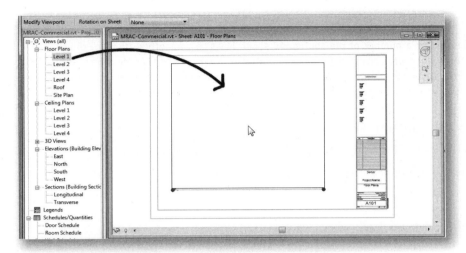

FIGURE 4.46 *Drag the Level 1 floor plan view and drop it on the sheet*

- Zoom in on the view title beneath the floor plan on the sheet.

Notice that the name of the drawing is "Level 1." While this is logical, and in some cases desirable, chances are we would like the name of this drawing on the printed Sheet to be something else. This is easily changed.

3. On the Project Browser, select the *Level 1* floor plan view.

- On the Properties palette, scroll down to the Identity Data grouping, and then type **First Floor Plan** in the "Title on Sheet" parameter field (see the left side of Figure 4.47).

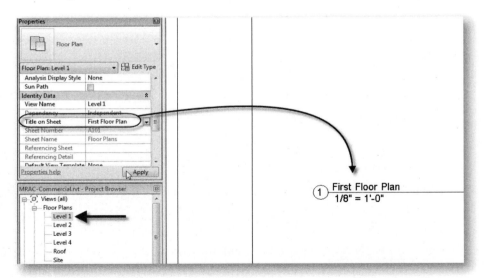

FIGURE 4.47 *Edit the value of the Title on the Sheet*

- Click Apply to complete the change (see the right side of Figure 4.47).

Notice the change to the view on the sheet. The Title on Sheet parameter overrides the view name. If you delete the value for Title on Sheet, it will revert back to the view name on the sheet. Also notice that the number in the view title tag has automatically filled in with the number 1. This is because this is the first (and currently only) view that we dragged to this sheet. Revit will automatically enumerate the annotation of all views as you drag drawings onto sheet views.

4. Save the project.

Take note of the Sheet on the Project Browser. A small plus (+) sign will appear next to a sheet when there are views placed on it. You can click the plus sign to see each view listed beneath the sheet. You can double-click on views from here or the standard location in Project Browser to open them.

Create a Sheet List

Revit makes it easy to create a list of all the drawing sheets in our project. A sheet list will resemble the other schedules that we already have, and it can even be placed on a sheet as a sheet index. Furthermore, it can actually be used to help us create and edit the sheets in our project. Let's create a sheet list and then use it to add more sheets.

5. On the Project Browser, right-click on the *Schedules/Quantities* branch and choose **New Sheet List.**

    - In the "Sheet List Properties" dialog, select Sheet Number and then click the Add -> button.
    - Add additional fields like Sheet Name, Drawn by, Checked by, Sheet Issue Date and Guide Grid.
    - Click the Sorting/Grouping tab.
    - For Sort by, choose Sheet Number and then click OK.

Fields like drawn by, checked by, and other similar text fields must be edited individually on each sheet. However, one of the benefits of the Sheet List schedule is that you can edit those values directly in the table, which is usually more efficient than doing them individually. Another key advantage to the Sheet List Schedule is that we can use it to create Sheets. Think of it as queuing up list of available sheets, complete with pertinent information filled in and ready to add to the project.

6. On the Modify Sheet List tab, on the Rows panel, click the New button.

A new row will appear in the Schedule. It will automatically number as sheet A102. Using this method, you can create several "placeholder" sheets and fill in their values in the other columns of the Schedule.

7. Change the name of A102 to Floor Plans (see Figure 4.48).

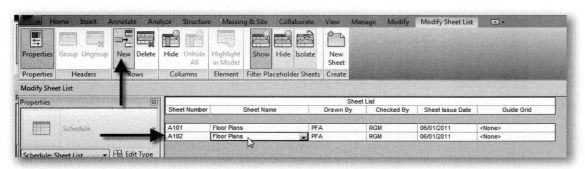

FIGURE 4.48 *Create a Drawing List and use New Rows to add a placeholder sheet*

You can input your initials in the Drawn By and Checked By columns if you wish. You can also input a date for sheet issuance if you like. Using Table 4.2, we'll set up the rest of the floor and ceiling plan sheets.

Create the Remaining Floor Plan Sheet Views

Let's create the remaining floor plan sheets.

8. On the Project Browser, select the *Level 2* floor plan.

- On the Properties palette, change the "Title on Sheet" parameter to **Second Floor Plan**.
- Repeat this process for each view listed in Table 4.2.

TIP	If you prefer to input the Title on Sheet parameter in a schedule, try making a View List. The process is the same as adding a Sheet List and you can you add the Title on Sheet parameter to the View List and edit it from the Schedule.

9. Repeating the process outlined above, from the Sheet List Schedule, click the New Row button for each of the sheets listed in Table 4.2 below.

- Create a new Row in the Schedule for each of the additional Sheets listed in the "Drag to Sheet" column of Table 4.2.

TIP	To reuse existing values in the Schedule, click the drop-down arrow in the text field.

TABLE 4.2 *Titles on Sheets for Plan Views*

View Name	Title on Sheet	Drag to Sheet
Floor Plan Views		
Level 2	**Second Floor Plan**	A102 – Floor Plans
Level 3	**Third Floor Plan**	A103 – Floor Plans
Level 4	**Fourth Floor Plan**	A104 – Floor Plans
Roof	**Roof Plan**	A105 – Roof Plan
Site Plan	**Site Plan**	A100 – Site Plan
Ceiling Plan Views		
Level 2	**First Floor Reflected Ceiling Plan**	A106 – Reflected Ceiling Plans
Level 2	**Second Floor Reflected Ceiling Plan**	A107 – Reflected Ceiling Plans
Level 3	**Third Floor Reflected Ceiling Plan**	A108 – Reflected Ceiling Plans
Level 4	**Fourth Floor Reflected Ceiling Plan**	A109 – Reflected Ceiling Plans

NOTE	The sheet numbers suggested here are based on the US National CAD Standard recommendations. If you prefer a different sheet numbering scheme, feel free to use it instead.

Creating Sheet Views from Placeholder Sheets

We now have nine new placeholder sheets in our sheet list, but if you look at the Project Browser, we still have just the one sheet. The purpose of adding placeholder sheets is that it allows you to configure a list of standard sheets and to input all the fields before you actually need the sheet in the set. The other purpose of Placeholder Sheets is to allow the addition of consultant sheets to the Sheet List without actually having those sheet views in your project. When you are ready, you simply add a new sheet as we did before, but this time we will be able to choose from the list of available placeholder sheets.

10. On the Modify Sheet List tab, click the New Sheet button. (This is the same as right-clicking the Sheets (all) branch on the Project Browser as we did above).

    Notice the list of placeholder sheets.

11. Select A100 in the list. Hold down the SHIFT key and then select A109 (see Figure 4.49).

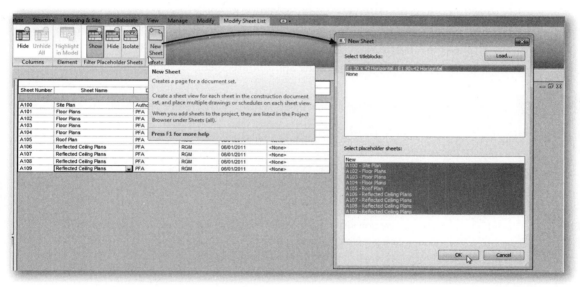

FIGURE 4.49 *Create several sheets at once using placeholder sheets*

12. Click OK.

Nine new sheets will appear on the Project Browser ready for you to drag and drop views. Zooming in on the titleblocks will reveal that all the data input in the Schedule is already filled in to the appropriate fields.

Add Plan Views to Sheets

Now that we have all our plan sheets created, let's add the views to them.

With the Guide Grid feature you can overlay a grid on top of a titleblock sheet and use it to keep similar views lined up across multiple sheets. However, to be most useful, you need to have Structural Grids in your project to assist in the alignment. Since adding Structural Grids is one of the primary topics of the next chapter, we will postpone discussion of the Guide Grid feature to Chapter 5. Meanwhile, we will simply drag the views onto the appropriate sheets and place them in approximately the same location. Once we have the Structural Grids in Chapter 5, we can adjust the position of the manually placed views to make sure that they all align with one another using the Guide Grid feature.

13. Using Table 4.2 as a guide, drag each of the floor and ceiling plan views onto their respective sheets.

    • Place each one in the same general area on the sheet.

14. Save the project.

Create Elevation and Section Sheets

We will need a couple of Sheets for our building elevations and building sections.

15. Repeating the process outlined above, create the sheet views indicated in Table 4.3.

Two methods were covered above. Create the sheet directly (using the tool on the ribbon or right-click on the Sheets branch of the Project Browser) or you can first create placeholder sheets in the Sheet List Schedule and then create the sheets from those placeholders. You can use either method here.

    • Repeating the process outlined above, edit the "Title on Sheet" parameter of each of the elevation and section views as indicated in Table 4.3.

    • Drag each of the views to the appropriate Sheets as indicated in Table 4.3.

NOTE You are only creating three additional sheets here and there are two views being placed on each of these sheets.

TIP If your elevation or section views are too wide for the sheet, you can use the crop region handles to resize them. The crop region is already on by default for section views. For elevation views, you must repeat the process outlined earlier in the chapter to edit.

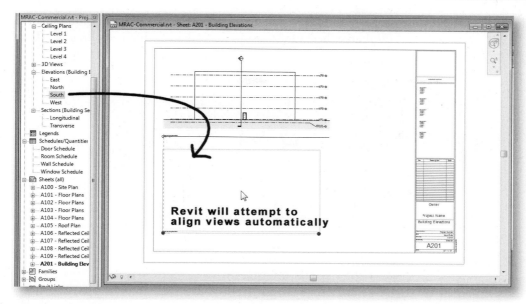

FIGURE 4.50 *Create elevation and section sheets and drag the appropriate views*

When you create your first elevation sheet, you will notice that the second view you drag to the sheet attempts to align automatically with the first as you drag. This will help you keep the sheet neatly organized (see Figure 4.50).

TABLE 4.3 *Titles on Sheets for Elevation and Section Views*

View Name	Title on Sheet	Drag to Sheet
Elevation Views		
North	**North Building Elevation**	A201 – Building Elevations
South	**South Building Elevation**	A201 – Building Elevations
East	**East Building Elevation**	A202 – Building Elevations
West	**West Building Elevation**	A202 – Building Elevations
Section Views		
Longitudinal	**Longitudinal Building Section**	A301 – Building Sections
Transverse	**Transverse Building Section**	A301 – Building Sections

16. Create one more sheet numbered **A601** and titled **Schedules**.

   • Drag each of the schedule views and drop them on this sheet.

If you zoom in on the schedules on the sheet, in some cases you will notice that the data in columns might wrap to a second line. You can leave it like this, or use the small triangular control handles to drag and resize the columns. The choice is yours.

When you have completed these Sheets, your Project Browser should resemble Figure 4.51. Again take note of the elevation and section tags in the various views.

All of them now properly indicate the Sheet references and drawing numbers. All annotations were automatically enumerated in the order in which you dragged them to the Sheets.

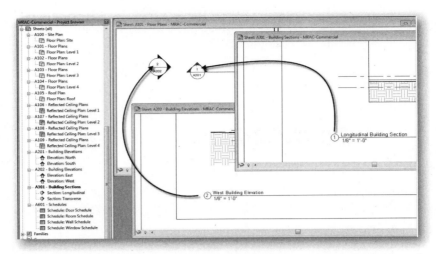

FIGURE 4.51 *Project Browser showing all sheets and the views they contain and examples of cross referenced callouts*

At this point, every sheet that we have created is likely still open. You can see which views and sheets are open using the Switch Windows drop-down button on the QAT. In a project this size, it is probably OK to have several windows open at once; however, as a matter of best practice, it is a good idea to periodically close unused windows. The easiest way to do this is with the Close Hidden tool on the View tab. Views must be maximized first. If your screen shows view windows cascaded or tiled, please maximize before using the Close Hidden command.

17. Make sure that windows are maximized, and then on the View tab of the ribbon or the QAT, click the Close Hidden icon.
18. Save the project.

In some cases, when you increase the length of the title on the sheet parameter, the titlebar under the viewport will wrap the title onto two lines. If you prefer to have it not wrap, you can edit the View Title Family. To do this, expand *Families* on the Project Browser and then expand *Annotation Symbols*. Right-click on **View Title [M_View Title]** and then choose **Edit**. In the dialog that appears, click Yes. Click on the View Name label on screen, and use the Drag handle on the right to make it wider. On the ribbon, click the Load into Projects button. If a dialog appears showing multiple projects, check the box next to your *MRAC-Commercial* project and then click OK. In the Reload Family dialog, choose the "Override parameter values of existing types" item. Any sheets that previously wrapped the title to two lines should now be displaying on a single line.

Create a Custom Titleblock Family

The only Sheet we still need at this stage is a cover sheet. To create it, we will build a custom titleblock Family. Many times, firms have titleblocks already drawn in other CAD programs that they wish to import into Revit. You can do this if you wish.

The exact steps will not be covered here, but you can create a new titleblock Family using the procedure covered here, and then import a CAD file into it. The approach we will use here will be to create a very simple titleblock Family from scratch using Revit tools.

> **NOTE** If you prefer to skip this exercise, you can instead load the custom titleblock Family file provided with the dataset files from the student companion. To do so, skip to the "Create a Cover Sheet" topic below.

19. From the Application menu, choose **New > Title Block**.

   - In the "New" dialog, select the *E1 – 42 × 30.rft* [*A1 metric.rft*] template and then click Open.
   - The ribbon will change to reflect that we are now in the Family Editor.

20. On the Home tab of the ribbon, click the **Line** tool.

   - On the Modify | Place Lines tab, on the Line Style panel, choose Wide Lines from the Style list.
   - On the Draw panel, click the rectangle icon, and on the Options bar, type **3/4" [18]** in the Offset field.

21. Draw a rectangle by snapping to the corners of the one already in the file.

   - After you click the first corner, tap the SPACEBAR to flip the offset to the inside and then click the other corner.
   - The new rectangle will appear 3/4" [18] smaller than the original one due to the offset.
   - On the ribbon, click the **Modify** tool or press the ESC key twice.

22. Copy the bottom line up **6" [150]**.

   To Copy, select the line, on the Modify panel, click Copy, click a start point on the line, move straight up, type in the distance, and then press ENTER.

23. Draw some smaller vertical lines within this space to separate it into a series of boxes (see Figure 4.52).

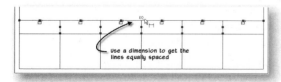

FIGURE 4.52 *Draw a horizontal bar at the bottom and divide it into boxes*

> **TIP** Draw five vertical lines in approximately the correct locations. Use the Aligned dimension tool to add a dimension starting with the vertical edge on one side, including all five vertical lines, and ending with the vertical edge on the other side. Click the EQ toggle. This will space all the lines equally. Then delete the dimension.

24. On the Home tab of the ribbon, click the **Text** tool.

   - On the Properties palette, click the Edit Type button and then click the Duplicate button.

- Name the new Text Type: **Title Text** and then click OK.
- For the Text Size, type: **1/2" [12]**.
- Select the Bold checkbox and then click OK.

25. In the first box that you drew above, click a point in the upper left corner to place the text.

  - Type the following: (press ENTER after each line)

    Consultant Name

    Address 1

    Address 2

    Phone

  - On the ribbon, click the **Modify** tool or press the ESC key twice.

26. Copy this block of text to a few of the other boxes along the bottom of the titleblock.

  - Leave a couple of boxes empty for a company logo and an Architect's seal.
  - You can add a new text note to one of them that says: **Drawing Index** and another that says: **Seal**. You might want to use the smaller text size (see Figure 4.53).

FIGURE 4.53 *Add text blocks for consultants and a location for the Architect's seal*

You can either leave the logo box empty for now or use the Image tool on the Insert tab to import your company logo. The file must be saved in JPG, JPEG, BMP or PNG image file formats. Once you place the image, use the handle at the corner to resize it.

27. On the Home tab, click the **Label** tool.

  - Use the same procedure as above to create a new Label Type called **1" Arial Titles [25mm Arial Titles]**. Make the text size **1" [25]** and Bold.
  - Click OK after creating the Type and then choose the Center icon on the Format panel before placing the label.

28. Click a point near the top middle of the titleblock.

29. In the "Edit Label" dialog, double-click Project Name to add it to the Label Parameters list Project Name and then click OK.

  - Drag the handles to make the title wider.
  - Create another Label Type with smaller text size and add additional centered labels beneath the title such as Project Address and Issue Date (see Figure 4.54).

 NOTE In the "Edit Label" dialog, you can add more than one parameter to a single Label. So for example, Project Address, Project Issue Date and Client Name can all be added to the same Label. On the far right, check the box in the break column to make each parameter fall on its own line.

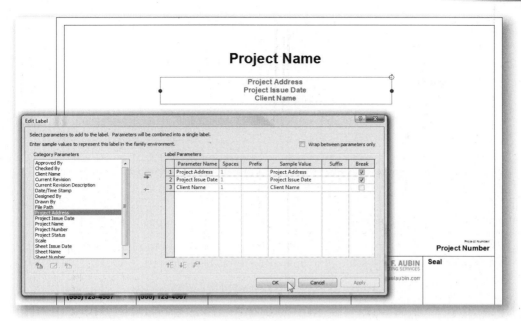

FIGURE 4.54 *Add Labels to report the project information*

Feel free to make any other embellishments you wish. You can draw additional lines and add more labels and text. Use labels when you wish to link the text to project data. When this titleblock is used in the project, the Project Name and date will automatically fill in. The consultants on the other hand will need to be edited manually in the titleblock later. While it is possible to make the consultant items link to labels as well, the process to do so is beyond the scope of this short tutorial. Consult the online help for more information on creating titleblocks if you wish to learn more.

30. Save the titleblock Family to the *Chapter04* folder.

    • Name it: **MRAC E1 30 × 42 Cover Sheet.rfa [MasterRAC A1 Cover Sheet.rfa]**.

31. On the ribbon, click the **Load** into Project button.

Create a Cover Sheet

Now that you have built your own custom titleblock Family for the cover sheet, we can use it in the project and create the cover sheet.

32. On the Project Browser, right-click *Sheets* and choose **New Sheet**.

    • In the "New Sheet" dialog, choose your new cover sheet titleblock and then click OK.

NOTE

If you decided to skip the titleblock Family creation exercise above, click the Load button and browse to the *Chapter04* folder with the other files for this chapter. Select the *MasterRAC E1 30 × 42 Cover Sheet.rfa* [*MRAC A1 Cover Sheet.rfa*] file and then click Open.

    • Right-click on the new sheet (currently *A602 – Unnamed*) and choose **Rename**.
    • Re-number it to **G100** and rename it to **Cover Sheet**.

Why don't we add a 3D axonometric of the project to our Cover Sheet? We can use the default *{3D}* view, but it might be better to create a copy of it first.

33. On the Project Browser, expand *3D Views*, right-click the *{3D}* view and choose **Duplicate View > Duplicate**.

- Rename the view **Cover Axonometric**.
- Use the ViewCube and Steering Wheels to adjust the view to a pleasing vantage point.

TIP	If you have a wheel mouse, hold down the SHIFT key and drag with the wheel to quickly spin the model.

34. Drag the *Cover Axonometric* view onto the Cover Sheet and position it.

When you drop the view on the Cover Sheet, it is probably too large for the Sheet. We can easily adjust the scale of a view after it has been placed.

35. On the Project Browser, select the *Cover Axonometric* view.

- On the Properties palette, from the view Scale list, choose a smaller scale.

We also do not want a title under this viewport on the cover. To do this, we need to duplicate the viewport Type.

36. Select the viewport on the sheet (click directly on it).

The Properties palette should read Viewport: Viewport 1 on the Type Selector. A

- On the Properties palette, click the Edit Type button (see Figure 4.55).

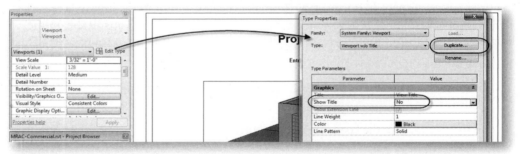

FIGURE 4.55 *Duplicate the Viewport Type and turn off the View Title*

- Click the Duplicate button.
- Name the new Type: **Viewport w/o Title** and then click OK.
- For Show Title, choose **No** and then click OK.

NOTE	If you are using the Metric files, you can use the *No Title* Type already in the project instead of creating a new one.

- Reposition the view as necessary.

TIP	If the scale is still too large for the Sheet, you can choose **Custom** from the view Scale list and type in a value. In the Metric files, try a scale 1:250, for example.

Edit Project Information

When we built the titleblock Family, we noted that the labels would automatically fill in with the appropriate project data. You may have noticed that most fields have not yet changed. The information that these labels reference is global to the entire project and is stored in a common System Family named "Project Information." We can edit its values at any time, and the new values will immediately populate *all* Sheets in the project. Let's take a look.

37. On the Manage tab, click the Project Information button.

   - In the "Instance Properties" dialog, input values for the various fields (see Figure 4.56).

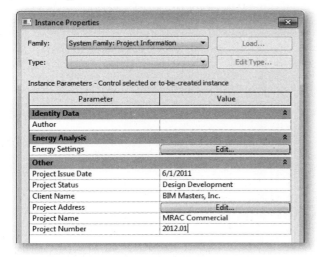

FIGURE 4.56 *Edit Project Information*

38. Open any Sheet and view the Titleblock fields.

Notice the change to the values.

39. Save the project.

Assign a Starting View

New in this release, we can assign a starting view to our project. This view will open whenever you open the project regardless of which view was active when the project was last saved. In this case, we will use this feature to set our new cover sheet as the starting view for the project.

40. On the Manage tab, on the Manage Project panel, click the Starting View button.

   - In the "Starting View" dialog that appears, choose **Sheet: G100-Cover Sheet** and then click OK.

To see the effect of this change, you would need to close the project and then reopen it. Feel free to do this now if you wish, or you can simply wait till later to test it out.

Working with Browser Organization

As the quantity of views and sheets in your projects increases, you may find it useful to explore other ways to organize your Project Browser. First, as noted above, you can open a view directly from the sheet entry in the browser by expanding the plus sign next to the sheet and then double-clicking the view indented beneath. You can see what this looks like above in Figure 4.51. You can also open the sheet and then right-click the viewport and choose **Activate View**. This allows you to work directly in the view while maintaining the context of its placement on the sheet for reference. If you take advantage of either of these methods, you can choose to hide all views that are already on sheets from the *Views* branch of the Project Browser.

41. On the Project Browser, scroll to the top and then select Views (All).

    • On the Properties palette, choose "not on sheets" from the Type Selector and then click to see the change.

This will filter the view list to show only those views that have not yet been placed on sheets. In our case, only the default *{3D}* view will remain (see Figure 4.57).

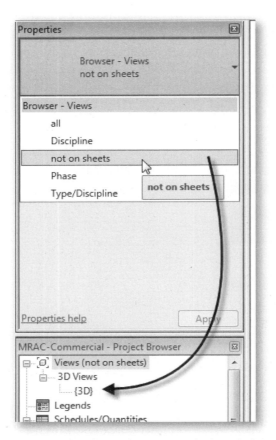

FIGURE 4.57 *Change the Browser to display only the views not already on sheets*

There are other Browser organization types as well. One sorts by discipline, another by phase, and so on. The not on sheets type gives the most dramatic difference at the current stage of our project. Similar filtering types are available for the *Sheets* branch. To see them, select the Sheets (All) branch of the Project Browser and look to the choices on the Type Selector. For example, try **Sheet Prefix**. This organization will give us two groupings, one for the "A" sheets and another for the "G" sheets. Often architects have sheet numbering schemes where the first two or three characters are codes for drawing types. With the NCS-derived numbering scheme suggested above, the first character is the discipline code and the number immediately following it is a drawing type code where 1 equals plans, 2 equals elevations, 3 equals sections, and so on. In this case, it might be useful to filter the list by the first two characters rather than just the first. Using Browser Organization, we can create our own types.

42. On the Project Browser, right-click the Sheets (Sheet Prefix) branch and choose **Type Properties**.

    • In the "Type Properties" dialog, click the Edit button next to Folders.

    • On the Folders tab, change the number of leading characters to 2 (see Figure 4.58).

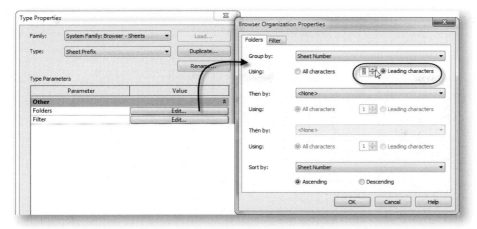

FIGURE 4.58 *Edit the Sheet Prefix organization to group by 2 characters*

- Click OK twice.

Explore the results in the Project Browser. It should look like Figure 4.59.

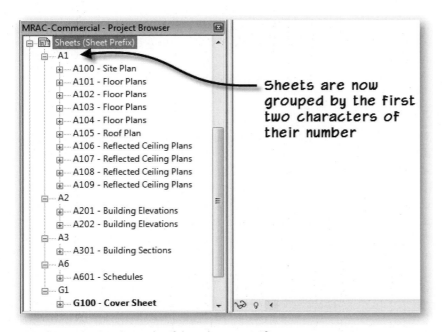

FIGURE 4.59 *View the results of the 2 character prefix*

Create a Custom Parameter

Return to the Sheet List that we created above. All of the latest sheets including the Cover Sheet should now appear on the list. You can make any modifications to the parameters listed that might be necessary. You may notice one small problem with our sheet list however. While all the "A" sheets sort nicely, our Cover Sheet appears last in the list. This is an unfortunate limitation in the sorting options we have available. Let's look at a work-around solution to address this problem.

43. On the Properties palette, click the Edit button next to Fields.

- On the Fields tab, click the Add Parameter button.
- In the "Parameter Properties" dialog, input **Sheet Sort Number** for the Name.
- For the "Type of Parameter" choose **Integer** and then click OK (see Figure 4.60).

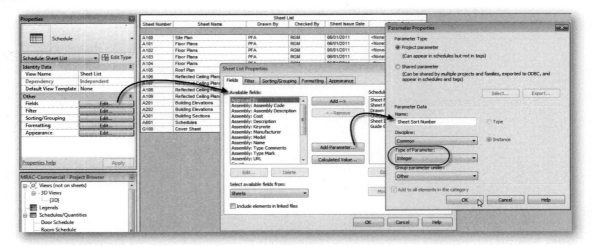

FIGURE 4.60 *Create a new Project Parameter for sorting sheets*

What we have created is a custom parameter that can receive a numeric value. Next we'll change the sorting of the table to sort on our new field so that sheet G100 can come before sheet A100.

44. Still in the "Sheet List Properties" dialog, click the Sorting/Grouping tab.

- Change Sort by to Sheet Sort Number and then. Select the Descending radio button.

- For the next sort criterion (next to "Then by"), choose Sheet Number and leave it set to Ascending.

- Click OK to finish.

An empty Sheet Sort Number column will appear in the table. Simply input numbers into these fields to indicate how you want the sheets to sort. However, since we only needed to force the cover sheet to the top of the list, using two sort criteria as indicated will spare us the effort of inputting values in this field for all sheets. When you sort in ascending order, any blank fields will come first. This is why the list currently looks the same. However, inputting any value in the cover sheet will force it to the top (since we told the first sort to go in reverse – descending). If you find this "trick" confusing, feel free to input an actual number in the Sheet Sort Number field for every sheet.

45. Input **1** in the Sheet Sort Number field for the G100 sheet (see Figure 4.61).

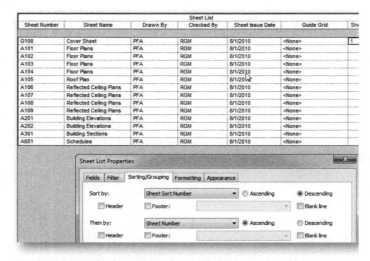

FIGURE 4.61 *Edit the sorting values in the Drawing List table*

46. Save the project.

MANAGER NOTE BIM

You can customize the Browser Organization even further. For example, you can add custom parameters to Views as well as Sheets. So if you prefer to organize the view list by something other than the defaults like Family and Type, try making a custom text parameter called "Folder Name" for views. This will allow you to edit the properties of any view and type in a user-defined designation for the view's "folder." Then edit the Browser Organization to sort and group based on the custom Folder Name property. Some examples are shown in Figure 4.62. The top example shows a custom View Folder parameter added to Views and then used for grouping. The bottom example is in our Commercial Project and simply sorts views by associated Level to make the Site Plan list before Level 1. No custom parameter is required. Give it a try.

The basic steps are easy. Edit the Type Properties of the current Browser Organization (for example, not on sheets). Duplicate the Type and give it a new name such as MRAC. On the Folders tab, simply change the Sort by criterion to Associated Level. If you wish to show all views again, click the Filter tab and set the Filter by to <None>.

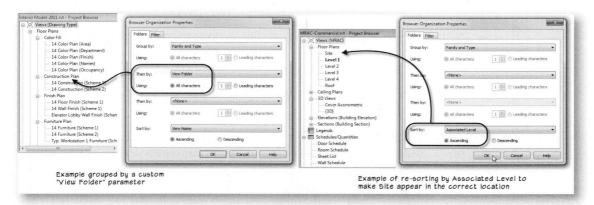

Example grouped by a custom "View Folder" parameter

Example of re-sorting by Associated Level to make Site appear in the correct location

FIGURE 4.62 *Examples of custom Browser Organization settings*

TIP

If you are displaying all views in the Views branch, consider colapsing the Ceiling Plans category until you need it. Many users will inadvertently open a ceiling plan when they intend to open a plan since these two views often have the same name.

DRAFTING VIEWS

One other type of view bears mention in this discussion: the Drafting view. A Drafting view is unlike other views in that it is not linked to anything in the model. The purpose of a Drafting view is to create details, diagrams, or sketches that are not easily modeled or that you and your project team have deemed not *worth* modeling. Remember that Building Information Modeling is as much about information as it is about graphics. In some cases, a simple diagram or some sketched embellishment on a detail is all that is required to convey design intent. In these cases, a Drafting view is often appropriate. Drafting views provide a blank sheet of paper suitable for importing legacy CAD details, image files or natively drawn sketches.

While we are not creating any Drafting views in this chapter, Chapter 11 is devoted to the subject of detailing. We will have the opportunity to add Drafting views to our project at that time.

PRINTING A DIGITAL CARTOON SET

Going though the upfront process of creating sheet views gives you a few benefits. First, this task can be handled by a single individual early in the project without the pressure of a looming deadline. If one person sets up all of the necessary sheets, there is a greater chance for consistency from sheet to sheet in terms of naming and drawing placement. Following this process also gives you a digital cartoon set. Just like the traditional cartoon set, the digital version will help make good decisions about project documentation requirements and the impact on budget and personnel considerations. One extra advantage of the digital cartoon set over the traditional one is that the digital one becomes the real building model and document set! Don't be concerned with the finality that this seems to imply. The model remains completely flexible and editable (as we have seen).

Publish a 2D DWF

As the final task in this chapter, go ahead and plot out the project sheets. You could print to a physical plotter, but to save paper, let's create a digital plot instead of using the Publish to DWF function.

1. On the Project Browser, double-click G100 – Cover Sheet.
2. From the Application menu, choose **Export > DWF/DWFx**.

    • In the "DWF Export Settings" dialog, choose **<In session view/sheet set>** from the Export list on the right.
    • From the Show in list menu, choose **Sheets in the Model**.
    • Click the Check all button (see Figure 4.63).

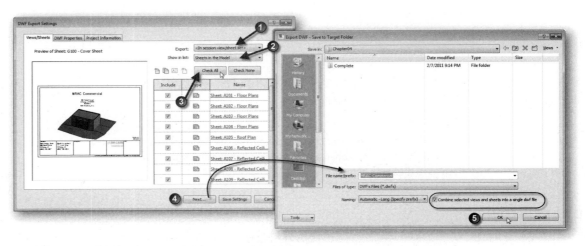

FIGURE 4.63 *Select the sheets to export to DWF*

3. Click the Next button.

    • In the "Export DWF" dialog, browse to the *Chapter04* folder, accept the other defaults (make sure that "Combine views and sheets into a single dwf file" is checked) and then click OK.

Revit will begin processing your request. A multi-sheet DWF file will be generated that you can open in Autodesk Design Review. This is a free application that you can download from the Autodesk web site. It allows you to view and redline DWF files that you share with your consultants and clients. DWF files are very small and read-only while maintaining full vector-based quality of the original model. In this way, you can share your designs with your extended team without fear of file size limitations or unauthorized editing.

After the export is complete, browse to the folder where you saved it and double-click the file. Autodesk Design Review will open the file, and you will see thumbnails of each sheet on the left side. You can view sheets, add comments, redlines, and save them for later coordination back in Revit. If you CTRL click on the section and elevation tags and level heads, you will be taken to the corresponding view instantly. You can email the DWF file to a recipient who does not need to have Revit to open, view and print the DWF. All they need is the free Design Review application.

Now that we have thoroughly explored Templates and views and have created a project that contains several views, we might be looking for ways to save time on the next project we begin. You can take any Revit Architecture project and save it as a template. This is very easy to do and allows you to reuse your work on similar projects in the future. To save the current project as a template, simply choose Save As > Template from the Application menu. In the Save As dialog, browse to a location to save the template file. The best location is a central location on the network server where all users have ready access. Give the new template file a name, change the "Save as type" to Template Files (*.rte) and then click Save. The template file will be saved to the location you specified and be ready for use.

SUMMARY

- Template files provide a means to start new projects with consistent content and setup.
- All new projects should be created from an agreed upon office template file.
- Levels establish horizontal datum elements positioned vertically in the Z direction for use as floor levels and other reference points.
- Walls heights can be constrained to Levels and will thus change height automatically if a Level changes.
- Views are used to study, create, and manipulate model data.
- Views can be graphical like plans, sections and elevations or non-graphical like schedules and drawing lists.
- Edits made in one view are immediately seen in all appropriate views.
- Sheets are used to create drawing sets for printing.
- You can create your own custom titleblock to use in your Revit projects.
- Generate a Drawing List to plan your set or place on your cover page.
- Before printing to a paper plotter consider generating a digital plot instead—DWF files are small, high quality "digital plots" that you can use to share design data among the extended project team.
- Anyone with a free copy of Autodesk Design Review software can open, view, redline, and print a DWF file.

CHAPTER 5

Column Grids and Structural Layout

INTRODUCTION

In this chapter, we will explore the layout of basic structural components. We will begin with the layout of the column grid lines for the commercial project started in the last chapter. We will also add columns and framing members to this project and begin to explore strategies related to separating and sharing data among the various disciplines involved in a project team. Finally, we will revisit the residential project to create a foundation plan.

OBJECTIVES

In this chapter, we will begin by adding column Grid lines and Columns. The goal will be to understand the features and behaviors of Grids and Columns in Revit. We will also look at Beams and Joists to complete the framing. Footings will also be explored. After completing this chapter you will know how to:

- Add and modify column Grids
- Add and modify Architectural and Structural Columns
- Load Steel Shape Families
- Create Structural Framing elements
- Copy elements to Levels
- Create foundation Wall footings

WORKING WITH GRIDS

In Chapter 1, Figure 1.1 we explored the hierarchy of Revit Architecture elements. In the Datum elements branch of the hierarchy, we find Levels, Grids, and Reference Planes. In the last chapter, we worked with Levels. These are horizontal planes (expressed graphically as lines in views that show them with level head annotation). They typically define floor levels or stories of the building but can also be used to define other horizontal datum elements such as "top of footing" or "bottom of steel." Grids are very similar to Levels in almost every way except that they define vertical planes through the building. The most common use of Grids is to define the locations of structural columns in the building. As such, they are typically adorned with grid bubble annotation. Grids are typically added in two directions and their intersections are used to label unique locations for each column. Like Levels and Section lines, Grids exhibit intelligent annotation characteristics and will appear automatically in all appropriate views such as plans, elevations, and sections at the proper scale and location. The third type of datum, the Reference Plane, also shares many characteristics of Levels and Grids, but is more generic in nature and does not have any annotation automatically associated with it. Reference Planes will be explored in later chapters.

Install the Dataset Files from the Student Companion Files and Open a Project

The lessons that follow require the dataset included on the *Aubin Academy Master Series: Revit Architecture 2012* dataset files from the student companion. If you have already installed all of the dataset files from the student companion, simply skip down to step 3 below to open the project. If you need to install the dataset files from the student companion files, start at step 1.

1. If you have not already done so, install the dataset files from the student companion.

   Refer to "Files Included on the dataset files from the student companion" in the Preface for instructions on installing the dataset files from the student companion.

2. Launch Revit Architecture from the icon on your desktop or from the *Autodesk ≥ Revit Architecture 2012* group in *All Programs* on the Windows Start menu.

You can click the Start button and then begin typing **Revit** in the Search field. After a couple letters, Revit Architecture should appear near the top of the list. Click it to launch to program.	**TIP**

3. On the Recent Files screen, click the Open link beneath Projects.

The keyboard shortcut for Open is CTRL+O. You can also click the Open icon on the Quick Access Toolbar (QAT).	**TIP**

   • In the "Open" dialog box, browse to the location where you installed the *MasterRAC 2012* folder, and then double-click the *Chapter05* folder.

4. Double-click *05 Commercial.rvt* if you wish to work in Imperial units. Double-click *05 Commercial Metric.rvt* if you wish to work in Metric units

   You can also select it and then click the Open button.

Add Grid Lines

Let's begin laying out a column grid in this project. A Grid is a vertical datum element used mostly to represent column locations in the building. Grids help with locating objects in plans and elevations.

5. On the Project Browser, double-click to open the *Level 1* floor plan.

This view shows the first floor plan as we left it at the end of the previous chapter (see Figure 5.1).

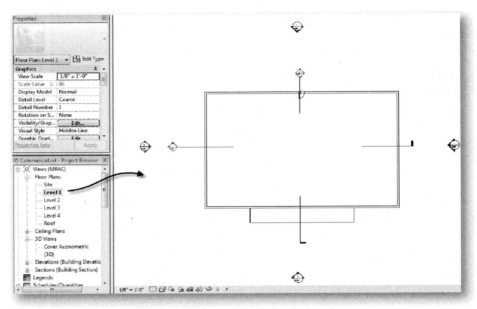

FIGURE 5.1 *Open the Level 1 floor plan view*

6. On the ribbon, click the Home tab; on the Datum panel, click the **_Grid_** tool.

TIP	The keyboard shortcut for Grid is **GR**.

Like many elements we have already seen, a new tab will appear on the ribbon. Some common tools appear on the Draw panel like the Line and Pick Lines icons. Line is the default and allows you to click any two points to specify the extent of the grid line. Grid lines can be drawn at any angle: they can even be curved. You can also specify the grid lines based upon existing elements in the model. To do this, use the Pick Lines icon.

- Click any two points within the building footprint to add a Grid line.
- On the ribbon, click the Modify tool or press the ESC key twice.
- Click on the Grid line just created.

Notice that many of the same handles and controls appear on the Grid line that appeared on the Level lines in the last chapter (see Figure 4.14 in the previous chapter). Place your pointer over each control to see a screen tip indicating its function (see Figure 5.2). Zoom in as required.

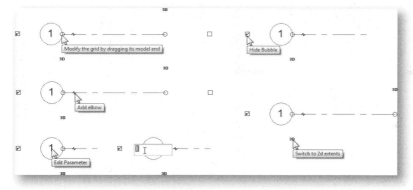

FIGURE 5.2 *Grids have many control handles and drag points*

Moving left to right in Figure 5.2, the following briefly describes each control.

- **Modify the grid by dragging its model end**—This round handle is used to drag the extent of the Grid. If the length or alignment constraint parameter is also active, dragging one Grid will affect the extent of the other constrained Grid lines as well. Grids are constrained to one another by default.

- **Hide/Show Bubble**—Use this control to hide or show the Grid bubble at either end of the Grid line.

- **Add Elbow**—This small handle creates an elbow in the Grid line. This is useful when the annotation of two Grid lines overlap one another in a particular view. Click this handle to create the elbow, and then drag the resultant drag handles to your liking to make the annotation legible.

- **2D/3D Extents Toggle**—When 3D Extents are enabled, editing the extent of the Grid line in one view affects all views in which the Grid appears (and is also set to 3D Extents) and literally changes the extent of the virtual plane representing the Grid. If you toggle this to 2D Extents, dragging the extent of the Grid line affects only the graphical representation in the current view.

- **Edit Parameter**—Click the blue text to re-number the Grid bubble.

- **Length and Alignment Constraint** (not shown in the figure)—A padlock icon used to constrain the length and extents of one Grid line to the others nearby. This is useful to keep all of your Grid lines lined up with one another.

7. Undo the Grid line (Undo is on the QAT or press CTRL+Z).

NOTE

You can also delete the Grid, but the automatic numbering of the next Grid you add will start at number 2 since it is the next number in the sequence. If you undo the Grid instead, it returns the starting number to 1.

Sometimes it is easier to use existing geometry to assist in the creation of the Grid lines. This can speed up the creation process and ensure that design relationships are maintained.

8. If you canceled the command, on the ribbon, click the Grid tool again.

- On the Place Grid tab, on the Draw panel, click the Pick Lines icon.

- On the Options bar, in the Offset field, type **4" [100]** and then press ENTER.

- Position the mouse pointer over the inside edge of the bottom horizontal Wall.

- When the dashed line appears on the inside of the building, click to add the Grid line (see Figure 5.3).

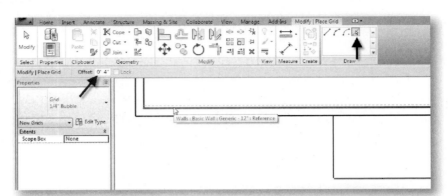

FIGURE 5.3 *Create a Grid line based on a picked Wall edge*

Depending on whether you deleted the first Grid line or used Undo, the number parameter of this Grid line may be "1" or it may be "2." If it is 2, we can edit it.

- If necessary, click on the Grid bubble text parameter (blue text) and then change the value to **1**.
- Make sure the Grid bubble shows on the left. If it is on the right, use the Show/Hide Bubble controls (small checkboxes) on each side to switch it.

You should now have Grid line 1 offset slightly to the inside edge of the bottom horizontal Wall with a bubble on the left. The extent of the Grid line matches the length of the Wall since we used the Wall to create it. Let's stretch it longer.

9. Click on the model end handles of Grid 1 (see above) and stretch each end longer away from the outside edges of the model (see Figure 5.4).

FIGURE 5.4 *Lengthen the Grid lines by dragging the model end handles*

It is a good idea to get the extent of the first Grid line to your liking before you continue, as the additional Grid lines that you add will snap to this length automatically. However, it is not critical as new Grid lines will automatically constrain to their neighbors, so if you do not stretch it longer first, you can stretch any Grid line later and they will all lengthen together.

You should still be in the Place Grid command. If you canceled the command, click the *Grid* tool again.

10. On the Draw panel, click the Line icon, and on the Options bar change the Offset back to **0**.

- Move the pointer to the right side, just above the previous Grid line.

When your cursor is above the previous Grid line end, a guideline will appear aligned with the previous Grid line and a temporary dimension will also appear.

- When the temporary dimension reads approximately 20'-0" [6000] click to set the first point—make sure it is still aligned.
- Move the pointer to the left and when the alignment guidelines (horizontal and extension) appear, click to place the new bubble aligned to the first (see Figure 5.5).

FIGURE 5.5 *Draw a second Grid line aligned to the first one*

Notice the closed padlock icon that appears connected to the new Grid line at each end. This indicates that the new Grid line is constrained to the previous one. Like Levels in the previous chapter, if you edit the extent of one Grid, they will both change together. Also take a careful look at the temporary dimensions. The witness lines are actually attached to the Walls, *not* the previous Grid line. We can adjust this and then edit the value to set the desired spacing between the two Grid lines.

11. With the Grid line command still active, move the pointer over the lower "Move Witness Line" handle (see the left panel in Figure 5.6).

- Drag the Witness Line up until it highlights Grid line 1 and then release (see the middle panel in Figure 5.6).
- Edit the temporary dimension value to **20'-0" [6000]** (see the right panel in Figure 5.6).

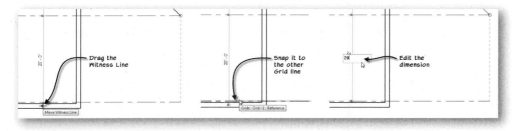

FIGURE 5.6 *Move the Witness line and then edit the value of the temporary dimension*

At first it might be a bit tricky to do this operation with the command still active. If you prefer, you can press ESC twice to cancel out, select Grid line 2, and then perform the steps above. The result will be the same. The motion should be fluid. Move the pointer over the Witness Line handle, click and drag it to Grid line 1 and release. Grid line 1 will highlight as you drag it. When it highlights, you should release the mouse. At this point, we could continue adding Grid lines using the same process. Instead, let's test the alignment constraint and then explore alternate techniques to create the additional Grid lines.

12. On the ribbon, click the Modify tool or press the ESC key twice.
13. Select Grid line 2.

- Using the "Model end" drag handle, drag the bubble horizontally a bit.

Notice that both Grid bubble 1 and 2 move together.

- Position the bubbles where you want them.

14. Save the project.

Copy and Array Grid Lines

We could draw each additional Grid line as noted above using either the *Line* or *Pick Lines* tools. We can also copy the existing ones using either the *Copy* tool or the *Array* tool. The *Copy* tool is basically a Move command that moves a copy instead of the original. The Array command can create several equally spaced copies in one action. It also has the option to "group and associate" the copies so that the parameters assigned to the array are maintained and remain editable. Let's explore a few options.

15. Select Grid line 2.

The Modify | Grids tab will appear on the ribbon.

- On the Modify | Grids tab, click the **Copy** tool.

TIP	The keyboard shortcut for Copy is **CO**. (**Note:** choose *Copy* on the Modify panel, not *Copy* on the Clipboard panel.)

On the Options bar, there are three available checkboxes. The "Disjoin" box is grayed out. ("Constrain" and "Multiple" are not currently selected).

- On the Options bar, place checkmarks in both the "Constrain" and "Multiple" checkboxes.

The "Constrain" checkbox limits the movement of the copy in the vertical or horizontal direction and the "Multiple" option makes more than one copy in the same operation.

- For the start point, click at the end of the Grid line (snap to the endpoint).
- Move the pointer straight up (you don't have to be too careful, this is what Constrain is for).
- Using the temporary dimensions, place two copies above Grid line 2 spaced at **20'-0" [6000]** (see Figure 5.7).

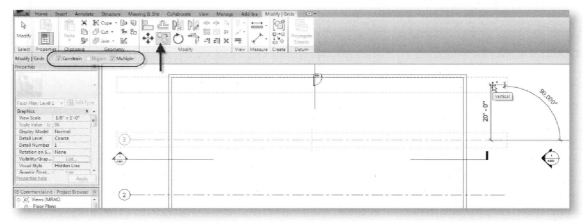

FIGURE 5.7 *Make two copies of the Grid line above Grid line 2*

- On the ribbon, click the Modify tool or press the ESC key twice.

Notice that the two new Grid lines numbered automatically as "3" and "4." Revit will always number the new Grid lines in sequence after the last one placed. If you were to delete one Grid (Grid line 4 for example), and then add another Grid line, it would automatically number to "5." However, if you were to undo the placement of Grid line 4, then the next one would be "4" instead.

16. Click on either of the new Grid lines.

Notice that again, the temporary dimensions reference the Walls and not the other Grid lines. If you wish to move them with the temporary dimensions, you can first move the Witness Lines as we did above. You can also use the Move tool if you prefer. Let's move both Grid lines up slightly to make the middle bay larger.

17. Select Grid lines 3 and 4. (You can use the CTRL key or a Crossing selection to do this—be certain to select only the Grid lines.)

   - On the Options bar, click the Activate Dimensions button.
   - Use the technique outlined above to move the Witness Line of the lower vertical dimension up to Grid line 2 (see the left side of Figure 5.8).

     When you have moved the Witness Line, the temporary dimension should read 20'-0" [6,000] (see the middle of Figure 5.8).

   - Edit the value of the lower temporary dimension to **21'-2" [6,480]** (see the right side of Figure 5.8).

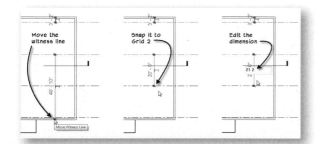

FIGURE 5.8 *Move Grid lines 3 and 4 using the temporary dimensions*

18. On the ribbon, click the Modify tool or press the ESC key twice.
19. On the Home tab, click the Grid tool.
   - On the Place Grid tab, click the Pick Lines icon.
   - On the Options bar, in the Offset field, type **4" [100]** and then press ENTER.
   - Position the pointer so that the light blue dashed line appears at the inside edge of the left vertical Wall and then click to add the Grid line.
   - On the ribbon, click the Modify tool or press the ESC key twice.

Notice that this Grid line became number 5. For the vertical bays, we want to use letters.

20. Click on the new vertical Grid line.
   - Adjust its length so that it projects above and below the building footprint (like we did for the others) and then edit the bubble designation to **A**.
21. Select Grid line A.
   - On the Modify | Grids tab, in the Modify panel click the *Array* tool.

The Array command has many options. First, you can choose between a linear or radial array on the Options bar. Next there is the "Group and Associate" checkbox. With this option enabled, all of the arrayed items (the original and the copies) will be grouped together. When thus grouped, you can select them later and edit the quantity of items in the array. The remaining items will re-space to match the new quantity either adding or deleting elements in the group. This can be a very powerful and useful functionality that has many applications in building design. This being our first opportunity to explore this tool, we will test it out here with our Grid lines. Ultimately, we will want an ungrouped array because the final spacing of our column grid is not equal (as is typical in most buildings), however it will still be educational to explore briefly the "Group and Associate" option of array nonetheless.

- On the Options bar, verify that the "Linear" icon is chosen and that "Group and Associate" is selected.

Array is much like the move and copy commands in that it requires you to pick two points on screen to indicate the spacing of the array. We have two options for what these two points can represent in the array: they can be the spacing between each element, or the total spacing of the entire array. To set the spacing between each element, choose the "2nd" option next to "Move To." To set the total spacing, choose the "Last" option.

- Choose the "2nd" option.
- For the "Number," type **6** and make sure that the "Constrain" box is selected.

The number indicates the total quantity of elements in the final array. This quantity includes the original selection. The "Constrain" option works like the same option in the Copy command (see above) by limiting movement to horizontal and vertical only.

- Watch the Status bar for prompts. Click the first point on the endpoint of the selected Grid line.
- Move the pointer to the right about 18'-0" [5400] and then click the second point (see Figure 5.9).

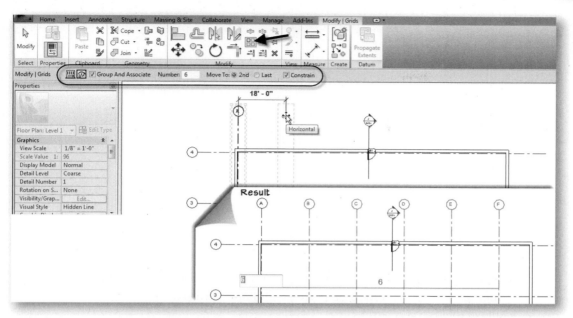

FIGURE 5.9 *Array a total of 6 Grid lines horizontally*

The last element of the array is not quite close enough to the inside edge of the right Wall. Using the temporary dimensions like we did above, we can adjust it to the exact location we need. Since the array is grouped, all items will adjust to maintain an equal spacing as we do this.

22. On the ribbon, click the Modify tool or press the ESC key twice.
23. Click the last item (Grid line F on the right).

    Notice that a dashed box appears around the Grid line. This indicates that the element is part of a Group.

    • On the Options bar, click the Activate Dimensions button.

    You may need to zoom out a bit to see them.

    The last Grid line (labeled F) needs to be 4" [100] from the inside face of the right Wall (like its counterpart on the left).

    • Move Witness Lines and edit the temporary dimension value to **4" [100]** from the inside face of the Wall.

Instead of dragging the Witness Line handle, simply click it. Each time you click it, it will shift to a new reference point in the Wall. It will cycle from inside to center to outside and then back again.

TIP

The Wall is 12" [300] thick. So you can place the Witness Lines in any convenient location and take this value into account to achieve the correct location.

NOTE

Notice that all the Grid lines between A and F adjust with this change and remain equally spaced. This is because of the grouped Array.

    • Deselect Grid line F and then re-select it (click on it again).

A temporary dimension will appear on the arrayed items with a single text parameter that indicates the number of items in the array (see Figure 5.10).

FIGURE 5.10 *Edit the quantity of items in the array with the Edit Text parameter*

- Click to select this text parameter.
- Type **7** and then press ENTER.

Notice that a new Grid line is added and lettered automatically.

- Repeat the process typing **5** this time.

Notice that two of the Grid lines are deleted. Notice also that the spacing between the items remains unchanged. This is because we used the "Move To 2ⁿᵈ" option above to create the array.

- Repeat the process one more time setting the value back to **6**. (You can also Undo twice if you prefer.)

This is the benefit of using the "Group and Associate" parameter when you create the array. However, as we mentioned above, the spacing of our commercial building's column grid is not actually equal. Therefore, we will need to ungroup this array. Before we do, feel free to experiment further with the array. For example, you can select any one of the Grid lines and move it. All of the others will move accordingly. This can be tricky, as the one you select actually stays stationary and all the others move relative to it. Be sure to undo after you try this.

You can also choose the Edit Group icon on the Modify Model Groups ribbon with one of the elements selected. This will make an Edit Group mode active and a toolbar will appear. In this mode, you can manipulate the elements within the group. When you choose the Finish button, the change is applied to all group instances (all of the elements in the array in this case). In this case, appropriate edits might be resizing the Grid line or adding or removing the bubble at either end. We will explore Groups in more detail in Chapter 6. Be sure to undo any explorations you made returning to 6 Grid lines before continuing.

Dimension the Grid Lines

Next we'll ungroup the array. This will allow us to adjust the Grid line spacing independently.

24. Select all vertical Grid lines—A through F (use the CTRL key or a crossing selection).

    The Modify Model Groups tab will appear on the ribbon.

    - On the Modify | Model Groups tab, on the Group panel, click the Ungroup button.

The Grid lines will now be ungrouped and remain selected. Since they are no longer part of an array, they now move independently of one another. Try a few moves if you like, but be sure to undo before continuing. We are going to adjust several Grid lines now. This will be easier with some permanent dimensions rather than the temporary dimensions.

25. On the Annotate tab of the ribbon, on the Dimension panel, click the ***Aligned*** tool.

    • Starting with Grid line A, click successively on each lettered Grid line A–F (see Figure 5.11).

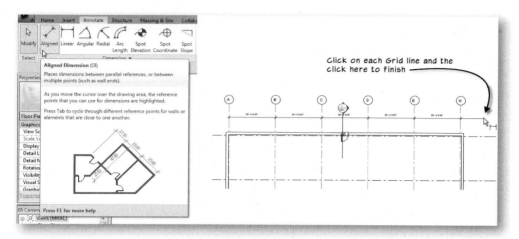

FIGURE 5.11 *Add a dimension string to the lettered Grid lines*

    • After you click on grid line F, move the mouse to a spot where you want to place the dimension string and then click to set the dimension string and end the command.

Make sure that you click in an empty spot to place the dimension, if you click on another model element, it will add a Witness Line instead.	**TIP**

26. On the ribbon, click the Modify tool or press the ESC key twice.
27. Select Grid line B

Notice that now that we have added a string of dimensions, in addition to the normal temporary dimensions we also can edit the dimension values of the string we just added.

    • Click on the blue dimension text between Grid line A and B, type **22'-2" [6740]** and then press ENTER.

Grid line B will move to the right slightly.

    • Select Grid line C next and edit the distance between it and B to **14'-4" [4400]**.

This time, Grid C will move. Remember; always click the element that you wish to move *before* editing dimension text.

    • Repeat moving left to right for the remaining vertical Grid lines as shown in Figure 5.12.

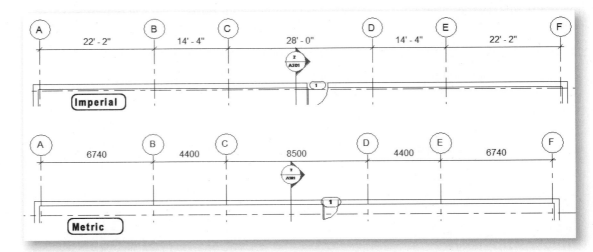

FIGURE 5.12 *Edit dimension values to move Grid lines*

Our feature façade on the front of the building (we created a Mass for this in the last chapter) will require some structural support. We can add a few additional Grid lines in the front for these.

28. On the Home tab of the ribbon, click the Grid tool.

- On the Place Grid tab, verify that the Line icon is selected.
- Move the pointer to the inside of the building above the lower Wall and click between Grid lines C and D.
- Move down past the building and click again to place the bubble.

The bubble will automatically enumerate as "G."

- Click in the text parameter of the new bubble and change the value to **C.3**.
- Using the temporary dimensions, move the Grid line so that it is **9'-0" [2700]** from Grid line C (see Figure 5.13).

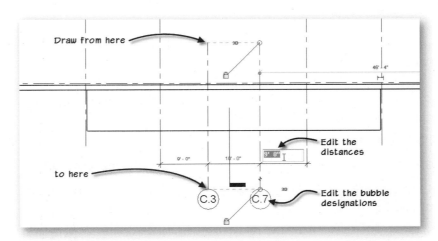

FIGURE 5.13 *Add Grid lines at the bottom of the plan for support of the front façade*

- Add Grid line C.7 to the other side of the C-D bay using the same **9'-0"** [2700] off-set (see Figure 5.13).

29. Save the project.

Add and Edit Dimensions

Let's add a dimension string to the vertical Grid lines too. When you add a dimension, a drop down on the Options Bar allows you to choose between dimensioning to Wall faces or Wall centerlines.

30. Start the Aligned Dimension tool and on the Options Bar, choose **Wall Faces** from the drop-down list.
    - Add a dimension string to the numbered bubbles on the left (running vertically) starting at and including the outside face of the Wall and ending at the outside face of the opposite Wall.

Remember to finish placing a dimension string, you click in empty space not on any geometry. If you wish to add the exterior Walls to the horizontal string we created earlier, you can edit the dimension. To do this, select the dimension. On the Modify | Dimensions ribbon tab, click the Edit Witness Lines button. With edit Witness Lines active, if you click on geometry that is not already part of the dimension, it is added. If you click on a witness line that is already part of the dimension, it is removed. To finish, click in empty space (not on any geometry).

Viewing Grid lines in Other Views

All of the work we have done on Grid lines so far has been in the *Level 1* floor plan view. However, like other datum elements, Grids will automatically appear in all orthographic views.

31. On the Project Browser, double-click to open the *West* elevation view.

Notice how the numbered Grid bubbles appear here running vertically on the elevation.

32. On the Project Browser, double-click to open the *South* elevation view.

This time only the lettered bubbles appear. Like the plan views, if you drag the end of one Grid, the others will stretch as well. If you find the bubbles in the middle to be too close together, you can use the handles to give them an elbow.

33. On the Project Browser, double-click to open the *Longitudinal* section view.

In this view, Grid lines C.3 and C.7 do not show since the section is looking toward the north. If the section were looking south, they would appear. If you want to experiment, you can return to the *Level 1* floor plan view, select the Section line for the *Longitudinal* section and then click the flip handle. Re-open the *Longitudinal* section view to see the Grids appear. Flip it back before continuing.

Our column Grid layout is now complete. We are ready to begin using the intersections between Grids to locate our Columns. Before we begin that process however, our Grid lines can also be used to enhance our sheet layouts. The next topic will look at doing so.

Add a Guide Grid

With the Guide Grid feature, you can overlay a grid on top of a sheet titleblock and use it to align similar views across multiple sheets. For example, using this tool, we can make all of the floor and ceiling plans appear on the same relative location from one sheet to the next. Sheets were set up for this project in the previous chapter. At that time, we simply dropped each floor plan view on the sheet in an approximate location. Using the Grids we have now added to the project and the new Guide Grid tool, we can align our floor plan sheets with one another precisely.

1. On the Project Browser, double-click to open the *A101 – Floor Plans* sheet.
2. On the View tab, on the Sheet Composition panel, click the Guide Grid tool.

    • In the "Guide Grid Name" dialog that appears, type **Plan Sheets** and then click OK.

A light blue grid object will appear across the sheet. If you select it, you can modify its grid spacing on the Properties palette and manipulate the overall extents with the control handles on the edges.

    • Select the Guide Grid onscreen.
    • On the Properties palette, change the Guide Spacing to **5" [125]** and then click Apply.

The grid of blue squares will increase in size, making it easier to work with. You can now move the viewport on the sheet and snap it to any of the Guide Grid's lines or intersections. Use the Move tool to do this. To move a viewport precisely, snap to column Grid lines for the move start point, and then snap them to the Guide Grid.

3. On the Modify tab, click the Move tool.

> **NOTE** This is an alternative to the method previously shown. You can select modify commands first, and then build your selection. Press ENTER to finish selection and continue with the command.

4. Click on the viewport to select it and then press ENTER to complete the selection.
5. Snap the first point of the move to the column Grid lines (such as grid intersection F1).
6. Snap the second point of the move to the Guide Grid (see Figure 5.14).

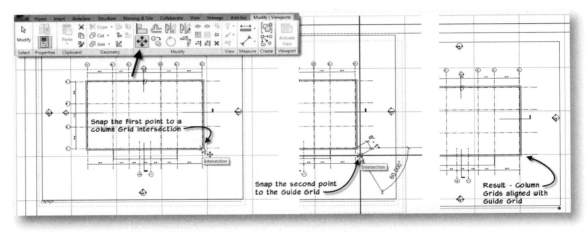

FIGURE 5.14 *Add a Guide Grid to the plan sheets and use it to line up the viewports*

7. On the Project Browser, double-click *A102 – Floor Plans*.

    • On the Properties palette, from the Guide Grid list, choose **Plan Sheets** and then click the Apply button.

Notice that the same Guide Grid will now appear on this sheet. You can now align the viewport in the same way as the previous sheet and the views will thereby be in the same relative location on each sheet.

8. Repeat the move steps to align the viewport on sheet A102 with the same grid location used on sheet A101.
9. Open each floor plan and ceiling plan sheet and repeat the process.

> **TIP** You can also open the Working – Sheet List Schedule and set the Guide Grid parameter there as an alternative to using the Properties palette.

You can use Levels, Grids, Reference Planes and view Crop boundaries to align with Guide Grids. Unfortunately, you cannot use Walls or other model geometry. This is why we waited until this chapter to add the Guide Grid and line up the viewports. If you wish to line up viewports on other kinds of sheet like elevations or sections, you can use the Guide Grid tool on the View tab to create another Guide Grid. You can create as many Guide Grids as necessary to set up all your sheets.

When you are finished aligning the viewports, you can set the Guide Grid parameter back to <None> if you no longer wish them to display. However, it should be noted that Guide Grids do not print, so you can leave them displayed indefinitely onscreen without worry that they will appear in final output.

If the views are not maximized, maximize them now.

10. On the Project Browser, double-click to open the *Level 1* floor plan view.

   • On the View tab of the ribbon, on the Windows panel, click the **Close Hidden** Icon.

   • Save the project.

WORKING WITH COLUMNS

A collection of structural tools is included in Revit Architecture. A larger collection of structural tools appears in the Revit Structure application. Users of Revit Architecture can seamlessly share models back and forth between structural consultants using Revit Structure. (The file format is the same between the three flavors of Revit, Architecture, MEP, and Structure.) You can use the structural tools provided with Revit Architecture to begin the layout of Columns and if desired, Beams and Braces as well. Like all Revit elements, structural elements remain parametric and editable after they have been added to the model. Therefore, when the structural analysis comes back from the engineer, you can swap in the correct types and sizes to meet the requirements of the design.

Disclaimer:

No structural analysis of any kind has been performed on the designs in this book. The shapes used in this book serve illustration purposes and are not to be construed as a recommendation of structural integrity or a design solution.

Add Columns

Revit Architecture includes two types of columns: Architectural Columns and Structural Columns. An Architectural Column provides the location and finished dimensions (including column wrap, furring, finishes, etc.) of a column in an architectural space. The Structural Column is the actual structural material responsible for supporting building loads typically without any enclosure or finish. Examples include steel or concrete columns. Let's take a look at both kinds.

Continue in the *Level 1* view opened above.

1. On the ribbon, click the Home tab and then click the drop-down button (bottom half) on the **Column** tool and choose Architectural Column.

   • From the Type Selector, choose Rectangular Column: 24" × 24" [M_Rectangular Column: 610 × 610mm].

   • Move the pointer to the intersection of Grid lines 4 and A (upper left corner of the building).

   • The pointer will snap to the intersection automatically (see the left side of Figure 5.15).

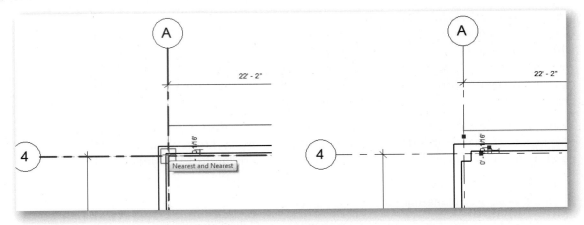

FIGURE 5.15 *Add an Architectural Column to the A4 Grid intersection*

- Click at the intersection to place the Column (see the right side of Figure 5.15).

Notice that the Architectural Column automatically interacts with the Walls at the intersection and graphically displays as an integrated column.

2. Continue placing Columns at all of the intersections around the perimeter of the building (see Figure 5.16). Do not place columns at the interior intersections.

TIP	You can simply click at each intersection or you can use the Copy tool. If you use Copy, be sure to select "Multiple" on the Options bar, and use the TAB key to select a good start point for the copy.

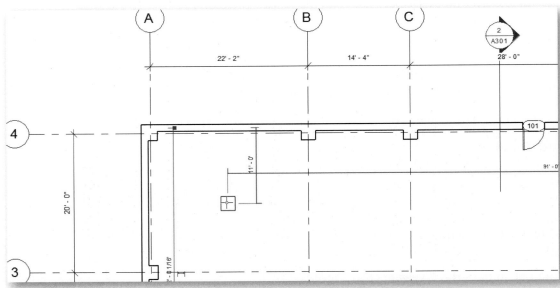

FIGURE 5.16 *Architectural Columns around the perimeter integrate with the Walls*

- On the ribbon, click the Modify tool or press the ESC key twice.

Let's try some Structural Columns next.

3. Click on the drop-down button on the **Column** tool.

   - Choose **Structural Column** from the pop-up.
   - Open the Type Selector.

Notice that there are only two Structural Column types loaded in the current project (there is only one in the metric project). Structural Columns, like other Families in Revit Architecture can be loaded as needed from external libraries. Close the Type Selector.

4. On the Modify | Place Structural Column tab of the ribbon, click the Load Family button.

If you are working in Imperial units, the *Imperial Library* folder should open automatically. If you are using Metric units, the *Metric Library* folder should open automatically. If the appropriate location fails to open automatically, browse there now. The easiest way to do this is to click the appropriate shortcut button on the left side of the dialog.

   - Double-click the *Structural* folder, then double-click the *Columns* folder and finally double-click the *Steel* folder.
   - Double-click the *W-Wide Flange-Column.rfa* [*M_W-Wide Flange-Column.rfa*] Family file.

The "Specify Types" dialog will appear on screen. This shows a list of industry standard steel-shape sizes. Scroll through the list to see all of the sizes. You can select one size or multiple items using the SHIFT and CTRL keys.

   - Scroll in the list, locate W12×87 [W310×97] and select it (see Figure 5.17).
   - Click OK to load the Family.

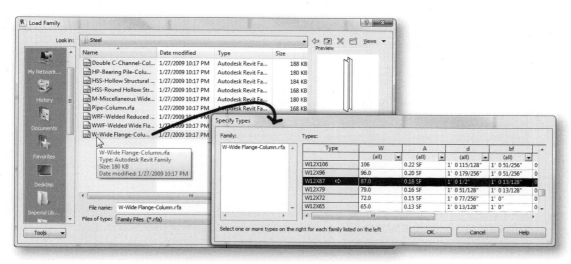

FIGURE 5.17 *Common Industry Steel Shapes are available to load from the Family file*

   - When the "Family Already Exists" dialog appears, click the "Overwrite the existing version" option. (This is the same Family that was already loaded, but we have loaded a new type.)
   - From the Type Selector, choose the newly loaded W12×87 [W310×97] type.

Structural Columns have many of the same parameters as Architectural Columns—and more. If you move the pointer around the screen before you click, you will note that like the Architectural Column, the Structural Column will pre-highlight Grid lines and Walls as possible hosts. (However, if you insert a Structural Column at a Wall, it will not merge with the Wall the same way that the Architectural one did.)

In addition to behaviors shared with Architectural Columns, there are some additional options on the ribbon. For example, the "On Grids" button (on the Multiple panel) allows you to place several Structural Columns at once on a selection of Grid lines, and the "At Columns" button allows you to place a Structural Column at the location of a selection of Architectural Columns. Let's give these both a try.

5. On the Modify | Place Structural Column tab, on the Multiple panel, click the "At Grids" button.

   The Modify | Place Structural Columns > At Grid Intersection tab will appear on the ribbon.

6. Click the pointer above Grid line 3 and to the right of Grid line E and then drag to the left of Grid line B and below Grid line 2 (see Figure 5.18).

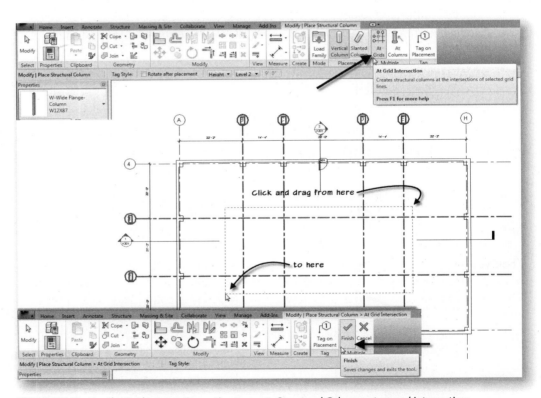

FIGURE 5.18 *Use the Grid Intersection option to create Structural Columns at several intersections*

The Grid lines will highlight and several ghosted columns will appear. If you are satisfied with this selection, you click the Finish Selection button on the ribbon. Otherwise, you simply select again until you are satisfied. If you decide not to add the Columns, you can click the Cancel button on the ribbon.

• On the ribbon, click the Finish button.

You will now have a steel column at each Grid intersection in the middle of the building. Let's try the "At Columns" option next. In addition, we will also adjust the height of the columns as we add them.

7. On the Options bar, from the button that currently reads "Level 2," choose **Roof** from the pop-up.

This will create a single continuous Structural Column from Level 1 (the current Level) up to the Roof.

- On the Modify | Place Structural Columns tab, click the At Columns button.
- Using a crossing or window selection (click and drag a selection window—either left to right or right to left will work in this case) surround the entire building to select all Architectural Columns.
- On the ribbon, click the Finish button.
- On the ribbon, click the Modify tool or press the ESC key twice.

Edit Columns

We now have Structural Columns at all Grid intersections including the ones at the perimeter Walls. We can modify Columns using many of the same techniques we use for other elements. If you wish to change the Type of column, you can select one or more and choose a new Type from the Type Selector. You can move, copy, or rotate them. For example, sometimes we need to rotate the flange of the steel.

8. Select all of the steel Columns in the center of the plan (the ones not enclosed by Architectural Columns).

- Tap the SPACEBAR to rotate them 90 degrees.

However, the ones in the center (not surrounded by Architectural Columns) span from Level 1 to Level 2, while the ones at the perimeter span the complete height of the building. We cannot see this in the current plan view. Let's take a look at the section.

9. On the Project Browser, double-click to open the *Transverse* section view.

Understanding Level of Detail

When you switch to this view, you will likely not notice much detail with regard to the Structural Columns. This is because structural elements in Revit Architecture display at various levels of detail. This allows for simplified graphics in smaller scale drawings, and more detail in larger scale drawings. You can easily modify the default settings for this behavior, or simply adjust the level of detail display for an individual view. For example, Figure 5.19 shows the variation in the three detail levels when viewing Structural elements such as the Columns we have in our model. As you can see, the Coarse detail level shows simple linework for the steel shape in plan and a single line for it in section (and elevation). This is the currently active display detail level in both the floor plan and section views in our project. If you switch to Medium or Fine detail level, the graphics for the steel shape will get progressively more detailed in plan. In elevation and section, display is more detailed than Coarse, but the same graphics display for both higher levels of detail (see Figure 5.19).

Lineweight display adjusts automatically when you vary the scale of the view. Note the change in scale in each panel of the figure and experiment on your own screen to make the medium and high displays more legible.	**NOTE**

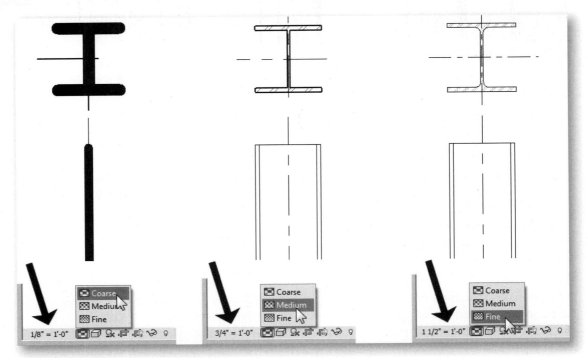

FIGURE 5.19 *Three levels of Display Detail Level*

10. In the *Transverse* section view, on the View Control bar, change the Detail Level to **Medium**.

The Columns will now display with the correct dimensional thickness in the section rather than the diagrammatic single-line display. Zoom around the section as needed to see clearly that the Architectural Columns stop at Level 2 while the Structural ones go all the way to the Roof. Naturally it is likely that these Columns would actually be constructed in sections and not be a single continuous four-story-tall Column. However, for design purposes, they function in the building as a single continuous column. Therefore, it is up to you decide in your own projects what the height of the Column object represents and build it accordingly. You can make the decision to create the Columns the actual height of the material that will be shipped to the site (two-stories tall for instance); this would be important if you use Revit Architecture to assist you in deriving construction quantities or when we send the model to our Structural Engineer for structural analysis and design. If you wish to create shorter Columns, select them, edit their properties and then simply choose a different Top Level parameter value. If you want them to end between levels, use the Base Offset and/or Top Offset parameters with positive or negative offsets as required to achieve the desired result. If you do choose to shorten the Columns, you will then need to open the plan view for the next Level and copy/paste aligned the Columns there.

In this project, we will leave this decision to the Structural Engineers. For now, we will make our Structural Columns span the full height of the building and leave our Architectural Columns (which in this project represent the finish materials wrapping the Columns) spanning a single Level. Later we will copy these Architectural Columns to the other floors.

11. Zoom in on Grid line 2 or 3 at Level 1. (The middle bay).

- Move your pointer over the Columns and note the screen tips that appear (see Figure 5.20).

> **TIP**
> If you have trouble highlighting the overlapping elements, use the TAB key to cycle through adjacent elements.

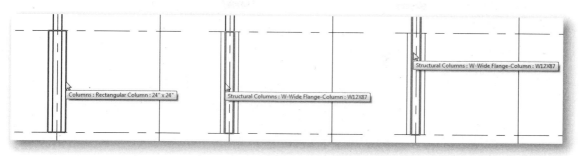

FIGURE 5.20 *In section view, we see columns in the middle of the building at different heights than those beyond*

In our section view, several of the Columns are in the same place. Some of these are the shorter ones in the center of the building and the tall ones appear beyond at the perimeter of the building. We can still select the ones we need to edit in this view, or if you prefer, we can return to the Level 1 plan view and select there. It makes no difference in which view you select them. If you select in section, you can easily avoid the full height Columns with a crossing or window selection; however, it will be difficult to avoid the Architectural Columns behind them. Furthermore, some of the Columns would not be selected as they are cropped away in the section view. Therefore, in many cases you will want to consider selecting in a view other than section. However, in this case we can make use of another tool to assist in this selection.

12. Click to select one of the steel Columns (any one, tall or short).

   • Right-click and choose **Select All Instances > In Entire Project**.

> **TIP**
> It is not actually necessary to select one element first. You can simply pre-highlight one of the Columns and then right-click to select all instances.

Be careful with this command. It literally selects all instances of a Family Type in the entire project—both seen and unseen. In this case, that is exactly what we want, but in many cases, it may not be. Keep this in mind before using this tool in your work. For example, if you choose to select all instances of a door while working in a plan view, realize that you are also selecting similar doors on other floors of the building! For this type of selection, you would use the other option; **Select All Instances > Visible In View**. This option functions the same way by selecting all instances of the selected Family and Type but limits selection on to those you can see in the active view. For our current selection needs, this would exclude those columns that are cropped out of the section view. So use of the "In Entire Project" option is better in this case.

   • On the Properties palette, for Base Level, choose **Street Level**.
   • For Top Level, choose **Roof** and then click Apply (see Figure 5.21).

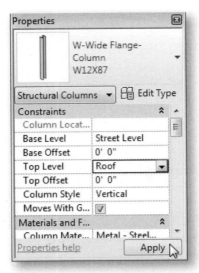

FIGURE 5.21 *Edit the Base and Top Level parameters of the selected Columns*

Notice the height of all of the steel Columns now spans from below the first floor to the roof.

13. On the QAT, click the Default 3D View icon.

   • Click on the Roof and then on the View Control bar, from the Hide/Isolate pop-up (icon looks like sunglasses), choose **Hide Element**.

   Orbit the model as required to get a better look. You can drag the ViewCube, use a steering wheel (or hold down the SHIFT as you drag with the mouse wheel).

You will notice that all steel Structural Columns were impacted by this change, not just the ones we could see in the section view. This is because as noted above, Select All Instances selects across the entire model, not just the elements visible in the current view.

Copy and Paste Aligned

Let's copy our Architectural Columns and add them to other floors.

14. On the Project Browser, double-click to open the *Level 3* floor plan view.

Notice that only the Structural Columns show in this view. The Structural Columns show because they span multiple levels and therefore intersect this view's cut plane. The Architectural Columns do not show because they occur on Level 1 and are a single story in height.

15. On the Project Browser, double-click to open the *Level 1* floor plan view.

We could use the same selection method that we just used to select all of the Architectural Columns, but let's explore another technique this time.

16. Make a window selection (from left to right) surrounding the entire building (don't include the Grid lines). (See item 1 in Figure 5.22.)

This will select all visible elements in the building model.

   • On the Modify | Multi-Select tab of the ribbon, click the Filter icon (see item 2 in Figure 5.22).

- In the Filter dialog, click the "Check None" button.
- Place a checkmark in the Columns checkbox and then click OK (see item 3 in Figure 5.22).

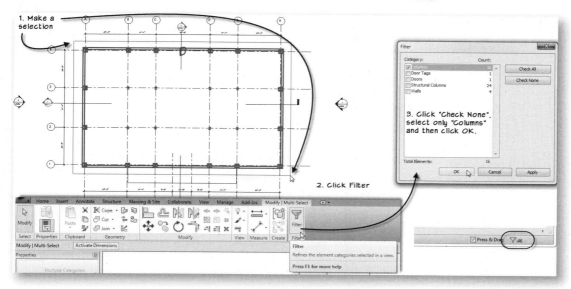

FIGURE 5.22 *Use Filter Selection to modify a selection and remove unwanted items*

Only the Architectural Columns should now be selected in light blue. (Note that Architectural Columns are referred to in the Filter dialog simply as "Columns".)

17. On the Modify | Columns tab of the ribbon on the Clipboard panel, click the **Copy to Clipboard** tool (or press CTRL+C).

    - On the Clipboard panel, click the drop-down button on the **Paste** tool and then choose **Aligned to Selected Levels**.
    - In the "Select Levels" dialog, select Level 2 and drag down to Level 4. (This should highlight Levels 2, 3 and 4.)

      You can also select Level 2 first and then use the CTRL key to select levels 3 and 4 as well.

    - Click OK to complete the paste operation.

18. On the Project Browser, double-click to open the *Level 3* floor plan view.

Notice that all of the Columns, both Structural and Architectural, show in this view. The Architectural Columns are the copies that we just added. Check other views like the section as well if you wish.

19. Save the project.

ADDING CORE WALLS

Now that we have a column grid with associated Columns, it is a good time to sketch in the building core. To do this, we will use Structural Walls. Structural Walls are essentially the same as other Walls except that the "Structural Usage" parameter for them is automatically set to **Bearing**. This parameter is used in the Revit Structure application to distinguish the function of Walls. We add Structural Walls in the same fashion as other Walls.

Add Structural Walls

1. On the Project Browser, double-click to open the *Level 1* floor plan view.

   - Zoom in on the two bays between column Grids C and E at the top of the plan (between Grid lines 3 and 4).

TIP	Right-click and choose Zoom In Region or type **ZR. Zoom In Region** is also located on the Navigation bar.

2. On the Home tab of the ribbon, click the drop-down button on the **Wall** tool.

   - From the pop-up that appears, choose the **Structural Wall** tool.
   - From the Type Selector, from the Basic Wall Family Types, choose Generic - 8" Masonry [Generic - 225mm Masonry].
   - On the Properties palette, set the Location Line to: **Wall Centerline**.
   - Set the "Base Constraint" to **Street Level**.
   - Set the "Top Constraint" to **Up to level: Roof**.

These two settings instruct our Structural Wall to go the full height of the building from the lowest level to the highest. We will also need the core Walls to continue past the Roof to allow stair access to the roof. So let's also add a "Top Offset" parameter. Keep in mind that all of these parameters can be edited later as the project progresses and as design needs dictate.

   - In the "Top Offset" parameter field, type **12'-0" [3600]** (see Figure 5.23).

Take note of the "Structural Usage" parameter—note that it is set to "Bearing" automatically.

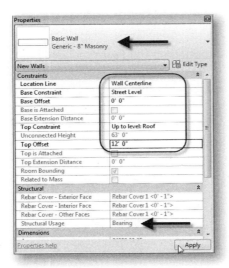

FIGURE 5.23 *Configure the parameters for Structural Wall the full height of the building*

   - Click Apply (or simply shift focus away from the palette) to continue.
   - On the Draw panel, click the Pick Lines icon.
   - On the Options bar, in the Offset field, type **1'-2" [350]** and then press ENTER.

With the Pick Lines option enabled, we will be able to select existing geometry from which to create the Walls. The Walls will be created at a distance of 1'-2" [350] from the selected geometry.

3. Highlight Grid line C and when the light dashed offset line appears to the right of the Grid line, click the mouse to create the Wall.

   - Repeat for Grid line 3 (horizontal) when the offset line appears below the Grid line (see Figure 5.24).

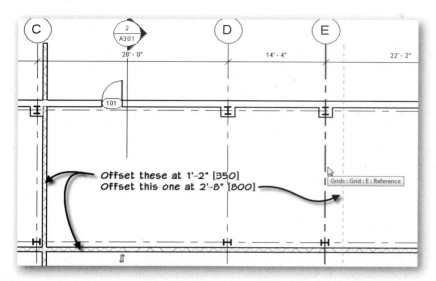

FIGURE 5.24 *Create Walls from Grid lines using the Pick Lines option with an offset value*

   - Change the Offset value on the Options bar to **2'-8"** [**800**] and then create one more Wall to the right of Grid line E.
   - On the ribbon, click the Modify tool or press the ESC key twice.

4. On the Modify tab of the ribbon, on the Modify panel, click the ***Trim/Extend to Corner*** tool.

   - Trim the two bottom corners of the Structural Walls to appear as in Figure 5.25.

Use the Trim/Extend to Corner option for the two bottom intersections (remember to select the side of the Wall that you want to keep).

5. On the Edit panel, click the ***Extend*** icon.

   - From the pop-up that appears, choose **Trim/Extend Multiple Elements**.
   - Trim the two top intersections to butt into the inside edge of the exterior horizontal Wall as shown in Figure 5.25.

Remember to click the side of the Wall you want to keep. For more information on the Trim/Extend tool, refer to the tutorials in Chapter 3 or search the online Help.

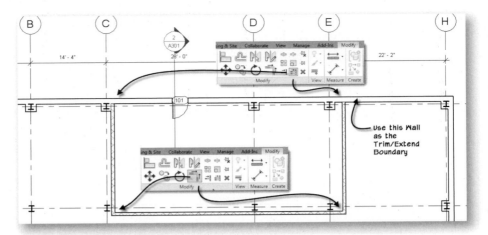

FIGURE 5.25 *Trim the Walls to form the core layout*

- On the ribbon, click the Modify tool or press the ESC key twice.

Convert a Wall to a Structural Wall

Let's view the model in section to see what we have so far.

6. Double-click the blue Section Head between Grid lines C and D.

The *Transverse* building section view will open.

7. Click on the Crop Box to select it. Click and drag the control handle at the top edge and drag it up enough to see the top of the core Walls (see Figure 5.26).

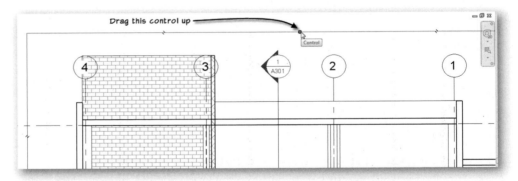

FIGURE 5.26 *Drag the Crop Box large enough to see the top of the new Walls*

Notice that the exterior Wall at the core only projects to the parapet height. In reality, this Wall would also need to be part of the core. Let's split this existing Wall and then change its parameters to match the rest of the core.

8. On the Project Browser, double-click to open the *Level 1* floor plan view.
9. On the Modify tab of the ribbon, on the Modify panel, click the **Split Element** tool.

The keyboard shortcut for Split is SL.

Move the split pointer near the intersection between the vertical core Wall and the horizontal exterior Wall. (Be sure that "Delete Inner Segment" on the Options bar is not selected.)

- When the pointer snaps to the intersection, click the mouse to split the Wall.
- Repeat on the other intersection (see Figure 5.27).

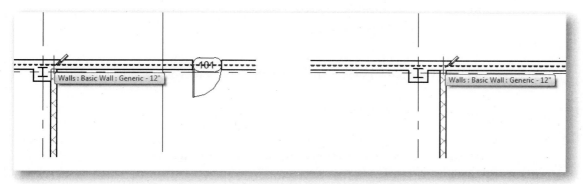

FIGURE 5.27 *Split the upper horizontal Wall to form the top Wall of the core*

- On the ribbon, click the Modify tool or press the ESC key twice.
10. Select the middle horizontal Wall segment (created by the splits).
    - On the Properties palette, from the Type Selector, choose Generic - 12" Masonry [Generic - 300mm Masonry].
    - Change the "Top Offset" to **12'-0" [3600]**.
    - Beneath the Structural grouping, check the "Structural" checkbox and verify that the Structural Usage parameter is set to **Bearing** and then click Apply.

Edit Wall Joins

Graphically, the Wall should now appear as the other core Walls do (the thickness varies). If you are unhappy with the way the Walls clean up at the corners, you can use the *Wall Joins* tool to edit them.

11. On the Modify tab of the ribbon, on the Geometry panel, click the **Wall Joins** tool.
    - Click at the first intersection that you wish to correct.
      A square will appear surrounding the intersection.
    - On the Options bar, click Next or Previous to see the various solutions. Stop at the one you like.
      The Edit Joins tool will stay active.
    - Click on the other intersection and repeat the process (see Figure 5.28).

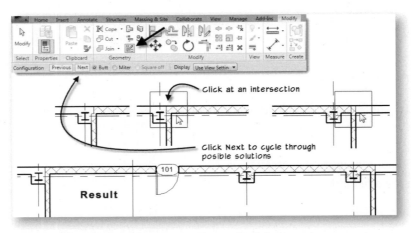

FIGURE 5.28 *Edit Wall Joins to inprove the graphical display at the corners.*

- When you are finished editing, click the Modify tool or press the ESC key twice.

TIP	In some cases, you can disjoin the Walls using the drag handles at the ends instead of Edit Joins. Drag the handles away from the intersection to disconnect the Walls from one another. Then use the *Trim* tool to re-join the Walls. Try Edit Joins first. You can also right click the handle and choose Disallow Join. This prevents the Walls from joining at all. This is useful when you want a Wall to butt into a curtain wall mullion for example.

12. Double-click the blue Section Head between Grid lines C and D again.

Notice that the outside Wall now matches the others in the core. You can check your work in the default 3D view as well if you wish.

13. Open the Longitudinal Section and expand the Crop Region there as well.
14. Save the project.

ADDING FLOORS

The most obvious omission that you might notice in our current sections is the lack of any floor slabs. Adding floors is easy to accomplish.

Create a Floor from Walls

The easiest way to create a Floor is by using the existing Walls that bound it.

1. On the Project Browser, double-click to open the *Level 1* floor plan view.
2. On the Home tab of the ribbon, click the **Floor** tool.

   The Modify | Create Floor Boundary ribbon tab appears. On the Draw panel, the Boundary Line and Pick Walls modes should be active.

   - Click first on each of the vertical exterior Walls, and then the bottom horizontal one, and one of the top horizontal ones (see Figure 5.29).

NOTE	Be sure to click on the inside edges. If you accidentally create sketch lines to the outside edge, use the Flip control at the middle of the sketch line to flip the sketch to the inside.

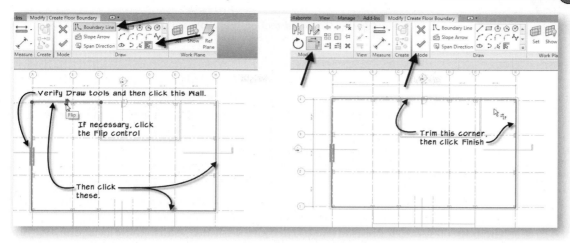

FIGURE 5.29 *Create sketch lines from Walls (Sketch lines enhanced for clarity)*

- Use the ***Trim/Extend to Corner*** tool to join the two open sketch lines.

Above, you were instructed to click the vertical Wall on the left first. Notice that this applies a special symbol to that sketch line only. This is the Span Direction of the Floor. Revit allows you to edit the Floor Type to apply Structural Deck to your design. The symbol on this first sketch line indicates the direction of the decking. You can use the Span Direction button on the ribbon to change the span direction later if required.

- **C**lick the Finish Edit Mode button.

A message will appear on screen that reads: "The floor/roof overlaps the highlighted wall(s). Would you like to join the geometry and cut the overlapping volume out of the Wall(s)?" If you answer yes to this message, Revit will use the Floor volume to cut away the overlapping portion in the various Walls. If you answer no, the two elements will remain overlapping and appear unfinished in sections and throw off volume calculations you may later generate.

- In the dialog that appears, click Yes.

You can always edit the automatic joins later, but in general, it is a good idea to allow Revit Architecture to apply joins when prompted at least at the early stages of design. This connects the various building elements in logical ways and does so more efficiently than we can do manually at this stage. As your model becomes more refined and the project progresses beyond design development, you may wish to manage the joins manually. You would use the Join Geometry tool on the Modify tab for this purpose.

Copy Floor Slabs to Levels

3. Double-click the blue Section Head between Grid lines C and D again.

You can see the new Floor slab at Level 1. If you look carefully at the intersections, particularly at the right side, you will see the effect of the join geometry message above.

4. Select the new Floor slab with the Modify tool, and then press CTRL+C.

If you prefer, click the **Copy to Clipboard** tool on the Clipboard panel of the Modify | Floors ribbon tab.

- On the Modify | Floors ribbon tab, on the Clipboard panel, click the drop-down button on the **Paste** tool and choose **Aligned to Selected Levels**.
- In the Select Levels dialog, select Level 2, Level 3, and Level 4 and then click OK.

Use the CTRL key to select more than one Level, or simply drag through the names.

The other approach you can take to create the upper level Floors would be to simply open each plan view and repeat the sketching steps. Naturally the copy/paste method is a little quicker. However, looking in the section view, you may notice that the upper Floors did not automatically join with the Walls. There is a Join tool on the Modify tab of the ribbon on the Geometry panel. You can find examples of how to use this tool in Chapter 8. While using the Join tool would give the desired result in this section view, the process can be tedious and would need to be repeated in the *Longitudinal* section as well.

More efficient would be to have Revit prompt us again. In order to be prompted by the message we received in the previous topic, you have to create the Floor manually or edit the Floor sketch of an existing Floor. Since we have copied the Floors here, and did not benefit from auto-prompting, we can simply edit the Floors, and without making any actual edits to the sketch, click the Finish Edit Mode button on the ribbon. This will trigger Revit to prompt us to join and cut volumes as above. To try this, simply click a Floor, on the Modify | Floors tab that appears, click the Edit Boundary tool. A "Go To View" dialog will appear (you cannot edit the sketch from a section view), select any floor plan view and click the Open View button. Make no changes. On the ribbon, click the Finish Edit Mode button. The message we saw above will appear again. Click Yes. Reopen the section and repeat the process on the other Floors. You may have to complete a few of the joins on the back Wall manually. This process is not required to complete the tutorials in this chapter and is left to the reader as an optional exercise.

Create a Shaft

You should now have a Floor slab at each Level and we already had a Roof at the top. However, these new floor slabs now interrupt our building core. We will need a shaft for the elevators and stairs that will come later. We will add a shaft now using approximate dimensions and then later, in Chapter 7, we will add the vertical circulation elements and fine-tune the size of the shaft to suit the design as required.

5. On the Project Browser, double-click to open the *Level 1* floor plan view.
6. On the Home tab, on the Opening panel, click the **Shaft** tool (see Figure 5.30).

FIGURE 5.30 *Create a Shaft Opening*

The Modify | Create Shaft Opening Sketch ribbon tab appears. On the Draw panel, the Boundary Lines and Line modes should be active.

- On the Draw Panel, click the Rectangle icon.
- Snap to the lower left outside endpoint of the core Walls.
- For the other corner, snap to the inside midpoint of the top exterior Wall (see Figure 5.31).

NOTE

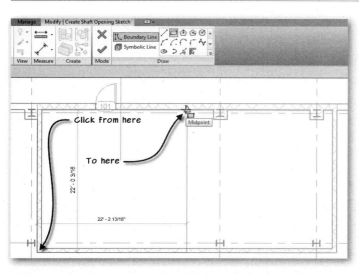

FIGURE 5.31 *Sketch a rectangular Shaft Opening*

- On the Properties palette, for the "Base Constraint" choose **Level 1** and for the "Top Constraint" choose **Up to level: Roof**.
- For the Top Offset, type **1'-0"** [**300**] and then click Apply.
- On the QAT, click the Default 3D View icon.
- Using the Orbit controls (or hold down the SHIFT key and drag with your wheel button, drag the ViewCube or use the Steering Wheel) orbit the model around so you can see down into the core.
- On the ribbon, click the Finish Edit Mode button and then deselect the Shaft (see Figure 5.32).

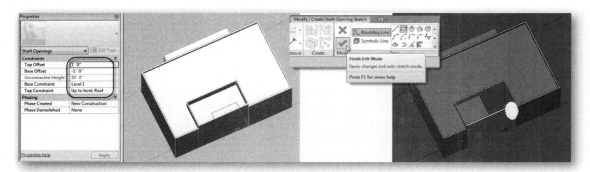

FIGURE 5.32 *The completed Shaft cuts through all of the Floors*

When you finish the sketch, you should see the Shaft Opening cut through all of the Floors to create a void.

7. On the Project Browser, double-click to open the *Level 1* floor plan view.

8. On the View tab, on the Create panel, click the click the **Section** tool.

   - Create a section (from left to right) through the top portion of the core.
   - Click anywhere to deselect the section, and then double-click the new Section Head to open the view (see Figure 5.33).

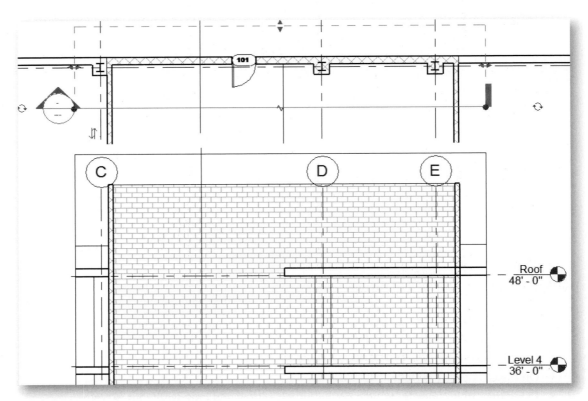

FIGURE 5.33 *View the Shaft in a new Section view*

Study the way that the shaft has cut the slabs in this and other views.

9. Save the project.

CREATING STRUCTURAL FRAMING

So far we have explored Columns, Grids, and Floor Slabs. In addition to these tools, Revit Architecture also includes tools to create Beams and Braces. If your primary job function is architectural, you may not be responsible for adding these elements to the model. This task might fall on the structural engineers on your team. However, like the Columns, even though they will ultimately be sized by the Structural Engineer, you can add the basic components to your model at any appropriate point in the design process and later re-size and reconfigure them as appropriate based upon your structural engineer's design and analysis. This is accomplished by simply selecting the elements and editing the Type applied in the Properties palette. In this topic, we will take a brief look at the tools available. Feel free to explore beyond what is covered here.

Working with Beams

You can add Beams to your model by sketching, or you can create them automatically based upon a column grid. In this exercise, we will use our grid to create Beams.

1. On the Project Browser, double-click to open the *Level 2* floor plan view.
2. On the Structure tab, on the Structure panel, click the **Beam** tool.

Like the Columns above, we will first load a Family to use for the Beam shape.

- On the Modify | Place Beam tab, click the Load Family button.
- Double-click the *Structural* folder, then double-click the *Framing* folder and finally double-click the *Steel* folder.

3. Double-click the *W-Wide Flange.rfa* [*M_W-Wide Flange.rfa*] Family file.

A list of industry standard steel-shape sizes will appear. Scroll through the list to see all of the sizes (similar to Figure 5.17 above).

- Scroll in the "Specify Types" dialog. Select **W18×40 [W460×52]** and then click OK.
- In the "Family Already Exist" dialog, click the Overwrite the existing version button to accept the reload.

4. From the Type Selector, choose the newly loaded Select **W18×40 [W460×52]** Type.

- On the Modify | Place Beam tab, toggle off Tag on Placement and then click the **On Grids** button.
- Dragging from right to left, select all Grid lines.
- On the Modify | Place Beam > On Grid Lines tab, click the Finish button.
- On the ribbon, click the Modify tool or press the ESC key twice.

After a short pause, it will appear as though the command is finished, but nothing will appear to have been created. This is because the Detail Level of our plan view is set to Coarse.

5. On the View Control bar, change the Detail Level to **Medium**.

Beams will appear. They have been created at each major Grid line between Columns (see Figure 5.34).

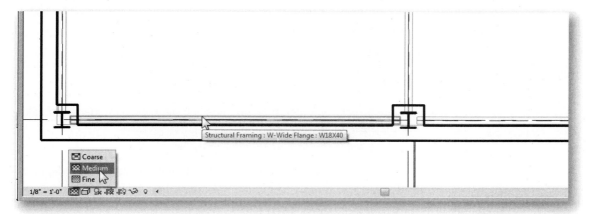

FIGURE 5.34 *Switch to Medium Detail Level to see the Beams in plan*

6. On the Project Browser, double-click to open the *Longitudinal* section view.

- On the View Control bar, change the Detail Level to **Medium**.
- Zoom in on Level 2 and view one of the Beams.

Notice that the Beams appear at the same height as the top of the slab.

7. Pre-highlight one of the Beams.

- Right-click and choose **Select All Instances > In Entire Project**.
- On the Properties palette, beneath the Constraints grouping, set the Start Level Offset and the End Level Offset parameters to **-4"** **[-100]** (negative values) and then click Apply.

Adding this negative offset moves the Beams below the level enough to allow the topping slab to cover them. (Right now, we have a generic slab, but in later chapters, we can change the composition of the slab to show more detail. When we do, the Beams will already be at the correct elevation.)

There are other Beam creation methods and options. You can sketch Beams and create them specifically as Girders, Joists, Purlins, etc. When you use the Grid option as we have here, Revit determines the usage automatically. In this case we have Girders. You can view the Structural topic in the online help for more details on this subject.

Feel free to experiment further with the various Beam options before continuing. For example, there is also a ***Beam System*** tool on the Structure tab. With this tool, you create an array of beams within a sketched shape—usually infilling a column bay. You can use the Automatic Beam System option to create the system by just clicking on one of the Beams we just created, or you can sketch the shape of the system manually. The shape that you sketch will be filled in with an automatically generated array of Beams. Before starting the Beam System routine, load an appropriate Beam Family such as a bar Joist. To do this, click the Insert tab of the ribbon and then click the Load Family button. Browse to the same folder as the girder above. Select a Family and Type that you want to load, such as a K-Series Bar Joist Family. To use the newly loaded Bar Joist Family in the Beam System, choose it on the Options bar. There is also an Elevation parameter on the Properties palette so you can give the system a negative offset relative to the Level like the Girders above. Give it try!

Working with Braces

To complete our preliminary structural layout of our commercial project, let's add a few cross braces at the building core.

8. On the Project Browser, double-click to open the *Level 1* floor plan view.
9. On the Structure tab, on the Structure panel, click the ***Brace*** tool.

- On the Modify | Place Brace tab of the ribbon, click the Load Family button.
- Double-click the *Structural* folder, then double-click the *Framing* folder and finally double-click the *Steel* folder.
- Double-click the *L-Angle.rfa* [*M_L-Angle.rfa*] Family file.

Once again, a list of industry standard steel-shape sizes will appear. Scroll through the list to see all of the sizes (similar to Figure 5.17 above).

- Scroll in the list, select **L6×6×3/8 [L152×152×9.5]** and then click OK.

10. Verify that **L6×6×3/8 [L152×152×9.5]** is chosen from the Type Selector.

- On the Options bar, from the "Start" list, choose **Level 1** from the drop-down list and then type **2'-0" [600]** in the offset field next to it.
- From the "End" list, choose **Level 2** and type **-2'-0" [-600]** in the offset field.

- Toggle off Tag on Placement.
- Snap the start point to the midpoint of the Column at Grid intersection 3C.
- Snap the end point to the midpoint of the Column at Grid intersection 4C.
- Repeat to create another Brace from Grid intersection 3D to 3E.

Feel free to add additional Braces or Beams as desired.

- On the ribbon, click the Modify tool or press the ESC key twice.

As with the Beams, Braces will not appear in Coarse Detail Level. You can change this in the *Level 1* plan if you wish. To view either Brace in elevation, you can create additional Section views as we did above (see Figure 5.35).

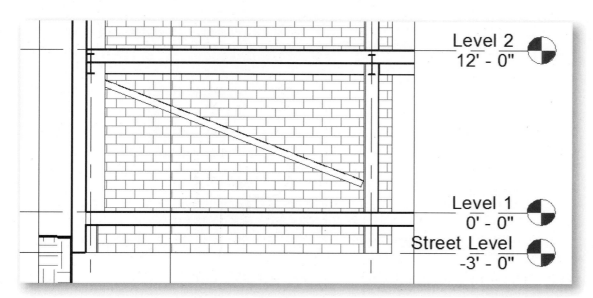

FIGURE 5.35 *Create a new Section view to see the Braces*

> Get in the habit of naming each new view (like the sections created here) with a good descriptive name. (Follow office standard naming guidelines as appropriate).

> **TIP**

Create a New Model Group

The same Framing layout will occur on all four floors of the project. We could copy it to each Level as we did with the other elements, but let's take advantage of Groups to make it easier to edit the framing later. Anytime that you have a repetitive (or "typical") portion of your building design—such as a typical Stair, Toilet Room or Framing layout as in this case—you can use Groups to manage them. We saw Groups briefly at the start of the chapter with the Array command. In that instance, we actually ungrouped the array. In this case, we will leave the items grouped. The process is simple. Select the elements, Group them, and then insert instances in the model. Whenever you need to make a change, edit any instance of the Group. When you are finished editing, the change will apply to all instances of the Group across the entire project.

11. On the QAT, click the Default 3D View icon.

- Dragging from right to left, select the entire model.
- On the Modify | Mulit-Select tab of the ribbon (or at the lower right corner of the Application Status bar), click the Filter icon.

- In the "Filter" dialog, click the Check None button.
- Place a checkmark only in the Structural Framing checkboxes (there should be two such items, unless you added joists, in which case there will be three) and then click OK. (Do not include the Structural Columns in the selection).

You may not be able to see the selected elements. It depends on the angle of your 3D view and what other elements obscure them. It is not important to see the elements. Choosing "Structural Framing" selected all of the Beams and Braces. If you do wish to see the elements, use the Temporary Hide/Isolate icon and the Isolate Element option.

12. On the Modify | Structural Framing tab of the ribbon, on the Create panel, click the Create Group icon.

- In the "Create Model Group" dialog that appears, type **Typical Framing** and then click OK (see Figure 5.36).

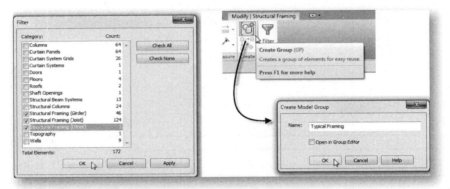

FIGURE 5.36 *Use Filter to help create and name a new Model Group*

13. On the Project Browser, double-click to open the *Level 1* floor plan view.
14. On the Project Browser, expand *Groups* (near the bottom of the list), then expand the *Model* branch.

Notice that "Typical Framing" is listed in the Model category.

- Right click on *Typical Framing* and choose **Select All Instances > In Entire Project**.

Currently there is only one instance of this Group in the model, so this just selects that one instance. We could have used any other selection method here as well. Notice the small "L" shaped icon with grip handles and "X" and "Y" labels that appear at the center of the Group. This indicates the insertion point of the Group. Using the handle at the intersection of the two "L" shapes, we can relocate the insertion point. The other round handles allow you to rotate the axes. The "X" and "Y" labels allow you to flip the Group about either the X or Y-axis. Refer to Chapter 6 for more detailed information on Groups.

- Click and drag the insertion point handle (at the intersection of the "L") and snap it to the intersection of Grid lines A and 1.

TIP

> If you have trouble doing this, drag it close first, then zoom in to finish. You can use Temporary Hide/Isolate to hide the column and make snapping to the Grids easier. Use Reset Temporary Hide/Isolate to re-display the Column when finished.

15. On the Project Browser, double-click to open the *Level 2* plan.
16. Right-click the *Typical Framing* Group on the Project Browser and choose **Create Instance**.

  • Snap the new Group instance to the intersection of Grid lines A and 1.

  • Repeat for Level 3 and Level 4.

    If you prefer, you can use the Copy and Paste Aligned method showcased elsewhere in this chapter instead.

  • Open any section view to see the result.

Editing Groups

Now that we have a Group of our typical framing, if the design should change, we can edit the Group and the change will propagate across the project to all instances of the Group. To edit a Group, select it (A dashed box appears around a Group when you select it) and then on the Modify | Model Groups tab of the ribbon, on the Group panel, click the Edit Group button. Edit Group mode grays out the rest of the model and you can only edit the elements within the Group using any tools or methods. When you are finished, you click the Finish button on the Edit Group panel. The changes you make will then appear in all instances of the Group throughout your model.

As you experiment with the Group, it is important to note that editing the Group and ungrouping it do not yield the same results. When you ungroup a Group, you end up with individual elements that are no longer associated as a Group. Therefore, if you make changes to the ungrouped Elements, such changes will apply only to the Elements themselves and not be propagated back to the other instances of the Group. If you choose to Group the Elements again, you will be creating a new Group, not replacing the existing one.

This completes the work on the commercial project for this chapter. You can experiment on your own with the Group if you wish. We will not cover the steps in detail now as the next chapter is devoted almost exclusively to working with Groups. Therefore, you can undo any changes you make when you are finished or you can leave them. It is up to you. Please refer to Chapter 6 for detailed information and tutorials on Groups and Group editing.

17. From the Application Menu, choose Close. When prompted to save the project, click Yes.

To complete our brief exploration of structural tools in Revit Architecture we will have a look at the Footing tools provided. To do this, we will switch to the Residential project. Some new Footings are required for the addition on the back of the house. If you wish, after completing the exercise in the Residential project, you can return to the Commercial project and add footings there as well.

Load the Residential Project

Be sure that the Commercial Project has been closed and saved.

1. On the QAT, click the Open icon.

| TIP | The keyboard shortcut for Open is CTRL + O. **Open is also located on the Application menu.** |

- In the "Open" dialog box, browse to the location of your *MasterRAC 2012* folder, and double-click the *Chapter05* folder.

2. Double-click *05 Residential.rvt* if you wish to work in Imperial units. Double-click *05 Residential Metric.rvt* if you wish to work in Metric units

   You can also select it and then click the Open button.

The project will open to a cover sheet showing a 3D view of the model. The *First Floor* view looks like we left it in Chapter 3 with the exception of some new Walls boxing out where the addition will be. The *Second Floor* and *Basement* floor plan views have been provided for you. If you wish to gain some additional practice modeling the elements on these levels yourself, refer to Appendix A for exercises covering the layout of these levels.

Add Continuous Footings

You can create three types of Footings in Revit: Wall Foundation, Isolated Foundation, and Slab Foundation. A Wall Foundation is very easy to add and is associated with a Wall. An Isolated Foundation is a Component Family that can be inserted anywhere a Footing is required, such as a pier footing, pile cap, etc. A Slab Foundation is basically a Floor element that you can use to sketch any shape that is required. Foundation tools are accessed from the Foundation panel of the Structure tab on the ribbon.

3. On the Project Browser, double-click to open the *Basement* floor plan view.
4. On the Structure tab, on the Foundation panel, click the **Wall Foundation** tool (see Figure 5.37).

FIGURE 5.37 *Create a Wall Foundation*

The Modify | Place Wall Foundation tab will appear on the ribbon.

- Choose Continuous Footing [Bearing Footing - 900 × 300] from the Type Selector. On the Status bar, a "Select Wall(s)" prompt will appear.

We are adding footings only to the new construction. The new construction for the addition is bold and black while the existing construction is gray in color.

- Click the vertical new construction Wall on the left of the addition (top left of the plan).

A "Warning" message will appear in the lower right corner of the screen. This is considered an "Ignorable" Warning message in Revit. It indicates a situation to which you may or may not be aware, but if you choose to ignore it, this condition does not prevent you from continuing your work. In this case, the Foundation that we just added is not visible in the current view (see Figure 5.38).

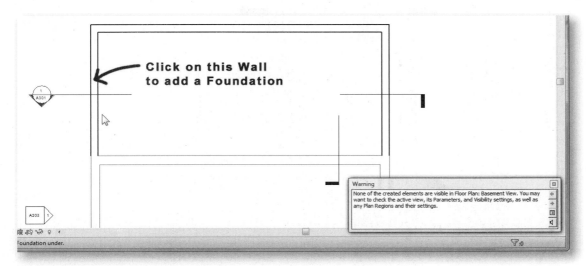

FIGURE 5.38 *Revit "Ignorable" Warning box*

We have to perform a few steps to correct this and display the Foundations.

- On the ribbon, click the Modify tool or press the ESC key twice.

Adjusting View Range

5. On the Properties palette, make sure the drop-down list near the top reads: Floor Plan: Basement.

   This tells us we are editing the view's properties not the properties of a selected object.

   - Scroll down to the Extents grouping and then click the Edit button next to "View Range."
   - In the "View Depth" area, type **-2'-0" [-600]** in the Offset field next to Level (see Figure 5.39).

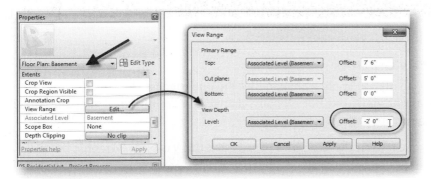

FIGURE 5.39 *Edit the View Range of the Basement Level Floor Plan*

- Click OK to see the results.

We now see the footing, but it is displaying in a continuous line style. To fix this, we need to understand a little more about the View Range settings. When you make the View Depth lower than the "Bottom" of the Primary Range, the elements falling within the range between Bottom and View Depth use a special Revit Line Style called <Beyond>. By default this style uses a continuous line, with a lineweight of 1 so the result is not very apparent in the current view. We can, however, edit this line style and make it use a hidden line pattern which will in turn make geometry within the View Depth range also display as hidden.

6. Click the Manage tab of the ribbon.

   • On the Settings panel, click the ***Additional Settings*** drop-down button.

   • From the pop-up menu that appears, choose **Line Styles**.

   • Click in the Line Pattern field next to <Beyond>.

   • From the pop-up menu that appears, choose **Hidden** (see Figure 5.40).

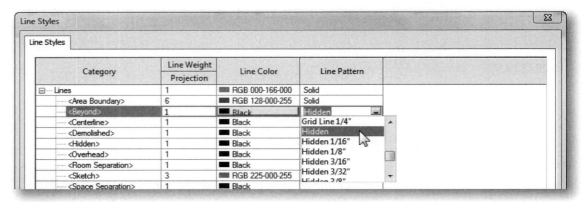

FIGURE 5.40 *Change the <Beyond> Line Style to use a Hidden Line Pattern*

   • Click OK to see the change.

The Foundation should now use a hidden line pattern. This is a global change. It will affect the <Beyond> line style in all views throughout the project. If you prefer not to globally change the <Beyond> Line Style as we have, you can apply this edit to just the current view instead. To do so, on the View tab, click the Visibility/Graphic button. We will learn how to do this next.

7. Undo the previous change.

Override View Graphics

It is sometimes better to apply a change like the one suggested in the previous step only to the current view, and not globally change the definition of the Line Style throughout the entire project.

8. On the View tab, on the Graphics panel, click the Visibility/Graphic button.

   Changes made in this dialog affect only the current view (*Basement* in this case).

9. On the Model Categories tab, scroll down and locate the Lines category.

   • Click in the Lines colunm next to <Beyond>.

   • Click the Override button.

   • In the "Line Graphics" dialog, change the Pattern to Hidden and then click OK (see the left side of Figure 5.41).

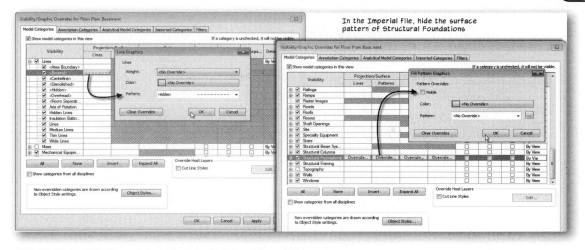

FIGURE 5.41 *Edit the surface pattern for Structural Foundations*

10. Scroll down and locate the Structural Foundations category.

   • Click in the Projection/Surface Patterns field next to Structural Foundations (see the right side of Figure 5.41).

   • Clear the "Visible" checkbox and then click OK twice.

You can make the same edit at the element level choosing **Override Graphic in View > By Element** instead when you right-click. However, since we did it by category, it will apply to all Structural Foundations even those we have not added yet. Only use the By Element option when you do not want the change to apply to all similar objects also in the view. In either case, By Element or By Category, the change applies *only* to this view.

11. Add two more Continuous Footings to the other two Walls of the addition.

Select the Footing already in the model and then click the ***Create Similar*** tool on the ribbon or type "CS". The Create Similar tool allows you to use the elements visible within the Canvas as an on-screen tool palette.

TIP

Edit the Footing (Edit Wall Profile)

Like many building model elements in Revit Architecture, the Footings will interact intelligently with the Walls to which they are associated. For example, if we edit the Profile of the Wall to make a step at the bottom, the Footing will adjust accordingly.

12. On the Project Browser, double-click to open the *North* elevation view.

In this view, the topography is concealing the foundation Walls and Footings from view. For now, we will simply hide the Terrain so that we can work with the Footings and Foundations Walls.

13. Select the Toposurface, and then on the View Control bar, choose **Hide Element** from the Temporary Hide/Isolate popup icon (looks like sunglasses).

You should now be able to see the Footings and Walls. Remember that Temporary Hide Objects is limited to this work session only.

14. Select the Wall facing this elevation. (Use TAB if necessary to select it).
15. On the Modify | Walls tab of the ribbon, on the Mode panel, click the Edit Profile button.

The Profile is a sketch that defines the elevation of the Wall. By default, this profile is a simple rectangular shape defined by the length and height of the Wall. We can edit the sketch of this profile to create a Wall with an alternate shape.

- Using the Offset tool, offset the bottom edge down **1'-6" [450]**.
- Draw a vertical sketch line (click the Line icon on the ribbon) at **11'-0" [3300]** from the right side.
- Using the *Trim* tool, complete the sketch as shown in Figure 5.42.

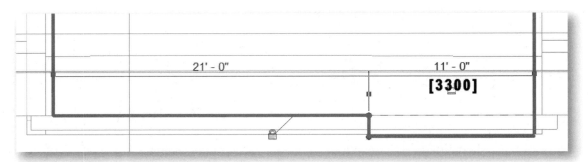

FIGURE 5.42 *Create a step in the Footing (sketch lines enhanced in the figure for clarity)*

16. On the ribbon, click the Finish Edit Mode button.

The Footing will automatically adjust to match the new configuration of the Wall.

17. Perform similar Steps on the adjoining Wall (right side in the current elevation).

Study the model in various views. Remember that you will need to hide the Toposurface temporarily in other views to see the foundations (see Figure 5.43).

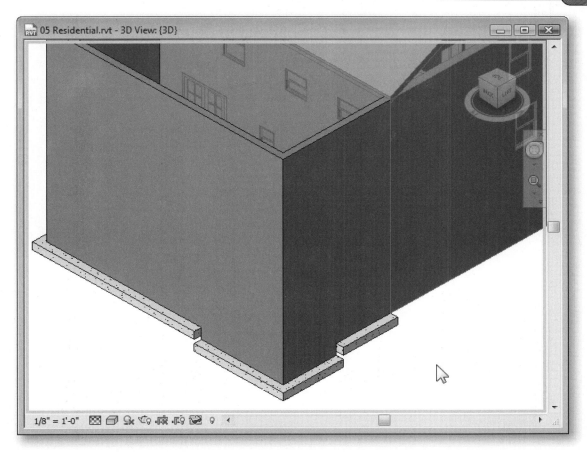

FIGURE 5.43 *The completed foundation in the {3D} view (with the Toposurface hidden)*

There is plenty more that we could do to finish up the foundation of the residential project. We could add existing Footings (remember to work in that phase) or we could add an Isolated Foundation to the chimney and the columns in the basement. We could also add isolated Foundations at each column in the Commercial project. When you choose the Isolated Foundation tool, you will be prompted to load a Family for its use. This is left to the reader as an additional exercise. Feel free to experiment in both projects with these and other structural tools.

SUMMARY

- Column Grid lines are datum elements that establish Column Grid intersections.
- Column Grid lines share many features with Levels and appear automatically in all orthographic views.
- Structural tools in Revit Architecture include Columns, Beams, Beam Systems, Braces, Walls and Slabs.
- Revit Architecture includes many Structural tools useful to Architects for basic structural layout.
- For complete Structural tools including interface with analysis packages, look into the Revit Structure application.

- Structural Columns represent the actual structural support members of the building.
- Architectural Columns are typically used to represent the finished Column as it would be seen in the building. Use them for Column wraps, etc.
- Structural Walls are Walls that have their Structural Usage parameter set to Bearing.
- Floor Slabs can be created from bounding Walls.
- You can create Framing layouts and plans using Beams and Braces.
- Group a collection of elements and reuse it elsewhere in the model. If the Group changes, all instances will update.
- A continuous Wall Foundation is associated to Walls.
- An Isolated Foundation is a Component Family that you place in your model.

CHAPTER
6

Groups and Links

INTRODUCTION

In the previous chapter, we briefly covered the use of Groups. In this chapter, we will make a more detailed exploration of this feature. Groups provide a mechanism to standardize typical design elements throughout the project. A Group consists of a collection of elements that can be placed into the model as a single unit. You can edit any single instance of the Group and the changes will propagate to all instances throughout the model. Groups have many other features as well including the ability to have overrides applied to individual instances.

OBJECTIVES

In this chapter, we will work with both Model and Detail Groups. We will explore how to create Groups, modify them and strategies for using them effectively in your projects. After completing this chapter, you will know how to:

- Create Groups
- Modify Groups
- Override elements in Group instances
- Create Attached Detail Groups
- Swap Groups with one another

CREATING GROUPS

The dataset for this chapter will deviate from our Commercial and Residential projects to explore Groups in a dataset that will be more suitable to conveying the critical concepts. Toward the end of the chapter, we will return to the Commercial project to put into practice what we have learned about Groups and also explore Links. Groups can be created in any project. Creating them is as simple as making a selection of elements in your project and then clicking the ***Create Group*** button. Groups appear on the Project Browser beneath the Families branch.

Install the Dataset Files and Open a Project

The lessons that follow require the dataset included on the Aubin Academy Master Series student companion. If you have already installed all of the files from this site, skip to step 3 to begin. If you need to install the files, start at step 1.

1. If you have not already done so, download the dataset files located on the CengageBrain website.

    Refer to "Accessing the Student Companion site from CengageBrain" in the Preface for information on installing the dataset files included in the Student Companion.

2. Launch Autodesk Revit Architecture from the icon on your desktop or from the *Autodesk* ≥ *Revit Architecture 2012* group in *All Programs* on the Windows Start menu.

TIP	You can click the Start button, and then begin typing **Revit** in the Search field. After a couple letters, Revit Architecture should appear near the top of the list. Click it to launch to program.

3. On the Recent Files screen, click the Open link beneath Projects.

TIP	The keyboard shortcut for Open is CTRL + O. **You can also click the Open icon on the Quick Access Toolbar (QAT).**

-   In the "Open" dialog box, browse to the location where you installed the *MasterRAC 2012* folder, and then double-click the *Chapter06* folder.

4. Double-click *Understanding Groups.rvt*.

    You can also select it and then click the Open button.

Please note that for this chapter, units are immaterial to the lessons covered and as such only one dataset has been provided rather than the customary separate Imperial and Metric datasets of other chapters.

Explore the Dataset

Groups are appropriate for nearly any repetitive (typical) design condition. Hotels, dormitories, apartment complexes, and condominiums give us plenty of opportunities to utilize Groups in very effective ways. In this example, we will work with a very simple hotel room layout.

5. On the Project Browser, double-click to open the *Architecture* floor plan.

This view shows the basic floor plan of a hotel guest room. Walls, Doors, Windows, and some basic fixtures are included.

6. Double-click to open the *Section 1* section view.

Here you will notice in addition to the items we can see in plan, there is also a multi-height ceiling plane in this model and some furniture items.

-   Open other views to explore the dataset further if you wish before continuing (see Figure 6.1).

7. On the Project Browser, double-click to return to the *Architecture* floor plan.

-   Maximize the view (if it is not already maximized).

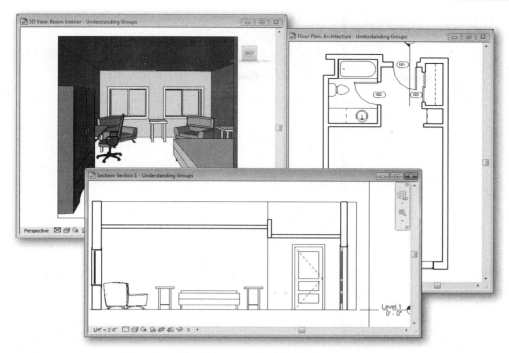

FIGURE 6.1 *The dataset represents a simple hotel guest room layout*

- On the View tab of the ribbon, on the Window panel, click the **Close Hidden** button.

Create a New Group

The first step to understanding Groups is to create one.

Continue in the *Architecture* floor plan.

8. Using a window selection, select all elements on screen. (Click above and to the left of the model and drag down to the right surrounding all elements).

The Modify | Multi-Select tab will appear on the ribbon.

- On the Create panel, click the **Create Group** button.

An error dialog will appear. When you made your selection window, the elevation view tag and possibly the section view tag were included in the selection. View tags cannot be included in a Group. Simply clicking OK in this warning will automatically exclude them from the selection set.

- Click OK in the warning dialog to dismiss it.

The "Create Model Group and Attached Detail Group" dialog will appear.

- In the Model Group Name field, type **Guest Room A**.
- Leave the Attached Detail Group name as **Group 1** and then click OK (see Figure 6.2).

Since we have selected both model and detail elements, Revit will actually create two groups. One will be a Model Group containing the Walls, Doors, Windows,

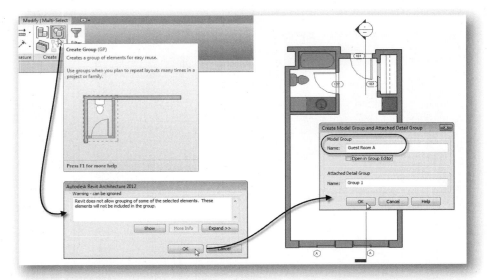

FIGURE 6.2 *Create a new Group and give it a name*

and fixtures. The other will be a Detail Group that contains the door and window Tags. It is not possible for detail elements and model elements to be in the same group. The Detail Group will actually be an "Attached Detail Group." This means that this Detail Group is associated to its parent Model Group. Later, we can have instances of the Attached Detail Group automatically applied to instances of the Model Group. To see each Group, simply move your mouse over them on screen.

- Move your mouse near the edge of one of the Walls.

  You will see a dashed box appear around the Model Group with a screen tip indicating its name.

- Move your mouse over one of the Tags.

- You will see a dashed box appear around the Detail Group with a screen tip indicating its name (see Figure 6.3).

Groups that you create will also appear in the Project Browser.

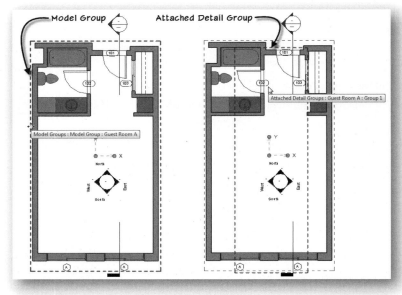

FIGURE 6.3 *Two Groups were created—A Model Group and an Attached Detail Group*

9. On the Project Browser, expand the *Groups* branch.

   This reveals the Detail and Model branch.

   • Expand the *Model* branch.

You will see the *Guest Room A* Group indented beneath the *Model* branch.

   • Expand the *Guest Room A* entry to reveal the Attached Detail Group named *Group 1* (see Figure 6.4).

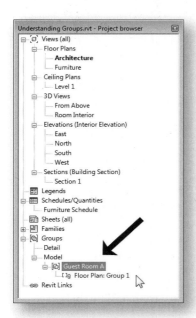

FIGURE 6.4 *Groups appear hierarchically on the Project Browser*

Each Model Group you create will appear beneath the *Groups > Model* branch in the Project Browser. Attached Detail Groups will always appear beneath the Group to which they are attached. (If you create a Group from detail elements by themselves, without associated model geometry, they will appear beneath the *Groups > Detail* branch).

Create a Group Instance

Now that we have created a Group, we can easily add additional instances of the Group in our project. You can do this from the **Model Group** button on the Home tab or the *Groups* branch of the Project Browser.

10. On the Project Browser, expand *Groups*, then *Model*, and then right-click *Guest Room A*.

   • Choose **Create Instance**.

     A dashed rectangle will appear on screen with the mouse pointer in the center.

   • Click on screen to place the new Group instance to the left side of the original.

   • On the Modify | Model Groups tab of the ribbon, click the **Finish** button.

Another guest room will appear. (It does not include any annotation because as we saw above, the annotation is included in a separate Detail Group). When you create a Group, the geometric center of the Group becomes the insertion point by default. This is why when we added this instance; our mouse pointer was positioned in the center of the Group. You can move the origin to a more useful location simply by dragging it.

11. Select the Group instance that you just created.

    In the center of the Group, a blue Group Origin icon will appear.

    • Click and drag the round handle at the intersection of the two axes.

    • Drop the icon on the Wall endpoint at the top left corner of the hotel room (see Figure 6.5).

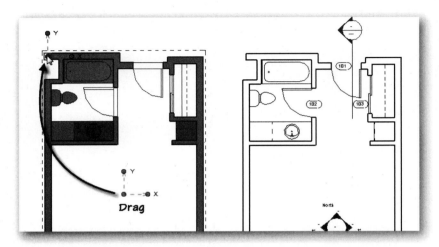

FIGURE 6.5 *Move the Group Origin by dragging*

12. On the Project Browser, right-click *Guest Room A* again and choose **Create Instance**.

    Notice the location of the mouse pointer relative to the Group outline this time.

    • Click a point to the right of the original to place the new Group instance.

    • On the Modify | Model Groups tab of the ribbon, click the ***Finish*** button.

If you select the original Group instance, you will notice that the insertion point for it is also at the upper left corner. Edits you make to one instance of a Group apply automatically to all instances.

Working with Attached Detail Groups

Take a look at the original guest room (the one that has Tags). Notice that there are three Doors, each with its own unique number. On the other hand, the Windows share the same designation of "A." The default Revit Door Tags show the instance "Mark" parameter of Doors, which is unique for each Door while window Tags show the "Type Mark" for Windows which is the same for all instances of a given Type. Keep these observations in mind as we perform the next several steps.

To add tags to the other Group instances, we could manually tag each item in the Group. A faster method is to use an Attached Detail Group. An Attached Detail Group can automatically be applied to any instance of its parent Model Group.

13. Select one of the Groups on the left or right (without annotation).

    • On the Modify Model Groups tab of the ribbon, on the Group panel, click the ***Attached Detail Groups*** button.

    • In the dialog that appears, place a checkmark in the box next to Floor Plan:Group 1 and then click OK (see Figure 6.6).

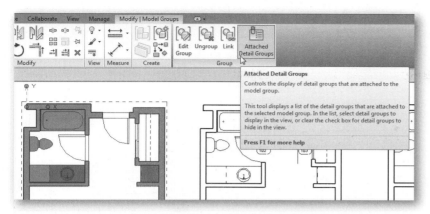

FIGURE 6.6 *Add an Attached Detail Group to the selected Model Group*

An instance of the Attached Detail Group will appear. Notice that each of the door Tags will have incremented sequentially to show unique numbers. The window Tags however will show the same designation that the originals did.

14. Repeat the process to add an Attached Detail Group to the other Model Group as well.

EDITING GROUPS

You can edit a Group at any time. When you do, changes you make to the Group will be applied to all instances when the edit is complete. This is one of the most powerful benefits of using Groups. Furthermore, edits to a Model Group can also have an automatic impact on Attached Detail Groups.

Edit a Group

To understand the value and potential of editing a Group, we can start with a very simple modification.

15. Select one instance of the *Guest Room A* Group onscreen.

- On the Modify | Model Groups tab of the ribbon, on the Group panel, click the **Edit Group** button.

This enables the Group Edit mode. The background of the canvas area is tinted yellow and the Edit Group panel appears at the upper left corner of the view. The elements that are members of the Group remain bold, and all of the other elements on screen become grayed out and cannot be selected or edited, but they can be added to the group as we will see below. For this example, we will make a very simple edit.

16. Select one of the Windows (one of the bold ones).

- On the Type Selector, choose **Slider with Trim:36" × 48"**.

In the Group editor, the selected Window will immediately reduce in size and the associated Tag will change from A to E—even though this Tag resides in a separate Detail Group!

- On the Edit Group panel, click the **Finish** button (see Figure 6.7).

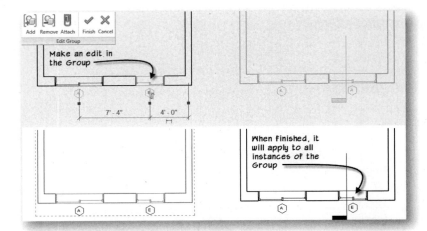

FIGURE 6.7 *Make a change in the Group Editor*

When you have finished, the edit will be applied to all instances of the Group. Notice that the Attached Detail Groups update as well. Let's try another edit.

17. Select one instance of the *Guest Room A* Group.

   • On the Modify | Model Groups tab of the ribbon, on the Group panel, click the **Edit Group** button.

   • Repeat the Window edit made above to the other Window.

   • Create a new Window in the space between the two existing Windows. (Try a **Fixed:36" × 48"**).

Notice that the window Tag is also created, but it comes in grayed out. This is because it is annotation and is therefore automatically excluded from the Model Group that we are currently editing.

   • On the Edit Group panel, click the **Finish** button.

18. Select the window Tag.

Notice that the Tag selects independently as a freestanding element in the project. It is not automatically added to the Attached Detail Group as you can see by examining the other two instances in this project (see the top half of Figure 6.8).

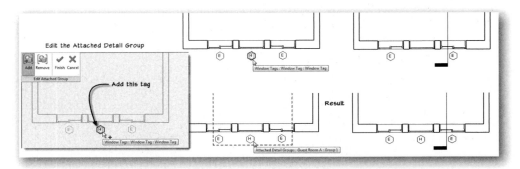

FIGURE 6.8 *Tags for newly added Group elements must be added manually to the Attached Detail Group*

19. Select the *Group 1* Attached Detail Group (the instance with the stray Window tag).

    - On the Modify | Attached Detail Groups tab of the ribbon, on the Group panel, click the **Edit Group** button.
    - On the Edit Attached Group panel, click the **Add to Group** button.
    - Select the window Tag (Tag H) and then click the **Finish** button (see the bottom half of Figure 6.8).

The Tag for Window Type H should now appear in all three instances of the Attached Detail Group. Again, since this Tag references a Type parameter (the Type Mark), the letter displays the same value in all instances of the Attached Detail Group. For adding a Tag, we had to manually edit the Attached Detail Group; however, if you were to edit the Model Group again and delete one of the tagged elements (a Door or Window) then the tag in the Attached Detail Group would also be deleted automatically even though it is in a different Group—you would not have to separately edit the Detail Group. Hosted elements like tags cannot exist without a host.

Duplicate and Edit a Group

Making a variation of a Group is simple to do. Once you have two or more variations, you can easily swap them out with one another.

20. Select the instance of the *Guest Room A* Model Group on the left.

    - On the Properties palette, click the Edit Type button.
    - In the "Type Properties" dialog, click the Duplicate button.
    - In the "Name" dialog, type **Guest Room B** and then click OK (see the left side of Figure 6.9).

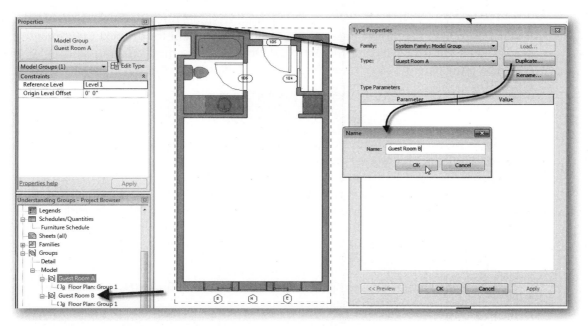

FIGURE 6.9 *Create a duplicate of the Model Group*

Notice the appearance of *Guest Room B* on the Project Browser. If you expand it, you will see that a copy of the *Group 1* Attached Detail Group has also been created and associated to the new Model Group.

21. Select the same instance onscreen (now *Guest Room B*).

    • On the ribbon, click the **Edit Group** button.

    • Delete the middle Window and make some other obvious change (such as enlarging the bathroom or flipping a Door).

    • Finish the Group (see Figure 6.10).

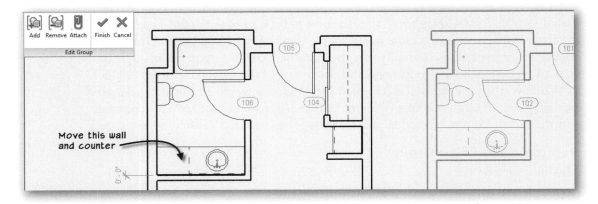

FIGURE 6.10 *Modify Model Group Guest Room B*

Not surprisingly, the change only affects the currently selected Group. This is because it is currently the only instance of *Guest Room B* in the project. At this point however, we can experiment with swapping Group instances and see the ease at which we can switch from one Group to another and also see another one of the benefits of the Attached Detail Group functionality. Take notice of the names of the Attached Detail Groups on the Project Browser. For both Model Groups, the Attached Detail Groups have the same name—*Group 1*. This is important for the next experiment. If you use the same name for the Attached Detail Groups, they will automatically swap when the parent Model Groups swap. Let's take a look.

22. Select one instance of the *Guest Room A* Group.

    • From the Type Selector, choose *Guest Room B*.

Since we tried to make the difference between the A and B Guest Rooms obvious, you should be able to spot the changes to the model right away. The most interesting change however is that the Attached Detail Group has changed automatically as well. Since the Attached Detail Groups for each Model Groups have the same name, Revit is able to swap them appropriately as well. This behavior works as long as the name of the Attached Detail Group is the same for each Model Group. In this example, we left the name "Group 1" but this name is not required. You can choose a more descriptive name if you wish.

23. On the Project Browser, right-click the Attached Detail Group (*Floor Plan: Group 1*) for *Guest Room A* and choose **Rename**.

    • Change the name to **Tags** and then click OK.

    • Repeat for the Attached Detail Group of *Guest Room B* and rename it to **Tags** as well.

    • Repeat the process above to swap one of the *Guest Room B* instances in the project back to *Guest Room A*.

Notice that the Attached Detail Group continues to swap as well. With this in mind, you should try to pick useful and descriptive names for both your Model Groups and your Attached Detail Groups in your own projects. Careful naming often helps reduce errors and limits the complexity of projects.

EDITING GROUP INSTANCES

Situations will often arise in the course of a project where one instance of a Group needs to be slightly different from the other instances in the project. In this case, we could certainly repeat the process covered in the previous sequence and duplicate and edit another Group. However, doing so could begin to dilute the usefulness of Groups and make management of the multiple potential variations cumbersome and time consuming. In scenarios such as this, Revit offers us the ability to create overrides to individual Group instances. To illustrate the point, a simple example is appropriate.

Excluding Group Members

Continue from the previous Exercise.

1. Delete the two Groups copied above leaving only the original one (in the middle) and its Attached Detail Group.

Notice that the Attached Detail Groups are deleted automatically when their hosts are deleted. Make sure the Group in the middle is Guest Room A. If it is not, select it and change it on the Type Selector.

2. Select the remaining Model Group instance on screen.

    *   On the Modify panel, click the ***Mirror - Pick Axis*** button.
    *   Click on the centerline of the vertical Wall on the right side of the Group as the mirror edge (see Figure 6.11).

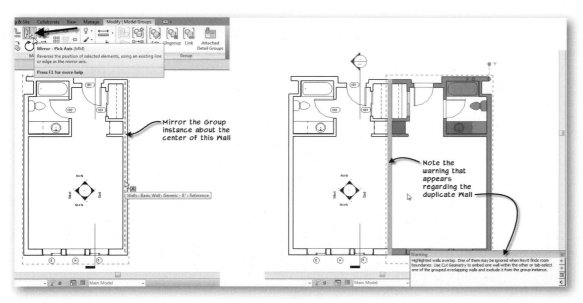

FIGURE 6.11 *Mirror a copy of the Guest Room Group*

When you complete the mirror command, a copy of the Group will appear, and a Warning dialog will also appear at the bottom right corner of the screen. This warning is not serious and can be ignored. All such "ignorable" warnings will appear in this location on screen and will have a yellow tint to the dialog in which they appear. It is still a good idea to read the warning message as there is some useful information conveyed in them. In this case, Revit is alerting us that we now have two Walls overlapping in the same spot. While we can ignore this situation, the message further explains that Room boundaries may be affected:

"Highlighted walls overlap. One of them may be ignored when Revit finds room boundaries. Use Cut Geometry to embed one wall within the other. Or tab-select one of the grouped overlapping walls and exclude it from the group instance."

The right side of Figure 6.11 shows the warning message. If you wish, you can expand the warning dialog to get more detailed information. Do this with the small icon on the right side of the warning dialog. If the warning has closed already, you can click the ***Warnings*** button on the Inquiry panel of the Manage tab of the ribbon to access it. When you fully expand the error, you can click on each of the overlapping Walls to highlight them on screen (see Figure 6.12).

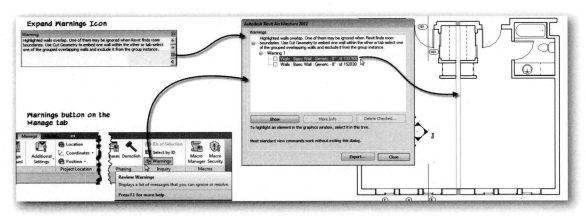

FIGURE 6.12 *Click the Expand icon to see a more detailed error dialog*

If you wish to see the element highlighted in other views, click the Show button. Each time you click Show, it will open another view window and highlight the element in question. When you are done reviewing the warning, click the Close button to dismiss it. If you did click show, make sure you return to the *Architecture* floor plan view and then close hidden windows.

In addition to potentially having an adverse effect on Rooms, you can also see that the overlapping Walls do not cleanup very nicely. The solution to both problems is simple: any element in any Group can be excluded from an individual instance of the Group. In this situation, we can exclude the duplicate Wall from one of the Groups.

3. Place your mouse over the double Wall.

Notice that the Group pre-highlights.

• Press the TAB key.

Notice that the other Group pre-highlights.

- Press the TAB key again.

This time, the Wall within one of the Groups pre-highlights.

- Click to select this Wall.

- Click the Group Member icon in the canvas area to exclude this Wall from the Group (see Figure 6.13).
- Deselect the Group.

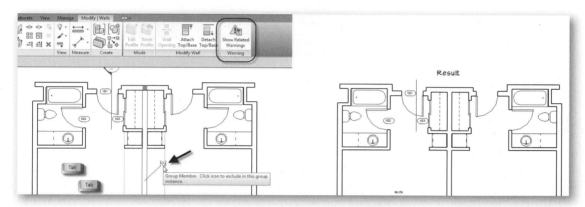

FIGURE 6.13 *Tab into the Group, select the Wall and then click the icon to exclude it*

Notice that the extra Wall has been removed and the cleanup is now correct. It is important to realize that this change is not simply graphical override—Revit has actually removed one instance of one of the Walls. For example, were we to have counted the Walls before we started and then re-count them now, there would be one Wall fewer in our model. Rather than count the Walls, which might prove tricky, let's do a similar experiment using furniture, which is easier to count.

4. On the Project Browser, double-click to open the *Furniture* floor plan view.

Furniture elements will appear in the original guest room.

5. Select all of the furniture elements.

- On the ribbon, click the **Create Group** button.
- In the Create Model Group Name field, type **King-01**.
- Using the technique covered in the "Create a Group Instance" topic above, move the origin point to the same location as the Guest Room Group.

6. Mirror the furniture Group to the other guest room (see Figure 6.14).

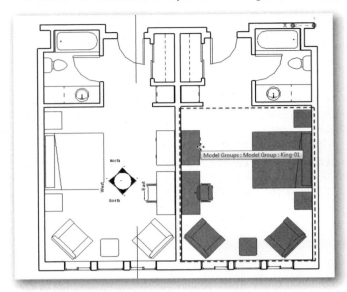

FIGURE 6.14 *Group the furniture, move the origin point, and then mirror a copy*

7. On the Project Browser, beneath Schedules/Quantities, double-click to open the *Furniture Schedule* view.

Study the table and take note of the totals in the "Count" column. In particular, notice that we currently have 4 Side Chairs (Chair-Viper: Chair). This is impressive; Revit gives us an accurate count even when the items it is counting are inside Groups! Let's exclude one and see the impact on the Schedule.

8. Return to the *Furniture* floor plan.

TIP On the View tab, choose Cascade or Tile Windows to see both plan and schedule side by side. The shortcut for Tile Windows is WT.

- Using the TAB key, select one of the Side Chairs (near the Windows).
- Click the Group Member icon to exclude this chair from the Group (see Figure 6.15).
- Deselect the chair after excluding it to see it disappear.

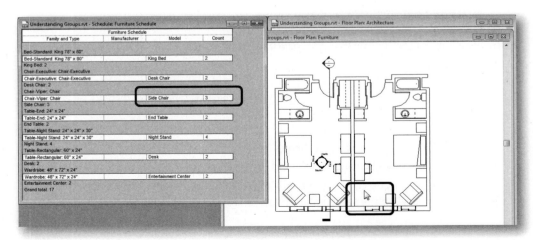

FIGURE 6.15 *Excluding an item from a Group removes it from the schedule as well*

Now that is even more impressive; the schedule accurately reflects the quantity shown in the model. This example illustrates that you can use the exclude from Groups feature with confidence, as Revit will accurately reflect the exclusions throughout the model.

You should try a few more experiments to become comfortable with the full behavior of this feature. For example, move your mouse over the missing chair and it will pre-highlight as if it were there. In this way, you can TAB back into the Group and bring the element back (include it). Be careful when editing a Group that has overrides applied. You can edit either instance of the furniture Group on screen. However, if you edit the one with the excluded chair, you will not be able to edit the chair at all. If you instead edit the one without exclusions, you will have the ability to edit all its elements including both chairs. You can even move or otherwise edit the chair that is excluded in one of the other Groups. The change will be visible in the Group (or Groups) that shows the chair and not visible in any that exclude it. Try some of these experiments now before continuing if you like.

9. Save the project.

ADDITIONAL GROUP DESIGN TECHNIQUES

As you refine your design using Groups, you will find some of the additional techniques covered here useful.

Creating Attached Detail Groups for Existing Model Groups

The Attached Detail Groups that we made earlier were created at the same time as the host Model Groups. You can also create them later even after the Model Group has been created.

1. Return to the *Furniture* floor plan.
2. On the ribbon, click the Annotate tab.

-   On the Tag panel, click the **Tag by Category** button.
-   Tag each piece of furniture in only the guest room showing both chairs.

All of the furniture numbers have already been input in this dataset. To learn more about tagging and editing tag parameters, see Chapter 12. You can move the tags around after placement and adjust the leaders with the handles as required.

-   Select all of the furniture Tags, and then on the Modify | Furniture Tags tab, click the **Create Group** button.

Remember you can use a window selection and the Filter button to get just Furniture Tags.	**TIP**

-   In the Attached Detail Group Name field, type **Tags** (see Figure 6.16).

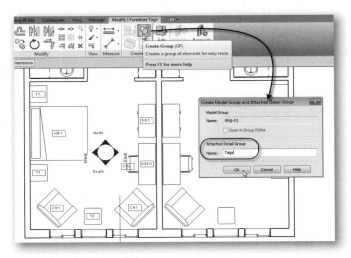

FIGURE 6.16 *Revit will automatically create an Attached Detail Group from Tags attached to a Model Group*

As you can see, Revit will automatically recognize that the items being grouped are Tags associated with model elements contained in a Model Group. As a result, an Attached Detail Group is created automatically. Now that we have an Attached Detail Group for our furniture, we could add it to the other instance of our furniture Model Group. However, we will do something a little different with it below.

Adding Detail Groups to Mirrored Groups

Attached Detail Groups can be very useful as we have seen. They do have limitations as well as we will see. If we return to the *Architecture* floor plan, we will notice that the mirrored Group has no Tags.

3. From the Project Browser, re-open the *Architecture* floor plan.

    • Select the mirrored Group (on the right side).

    • On the ribbon, click the **Attached Detail Groups** button.

    • In the "Attached Detail Group Placement" dialog, place a check in the box next to the *Floor Plan:Tags* Detail Group and then click OK.

Notice that despite the Model Group's being mirrored, the Detail Group remains "right-reading." This is true if you mirror in any direction. Furthermore, you can mirror a selection of both Model and Detail Groups together in the same operation and the Detail Groups will remain right-reading as the Model Group mirrors.

4. Select both Model Groups and both Detail Groups.
   Be careful *not* to select the Section or Elevation tags.

    • On the Modify | Multi-Select ribbon tab, click the **Mirror - Draw Axis** button.

This allows you to sketch the mirror line rather than pick an object. In this case, we will mirror the selection up at a distance above the selection to allow room for a corridor between the rooms.

- Using the temporary dimension as a guide, click the first point about 5'-0" above the top Walls to indicate the middle of the corridor.
- Drag the mouse horizontally and then click again to complete the mirror (see Figure 6.17).

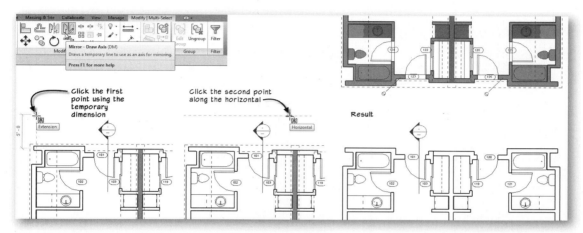

FIGURE 6.17 *Tags in mirrored Detail Groups remain right reading*

If you zoom in on the new rooms and study the tags, you will see that they are right-reading, yet they each display a unique door number sequentially incremented from where the previous door numbering left off.

5. Select all Groups on screen (4 Model and 4 Attached Detail Groups).

- Mirror the selection around the centerline of the right most vertical Wall.

As before, a warning will appear indicating that you have duplicate Walls again. Ignore this warning for now.

There are now eight total guest rooms each with its own Tags.

6. Select the four guest rooms in the middle. (Clicking with the CTRL key is the easiest method of selection in this case).

- From the Type Selector, choose *Guest Room B*.

As we saw earlier in the chapter, not only does the guest room geometry change, but the Attached Detail Group updates as well. Remember, the Attached Detail Groups swap out as well because they have the same names. If you do not keep the names the same ("Tags" in this case), they will not swap out. The duplicate Walls warning will appear again.

- Use the process covered in the "Excluding Group Members" topic above to exclude the extra Walls. (You may need to do this for six Walls total).
- Save the project.

Duplicate Groups on Project Browser

Returning to the Furniture view will reveal that while the Walls, Doors, and Windows contained in the duplicated Model Groups were copied to form additional guest rooms, the furniture was not. This is simply because the furniture is contained in a separate Model group. Let's duplicate our furniture Group and make an alternate for the other guest room type.

7. On the Project Browser, beneath Model Groups, right-click the *King-01* Group and choose **Duplicate**.

   • Right-click the new copy and choose **Rename**. Call it: **Queen-01** (see Figure 6.18).

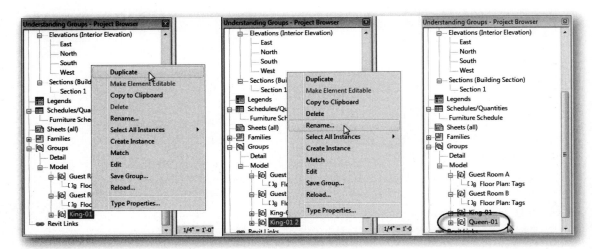

FIGURE 6.18 *Duplicate and rename a Group on the Project Browser*

Notice that the *Queen-01* Group also has its own attached Detail Group named *Tags*.

There are several commands on the right-click menu. Here is a brief description of each:

- **Duplicate**—Creates a copy of the Group and assigns it a default name.
- **Make Element Editable**—This command is only active in a project using Worksharing. Worksharing is a process enabling a team of people to work in the same Revit project. See Chapter 15 for more information.
- **Copy to Clipboard**—This copies the Group to the clipboard so you can paste it in other projects. This is a fast way to use the same Group in another project. If you paste it in the same project, it behaves like Duplicate.
- **Delete**—This deletes the Group definition from the project. You can only use this command if no instances of the Group are inserted in the project. Use with caution.
- **Rename**—Use to assign a new name to a Group definition.
- **Select All Instances > Visible In View**—Use this command to select all instances of the Group in the current view only.
- **Select All Instances > In Entire Project**—Use this command to select all instances of the Group throughout the entire project. Be careful as this command selects all instances on all levels, even the ones that may not show in the current view.
- **Create Instance**—This adds an instance of the Group.
- **Match**—Use this command to swap one Group with another on screen. You will be prompted to select the Group to use as the source and then the Groups to which to apply the source definition.

- **Edit**—This command will export the Group to a new Revit project and open it for editing. In this way, you can edit a Group independently of the current project and save it as its own file outside of the project. To use the Group saved this way in a project, click the Insert tab and then on the Load from Library panel, click the **Load as Group** button and follow the prompts.

- **Save Group**—This command will also export the Group to a new Revit project but it will not automatically open it. You will simply be prompted for the file name and location in which to save it.

- **Reload**—This is a shortcut to the **Load as Group** button on the Load from Library panel of the Insert tab. Use it to load an external file and replace the internal Group definition.

- **Type Properties**—Opens the Type Properties dialog for the selected Group.

A similar list of commands appears when you right-click a Detail Group.

Return to the *Furniture* floor plan.

8. Select the furniture Group with the excluded chair.

Notice that the excluded chair appears when the Group is selected.

- On the ribbon, click the **Restore All Excluded** button.
  The previously excluded chair will be restored.

- With the Group still selected, choose *Queen-01* from the Type Selector.

> **NOTE** If you skip the "Restore All Excluded" step, the excluded chair will still reappear when swapping Groups. Instance-level overrides are not retained when changing types.

9. On the Group panel, click the **Edit Group** button.

- Select the bed and from the Type Selector, choose **Bed-Standard : Queen 60" × 79"**.

- Delete one of the lounge chairs and move the bed and nightstands down to fit the room better (see Figure 6.19).

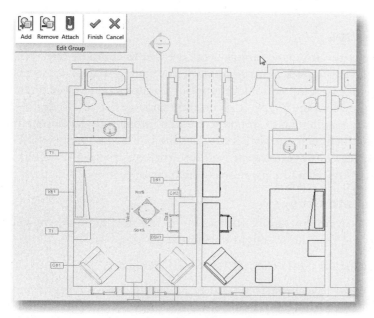

FIGURE 6.19 *Swap in the Queen-01 Group and then modify it*

10. On the Edit Group panel, click the **_Finish_** button to complete the edit.

11. Select the *Queen-01* Group on screen.

    • On the Group panel, click the **_Attached Detail Groups_** button.

    • In the Attached Detail Group Placement dialog, place a check in the box next to the *Floor Plan:Tags* Detail Group and then click OK.

Notice that the Tags have automatically adjusted to the new furniture layout of the Group.

12. Using any of the techniques covered so far; add furniture and Model and Detail Groups to the remaining guest rooms (see Figure 6.20).

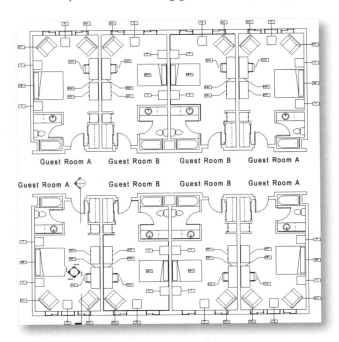

FIGURE 6.20 *Add furniture groups to the remaining rooms*

MAINTAINING GROUPS

As you can see, working with Groups so far has made it easier to compose our overall plan layout and quickly replicate a series of similarly configured spaces. After this initial design work, you may be tempted to ungroup your Groups to gain more direct access to the elements they contain. While it is certainly possible for you to do this, you may want to consider keeping your Groups active well into design development or even CDs. The reason for this is simply because despite our best efforts to minimize them, design changes continue to occur well into the construction document phase and even beyond. Groups can help you make such changes more efficiently.

Add missing elements to Groups

While design changes occur for any number of reasons, in this next example, we will consider a change resulting from an oversight during the design phase.

1. On the Quick Access Toolbar (QAT), click the Default 3D View icon.

Compare the original room that we started with to all of the copied versions and notice that the copies do not include the ceiling elements. Since we have been working exclusively in plan views, we did not notice that the ceilings were not included in the original selection set from which the Groups were created (see Figure 6.21).

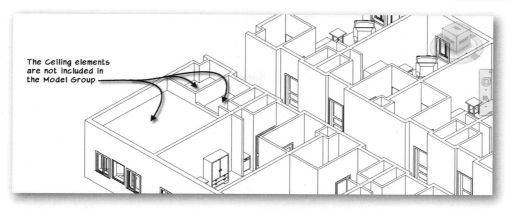

The Ceiling elements
are not included in
the Model Group

FIGURE 6.21 *Switching to 3D view reveals elements missing from the Groups*

The fix for Guest Room A is simple. For Guest Room B there is an extra step. In the original Guest Room A Group (shown in the figure) the ceiling elements are positioned in the proper location. All we have to do is add those stray ceiling elements to the Group and they will appear in all instances of Guest Room A. For Guest Room B, we first need to copy the ceiling elements into position relative to one of our Guest Room B Groups and then add them to the Group. While the 3D view was useful to identify the problem, a ceiling plan view is the best choice for making the required edits.

2. On the Project Browser, double-click to open the *Level 1* Ceiling Plan.

Zoom in on the original Guest Room A; in this case, it has an interior elevation tag within it, making it easy to locate. If you compare the elements in the original Guest Room A to the others, you will notice that there is a small Wall separating the main guest room from the entry foyer. We'll need to mirror this Wall from Guest Room A to Guest Room B to form the boundaries for the Ceiling elements.

3. Place your mouse over the Wall between the foyer space and the main room and then look at the Status Bar (at the bottom of the Revit screen).

   Press tab if necessary to highlight the Wall.

A message will appear reading "Walls : Basic Wall : Generic 5"." The format of this message is: *Category : Family : Type*. All Revit objects appear in this format when pre-highlighted on screen. (See the "Status Bar" heading in Chapter 2 for more information.)

Depending on your settings, the same information may appear in a tooltip on screen (see Figure 6.22).

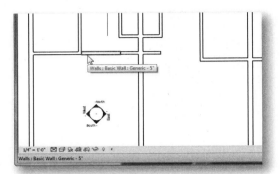

FIGURE 6.22 *The Status Bar reports the Category, Family, and Type of pre-highlighted elements*

- When the Wall between the foyer space and the main room highlights, click it to select it.

If your Wall is not in the same location, you may have inadvertently moved it during the previous exercises. Use the *Align* button on the Modify tab to position it back where it was shown in the figure before proceeding.

4. On the ribbon, click the *Mirror - Pick Axis* button.

- Using the center of the Wall between the two guest rooms, mirror the elements to the neighboring room (see Figure 6.23).

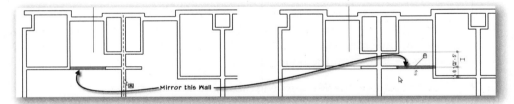

FIGURE 6.23 *Mirror the Wall required to enclose Ceilings*

While it would be possible to select and mirror the existing Ceiling objects to the other space as well, in this case it will be better to recreate them since the shapes of the rooms do not match.

5. Select the original instance of the Guest Room A Group.

- On the Group panel, click the *Edit Group* button.

TIP	The shortcut for Edit Group is EG.

- On the Edit Group panel, click the *Add* button.

TIP	The shortcut for Add to Group is AP.

This tool allows us to add items from the main model into the Group. When we are finished editing, these elements will appear in all instances of the Group.

6. Move your mouse near the edge of the toilet room in Guest Room A.
   The Ceiling object will pre-highlight.

- Click the Ceiling to add it to the Group (see Figure 6.24).

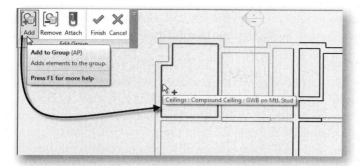

FIGURE 6.24 *Add Ceilings and the Wall to the Group*

Repeat for the foyer Ceiling, the closet Ceiling, and the small Wall.

7. Finally, add the Ceiling in the main guest room space to the group.
Press the tab key if necessary to assist in adding any of the elements.

 • Once you have added four Ceilings and the Wall to the Group, click the **Finish** button on the Edit Group panel.

If you return to the *{3D}* view, you should notice that the four Guest Room A Groups now have Ceilings. To add ceilings to the Guest Room B Groups, we will simply create new ones.

8. In the Ceiling Plan *Level 1*, select the instance of Guest Room B to which we mirrored the small Wall above.

 • Click the **Edit Group** button.

 • Add the small Wall to the Group.

9. Click the Home tab of the ribbon.

 • Click the **Ceiling** button.

 • From the Type Selector, choose **Compound Ceiling : GWB on Mtl. Stud.**

 • Click inside each enclosed space to add Ceilings.

If you wish to make the Ceiling in the main part of the guest room taller, select it and on the Properties palette, change the Height Offset From Level to 9'-0".

 • Click the **Finish** button on the Edit Group panel when done.

If you return to the *{3D}* view, you should now have Ceilings in all rooms.

NESTING GROUPS

As we have seen, most model elements can be added to Model Groups. We can also make a Group that contains other Groups. So-called "Nested Groups" can be useful, but can also present certain challenges. For example, in the dataset we have open, it might be useful to group all of the various guest rooms and their furniture into a single Group named something like "Typical Floor Layout." This approach may certainly prove valuable at the early stages of design where you stand to gain an advantage from the ease of being able to edit a Group instance and have the changes apply across the entire project. However, there are limitations. The most notable is that Attached Detail Groups only work one level deep. This means that you cannot make a Group containing both Model and Detail Groups as members. You will still be able to apply Attached Detail Groups to the nested instances of the Model Groups, but you will have to use the TAB key to select each instance before placement. With careful planning, you can certainly make a workable solution; the only caution is to plan your strategy carefully before execution.

To create a nested Group, you simply select objects (including other Groups) and then click the ***Create Group*** button as we have done already.

Creating a Nested Group

Let's do a quick example of a nested Group in the project we have open. Suppose we wanted to explore adding some more Levels to this project and reusing the layout we have devised here on those Levels. We can certainly use Copy and Paste in that scenario, but creating a Group of the entire layout affords us the opportunity to make edits to the Group later and have those edits apply automatically to all Levels.

1. On the Project Browser, double-click to open the *Section 1* Building Section view.
2. On the Home tab, on the Datum panel, click the ***Level*** button.

   - Using the Pick Lines draw icon, create a Level 12'-6" above the existing Level 1.
   - Rename it **Level 2** (see Figure 6.25).

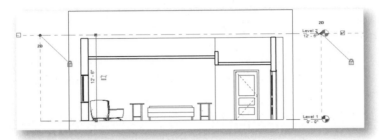

FIGURE 6.25 *Add and rename a new Level*

3. Return to the *Architecture* Floor Plan view.
4. Select all eight Model Groups on screen.

   - On the Modify Model Groups tab, on the Create panel, click the ***Create Group*** button.
   - Name the Group: **Typical Floor Layout** and then click OK.

Take notice of the Project Browser after you complete the Group. Typical Floor Layout will show Guest Room A and B indented beneath it. This indicates that these two Groups are nested within it (see Figure 6.26).

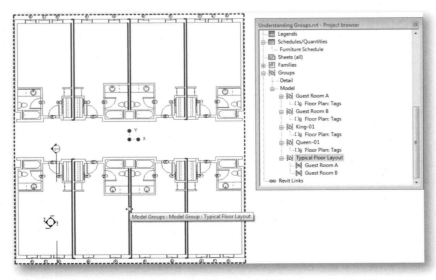

FIGURE 6.26 *Add and rename a new Level (continued)*

5. On the Project Browser, double-click to open the *Level 2* Floor Plan.

6. On the Project Browser, right-click Typical Floor Layout and choose **Create Instance**.

   • Snap to the Group Origin of the Group on the level below and then click the ***Modify*** tool (see Figure 6.27).

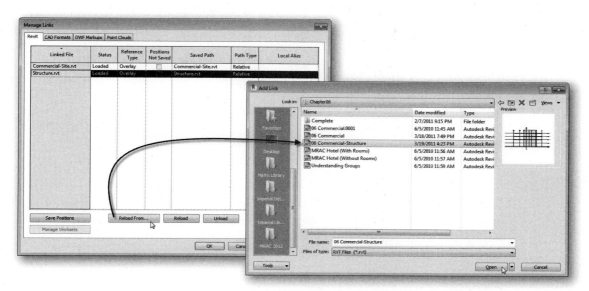

FIGURE 6.27 *Create a new instance of the Group at the Group Origin*

Adding Attached Detail Groups to Nested Groups

If you want to add the Tags Groups, you have to use the TAB key.

7. Place your mouse over one of the Groups.

Notice that the entire floor layout Group pre-highlights.

   • Press the TAB key.
   • Click to select the nested Group instance.
   • Click the ***Attached Detail Groups*** button, choose the *Floor Plan:Tags* Group and then click OK (see Figure 6.28).

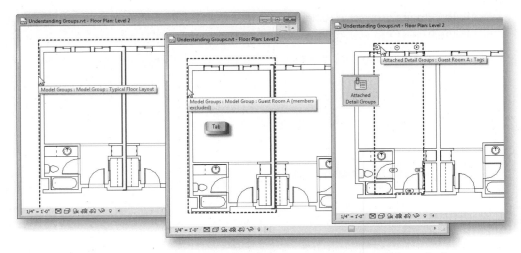

FIGURE 6.28 *Add an Attached Detail Group to the nested Model Group (using* TAB *to select)*

You can repeat the process on the other Groups if you like. Following any of the procedures covered so far, you can also edit either the nested Groups or the overall Group and see the results throughout the model and in the Attached Detail Groups. Feel free to experiment further before continuing.

GROUPS AND REVIT LINKS

We have explored many techniques and advantages of working with Groups so far in this chapter. You can begin to see the many advantages of including Groups in your workflow. While working with Groups directly in a project can prove a useful strategy for managing typical and repetitive design conditions, it is sometimes even more advantageous to export a Group to its own separate file. This can be achieved by saving the Group or converting it to a Linked file.

Saving a Group to a File

A Group can be saved to a separate Revit file. This enables you to work on the Group independently of the project. This can be useful if the project is particularly large and/or if you want to have another colleague working on the Group at the same time as you or someone else is in the project file. It also allows you to use the Group in other projects.

1. On the Project Browser, right-click the *King-01* Group and choose **Save Group**.

   • In the Save Group dialog, browse to the *Chapter06* folder, verify that the "Include attached detail groups as views" checkbox is selected, and then click Save (see Figure 6.29).

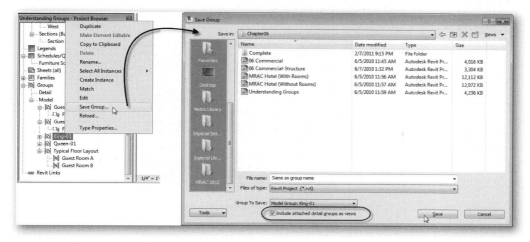

FIGURE 6.29 *Save a Group to a separate file*

The file name defaults to the same name as the Group: King-01 in this case. Once the save is complete, you can open the file to study the result.

2. From the *Chapter06* folder, open the *King-01* file.

   • On Project Browser, expand Views.

You should have two floor plan views: *Level 1* and *Tags*. *Tags* contains the annotation contained in the Attached Detail Group of the same name in the main project. The *Level 1* view contains the model geometry. You can make any edits here that you like. Upon saving those changes, we can reload them back into the main project.

3. In the *Level 1* floor plan view, make a change to the furniture layout.

   • Save and close the *King-01* file.

   • Back in the main project, right-click the *King-01* Group on Project Browser again and choose **Reload**.

   • In the dialog that appears, browse to the *Chapter06* folder and select the *King-01* file to reload.

   • Accept the defaults and click OK in any warnings.

     If you are not in a view that shows the furniture, switch to one now to see the results.

Convert a Group to a Linked file

Revit also provides the ability to embed other Revit files in your project as Revit links. A linked file provides many of the same advantages as Groups but remains a separate project file on your hard drive or server maintaining a live link for easy reloading. In this way, another individual can work in the linked file simultaneously. When the linked file is saved, you can capture the latest changes by reloading the linked file. The process is similar to the one just outlined, but the path to the link file is saved with the project so that we do not have to browse to each time we reload.

4. On the Project Browser, double-click to open the *Level 2* floor plan.

5. Select the *Typical Floor Layout* Group on screen.

   • On the Group panel, click the **Link** button.

   • If a warning appears, click OK.

   • In the "Convert to Link" dialog, click the "Replace with a new project file" option (see Figure 6.30).

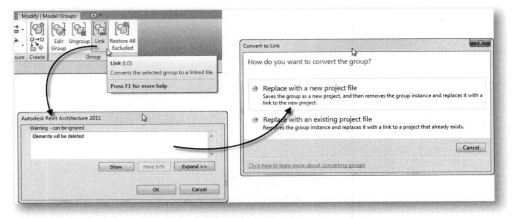

FIGURE 6.30 *Convert the Group to a separate linked file*

In the "Save Group" dialog, browse again to the *Chapter06* folder and then click Save.

The first dialog was a warning about elements being deleted. This is because we had previously applied tags (via an Attached Detail Group) to the original model Group. When you convert a Group to a Link, the annotation cannot remain applied. The second message allows us to create a new file from the Group we are replacing, or to point to an existing file already on our hard drive or server to swap in its place. In this case, creating a new project file was our obvious choice.

When the conversion is complete, you will see that the newly created file automatically appears beneath the Revit Links node of the Project Browser (see Figure 6.31). Here you can access features of the linked file via the right-click menu.

FIGURE 6.31 *Linked Revit files appear beneath Groups on the Project Browser*

Feel free to open the linked file, make a few edits, and then re-save and reload the file. When you try to open the linked file, Revit will warn you that it must be unloaded in the current project, and the unload can't be undone. This is normal. Click Yes to proceed. Once edits are complete in the *Typical Floor Layout.rvt* file and saved, you can close it and back in the *Understanding Groups.rvt* file, right-click the linked file on the Project Browser and choose **Reload**. You will not need to browse to it again. The path for links is saved with the project. You can also right-click on the Revit Links branch of the Project Browser and choose **Manage Links** to see a dialog listing paths and other information about linked files.

Binding Linked Files (to Groups)

The opposite of converting a Group to a linked file is "Bind," which converts a linked file into a Group.

1. Select the linked file.

   - On the Modify | RVT Links tab, on the Link panel, click the ***Bind Link*** button.
   - In the dialog that appears, choose Attached Details and then click OK.
   - Accept the default in the remaining dialog(s).
   - Remove the link when prompted.

The Revit link should now be removed and in its place the Group that we started with should have been restored.

2. Save the project.

Working with Rooms in Groups

When the time comes to add Room objects to our project we can choose to add them within the Groups or outside the Groups. If we add a Room to each Guest Room Group, they will appear in all instances like other objects. We can then tag them inside an Attached Detail Group or directly on the floor plan view.

Another approach is to simply add the Rooms outside of any Groups directly in the project. Since the Room object will conform automatically to the shape defined by the Walls, either approach is completely valid.

To compare methods, try both approaches in the current project.

3. In the *Architecture* floor plan view, select one of the Guest Room Groups (use the TAB key to assist in selection).

   - Click the ***Edit Group*** button on the ribbon.
   - Using the ***Room*** button on the Room & Area panel of the Home tab, add a Room in the main space.

You will see the Room object conform to the shape of the main room plus the entry foyer. If you like you can repeat the process to add additional Rooms for the closet and toilet room. However, in some cases, for a hotel room layout such as this, you may not want to have separate Rooms for each of these spaces, but might instead prefer a single Room that expands to include the closet and toilet rooms within it. To do this, we can select the Walls between the toilet room and main guest room area and make them "non room bounding."

4. Click the ***Modify*** tool and then select the Walls that separate the toilet space and closet from the main space (5 total).

   - On the Properties palette, turn off "Room Bounding" (see Figure 6.32).

     When you click Apply, the Room should now ignore the interior Walls and fill the entire guest room layout.

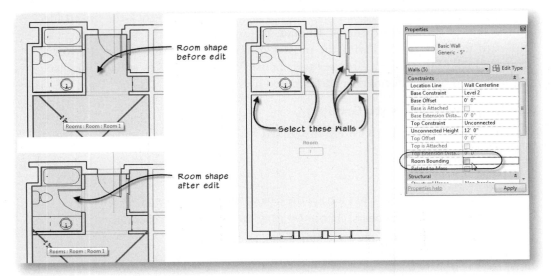

FIGURE 6.32 *Adding Rooms to the Group and varying the Room Bounding behavior*

When you finish the Group the Room will be added to all instances. Again, the Room Tag will need to be attached separately as explained in "Edit a Group" above. If you wish to try the alternative method, simply exit the edit Group mode and add Rooms directly to the project. Tags can also be free-standing or grouped in Attached Detail Groups.

There are certainly plenty of other equally useful applications of Groups including typical toilet room layouts, typical stair tower, office furniture layouts, etc. For example, in the previous chapter we used a simple Group to create a typical floor framing condition that was copied to multiple floors in the building. There are almost limitless applications for Groups.

For your further experimentation, a larger and more complete dataset similar to the one utilized in this chapter has been provided. You will find two versions of "MRAC Hotel" in the *Chapter06* folder. One version named *MRAC Hotel (With Rooms).rvt* has the Room objects embedded within the Guest Room Groups. The other version *MRAC Hotel (Without Rooms).rvt* has the Rooms placed directly in the project (not in the Groups). You are encouraged to open each of these files and experiment further with all of the techniques covered in this chapter (see Figure 6.33).

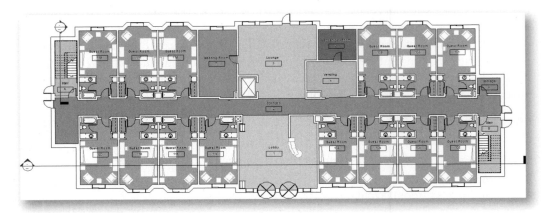

FIGURE 6.33 *The MRAC Hotel.rvt file is provide for your further experimentation*

LINKED PROJECTS

Throughout the course of this chapter, we have worked in a separate dataset and not re-visited our commercial and residential projects. While theoretically any project can make use of Groups and Linked files, our residential project has no need for either. However, our commercial project can make use of both. We have already added a Group to our commercial project at the end of Chapter 5 (for the structural framing) and we can also make use of Linked files for certain aspects of the project as well.

Many firms using Revit take advantage of Linked files as a way of splitting up larger projects into more manageable pieces. It is common to see separations made along various disciplines (such as architectural, structural and mechanical) and sometimes between major functional areas of the project (like core, shell, and interiors). These are of course suggestions and each firm and in fact each project can and often will implement some variation of these. The one strategy that is common is to separate a large project into smaller files so that different team members can readily work in separate areas simultaneously.

Now that we have explored most of the concepts of Groups and Links in the dataset files accompanying Chapter 6, let's return to our commercial project to see how some of these concepts might apply and allow the project to progress.

Using Linked files is only one way in which Revit teams collaborate. The other method involves a toolset called "Worksharing," "Worksets," and "Element Borrowing." The Worksets function of Worksharing provide a means to separate a single building model into discrete portions for purposes of facilitating multi-user access to the same model. The process involves the creation of a "Central" model stored on a common network server and individual "Local" files on each team member's workstation. Revit keeps track of changes that each user makes by enabling object locking at the Workset level and the level of the individual element (referred to as "Element Borrowing"). Worksets will be discussed more extensively in Chapter 15. If you are working in a team of Revit users, then Worksharing is a must. Please set aside time to read the material included in Chapter 15 before participating in your first team project. Worksets are used within the office local area network (LAN) and Linked files are typically used to collaborate with external consultants such as Civil, Structural, MEP and other design firms. Links are not limited to external consultants, and are certainly used for internal workflow management as well.

Load the Commercial Project

Be sure that the *MRAC Hotel.rvt* and *Understanding Groups.rvt* Projects are saved and closed.

1. On the QAT, click the Open icon.

The keyboard shortcut for Open is CTRL + O. **Open** is also located on the Application menu. **TIP**

- In the "Open" dialog box, browse to the location where you installed the *MasterRAC 2012* folder, and then open the *Chapter06* folder.

2. Double-click *06 Commercial.rvt* if you wish to work in Imperial units. Double-click *06 Commercial Metric.rvt* if you wish to work in Metric units.

You can also select it and then click the Open button.

The project is in much the same state as we left it at the end of the previous chapter. However, some important changes have been made since we closed it there. For this reason, be certain that you use the new dataset provided for Chapter 6 and do not attempt to continue in your own files from the previous chapter. The building still looks the same, but the toposurface is no longer in the file. The geometry for the site was removed and we will now walk through the process of creating a separate Revit project for the site data. Using techniques covered in this chapter, we will then link the Site project we create back into the Commercial project.

Create and Link a Site Project

Frequently you will receive site plan data from outside firms in AutoCAD DWG or Microstation DGN format. Revit Architecture readily imports files saved in either format (and others as well). The linework in those files can be used to create a Toposurface. (In order for this to work correctly, the linework in the file has to be drawn at the correct z-height corresponding to the actual contour level you wish to create).

 This means that if your Civil Engineer did not draw the contours at their actual elevations, you will have to open the file in the original application (AutoCAD or Microstation) and move the contour lines to their correct Z heights. If you do not own a copy of the application, you can ask your consultant to do this for you before they send the file.

Let's import some contour lines from a DWG file and generate a new Toposurface. We will create a new Revit project in which to do this. We could, of course, import the CAD file directly into our Commercial project, but as noted above, it is common "best-practice" for such data to be contained in a separate project and then linked back into our project. This makes it easier to coordinate the sometimes different needs and workflows of the different disciplines responsible.

NOTE If you prefer to skip this exercise, you can instead use the Commercial Site project file provided in the *Chapter06\Complete* folder. To do so, skip to the "Link the Site Project" topic below.

Create a New Project

1. From the Application menu, choose **New > Project**. (Access the Application menu by clicking on the big "R" button at the top left corner of the application frame).

    *   In the "Template File" area, be sure that default template file is selected: *default.rte* [*DefaultMetric.rte*] (dialog shown in Figure 4.1 in Chapter 4).

NOTE If your version of Revit Architecture does not include the template files cited here, both have been provided with the dataset files. Please browse to the *Templates* folder in the location where you installed the dataset files to find them.

    *   In the "Create New" area, verify that Project is selected and then click OK.

2. Double-click to open the *Site* floor plan view.

Link the Site Plan CAD File

3. On the Insert tab of the ribbon, click the **Link CAD** button.

This command creates a live link to a CAD file. If the original file should be changed in its host application, Revit will be able to reload the changes.

4. In the "Link CAD Formats" dialog, browse to the *Chapter06* folder.

- Select (do not double-click) the *Commercial-Site.dwg [Commercial-Site-Metric.dwg]* file (don't click Open yet).

Several options appear at the bottom of the dialog.

Current view only—this checkbox will import the file into the active Revit view only. This means that the CAD file will not display in any other view. (This can be useful in some cases, but if you wish to use the imported file to generate a Toposurface, as we do here, do not use this option. You will not be able to select the contours of the imported file while using this option.) If you recall the "Revit Architecture Elements" topic in Chapter 1, we learned there that Revit treats model elements differently than annotation elements. In particular, model elements appear in all views while annotation elements appear only in the view in which they are created. This checkbox basically tells Revit to treat the linked file like model elements when it is unchecked, or annotation elements when it is checked.

Layers—most CAD files use layers (or levels in DGN files) to organize the geometry they contain. These layers/levels can be interpreted in the incoming file. If you wish to import only certain layers in the CAD file, you can choose either the "Specify" or "Visible" options. Visible brings in only those Layers not turned off in the CAD file, and the Specify option will display the list of all Layers and let you select the ones you need. The default setting brings in all layers.

Colors—most DWG or DGN files are saved in multiple colors. The options here allow you to control how this color data is handled on import. If the CAD data was drawn in a black background, try the Invert option to make the colors read better.

Import units—Auto-Detect is usually the best option. However, in cases where Revit misinterprets the units in the CAD file, you can designate the proper unit manually.

Positioning—there are several options. "Auto - Center to Center" is the simplest option. It simply matches the geometric center of the imported file to the geometric center of your active Revit view ensuring that something will show up when you finish. If the file is a one-time import and you are reasonably certain that you will not need to import additional files, this can be the most convenient option. If the imported file has a known and meaningful origin, the "Auto - Origin to Origin" can be used. When you allow Revit to align the origin of the DWG or DGN file to the Revit model origin, you can later import additional DWG or DGN files based upon the same origin point and be certain that they will automatically align properly with the existing geometry. The only problem with the origin to origin option is that your Revit project may not be built to match the origin in the incoming file, or the origin might be far from the model geometry. In this case, you can use one of the other options, make adjustments and then establish shared coordinates between your project and the link. The benefit of this approach is that it does not force you to adopt the origin of the incoming file at the expense of the host project. For the CAD file we are importing here, we will use the origin option and see if that gives us acceptable results.

Several manual options are also available allowing you to use the mouse pointer to place the imported file in any location you like. If you intend to move the linked file into the proper position after import, the manual options can prove more convenient. The "Place at" option controls which level the link is imported to.

- From the "Colors" list, choose **Preserve**.
- Leave Layers set to **All** and Import Units at **Auto-Detect**.
- Leave "Current View Only" checked off.
- Leave "Orient to View" checked on.
- For "Place at" leave "Level 1"
- From the "Positioning" list, choose **Auto – Origin to Origin**, and then click Open (see Figure 6.34).

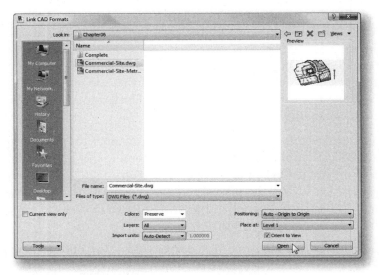

FIGURE 6.34 *Import the Site data from a DWG file*

5. On the Navigation Bar, choose Zoom to Fit.

TIP	The shortcut for Zoom to Fit is ZF.

The CAD file has been inserted into the Revit project relative to its own origin point, which occurs at the point where the Project Base Point and Survey Point icons (the blue icons in the center of the elevation marks) appear.

BIM MANAGER NOTE

In this particular CAD file, the origin is not too far from the site of the building. However in many real-life projects, the origin can actually be quite far from the building(s). If the origin of the CAD file is greater than 20 miles [32.18km] away from the origin of the Revit project, Revit may not be able to use the Origin to Origin option. In such a case, a warning will appear on screen during import and the Center to Center option will be substituted. If this occurs in your projects, use Shared Coordinates to keep the relative origins in synch with one another. This allows linking of data files (RVT, DWG, DGN, or other formats) where the origin point within the linked file may be very far away from the fixed Revit Base Point (origin). Shared Coordinates is covered below.

If you zoom in on the CAD file, you will notice labels on the contours and several spot elevations. These labels are in feet[meters]. They tell us the actual height at which each of the contour lines is placed in the CAD file.

6. On the Project Browser, double-click to open the *South* elevation view.

Notice that we have the two default Levels at 0'-0" and 10'-0" [4000] and that the contour lines from the imported DWG file appear above these Levels at the distance indicated by the plan labels we just studied. Returning to the plan view, we will note a blue rectangular shape in the file indicating the location of the building. Furthermore, interpolation of the contour labels puts the elevation at the front façade at approximately 81'-0"[24300]. We will use this number in the elevation view to adjust the height of Level 2. Then we will rename the two Levels to something more descriptive.

- In the *South* elevation view, change the height of Level 2 to **81'-0" [24300]**.
- Rename Level 1 to **Datum** and rename Level 2 to **Street Level**. When prompted about renaming corresponding views, choose Yes.

7. Save the project as **Commercial-Site**.

- On the QAT, click the **Default 3D View** icon.
- On the View Control Bar, choose **Shaded**.
- Orbit the model.

Notice that the contours in the CAD file are just lines and do not have any surface.

Build a Toposurface from imported Data

Now that we have imported the 3D contour line data from the DWG file, we can use it to create a more accurate Toposurface than the one created from manual points in Chapter 4.

8. On the Project Browser, open the *Site* view.
9. On the ribbon, click the Massing & Site tab and then click the **Toposurface** button.

The Modify | Edit Surface tab will appear with the **Place Point** button active.

10. On the ribbon, click the **Create from Import** button.

- From the pop-up that appears, choose **Select Import Instance**.
- Click on any of the objects in the imported DWG to select it.
- In the "Add Points from Selected Layers" dialog, click the "Check None" button.
- Place a checkmark in only the "C-Site-Cntr" and the "C-Site-Cntr-Intm" checkboxes to select only those two layers (see Figure 6.35).

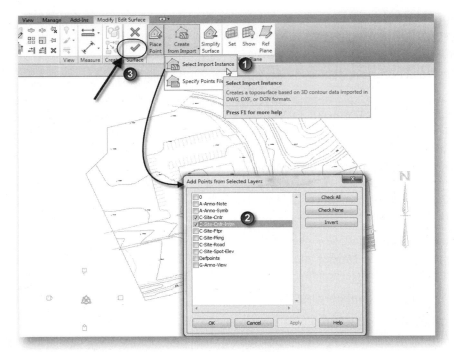

FIGURE 6.35 *Choose the Layers from which to create the Toposurface*

- Click OK to create the points.

Several points will be extracted from the geometry on the selected layers, and from those points a Toposurface will be created.

- On the ribbon, click the **Finish Surface** button to exit sketch mode and complete the Toposurface.
- On the QAT, click the **Default 3D View** icon.
- Orbit the model.

Notice that there is now a three-dimensional surface spread across the contours from the CAD file.

11. On the Project Browser, double-click the *East* elevation view.

If you zoom in, you'll see a sloping profile of the terrain similar to the manual surface we created in Chapter 4. This Toposurface has more points and does a better job suggesting the roads that surround the building site. The manual method of placing individual points that we used in Chapter 4 is effective when you do not have any civil engineering files to import. If you receive a site plan file, it is usually easier to use the linked file. Otherwise, you can quickly create a suitable site for your building model with the point sketching method as well. Either method is appropriate for creating Toposurfaces in your own projects.

Add a Building Pad

Let's add a Building Pad. A Building Pad adds a simple slab surface that cuts into the terrain model as appropriate to suggest the required excavation or other prepared surface of construction. It will be easier to do this without having the model shaded.

12. On the Project Browser, double-click to open the *Site* view.

The site plan data imported from the DWG file includes a rectangle that approximates the rough footprint of the building. We can use this to assist us in sketching the Building Pad. However, at the moment the Toposurface is concealing the linked file.

- On the View Control Bar, click the Model Graphics pop-up and choose **Wireframe**.

13. On the Massing & Site tab, click the ***Building Pad*** button.

- On the Modify | Create Pad Boundary tab of the ribbon (now in Sketch mode) click the ***Pick Lines*** tool.

- In the view window, position the pointer over one line of the building footprint in the middle of the site (the line should pre-highlight) and then press the TAB key.

- When all lines of the shape pre-highlight, click the mouse to create sketch lines (see Figure 6.36).

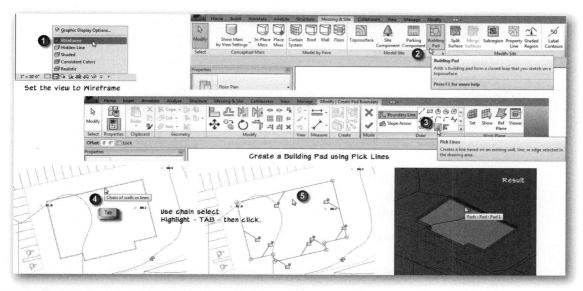

FIGURE 6.36 *Use the tab key to Chain Select and create Sketch Lines*

On the Properties palette, change the Level to **Street Level**.

For the "Height Offset From Level" parameter, input **-4'-0"** [**-1200**] and then click Apply.

On the ribbon, click the ***Finish Edit Mode*** button (big green checkmark).

14. On the QAT, click the Default 3D View icon.

- Zoom as required to see the Pad and its relationship to the Toposurface.

15. Save the Project.

There is plenty more work that we can do to the Site project. However, for the purposes of preparing the file for linking into the Commercial project (which is the primary goal for it in this chapter) we have completed enough work in the Site file for now. If you wish to go further with the Site project, refer to Appendix A for additional exercises on splitting the surfaces, applying materials and suggestions on adding trees and parking.

16. From the Application menu, choose **Close**.

Link the Site Project

Now that we have built a Site project, imported contours from the Civil Engineer and created a Toposurface, we are ready to link this project into the Commercial project. The process of creating a Revit link is nearly identical to the process used in the "Link the Site Plan CAD File" topic above. The Commercial project file should still be open. If you closed it, please reopen *06 Commercial.rvt* [*06 Commercial Metric. rvt*] now.

17. In the *06 Commercial.rvt* [*06 Commercial Metric.rvt*] project file, on the Project Browser, double-click to open the *Site* plan view.

18. On the Insert tab, on the Link panel, click the **Link Revit** button.

   - In the "Import/Link RVT" dialog, choose *Commercial-Site.rvt*, accept the default Center to Center positioning and then click Open.

     If you decided to skip the previous exercise and did not create the *Commercial-Site* file, you can instead link to the *06 Commercial-Site-Complete.rvt* [*06 Commercial-Site Metric-Complete.rvt*] file provided in the *Chapter06\Complete* folder.

When the site file is linked in, it does not align with the building properly. It is to the upper left of the building (and if you looked in elevation, it is also at the wrong height). The Site project was created from a CAD file using the CAD file's origin. Our building was created in the center of our project template's elevation view markers where the Revit file's Base Point (origin) is located (you can see this origin in the current view as the blue icons in the center of the screen). Therefore it is not surprising that things do not line up. While we could have started with the site plan data and built our model to match the orientation of the imported file (True North), it was more convenient to model our building relative to the project template setup (Project North) with the footprint of the building orthogonal to the screen edges.

One of the advantages of using linked files is that each model file can maintain its own coordinate system, called Project Coordinates, with its own Base Point (origin) without imposing it on the other files in the project. There is a bit of setup required to synchronize the different models' coordinate systems, but once complete, each model maintains its own internal coordinates (Project Coordinates) and also understands how it relates to the other files (Shared Coordinates).

The basic setup process for establishing shared coordinates is as follows:

- Gather all required project files and decide which one will be "primary." (This is the file with which the others will synchronize their coordinates).
- Link the files.
- Move and rotate linked files as required to establish the correct geographic relationships. (For example, move the site file link so that the site data is correctly oriented and located under the building).
- Save Shared Coordinates for each pair of files.

The overall process is straightforward. Let's walk through the process now with our two project files. For step 1, we have two project files, the Commercial project and the Commercial Site. The "Primary" file in this case will be the Site file. This just means that we will "acquire" coordinates from the Site file. More detail on acquiring coordinates is found in the "Set up Shared Coordinates" topic below. We have already accomplished step 2 by linking the Site file in the previous sequence. The next task is to move and rotate the linked file (Site project) to match the orientation and location of the host project (Commercial project). We can achieve this using the *Move* and *Rotate* tools or the *Align* tool. Let's look at both options.

Using Rotate

19. Select the linked site file. (You can click on it anywhere and the entire file will highlight.)

    • On the Modify | RVT Links tab, click the **Rotate** button.

The shortcut for Rotate is RO.

TIP

If you know how much you want to rotate, you can simply type in the angle in the field on the Options Bar. Otherwise, you can rotate graphically on screen. A small round "center of rotation" control will appear at the middle of the selection. Using this control, you can change the center of the rotation to the desired position. New in this release, this can be accomplished with two clicks rather than dragging. There is also now a button on the Options Bar for this purpose. You can either click the center of rotation onscreen and then click the new location, or simply click the Place button on the Options Bar and then click where you would like the center to be.

• Click the small blue circle handle indicating the center of rotation or click the Place button on the Options Bar (see panel 1 in Figure 6.37).

• Click at the lower right endpoint of the building footprint to place the center of rotation (see panel 2 in Figure 6.37).

• Move the mouse toward the opposite endpoint of the building edge and then click along the line (see panel 3 in Figure 6.37).

• Finally, move the mouse to the right. It will snap vertically. When it does, click to finish the rotation (see panel 4 in Figure 6.37).

The result should be that the building footprint is perfectly orthogonal to the screen as shown in the far right side of the figure.

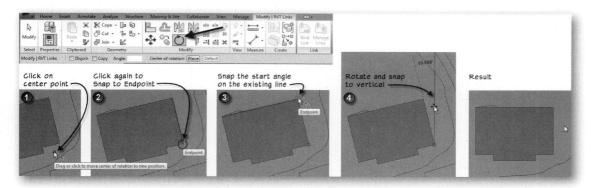

FIGURE 6.37 *Rotate the linked site file to make the building footprint horizontal*

Next we'll move the file and snap it to our building model.

The Link should still be selected, if it is not, click to select the linked file again.

20. On the ribbon, click the **Move** button. (Make sure that Constrain on the Options Bar is *not* checked.)

The shortcut for Move is MV.

TIP

• For the start point, use the same endpoint about which you rotated.

• For the move end point, snap to the corresponding endpoint on the building model geometry. (You may need to pan or zoom. Use your wheel mouse for this.)

| TIP | If you are having trouble snapping to the precise point, move it close, then zoom in and repeat. You can type the shortcut SE to force Revit to snap to the endpoint. |

The linked file will remain selected (highlighted in light blue). The footprint of the building should be shaded a little darker helping you determine if it is lined up properly. Repeat move or rotate if necessary to finetune the position or try the *Align* command as outlined next.

Alternative Positioning Technique

The *Rotate* tool is an important tool and you should be sure you are comfortable with the techniqe for changing the center of rotation. However, there is an easier way to position the linked file with the building model—use the *Align* tool.

> If you wish to follow the steps, undo the move and rotation. Otherwise, skip to the next topic: "Position the Link Vertically."
>
> • On the Modify tab, click the *Align* button.

| TIP | The shortcut for Align is AL. |

> • For the alignment reference, click the outside edge of the top horizontal Wall of the building model.
> • Click the top angled edge of the building footprint sketch in the linked file next (see the top of Figure 6.38).

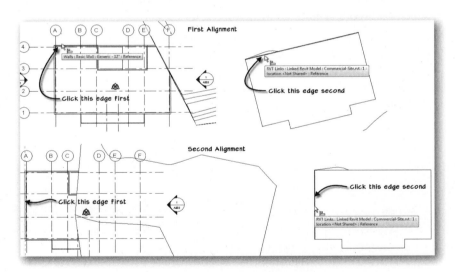

FIGURE 6.38 *Align can be used to rotate and move simultaneously*

> • Repeat the Align process to align the vertical edge (see the bottom of Figure 6.38).

Select the linked file when you are finished and as before, it will highlight and turn transparent so that you can visually verify that the alignment is correct. Both the align and rotate/move methods achieve the same result. The align method is a bit quicker, but there will be times when rotating and/or moving is preferred. Be sure you are comfortable with both methods before continuing.

Position the Link Vertically

Finally, if we look at one of the elevation views, it will become clear that there is one more adjustment required. Recall that in the "Link the Site Plan CAD File" topic above, we noted that the contour lines near the center of the building are at 81'-0". We adjusted one of the Levels to this height in the Site file in preparation for making the vertical adjustment here.

21. Open the *North* elevation view.

Currently we do not see the linked file in this view at all. This is because the elevation view has the crop region enabled. We must temporarily turn it off to make the required adjustment.

22. On the View Control Bar, click the Do Not Crop View icon.

    You can see the View Control Bar in Figure 4.36 in Chapter 4.

Notice that once you disable the cropping, the linked file appears and the Toposurface in that file covers the entire building. Zoom out as necessary.

23. On the Modify tab, click the **Align** button.

    - For the alignment reference, click on the Street Level line of the building model (in the current project).
    - For the entity to align, click the Street Level line in the linked file. (You may have to zoom to find it.)
    - Do not click the lock icon.
    - On the ribbon, click the **Modify** button.

The Toposurface should now appear at the base of the building model (see Figure 6.39).

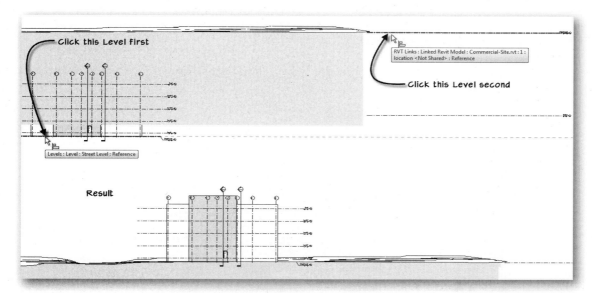

FIGURE 6.39 *Move the link file vertically in the elevation view using the Align tool*

24. On the View Control Bar, click the Crop View icon.

Check all four elevations. If necessary, click the Show Crop Region icon in each elevation, adjust the Crop Region, hide the Crop Region, and finally make sure that the Crop is turned on for each elevation.

25. Save the project.

Survey Points, Project Base Points and Shared Coordinates

The basic mechanics of linked files have already been discussed in the topics above. As we have seen in the previous sequence, when files authored by different parties are linked together, it is important to establish a common reference point for these files. This will ensure that common physical features maintain their proper geometric relationship and position relative to one another.

To help us in maintaining these relationships, Revit projects have the following tools:

Project Base Point—is the origin (0,0,0) of the project with respect to the *project coordinate system*. This coordinate system relates to the project itself, and by default is in the center of the default plan view.

Survey Point—represents a known point in the physical world. This might be some benchmark indicated by the Civil Engineer. The Survey Point is useful to correctly locate and orient your building project in another meaningful coordinate system such as the coordinate system used by a civil engineering application.

Shared Coordinates—reconcile the differences between the current project's coordinate system and the coordinate system used by a linked file. Setting up a Shared Coordinate system keeps all linked files in the correct relative positions to one another while allowing each to maintain its own internal coordinate systems.

Understanding Project Base and Survey Points

The default template from which we began our Commercial project has the Project Base Point and the Survey Point icons displayed in the Site plan view. Other views do not display them by default. If you wish to see these points, their visibility is easy to turn on.

1. Open a plan view.
    - On the View tab, click the **Visibility/Graphics** button.

TIP	The keyboard shortcut for Visibility/Graphics is V G.

The Project Base Point and the Survey Point icons are subcategories of the Site category.

- Expand the Site category.
- Place checkmarks in the boxes next to Project Base Point and Survey Point and then click OK (see Figure 6.40).

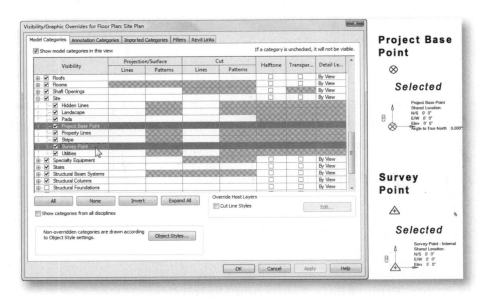

FIGURE 6.40 *Turn on Project Base Point and Survey Point*

Like other display settings, this visibility change must be performed in each view where you wish to have these points display. The Project Base Point is a round symbol with a cross through the middle. The Survey Point is triangular in shape with a small plus (+) sign in the center. Both points are displayed on the right side of Figure 6.40 as they appear when not selected and when selected. In our project, they are currently directly on top of one another. This simply means that both coordinate systems currently share the same origin.

2. Select the Project Base Point. (Use the TAB key if necessary.)

The coordinates of the point are listed as editable dimensions next to the icon. You can type new values into any of these dimensions to move the point. For now, do not make any changes.

3. Deselect the Project Base Point and select the Survey Point. (Use the TAB key if necessary.)

Notice that both points are currently located at the origin (0,0,0). Again, do not change the location at this time.

Saving Shared Coordinates

If the linked file(s) were to later change and require reloading, we want to be sure that they reload in the same relative location. More importantly, if we decide to also link the architectural file to the site file, we would not want to have to move, rotate or align the positions of the file again. Saving a shared location for the file basically makes each file (the host and the link) aware of the other and what offsets and rotations are required for correct orientation in both files. When a Shared Location is saved from the host file, Revit saves this information into the link file, even if the link file is not open. This is also true of linked .dwg files.

4. Open the *Site* view.
5. Select the linked (site) file on screen.

   *   On the Properties palette, beneath Identity Data, change the Name to: **Site**.
   *   Beneath Other, next to Shared Site, click the <Not Shared> button.

The "Share Coordinates" dialog lists two ways that the coordinate systems can be reconciled. The two methods are very similar and differ only in which file will be recorded as the "primary" file. In each case, the Shared Coordinate information must be saved to both the host file and the linked file. Publishing makes the host file predominant, while acquiring **mak**es the linked file predominant. While there is no "correct" choice, making the file that contains the site data the primary file usually makes sense and is a common practice. In this case, this means we will want to acquire the coordinates from our site file. After we do so, Revit will save the Shared (Location) Coordinates to both files the next time we save the current model. In other words, on the next save of the current model, the host model, i.e., the *06 Commercial.rvt* [*06 Commercial Metric.rvt*] project file, the "Location Position Changed" dialog will appear. In addition to saving the current model Revit also writes the Shared Coordinates back to the Site model. This happens any time a linked file in a host file with reconciled shared coordinates moves in any direction.

   *   Click the "Acquire the shared coordinate system of..." radio button.
   *   At the bottom of the dialog, click the Change button.

Revit projects can have one or more saved locations within them. This is useful when the same building model must be repeated on a site, such as a multi-building campus of condominium buildings. In our case, we have only one building model and could simply accept the default location name. However, it is good practice to get in the habit of renaming the default location so that later you can use this to verify that you have, in fact, reconciled the coordinates.

- In the "Location Weather and Site" dialog, click the Rename button and change the name to **MRAC Commercial Site** (see Figure 6.41).

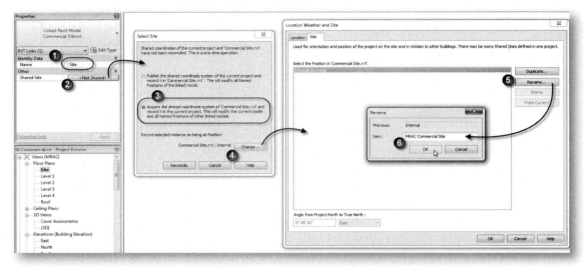

FIGURE 6.41 *Acquire the Shared Coordinates system and rename the default location*

- Click OK two times.
- In the "Select Site" dialog, click the Reconcile button.

Take notice of the new location of the Survey Point icon. You may need to Zoom. When we acquired the coordinates of the site project, the Survey point moved to the origin of that file.

6. Save the project.

- In the "Location Position Changed" dialog, choose the Save option (see Figure 6.42).

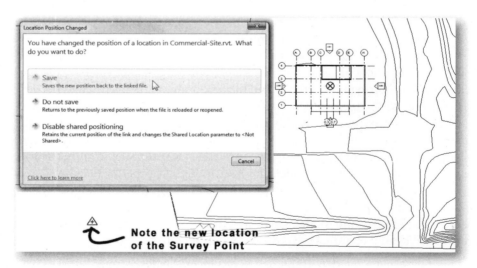

FIGURE 6.42 *Save the shared location in the Commercial Site file*

7. Select the Project Base Point (the round icon).
8. Deselect the Project Base Point and select the Survey Point (the triangular icon).

Notice that the Survey Point is still located at (0,0,0). However, this origin is now reading the origin within the site file which we acquired. Furthermore, the Project Base Point's coordinates are now relative to this origin point. Note also the angles of both points (see Figure 6.43).

FIGURE 6.43 *Now that coordinates are shared between the host file and the linked file, the Project Base Point origin is relative to the Survey Point*

Next to each icon is a paper clip icon. The Project Base Point icon is not clipped and the Survey point is clipped. If you move the Survey Point while it is clipped, the entire linked file will move with it. You are not really moving the Survey Point but rather the location of the host project relative to the linked file. This would be very similar to the move, rotate and/or align steps performed in the "Link the Site Project" topic above. If you unclip the Survey Point and then move it, you will be changing the position of the Survey Point only. You would do this if there were a more meaningful benchmark rather than the linked file's origin point. This might be a location designated by your Civil Engineer such as some known site feature.

Similar behaviors are exhibited by the Project Base Point. If its paper clip icon is clipped, moving it will relocate the entire project relative to the Survey Point and shared coordinate system. Moving an unclipped Base Point simply changes the reference point itself. You might wish to do this to make the Project Base Point reference a more meaningful point, such as the corner of the building. Feel free to experiment with moving these points both clipped and unclipped. However, be sure to save the project before experimenting. In most cases you can undo, but in cases where you cannot, you can close without saving and then reopen the saved file.

> Be sure the Project Base Point is unclipped (shows a red line slash through the paper clip icon).

9. Zoom in as necessary and drag the Project Base Point to the intersection of Grids 1 and A and then clip it again (see the right side of Figure 6.43).

10. Save the project.

Rotate a View to True North

Floor plan views can be oriented either to "Project North" or to "True North." Every plan view has this parameter. True North is the geographic direction of North given to us by the Civil Engineer or Land Surveyor. Project North is typically parallel to the predominant geometry in the building and is oriented for convenience when composing sheets. By default, floor plan views are oriented to Project North. Changing a view to display True North is easy once shared coordinates are set up.

> Continue in the *Site* view. Make sure that no objects are selected and that the drop-down on the Properties palette reads: Floor Plan: Site.
>
> • On the Properties palette, beneath Graphics, next to Orientation, choose **True North** and then click Apply.

The view will rotate to make True North point straight up. If you click either the Survey or Project Base Points, you will see the new orientation reflected in these icons as well.

11. On the Project Browser, open any other floor plan view.

Notice that the orientation in the other plans is still Project North.

Location Weather and Site

We touched briefly on project location in the previous sequence. Using the "Location Weather and Site" dialog, we can configure two important pieces of information for our projects: the actual geographical location in the World and the site-specific information relative to the project and its links. In the previous sequence, we renamed the default "Internal" Site for the linked topography file. Let's take a look at this dialog again from the vantage point of our current project.

1. On the Manage tab, on the Project Location panel, click the **Location** button.

On the Location tab, you can input the address of the project in the "Project Address" field and press enter. Revit will use an Internet mapping service to set your project's location. If you require more precision, you can drag the icon in the map or type in an exact longitude and latitude for your project location. This is important to get proper lighting and shadows. Refer to Chapter 17 for more information.

- Click on the Site tab (see Figure 6.42).

This is the dialog that we saw above. Notice that even though we renamed the Site while saving Shared Coordinates, it is still listed as Internal here. Remember, there are two files: the Site we renamed is in the *Commercial Site.rvt* file. We are currently in the *06 Commercial.rvt* [*06 Commercial Metric.rvt*] project file. As noted above, it can be easier to verify the shared coordinates if you rename the Site to something other than Internal. Also notice the angle from Project North to True North corresponds to the rotation that we have between the two files.

- Click on the Rename button and change the name to **MRAC Commercial Arch** and then click OK twice (see Figure 6.44).

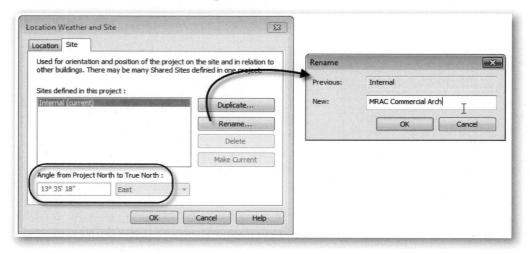

FIGURE 6.44 *Filter the selection to just framing elements*

Linking by Shared Coordinates

For each pair of files, you only need to establish the shared coordinates once. In other words, if the owner of the Commercial Site project wishes to link the architectural file, they can do so without repeating the steps in the "Saving Shared Coordinates" topic above. If you wish to try this, you must first close the Commercial project. You cannot have both a host file and a link file open at the same time in the same session of Revit.

First save the current project and then close it. Open the *Commercial Site* project. Open the 3D view or Site plan. On the Insert tab, choose the **Link Revit** button. Browse to the *06 Commercial.rvt* [*06 Commercial Metric.rvt*] project file. Before clicking Open, choose **Auto – By Shared Coordinates** from Positioning. Click

Open to complete the linking. The Commercial project will appear in exactly the correct spot on the site. If you edit the properties of the linked file, you will see that the Shared Site was assigned to the location named "MRAC Commercial Arch".

There are other important features of Shared Coordinates that are worth your time and exploration. For example, the same linked file can be copied multiple times in a host project and assigned to different named locations. This is useful in a campus situation with multiple identical buildings on the same site. To do this, you reopen the Location Weather & Site dialog, click the Site tab, and Duplicate one of the named locations rather than rename it. Each named location can have its own saved coordinates.

> If you opened the site project, save and close it now. Then reopen the commercial project before continuing.

Create and Link a Structural Project

Now that we have completed setup of our site model, we will next isolate the structural elements and create a separate linked file from them. This will leave only architectural elements (Walls, Doors, Windows, Column enclosures (Architectural Columns), Floors, and Roofs) in the *06 Commercial.rvt* [*06 Commercial Metric.rvt*] project file. You may recall that in the previous chapter, we created Columns, Beams, Beam Systems, and Braces in our commercial project. These elements and the core Walls are the ones that we will separate out to their own structural model. However, some of these elements like the core Walls actually need to appear in both files. We'll look at a special way to achieve that as well.

Convert a Group to a Link

The task of separating the structural elements into their own file can be accomplished in a few ways. We could create a separate file, and then select the required elements and copy and paste them from our project to the new one. We can also use the techniques already covered in this chapter and isolate the required objects using Groups. The process is as follows: select the desired objects, make a Group from them and then convert that Group to a Link. (We could also save the Group without converting it to a Link as well.) Once we have the Group saved as a separate file, we could open it directly, or import into another project. If you prefer to use copy and paste in the next sequence, feel free to do so. The steps that follow will highlight the approach using Groups to reinforce the skills we learned at the start of the chapter.

2. In the *06 Commercial.rvt* file, on the QAT, click the Default 3D View icon.

   The view named {3D} will re-open.

3. Using a Window selection, select all elements on screen.

   • Click the *Filter* button, deselect everything except Model Groups (similar to Figure 5.36 in Chapter 5).

   If there are Structural elements shown, choose those as well, but do *not* include Structural Columns.

At this point, we can go directly to the Group step; however, it is often good practice to use the Temporary Hide/Isolate icon (on the View Control Bar) to isolate the selected elements first to be sure you have the desired selection.

4. On the View Control Bar, click the Temporary Hide/Isolate icon and choose **Isolate Element** from the pop-up.

   You should only have Beams, Braces, and Joists selected on screen.

   • On the ribbon, click the *Create Group* button.

   • In the "Create Model Group" dialog, type: **Structure** for the Name and then click OK (see Figure 6.45).

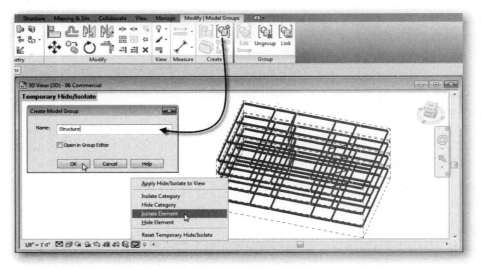

FIGURE 6.45 *Group all the Structural framing*

At this point we have two options: we can simply save the Group to a file (as seen in the "Saving a Group to a File" topic above) or we can convert it to a Link (using the procedure covered in the "Convert a Group to a Linked file" topic above). Let's use the Link option.

5. With the Structure Group still selected, click the **Link** button on the Group panel of the ribbon.

   • In the dialog that appears, choose the "Replace with a new project file option".
   • In the "Save Group" dialog, browse to the *Chapter06* folder and then click Save.

Using this method, we have quickly and efficiently gathered all of the structural framing and moved it to a separate linked Revit file. If you were to unload the Link now, none of the Structure would remain in the *06 Commercial.rvt* [*06 Commercial Metric.rvt*] project file. At this point we could open the new Structure.rvt file and continue the exercise in there. However, one limitation of either the Save to Group or convert to Link method is that the resulting file is not based on the default Revit template. This means the resulting file has no levels, few annotations, and only the bare minimum of views. As a result, it will prove better in practice to create a new file using your preferred template first, and then insert the newly created file into it as a Group. In this way, we can ensure that the Structural file (or any file created this way) benefits from the office standards embedded in a template project. The process would be very similar to the steps covered above in the "Binding Linked Files (to Groups)" topic.

Working with Copy/Monitor

To save a few steps, a file has already been created from the standard template and included with the other Chapter 6 files. We'll open this file, make a few preparations and then insert our Group into it. The preparations that we need involve copying the Levels and Grids from our main commercial project over to the structural file. While it is possible to simply copy these items to the structural file using Groups or copy and paste, a better approach is the *Copy/Monitor* tool which is specifically designed for this purpose. The *Copy/Monitor* tool allows you to copy certain elements (Levels, Grids, Walls, Floors and Columns) from a linked file and keep them associated back to the originals. In this way, you can monitor changes as they occur and update the copies to match. This provides a very practical way for project teams to collaborate on shared elements even when they are not physically located in the same office.

6. Save and Close the *06 Commercial.rvt* [*06 Commercial Metric.rvt*] project file.

7. On the QAT, click the Open icon.

TIP

The keyboard shortcut for Open is CTRL + 0. **Open** is also located on the Application menu.

- In the "Open" dialog box, browse to the *Chapter06* folder.

8. Double-click *06 Commercial-Structure.rvt* if you wish to work in Imperial units. Double-click *06 Commercial-Structure Metric.rvt* if you wish to work in Metric units.

The project will open with the *South* elevation view visible on screen. We had to close the commercial project above because Revit will not allow you have both projects (the host and the Link) open at the same time.

9. On the Project Browser, right-click the *Revit Links* node and choose **New Link**.

NOTE

This is simply an alternative to the method covered above. If you prefer, you can continue to use the *Link Revit* button on the Insert tab instead.

- In the Import/Link RVT dialog, choose 06 Commercial.rvt [*06 Commercial Metric. rvt*].
- For Positioning, choose Auto – Origin to Origin and then click Open. In the "Nested Links Invisible" dialog that appears, click Close.

MANAGER NOTE BIM

The warning simply informs us that the file has Links of its own that will not carry through to the current host. If we wanted them to, we could edit the Link type in the commercial project to be Attachment rather than Overlay. This is done in the Manage Links dialog. Overlaid reference file are direct links. Attachments can nest several levels deep.

10. On the Collaborate tab, click the **Copy/Monitor** button and then choose the **Select Link** option from the pop-up.

- Select the Commercial project onscreen.

  The Copy/Monitor tab will appear on the ribbon.

11. On the Copy/Monitor tab of the ribbon, click the **Copy** button.

- On the Options Bar, click the Multiple checkbox.
- Select Levels 1 through 4 and "Roof" in the linked file. (Use the CTRL key or a crossing selection).
- On the Options Bar, click the **Finish** button (see Figure 6.46).

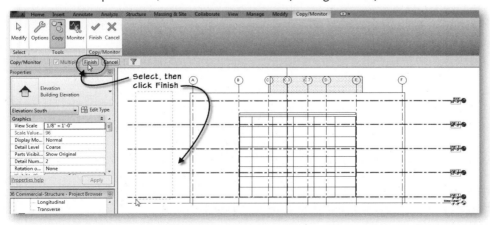

FIGURE 6.46 *Use Copy/Monitor to copy the Levels*

Be sure to click the Finish button on the Options Bar, *not* the one on the ribbon. This finishes selection. The one on the ribbon finishes the command.

A small monitor icon will appear next to each monitored item. If a Level is changed in the main Commercial project, the structural engineer can re-enter the Copy/Monitor mode and use the Coordination tool on the Ribbon to synchronize the changes.

12. Return to the Copy/Monitor tab of the ribbon and then click the **Finish** button.

   This finishes the Copy/Monitor mode.

Create Structural Plans

13. On the View tab, click the **Plan Views** button.

   - From the pop-up that appears, choose **Floor Plan**.
   - In the dialog that appears, select all of the Levels and then click OK.

14. On the Project Browser, expand Floor Plans, select *Level 1*, hold down the SHIFT key and then select *Roof*.

   This will select all the plan views, not including the Site view.

   - Right-click the selected plans and choose **Apply View Template**.
   - Select Structural Framing Plan and then click OK.

Having assigned this view template we now see only structural components. This will make selecting the remaining items for copy/monitor much easier.

Copy/Monitor Grids, Walls and Columns

Now that we have copied the Levels and set up framing plans we are ready to copy and monitor the remaining structural items.

15. On the Project Browser, double-click to open the *Level 2* plan view.

   If necessary, zoom to fit (type ZF).

   - Use the Copy/Monitor steps above (with the multiple option) to copy all of the Grids, all of the structural (steel) columns and the four core Walls.

| TIP | Make a crossing window selection, click the Filter icon on the Options Bar, deselect Columns and Floors and then click OK. |

   - Click the **Finish** button on the Options Bar.
   - On the Copy/Monitor tab of the ribbon click the **Finish** button.

Insert a Group

To complete our structural model, we'll import the Group we created from the framing members.

16. On the Project Browser, double-click to open the *South* elevation view.

   - Use Temporary Hide/Isolate to hide the linked model.

17. On the Insert tab, click the **Load as Group** button.

   - In the "Load File as Group" dialog, browse to the *Chapter06* folder.
   - Select the *Structure.rvt* file and then click Open. (If a message regarding duplicate Types appears, click OK.)

After a short pause, you will note that the *Structure* Group is now available on the Project Browser beneath the Model Groups branch.

18. On the Project Browser, beneath the Model Groups branch, right-click *Structure* and choose **Create Instance**.

   - To insert it in the correct location, simply type 0 (zero) and then press ENTER.
   - On the Edit Pasted panel of the ribbon, click the ***Finish*** button.

19. On the Project Browser, double-click to open the *Longitudinal* section view.

   - Select the Group instance on screen, and then on the ribbon, click the ***Ungroup*** button. We will leave the nested *Typical Framing* Groups alone. This way , we can still benefit from their being grouped should we need to edit the framing later.

20. Save and close the Structural model.

Using Reload From to Swap a Link with another File

Now that we have completed the setup of our structural model, we are ready to load it into our main Commercial project. Since we already have a link to the Structural Group created above, the process will involve simply swapping the file referenced by this Link. This is done in the Manage Links dialog.

21. Re-open the *06 Commercial.rvt* [*06 Commercial Metric.rvt*] file.

22. On the Insert tab, click the ***Manage Links*** button.

   - In the "Manage Links" dialog, be sure that the Revit tab is active.
   - Click on the *Structure* entry at the left.
   - Click the "Reload From" button at the bottom (see Figure 6.47).

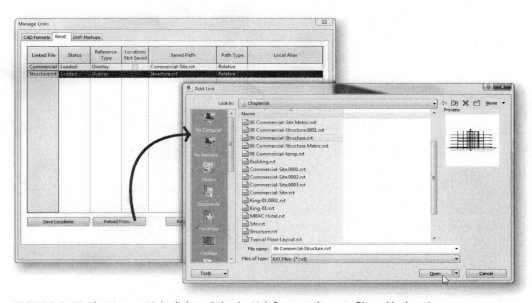

FIGURE 6.47 *The Manage Links dialog—Reload a Link from an alternate file and/or location*

23. Select *06 Commercial-Structure.rvt* [*06 Commercial Metric-Structure.rvt*] and then click Open.

   - Click OK to close the Manage Links dialog.

In this example, we used "Reload From" because we wanted to point the linked file to a different project file than the one originally used. In normal circumstances, you will want to use the Reload button that simply loads the latest saved changes from the same RVT file. If you no longer want the Linked file, you can use the Remove button.

MANAGER NOTE

If you try to open a project that is actively linked by the currently open project, Revit will display a warning indicating that the Linked file must first be unloaded before it can be opened. In other words, you cannot have both projects open at the same time in the same Revit session. This limitation does *not* prevent two different users from working simultaneously in each of the projects since each team member will be working on a different system. However, if both users are actively changing their respective models, you should save frequently and use the "Manage Links" dialog periodically to reload the linked project(s).

The best place to see the results of the work we have done here is in one of the section views. Open a section view and select the linked structural file. (Use the TAB key if necessary.) You will see it highlight on screen and be able to see clearly the elements that have been moved and copied to this linked file in comparison to those that remain in the host commercial project.

If you should ever wish to "turn off" a linked file, you can unload it. Do this with the *Manage Links* tool. When a linked RVT file is unloaded it will not be visible in any view. At a later point in time it can be reloaded and therefore visible. Unloading it also has the benefit of removing the data from the computer's memory, which reduces the burden on the computer's resources. If you wish to try it out, follow these steps:

24. On the ribbon, click the Manage tab.

- On the Manage Projects panel click the *Manage Links* button.

TIP — The Manage Links button can also be found on the Insert tab.

- Click the Revit tab, select the *06 Commercial-Structure* [*06 Commercial Metric-Structure*] file, and then click the Unload button.

NOTE — Revit will warn you that this cannot be undone. While this is true, all it really means is that the undo command will not work with this action; you can always return to the Manage Links dialog and reload it.

- When asked to confirm, click Yes and then click OK.

In the next chapter, we will continue to refine the Commercial project and its linked files. For now however, our work with Groups and Links in this chapter is complete.

25. Save and close all project files.

SUMMARY

- Groups offer a powerful means to create and manage typical design conditions and keep all instances of them coordinated throughout the project.
- Any selection of objects can be grouped. Model element and Annotation elements however cannot occupy the same Group.
- When creating a Group from a selection of Model elements and attached Annotation elements (such as Tags) the Annotation becomes a separate Attached Detail Group.

- Attached Detail Groups contain annotation that is linked to the model elements in the corresponding Model Group.

- The insertion point of a Group can be adjusted by dragging the control on screen. Subsequent instances of the Group will insert relative to this location.

- When you edit any instance of a Group, the changes are applied to all instances.

- Elements can be excluded from individual instances of Groups making a unique condition. Changes to the Groups still apply to the other elements in the Group.

- Groups can contain other Groups creating so-called "nested" Groups.

- Groups can be saved to files thereby becoming independent Revit projects.

- Groups can be converted to linked Revit projects.

- A separate Revit project can be linked to your current project. If the external project changes, the link will update to reflect the changes.

- Linked projects can be converted to Groups.

- You can link AutoCAD or Microstation files to Revit projects.

- Contour lines in linked DWG files can be used to create points in a Revit Toposurface element.

- Maintain proper positioning of linked projects using Shared Coordinates.

- Project Base Points and Survey Points give on-screen icons for easy understanding and manipulating of project coordinates.

- Use Copy/Monitor to copy certain kinds of elements from a linked file and keep the copies synchronized with the original linked file.

- You can swap instances of Groups and Links with other Groups or Links respectively.

Vertical Circulation

INTRODUCTION

In this chapter, we will look at Stairs, Railings, and Ramps. We will explore these elements in both of our projects. The Residential Project contains existing Stairs on the interior and an existing exterior Stair at the front entrance. The core plan of the commercial building will include Stairs, Railings, and Elevators. The exterior entrance plaza leading up to the commercial building calls for Stairs, a Ramp, and Railings.

OBJECTIVES

We will add Stairs for the existing conditions of the residential project and lay out the core for the commercial project. Our exploration will include coverage of the Stairs, Railings, Ramps, and Elevators. After completing this chapter you will know how to do the following:

- Add and modify Stairs
- Add and modify Railings
- Add and modify Floors and Shafts
- Add and modify Ramps
- Add Elevators

STAIRS AND RAILINGS

Revit Stairs are "sketch-based" objects. Sketch-based objects require that you sketch out their basic form with simple 2D sketch lines. This linework is then used to generate the 2D and/or 3D form of the final object. We have seen other examples of sketch-based objects such as the Roof and Floor Slabs in the previous chapters. Stairs (like Walls) are a System Family. This means that the Stair Family is predefined in the software and is therefore part of the original template from which our project was built. The Stair Family includes a couple basic Types to get us started. If we want to use a Stair Type that is not included in the original template, either we have to create a duplicate of one of the existing types within the project, or we have to import a Type from another project (Transfer Project Standards or Copy and Paste can be used for this purpose). Stairs, like most Revit elements, have both instance and type parameters. Type parameters include riser and tread relationships, stringer settings, and basic display settings. Width and height parameters and clearances belong directly to the Stair object (instance parameters). Railings are separate elements but the Stair element will create Railings by default. Railings can also be edited or added independently later. For situations where the Railing is not required, we simply delete the auto-created one after we build the Stair.

Install the Dataset Files and Open a Project

The lessons that follow require the dataset included on the Aubin Academy Master Series: Revit Architecture student companion. If you have already installed all of the files from this site, skip to step 3 to begin. If you need to install the files, start at step 1.

1. If you have not already done so, download the dataset files located on the CengageBrain website.

    Refer to "Accessing the Student Companion site from CengageBrain" in the Preface for information on installing the dataset files included in the Student Companion.

2. Launch Autodesk Revit Architecture from the icon on your desktop or from the *Autodesk > Autodesk Revit Architecture 2012* group in *All Programs* on the Windows Start menu.

You can click the Start button, and then begin typing **Revit** in the Search field. After a couple letters, Revit Architecture should appear near the top of the list. Click it to launch to program.	**TIP**

3. On the QAT, click the Open icon.

The keyboard shortcut for Open is CTRL + O. **Open** is also located on the Application menu.	**TIP**

- In the "Open" dialog box, broswe to the location where you installed the *MasterRAC 2012* folder, and then double-click the *Chapter07* folder.

4. Double-click *07 Residential.rvt* if you wish to work in Imperial units. Double-click *07 Residential Metric.rvt* if you wish to work in Metric units.

    You can also select it and then click the Open button.

Add a Stair to the Residential Plan

This is the Residential project that was begun in Chapter 3. (We added some footings to it in Chapter 5, and the second floor existing conditions have also been laid out for you. You can find an additional exercise for the layout of the second floor in Appendix A. Feel free to complete that exercise first before continuing if you have not completed it already.) In Chapter 3, you may recall that we completed the existing conditions on the first floor without adding any Stairs. We will add Stairs to the Residential project in a few locations.

1. On the Project Browser, double-click to open the *First Floor* plan view.

An additional view of the first floor named *First Floor (Chapter 3)* also appears in the Project Browser. This view contains the dimensions that were used in Chapter 3 to set the locations of the Walls. While you can delete the dimensions without removing the associated constraints, it is more convenient to modify those constraints later if required if you do not delete them now. The recommended approach (used here) is to duplicate the view and apply dimensions to the copied view. Think of this as a "working view" as opposed to the original from which it was copied. The original is more likely to be presented and printed on Sheets; the "working" view simply provides a convenient place for us to work. Remember, in Revit views are live windows looking in on the total model. Edits you make to model geometry in one view affect all other appropriate views automatically. Therefore, even though the dimensions appear in only one of the floor plan views, the constraints apply to the elements in the model in *every* view.

We'll start with a very simple Stair—the existing front entrance stairs.

2. Right-click and choose **Zoom In Region**.

TIP

The keyboard shortcut for Zoom In Region is **ZR**. **Zoom In Region** is also located on the Navigation Bar.

- Click and drag a region around the front door at the bottom of the plan (see Figure 7.1).

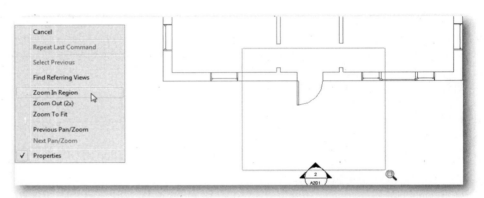

FIGURE 7.1 *Zoom in on the front entrance*

3. On the Home tab, on the Circulation panel, click the ***Stairs*** tool.

   The view window will gray out and the Modify | Create Stairs Sketch tab will appear on the ribbon.

Stairs have two sketch creation modes: "Boundary and Riser" and "Run." (Note that Boundary and Riser are two different buttons that are used together as parts of the same sketch.) You basically sketch out the plan view of the Stair in simple 2D linework and when you finish the sketch, Revit creates a Stair element from it that includes all of the correct parameters. A Stair sketch requires at least two boundary lines (representing the edges or stringers of the Stair) and several riser lines. If you use the "Run" option, you simply draw the path of the Stair, and Revit will automatically create the required Boundary and Riser lines. If you prefer, you can choose either the "Riser" or "Boundary" mode to draw these items manually. In general, you can create most common stair configurations including straight runs, U-shaped, L-shaped, etc. with the "Run" option. Use the other modes when you wish to modify the automatically created graphics to add a special feature to the Stair or when you want to build a sculptural Stair. We will explore these options later in the tutorial. For now, let's stick with the "Run" option (see Figure 7.2).

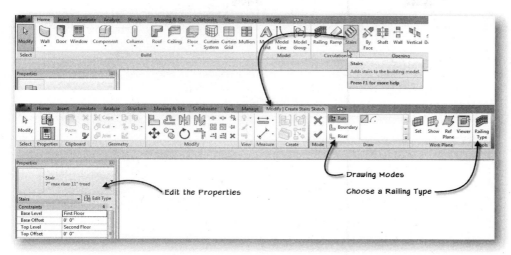

FIGURE 7.2 *The Sketch-mode Design Bar for Stairs contains some unique tools*

Also like other sketch-based Revit elements, you can use the Properties palette to access the instance and type parameters of the Stair that you are sketching. Another useful button (on the Tools panel) is the "Railing Type" button. When you create a Stair, Revit automatically adds a Railing, which is a separate element. Most stairs require railings, so Revit Architecture adds them automatically to save you time. Use the ***Railing Type*** tool to choose the Railing Type you wish it to add, or if you prefer, choose "None" to instruct Revit not to add a Railing. If you forget to use this tool to choose a Railing, you can always select the Railing(s) in the model window and edit it or delete it later.

- On the Draw panel, verify that the ***Run*** tool is selected.
- On the Properties palette, from the Type Selector (at the top) choose **Monolithic Stair**.
- Beneath the "Constraints" grouping, for the "Base Level" choose **Site**, and for the "Top Level" choose **First Floor**.
- Beneath the "Graphics" grouping, uncheck all the boxes.

These boxes control where and when arrows and annotation will appear indicating the direction of a Stair in plans. In this case, this is a simple existing entrance stair. It does not require any graphics in plan.

- Beneath the "Dimensions" grouping, set the "Width" to **4'-0" [1200]**.
- Beneath the "Phasing" grouping, from the "Phase Created" list, choose **Existing** (see Figure 7.3).

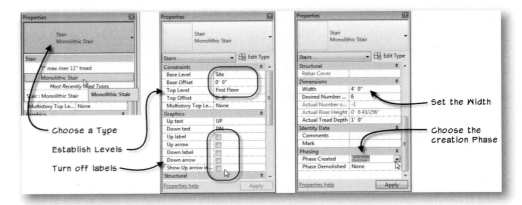

FIGURE 7.3 *Configure the Stair parameters*

4. Move your pointer to the middle of the front Door.

  • When the guideline appears at the midpoint, click the mouse to set the first point of the Stair run (see the left side of Figure 7.4).

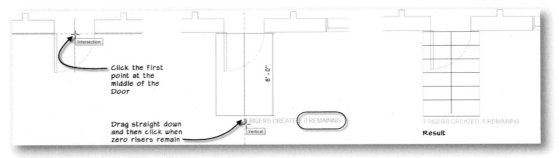

FIGURE 7.4 *Create a straight run of Stairs with two clicks*

When you create the Stair, the first point of the run is at the bottom of the run and the last point you click is at the top. If you click just two points, you get a straight run. As you drag the mouse (after the first pick) a note in grey text will appear indicating how many risers you have placed. If you click again before you use up all the risers, you create a landing. In this case, we only have 6 risers and want a straight run, so we will click only one more point.

  • Move the pointer straight down and when the number of risers reads "0 Remaining" click the second point (see the middle of Figure 7.4).

A sketch of black, blue, and green lines will appear indicating the Stair (see the right side of Figure 7.4). The green lines are the boundary lines of the Stair, the blue line is the Stair path and the black lines are the risers. If necessary, you can edit this sketch before finishing the sketch. In this case, we will leave the sketch as is. However, let's modify the Railing before we finish the sketch.

5. On the Tools panel of the ribbon, click the ***Railing Type*** button.

  • In the "Railings Type" dialog, choose ***Handrail – Rectangular*** [***900mm***] and then click OK.

6. On the ribbon, click the **Finish Edit Mode** button (large green checkmark).
7. On the Project Browser, double-click to open the *East* elevation view.

Notice that the Stair is oriented the wrong way. This is easy to fix (see the left side of Figure 7.5).

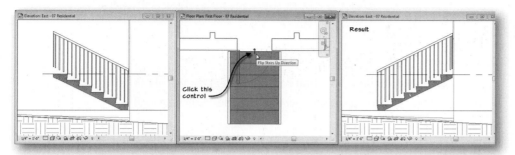

FIGURE 7.5 *Flip the Stair to orient it properly*

8. On the Project Browser, double-click to open the *First Floor* plan view.

  • Select the Stair.
  • Click the small flip control at the top of the Stair.

    Note the change in the elevation view (See Figure 7.5).

You can view it in other views as well, to see that it was created properly. Try any section, elevation, or 3D view for this.

Modify the Railing

Sometimes when you create an element, it looks fine in plan view, but then you look at it in another view and notice that something is not correct. This is the case here.

9. Return to the *East* elevation view.

  • Use Zoom In Region again, to zoom in on the Stair (see Figure 7.6).

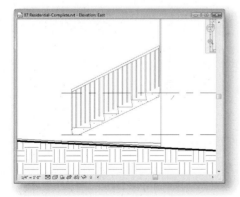

FIGURE 7.6 *Take note of the balusters in the elevation view*

Notice that the balusters do not attach to the Stair treads. There is an easy fix for this.

10. On the Project Browser, double-click to open the *First Floor* plan view.

- Move your mouse over the Stair to pre-highlight it, and then over one of the Railings.

Notice that the Railing is actually next to the Stair rather than overlapping it. This is why the balusters terminate the way they do. If we flip the Railings to make them sit on top of the Stair, then the balusters will project down to the treads properly.

11. Select one of the Railings.

- Click the Flip Railing Direction control. Repeat for the other Railing.
- Return to the *East* elevation view to verify the result (see Figure 7.7).

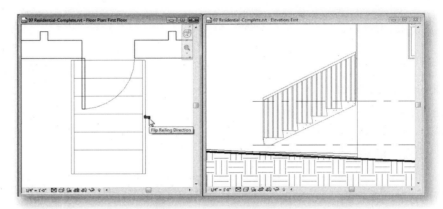

FIGURE 7.7 *Flip the Railings to make the balusters interact properly with the Stair*

12. Select both Railings and on the Properties palette, set their Phase Created parameter to **Existing** and then click Apply.

Modify the Toposurface

As we continue to view our Stair in the *East* elevation view, we discover that there is another perhaps even more obvious issue; the Stair does not actually sit on the grade. It appears to "float." We could edit the Stair and adjust it accordingly (using Base and Top Offsets), but upon further consideration, it is typically not desirable to have the terrain slope toward the building structure. Therefore, a better approach in this situation would be to adjust the Toposurface to provide better grading.

13. On the Project Browser, double-click to open the *Site* plan view.

- Select the Topography: Surface element.
- On the Modify Topography tab, click the **Edit Surface** button.

If you receive a warning about the possibility of points not being visible due to the current view's clipping, you can ignore it as it will not affect the edits that you need to make.

14. On the Tools panel, click the **Place Point** tool.

  • On the Options Bar, type **-3'-1" [-940]** in the Elevation field.
  • Snap a point to each of the lower corners of the building and to the two lower corners of the Stair (see Figure 7.8).

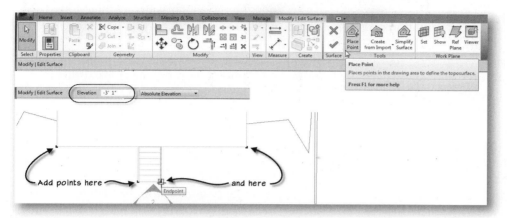

FIGURE 7.8 *Add points along the front of the house and Stairs*

15. On the ribbon, click the **Finish Surface** button (large green checkmark).
16. On the Project Browser, double-click to open the *East* elevation view.

Zoom in again if necessary. Notice that the Stair is now touching the ground instead of floating in space and the terrain now has positive drainage at the front of the house. It will be easier to see this result if you change the view to Shading (see Figure 7.9).

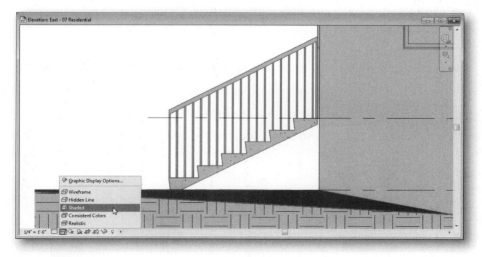

FIGURE 7.9 *After the edit, the Stair sits properly on the grade*

17. Save the project.

CREATE A GRADED REGION INSTEAD

In this particular project, we are still modeling existing construction, so the procedure covered here is appropriate since it is likely that the topography was properly graded in the existing conditions. However, if you wish, you can treat the grading as a new improvement, rather than an existing condition and use the *Graded Region* tool in this situation instead. The *Graded Region* tool works with the project's phases by "demolishing" the original Toposurface and then creating a new one assigned to the current phase in its place. For this to work properly, your existing Toposurface must be assigned to the previous phase. When you edit the grading of the new Topography element, Revit is then able to calculate cut and fill volumes in your project.

If you wish to try this, undo the previous steps. The Toposurface element is already assigned to the Existing phase in the dataset. Click on the Massing & Site tab of the ribbon and then click the *Graded Region* tool. The "Edit Graded Region" dialog will appear, offering you two options. The first option is useful in situations like this where there is a small amount of new grading. You will get an exact copy of the Topography, which you can then edit. The second option is useful when you are doing major site work. In that case, only the perimeter is copied and you re-grade the entire site. Choose the first option and then follow the same steps outlined in the "Modify the Toposurface" topic. When you finish the surface, you will have two topography elements: the existing one and a new one with the new grading assigned to the New Construction phase.

To see the cut and fill calculations, select the Toposurface element and edit its properties. The cut, fill, and Net cut/fill will be listed on the "Other" grouping. You can also create a Schedule for Topography elements and include the cut and fill fields there.

Copy and Demolish a Stair

At the back door of the existing house is another Stair like the one in the front. However, this one will be demolished to make way for the new addition. Feel free to create a new Stair using the procedure covered above. However, another way to create this Stair is to copy, rotate, and modify it from the one we just created at the front.

18. On the Project Browser, double-click to open the *First Floor* plan view.
19. Select both the Stair and its Railings. (Use a crossing selection box or the CTRL key.)

    - On the Modify tab, click the **Copy** tool.
    - Click any start point and then click to place the copy near the back of the house within the space of the new addition (the exact location is not important yet).

20. With the Stair and Railings still selected, click the **Rotate** tool.

    - On the Options Bar, type **180** in the Angle field and then press ENTER.

21. Move the Stair (and Railings) so that it touches the house centered on the existing back Door (see Figure 7.10).

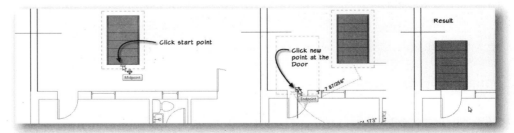

FIGURE 7.10 *Copy, Rotate, and then Move the Stair and Railings into place at the back Door of the existing house*

22. Click the **Modify** tool on the ribbon or press the ESC key twice.

You can use other methods to achieve the same end result. For example, we could have used object snaps when copying the Stair to snap it directly into the correct place, and then use the flip control as we did above to flip the Stair instead of rotating it. You can also do the rotation first with the "Copy" option on the Options Bar to copy and rotate it in one step in the original location, and then move the copy to the new location with object snaps. The exact procedure you follow is not important as long as you arrive at the desired result.

23. Double-click the blue vertical Section Head (cutting through the new addition) or open the *Transverse* section view from the Project Browser.

You will note, similarly to above, that the Stair floats above the terrain at the back of the existing house. This is because the grade slopes from the front of the house down toward the back. So by the time we reach the back of the house, the Stair requires a few additional risers.

24. On the ribbon, click the Modify tab.

25. On the Measure panel, click the **Measure Between two References** tool.

- Snap to the bottom corner of the Stair for the first point.
- Click on the terrain directly below the Stair for the second point (see Figure 7.11).

Change the view to Wireframe before measuring to more accurately measure to the topography.	**TIP**

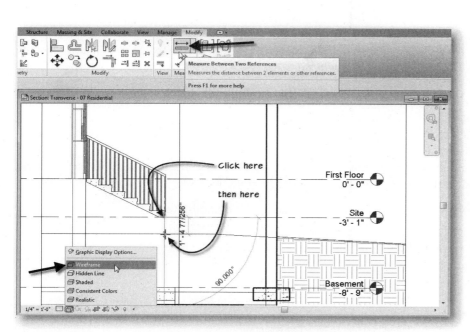

FIGURE 7.11 *Measure the distance between the bottom of the Stair and the terrain*

The distance rounded off should be about 1'-4" [410]. We will use this value to adjust the Stair height, which will add risers to it.

26. Select the Stair, and then on the Modify | Stairs tab, click the **Edit Sketch** button.

A dialog will appear indicating that you need to choose a view better suited to the edit than the currently active section view. The sketch was created in a plan view and it would be difficult or impossible to edit it from a section view.

27. From the "Go To View" dialog, choose *Floor Plan: First Floor* and then click Open View.

- On the Properties palette, In the Base Offset field, type **-1'-4" [-410]**.
- Change the "Desired Number of Risers" to **8**.
- Change the Phase Created to **Existing**.

Notice the gray message that indicates that 2 Risers are remaining (see the left side of Figure 7.12).

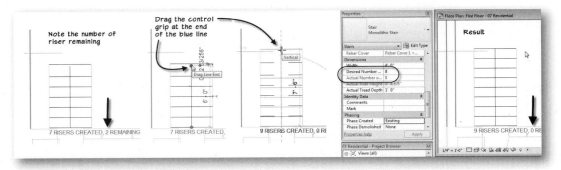

FIGURE 7.12 *Add two new risers to the Stair sketch*

28. Click on the blue path line in the center of the Stair sketch.

- Drag the control grip at the top and release it when the temporary dimension reads 7'-6" [2250] (see the right three panels of Figure 7.12).
- On the ribbon, click the **Finish Edit Mode** button.

29. Click the **Modify** tool or press the ESC key twice.
30. Click on the railings and on the Properties palette verify that Phase Created is set to Existing.

If you still have the section view from above set to wireframe, you should see that the bottom of the Stair now extends down to grade level.

31. On the Manage tab's Phasing panel, click the **Demolish** tool.

- Click on the Stair and each of the Railings to mark them as demolished in the current phase. (You can do this in any view.)
- Click the **Modify** tool or press the ESC key twice.

NOTE As an alternative, you can also select the Stair and Railings, edit their Properties, and then change the "Phase Demolished" setting to New Construction.

The Stair and Railings will turn dashed to indicate that they are to be demolished.

Create an Existing Conditions View

The next Stair that we will build is the main stair in the existing house. Before we do, let's explore a technique to make working with phases a bit simpler. In Chapter 3, we temporarily changed the current Phase of the *First Floor* plan view (named *Level*

1 in that chapter) to "Existing" to add all of the existing construction. At the end of the lesson, we set the Phase back to "New Construction." We could repeat that process here. As an alternative approach, it is easier to maintain two first floor plan views—one set to the Existing phase, and the other set to the New Construction phase. In this way, you can simply open the view for the phase in which you wish to work. This will prevent much of the manual phase assignments like we have done on the Stairs so far.

32. On the Project Browser, right-click the *First Floor* plan view and choose **Duplicate View > Duplicate**.

    • Right-click the new *Copy of First Floor* view and choose **Rename**.

    • For the new Name type **Existing First Floor** and then click OK.

      The *Existing First Floor* view should have opened automatically (it should be bold on the Project Browser). If it did not, double-click to open it now.

    • Make sure there is nothing selected, the Type Selector on the Properties palette should read "Floor Plan."

    • On the Properties palette, from the Phase list, choose **Existing**.

This will set the current Phase of the copied view to Existing. This means that any elements we add while working in this view will be assigned to the Existing construction phase automatically. We did not *need* to create this view, but sometimes it is easier than remembering to set the Phase parameter of each object we create or changing the Phase of the view back and forth. With Existing set as the current Phase, notice that the rear Stair is no longer dashed. It is still set to be demolished, but since we are now viewing the plan as it looks during the Existing Phase, these Stairs are not yet demolished at this point in time. The Walls of the addition have also disappeared. Again, during the Existing Phase, those Walls do not yet exist. Remember, as we saw in Chapter 3, when you set a view to a different phase, the relative "definition" of "Existing," "New" and "Demolished" shift accordingly.

33. Double-click to open the *First Floor* plan view.

Notice that this view, still set to New Construction Phase, continues to show all of the new construction Walls and the demolished Stair.

34. Double-click to open the *Existing First Floor* plan view.
35. Save the project.

Create a Straight Interior Stair

The existing interior Stair is also a simple straight-run stair going from the first to the second floor. However, there are a few custom details on the first few treads and some Railing treatments that will make the Stair a bit more interesting.

36. Working in our newly created *Existing First Floor* plan view, Zoom and Pan to the middle of the existing house layout.
37. On the Home tab, on the Circulation panel, click the *Stairs* tool.

    We return to sketch mode and the Modify | Create Stairs Sketch tab appears again on the ribbon.

Sometimes it is difficult to set the precise location of the element that you are placing. In the case of Stairs, the Run option adds the Stairs from the midpoint at the bottom of the run. We do not currently have a convenient point to snap to at that location. We could simply draw the sketch at a random location and then move it into position relative to the Walls. In many cases, you will find this to be the easiest approach. An alternative approach that we will explore here is to use Reference Planes. Reference Planes are simple datum elements. They share many common characteristics with the Levels and Grids we explored in previous chapters, but serve less specific functions. You can add them anywhere that you need assistance in achieving alignments or reference points. Think of them as analogous to blue pencil lines in hand drafting.

- On the Work Plane panel, click the **Ref Plane** tool and then from the Draw panel choose the **Pick Lines** tool.
- On the Options Bar, in the Offset field, type **1'-6 ½" [471]**. (This is half the width that the Stair needs to be.)
- Highlight the right side of the vertical Wall on the left of the stair corridor. When the dashed guideline appears in the middle of the stair space, click to create the Reference Plane (see the left side of Figure 7.13).

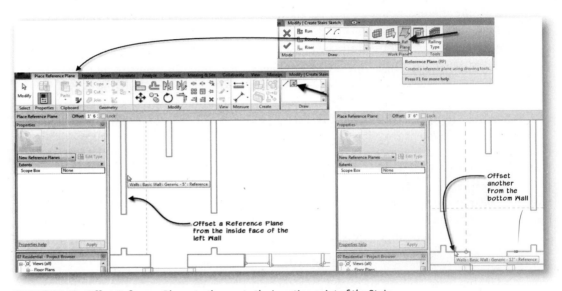

FIGURE 7.13 *Offset Reference Planes to the create the insertion point of the Stair*

38. On the Options Bar, change the Offset parameter to **3'-6" [1100]**.

- Offset a Reference Plane up from the inside face of the bottom exterior Wall (see the right side of Figure 7.13).
- Click the **Modify** tool.

The way in which Reference Planes are presented here is one acceptable use for them. However, they are also part of the constraint mechanism in Revit. This means that they are very similar to Levels & Grids (see previous chapters). If an element in the model is constrained to a Reference Plane, the location of the Reference Plane will "drive" the position and/or shape of the constrained geometry. This important feature can be exploited directly in the project, but is most often seen in the Family Editor. Detailed coverage of the Family Editor and Reference Plane usage within Families can be found in Chapter 10.

39. On the Modify | Create Stairs Sketch tab, on the Draw panel, click the **Run** tool.

On the Properties palette, notice that Revit remembers the settings of the previously created Stair. Several settings must be adjusted to make this Stair match the interior Stair occurring in the existing house.

- In the Properties palette, from the Type Selector at the top, choose **7" max riser 11" tread** [**190mm max riser 250mm going**].
- On the Properties palette, verify that the "Base Level" is set to **First Floor** and the "Top Level" is **Second Floor**.
- Beneath the "Graphics" grouping, place checkmarks in only the "Up label" and "Up arrow" boxes; be sure the others are cleared.
- In the "Dimensions" grouping, change the Width to **3'-1"** **[942]** (see Figure 7.14).

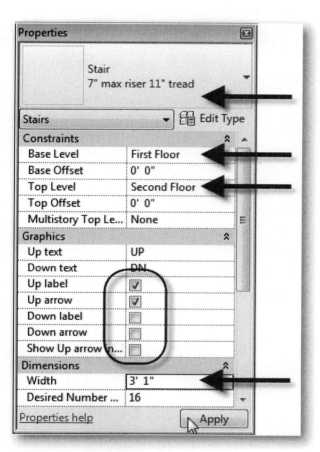

FIGURE 7.14 *Set up the parameters for the Stair*

40. Click the first point of the Stair run at the intersection of the two Reference Planes.

- Move the pointer straight up far enough to place all of the risers in a single run and then click again.

The sketch created will be too long for the space. The parameters configured above were all "instance" parameters. Instance parameters apply directly to each individual object in the model. To adjust our Stair to fit the space, we need to modify the "type" parameters. Type parameters apply to all objects sharing the same type. We can edit them directly, but any changes would affect all Stairs in the model with type: **7" max riser 11" tread** [**190mm max riser 250mm going**]. A more prudent practice is to first create a new type and then adjust its parameters as appropriate. In this way, we preserve the existing type for use on other Stairs while creating a variation suited to the Stair we are currently creating.

Create a New Stair Type

In this sequence, we will create a new Stair type and edit the parameters to help it conform to the existing conditions. This will involve applying less stringent tread and riser rules on our Stair. Since the stairs existing in this house were built before current code requirements were in place, an alternate Stair Type needs to be created allowing more freedom.

41. On the QAT, click the **Undo** icon (or press CTRL + Z).

This will remove the sketch so that we can adjust the parameters and re-create it.

42. On the Properties palette, click the Edit Type button.

    - Click the Duplicate button.
    - In the "Name" dialog, type **Existing House Stair** and then click OK.

Here you can enter a range of values for minimum and maximum tread and riser. You can also enter building code values for your jurisdiction in a rule-based calculator (click the Edit button next to "Calculation Rules" to do this). Values assigned to these rules will constrain the parameters of the Stair as it is being sketched. Again, because the Stair we are building is an existing Stair, we will set the limits very broadly so that we can enter actual values found in the field without restriction.

43. Scroll down to the "Treads" grouping, type **8" [200]** for the "Minimum Tread Depth."

    - From the "Nosing Profile" list, choose **Default**.
    - Beneath the "Risers" grouping, type **12" [300]** for the "Maximum Riser Height" (see Figure 7.15).

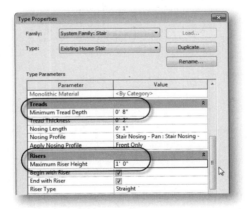

FIGURE 7.15 *Edit the Type parameters of the new Stair type*

44. Click OK to accept the values.

    - On the Properties palette, beneath the "Dimensions" grouping, verify that the Width is **3'-1" [942]** and the Actual Tread Depth value is **8" [250]**.
    - Set the Desired Number of Risers parameter to **14** (see Figure 7.16).

FIGURE 7.16 *Set the actual Tread and Riser settings of the existing house Stair*

45. On the Draw panel, click the **Run** tool.

- Click the first point of the Stair run at the intersection of the two Reference Planes.
- Move the pointer straight up far enough to place all of the risers in a single run and then click again.

Notice that the Stair sketch now fits the available space (see Figure 7.17). As an alternative to undoing the first Stair and redrawing it, you could grip stretch the blue path line like we did previously. It is a little tricky, however, to get the Stair to rescale properly when using this method. Feel free to try it out if you wish.

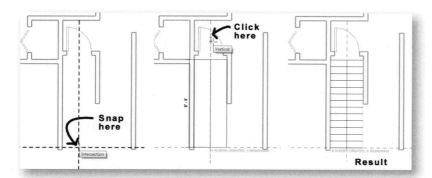

FIGURE 7.17 *The Stair sketch based on the newly created Stair Type properly fits the space*

46. On the ribbon, click the **Finish Edit Mode** button.

The Stair will appear with its railings. Notice that a cut line appears automatically. Also, since we asked only for "up" annotation, we only get an arrow and label pointing up, not up and down. We will add the Stair going down to the basement later.

Edit the Stair Type

When you look at the Stair we created, it appears to be too wide—part of it overlaps the neighboring Walls. This overlap is actually the stringer of the Stair and the Railing that sits on top of the stringer. We will need to make some adjustments to the Stair Type to address this.

47. Select the Stair.

- On the Properties palette, click the Edit Type button.
- Beneath the "Stringers" grouping, choose **Open** for both the Left Stringer and the Right Stringer.
- For the Trim Stringers at Top option, choose **Match Level**.

An "Open" stringer is notched in the shape of the treads and risers and supports the treads from underneath. The "Closed" stringer occurs at the edges of the Stair with the treads and risers spanning in between. A typical wooden stair would use an open stringer, while a steel pan stair typically would be closed. The tops of the stringers can be cut to match the Level or they can be left uncut. Illustrations of several of stringer settings are shown in Figure 7.18.

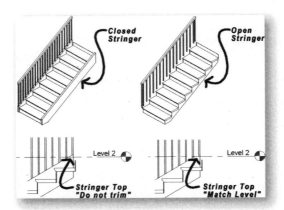

FIGURE 7.18 *Understanding some of the Stringer settings*

In addition to these settings, we can also edit the dimensions of the stringer material such as the height and thickness. In this case, we will leave those settings as they are.

48. Click OK to apply the changes and return to the model.

It appears as though nothing has changed in our plan view. However, if you slowly pass your mouse over the Stair and its edge, you will note that when the Stair pre-highlights, it has indeed reduced in width. The overlapping portion that we still see is actually the Railing as on the front door Stair we worked with earlier.

49. Click on one of the Railings.

- Click the small "Flip Railing Direction" control to flip the Railing to the other side of the stringer.
- Repeat on the other side. (This is similar to the situation pictured above in Figure 7.7.)

Create the "Down" Stair

In addition to the Stair we have added here, the house also has an existing Stair going down to the basement. The simplest way to create this one is with Copy and Paste.

50. On the ribbon, click the View tab and then on the Create panel, click the **Section** tool.

- Create a section line running vertically through the Stair. (Click from bottom to top.)
- Deselect the Section Line (click away from it) and then double-click the Section Head to open the associated view.

In this view, you can see very clearly that we have a single Stair spanning from first to second floor.

51. Select the Stair.

- On the Modify Stairs tab, on the Clipboard panel, click the **Copy** tool (or press CTRL + C). Make sure you click Copy on the Clipboard panel, *not* the Modify panel.
- On the Clipboard panel, click the **Paste Aligned** drop-down button and then choose the **Aligned to Selected Levels** tool.
- In the "Select Levels" dialog, choose Basement and then click OK (see Figure 7.19).

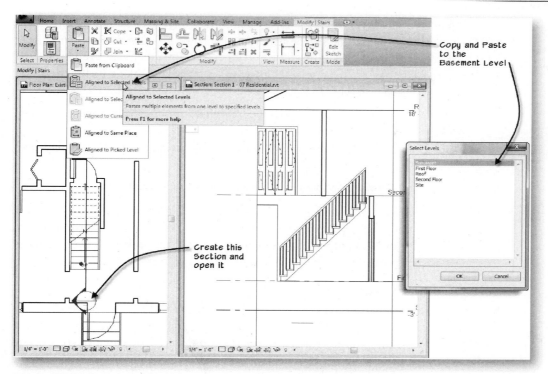

FIGURE 7.19 *Add a section view and then copy and paste the Stair to the Basement Level*

52. With the new Stair still selected, on the Properties palette verify that the "Base Level" is Basement and that the "Top Level" is First Floor.

When you paste a Wall, Stair, or Column to a different level like this, Revit will adjust the base and top offsets to the new levels, but will try to maintain the original height of the element by adding a Top Offset. In this case, the height between the Basement and First Floor is 3" [150] shorter than the height between the First and Second floors.

- Remove the "Top Offset" by typing a value of **0** (zero) (see Figure 7.20).

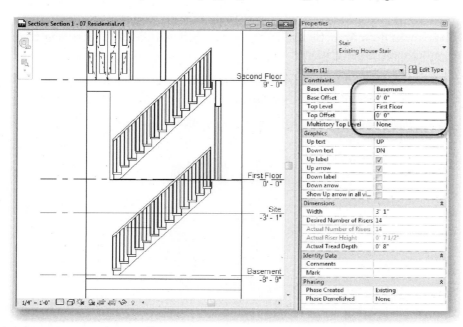

FIGURE 7.20 *Set the Top Offset back to zero*

53. On the Project Browser, double-click to open the *Existing First Floor* plan view.

54. The basement Stair should still be selected. (If you need to reselect it, you may need to use the TAB key to select the correct stair because the other stair overlaps.)

    You can verify that you have selected the correct stair by checking its Properties on the Properties palette for the Levels settings.

    • On the Properties palette beneath the "Graphics" grouping, clear the "Up arrow" and "Up label" checkboxes.

    • Place checkmarks in the "Down arrow" and "Down label" checkboxes and then click Apply.

55. On the Project Browser, double-click to open the *Basement* floor plan view.

Notice only the basement Stairs appear here. If you open the Second Floor plan view as well, you will note that only the upper Stair appears. However, since we have not yet added any Floor elements to our residential model, they will show even beneath the closet at the front of the house. We'll adjust this in later chapters.

Edit the Stair Sketch

So far we have used the "Run" option to automatically create very simple Stair sketches based upon our riser and tread dimensions. However, you can edit these simple sketches to add architectural detailing to your Stairs. You can also completely create your own custom sketch to form very complex Stair configurations. We'll look at such an example later. In this example, we will widen the lower portion of the Stair and add a bull nose to the bottom two treads.

56. On the Project Browser, double-click to open the *Existing First Floor* plan view.

    • On the View tab, click the **Close Hidden** tool. (This tool only works if the current view is maximized first.)

57. Select the basement Stair. (Use the TAB key if necessary to assist in selection.)

One way to be sure that you are selecting the correct Stair is to open the section view and tile it next to the floor plan. Select the Stair in section and then switch to the floor plan where it will still be selected. Another way to be sure is to remember that the upper Stair only has an "UP" label and the basement Stair only has a "DN" label. We want to temporarily hide the basement Stair to make it easier to work on the sketch of the main floor Stair. So use whatever method you prefer to be sure you are selecting only the basement Stair.

    • On the View Control Bar, click the Temporary Hide/Isolate icon (sunglasses) and choose **Hide Element** (see Figure 7.21).

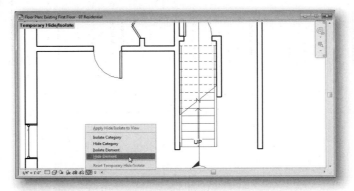

FIGURE 7.21 *Stairs include three types of sketch lines*

Temporary Hide/Isolate allows you to either hide selected elements or isolate the selected elements (hide anything not selected). This is a temporary mode for the active view window. While the mode is active, a cyan-colored border will surround the view window. New in this release, there is also a descriptive label that appears indicating the display mode. When you wish to restore the normal display settings of the view, choose the sunglasses icon again and then the Reset Temporary Hide/Isolate option.

58. Select the "Up" Stair. (With the other one hidden, the selection is now easy.)

    • On the Modify | Stairs tab, click the ***Edit Sketch*** tool.

This returns you to sketch mode and reveals the sketch lines created when we added the Stair. By adjusting these sketch lines, we can change the shape of the Stair. The Stair sketch uses three colors on screen to indicate the function of each line. Green indicates a boundary line. The Stair must have two boundary edges. The blue line at the center indicates the Stair path of a Stair sketch created with the ***Run*** tool. Editing this line has an effect on the overall Stair as we saw above. The black lines are the individual risers. A gray text note will also appear near the sketch indicating how many risers are created and required.

If you position your pointer over one of the black riser lines or green boundary lines to pre-highlight it, you will notice in the screen tip that appears that the sketch lines of the Stair are actually "Model Lines." There are two types of line elements in Revit: they are Model Lines and Drafting Lines. A Model line, while two-dimensional, is a part of the building model. Therefore, like other elements of the model, it will appear in *all* appropriate views. Imagine it as if the line was actually painted on the floor or wall much like the stripping on roads. A Drafting line on the other hand appears *only* in the view in which it is drawn. Model Lines are used to represent real things, while Drafting Lines are like annotation, used to embellish a particular view to convey intent. Model Lines are used in the Stair sketch because they indicate the actual form of the Stair in the model.

When you are working on a Stair sketch, there are three sketching tools on the ribbon: Run, Boundary, and Riser. We have seen Run already in each of the Stairs we built earlier. If you have placed all of the required risers, the Run option will be unavailable. To edit the existing run, you would edit the blue sketch line. When you wish to customize an existing Stair, or when you wish to draw a Stair free-form, you use the Riser and Boundary tools. In this case, we'll edit the boundary.

59. On the Modify | Stairs > Edit Sketch tab, on the Draw panel, click the ***Boundary*** tool.

    • On the Options Bar, place a checkmark in the "Chain" checkbox.
    • Click the first point of the new boundary Line at the intersection of the right-hand Boundary Line and the bottom of the existing Wall (see "Point 1" in Figure 7.22).
    • Click the next point at the other side of the Wall's width (see "Point 2" in Figure 7.22).
    • Place the last point aligned with the original boundary Line's bottom end and the Wall's right face (see "Point 3" in Figure 7.22).

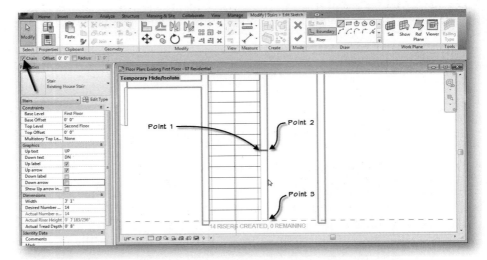

FIGURE 7.22 *Sketch two new Boundary Lines snapping to the existing Wall*

60. On the ribbon, click the ***Modify*** tool or press the ESC key twice.

- Use the ***Trim/Extend to Corner*** tool to remove the unneeded segment (see Figure 7.23).

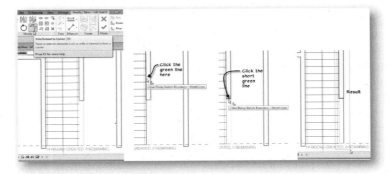

FIGURE 7.23 *Trim the existing Boundary Line to join it to the new sketch*

- On the Modify panel, click the ***Trim/Extend Multiple Elements*** tool.
- For the Trim/Extend boundary, click the newly drawn boundary Line (vertical one).
- Click each of the riser Lines to extend them to this boundary (see Figure 7.24).

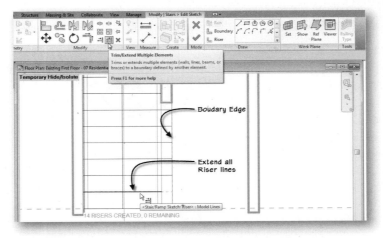

FIGURE 7.24 *Extend the riser Lines to the new boundary Line*

- Click the *Modify* tool or press the ESC key twice.

On the lower portion of the Stair, the risers will now come out flush with the Wall. Next we'll add the bull nose to the lower two steps. The first step we need to do is stretch the Boundary Line back a bit. Otherwise, the stringer will follow the shape of the bull nose.

61. Click the right Boundary Line (the new vertical one).

- Using the drag handle at the bottom, drag the endpoint up and snap it to the end of the third Riser Line.
- When the error dialog appears, click the Unjoin Elements button (see Figure 7.25).

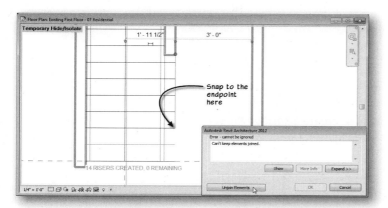

FIGURE 7.25 *Reduce the length of the boundary Line by two treads*

The boundary line is joined to each of the riser lines. This is required to create a valid Stair. When we stretch the boundary line as indicated above, it breaks this relationship with the lower risers. Before Revit will allow us to finish the sketch, we must join them back to the boundary. Next, we will edit the shape of the riser lines to make the bull nose condition which will also connect the riser back to the Boundary.

Add a Bull Nose Riser

To add the bull nose riser, we simply sketch them using the *Riser* tool on the ribbon.

62. Deselect the Boundary line and then on the Draw panel, click the *Riser* tool.

- On the Options Bar, clear the "Chain" checkbox.
- Click the "Start-End-Radius Arc" icon.
- Place the first point of the arc at the right endpoint of the bottom Riser Line (see "Point 1" in Figure 7.26).
- Place the second point at the endpoint of the third Riser line from the bottom (see "Point 2" in Figure 7.26).

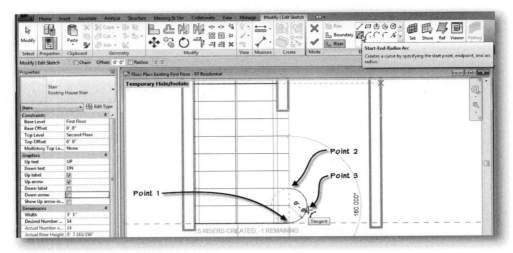

FIGURE 7.26 *Add an Arc to the lower Riser*

- Move the pointer to the right and when the shape snaps to a half-circle, click to place the last point (see "Point 3" in Figure 7.26).
- Repeat the process to create a smaller Arc (one tread deep) between the second and third risers (see Figure 7.27).

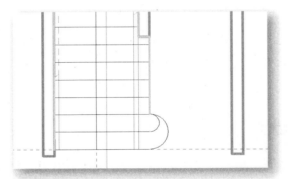

FIGURE 7.27 *Add an additional Riser Arc*

63. On the ribbon, click the **Finish Edit Mode** button.

View the Stair in 3D

At this point, it might be easiest to understand what we have created if we view the edited Stair three-dimensionally. We have used the "Default 3D View" icon several times already. However, this tool will show the entire model in 3D. With all of the exterior Walls visible, it will be impossible to see the Stairs inside the building from such a vantage point. There are a few tricks we can use to quickly view just the Stairs in 3D.

64. On the View tab, click the **Close Hidden** tool.

NOTE If this command is not available, then you either have only the current view window open or your view is not maximized. If this is the case, maximize the current view and then try again.

- On the Create panel of the View tab, click the **3D View** tool. (This is a split button; be sure to choose the top part.)

- On the View Control Bar, click the Visual Style pop-up and choose **Hidden Line**.
- On the View tab, on the Windows panel, choose the *Tile* button.

65. Click in the *First Floor* view window to make it active.

- Select the Stair, Railings, and their immediately adjacent Walls (see the left side of Figure 7.28).

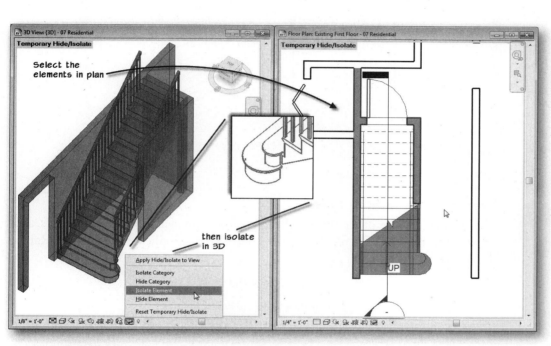

FIGURE 7.28 *Select elements in the plan view, isolate them in the 3D view*

- Click the titlebar of the 3D view window to make it active.
- From the View Control Bar in the 3D view window, choose **Isolate Element** from the Temporary Hide/Isolate pop-up icon (see the right side of Figure 7.28).

Feel free to use the ViewCube or the Steering Wheels controls to spin the isolated selection around to an alternative viewing angle. Notice the way the two treads at the bottom of the Stair have the bull nose applied to them. Notice also the way that the stringer stops before these two treads. These are direct effects of the way that we sketched the Stair above.

Adjust a Railing Sketch

Having completed the edit of the Stair above, it is now apparent that the Railing could use a bit of adjustment as well. You edit Railings the same way as Stairs—by editing the sketch.

66. In the plan view, zoom in on the Railing on the right side.

As you can see, this Railing has matched the shape of the Boundary Line that we sketched above and makes a slight jog as it goes up the run of Stairs. Let's eliminate the portion of the Railing above the jog (adjacent to the Wall).

67. Select the Railing on the right side (use TAB if necessary to select it. Select in either view).

- On the Modify | Railings tab, click the **Edit Path** button.
- Delete the top two sketch lines (the vertical one and the short horizontal one that are adjacent to the Wall).
- On the ribbon, click **Finish Edit Mode** (see Figure 7.29).

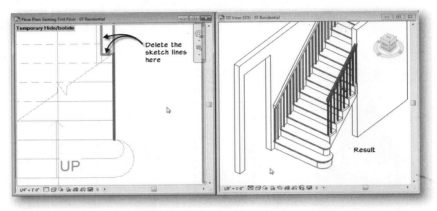

FIGURE 7.29 *Edit the Railing sketch*

Create a New Railing Type

The Railing on the other side of the Stair does not need balusters or posts. It is attached to the Wall. To represent this correctly, let's duplicate the existing Railing Type and then modify it.

68. Select the long Railing on the left side of the Stair.

- On the Properties palette, click the Edit Type button.
- In the "Type Properties" dialog click the Duplicate button.
- In the "Name" dialog, type **Existing House Railing – No Balusters** and then click OK.
- In the Type Properties dialog, beneath the "Construction" grouping, click the Edit button next to "Baluster Placement."
- In the "Edit Baluster Placement" dialog, in the "Main Pattern" area (at the top), choose **None** from the Baluster Family list for element 2 (see the top of Figure 7.30).

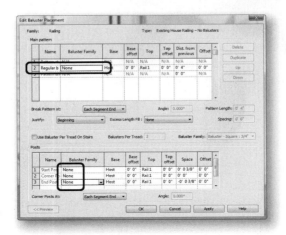

FIGURE 7.30 *Remove balusters and posts from the new Railing Type*

- In the "Posts" area (at the bottom), choose **None** for all three components (see the bottom of Figure 7.30).

- Click OK twice to return to the model window.

Import a New Railing Type

Railings (like Stairs, Walls, Floors, and Roofs) are System Families. System Families are so called because they are "built in" to the system. This means that the parameters, behavior, and overall geometric characteristics of System Families are predefined in the software and not editable by the user. In the case of Railings, there is a single Railing System Family. Each System Family can, however, have multiple Types. A Type is a collection of parameters that are applied to all elements assigned to the Type. If a Type parameter is modified, it affects all instances of that Type in the entire project.

The item we just created, ***Existing House Railing – No Balusters***, was a Type. In some cases, you will build a Type like this and later wish to use it in another project. However, unlike Component Families (Doors, Windows, Furniture, etc.) we cannot save and load Railings (or other System Families) in individual RFA files. You can however copy and paste them from other projects or use the Transfer Project Standards command (on the Manage tab) to copy Types between various projects. To illustrate this approach, files containing several Railing and Stair examples have been provided with the dataset files from the student companion. These files are also available on the Autodesk Seek website, which is an online library of free content. They are provided with the book dataset for your convenience. We will be exploring how to use Seek later in this chapter.

69. From the Application menu, choose **Open > Project**.

- Browse to the location where you installed the dataset files.

- Browse to the *MasterRAC 2012\Library\Stair Samples* folder.

70. Double-click the *Railing Samples* [*Railings*] project file to open it (see Figure 7.31).

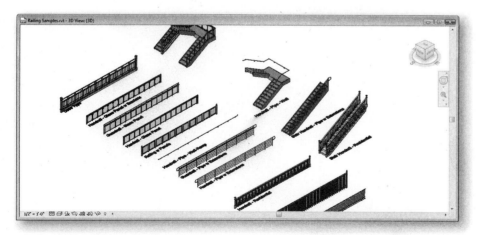

FIGURE 7.31 *Open the Railing Samples file*

When the file opens on screen, you will see a three-dimensional view that has several Railing samples placed in it. Each has a text label next to it indicating the name of the Railing Type. To use one of these Railings in your project, simply select and copy it to the Windows Clipboard from this project and then paste it into your project.

Zoom and pan around the file and locate the *Handrail – Residential [Residential Timber Newel and Spindles]* Type.

71. Select one of the *Handrail – Residential [Residential Timber Newel and Spindles]* Railings and then press CTRL + C.

- On the QAT, click the Switch Windows drop-down button and choose *07 Residential. rvt – Floor Plan: Existing First Floor* plan view from the list.
- Press CTRL + V and then in the dialog that appears, click OK.
- Click a point anywhere on screen to place the pasted Railing.
- On the Modify | Model Group tab, on the Edit Pasted panel, click the **Finish** button.
- Select the newly pasted Railing and then press the DELETE key.

You may be wondering why you should go through all the trouble to paste the Railing only to delete it. You must actually place the pasted element to import the Type into the project. Once you have done this however, you can safely delete the extraneous Railing instance without fear of deleting the newly imported Type. The *Handrail – Residential [Residential Timber Newel and Spindles]* Type is now part of the Residential project and we can assign it to our existing Railing element.

72. On the Project Browser, double-click to open the *{3D}* view (or use Switch Windows again).
73. Select the short Railing (the one with balusters).

- On the Properties palette, choose *Handrail – Residential [Residential Timber Newel and Spindles]* from the Type Selector.
- With the Railing still selected, choose **the Edit Type button on the Properties palette**.
- Edit the Baluster Placement and change the End Post to None.
- Click OK twice to return to the model and view the results (see Figure 7.32).

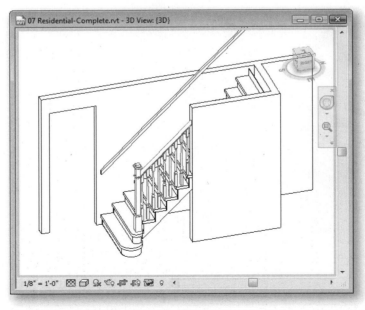

FIGURE 7.32 *After pasting in the new Railing Type, apply it to the Railing and remove the top post*

Examine the Railing in both 3D and plan after placement. It may be necessary to flip it to seat the balusters on the treads properly. You may also want to fine-tune the Railing sketch to get the look you like. Feel free to experiment with it further. For example, if you edit the path and stretch it slightly, you can get the post to sit comfortably on the tread with the small bull nose.

Adjust Wall Profile

Although our Stair looks fine, the 3D view reveals that the Wall on the right does not continue under the Stair.

> Continue working in the 3D view window. If you shaded the view, on the View Control Bar, choose **Hidden Line** from the Visual Styles pop-up.

The keyboard shortcut for Hidden Line is **HL**.	**TIP**

- On the ViewCube, click the Right face.

This will orient the view to look directly at the Wall. If the Railing is still selected, it will zoom in on it.

74. Select the Wall to the right of the Stair.

    It is to the right in plan; it is in front of the Stairs in the 3D view.

    - On the Modify | Walls tab, on the Mode panel, click the **Edit Profile** button.
    - Edit the sketch lines as indicated in the left two panels of Figure 7.33.

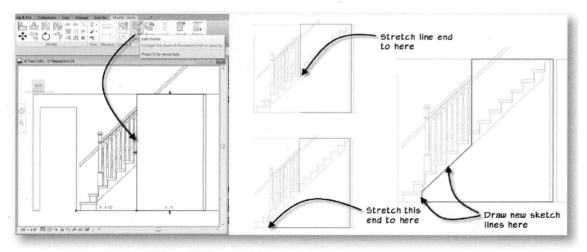

FIGURE 7.33 *Edit the sketch of the Wall profile to make it pass under the Stair*

- On the Draw panel, click the **Line** tool.
  On the Options Bar, verify that the "Chain" checkbox is selected.
- Draw two sketch lines to close the Wall profile shape (see the right panel of Figure 7.33). Clean up with Trim/Extend if necessary.

75. On the ribbon, click the **Finish Edit Mode** button.

    - Deselect the Wall.
    - Click the top-left corner of the ViewCube to orient the view to a southeast vantage point.

- From the View Control Bar in the 3D view window, use the pop-up icons to choose **Shading** and **Shadows On** (see Figure 7.34).

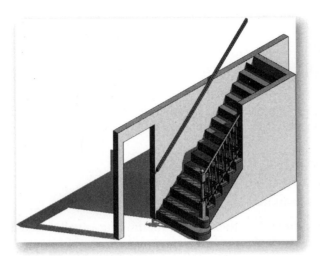

FIGURE 7.34 *The completed residential Stair hallway*

This view shows the results of our efforts nicely. Note, however, that the view window is still surrounded by a cyan border. This indicates that we are still in Temporary Hide/Isolate mode. When we close Revit, the mode will be reset automatically and all of the building geometry will redisplay in this view concealing the Stair again. If you like, we can make this display permanent in this view. Let's do that now.

76. From the Temporary Hide/Isolate pop-up, choose **Apply Hide/Isolate to View**.

- On the Project Browser, right-click on the {3D} and choose **Rename**.
- In the "Rename View" dialog, type: **3D Existing Stairway Hall** and then click OK.

We now have a permanent 3D view of the Stair Hallway. This completes our work in the residential project. We have plenty more vertical circulation explorations to make. We will now switch to the commercial project to continue those explorations.

77. Save and Close the Residential project.

COMMERCIAL CORE LAYOUT

While we have done much with the Stair tools in the residential project, there is still more we can explore. To continue our exploration of vertical circulation, we will switch to the Commercial Project. In this project, we will add a multi-story Stair and some elevators in the building core and we will add ramps to the front entrance of the building. We need to adjust a few issues with our linked structural file first.

Load the Commercial-Structure Project

Be sure that the Residential Project is saved and closed.

1. On the QAT, click the Open icon.

TIP	The keyboard shortcut for Open is CTRL + O.

- In the "Open" dialog box, browse to the *Chapter07* folder.

2. Double-click *07 Commercial-Structure.rvt* if you wish to work in Imperial units. Double-click *07 Commercial-Structure Metric.rvt* if you wish to work in Metric units.

> Please start with the files provided for Chapter 7 rather than attempting to continue from your own saved version from the previous chapter. A few changes have been included here that were not detailed in the previous chapter.

NOTE

A warning dialog will appear on screen alerting you that your file requires coordination review. This warning is informational and can be ignored.

3. Click OK to dismiss the warning dialog and finish loading the file.
4. On the Project Browser, double-click to open the *Level 2* floor plan view.

- Right-click in the workspace and choose **Zoom In Region**.

> The keyboard shortcut for Zoom In Region is **ZR. Zoom In Region** is also located on the Navigation Bar.

TIP

- Drag a box around the core to zoom in on it (see Figure 7.35).

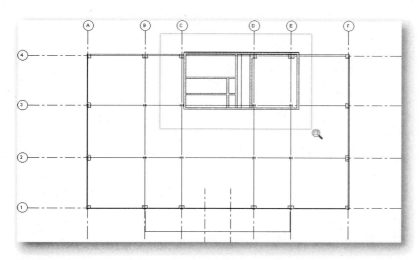

FIGURE 7.35 *Zoom in on the core and note the new Walls and Doors added since the previous chapter*

Perform a Coordination Review

In the previous chapter, we established links between the structural model and the commercial project. We used Copy/Monitor to create monitored copies of the geometry from the commercial project that is pertinent to the structural file. As you can see, some additional geometry has been added to the building core area in the linked commercial project since we established the Copy/Monitor link. In addition, there are other less obvious changes that require coordination. This was what the warning during file open was telling us. To address these issues, we can perform a coordination review and reconcile the differences between the two files. An additional exercise has been provided in Appendix A covering the creation of the Walls provided here in the core. Feel free to do that exercise before continuing if you like.

5. On the Collaborate tab, click the **Coordination Review** tool and then choose the **Select Link** option from the pop-up.

- Click on the commercial project linked file on screen to select it (it is easiest to click on one of the core Walls).

In the "Coordination Review" dialog, various messages will appear in a column on the left. You can choose an Action for how you wish to deal with the differences and, if desired, add a comment. Typically there will be a "Postpone" option which just puts off your decision until next time. The "Reject" option keeps your element as is and flags the change from the linked file as rejected. If you choose "Accept Difference," you are allowing your file and the linked file to be out of synch. Basically, this is like agreeing to disagree. The last option describes what action you must take to make your model match the linked file.

In this case, we have two issues. The first warning notices that all of the Structural Columns have been deleted in the parent file. Expand the "Check whether an Element exist" option to see this (see the left side of Figure 7.36). At this point in the project, the team has decided that the Structural Columns should exist only in the structural model. Since this model is linked back to the architectural file, the architectural team will still see Structural Columns. You can review this warning and its sub-warnings here, but we will deal with it outside of this dialog, so please do not select any action for it at this time.

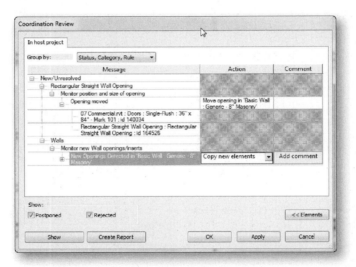

FIGURE 7.36 *Choose an action for how to deal with the difference in coordination review*

The second issue we will deal with here. It informs us that there are new Wall openings/inserts. Openings have been added to the commercial core. (This was not part of the lessons in Chapter 6, so be sure you are using the provided files for Chapter 7.) With the "Coordination Review," we can copy these openings to the structural model.

- Expand Walls, then Monitor new Wall openings/inserts.

- From the Action list, choose Copy new elements and then click OK (see the right side of Figure 7.36).

The "Elements >>" button expands each item to show each individual element affected by the message. The Show button will zoom to the affected element in the model. Revit may need to open other views to do this. The Create Report button will generate an HTML file detailing the results of the coordination review. Feel free to experiment in this dialog if you wish. As you can see, this tool allows you to synchronize changes between the two linked files for elements that are monitored.

6. Click OK to dismiss the "Coordination Review" dialog.

   • If an alert appears, simply click OK.

To deal with the missing columns, we will simply stop monitoring them. They are deleted in the architectural file, but we want them to remain here in the structural file.

7. Select one steel column, right-click and then choose **Select All Instances › In Entire Project**.

   • On the ribbon, click the Stop Monitoring button (see Figure 7.37).

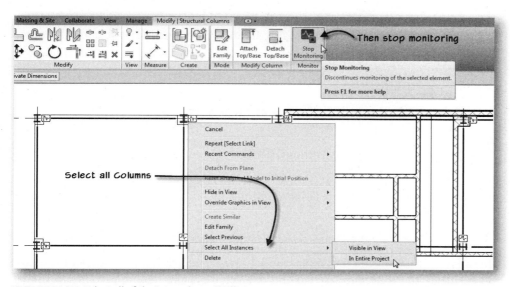

FIGURE 7.37 *Select all of the internal core Walls to copy*

If you return to the "Coordination Review" dialog, you will see that the message about the columns no longer appears. There may still be a message about the openings, but since we copied them over, it is safe to ignore this.

8. Save and close the *07 Commercial-Structure.rvt* [*07 Commercial-Structure Metric.rvt*] file.

Using Visibility/Graphic Overrides

Now that we have resolved any coordination issues in the core area, let's open the commercial project and begin adding some Stairs. We will add a U-shaped egress Stair to the space in the upper left corner of the core.

9. On the QAT, click the Open icon.

The keyboard shortcut for Open is CTRL + O.

TIP

   • In the "Open" dialog box, browse to the *Chapter07* folder.

10. Double-click *07 Commercial.rvt* if you wish to work in Imperial units. Double-click *07 Commercial Metric.rvt* if you wish to work in Metric units.

11. On the Project Browser, double-click to open the *Level 1* floor plan view.

- Zoom in to the same area of the plan as above.

Some of the Walls or columns might appear double in the core area. This is the linked structural file. We can adjust the visibility settings of the view to address this.

- On the View tab, click the Visibility/Graphics button (or type VG).
- In the dialog that appears, click the Revit Links tab.

The possibilities in the "Visibility/Graphic Overrides" dialog are nearly limitless. On Model Categories, we can turn on or off individual categories of Model Elements. We can also override their line styles, patterns, halftone settings, and new to this release, we can use a feature called Ghost Surfaces which makes the surfaces of elements appear semi-transparent. This can be a nice effect in 3D views. Several other tabs also appear with similar options for Annotation elements, analytical elements, imported categories, etc. In this example, we will make changes only on the Revit Links tab. We want to turn off certain redundant elements in the structural model.

12. Click on the Revit Links tab.

Two files will be listed.

- Select *07 Commercial-Structure.rvt* and then in the Display Settings column, click the By Host View button.

- In the "RVT Link Display Settings" dialog, on the Basics tab, click the Custom radio button.

13. Remaining in the "RVT Link Display Settings" dialog, click on the Model Categories tab.

- At the top, from the Model categories drop-down list, choose **<Custom>**.

This allows us to edit the individual model categories just like we could for the host model itself. The difference is that the choices we make here apply *only* to the 07 Commercial-Structure linked file.

14. In the Visibility column, uncheck Walls and then click OK twice (see Figure 7.38).

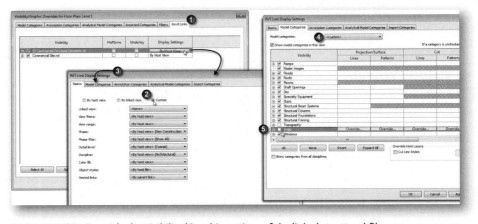

FIGURE 7.38 *Override the Visibility/Graphic settings of the linked structural file*

The redundant Walls should no longer be displayed in this view. When you make a change like this, you likely want similar results to apply on the other floor plans. Visibility/Graphic Overrides are view-specific. This means that they apply only to the view where you apply them. However, we can save our changes to a View Template and then apply this View Template to other views.

Create and Apply a View Template

View Templates offer as much power and flexibility as Visibility/Graphic Overrides. In this example, we will create a View Template containing only the customizations to our linked structural file's display.

15. On the View tab, on the Graphics panel, click the **View Templates** drop-down and choose **Create Template from Current View**.

   • Name the new template **Structural Model Overrides** and then click OK.

   • In the "View Templates" dialog, in the Include column, uncheck everything except V/G Overrides RVT Links and then click OK (see the left side of Figure 7.39).

   • On the Project Browser, select *Level 2*, hold down the SHIFT key and then click *Level 4* to select all three.

   • Right-click and choose Apply View Template. In the dialog that appears, choose Structural Model Overrides and then click OK (see the right side of Figure 7.39).

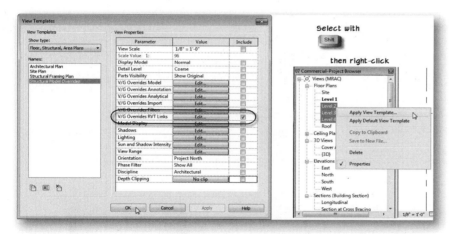

FIGURE 7.39 *Create a View Template and apply it to all plan views*

The other plans now hide the Walls in the structural file as well. If necessary, this View Template can be applied to other types of views as well. For now, we will stick to just the plans and move on to creating the core Stairs.

16. Save the file.

Add a New Stair to the Commercial Plan Core

Previously, in the "Create a Straight Interior Stair" heading, we used the **Run** tool to create a straight Stair for the residential project. To create a Stair with a landing, you start the same way but create more than one run within the same Stair sketch. To do this, simply click short of the complete run. This leaves some of the risers unplaced.

You then start another run nearby. When you do this, a landing will be created in your sketch automatically. We will use this approach to create a U-shaped Stair.

17. On the Home tab, on the Circulation panel, click the **Stairs** tool.

• On the Properties palette, verify that the Type is set to: **7" max riser 11" tread** [**190mm max riser 250mm going**].

• In the "Dimensions" grouping, change the Width to **3'-8" [1100]**.

• Set the "Desired Number of Risers" to **21**.

18. Locate the first point on the inside edge of the right vertical Wall of the Stair core.

• Drag the mouse horizontally to the left. When the message reads "11 Risers Created, 10 Remaining" click the mouse to set the next point.

• Move the pointer above the point just clicked, and then click again.

The point is approximately 2'-0" [600] above the previous one, but the exact location is not important right now as we will adjust it below.

• Drag horizontally to the right and click to place the remaining risers (see Figure 7.40).

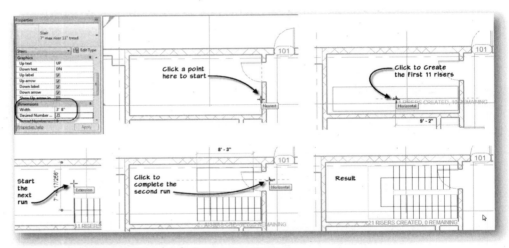

FIGURE 7.40 *Sketch the Stair by placing two separate runs*

A sketch of the U-shaped Stair will appear. The size is not exactly correct and it is not in the right spot. We can adjust both of these issues easily.

19. Select the green vertical sketch line on the left.
This is part of the Stair Boundary.

• On the Modify panel, click the **Move** tool.

• Move the Boundary line to the left **1'-0" [300]**.

Notice that the other Boundary lines remain attached to the moved line.

20. Using a window selection box, select the entire Stair sketch.

NOTE Don't worry about selecting the neighboring elements. When you are in sketch mode, you can only select elements of the sketch.

- Using the **Move** tool again, snap the start point of the move to the lower left endpoint.
- Snap the end point of the move to the endpoint of the Wall at the lower left inside corner of the room (see Figure 7.41).

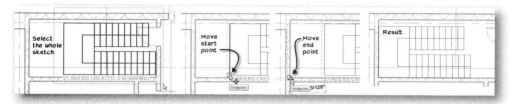

FIGURE 7.41 *Move the entire sketch to the inside corner*

This positions the Stair in the correct general location. However, the Stair Type that we are using here uses Closed stringers instead of open ones like the Stair above. Therefore, we need to take the thickness of our stringers into account when establishing our final position.

21. Using Move again, move the entire sketch up vertically **2″ [50]** and then to the right horizontally **2″ [50]** (see Figure 7.42).

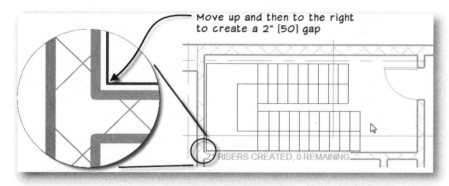

FIGURE 7.42 *Move the entire sketch to create a gap for the stringer*

This will take two moves.

22. Using a window selection box, surround just the top run of the Stair sketch (see the top left corner of Figure 7.43).

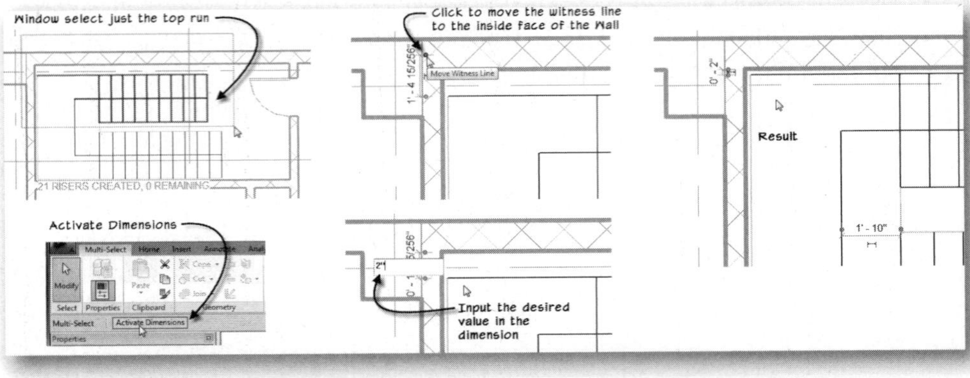

FIGURE 7.43 *Activate Dimensions to move the top run the required gap from the Wall face*

- On the Options Bar, click the Activate Dimensions button.
- Locate and click the small round "Move Witness Line" grip control (zoom as required).

This will move the Witness line to the inside face of the Wall.

- Click on the temporary dimension value, type **2" [50]** and then press ENTER (see the middle of Figure 7.43).
- Deselect the lines.

23. On the Modify | Create Stairs Sketch tab, click the ***Railing Type*** button.

- In the "Railing Type" dialog, choose ***Handrail – Pipe*** [***900mm Pipe***] and then click OK.
- On the Stairs panel of the ribbon, click the ***Finish Edit Mode*** button.

24. Save the project.

Notice that the Stair fills the core space nicely with just the right amount of room for the Railings and stringers.

Create a Multi-Story Stair

Let's take a look at our Stair in a section view.

25. A section line runs horizontally through the Stair. Double-click the blue section head to go to that view.

As you can see, the Stair that we built only occupies the first floor. We can edit the properties of the Stair to make it apply to multiple stories.

26. Select the Stair element in the current Section view.

Be sure to select the Stair and not the Railing for this step.

- On the Properties palette, beneath the "Constraints" grouping, from the "Multistory Top Level" list choose **Roof** (see Figure 7.44).

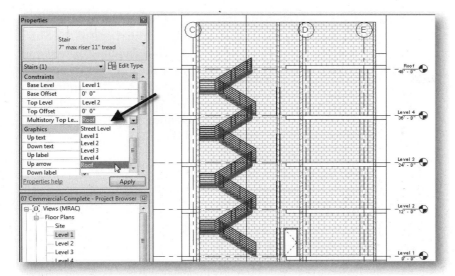

FIGURE 7.44 *Change the Stair to a multistory Stair*

27. On the Project Browser, double-click to open the *Level 1* floor plan view.

Notice that only the UP Stair annotation displays on the first floor and that the Stair above the cut plane is shown dashed.

28. On the Project Browser, double-click to open the *Level 3* floor plan view.

Notice that in this view (and in *Level 2* and *Level 4* as well) that both the UP and DOWN Stair annotations display.

29. On the Project Browser, double-click to open the *Roof* plan view (see Figure 7.45).

Notice that here, only the DOWN annotation shows.

The UP and DOWN annotations can be moved. First, select the Stair. A blue control dot appears next to the text. Click and drag the text to a better location.

TIP

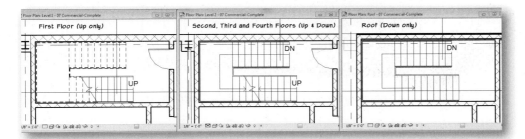

FIGURE 7.45 *Stairs automatically display directional annotations appropriate to the level*

30. Save the project.

LANDINGS AND SHAFTS

In the previous chapter, we created some simple Floors and a Roof for the project. We used a Shaft element to cut these horizontal surfaces through all floors at the stair and elevator shafts. Another glance at the section in our current model will reveal that our Shaft requires modification.

We need to accommodate the Stair elements that we have added and add landings on the entrance side to meet the Stairs. To do this, we can edit the shape of the Shaft, we can remove the Shaft completely and instead edit the Slab sketches, or we could do a little bit of both. The exact approach that we take is really a matter of personal preference. Since the shape of the void is the same on all levels, the shaft approach is easier. If the shape of each landing varies, it is better to edit the sketch of each Floor. If the thickness of the landing is different than the thickness of the floor, creating separate Floor elements for the landings might be best. For our purposes, we'll edit the Shaft.

Edit the Shaft

Editing the Shaft is easy. We simply select it and then edit the sketch.

1. On the Project Browser, double-click to open the *Section at Stair* section view.
2. Move the pointer near the outside edge of the Shaft where it cuts the Floors.

   You should see it highlighted when you move the mouse nearby.

   • When the screen tip reads "Shaft Opening: Opening Cut" and the Shaft pre-highlights, click the mouse to select it.

   The Shaft will turn semi-transparent blue while it is selected.

   • On the Properties palette, change the Base Offset to **0** (zero) (see the left side of Figure 7.46).

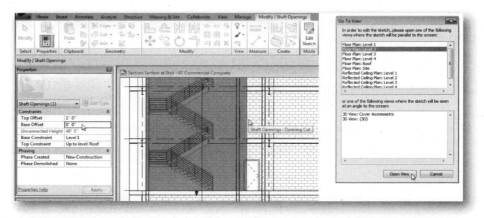

FIGURE 7.46 *Move sketch lines of the Shaft to fit the elevator core (sketch lines in the figure enhanced for clarity)*

3. On the Modify | Shaft Openings tab, click the **_Edit Sketch_** button.

   • In the "Go To View" dialog, choose *Floor Plan:Level 2* and then click Open View (see the right side of Figure 7.46).

   • Move the top horizontal sketch line up to the outside of the top exterior Wall (at the Stair).

   • Move the bottom line up to the outside bottom edge of the elevator shaft.

   • Move the right line to the outside right edge of the elevator shaft (see Figure 7.47).

   You can drag the lines; use the Move tool or the Align tool for this.

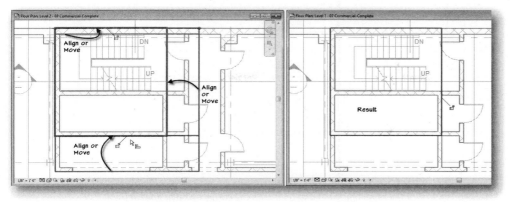

FIGURE 7.47 *Add sketch lines around the Stair tower space (sketch lines in the figure enhanced for clarity)*

You can make your sketch lines appear bolder on screen as you work by editing their lineweight in the Line Styles dialog (Manage tab, Additional Settings drop-down button).

4. On the Draw panel, click the Pick Lines icon.
    *   At the top of the Stair, click the top right tread line near the DN label.
    *   Click the lower inside face of the Wall in the stair core (see Figure 7.48).

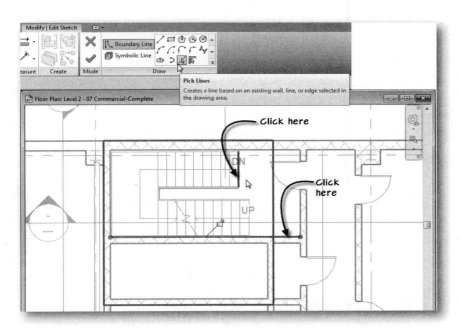

FIGURE 7.48 *Use the Pick Lines option to create sketch lines from existing Wall edges (sketch lines in the figure enhanced for clarity)*

5. On the Modify panel, click the **Trim/Extend to Corner** button.
    *   Using Figure 7.49 as a guide, trim/extend the three corners indicated. (Remember, click the side of the line you wish to keep.)

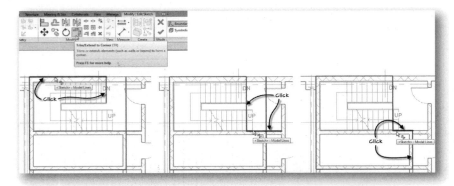

FIGURE 7.49 *Pick the remaining lines and then Trim the corners (sketch lines in the figure enhanced for clarity)*

6. On the Modify | Edit Sketch tab, click the **Finish Edit Mode** button.

   • Deselect the Shaft.

The result of this change in plan is noticeable at the middle of the Stair between the two runs. The section will show the results much more clearly.

Join Geometry in Section

When you open the section view, you will notice that the Floor slabs now extend to the Stairs providing a nice landing at the right side. However, we also notice that the Walls in the core area pass right through the Floor slab. We can make these intersections clean up nicer by using the Join Geometry command.

7. (If it is not already open) On the Project Browser, double-click to open the *Section at Stair* view.

   • On the Modify tab of the ribbon, on the Geometry panel, click the **Join Geometry** button.

   • On the Options Bar, check the Multiple Join checkbox.

   • Click the Floor slab at Level 1 and then click each of the Walls that intersects it (see Figure 7.50).

   • Click the **Modify** tool or press the ESC key twice.

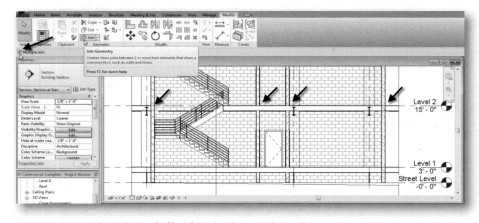

FIGURE 7.50 *Adjust the Roof offset from level to match the Floors*

8. Repeat the process for each of the other Floor slabs shown in the section.

The Floors will now clean up much nicer with the core Walls. Multiple Join is handy and reduces the number of clicks required to complete the task, but you must remember to cancel out before moving on to the next Floor.

9. Save the Project.

Edit Railings

We need to make one more adjustment in our Stair tower. The Railings should wrap around at the landings. We can accomplish this by editing the Railing sketch.

10. On the Project Browser, double-click to open the *Level 2* floor plan view.
11. Select the inside Railing.

    Make sure you select the Railing and not the Stair; use the TAB key if necessary.

    - On the ribbon, click the **Edit Path** button.
    - On the Draw panel, click the **Line** icon.
    - On the Options Bar, select the "Chain" checkbox.
    - Draw the two segments indicated in Figure 7.51.

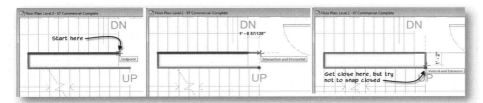

FIGURE 7.51 *Draw two segments in the Railing sketch*

12. On the ribbon, click the **Finish Edit Mode** button.

The best way to view the result is in a 3D view. If we use the default 3D view, we will have to isolate the Stair (as we did above in the residential project) in order to see the Railing. Another option is to create a new 3D view of just the stairs. In this case, creating a 3D view from the *Section at Stair* view would be the best choice. It will give use a three-dimensional section view in which to study the Stairs and Railings.

13. On the Project Browser, right-click the *{3D}* view and choose **Duplicate View > Duplicate**.

    - Right-click the *Copy of {3D}* view and choose **Rename**.
    - In the "Rename View" dialog change the name to **3D Stair Section** and then click OK.

14. Click anywhere in the 3D window to shift focus from the Project Browser to the model window.

    - Right-click the ViewCube and choose **Orient to View > Sections > Section: Section at Stair** (see Figure 7.52).

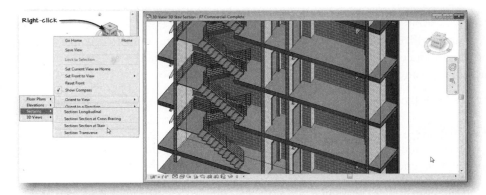

FIGURE 7.52 *Orient the view to the Section at Stair Tower then spin it in 3D*

15. Click on the small corner of the ViewCube between the Top, Front, and Right sides.
    • Hold your SHIFT key down and drag with the wheel on your mouse to orbit the model around and see the effect.

NOTE If you prefer, you can drag the ViewCube or use the Steering Wheel instead.

16. Zoom in on the second floor at the point where we edited the Railing.

    • Pan to other floors as well.

Notice that the Railing now wraps around the gap between the two runs and this change has occurred on all Levels (see Figure 7.53).

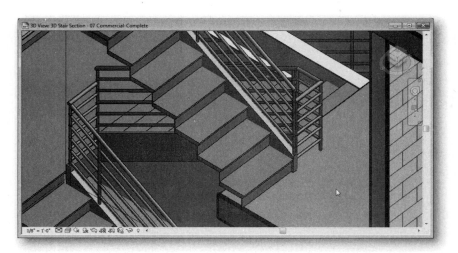

FIGURE 7.53 *The Railing edit we made is visible on all Levels*

For the top and the bottom of the Stair, you can add additional Railings if you like to complete the layout. The easiest way to do this is to select the Railing, and then on the ribbon click the ***Create Similar*** tool. Sketch the new Railing and use the ***Set Railing Host*** tool on the Tools panel to attach it to the Landing.

If you witness a strange artifact for the balusters in the 3D view, you can edit your sketch and pull the vertical sketch line back slightly from the horizontal segment so they don't join. Leave the vertical segment about an inch or two [few millimeters] short.

Adjust Outside Railing

Previously in the "Create a New Railing Type" heading of the residential project, we created a new Railing Type that did not have any balusters. A Railing Type similar to this has been provided in the dataset that we can use for the Railing along the Stair core Walls.

17. Select the outside Railing (the one adjacent to the Wall).

    - From the Type Selector on the Properties palette, choose **Handrail – Pipe – Wall** for the Type.
    - Study the results in other views.

18. Save the project.

RAMPS, CUSTOM STAIRS, AND ELEVATORS

In this topic, we will add some Ramps, a custom Stair, and elevators. While elevators are included here to round out our look at vertical circulation, the approach to working with them is quite different. Creating a ramp in Revit is much like adding a Stair. Elevators on the other hand are simply Component Families that we insert in our models like other components. Creating a custom Stair will use the Boundary and Riser tools.

Create an Enlarged Plan View

1. On the Project Browser, double-click to open the *Level 1* plan view.

Let's create an enlarged floor plan view showing just the front portion of the building.

2. On the View tab of the ribbon, on the Create panel click the ***Callout*** tool.

    - Drag a box around the patio slab at the front of the building.
    - Deselect the Callout boundary and then double-click the Callout head to open the view (see Figure 7.54).

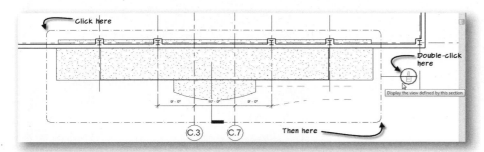

FIGURE 7.54 *Create the Callout View of the patio slab at the front of the building*

3. On the Project Browser, right-click *Callout of Level 1* and choose **Rename**.

    - Name the view: **Entrance Plan** and then click OK.

Add a Ramp

The *Ramp* tool is next to the Stairs tool on the Circulation panel of the Home tab. Ramps are very similar to Stairs. To save a bit of time and effort, the dataset includes a floor slab at the front of the building as an entrance patio. Some Reference Planes have been included here as well. These Reference Planes are provided to make it easy for us to the sketch the ramp in the following example. When building your own Ramps and Stairs, you can also use Reference Planes (as we did in the residential project above), or any other technique you find helpful in laying out the sketch (such as the method we used in the Stair core).

- Zoom in on the lower right corner at the front of the building.

You will notice four reference planes in this location. We will use these to help us sketch the ramp.

4. On the Home tab, on the Circulation panel, click the **Ramp** tool.

Notice that the Create Ramp Sketch tab that appears is nearly identical to the Stair Sketch tab.

- On the Properties palette, beneath the "Dimensions" grouping set the Width to **3'-0" [900]**.
- Change the "Base Level" to **Street Level** and the "Top Level" to **Level 1**.
- Starting at the lower left endpoint of the inclined Reference Plane, click on each Reference Plane endpoint moving first left to right, then up, then right to left (see Figure 7.55).

NOTE As when we created the U-shaped Stair previously, you only click the start and end of the runs. The landings fill in automatically.

Also, note in the figure, that the first click in the run will show a tooltip for the Endpoint, but not highlight the point with the customary square snap indicator. The square snap indicator does appear on the second click. This anomaly should not prove detrimental to completing the exercise.

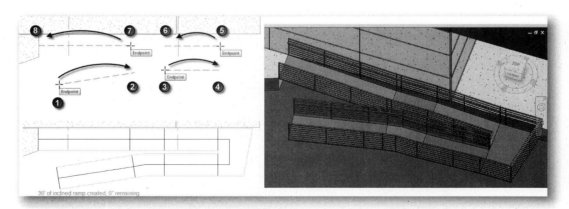

FIGURE 7.55 *Sketch the ramp using the provided Reference Planes and then study it in other views*

If you wish, you can edit the shape of the green outline on the right side at the landing to make the ramp match the building Pad in the linked Site model. Try moving the vertical green boundary line to the right about 1'-0" [300] to widen the landing, or even change the shape to angle or curve it if you like.

While we have provided the Reference Planes in this exercise to make the process of sketching the ramp go smoothly, you can use any of the sketching techniques that we covered above on Stairs when adding ramps in your own projects. Remember that you can sketch the Run, or edit the Risers or Boundaries directly with the appropriate tools on the ribbon.

5. On the Modify | Create Ramp Sketch tab, click the **Finish Edit Mode** button.

   • View the ramp in various views to see the results (shown in 3D on the right side of Figure 7.55).

6. Return to the *Entrance Plan* view and then Mirror the ramp and its Railings to the other side.

   Use any selection technique, but be sure to select only the Ramp and Railings.

   • On the Modify | Multi-Select tab, click the **Mirror – Draw Axis** tool.

   • Use the midpoint of the patio curve to draw the mirror line.

     Be sure that "Copy" is selected on the Options Bar so that you maintain the original when mirroring.

7. Save the project.

Add a Custom Shaped Stair

We can add a few steps up to the patio leading to the main building entrance. To do this, we simply add another Stair. However, we'll sketch it a bit differently than the ones above.

8. On the Home tab, click the **Stairs** tool again.

   • On the Properties palette, from the Type Selector, choose **Monolithic Stair**.

   • Set the Base Level to **Street Level** and the Top Level to **Level 1**.

9. On the Draw panel, click the **Boundary** tool.

   • On the Options Bar, clear the "Chain" checkbox and set the Offset to **6'-0"** **[1800]**.

   • Click the first point at the midpoint of the curved edge of the patio (see Figure 7.56).

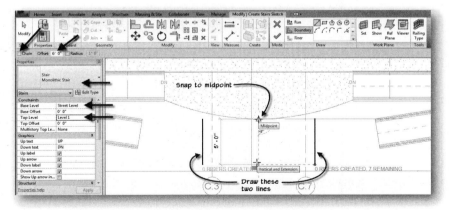

FIGURE 7.56 *Add two Boundary lines drawn with the Offset option*

- Draw straight down and click below the bottom edge of the Ramps about 5'-0" [1500].

  If you forgot to clear "Chain" above, press ESC once.

  Repeat the process using the same settings and start at the same midpoint. Before you click the other end, tap the SPACEBAR to flip the sketch to the other side.

You should have two vertical green lines centered on the patio and spaced 12'-0" [3600] apart. These are the Boundaries of the Stair (this is a monolithic Stair, but in other Stairs, these lines would become the stringers).

10. On the ribbon, click the **_Riser_** tool.

    - On the Draw panel, click the Pick Lines icon.
    - On the Options Bar, set the Offset to **5'-0" [1500]**.

11. Highlight the curved edge of the patio. Move the mouse slightly to make a dashed guideline appear below, and then click to create the Riser line (see top left panel of (Figure 7.57).

    - Change the Offset to **1'-0" [300]**.
    - Highlight the Riser line you just added, and offset a new line up from it. Repeat until you have created 5 more (6 total) (see top right panel of Figure 7.57).

      The last one will be right on top of the patio edge.

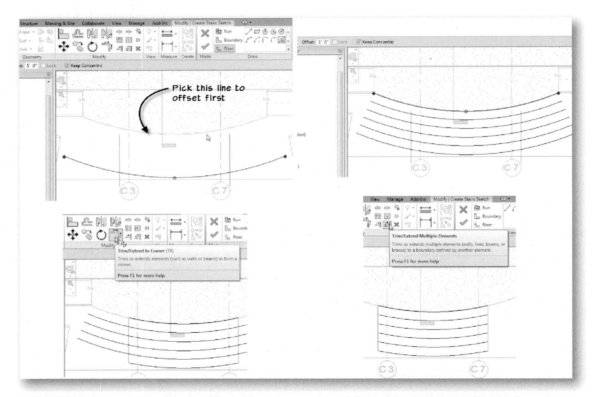

FIGURE 7.57 *Create the Riser lines by offsetting from the curved patio edge and then Trim to complete sketch*

12. Use the **_Trim/Extend_** tools to clean up the sketch.

    - Use the **_Trim/Extend to Corner_** to trim and extend the green Boundary lines to the curved black Riser lines.

- Use the Trim/Extend Multiple Elements to trim the curved lines back to the green Boundary lines (see bottom right panel of Figure 7.57).

  Remember to click the green Boundary line as the trimming edge first, and then pick the side of the curve you wish to keep when trimming.

13. On the ribbon, click the Finish Stairs button.

An ignorable warning will likely appear. (Warnings that appear in the lower right corner of the screen can sometimes be ignored, either because they are benign, or as in this case when to resolve them would sacrifice design intent.) Normally you would include the landing at the top of the stair flight as part of the Stair and therefore include one more Riser. In this case, we have the patio slab instead and will be fine ignoring this message for now.

You should always read the warnings and address them when you can, as a large number of unresolved warnings can seriously degrade model performance. You can check for any unresolved warnings on the Manage tab on the Inquiry panel. Click the Warnings tool to display a dialog and review any warnings in your model. If this tool is greyed out, congratulations, this means that you have no unresolved warnings in the model, which would be very good indeed! In this case, the warning tells us that we do not have the desired number of risers.

One problem not indicated by the warning or obvious from the plan view is that the stair direction is wrong. The highest tread is at the bottom of the screen and lowest by the patio. We can fix this with the flip control like we did for the residential Stair at the start of the chapter.

14. Select the Stair and click the Flip Stair's Up Direction control (refer back to Figure 7.5 above for an example).

If your Stair disappears after clicking the flip control, undo and then edit the Stair sketch. Zoom in closely on the Stair. Select the green Boundary on one side of the Stair and nudge it slightly. (Press the left or right arrow keys on the keyboard one time.) Finish the sketch and try again.

TIP

- Flip both Railings as indicated in the "Modify the Railing" topic earlier in this chapter.
- Choose a more suitable Railing Type.
- Study the Stair in section or 3D views before continuing.

Add Railings to the Entry Stair

If you want to add railing extensions to the Railings, just edit the sketch.

15. Return to the *Entrance Plan* view.
16. Select one of the Stair Railings for the Stair we just created.

- On the Modify | Railings tab, click the ***Edit Path*** button.

A single sketch line appears.

- On the Modify | Railings > Edit Path tab, on the Draw panel, click the **Lines** tool.
- At the bottom endpoint of the line, draw a new line **1'-0" [300]** long.
- At the top, draw a **2" [50]** long vertical line and then use the pick option to offset a curved line **2" [50]** from the patio edge.
- Trim to clean up.
- Select one of the new lines and on the Options Bar, choose Flat for the Slope (see Figure 7.58).

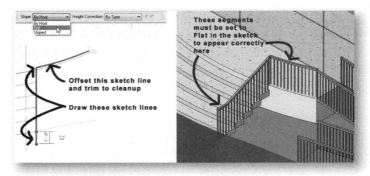

FIGURE 7.58 *Add lines to the Railing Sketch*

- Repeat for the other two lines added in this sequence.
- On the ribbon, choose Finish Edit Mode.

17. Return to the *Entrance Plan* view.

- Edit the other Railing and repeat the process. (Or you can delete the other Railing and mirror the one we just edited.)

If you wish, you can add a center Railing on the Stairs. To do this, click the ***Railing*** tool on the Home tab. Sketch a line down the center of the Stairs (snapping to mid-points). Add an additional line at the top and another at the bottom each 1'-0" [300] long. You will have three vertical lines running end to end. Set the two short ones' Slope option to Flat. Finally, click the ***Pick New Host*** tool on the Tools panel, and then select the Stair as the host. Finish the sketch.

Add Railings to the Patio

Let's add guardrails to the Patio.

Remain in the *Entrance Plan* floor plan view.

18. On the Home tab, on the Circulation panel, click the ***Railing*** tool.

- On the Draw panel, click the Pick Lines icon.
- On the Options Bar, set the "Offset" value to **4" [100]**.

19. Using Figure 7.59 as a guide, place Railing sketch lines around the patio edges.

- On the ribbon, click the ***Pick New Host*** tool and then click the patio slab.
- Finish the sketch.

NOTE The figure shows all the required sketch lines, however, Revit does not allow disconnected Railing segments. Therefore, you will need to create two separate Railing elements.

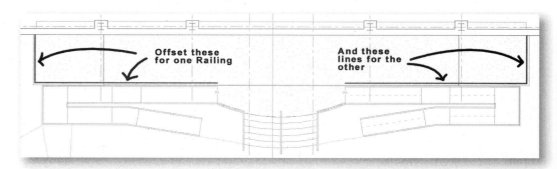

FIGURE 7.59 *Create Railings from the patio slab (sketch lines in the figure enhanced for clarity)*

20. Save the project when finished.

If you open the default 3D view, your model should look something like Figure 7.60. As we study this view, it now becomes obvious that the curtain wall at the entrance needs revision. That task will be completed in a later chapter. For now, we are finished with the commercial building's front entrance.

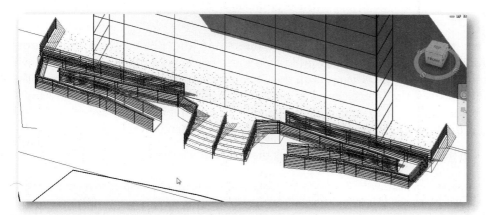

FIGURE 7.60 *The completed Ramps, entry Stairs, and Railings*

Add Elevators

Unlike Ramps and Stairs, we do not have a dedicated tool or element category for elevators in Revit. Elevators are like other component Families and we use the ***Component*** tool to add them in the same way that we added plumbing fixtures in Chapter 3. Elevator Families can be created from scratch or derived from manufacturers drawing files. Some sample elevator Families are provided with the files installed from the student companion dataset files in the *\MasterRAC 2012\Library* folder. Feel free to use these samples if you wish. In addition, Revit has the Autodesk Seek functionality built in. Seek is an online repository of manufacturer content. We can run a search on Seek directly from the Insert tab in the Revit application.

21. On the Project Browser, double-click to open the *Level 1* floor plan view.
22. Click the Insert tab of the ribbon.

   • On the Autodesk Seek panel of the ribbon, type **elevator** in the search field and then press ENTER.

Your default web browser will open and the results of the search will appear. You can also visit Seek manually by typing: **http://seek.autodesk.com/** into your web browser. You will notice that Seek contains content in many formats. You can use the links on the left-hand side to filter to just RFA files. Also, content is being added to Seek all the time. At the time of this writing, the search yielded some elevator files, along with a variety of supplemental content related to elevators.

For this exercise, we will go directly to the component we need for our project, but please feel free to spend some time exploring in this website and download any of the components you wish and add them to the current project. A new version of the dataset will be provided in the next chapter, so you do not have to worry about straying from the exercise. Exploration time is highly encouraged and is often a great way to learn more about Revit Architecture!

   • In Seek, on the left side, beneath Product Libraries, click on Revit Architecture.
   • Click on the Elevator-Hydraulic item.

- On the page that appears, download files will appear on the right. Check the most recent RFA item, and then click the Download Selected button.

  Make sure you check the box next to the latest version. At the time of this writing, the latest version of the elevator was Revit 2010—when you load it Revit it will automatically be upgraded to Revit 2012.

- In the dialog that appears, click Open. (If a further warning appears, click Allow.)

  You may also be required to accept the download conditions before being allowed to download.

When you open the file, it will open directly in the Revit Architecture workspace in "Family Edit" mode. Family Edit mode is a special mode for editing Revit Architecture Families. The tools on the ribbon will be unique to this mode, but otherwise most of the interface will remain the same. Here you could edit the Family file before adding it to your project. We will not edit it, but simply save the file and add it to our project. The Family Editor is covered in detail in Chapter 10.

23. On the Create tab of the ribbon, click the **Load into Project** button (see Figure 7.61).

NOTE If you have more than one Revit Architecture project open, you will be prompted to select the project in which you wish to load the Family. Choose the *07 Commercial* [*07 Commercial Metric*] project and then click OK. If you receive a warning that the 3000 lb Type elevator cannot be created, click OK to continue—the type we need will still be available.

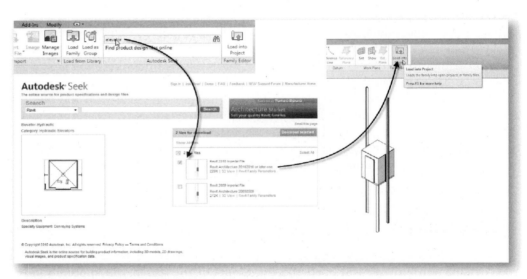

FIGURE 7.61 *Access the Web Library from any "Open" dialog*

The elevator Family is now loaded into your project file and ready to be placed in the model.

NOTE If you do not have Internet access, or are having trouble locating the files on Autodesk Seek, open the file from the *Library* folder of the dataset installed from the student companion instead.

- To load from the dataset folder instead, click the Insert tab on the ribbon.
- Click the **Load Family** tool.
- In the dialog that appears, browse to the *Library/Seek* folder in the location where you installed the dataset files.

Once you have loaded the Family using either method described here, you can continue to the next steps.

24. On the Home tab, on the Build panel, click the **Component** tool.

    The Elevator should now be loaded and already selected on the Type Selector.

    - If necessary, from the Type Selector, choose **Elevator – Hydraulic – 2000 lbs**.
    - Place the elevator anywhere in the project first.
    - Use Move or Align to position it in elevator core space. Repeat the steps or make a copy.
    - Also on Seek is a **Elevator Door - Center** Family. Load and place this one in your project too if you like (see Figure 7.62).

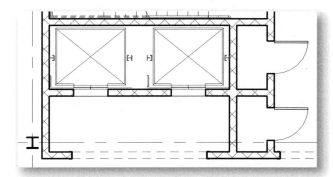

FIGURE 7.62 *Place two elevators in the core*

    - On the QAT, click the **Modify** tool or press the ESC key twice.

25. Save and close the project.

SUMMARY

- Stairs and Railings offer flexible configuration with Type-based parameters and Element-based variations.
- Stairs are "sketch-based" objects—create them with a 2D footprint sketch that includes risers and boundary edges.
- Like other Revit elements, Stairs and Railings can be demolished or added to any phase.
- Edit a Stair Type to apply changes to all Stair elements sharing that Type.
- Edit the Stair Sketch to make customizations such as bull nosed treads.
- Edit the Railing Sketch to modify the extent of the Railing or add railing extensions.
- Adjust Wall Profiles to cut away the space where they intersect Stairs.
- Create Stairs with integral landings by drawing short runs in the sketch. Revit automatically fills in the landing in between runs.
- Stairs can be made multiple stories tall by assigning the height constraints and specify a "Multi-Story Top Level."
- You can use Floor elements and Shafts to model landings where necessary.
- Ramps are similar to Stairs and are also sketch-based.

- Custom shaped Stairs can be sketched using the Boundary and Riser tools instead of the Run tool.
- Railings are added automatically to newly drawn Stairs. However, you can use the Set Host tool to sketch a custom Railing and apply it to an existing Stair, Slab, or Ramp.
- Autodesk Seek is an online content repository where manufacturers post content in a variety of file formats including Revit files.
- Pre-made elevator component families can be loaded into the project and placed in the model.

INTRODUCTION

In this chapter we will focus on Floors and Roofs in both our residential and commercial projects. Roofs and Floors in Revit are "sketch-based" objects. This means that, as we saw with Stairs in the previous chapter, a simple two-dimensional sketch is utilized to indicate the shape and form of the Floor or Roof you create. We will begin in the residential project with gable Roofs and some Floor elements. We will understand how to make Walls attach to Roofs, edit Roof structure, and how to apply edge conditions and gutters. The commercial project will give us the opportunity to explore flat Roofs and extrusion Roofs.

OBJECTIVES

We have already seen examples of Floors in previous chapters. You can construct Roofs in a variety of ways. The simplest method is similar to Floors. Roofs can interact with the Walls of the building and we can apply custom treatment to their edges. After completing this chapter you will know how to do the following:

- Build Roofs
- Create a Custom Roof Type
- Add and modify Floors
- Work with Roof Edges and Gutters
- Attach Walls and Join Roofs

CREATING ROOFS

Since we are working concurrently on two different projects in this book, we will get an opportunity to look at both traditional sloped residential roofs and "flat" commercial roofs. The only real difference between the two in Revit is the slope parameters that we assign and the way that we treat the edges. To get started, we'll begin with the residential project.

Install the Dataset Files and Open a Project

The lessons that follow require the dataset included on the Aubin Academy Master Series student companion. If you have already installed all of the files from this site, skip to step 3 to begin. If you need to install the files, start at step 1.

1. If you have not already done so, download the dataset files located on the CengageBrain website.

   Refer to "Accessing the Student Companion site from CengageBrain" in the Preface for information on installing the dataset files included in the Student Companion.

2. Launch Autodesk Revit Architecture from the icon on your desktop or from the *Autodesk > Revit Architecture 2012* group in *All Programs* on the Windows Start menu.

TIP	You can click the Start button, and then begin typing **Revit** in the Search field. After a couple letters, Revit Architecture should appear near the top of the list. Click it to launch to program.

3. On the QAT, click the Open icon.

TIP	The keyboard shortcut for Open is CTRL+O. **Open** is also located on the Application menu.

- In the "Open" dialog box, browse to the location where you installed the *MasterRAC 2012* folder, and then double-click the *Chapter08* folder.

4. Double-click *08 Residential.rvt* if you wish to work in Imperial units. Double-click *08 Residential Metric.rvt* if you wish to work in Metric units.

   You can also select it and then click the Open button.

Create an Existing Roof Plan View

In the previous chapter, we introduced the concept of creating a separate view for the existing construction and setting its parameters accordingly. We did this while adding Stairs in the first floor. Let's use the same technique now to create an "Existing Conditions Roof Plan" view.

5. On the Project Browser, right-click the *Roof* plan view and choose **Duplicate View > Duplicate.**

- On the Project Browser, select *Copy of Roof* and then on the Properties palette, change View Name to **Existing Roof**.
- Scroll back up to the top and take note of the Underlay setting which currently reads Second Floor.
- Scroll to the bottom and for the Phase, choose **Existing** and then click Apply.

The new construction will disappear. The existing second floor will show in gray since, as we just discovered, it is set as the underlay to the current view. Recall that in Chapter 3 we assigned the Second Floor as an underlay to the *Roof.* Copying the view here also copies this underlay parameter to our *Existing Roof* view. The underlay is useful in helping us initially build the Roof.

Add the Existing Roof

You may note that the temporary Roof provided in previous chapters has not been included for this chapter project. There is valuable experience to be gained in creating the Roof from scratch, as we will do in the tutorial that follows. We will begin with a simple gable Roof on the existing house and then add a slightly more complex double gable on the new addition.

6. On the Home tab, on the Build panel, click the ***Roof*** tool.

   The ***Roof*** tool's default is ***Roof by Footprint***. The ***Boundary Line*** and ***Pick Walls*** tools on the Draw panel should be active by default.

   - On the Options Bar, verify that the "Defines slope" checkbox is selected.
   - In the Overhang field, type **6" [150]**.
   - Verify that "Extend to wall core" is *not* selected.
   - Move the pointer over the topmost horizontal Wall and when the dashed line appears above and to the outside, click the mouse (see Figure 8.1).

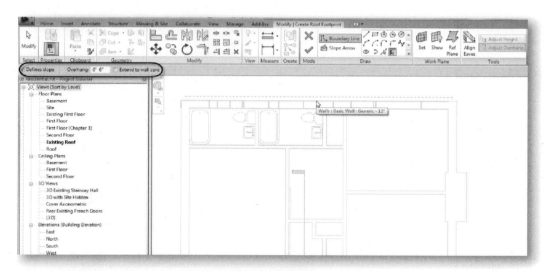

FIGURE 8.1 *Set the parameters for the first edge of the Roof on the Options Bar*

   - Repeat this on the lower horizontal Wall to create another sketch line below and to the outside of that Wall.

     Near both of these sketch lines, a triangular slope icon will appear.

7. On the Options Bar, clear the "Defines slope" checkbox and then click the outside vertical Wall on the left.

   - Also click to the outside of vertical Walls as shown in Figure 8.2.

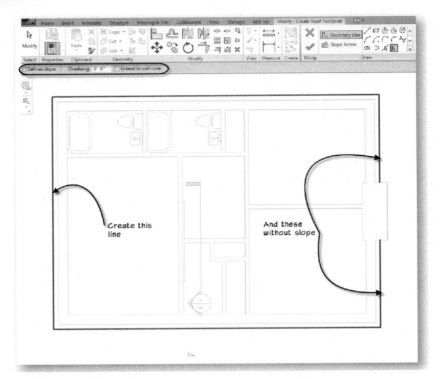

FIGURE 8.2 *Create the gable ends by clearing "Defines slope"*

Since we have the chimney on the right side, we need to make the Roof cut around it. We can accomplish this by drawing the remaining sketch lines relative to the chimney.

8. On the Draw panel, click the **Line** tool.

   • On the Options Bar, make sure that the "Chain" checkbox is selected and that there is no offset (set to zero).

   • Using Figure 8.3 as a guide, draw the remaining three sketch lines.

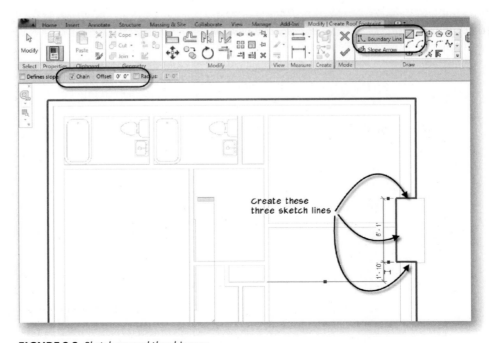

FIGURE 8.3 *Sketch around the chimney*

- On the ribbon, click the **Modify** tool or press the ESC key twice.

Before we complete the sketch, let's adjust the slope of the roof.

9. Click on the top horizontal sketch line, hold down the CTRL key and then click the bottom horizontal line as well.

   - On the Properties palette, beneath the "Dimensions" grouping, change the Roof slope. If you are working in Imperial units, type **6"** for the Rise/12" parameter. If you are working in Metric, set the Slope Angle to **26.57**.

> **TIP** As an alternative, you can select the sketch line and then edit the slope directly with the temporary dimension that appears next to the slope indicator of the slope defining line. However, you must do it one sketch line at a time.

10. On the ribbon, click the **Finish Edit Mode** button.
11. On the View tab of the ribbon, on the Create panel, click the **Default 3D View** button.

Notice that the two Roof edges that we designated as "Defines slope" have a pitch sloping up to a single gable ridge down the middle of the existing house. Feel free to orbit the model around and see it from different angles.

Adjust the Chimney

Notice also that the Roof has a cutout for the space of the chimney. (However, currently the chimney is too short). We can adjust the height of the chimney by simply dragging the control handles. To do this with a bit of accuracy, let's first lay down a reference plane.

12. On the Project Browser, double-click to open the *East* elevation view.
13. On the Home tab, on the Work Plane panel, click the **Ref Plane** tool.

   - Click the first point above the roof to the left of the chimney.
   - Click the other point in a horizontal line above the top of the Roof to the right of the chimney.

Depending on how high above the roof you drew your Reference Plane, an "ignorable warning message" will most likely appear when you place the second point of the Reference Plane (see Figure 8.4). This particular warning is ignorable, because it is merely informing you that the object that you created is not visible in the current view. Other warnings that may appear here could be more important and should be resolved as soon as possible, such as warnings about walls being created coincident with other walls, items that are unable to remain joined, etc. An ignorable warning message is simply an informational message that alerts you of a situation that may not otherwise be apparent. It can be ignored, thus the name. This type of dialog appears in the lower right corner of the screen and is tinted in color (which can be customized in the Options dialog). Simply continuing to draw or picking other tools closes this warning dialog. You do not need to click the close button to close it.

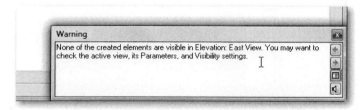

FIGURE 8.4 *"Ignorable Warning" dialogs report information that may be useful, but can be safely ignored*

This message appeared because the elevation view has a crop boundary enabled. This crop region is visible by default in the section views, but invisible in elevation views. We could turn on the visibility of Crop Region in this view so that the Reference Plane would become visible. However, it is likely that we sketched it too high above the building anyway. Let's use the temporary dimension to move it first and see if it comes into view before we take any action on the Crop Region. Incidentally, the display of the Crop Region is easily toggled on or off from the icon on the View Control Bar.

14. With the Reference Plane (the one you just drew) still active, click in the blue text of the temporary dimension.

   • Type a value of **8'-6" [2600]** and then press ENTER (see Figure 8.5).

NOTE — Be certain that the temporary dimension measures from the Roof Level to the Reference Plane. If it does not, adjust the witness lines first.

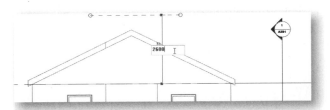

FIGURE 8.5 *Edit the temporary dimension to move the Reference Plane*

   • On the QAT, click the **Modify** tool or press the ESC key twice.

If you deselected the Reference Plane already, you can click the "Do Not Crop View" icon on the View Control Bar to toggle off the Crop Region. This will disable cropping in this view and the Reference Plane will appear above the Roof. Click to select the Reference Plane and then adjust the temporary dimension as noted previously. Toggle the Crop Region on again when finished.

15. Click on the chimney to select it (use TAB if necessary).

You may recall that the chimney was constructed in Chapter 3 from an In-Place Family.

   • Drag the control handle at the top and snap it to the Reference Plane.
   • Click the small padlock icon (close the lock) to constrain the top edge of the chimney to the Reference Plane (see Figure 8.6).

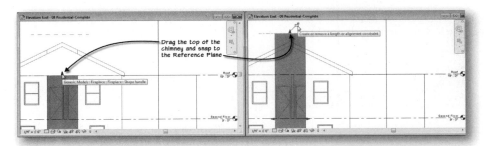

FIGURE 8.6 *Drag the top of the chimney and lock it to the Reference Plane*

Constraining the geometry to the Reference Plane is not required, but it makes it easier to adjust the height of the chimney later if required. Simply move the Reference Plane and the top of the chimney will follow. Try it now if you like.

> You can create a permanent dimension between the Reference Plane and the top Level and lock it to a height required by the building code. In this way, if the Roof level changes, the chimney height will change accordingly.

TIP

16. Save the project.

Attach Walls to the Roof

It may not be apparent from the elevation view, but the Walls do not project all the way to the Roof. In some cases when you draw a Roof, you will be prompted to automatically attach the Walls to the Roof. Usually for this to occur, the Walls must intersect the Roof. In this case, the Walls stopped beneath the Roof. However, we can still manually attach them to the Roof. It is a good idea to get in the habit of checking for this condition.

17. On the View tab, on the Windows panel, click the **Close Hidden** tool.

   • On the View tab, click the **Default 3D View** tool (or open the *{3D}* view from the Project Browser).

   • On the View tab, on the Windows panel, click the **Tile** tool.

> The keyboard shortcut for Tile is **WT**.

TIP

18. Pre-highlight one of the exterior Walls of the existing house.

   • Press the TAB key to highlight a chain of connected Walls.

   • With the existing house exterior Walls pre-highlighted, click the mouse to select them (see Figure 8.7).

> If you have trouble chain-selecting the existing Walls, you can try selecting in another view, like a floor plan.

NOTE

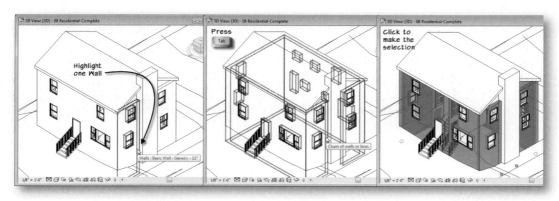

FIGURE 8.7 *Select a chain of Walls*

19. On the Modify | Walls tab, on the Modify Wall panel, click the **Attach Top/Base** button.

    • On the Options Bar, verify that "Top" is selected, and then click the Roof (see Figure 8.8).

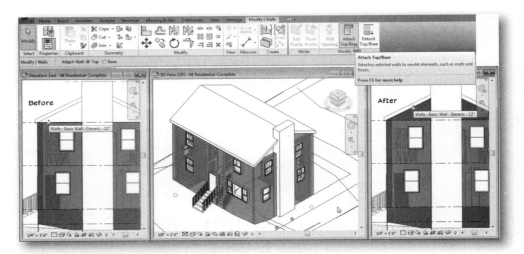

FIGURE 8.8 *Attach the Walls to the Roof*

    • Deselect the Walls.

Edit the Second Floor Wall Layout

There are many other Roof options to explore. The best place to do this in this project is on the new construction in the addition rather than the existing construction. The second floor of the addition will extend the two existing bathrooms and have an outdoor patio on the left side of the plan. On the north and west sides of the patio, we will have low-height, parapet-type Walls. The Roof will cover the other portions of the addition, but the patio will be uncovered. Before we add the Roof, we need to modify the Walls a bit to reflect these design features.

20. On the Project Browser, double-click to open the *Second Floor* plan view.

21. Using techniques covered in Chapter 3, add two Walls as shown in Figure 8.9. Use the same Type as the other three Walls in the addition.

    • Set the Location Line to **Finish Face:Exterior.**

    • Set the "Base Constraint" to **Second Floor** and the "Top Constraint" to **Up to level: Roof.**

TIP	Select one of the existing Walls, and then on the ribbon choose the *Create Similar* tool to easily match all the parameters.

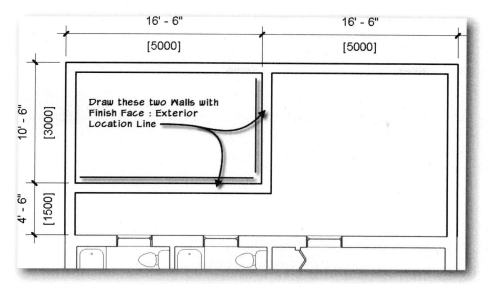

FIGURE 8.9 *Add two new Walls to the Second Floor*

The space that we have just described in the top left corner is the outdoor patio. As noted previously, it will have low-height Walls on two sides. Now let's edit the original exterior Walls to reflect this condition. To do this, we are going to edit the Profile of the Walls.

22. Select the top horizontal Wall.

- On the Modify | Walls panel, click the **Edit Profile** button.

  The "Go To View" dialog will appear.

- In the "Go To View" dialog, choose *Elevation:North* and then click Open View (see Figure 8.10).

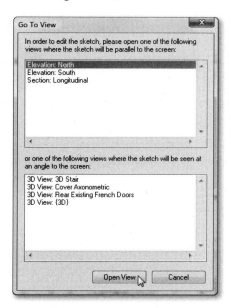

FIGURE 8.10 *Certain edits prompt for a more appropriate view in which to work*

When you begin an operation that cannot easily be performed in the current view, Revit will prompt you to open a more appropriate view. The "Go To View" dialog suggests all appropriate views in the current project in which to perform the operation. You should now be looking at the back of the house with the selected Wall in sketch mode. In this mode we can edit the shape of the Wall to "sculpt" it to meet the needs of the design.

23. On the Modify I Walls > Edit Profile tab, on the Draw panel, click the **Pick Lines** tool.

   • Click on the right vertical edge of the intersecting Wall in the middle of the second floor (see Figure 8.11).

NOTE This is the outside edge of the Wall we drew earlier.

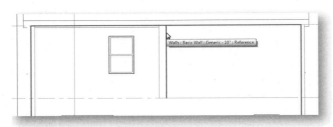

FIGURE 8.11 *Add a sketch line to the Wall Profile Sketch relative to the intersecting Wall*

   • On the Options Bar, type **3'-6" [1050]** in the "Offset" field.
   • Highlight the Second Floor Level line.
   • When the guide line appears above the Level line, click to create the sketch line (see Figure 8.12).

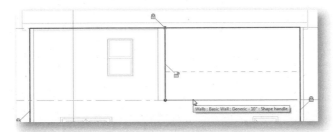

FIGURE 8.12 *Add another sketch line relative to the Second Floor Level Line*

24. Use the **Trim/Extend to Corner** tool to clean up the sketch and close all of the corners (see Figure 8.13).

NOTE Remember to click the side of the line that you wish to keep.

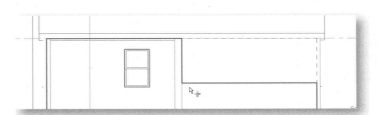

FIGURE 8.13 *Complete the Wall Profile Sketch (sketch lines enhanced for clarity)*

25. On the ribbon, click the **Finish Edit Mode** button.

26. Select the other Wall (the vertical one on the right in the current view—on the left in the plan view) and repeat the process.

    - When prompted, open the *Elevation:West* view.

    - Again use the outside edge of the Wall we drew above to create the vertical sketch line and then offset a line up from the Second Floor Level Line as before.

    - Trim/Extend to complete the sketch (see Figure 8.14).

    - On the ribbon, click the **Finish Edit Mode** button.

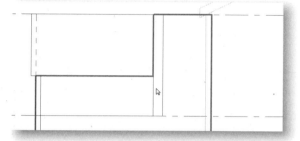

FIGURE 8.14 *Edit the Profile Sketch of the opposite Wall (sketch lines enhanced for clarity)*

The new addition will be brick veneer on a stud wall backup. We can apply a Wall Type that represents this type of construction to our new construction Walls.

27. Open the *Second Floor* plan view.

    - Select all of the Walls in the addition (the three original ones and the two new ones we just added).

    - On the Properties palette, from the Type Selector, choose **Basic Wall : Exterior - Brick on Mtl. Stud**

28. Open the *{3D}* view.

    - Hold down the SHIFT key and drag the wheel on your mouse to spin the model around and see the edits to the Walls (see Figure 8.15).

Remember: You can drag the ViewCube or use the Steering Wheel to orbit instead.	**TIP**

If the {3D} view does not show the new construction, change the Phase to New Construction on the Properties palette.	**NOTE**

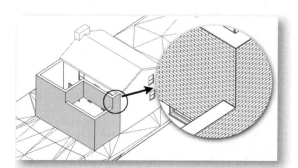

FIGURE 8.15 *View the model in 3D to see the completed Wall edits*

29. On the Project Browser, right-click the *{3D}* view and choose **Duplicate View >
Duplicate.**

   • Name the new view **New Addition Axon**.

30. Save the project.

Add the New Roof

Now that we have prepared the second floor Wall layout, we are ready to begin roofing the addition.

31. On the Project Browser, double-click to open the *Roof* plan view.

NOTE	Be sure to open *"Roof"* this time, not *"Existing Roof."*

32. On the Home tab, on the Build panel, click the ***Roof*** tool.
   The tools on the Draw panel should default to Boundary Line and Pick Walls.

   • On the Options Bar, select the Defines slope checkbox and in the Overhang field, type **6" [150]**.
   • Place the two sketch lines indicated in Figure 8.16.

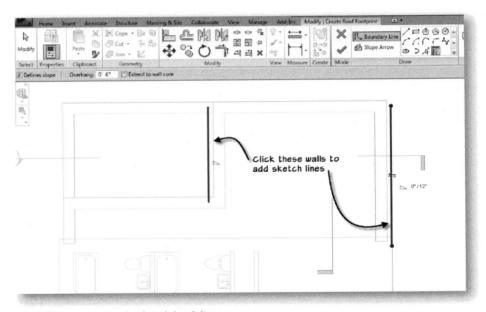

FIGURE 8.16 *Create the sloped sketch lines*

33. On the ribbon, click the ***Modify*** tool or press the ESC key twice.

34. Select both of these sketch lines.

   • On the Properties palette, beneath the "Dimensions" grouping, change the Roof slope. If you are working in Imperial units, type **6"** for the Rise/12" parameter. If you are working in Metric, set the Slope Angle to **26.57**.
   • Deselect the sketch lines.

TIP	As an alternative, you can select the sketch line and then edit the slope directly with the temporary dimension that appears next to the slope indicator.

35. On the Draw panel, click the **Line** tool.

- Clear both the "Defines slope" and "Chain" checkboxes.
- Draw a horizontal line aligned to the edge of the existing house Roof and the width of the two sketch lines we already have.
- Use the Pick Lines option and a **6" [150]** offset to create the top sketch line (see the left side of Figure 8.17).

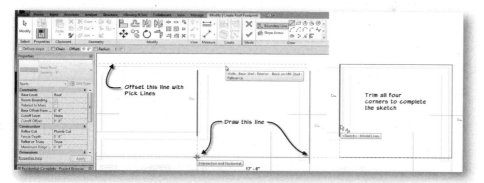

FIGURE 8.17 *Create two horizontal sketch lines that do not define slope*

36. Using the **Trim** tool, clean up the rectangular sketch shape (see the right side of Figure 8.17).

- On the ribbon, click the **Finish Edit Mode** button.
- On the Project Browser, double-click to open the *New Addition Axon* 3D view.
- Orbit the model to see the interaction of the two Roofs clearly.

Join Roofs

As you can see in the 3D view, the new Roof does not intersect with the existing one. This is easily corrected.

37. On the Modify tab, on the Geometry panel, click the **Join/Unjoin Roof** button.

Take a look at the Status Bar (lower left corner of the screen) and notice the message. You are prompted to: "Select an edge at the end of the roof that you wish to join or unjoin." We want to select one of the edges of the new construction Roof. The next prompt will ask us to select a face to which to join. In that case, we will select the face of the existing Roof.

- Click on the edge of the new construction Roof as indicated on the left side of Figure 8.18.
- Click on the face of the existing construction Roof as indicated on the right side of Figure 8.18.

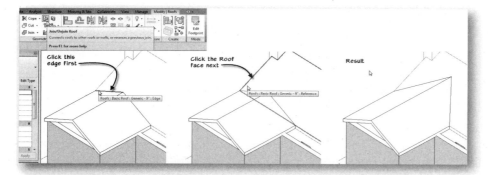

FIGURE 8.18 *Select an edge of the Roof to join to the face of the other Roof*

The new construction Roof will now extend over the existing Roof and form nicely mitered intersections. The same tool can be used in reverse if you ever need to unjoin a Roof.

38. Repeat the entire process to create the shorter new Roof on the other side of the addition (see Figure 8.19).

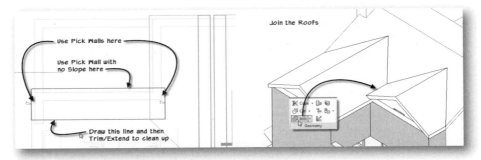

FIGURE 8.19 *Create the Roof on the other side of the addition*

Join Roofs works when the two Roofs meet perpendicular to one another. We will use a different technique to join the two new Roofs together at the small valley between them. For that we will use the more generic ***Join Geometry*** tool.

39. On the Modify tab, on the Geometry panel, click the ***Join*** button.

    This is a split button. The default is Join Geometry. The drop down button on the lower half gives access to unjoin geometry.

Again, watch the Status Bar prompts.

- Click on the first new construction Roof, and then click the other to join them.
- On the ribbon, click the ***Modify*** tool or press the ESC key twice.

40. Using the technique covered previously, Attach the Walls to the new Roofs (see Figure 8.20).

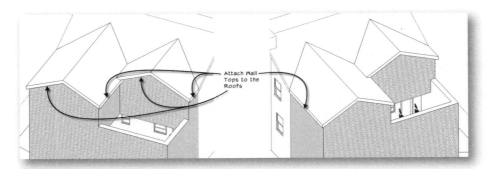

FIGURE 8.20 *Join the two new construction Roofs with Join Geometry*

41. Save the project.

EDITING ROOFS

Now that we have created the Roofs for the Residential project, let's turn our attention to Roofs editing techniques. In your own projects, you will work with Roofs in much the same way as other Revit Architecture elements—start by laying them out with simple generic parameters and then over the course of the project, layer in additional details and parameters as design and project needs dictate. In this sequence, we will explore some of the possibilities available for editing Roof elements.

Modify the Roof Plan View

You may have noticed that the *Roof* plan view looks a bit odd. The Roof elements are being cut in plan which shows us only part of the sloped surface. While appropriate for an attic space, this kind of display is not how we would represent a typical Roof Plan. To show the roof correctly, we want to see the entire Roof element uncut. To achieve this, we need to modify the view range of the *Roof* plan view.

> Continuing in the Residential Project.

1. On the Project Browser, double-click to open the *Roof* plan view.

   - Make sure you have nothing selected, then on the Properties palette, beneath the "Extents" grouping, click the Edit button next to "View Range."
   - In the "Primary Range" area, change the "Top" Offset to **15'-0" [4500]**.
   - Change the "Cut Plane" Offset to **10'-0" [3000]** (see Figure 8.21).

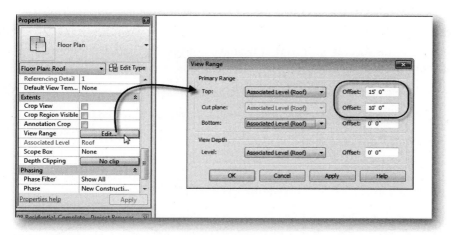

FIGURE 8.21 *Adjust the Cut Range of the Roof Plan view*

2. Click OK to return to the Properties palette.

   - Beneath the "Graphics" grouping, choose **None** for the "Underlay" and click Apply to see the results (see Figure 8.22).

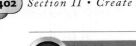

BIM *MANAGER NOTE* Roof plans provide an excellent case for View Templates. We learned how to apply a View Template to a view in the "Renaming Levels" topic in Chapter 4 and again in Chapter 7. Once you have established the settings for a standard roof plan, consider creating a Roof Plan View Template from it. You can do this on the View tab. On the Graphics panel, click the View Templates drop-down button and choose **Create Template from Current View**. If you want to use the View Template in other projects, or add it to your office standard project template (RTE) file, use the Transfer Project Standards tool on the Manage tab.

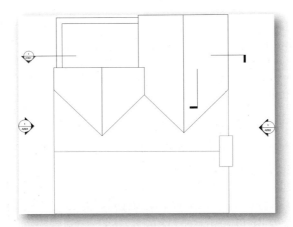

FIGURE 8.22 *Turn off the Underlay of the Second Floor and view the Roof Plan*

Understand Roof Options

The Roof element has several options that we have not yet explored. Let's take a look at some of them now.

3. On the Project Browser, double-click to open the *Longitudinal* section view.

- Zoom in Region around the eave at the left side of the section (see Figure 8.23).

Notice the way the Roof intersects the attached Wall.

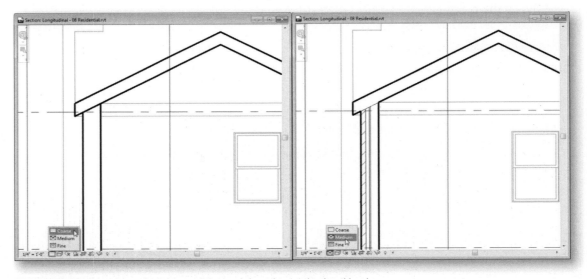

FIGURE 8.23 *Zoom in on the eave condition and then change the detail level*

Previously we swapped in a more detailed Wall Type for these Walls. However, since we are currently viewing the model in "Coarse" display mode, we do not see any difference in the Wall structure. To see the more detailed Wall structure, we need to adjust the detail display level of the current view.

4. On the View Control Bar, click the Detail Level icon and choose **Medium** (see the right side of Figure 8.23).

The Layers that make up the Wall Type's structure will appear including brick, a stud layer and air gap. This Detail Level will make it a little easier to understand the various Roof options that we are about to explore.

5. Select the Roof (the one sectioned in this view).

   • On the Properties palette, beneath the "Construction" grouping, change the "Rafter Cut" to **Two Cut – Plumb**.

The results will probably not be satisfactory.

   • On the Properties palette, set the "Fascia Depth" to **6" [150]**.
   • Repeat the process and choose **Two Cut – Square** (see Figure 8.24).

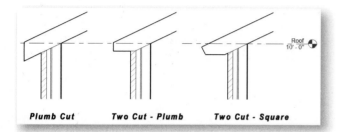

FIGURE 8.24 *Comparing the Rafter Cut options (Illustrations simplified for clarity)*

The section shown in the figure has been simplified to remove any of the beyond components (the Far Clip Offset of the section was adjusted to achieve this). However, the cut components of your model should closely resemble the figure. The panel on the left shows the original condition—which is the default. The middle and right panels show the conditions that we just tried.

6. Return to the Properties palette once more and change the "Rafter Cut" to **Two Cut – Plumb**.

When we were adding the Roof above, you may have noticed the "Extend to wall core" checkbox on the Options Bar. (We did not use this option when creating the Roof.) When you choose this option, the Overhang setting will be measured relative to face of the core layer rather than the finish face of the Wall. Figure 8.25 shows this option used with and without attaching the Walls to the Roof. (The Core layer of the Wall has been shaded gray in the figure for clarity). If you wish, you can experiment with these options. Remember, if you choose to edit the Roof while the section view is open, you will be prompted to open a more appropriate view—choose *Floor Plan:Roof.* Cancel or undo any edits made if you do experiment with this setting.

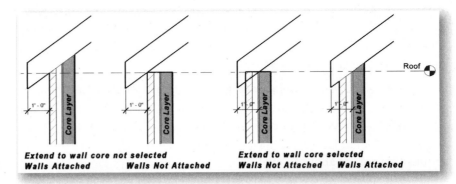

FIGURE 8.25 *Understanding the effect of the "extend to wall core" and Attach options*

There is one other setting worth exploring at this point. When you create your Roof, Revit offers two modes of construction: Rafter or Truss. The difference between these two settings is simply the point that is used as the spring point for the Roof.

7. Select the Roof.

- On the Properties palette, beneath the "Construction" grouping, change the "Rafter or Truss" to **Rafter**.

Notice how the entire Roof appears to move down. Rafter measures the plate of the Roof from the inside edge of the Wall. Truss measures from the outside edge. This option is only available for Roofs created using the "Pick Walls" option (see Figure 8.26).

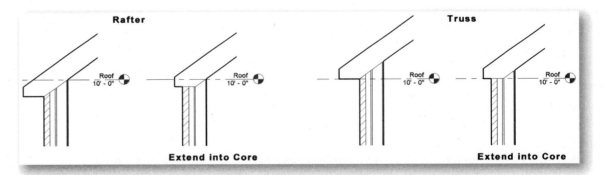

FIGURE 8.26 *Truss and Rafter settings impact how the Roof attaches to the Wall*

8. Back on the Properties palette again, experiment with changing the "Rafter or Truss" to **Rafter**, then back to **Truss** – finally leaving it at **Rafter**.

9. Select all Roofs (two new construction, one existing).

- On the Properties palette, change the "Rafter Cut" to **Two Cut – Plumb.**
- Set the "Fascia Depth" to **6" [150]**.
- Set "Rafter or Truss" to **Rafter** and then click Apply.
- If a warning dialog appears, click the Unjoin Elements button.

  Because of the change in the relationship between the Roof and the Wall on the left, Revit is "confused" and cannot maintain the join condition – we will fix this later.

10. Finally, select the wall on the left and use the ***Attach Top/Base*** tool to re-attach it to the roof. Attach it first to the existing Roof, then repeat for the Roof on the addition.

11. Save the project.

Create a Complex Roof Type

Much like Walls, Roof structure can be composed of several Layers. You can create your own Roof Type that contains the structure you need, or transfer an appropriate Type from another project using the Transfer Project Standards feature. In this example, we will build a new Roof Type from scratch.

12. Select all of the Roofs.

- On the Properties palette click the Edit Type button.

The Type Properties dialog will appear.

- Next to the Type list, click the Duplicate button.

A shortcut for this is to press ALT + D.	**TIP**

A new Name dialog will appear. By default "(2)" has been appended to the existing name.

- Change the name to **MasterRAC Wood Rafters with Asphalt Shingles** and then click OK (see Figure 3-43 in Chapter 3 for an example).

13. In the "Type Properties" dialog, next to the "Structure" click the edit button.

In the "Edit Assembly" dialog, you can see that the Roof Type currently contains only a single generic Layer. This is because we originally duplicated it from the generic Roof Type. You can add, edit, and delete Layers to the Roof structure in this dialog. We will keep the structure of our custom Roof Type simple. We need a structure layer, which will be wood rafters, a plywood substrate, and asphalt shingles for the finish layer. When you build a new Type in Revit, there are several things to consider. We can add as much or as little detail to the structure of the Roof Type as we wish. In some cases, it will prove valuable to represent each piece of the Roof's actual construction. However, also consider the potential negative impact that highly detailed Types can have on drawing legibility and computer performance. As a general rule of thumb, you should seek to build your models as accurately as possible while remembering that any architectural drawing includes a certain degree of abstraction as a matter of industry convention and the facilitation of clarity. All of these points hold true in other areas of Revit as well, such as creating and editing Wall Types. We will see more on this in coming chapters. With these issues in mind, we will abstract our Roof construction to just the three Layers noted previously, thereby excluding building paper, insulation, or interior finish. These items can be added to the Roof Type later (which will automatically apply to all Roof elements that reference the Type) or we can apply these items graphically as drafting embellishment in a Detail view (see Chapter 11 for more information on Detail views).

Edit Roof Structure

The "Edit Assembly" dialog lists each Layer of the Roof Type in a list with a numeric index number next to each one. There are four columns next to each item.

- **Function**—Click in this field for a list of pre-defined functions. The functions include "Structure," "Substrate," "Thermal/Air Layer," (2) "Finish" layers and

"Membrane." The number next to the function name indicates the priority of the Layer with regard to material joins. In this way, the Structure Layer of one Wall or Roof will attempt to join with the Structure Layer of another. One [1] is the highest priority, while five [5] is the lowest. "Membrane Layers" have zero thickness and thus do not have priority nor do they join.

- **Material**—Materials designate what the Layer is made from. Material properties include patterns, render material, shading, and even structural characteristics to represent real-life materials.
- **Thickness**—This is the dimensional thickness of the Layer.
- **Wraps**—Controls if the Layer wraps around corners at the ends or at openings. If this is not selected, the Layer simply cuts perpendicular at the ends and openings.
- **Variable**—When the Type is applied to a flat Roof, one component on the list can be given a variable thickness. Editing tools for manipulating flat Roofs (and Floors) provide a means to add high and low points to the surface of the Roof element. When the variable box is selected, such height variations are applied only to the variable component. (This is used to represent tapered insulation on flat roofs for example—see below.) When no layers use the variable option, thickness variations are applied to the entire Roof essentially warping the surface.

14. In the "Edit Assembly" dialog click in the Material cell for the Structure Layer (Layer number 2).

    A small browse icon will appear at the right side of the cell.

- Click the small browse icon to open the "Materials" dialog.
- From the "Name" list, select ***Structure - Wood Joist/Rafter Layer [Structure - Timber Joist/Rafter Layer]*** and then click OK.
- In the Thickness field, type **7 1/4" [190]** and then press ENTER (see Figure 8.27).

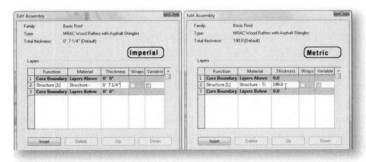

FIGURE 8.27 *Edit the existing Structure Layer to become Wood Rafters*

15. Beneath the list of Layers, click the Insert button.

    A new zero thickness Structure Layer will appear.

Notice that the new Layer appears within the "Core Boundary." Roofs, Walls, and Floors have a Core that contains the structural Layers. You can have additional Layers on either side or both sides of the Core. If a new Layer that you insert does not appear in the desired location, select it and then use the Up and Down buttons to adjust its position in the overall structure of the Roof.

- With the new Layer highlighted, click the Up button to place it above the Core Boundary.
- Change the Function of the new Layer to **Substrate [2].**

- Change the Material to ***Wood - Sheathing – plywood***.
- Set the Thickness to **5/8" [16]**.

16. Click the Insert button again.

- Change the Function of the new Layer to **Finish [4].**
- Change the Material to ***Roofing – Asphalt Shingle***.
- Set the Thickness to **¼" [6]**.

If you wish to see how the Type looks so far, click the Preview button at the bottom left corner of the dialog (see Figure 8.28).

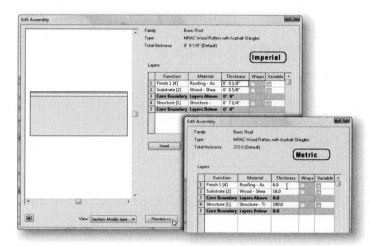

FIGURE 8.28 *Open the Preview window to see the completed structure graphically*

17. Click OK twice to return to the view window.

Notice that the new Roof Layers appear in the section view—if you switched to a different view, please return to the *Longitudinal* section view now. Also note that the new Layers will only appear if you left the section view in the Medium Display mode from above. If you set it back to Course, the graphics will simplify to show the outer edge of the Roof only. If you have difficulty seeing the sheathing and asphalt shingle Layers, try changing the scale of the view. The figure is shown at 1/2" = 1'-0" [1:25] (see Figure 8.29).

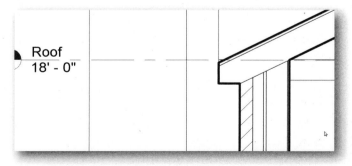

FIGURE 8.29 *Completed Roof Structure (shown at 1/2" = 1'-0")*

Apply a Host Sweep to Roof Edges

Creating the new Type and editing its structure provides a satisfactory representation of the overall Roof construction. However, the edges of the Roof could use some further embellishment. A Host Sweep allows us to apply a detailed profile condition to the edges of a Roof or along the length of a Wall. In this example, we will explore the use of Host Sweeps to apply fascia boards and gutters to our Roofs.

18. On the Project Browser, double-click to open the *New Addition Axon* view.

19. On the Home tab, click the drop-down button on the **Roof** tool and choose the **Fascia** tool.

There are some Fascia Types already in the project. However, as with the Roof Type above, we will create our own. A Fascia is a simple element whose primary parameter is the assignment of a Profile shape. The Roof here has a 6" [150] roof edge as defined earlier so we will want a Profile close to this depth.

20. On the Properties palette, click the Edit Type button.

   - Next to the Type list, click the Duplicate button.
   - Change the name to **MasterRAC Simple Fascia** and then click OK.
   - From the "Profile" list, choose **Fascia-Flat : 1 × 6 [M_Fascia-Flat : 19 × 140mm]**.
     If you like you can assign a material such as Finishes - Interior - Paints and Coatings – White.
   - Click OK (see the top of Figure 8.30).

A glance at the Status Bar will reveal the following prompt: "Click on edge of Roof, Soffit, Fascia, or Model Line to add. Click again to remove." With the *New Addition Axon* view open, it is easy to accomplish this.

21. Move the pointer over the various edges of the Roof elements on screen.

Notice that both the top and bottom edges of any given edge will pre-highlight (see the bottom of Figure 8.30).

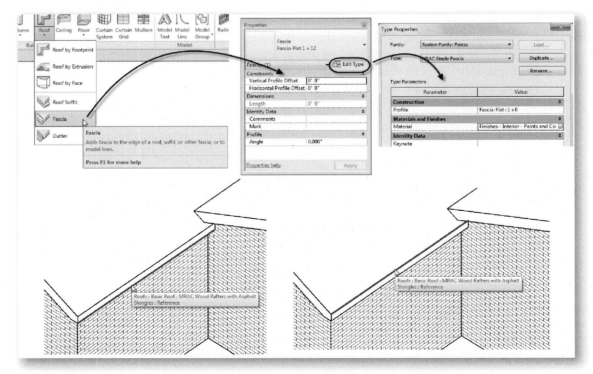

FIGURE 8.30 *Create a custom Fascia. You can Pre-Highlight either the top or bottom edge to place it*

- Click on a top edge to apply the Fascia.
- Place Fascia boards on the top edges of all of the horizontal (fascia) edges of the new Roof and all edges of the existing Roof as shown in Figure 8.31.

Be sure to click the top edge of each Roof edge, not the bottom.

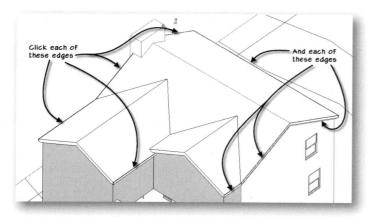

FIGURE 8.31 *Attach the Fascias to the horizontal edges of the Roofs.*

- On the ribbon, click the **Modify** tool or press the ESC key twice.

We did not apply a Fascia to the sloped (rake) edges of the new construction Roof yet because we are going to use a different Type for these. However, before we create the Type, we need a more complex Profile Family from the library.

22. On the Insert tab, click the **Load Family** button.

Either the *Imperial Library* or *Metric Library* folder should open automatically. If this has not occurred, you can use the shortcut icons on the left side of the dialog to jump to those locations. If you do not have these icons, the appropriate files have been provided with the dataset files installed from the *Aubin Academy Master Series: Revit Architecture 2012* student companion in the *Library* folder. You can navigate manually to that location and access the Families from there if necessary.

- Double-click on the *Profiles* folder and then double-click the *Roofs* folder.
- Select the *Fascia-Built-Up.rfa* file and then click Open.

This loads the new Profile Family into the current project. We now need to repeat the previous steps to create a new Fascia Type using this Profile.

23. On the Home tab, from the **Roof** tool, choose the **Fascia** tool.

- On the Properties palette, click the Edit Type button.
- Next to the Type list, click the Duplicate button.
- Change the name to **MasterRAC Built-up Fascia** and then click OK.
- From the "Profile" list, choose **Fascia-Built-Up : 1 × 8 w 1 × 6 [M_Fascia-Built-Up : 38 × 140mm × 38 × 89mm]** and then click OK.

24. Click each of the remaining Roof edges (remember to click the top edge).

- On the ribbon, click the **Modify** tool or press the ESC key twice.

Zoom in on one of the intersections between the Rake and Fascia conditions. If you select one of the Fascia boards, you will notice a drag control at the ends. You can stretch this control to modify the way the two Fascia boards intersect. Also, with one of the Fascia or Rake boards selected, you will notice that all of the boards select as a single element. To remove or add segments, use the ***Add/Remove Segments*** tool on the Modify Fascias tab. There is also a ***Modify Mitering*** tool. The options are "Horizontal," "Vertical," and "Perpendicular." Try them out if you wish to see how each option behaves. You can also use Join Geometry at the valley condition.

25. Save the project.

Add Gutters

Another type of Host Sweep available to Roofs is a Gutter. These are conceptually the same as Fascia boards. They use a Profile to determine the cross-section shape and sweep it along the path of the Roof edge(s) or the edges of other Host Sweeps.

26. Zoom out to see the whole Roof.

27. On the Home tab, from the ***Roof*** tool, choose the ***Gutter*** tool.

   • Click one of the top horizontal edges of the newly placed Fascia of the new construction Roof.

If the gutter is not visible, it needs to be flipped. There are flip controls on the selected Gutter.

   Click the Flip control if necessary to flip the gutter to the outside.

   • Before clicking the next segment, click the ***Restart Gutter*** button on the Modify | Place Gutter tab.

When you create Host Sweeps, multiple segments can be added to a single Sweep. If you later click to select it, you will notice that they all select as one. In this case, we will have more flexibility to select and flip our gutters if we start each segment as a new Sweep.

28. Add Gutters (and flip as required) to the remaining horizontal new construction Roof edges (see Figure 8.32).

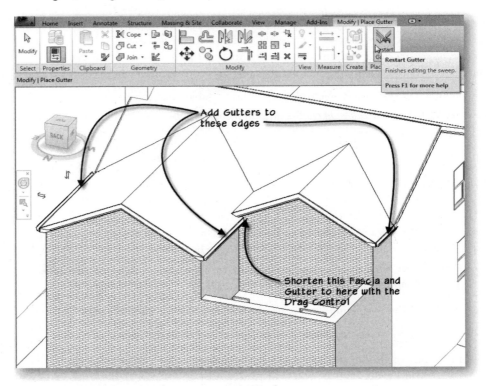

FIGURE 8.32 *Add Gutters to the new construction Roof*

After you place the Gutters, pre-highlight them and notice the one in the valley between the two new Roofs is too long. You can use the Drag control on the Fascia board (and the Gutter if necessary) to shorten it.

29. On the Project Browser, double-click to open the *Longitudinal* section view.

30. Zoom in on one of the Gutters.

   • On the View tab, on the Graphics panel, click the **Thin Lines** button.

Depending on which line you clicked (the Roof edge or the Fascia edge) when adding the gutters, you may notice that the Gutter overlaps the Fascia board. We can delete the gutter and re-add it, or we can apply an offset to the Gutter equal to the thickness of the Fascia profile to compensate for this.

31. Select any of the Gutters that require adjustment.

   • On the Properties palette, change the "Horizontal Profile Offset" to ¾" **[19]** and then click OK (see Figure 8.33).

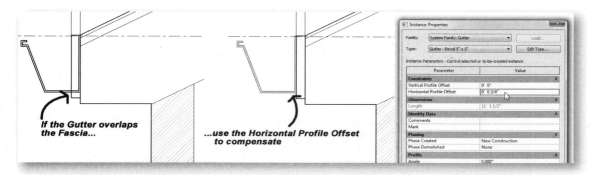

FIGURE 8.33 *Adjust Gutter Horizontal Offset as required*

32. Turn off **Thin Lines**.

The lineweight for the Fascia boards and Gutters in the section view is the same as that used for the Walls. Since both of these items have a very small thickness, we have two very bold lines right next to each other and the result is not very legible. You can override the lineweight (and other display characteristics) of elements directly in the view. Overrides can be applied to most elements at the category or element level. In the case of host sweeps, the cut lineweight can only be overridden at the category level. If you wish to try this, simply select one of the Gutters or one of the Fascia boards. Right-click and choose **Override Graphics in View > By Category**. (This command is also on the View panel of the ribbon.) In the "Visibility/Graphic Overrides" dialog that appears, the element category for the item you selected will already be highlighted (scroll down to see). In the Cut column, click in the Lines field. An "Override" button will appear—click it. In the "Line Graphics" dialog, change the Weight to 1 or 2 and then click OK twice to return to the view window and see the results. Repeat as necessary on other elements.

You can perform such edits to nearly any element. In many cases, you can also choose **Override Graphics in View > By Element**. In general, a category-level edit should be attempted first. It will apply to all similar elements in the current view. If you want to override just a specific element(s), use the "By Element" option instead.

It is important to get the overall defaults and global settings to your liking first. For example, if you find yourself applying the same override repeatedly, consider making the change "globally" in the "Object Styles" dialog. Object Styles establish the default graphical settings for all elements in all views throughout the project. Overrides, both category and element, modify the global settings only in current view. In some cases, what you may actually need is to edit the global setting making overrides unnecessary. To edit Object Styles, click the Manage tab, and then choose **Object Styles** from the *Settings* drop-down button. Remember, after applying category overrides, you can reuse them in other similar views by creating and applying a View Template.

Skylights

As a finishing touch to the residential Roof, let's add a skylight in one of the new Roofs. To do this, we must load another Family into the project.

33. On the Project Browser, double-click to open the *Roof* plan view.

34. On the Insert tab, click the **Load Family** tool.

   • Double-click on the *Windows* folder.

   • Select the *Skylight.rfa* [*M_Skylight.rfa*] file and then click Open.

35. On the Home tab, click the **Window** tool.

   • From the Type Selector, choose **Skylight : 28" × 38" [M_Skylight : 0711 × 0965mm]**

36. Place it approximately as indicated in Figure 8.34.

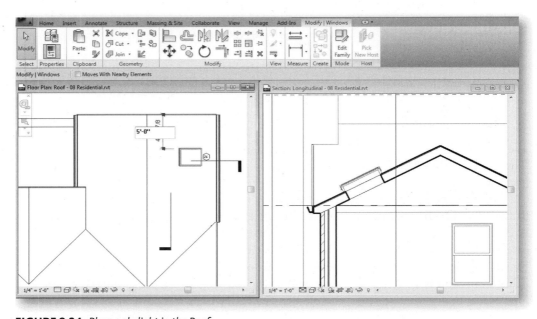

FIGURE 8.34 *Place a skylight in the Roof*

   • Click the **Modify** tool and then select the Window.

   • Adjust the witness lines as necessary and edit the value of the temporary dimension to **5'-0" [1500]** from the exterior (top) Wall as shown in Figure 8.34.

37. On the Project Browser, double-click to open the *Longitudinal* section view.

The skylight should appear cutting through the Roof. If the section does not intersect the roof window, you can move the section line slightly in plan.

38. Save the project.

Fine-tuning

More work could be done on the Roofs, to refine them further. Feel free to experiment further and make your own modifications if you wish.

CREATING FLOORS

Use Floor elements to model the floor platforms or slabs in your projects. Most often, the Floor element is a simple horizontal structure that you create via a closed sketch of its plan boundary. In some cases, the Floor may slope like in parking garages or theaters. In this topic, we will add and modify some Floor elements.

Add Floors

We have already worked with Floors a little in our commercial project. However, you have probably noticed that our residential project currently has no Floors.

1. On the Project Browser, double-click to open the *First Floor* plan view.
2. On the Home tab, click the **Floor** tool.

   On the Draw panel, the **Boundary Line** and **Pick Walls** tools should be active by default.

   • On the Options Bar, place a checkmark in the "Extend into wall (to core)" checkbox.

3. Click on the vertical Wall of the addition on the left.

   • Click on each of the other two exterior Walls of the new addition.
   • Click the existing Wall between the house and the addition.
   • On the ribbon, click the **Modify** tool or press the ESC key twice.

Sketch lines will appear at all four Walls. The one for the existing Wall is currently on the inside edge, but we need it to be on the exterior side adjacent to the new construction.

4. Drag the horizontal sketch line between the existing house and the addition to the other side of the existing Wall (see Figure 8.35).

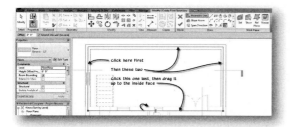

FIGURE 8.35 *Modify the sketch lines*

In this case, we can not use the flip control as it would flip all four lines. This is why we are dragging only the one line manually.

5. Deselect the line and then on the Properties palette, click the Type Selector.

   • From the Type list, choose **Wood Joist 10" - Wood Finish [Standard Timber-Wood Finish]**.

The structure of the Floor is similar to the structure of the Roof that we explored earlier. Feel free to edit the Type Parameters to study the structure but be sure to not make any changes at this time.

6. On the ribbon, click the ***Finish Edit Mode*** button.

   • In the dialog that appears, click Yes to join the geometry (see Figure 8.36).

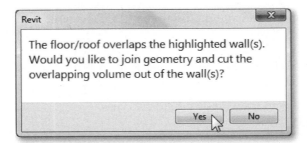

FIGURE 8.36 *Allow Revit Architecture to join the Floor to the neighboring Walls*

7. Repeat the entire process on the second floor.

Create the sketch the same way initially, however, modify the sketch to conform to the "L" shape (excluding the outdoor patio) of the interior space of the addition. Use the ***Trim/Extend*** tool to do this (see Figure 8.37). Remember to edit the Floor Properties and choose the same Type as the first floor.

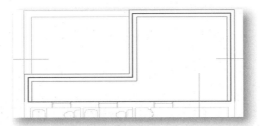

FIGURE 8.37 *Create the sketch for the second floor*

Create a Floor Type

The last Floor that we need to make is the one for the patio space on the second floor. The process is similar to the above. We are creating it separately because it will use a different type.

8. Click the ***Floor*** tool and use the same options.

9. Create sketch lines for each of the four Walls that make up the patio and use the Trim/Extend tool to complete the shape (see Figure 8.38).

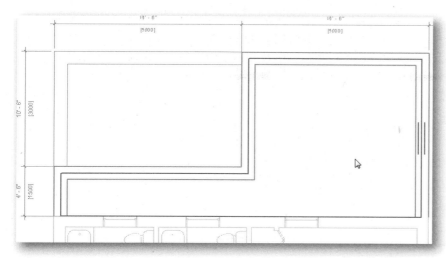

FIGURE 8.38 *Create the sketch for the patio floor*

10. On the Properties dialog, click the Edit Type button.

    - From the Type list, choose **Wood Joist 10" - Wood Finish [Standard Timber-Wood Finish]**.
    - Click the Duplicate button.
    - Change the name to **MasterRAC Patio Floor** and then click OK.

The patio will have a wood deck built up on top of sleepers to provide drainage below. Therefore, we can use nearly the same component makeup as the other Floor system and simply insert the sleeper Layer. While this Floor Type would have a membrane Layer and flashing, we will not include those in the model but rather show those components later in details (see Chapter 11 for more information on detailing).

11. Click the Edit button next to Structure, select Layer 1, and then click the Insert button to add a new Layer.

    - Move it to down to beneath the Finish Layer (but above the Core Boundary).
    - Set the new Layer Function to **Substrate [2].**
    - Change the thickness of the new Layer to **1 ½" [40]** and change the Material to **Wood - Stud Layer**.
    - Click OK to return to the "Type Properties" dialog.

12. Change the Function to Exterior, and then click OK.

    - On the ribbon, click the **Finish Edit Mode** button.
    - In the dialog that appears, click Yes to accept joining with the Walls.

Adjust Floor Position and Joins

A couple of issues remain with the two Floors on the second floor. First, the Walls that highlighted automatically for join did not include the two Walls we added at the start of the chapter. Second, the patio Floor is too low relative to the other one. We can see these issues best in section.

13. On the Project Browser, double-click to open the *Longitudinal* section view.

Examine the section to see both conditions noted here.

14. Select the patio Floor (it is on the right in the section and may still be selected).

- On the Properties palette, for the "Height Offset From Level" value, type **1½" [40]**.

This is the same amount as the thickness of the sleepers. By default, the entire thickness of Floors is set below the associated level. This is why we need to shift the Floor to properly represent that the sleepers are built up on top of the floor structure.

15. On the Modify tab of the ribbon, click the ***Join*** tool.

- For the first pick, click the patio Floor.
- For the second pick, click the vertical Wall (see Figure 8.39).

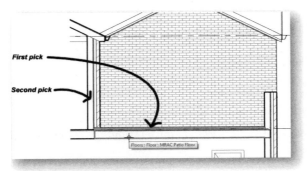

FIGURE 8.39 *Join the Floor to the Wall*

The Join command will remain active. You can perform several Join operations in a row.

16. Join the same Wall to the other Floor.
17. Join the two Floors to one another (see Figure 8.40).

TIP	Note the "Multiple Join" option on the Options Bar. Place a checkmark in this box to Join several elements in one operation.

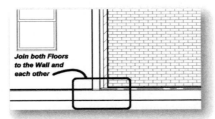

FIGURE 8.40 *Join the Floors and the Wall*

18. On the Project Browser, double-click to open the *Second Floor* plan view.

- Create a section cutting vertically (parallel to *Transverse*) through the patio (see Figure 8.41, left side).

  The ***Section*** tool is on the View tab of the ribbon.

- Rename it **Patio Section**.
- Open this section view, change the Detail Level to **Medium** and then repeat the steps here to join the Floors and Wall (see Figure 8.41, right side).

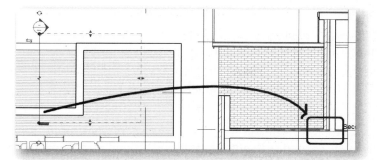

FIGURE 8.41 *Cut a section to assist in joining the remaining Floors and the Wall*

| Some joins may already be complete requiring no further action. Simply complete the joins that are necessary. | NOTE |

19. On the ribbon, click the **Modify** tool or press the ESC key twice.

Previously we mentioned the priorities of the Layers within the Roof (also Wall and Floor) structure. In these two sections, you can see this interaction very clearly. Notice the way that the structural Layer of the Floor cuts into the Core of the Walls.

20. Save and close the project.

COMMERCIAL PROJECT ROOF

Our commercial project already has a Roof. However, it is currently just a flat slab. Also, the Stair tower does not yet have a Roof. Our aim in this section will be to refine the Roof element already in the commercial project and to add additional required Roof elements. We will also begin work on our commercial project's Roof plan view.

Load the Commercial Project

Be sure that the Residential Project has been saved and closed.

1. On the QAT, click the Open icon.

| The keyboard shortcut for Open is CTRL + O. **Open** is also located on the Application menu. | TIP |

• In the "Open" dialog box, browse to the location where you installed the *MasterRAC 2012* folder, and then the *Chapter08* folder.

2. Double-click *08 Commercial.rvt* if you wish to work in Imperial units. Double-click *08 Commercial Metric.rvt* if you wish to work in Metric units

   You can also select it and then click the Open button.

| Some minor modifications have been made to the model since the last chapter. The layout of Walls on the Roof level is slightly different than it was in the previous chapter. As such, the core Walls were shortened to the level below and a new set of Walls added on the Roof. For this reason, please be sure to use the dataset provided with this chapter rather than attempting to continue in your previous files. | NOTE |

Create a Roof by Extrusion

Until now we have created our Roofs with the footprint option. It is also possible to create a Roof by sketching the profile of it in section and extruding this profile to form the Roof. In the Quick Start chapter we looked briefly at creating a Roof by Extrusion. Let's look at that process in more detail for the Roof at the top of our Stair tower.

3. On the Project Browser, double-click to open the *South* elevation view.

To assist us in placing the Roof, we will add some new Levels. You may recall from Chapter 4 that a Level can be created with automatically associated plan views, or it can be created without them and simply used for reference. We do not need a separate plan for the Roof of the Stair Tower, so the Levels we add here will not have associated plan views.

4. Select the Roof Level line.

    *   On the Modify Levels tab, click the **Copy** tool.
    *   Click anywhere to set the start point.
    *   Move the mouse straight up, type **8'-0" [2400]** and then press ENTER.
    *   Repeat the process and create another copy **4'-0" [1200]** above the previous copy (or **12'-0" [3600]** above the Roof Level).

Notice that when you deselect the Levels, both copies have black-colored Level heads. A Level head will be blue if it has an associated floor plan view (or ceiling plan), and black if it does not. Take a look at the Floor Plans on the Project Browser to confirm that no new floor plan views have been created. (If you wish to create a Level using the **Level** tool instead, it defaults to adding plan views: however, you can turn this option off on the Options Bar before drawing the Level.) You can also add plans later using the **Plan Views** tool on the View tab of the ribbon.

5. Click on the blue text of the new Level heads and rename the lower one to **Stair Roof Low** and the upper one to **Stair Roof High** (see Figure 8.42).

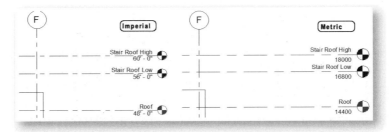

FIGURE 8.42 *Copy two new Levels without Associated Floor Plan views*

6. On the Home tab, click the drop-down button on the **Roof** tool and then choose the **Roof by Extrusion** tool.

A "Work Plane" dialog will appear. The Work Plane is the plane in which we will sketch. In this case, because we are making a Roof by extrusion, an effective Work Plane will be perpendicular to the Roof Levels. Any of our numbered Grid Lines can serve this purpose—the numbered Grid Lines form planes parallel to the screen in the current elevation view. Once we have chosen a plane, we will be able to sketch the shape of our Roof. When we finish the sketch, it will extrude perpendicular to the selected plane.

*   In the Roof "Work Plane" dialog, choose Grid 4 (from the "Name" list) and then click OK (see Figure 8.43).

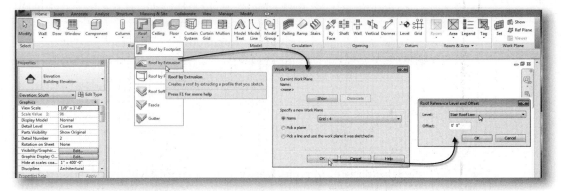

FIGURE 8.43 *Choose a Column Grid Line as the Work Plane*

As you can see in Figure 8.43, it is also possible to pick the face of some geometry such as a Wall in the model to set the Work Plane. (We took this approach in the Quick Start chapter.) In this case the named Plane associated with the Grid line works best. It is also common practice to create a Reference Plane to use as the Work Plane for the Roof.

- In the "Roof Reference Level and Offset" dialog that appears, select **Stair Roof Low** and then click OK.

The Roof must be associated with a Level. As you can see in the dialog, you can choose to create the Roof at any Level in the project and add an offset above or below the Level if appropriate. In this case, we specifically created Stair Roof Low for the task at hand, so no offset is required.

The Modify | Create Extrusion Roof Profile tab appears on the ribbon showing common sketching tools.

7. On the Draw panel, click the "Start-End-Radius Arc" icon.

- For the Arc start point, click the intersection of the Stair Roof Low Level and the left edge of the core Wall (see the top panel of Figure 8.44).

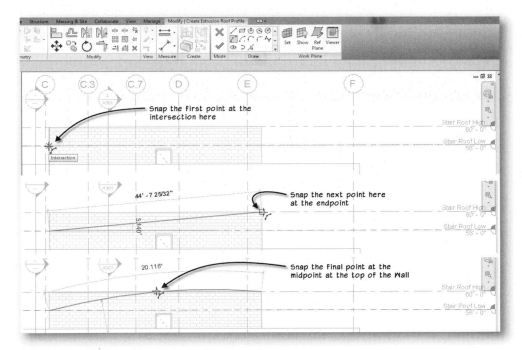

FIGURE 8.44 *Sketch an Arc Profile for the Extruded Roof*

- For the Arc end point, click the endpoint of the right edge of the core Wall at the Stair Roof High Level (see the middle panel of Figure 8.44).
- For the Arc intermediate point, click at the midpoint of the top edge of the core Wall (see the bottom panel of Figure 8.44).
- On the ribbon, click the **Modify** tool or press the ESC key twice.

When you draw a Roof by extrusion, you create an open shape, *not* a closed shape. The thickness of the Roof material will be determined by the Roof Properties and the Type assigned to it just like the other Roofs; therefore, we *do not* need to sketch the thickness of the roof.

8. On the Properties palette from the Type Selector, choose **Steel Truss - Insulation on Metal Deck – EPDM [Steel Bar Joist - Steel Deck - EPDM Membrane]**.

In the "Instance Parameters" area, notice that the Work Plane parameter is set to Grid 4. (This is unavailable for edit—to change the Work Plane after creation, select the Roof, and then click the **Edit Work Plane** button on the Work Plane panel of the ribbon.)

- For the "Extrusion Start" type **-2'-0" [-600]**.
- For the "Extrusion End" type **22'-0" [6600]**.

TIP	An alternate way to set the extrusion distance is to wait until after you have finished creating the Roof. It will be extruded an arbitrary distance initially. You can then move to a plan view and use the shape handles to adjust the extrusion depth graphically. This technique was used in the Quick Start chapter.

You probably noticed that there were no overhang parameters in the "Instance Properties" dialog or on the Options Bar. To create an overhang, you simply edit the sketch line, or add additional segments.

9. On the Draw panel, click the **Line** tool.

- Add a 1'-6" [450] long horizontal line at each end of the Arc (see Figure 8.45).

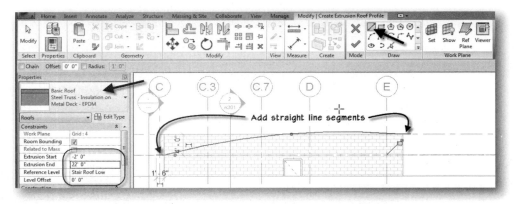

FIGURE 8.45 *Sketch overhangs by adding additional Line segments to the sketch*

10. On the ribbon, click the **Finish Edit Mode** button.

The Roof should appear with its thickness determined by the Type that we chose.

11. On the Project Browser, double-click to open the *{3D}* 3D view (see Figure 8.46).

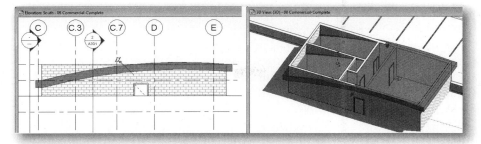

FIGURE 8.46 *The new Roof intersects the core Walls*

It appears as though our Walls could use some adjustment.

12. On the Project Browser, double-click to open the *Roof* plan view.

   • Dragging from left to right, surround the entire core.

Walls, Stairs, and other objects will be selected by this action.

   • On the Modify | Multi-Select tab of the ribbon, click the **Filter** tool.
   • In the Filter dialog, click the Check None button, click the checkbox next to Walls, and then click OK.

This will remove all other elements from the selection leaving only the core Walls on the Roof level selected.

13. On the Project Browser, double-click to open the *South* elevation view.

If you prefer, you can tile the two windows with the command on the Window menu instead. The selection of Walls will remain active. If the selection does not remain active, right-click in the *South* elevation view window and choose **Select Previous**.

   • On the Modify Walls tab, on the Modify Wall panel, click the **Attach Top/Base** button.
   • Verify that Top is selected on the Options Bar and then click on the Roof (see Figure 8.47).

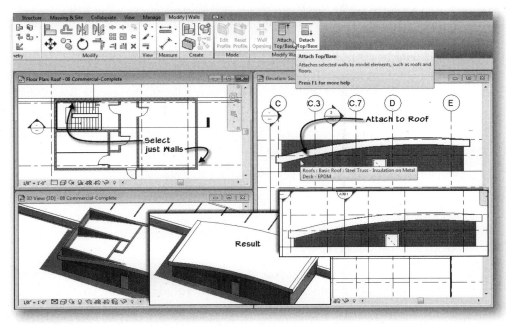

FIGURE 8.47 *Attach the core Walls to the imported Roof*

STRUCTURAL CORE WALLS

You may recall that in previous chapters we copied the Walls to the Structural model using the Copy/ Monitor tool. In the last chapter, the Walls from the Structural model were hidden using Visibility/ Graphic Overrides (VG) and a custom View Template (see the "topic" in Chapter 7). This means that we are not seeing the redundant Walls that would otherwise appear here from the Structural file. You can choose to return to the Structural model later, and update the Copy/Monitor Walls if you wish. Refer to the "Perform a Coordination Review" topic of the previous chapter for more details on the process.

You can study this change in other views as well, if you like, such as the *{3D}* view.

Working With Roof Sub-Elements

The final exercise in this chapter will be to add drainage sloping to the commercial project flat roof. In an early edition of this book, this task was accomplished with Slope Arrows. The Roof sub-element editing tools are largely considered a better way to represent the slope on flat Roofs and slabs. However, since Slope Arrows are still quite useful for certain roof modeling tasks, the original tutorial is included in PDF format in the *Chapter08* folder with the dataset files installed from the student companion. If you wish to perform that exercise, please perform a Save As (from the Application menu) to create a copy of your project in its current state. You can perform the steps here in one copy, then re-open the saved copy and perform the steps in the PDF. Use the PDF exercise as an opportunity to learn about slope arrows, but again realize, that in most cases, the sub-element editing process would be considered a better approach to drainage sloping on a flat roof. If you do not wish to do the tasks in the PDF, you can skip the Save As step.

Edit View Range

Before we can begin making modifications to our Roof, we need to make a few adjustments to the Roof and the *Roof* plan. If you study a section, you will note that the Roof element's thickness projects above the Level. This is the default behavior.

1. On the Project Browser, double-click to open the *Section at Stair* view.

Notice that the Roof sits higher than the Floor slab in the Stair tower.
- Select the Roof element and on the Properties palette, change the Base Offset From Level to **-1'-0"** [**-300**] .

There is a small problem with making this change. The Roof is now outside of the View Range in the *Roof* plan view. Let's fix that next.

2. On the Project Browser, double-click to open the *Roof* plan view.

- Make sure that you have no objects selected, then on the Properties palette, edit the View Range in the "Extents" grouping.
- Change offsets of both the Primary Range Bottom and the View Depth to **-1'-0"** [**-300**] and then click OK.

Another issue that now becomes apparent is that the height of the Architectural Columns protrudes into the thickness of the Roof. To address this, we will attach them to the underside of the Roof similar to how we edited the Core Walls in the previous sequence.

3. Dragging from left to right, make a selection box surrounding the entire building.

   • Click the Filter tool, click the Check None button and then select only the Columns checkbox near the top of the list.

   • Click OK to complete the selection.

   • On the Project Browser, double-click to open the *Longitudinal* section view.

   The selection of Architectural Columns should still be active. If it is not, right-click and choose **Select Previous**.

4. On the Modify l Columns tab, click the **Attach Top/Base** button.

   • Following the prompt on the Status Bar, click to select the flat Roof element in the model.

   • Return to the *Roof* plan view and check the results.

The Architectural Columns should no longer show through on the *Roof* plan.

Using Shape Editing Tools

When you have a flat Roof or Floor element in your model (no edges sloped and no slope arrows), you will see the Shape Editing panel on the Modify Roofs tab of the ribbon when the Roof or Floor is selected. If even one edge of the Roof or Floor is set to slope defining, the tools will not appear (see Figure 8.48).

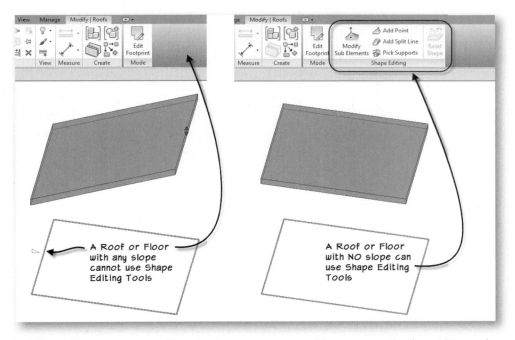

FIGURE 8.48 *Roofs or Floors with no sloping edges in their sketch have access to the Shape Editing tools (the sketch lines have been superimposed and enhanced for clarity)*

Continue in the *Roof* plan view.

5. Select the flat Roof element. (Use the TAB key as necessary.)

   Notice the collection of Shape Editing tools appears on the ribbon as shown on the right side of Figure 8.48.

Moving left to right, the tools are as follows:

- **Modify Sub-Elements**—This tool can be used to adjust the height of any points or edges drawn with the Draw Points and Draw Split Lines tools.
- **Add Point**—Use this tool to add points to the surface of the Roof or Floor. Each point has a height that you can adjust to either a negative or positive offset from the Roof or Floor level.
- **Add Split Line**—This tool adds elevation changes using lines instead of points. Like the Draw Point tool, each line has a height that you can adjust to either a negative or positive offset from the Roof or Floor level. You can also edit the height of the line's endpoints independently.
- **Pick Supports**—If you have structural supports set at accurate levels, you can use them to indicate the level changes of the Roof or Floor.
- **Reset Shape**—The button (grayed out in Figure 8.48) is used to remove all edits and return the shape of the Roof or Floor to flat with no slopes.

6. On the Shape Editing panel, click the **Add Split Line** tool.

- Click a point on the outside edge of the core Wall just to the left of the Door.
- Drag straight down and click a point on the inside edge of the parapet Wall (green dashed line) (see Figure 8.49).

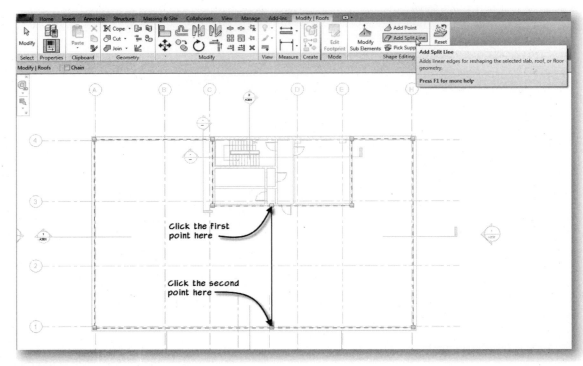

FIGURE 8.49 *Add a Split Line down the middle of the Roof*

7. Draw another vertical Split Line along column grid line B.

- Start just above intersection 3B and end just below intersection 2B.

The Grid lines will be grayed out and might be tough to see. As you pass your cursor over them, however, they will pre-highlight.

8. Draw a final vertical Split Line on the right side of the plan parallel to column grid line E.

   • Start at the corner of the core Walls and draw straight down to just below and to the right of intersection 2E (see Figure 8.50).

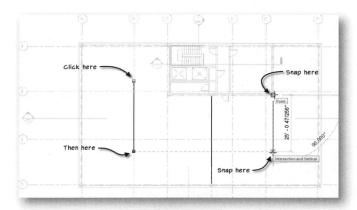

FIGURE 8.50 *Add two additional Split Lines*

9. On the Shape Editing panel, click the **Modify Sub-Elements** button.

   • Select the first Split Line drawn (the one in the middle).
     An elevation label will appear in the familiar blue temporary dimension color.

   • Click on the label (currently reading 0) and change it to **4" [100]**.

10. Select one of the other two Split Lines drawn and change the elevation to **-2" [-50]**.

   • Repeat for the remaining Split Line (see Figure 8.51).

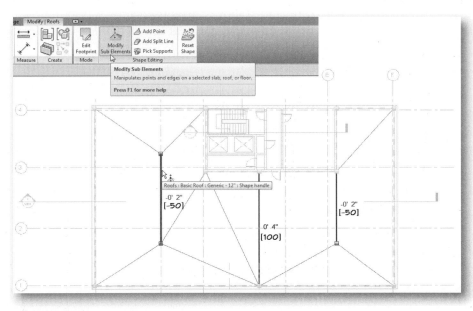

FIGURE 8.51 *Edit the elevation of the Split Line edges*

11. On the Project Browser, double-click to open the *{3D}* 3D view.

Changing views will terminate the sub-element editing mode. You can also click the *Modify* tool first if you prefer. Notice that the Roof now displays the edges of the ridges and valleys for the sloping planes (see Figure 8.52).

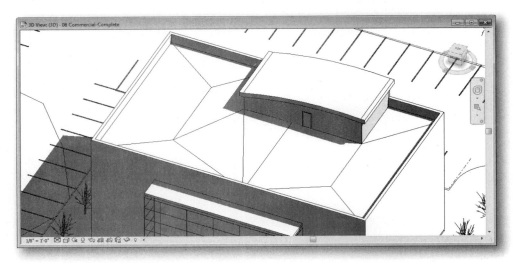

FIGURE 8.52 *Study the results in 3D*

The default behavior in the sub-element editing mode is for the entire Roof slab to be affected by the slope. If you prefer, you can edit the structure of the Roof Type applied to the Roof element and make one of its material layer thicknesses variable. When doing so, the bottom surface of the Roof will remain flat, while the top surface slopes according to the Split Lines and elevation points added above. This is an effective way to represent tapered rigid insulation in the construction (see the next few steps). The best way to see these sometimes subtle variations is in a section view.

12. On the Project Browser, double-click to open the *Longitudinal* section view.

Take a close look at the Roof element. Notice that both the top and bottom surfaces are sloped (maintaining a uniform thickness). We can designate one of the layers in the Roof Type structure as a variable thickness. When doing so, the bottom layers will remain flat, the variable layer will have a flat bottom and sloping top surface, and any layers on top of the variable one will follow the slope (with uniform thickness).

13. Select the Roof.

- On the Properties palette, click the Edit Type button.
- Click the Edit button next to Structure.
- Place a checkmark in the "Variable" column next to layer 2 and then click OK two times (see the top half of Figure 8.53).

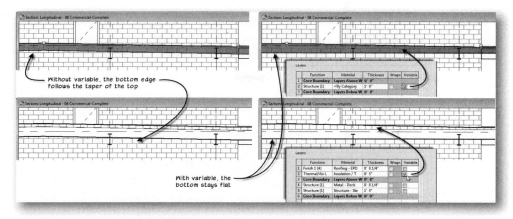

FIGURE 8.53 *Comparing the Roof with and without a variable thickness layer*

Currently we are using a Generic Type with only a single layer. To appreciate fully the effect of a variable component, we should assign a complex Roof Type.

14. With the Roof still selected, from the Type Selector, choose **Steel Truss - Insulation on Metal Deck – EPDM [Steel Bar Joist - Steel Deck - EPDM Membrane]**.

    - Edit the Type Properties again and then click the Edit button next to Structure.
    - Place a checkmark in the "Variable" column next to layer 2 (the insulation this time) and then click OK two times (see the bottom half of Figure 8.53).

Now that we have a more detailed structure in place you can see that we have effectively represented tapered insulation and the support structure beneath it remains level. However, this Roof Type's lowest layer is meant to represent the steel bar joists. This means that the entire roof structure sits too high in the model.

15. Select the Roof, and on the Properties palette, in the "Base Offset From Level" field, type **-1'-6" [-450]**.

Finally, the Door from the stair tower to the roof is still too low. We can move it up using nearly the same procedure (see Figure 8.54).

16. Select the Door, and on the Properties palette, in the "Sill Height" field, type **1'-0" [300]** and then click OK.

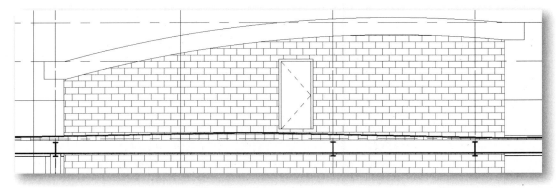

FIGURE 8.54 *Move the Roof element down with a negative offset from the level and move the Door up with a positive sill height offset*

If you wish, you can follow procedures covered in the previous chapter to add a few steps up to the Door inside the building on the *Roof* plan. Feel free to study the model in other views and experiment further with the Roofs and Floor slabs.

17. Save and Close the commercial project file.

SUMMARY

- Roofs are sketch-based elements that can be generated from existing Walls or manually drawn sketch lines. A single Roof can use a combination of each.
- Walls can be attached to Roofs and remain attached as the model is modified.
- Join Roofs together to resolve the intersection of complex Roof planes.
- Roofs have many options for their construction and how they interact with neighboring elements.
- You can apply edge conditions to each Roof Fascia and Rake.
- Gutters and other sweep profiles and Host Sweeps can be applied to Roof edges.
- Skylights interact with and cut holes in Roofs in the same way Windows interact with Walls.
- Create complex Roof Types that include layers of structure that share many features with complex Wall Types.
- Floors are added and modified via sketch mode similar to Roofs.
- Floors may also have complex structure like Roofs and Walls.
- Edit a Roof footprint or extrusion at any time by editing the sketch of the Roof.
- Roofs can be created with a plan sketch like Floors or via an extruded profile shape, usually drawn in elevation or section views.
- Flat Roofs can accurately represent subtle drainage slope using the sub-element editing tools.
- You can represent variable thickness materials (such as tapered rigid insulation) by enabling variable thickness for the layer in the "Roof Type Properties" dialog.

INTRODUCTION

In this chapter, we will enclose our commercial project with a building skin. The skin will be comprised of a masonry enclosure on three sides, with various curtain wall elements on the front and sides of the building. The front façade curtain wall begins on the second floor and spans the height of the third and fourth floors. In Chapter 4, we created a massing element to suggest this design element and applied a temporary Curtain System to it. We will now replace this with a more refined Curtain Wall.

OBJECTIVES

In order to complete the shell of the commercial project we will apply a detailed Wall type to the exterior Walls already in the project. We will also build Curtain Walls and Curtain Systems for the front and side façades. Upon completion of this chapter, you will be able to:

- Create and swap Wall types
- Add Curtain Walls to the model
- Modify a Curtain Wall
- Build a Curtain System type
- Build and add a Stacked Wall

CREATING THE MASONRY SHELL

The majority of the skin of the commercial building is comprised of masonry Walls. We will perforate portions of this masonry skin with Curtain Systems later in the chapter, but we will begin by swapping out the simple generic Wall types used in early chapters with a more detailed Wall type appropriate to the design at this stage.

Install the Dataset Files and Open a Project

The lessons that follow require the dataset included on the Aubin Academy Master Series student companion. If you have already installed all of the files from this site, skip to step 3 to begin. If you need to install the files, start at step 1.

1. If you have not already done so, download the dataset files located on the CengageBrain website.

   Refer to "Accessing the Student Companion site from CengageBrain" in the Preface for information on installing the dataset files included in the Student Companion.

2. Launch Autodesk Revit Architecture from the icon on your desktop or from the *Autodesk > Autodesk Revit Architecture 2012* group in *All Programs* on the Windows Start menu.

TIP	You can click the Start button, and then begin typing Revit in the Search field. After a couple letters, Revit Architecture should appear near the top of the list. Click it to launch to program.

3. On the QAT, click the Open icon.

TIP	The keyboard shortcut for Open is CTRL + O. **Open is also located on the Application menu.**

   • In the "Open" dialog box, browse to the location where you installed the *MRAC 2012* folder, and then the *Chapter09* folder.

4. Double-click *09 Commercial.rvt* if you wish to work in Imperial units. Double-click *09 Commercial Metric.rvt* if you wish to work in Metric units.

   You can also select it and then click the Open button.

Creating a Wall Type

As you can see, the project is largely unchanged from the previous chapter. We still have the very simple generic Wall Type used for the building skin. The first thing we'll do is refine that a bit.

5. On the Project Browser, double-click to open the *Level 1* floor plan view.

6. Pre-highlight one of the exterior Walls (try the vertical one on the left).

   • Press the TAB key once to pre-highlight a chain of Walls.

   All of the exterior Walls should pre-highlight including the one at the core.

NOTE	If your cursor is closer to the outside edge when you press the TAB key, all of the outer Walls will pre-highlight. However, if you move slightly to the inside edge of the initial Wall, the chain selection will shift to go around the inside Walls of the core. Be sure that you are highlighting the outside Walls before you click to make the chain selection.

   • Be sure the outer Walls are highlighted and then click the mouse to select the chain of exterior Walls.

   • Hold down the SHIFT key and then click to remove the masonry Wall(s) at the core from the selection.

   All of the exterior Walls except the masonry Wall at the core (the horizontal one at the top middle of the plan) should now be selected (see the left side of Figure 9.1).

NOTE	The Modify	Walls tab should appear on the ribbon and the Type Selector should read *Generic – 12"* [*Generic - 300mm*]. If it does not, then you still have a core Wall selected. Use SHIFT click to remove it.

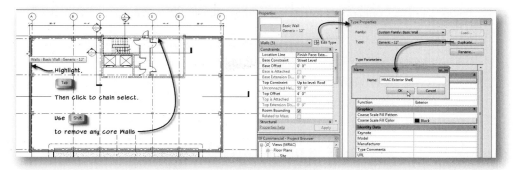

FIGURE 9.1 *Select a chain of Walls, remove the core Wall(s), and then use Type Properties to duplicate the type*

7. On the Properties palette, click the Edit Type button.

    • Click the Duplicate button.

    • For the name type **MRAC Exterior Shell** and then press OK (see the right side of Figure 9.1).

Edit Wall Type Structure

8. In the Type Properties dialog next to "Structure," click the Edit button.

If you completed the previous chapter and worked through the roof tutorials, you saw a very similar dialog there. The structures of Wall Types and Roof Types have much in common. Both allow you to configure the individual material layers of the structure, but Wall Types have also available additional parameters beyond this.

9. Highlight Layer 2 (Structure [1]).

    • In the Thickness column, type **7 5/8" [190]**.

To type a value of inches and fractions in Imperial units, you need to type a zero placeholder for feet first. Type the value as: **0 7 5/8** for example. That is zero space, seven space, five forward-slash eight. As an alternative, you can type just **7 5/8"** including the inch " symbol.	**TIP**

    • In the Material column, click where it says <By Category> and then click the browse icon that appears to open the Materials dialog to choose a Material.

    • In the Materials dialog, choose **Masonry - Concrete Masonry Units [Masonry - Concrete Blocks]** and then click OK.

10. Beneath the Layers list, click the Insert button.

    • Click the Up button to move the new Layer up above the Core Boundary.

    • From the Function list, choose **Thermal/Air Layer [3]**.

Functions determine the way that Walls join with other Walls, Floors, Ceilings, and Roofs. A pre-defined list of Functions is built into Revit. Structure layers will clean up with other Structure layers and interrupt layers of other functions. Substrate layers have the next highest priority and will clean up with other Substrate layers, will be interrupted by Structure layers, and will interrupt layers of all lower priorities, and so on. Materials also play a role in cleanup behavior. For example, you may still see a line between the Structure layers in a particular Wall join if they have different materials. To get a clean joint, both layers need the same function and material.

    • Change the Material to **Air Barrier - Air Infiltration Barrier**.

    • Change the Thickness to **2" [50]**.

TIP

Remember the inch " symbol when working in Imperial. Or you can type zero feet first: zero space two.

11. Insert one more Layer above the air gap Layer.

   - Set its Function to **Finish 1 [4]**.
   - Set its Material to ***Masonry – Brick***.
   - And set its Thickness to **3 5/8" [90]**.

TIP

Remember, zero space three space five forward-slash eight.

12. At the bottom of the Edit Assembly dialog, click the << Preview button (see Figure 9.2).

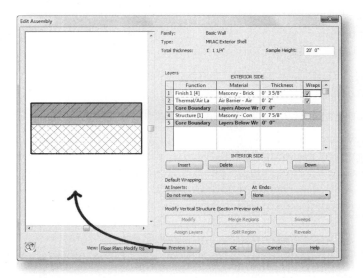

FIGURE 9.2 *Edit the Structure of the Wall type and preview it on the left*

13. Click OK twice to return to the view window.
14. On the View Control Bar (bottom left corner), change the Detail Level to Medium.

   - Study the results in the floor plan view.
   - Open the *{3D}* view and study the results there as well.

Zoom and pan around the two views to study the results. As you zoom in closer in the 3D view, you will see the brick pattern appear when you are close enough for it to display legibly. When studying the 3D view, you will note that we still have the simple "stand-in" curtain wall feature on the front of the building. (This was created in Chapter 4.) We are going to re-design this façade below. However, it should be clear from looking at the current state of the façade that we need to modify the Wall at the front of the building to at least allow for an entry lobby to the building. To do this, we'll simply split the front Wall and add a few wing Walls to add some depth to the main façade.

We could add more components to this Wall Type such as rigid insulation, membranes, and interior furring. Wall Types with this level of detail are provided in the out-of-the-box template used to create this project file. You can simply choose one from the Type Selector to compare it to the simpler Wall created here. Deciding how many material layers to represent within the structure of the Wall (and therefore "build in" to the model) is a topic occupying many a CAD/BIM Standards committee. The exact choice upon which you and your firm ultimately settle will be influenced by a variety of factors, including preference for graphical display, estimating and quantity take-off needs, Green Building analysis, and potentially several other factors. Ultimately when making any such decisions, consider all factors and decide what approach gives the firm and the project team the best balance between effort expended and benefits gained throughout the life of the project.

15. On the Project Browser, double-click to open the *Entrance Plan* floor plan view.
16. On the Modify tab, on the Edit panel, click the **Split Element** tool.

   - On the Options Bar, place a checkmark in the "Delete Inner Segment" checkbox.

We'll split the middle portion of the Wall out from one edge of the patio to the other.

   - Click the first point of the split on the left side of the plan (use the edge of the patio as a guide).
   - Click the second point of the split on the right side of the plan (again using the edge of the patio as a guide).
   - On the ribbon, click the **Modify** tool or press the ESC key twice.

Recall that the exterior Walls use a Finish Face: Exterior Location Line. If you click to select one of the exterior Walls (there are now two after the split), you will notice that the control handle (small round blue dot) is on the outside edge of the Wall.

17. Select the Wall on the left side of the plan.

   - Drag the control handle to snap at the intersection formed in the notch of the patio (see the left side of Figure 9.3).

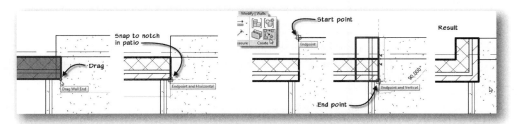

FIGURE 9.3 *Edit the end of the split Wall and add a new short segment*

   - Right-click the Wall and choose **Create Similar** (or click the tool on the ribbon).
   - Draw a new Wall as shown in Figure 9.3.

The figure shows drawing from top to bottom. If you draw from bottom to top, you will need to flip the Wall after you finish it. Or you can flip the Wall after the first click by pressing the SPACE bar on the keyboard. Then click the second point to finish the Wall.

TIP

18. Repeat the process on the other side of the plan.

   When drawing the second Wall, reverse the order of the first and second clicks. Click from bottom to top (or flip as indicated).

19. Open the *{3D}* view and study the results there as well.

Make sure that the height of the new Walls matches the height of the original Walls. If they do not, select the two new (short perpendicular) Walls and on the Properties palette change the Base Constraint to Street Level and the Top Constraint to Up to level: Roof with a 4'-0" [1200] Top Offset.

20. On the Project Browser, double-click to open the *Entrance Plan* floor plan view again.

Notice the way that the Wall layers terminate at the freestanding end of the Wall. The air gap is exposed on the end of the Wall. We can wrap the ends of the materials to close such gaps. We have the ability to do this both at the ends of Wall and at openings created by Doors and Windows (inserts).

21. Select one of the exterior Walls.

    You do not need to select them all. The following will be a Type-level edit, and Type-level edits will apply to all elements of that Type automatically.

    - On the Properties palette, click the Edit Type button.

22. Next to Structure, click the Edit button.

    - From the Wrapping at Ends drop-down, Choose **Exterior** from the "At Ends" drop-down.

This does not fully solve the problem. As you can see in the preview, the brick now wraps over the end of the Wall, but so too does the air gap.

    - In the Layers list at the top, remove the checkmark from the "Wraps" column next to layer 2 (Thermal/Air Layer [3]).

You should now see that only the brick wraps and the air gap is covered over by the wrapping brick layer (see Figure 9.4).

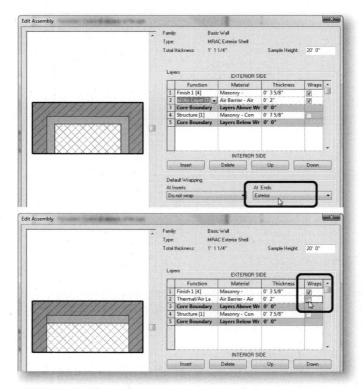

FIGURE 9.4 *Edit the wrapping of the Wall type layers*

23. Click OK twice to return to the view window and study the results.

24. Save the Project.

Wall System Families

Walls are System Families. A System Family is a Family that is built-in to the Revit system. They are predefined and cannot be modified at the Family level. To manipulate a System Family, we edit one or more of their Types. There are three Wall Families: Basic Wall, Curtain Wall, and Stacked Wall. All of the Walls that we have created so far are Basic Walls. Basic Walls can have one or more Layers in their Type's structure (as seen in the previous exercise). "Basic" does not necessarily mean simple or generic. A Basic Wall is defined by having a single continuous set of material layers (as seen when cut either horizontally or vertically). The "Generic" Type Walls we used in the early chapters are among the most "basic" Wall Types available, but the Wall Type we just created above with three Layers is still a Basic Wall. This is because all of the Layers run the full length and height of the Wall. A Curtain Wall, as we'll see next, defines a panel system along the length and/or height of the Wall. A Stacked Wall actually "stacks" two or more Basic Wall types on top of one another. We will look at Stacked Walls later in this chapter. The major focus of this chapter will be on Curtain Walls.

WORKING WITH CURTAIN WALLS

Curtain Walls in Revit Architecture are panelized wall systems. They come in two varieties: Curtain Walls and Curtain Systems. Curtain Walls are drawn the same way as Walls and using the same tool. Curtain Walls are always vertical walls. To create a Curtain Wall, you simply choose the appropriate type from the Type Selector as you draw the Wall. A Curtain System is typically created from the faces of other geometry such as a Mass. Curtain Systems allow for more complex geometry and can be vertical or sloped, curved (in section), etc.

A Curtain Wall Type sets up a panel modulation along the length and/or height of the Wall. Each panel can be assigned specific Family elements such as glass or stone panels or even other Wall Types; for example, brick or metal panel exterior with CMU, metal stud or metal frame backup. The edges between the panels are mullions. You can decide which edges should receive mullion elements and which type of mullion each edge should use. You can even control the way that the mullions intersect with one another.

Draw a Curtain Wall

The simplest way to create a Curtain Wall is to draw it using the Wall tool. Let's create our first Curtain Wall at the front entrance to the building on the first floor. We'll draw it in the place where the Wall we split out used to be.

1. On the Project Browser, double-click to open the *Entrance Plan* floor plan view.

2. On the Home tab, click the **Wall** tool.

    - From the Type Selector, choose **Curtain Wall: Curtain Wall 1 [Curtain Wall]**.
    - On the Properties palette, verify that the Top Constraint is set to: **Up to level: Level 2**.
    - Change the "Top Offset" to **0** (zero).

3. Click the first point of the Curtain Wall at the intersection of column Grid 1 and the small wing Wall on the left that we drew above (see the left side of Figure 9.5).

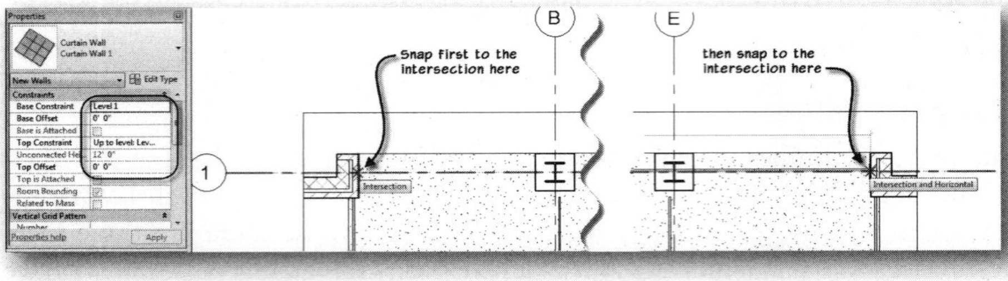

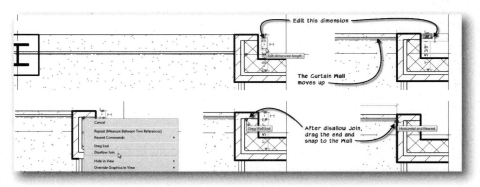

FIGURE 9.5 *Draw a single segment of Curtain Wall snaped to the intersection of the wing Walls and column line 1*

Use the TAB key if necessary to find the intersection. You can also use the keyboard shortcut **SI** to find it.

4. Pan to the right and then snap the endpoint of the Curtain Wall at the intersection of Grid 1 and the other small wing Wall (see the right side of Figure 9.5).

   • On the ribbon, click the **Modify** tool or press the ESC key twice.

The new Curtain Wall will appear as a thin plane of glass slightly offset from the column Grid line and passing through the Columns.

5. Select the new Curtain Wall.

   Locate the temporary dimension that appears (it should be at the right side).

   • Change the value of the upper temporary dimension to **4" [100]** (see the top of Figure 9.6).

FIGURE 9.6 *Adjust the location of the Curtain Wall with the temporary dimension*

When you draw Walls (including Curtain Walls), Revit attempts to join them automatically to other Walls. This is often advantageous. In this case, however, we'll be subdividing the Curtain Wall into bays and the automatic join might cause us issues as we proceed. When you click on the Curtain Wall for example, notice that the control grip at the ends is actually within the thickness of the neighboring Wall. We can use the "Disallow Join" command to turn off the automatic joining behavior for a single intersection.

6. With the new Curtain Wall still selected, right-click directly on the control handle grip at the end (the small blue dot).

   • From the menu that appears, choose **Disallow Join**.

- Drag the grip handle until it snaps to the face of the Wall (see the bottom three panels of Figure 9.6).

- Repeat the process on the other end of the Curtain Wall.

- On the ribbon, click the **Modify** tool or press the ESC key twice.

Hide in View

We will be working on this Curtain Wall from the vantage point of a few views. Let's take a look at what we have so far in 3D.

7. From the Project Browser, open the *{3D}* view.

   If necessary, orbit the model around to the front so that you can see the new Curtain Wall. You may also want to choose Shaded from the View Control Bar.

The new Curtain Wall will appear at Level 1 as a continuous pane of blue glass. What we also see is that the stand-in Curtain Wall created in Chapter 4 is obscuring our clear view of the new one. Let's hide the Curtain Wall (actually a Curtain System) from Chapter 4 to make the view easier to read. The use of the Temporary Hide/Isolate method discussed in the "Edit the Stair Sketch" topic in Chapter 7 would be appropriate here since we are going to later redisplay and unhide the Curtain System. However, as you may recall, the Temporary Hide/Isolate mode applies only during the current work session. If you decide to leave the tutorial and return later, you will have to reapply the hide later. For this reason, let's use this opportunity to discuss permanent hide instead. Permanent hide allows us to hide selected elements in the current view permanently until we decide to unhide them.

8. Select the Curtain System modeled in Chapter 4 (click near one of the corners).

- On the Modify | Curtain Systems tab, on the View panel, click the Hide in View drop-down button and choose **Hide Elements** (see Figure 9.7).

   The Curtain Wall drawn above should now be clearly visible and easy to select.

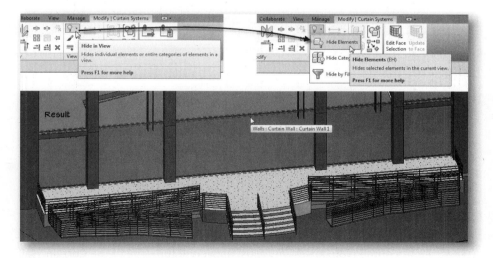

FIGURE 9.7 *Hide the Curtain System in the 3D view*

Understanding a Curtain Wall

A Curtain Wall is comprised of a Curtain Grid pattern in both the horizontal and vertical directions. Within each of the cells defined by these grids, is a panel. Panel Families can be made to represent anything from glass to stone panels to louvers to Doors. The number of grid divisions in the pattern is either a parameter of the Curtain Wall type or can be defined on the Curtain Wall instance directly. You can select the grid lines and panels independently. By default, the Curtain Wall itself pre-highlights first. To select the internal elements like grid edges, mullions, and panels, you can use the TAB key. Each time you TAB, a different portion of the Curtain Wall will pre-highlight. With Curtain Walls, you can also select a single element and then right-click to get additional selection options such as selecting all panels or mullions along a continuous line horizontally or vertically. We'll try all of these techniques below. Currently this particular Curtain Wall is one large panel of glazing because we haven't superimposed a Curtain Grid pattern on it yet. Therefore tabbing (to experiment with selection options) will not yield very interesting results. Naturally, it would be unlikely to have a single continuous panel of glass across the entire front of the building. We will learn how to sub-divide the Curtain Wall next.

Create a Working 3D Section View

We can sub-divide the Curtain Wall in any view using the Curtain Grid tool. However, it will be easiest to create a new view to see and work on the Curtain Wall isolated from the rest of the model.

9. On the Project Browser, double-click to open the *Level 1* floor plan view.

   - On the View tab, on the Create panel, click the **Section** tool.
   - Draw a Section line in front of (below in plan) and parallel to the Curtain Wall just a little wider both left and right.
   - Using the Control Handle, drag the "Far Clip Offset" of the Section back to just the inside of the Curtain Wall behind the Columns (see Figure 9.8).

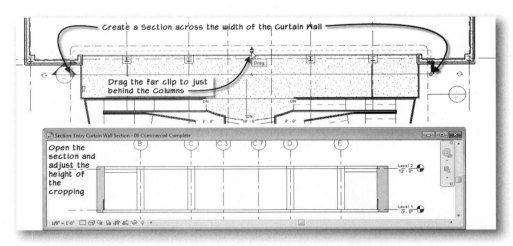

FIGURE 9.8 *Create a Section around just the Curtain Wall*

10. Deselect the Section line and then double-click the blue section head to open the section view.

    - Drag the top of the crop box down to just above the top edge of the Curtain Wall (slightly above Level 2).

The Curtain System that we hid above will be visible here again because hides are view specific.

    - Select the Curtain System (you may need to move the mouse around near column line B or E to find it).

- On the ribbon, choose the ***Hide in View > Hide Elements*** command again. You can use the same procedure to hide the Section line that appears in this view.

11. On the Project Browser, right-click *Section 1* and choose **Rename**.

   - Change the name to: **Entry Curtain Wall Section** and then click OK.

12. On the Project Browser, right-click the *{3D}* view and choose **Duplicate View > Duplicate**.

   - Right-click *Copy of {3D}* and choose **Rename**.

   - Change the name to: **Entry Curtain Wall 3D** and then click OK.

As you can see, this view currently shows the entire extent of the model. We can orient and simultaneously crop the view to match any other view on the Project Browser list, such as the section view we just created.

13. In the 3D view window, right-click on the ViewCube.

   - Choose **Orient to View > Sections > Section: Entry Curtain Wall Section** (see Figure 9.9).

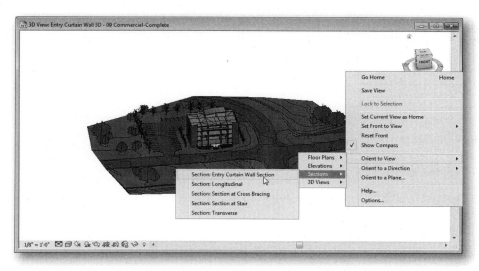

FIGURE 9.9 *Orient the copied 3D view to the Section at Curtain Wall and view it as Hidden Line*

14. On the View Control Bar, change the Model Graphics Style to **Hidden Line**.

   - Save the project.

We now have a section and a 3D view that are cropped to show only the Curtain Wall across the front entrance. In the 3D view, if you hold down the SHIFT key and drag the model to orbit the view (or drag the ViewCube), you will notice that the view is cropped to match the section in all directions, including the depth. To return to the head-on (elevation) viewpoint, click on **Front** on the ViewCube. This view will make it easier to work on the Curtain Wall. The 3D view is good for checking your progress, but you will get some additional editing functionality in the section view. We will use both views below as we work.

Add Curtain Grids

Now that we can see the Curtain Wall clearly without the rest of the model cluttering our view, let's sub-divide the Curtain Wall with Curtain Grids.

15. On the Project Browser, double-click to open the *Entry Curtain Wall Section* view.
16. On the Home tab, on the Build panel, click the **Curtain Grid** tool.

- Move the pointer near the top edge of the Curtain Wall.

  A Grid line and temporary dimensions will appear.

TIP	So you don't get confused about where the top of the Curtain Wall is in the hidden line display, click the Hide Crop Region toggle icon on the View Control Bar.

- Click to create a grid line in the middle bay between column C and C.3.
- Click to create another in the bay between column C.7 and D (see Figure 9.10).

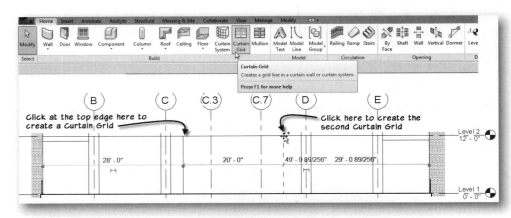

FIGURE 9.10 Add two Grid lines near the middle columns

- On the ribbon, click the **Modify** tool or press the ESC key twice.

17. Select the new Grid Line on the left (you should be able to just click it, but use the TAB key if necessary).

    Beneath the temporary dimension a small dimension icon will appear.

- Click on the dimension icon beneath the temporary dimension on the left (see Figure 9.11).

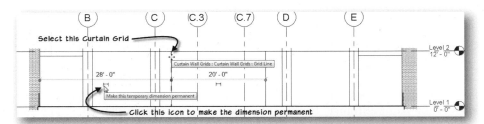

FIGURE 9.11 *Make the temporary dimension permanent for the selected Grid Line*

- Repeat by selecting the Grid line on the right and then making the temporary dimension on the right permanent as well.

  Do not make the temporary dimension in the middle permanent.

18. Click away from the selected element to deselect it (or click the **Modify** tool or press the ESC key twice).

- Select the new dimension on the left.

- Drag the leftmost witness line of the dimension to the right face of the column.
- Drag the rightmost witness line of the other dimension on the right to the left face of the column (see Figure 9.12).

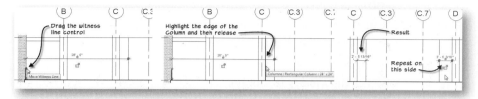

FIGURE 9.12 *Move the Witness Lines of the Dimensions to the faces of the Columns*

19. Select the new Grid Line on the left (you should be able to just click it, but use the TAB key if necessary).

Notice how the temporary dimensions now include the dimension that we just created and modified. We can now edit the value of this dimension to place the Grid Lines precisely relative to the Columns.

- Click on the temporary dimension between the Grid Line and the Column (the one edited above).
- Input a value of **1'-0" [345]**.
- Repeat on the other side.

You should now have a middle bay that has a Grid Line 1'-0" [345] away from the Columns on each side. The distance between the Grid Lines should be 24'-0" [7200]. If this is not the case, make any required adjustments (see Figure 9.13).

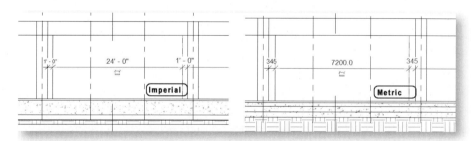

FIGURE 9.13 Move both Grid Lines relative the neighboring Columns

---

Rather than making Temporary dimensions permanent, we could have simply moved the Witness Line of the temporary dimension. Revit will "remember" the positions that Witness Lines have been moved to on subsequent selections. This lasts for the current editing session, however. When the project is saved and re-opened in the future, the Witness Lines will return to their default locations. Using the permanent Dimension technique showcased here will keep them available for editing in future sessions if necessary. The exact procedure that you employ in your own projects is a matter of personal preference.

TIP

We can add additional Curtain Grid lines by sketching them like the ones above, or we can copy them from the existing ones. Let's copy the ones we have to create three wide bays and two small bays for the entrance to our building.

20. Select the Grid Line on the left.
21. On the Modify Curtain Wall Grids tab, click the **Copy** tool.
    - On the Options Bar, place a checkmark in the "Multiple" checkbox.
    - Pick any start point and then drag to the right.
    - Type **6'-0" [1800]** and then press ENTER to create the first copy. Continue copying to the right.
    - For the next copy, type **3'-0" [900]** and then press ENTER.
    - Create another one at an offset of **6'-0" [1800]** and then one more at **3'-0" [900]**.
    - On the ribbon, click the **Modify** tool or press the ESC key twice (see Figure 9.14).

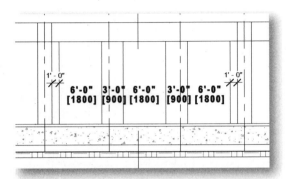

FIGURE 9.14 *Create several copies of the Grid Line making an A B A B A rhythm*

22. Zoom out so you can see the entire Curtain Wall.
23. On the Home tab, click the **Curtain Grid** tool.
    - Move the pointer near the left edge of the Curtain Wall (left of column Grid B). A Grid line and temporary dimensions will appear.
    - Click to create a horizontal Grid Line 4'-0" [1200] from the top (use the temporary dimensions to position it if necessary).
    - On the ribbon, click the **Modify** tool or press the ESC key twice.

At this point we have created several Grid Lines in our Curtain Wall. We can now use the tabbing technique mentioned above to see that this procedure has also cut the one large Curtain Panel into several smaller ones at each division we created. We can select any or all of these panels and assign them to other types. Some of the bays could be made solid construction like stone or brick while some could remain glass. Give it a try if you like, but don't make any permanent changes yet.

24. Save your project.

Assign Curtain Panel Types

Now that we have several Curtain Grid lines dividing our Curtain Wall into individual panels, let's assign some panel types to these bays. Our building needs an entrance. The three bays that we have roughed out in the front need some Doors. A Curtain Wall Door is not a regular Door Family. It is actually a Panel Family that looks like and schedules like a Door.

25. Zoom in on the bay between column line C and D.

- Place your Modify tool (mouse pointer) over the edge of one of the wider bays created above.

- Press TAB until the panel in this bay pre-highlights—when it does, click to select it (see Figure 9.15).

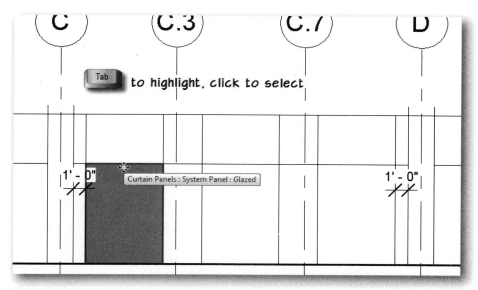

FIGURE 9.15 *Select the larger bays at the entrance*

- Pre-highlight the next wide panel using the same technique.
- Hold down the CTRL key and click to select it.

You should have two wide panels selected after this action.

- Repeat once more to select the remaining wide panel.

26. With the three panels selected, on the Properties palette click the Edit Type button.

- In the "Type Properties" dialog, click the Load button.

- In the Open dialog, double-click the *Doors* folder.

If you do not see a *Doors* folder, the Families that are referenced below have been provided with the dataset files installed from the *Aubin Academy Master Series: Revit Architecture 2012* student companion. You can browse first to the location where you installed the dataset files from the student companion and then double-click the *Imperial Library* [*Metric Library*] folder, then the *Doors* folder in that location.

- Select the *Curtain Wall-Store Front-Dbl.rfa* [*M_Curtain Wall-Store Front-Dbl.rfa*] file and then click Open.

Please note that even though the files referenced above are stored in a *Doors* folder, they are actually Curtain Panel Families and not Door Families. Normal Doors cannot be placed in a Curtain Wall. So if you create your own custom Curtain Wall doors, be sure to choose the dedicated Family template (*Door - Curtain Wall.rft* [*Metric Door - Curtain Wall.rft*]) for that purpose. Family templates and Family creation techniques are covered in the next chapter.

27. In the "Type Properties" dialog, click OK to see the results (see Figure 9.16).

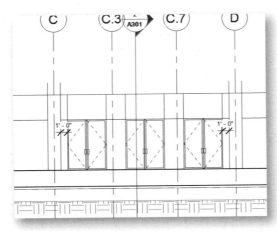

FIGURE 9.16 *Swap in Storefront Doors for the three wide bays*

A revolving Door Family has also been provided with the dataset files from the student companion. Feel free to use this Family instead of the double-door Family. A selection of Curtain Panel revolving Door Families (and many others as well) can be found by searching on the Autodesk User Group International (www.augi.com).

Assign Mullions

Our next task is to apply some mullions to the Curtain Grid Lines.

28. On the Home tab, on the Build panel, click the **Mullion** tool.

  - From the Type Selector, verify that ***Rectangular Mullion : 2.5" ✕ 5" rectangular [Rectangular Mullion : 50 ✕ 150mm]*** is selected.

On the Modify | Place Mullion tab of the ribbon, on the Placement panel, there are three methods to create Mullions: a Grid Line, a single Grid Line Segment, or All Grid Lines. Grid Line is the default choice.

  - With the "Grid Line" button selected, click on the horizontal Grid Line (the one 4'-0" [1200] from the top).

If you are zoomed in close enough, you should notice that the Doors adjust in size to accommodate the mullion.

  - Try the "Grid Line Segment" option next on any Grid segment.
  - Use the "All Grid Lines" option to complete the process of adding mullions (see Figure 9.17).

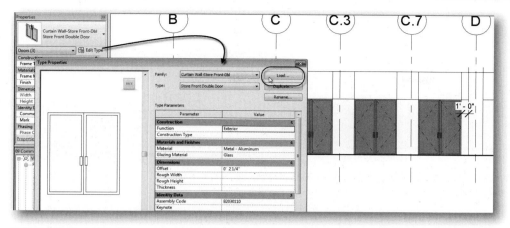

FIGURE 9.17 *Add Mullions to the Curtain Grid Lines*

Looking closely at the Door Panels, note that the last step added a Mullion at the sill. We can delete these individual Mullion segments to finalize the Door Panels.

29. Using the TAB and CTRL keys if necessary, select the Mullion segments indicated in Figure 9.18 and delete them.

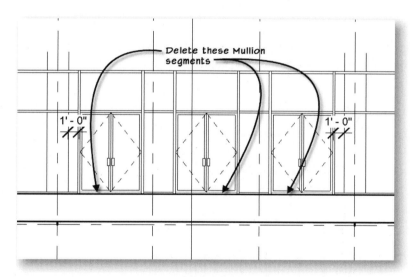

FIGURE 9.18 *Delete the Mullions under the Door Panels*

Edit Mullion Joins

You will notice that the vertical mullions' joins have been given priority over the horizontal ones. If you prefer to have the horizontal mullions continuous and have the verticals stop and start at each intersection, you can toggle the Mullion Join behavior for each join.

30. Select one of the vertical mullions at the Doors.

Notice the Join icons that appear at each end of the Mullion and on the right-click menu (see Figure 9.19).

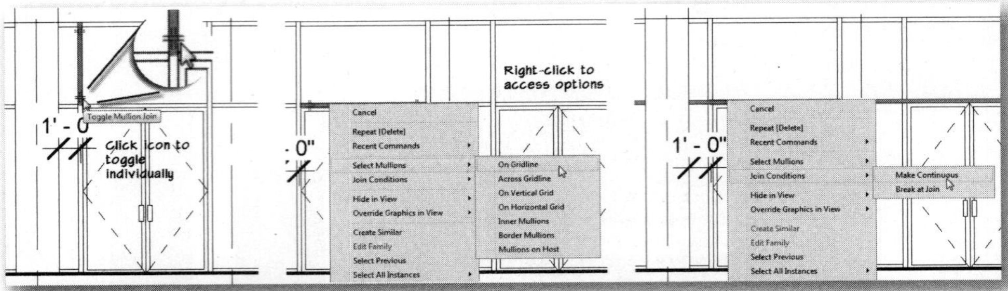

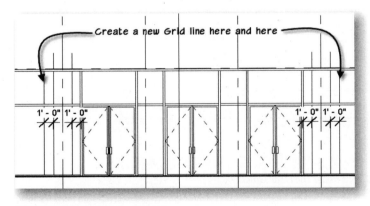

FIGURE 9.19 *Edit Mullion Joins using the control or the right-click menu*

- Click the Toggle Mullion Joins icon to model the join to your liking.

In addition to the control icons, you can right-click to access the same options. To use the right-click options effectively, select a single mullion, right-click, and choose **Select Mullions > On Gridline**. Once the entire gridline is selected, right-click again and choose **Join Conditions > Make Continuous** as shown in Figure 9.19. This would be the same as clicking the join icon on several separate Mullions.

Create Additional Grid Lines and Mullions

The middle entrance bay is complete, but the other bays need attention.

31. Using the techniques covered above, create a new Grid Line **1'-0"** **[345]** from the edges of the two middle columns as shown in Figure 9.20.

FIGURE 9.20 *Add more Grid Lines*

32. Add Mullions to each of these new Grid Lines.
33. Repeat the entire process to create Grid Lines and Mullions on each side of the remaining Columns (see Figure 9.21).

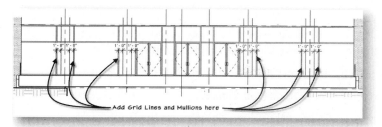

FIGURE 9.21 *Add more Grid Lines and Mullions at the remaining Columns*

Toggle the Mullion Joins as appropriate.

Using a Wall Type for a Panel

At each of the Columns, we want to remove the horizontal Mullions and swap out the glazing for a solid Wall panel.

34. Zoom in on one of the Columns.

  • Using the CTRL key, select all three of the horizontal mullions crossing through the Column (one at the top, one at the door head height, and one at the bottom).

If you have trouble with the selection, remember your TAB key. Also remember that they actually pass behind the Columns.	**TIP**

  • Delete the three selected Mullions (press the DELETE key) (see the left panel of Figure 9.22).
  • Select the horizontal Grid Line.
  • On the Modify Curtain Wall Grids tab, click the ***Add / Remove Segments*** button.
  • Click the Grid Line in-between the two vertical Mullions (see the middle panel of Figure 9.22).

The Grid Line will turn dashed between these Mullions to indicate that a portion of it has been removed.

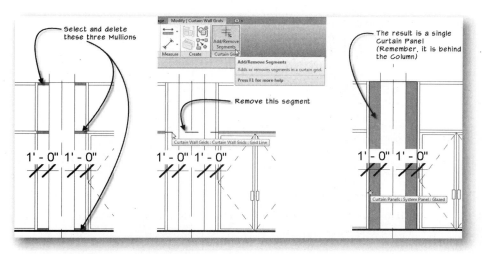

FIGURE 9.22 *Delete Mullions and Remove a Grid Line*

You will now have a continuous vertical panel at the Column as shown in the right panel of Figure 9.22. You may need to use the TAB key to pre-highlight it to see this. When you select it, remember that the Column is actually in front of it.

35. Using the TAB key, pre-highlight and then select the new full height panel at the Column.

   • From the Type Selector, choose **Basic Wall : Generic - 8"** [**Basic Wall : Generic - 200mm**].

36. Repeat the entire process for the other three Columns on the front façade.

   • On the Project Browser, double-click to open the *Entrance Plan* floor plan view.

   • Zoom in on the Columns at the front of the façade to see the result.

   • Select each Door (use the TAB key), and using the flip controls, flip the swing outside (see Figure 9.23).

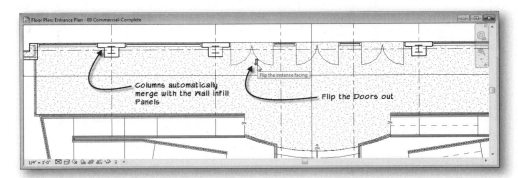

FIGURE 9.23 *Swap in a Wall type in place of the Curtain Panels*

This procedure illustrates that you can use Basic Wall types as panels within a Curtain Wall. This adds a whole range of possibilities to your design potential with this tool. Notice also how nicely these infill Wall "panels" interact with the Columns.

37. On the Project Browser, double-click to return to the *Entry Curtain Wall Section* view.

38. Using the TAB and CTRL keys, select each of the four lower panels (two to the right of the entrance, two to the left).

   • From the Type Selector, choose **Curtain Wall : Storefront** (see Figure 9.24).

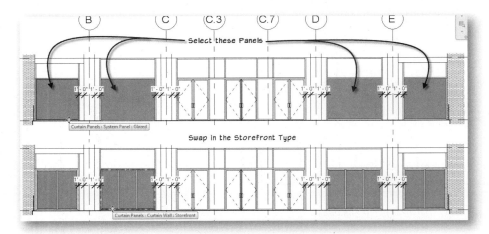

FIGURE 9.24 *Swap in a Curtain Wall type for the selected Panels*

There are two interesting points worth mention on this step: first, we are able to use one Curtain Wall type as a Panel in another. Second, the ***Storefront*** Curtain Wall type has built-in subdivisions of Curtain Grid lines and Mullions pre-assigned to it. We will explore this further in the remainder of the chapter.

39. Make one last substitution by selecting the two square panels between and above the Doors.

    • From the Type Selector, choose ***System Panel: Solid***.

40. On the Project Browser, double-click to open the *Entry Curtain Wall 3D* view.

    • Drag the ViewCube to orbit the model and study the results (see Figure 9.25).

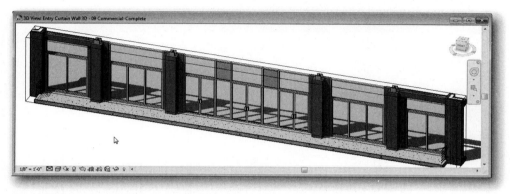

FIGURE 9.25 *Orbit the model to study the results*

If you wish to see the Curtain Wall in the context of the rest of the building, open the *{3D}* view and study it there. If you look closely, you will notice that the "Storefront" Curtain Walls that were used as panels include additional Mullions around their perimeters. For purposes of our exercise, we won't worry about this; however, if you wanted to remove them, you could do so. These objects are "pinned" however, which in the case of a Curtain Wall, indicates that the element is defined at the Type level. If you choose to delete them you will first have to click the blue pin icon that will appear to unpin them. This essentially adds an override to the Curtain Wall allowing you to remove the Type driven Mullions. We will learn more about Curtain Wall Type Properties in the coming exercises.

41. Save the project.

CREATING HOSTED CURTAIN WALLS

The Curtain Wall we created above filled a space that had no previous enclosure. We drew it just like any other Wall and then edited it. You can also create Curtain Walls from existing Walls or even embed them within existing Walls. In this way the Wall will "host" the Curtain Wall. In this sequence we will use one of the types already provided in the current file and draw a Curtain Wall hosted in the exterior shell Walls of the building.

Draw a Curtain Wall

The first part of this exercise is similar to the previous one. We will draw a Curtain Wall using a Type already in the file.

1. On the Project Browser, double-click to open the *Level 2* floor plan view.

2. On the Home tab, click the **Wall** tool.

- From the Type Selector, choose **Curtain Wall : Storefront**.
- On the Options Bar, from the "Height" list, choose **Level 4**.

- On the Properties palette, verify that the "Top Offset" is set to **0** (zero).
- On the Options Bar, clear the "Chain" checkbox.

   We will add the Curtain Wall to the vertical Wall on the left side of the plan.

3. Click the first point of the Curtain Wall halfway between column lines 1 and 2 on the inside brick face of the vertical masonry Wall (see Figure 9.26).
   Use the TAB as necessary to assist you in selection of the proper point.

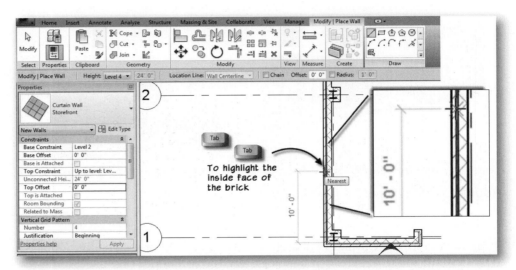

FIGURE 9.26 *Draw a new Curtain Wall on the second floor directly on top of the existing Wall on the face of the brick*

Draw the Curtain Wall **45'-0" [13500]** long (a little more than half way between Column lines 3 and 4).

The easiest way to do this is to move the mouse vertically, type the desired length, and then press ENTER (see Figure 9.27).

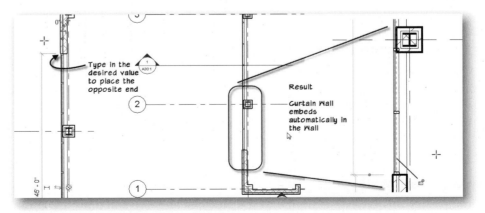

FIGURE 9.27 *Draw an embedded Curtain Wall*

On the ribbon, click the **Modify** tool or press the ESC key twice.

The Curtain Wall should cut the host Wall automatically. In some cases it will not (for example when using the Curtain Wall 1 Type). If it does not cut automatically, we can use the ***Cut*** tool on the Modify tab to cut the Wall manually with the Curtain Wall.

4. Zoom in on the column at column line 2.

Because we chose the ***Curtain Wall : Storefront*** Curtain Wall type above, this Curtain Wall already has Curtain Grid lines and Mullions assigned to it. However, as you can see, the Curtain Wall is too close and the Mullions intersect the Columns.

5. Select the Curtain Wall.

    Be sure to select the Curtain Wall itself, not the Mullions or glazing panels.

    The glazing should appear on the outside of the Curtain Wall. If it does not, click the "Flip Wall Orientation" control.

    • Edit the temporary dimension in the thickness of the Wall (it may be up near the top) to move the Curtain Wall out (away from the interior).

    • Experiment to find the right value. Try about **7"** **[175]**.

6. On the Project Browser, double-click to open the *{3D}* view.

    • Orbit the model as needed to gain a clear view of the new Curtain Wall.

Adjust the Curtain Grid Spacing

As you can see, this Curtain Wall type defines a large vertical spacing of grid bays. In addition, the floor element between the second and third floors is clearly visible. Let's make a few adjustments to address both of these issues.

7. Select the Curtain Wall.

    • On the Properties palette, click the Edit Type button.

    • In the "Type Properties" dialog, click the Duplicate button.

    • Name the new type: **MRAC Storefront** and then click OK.

Below, in the "Create a Custom Curtain Wall/System Type" topic, we will explore many of the settings in this dialog in detail. For now, we'll focus on just a few settings required by the current task.

    • Beneath the "Horizontal Grid Pattern" grouping, change the Spacing to **4'-0"** **[1200]** and then click OK.

        Notice the change in the overall grid spacing. Keep the Curtain Wall selected.

8. In the center of the Curtain Wall, click the Configure Grid Layout control.

Two dimension lines appear through the middle of the grid: one vertical and the other horizontal. To shift the horizontal grid lines up or down, edit the dimension on the horizontal origin line.

    • On the left side of the Curtain Wall, click the Horizontal Curtain Grid Origin temporary dimension.

        There is also one for the angle, so pause your mouse over the dimensions for a tooltip to help you select the correct one.

    • Type a new value of **2'-0"** **[600]** and then press ENTER (see Figure 9.28).

There are two temporary dimensions, both the horizontal and vertical origin lines. One moves the grid, the other rotates it. Be sure to use the tooltip to find the correct one before editing. If you want to experiment with the rotation as well, feel free to do so, but undo before continuing.

TIP

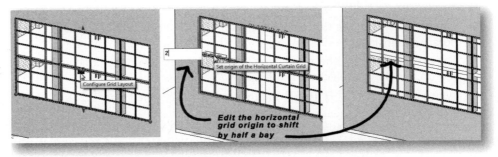

FIGURE 9.28 *Flip the Curtain Wall (if necessary) and move it toward the outside*

We now have a short bay at the top and the bottom of the overall Curtain Wall and a full grid centered at the floor line between second and third floors. Let's replace those panels with spandrel glass.

9. Select one of the panels in the middle of the Curtain Wall height (one that occurs at the floor line).

TIP	Place your mouse near the edge of the panel you want to select and then press TAB until it pre-highlights, then click.

- Right-click and choose **Select Panels > Along Horizontal Grid**.

Notice the small pushpin icons attached to each Panel—you may need to zoom to see them all. As with the Storefront Curtain Walls that were embedded into the larger Curtain Wall at the entry, since the Curtain Wall Type that we used here assigned the Curtain Grids, Panels, and Mullions automatically, these Panel definitions are pinned to the Type. If you look at the Type Selector, you will notice that you cannot change the Type (it is grayed out). To change a Panel, you must unpin it first. You can simply click the pin icon to unpin it, but since there are many selected, this would not be the most efficient approach.

10. On the Modify Curtain Panels tab, click the ***Unpin*** tool.

TIP	The keyboard shortcut for Unpin is UP.

- Keeping the same selection of Panels, from the Type Selector choose ***System Panel: Solid***.

Deselect the Panels to see that the entire row is now solid (see Figure 9.29).

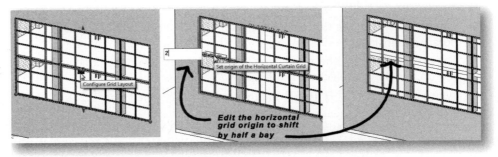

FIGURE 9.29 *Shift the grid pattern origin by half a bay*

Mirror the Curtain Wall

11. On the Project Browser, double-click to open the *Level 2* floor plan view.
12. Select the Curtain Wall.

Keep in mind that the Curtain Wall segment is made up of many subparts. The whole curtain wall is represented by a blue dashed that will appear when you select it or pre-highlight it.

13. On the Modify | Walls tab, click the ***Mirror – Draw Axis*** tool.

   • On the Options Bar, verify that the "Copy" checkbox is selected.
   • Snap to the midpoint of the Floor slab curve (in the center of the plan) and then drag straight up and click again (see Figure 9.30).

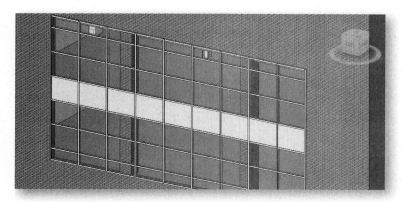

FIGURE 9.30 *Change the middle row of Panels to solid*

The Curtain Wall will mirror over to the other side of the plan. If it automatically cuts the Wall, then skip the next steps; otherwise continue.

Using Cut Geometry

You can perform the next action in either a plan or 3D view.

14. With the new Curtain Wall still selected, on the Modify | Walls tab, click the ***Cut*** tool.

   • At the "First Pick" prompt, click the vertical Wall on the right.
   • At the "Second Pick" prompt, click the mirrored Curtain Wall.

15. On the Project Browser, double-click to open the *{3D}* view.

   • Orbit the model and study the results.

Add Windows

Some basic punched Windows will help complete the masonry portions of the façade.

16. Add ordinary punched Windows to the masonry Walls on the first and fourth floors above and below the Curtain Walls.

   • Add Windows on all floors in the north and south Walls.

Simply open the *Level 1* floor plan, add some Windows to one side, mirror them to the other side, and then copy them to the *Level 4* plan using paste aligned. As we have done with other layout tasks, place the Windows in general locations first, and then use dimensions to fine-tune their placement. The easiest way to do this is to create a continuous string of dimensions. This will make moving individual Windows quick and easy.

17. On the QAT, click the ***Aligned Dimension*** tool.
    - On the Options Bar, from the Pick list, choose **Entire Walls** and then click the Options button next to it.
    - In the "Auto Dimension Options" dialog, check Openings and choose Widths and then click OK.
    - Select one of the exterior Walls to dimension it and click a point outside the building to place it (see Figure 9.31).

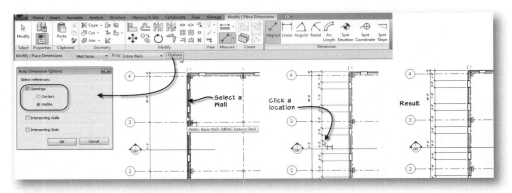

FIGURE 9.31 *Mirror the Curtain Wall to the other side of the model*

- Repeat for the other exterior Walls.
- Select a Window, edit the dimension value, and then move to the next.

18. Study the results in the {3D} view (see Figure 9.32).

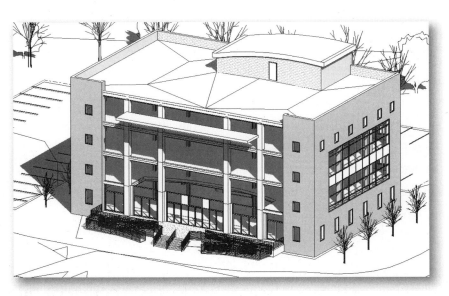

FIGURE 9.32 *Add Windows to the other floors*

19. Save the project.

CREATE A CUSTOM CURTAIN WALL/SYSTEM TYPE

We have now worked with a simple Curtain Wall with no divisions (in which we added all of the Curtain Grid lines and Mullions manually) and a Curtain Wall Type that included Curtain Grid spacing and Mullions pre-assigned within its Type. In this sequence, we will combine techniques starting with the creation of our own Curtain Wall Type. The goal is to try to include as many of the divisions and Mullions within the Type properties so that they occur automatically. We can then finish the design using the manual Curtain Grid line and Mullion techniques already covered.

Explore Curtain Wall and Curtain System Type Properties

Before we build our own Curtain Wall Type, let's take a closer look at the one we used in the previous sequence. There are three Curtain Wall Types provided in the out-of-the-box templates. The simplest is named ***Curtain Wall 1 [Curtain Wall]***. It has no built-in divisions and defaults to a single continuous panel of glass. We used this one in the first exercise at the start of the chapter. The next Type is slightly more complex including a default grid spacing in both the horizontal and vertical dimensions. This Type is named ***Exterior Glazing***. Finally we have the Type used in the previous sequence: ***Storefront***. As we have seen, not only does this one include pre-defined grid spacing but also adds Mullions on all edges automatically (see Figure 9.33).

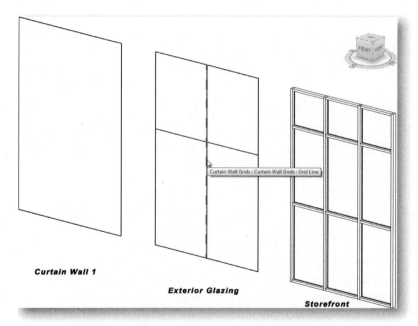

Curtain Wall 1

Exterior Glazing

Storefront

FIGURE 9.33 *Three Curtain Wall types are included in the out-of-the-box template files*

1. On the Project Browser, locate the *Families* branch and click the plus (+) sign icon next to it to expand it.

   • Click to expand *Walls* next, then *Curtain Wall* (see Figure 9.34).

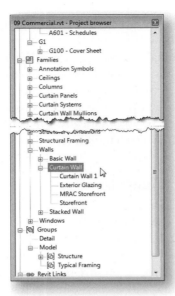

FIGURE 9.34 *Use the Project Browser to access the existing Curtain Wall types in the current project*

Each of the types currently resident in the existing project will appear in the tree listing. Notice that the three shown in the previous figure are included as well as the one we created in the previous exercise. You can duplicate and modify new Types directly from the Project Browser, or using the steps that we followed in the "Adjust the Curtain Grid Spacing" topic above. You can also add an instance of a particular Type to the model directly from here by dragging and dropping or using the right-click options.

2. Right-click on **Storefront** and choose **Type Properties**.

This takes you directly to the "Type Properties" dialog for this type. This dialog is similar to any other "Type Properties" dialog. Following is a brief description of some of the important parameters seen here.

Under the "Construction" grouping are the following parameters:

- **Function**—Since Curtain Walls are a Wall Family, they have the same list of "Functions" as other Walls. We can create Curtain Wall types that fulfill a variety of functions such as "Exterior" and "Interior." (Some of the other Functions like "Foundation" or "Retaining Wall" would not make sense for Curtain Walls.)

- **Automatically Embed**—When this checkbox is selected, the Curtain Wall will attempt to embed itself automatically within another Wall like the one that we created in the previous exercise. When this function is off you can still use the Cut tool to embed a Curtain Wall within another Wall manually.

- **Curtain Panel**—This parameter gives a list of all possible Panel Types that are loaded into the project. Choose the default Type for the Curtain Wall here. All Panels will default to this Type and they will appear pinned (as we saw above) when selected. You can always unpin them and override the Panel Type used for one or more Panels in the specific instance of the Curtain Wall element in the model (we also did this above).

- **Join Condition**—Early in this chapter in the "Edit Mullion Joins" topic, we edited the default join condition used by Mullions. You assign the default condition to a Curtain Wall Type here. Several options appear. You can make the horizontal or the vertical continuous, make the border continuous, or choose not to assign an option.

Beneath the "Vertical Grid Pattern" and the "Horizontal Grid Pattern" groupings are three parameters each:

- **Layout** and **Spacing**—Layout choices include "None," "Fixed Distance," "Fixed Number," and "Maximum Distance." When None is chosen the Curtain Wall will include a single Panel across the entire horizontal or vertical dimension as appropriate. Fixed Distance will create Curtain Grid Lines at the spacing indicated in the "Spacing" parameter. There may be panel space left over when using this option. Maximum Distance is similar in that it specifies the *largest* that the spacing can become but will size all bays equally up to the size indicated. In this case, there will not be any left over. The Spacing parameter is used for both of these settings. Choosing the "Fixed Number" option will disable the "Spacing" parameter. The number of bays is an instance parameter. This means that when you choose the "Fixed Number" option, you must then assign the number of bays individually to each instance of the Curtain Wall. You cannot do it globally in the "Type Properties" dialog.
- **Adjust for Mullion Size**—This parameter adjusts the location of Curtain Grid lines to maintain equal Panels even when Mullion sizes vary.

Beneath the "Vertical Mullions" and "Horizontal Mullions" groupings, you can choose from a list of available Mullion Families to use for the Borders and the Interior Mullions. These Mullions will be created automatically and will be pinned to the Curtain Wall as it is created (see above). You can unpin any Mullion, as we did in the exercise above, and assign an alternate type as design needs dictate.

The remaining parameters are the standard identity data properties that we have seen in other object Types. These are mostly used when doing quantities, schedules, and takeoffs.

- Click Cancel to dismiss the dialog without making changes.

3. On the Project Browser, beneath the *Families* node, expand *Curtain Systems*.

   A single Curtain System Family appears named simply: ***Curtain System***.

4. Expand Curtain System.

   A single Type appears named: ***5' × 10' [1500 × 3000mm]***.

- Right-click ***5' × 10' [1500 × 3000mm]*** and choose **Type Properties**.

Notice that the parameters here are nearly identical to the Curtain Wall. The only differences are that a Curtain System does not have the "Wall Function" or "Embed" options and the two sets of grid parameters are named more generically: "Grid 1" and "Grid 2." You are also limited to Curtain Panel Families and cannot use Wall Types as Panels like you can with Curtain Walls. In all other ways, Curtain Systems share the functionality and behavior of Curtain Walls. Curtain Systems however allow for more "freeform" designs than Curtain Walls. At the start of this book, in Chapter 4, we created a Massing element to represent the front façade of the building. To the faces of that Mass, we applied a simple generic Curtain System (review the "Add a Mass" topic in Chapter 4 for a refresher). At the start of this chapter, we hide that Curtain System to make it easier to work on the other portions of the façade. At this point, we are ready to redisplay this element and begin refining its design. To get started, we'll unhide it and then build a custom Curtain System Type to finish up the front façade of the building.

- Click Cancel to dismiss the dialog without making changes.

Editing Massing Elements

You create a Curtain System by picking faces of a Conceptual Mass object. The Curtain System will follow the shape of the selected face(s). We already have such an object from the Chapter 4 exercise. Let's redisplay it now.

5. On the Project Browser, double-click to open the *{3D}* view.

6. On the Massing & Site tab, on the Conceptual Mass panel, click the **Show Mass by View Settings** button.

This will toggle the button to the **Show Mass Form and Floors** option. The Mass object will reappear on screen. It will be a little difficult to see since Masses default to a transparent material. The show/hide Mass command is the only display command that affects *all* views, not just the active one. Therefore, if you switch to another view, the Mass will be visible there as well. You may find it easier to see in plan for example, but you will likely prefer to edit it in 3D. Therefore, you may wish to tile both the plan and the 3D view on screen together.

7. Close all views except *{3D}* and then on the Project Browser, double-click to open the *Entrance Plan* view.

    *   Tile the two windows and arrange and zoom them so you can see the front facade clearly in both view windows.

8. Select the Mass at the front of the building (it is the transparent box in 3D and appears as a box between Grid B and E in plan).

    *   On the Modify | Mass tab, click the **Edit In-Place** button (see Figure 9.35).

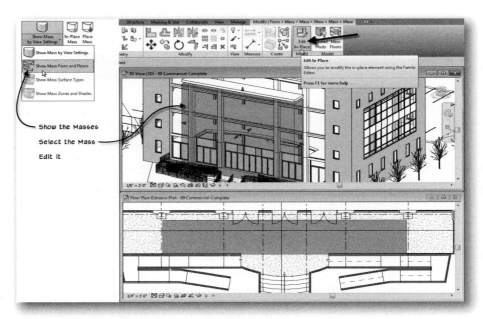

FIGURE 9.35 *Show Masses and then Edit the In-Place Mass*

The ribbons will change to the Massing Family Editor arrangement. This will revert back to the Project ribbon layout once we have finished editing the Mass. We will look closer at the Massing Environment in Chapter 15 and the Family Editor in the next chapter. The first series of edits will be easiest to accomplish in plan.

9. Make the *Entrance Plan* view active. Make edits to the sketch as shown in the top portion of Figure 9.36.

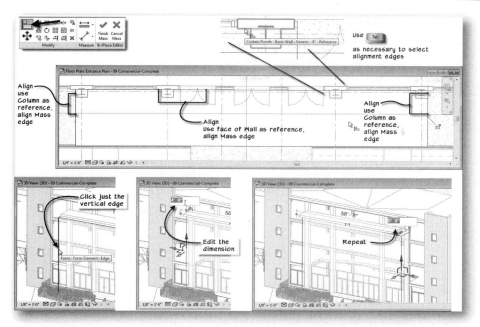

FIGURE 9.36 *Select the Mass using the tab key and the click the Edit In-Place button*

- Use the Align tool to align the left edge of the sketch to the left edge of the column.
- Repeat for the right side.
- Align the top edge of the sketch to the outside face of the Wall Curtain Panel.
- Use the control handles to change the shape of the sketch as shown. (Use temporary dimensions or guidelines to help.)

10. In the *{3D}* view, click on the vertical corner edge of the form on the left side.

Be careful how you select. In the Massing Environment, you can select the edges, the faces, or the entire form. Move the mouse around to highlight different possibilities and you can also use the TAB key to cycle through various selection options. Be sure you have the front left corner edge selected as shown in the figure.

- Edit the temporary dimension and change the value to **4'-0" [1200]**.
- Repeat for the right side using a dimension value of **8'-0" [2400]** (see the bottom portion of Figure 9.36).

11. On the Project Browser, double-click to open the *South* elevation view.

- Use the Align tool again to align the bottom of the Mass to Level 2.
- Align the top of the Mass to Level Roof.
- On the Modify tab, click the **Finish Mass** button.

Since we moved the bottom edge up to Level 2 (or 12'-0" [3600]), the Mass will no longer be visible in the plan view. We'll be able to see it in the 3D view, however.

12. On the Project Browser, double-click to open the *{3D}* view.

Notice that the ribbons have returned to normal now that we have finished editing the Mass. Your newly shaped Mass will appear as we edited it, but the existing Roof element (created from the Mass surfaces back in Chapter 4) no longer matches the shape of the Mass. Furthermore, the Curtain System that we hid at the start of the chapter was also originally based on the shape of the Mass. Both elements can be easily updated to match the updated shape of the Mass. But first, we need to restore the visibility of the Curtain System.

13. On the View Control Bar, click the Reveal Hidden Elements icon (small light bulb).

A dark maroon border will appear around the view window and the hidden elements (the Curtain System in this case) will reappear in dark maroon as well. New in this release, a label describing the mode also appears in the upper corner of the view.

- Select the Curtain System and on the ribbon, click the **Unhide Element** button (see Figure 9.37).

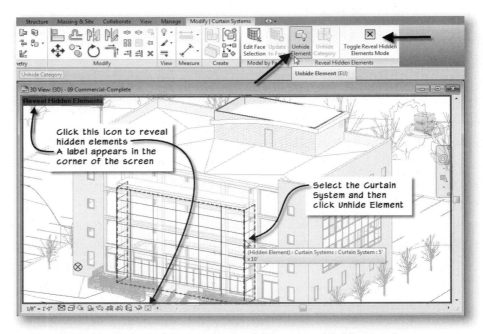

FIGURE 9.37 *Edit the shape of the Extrusion sketch (Dimensions added for clarity)*

14. On the View Control Bar, click the Exit Reveal Hidden Elements icon or click the **Toggle Reveal Hidden Elements Mode** button on the ribbon.
15. Select the small Roof at the top of the façade element.

- On the Modify | Roofs tab, in the Model by Face panel click the **Update to Face** button (see Figure 9.38).

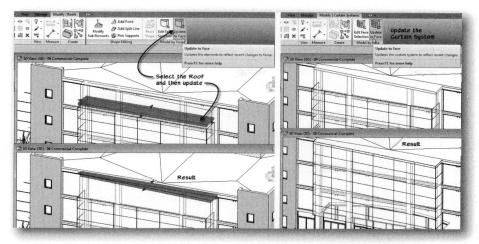

FIGURE 9.38 *Use the Update to Face button to update the Roof and Curtain System to match the edited Mass*

16. Repeat for the Curtain System.

17. On the Massing & Site tab, click the **Show Mass Form and Floors** button.

This will toggle the display of Masses off again. In this model we used a very simple Mass to illustrate the process. While you can model Masses inside the Project environment like we have done here, you can also work outside the project environment and create massing forms in nearly any shape you wish in the dedicated Massing Environment. Once created, they can be loaded into projects and Walls, Floors, Roofs, and Curtain Systems can be applied the same way as the simpler masses shown here. Later in Chapter 15, we will make a detailed exploration of the Conceptual Massing features.

Create a New Curtain System Type

Now that we have modified the shape of the facade bay, we are ready to create a new Curtain System type and apply it to the front façade.

18. Select the Curtain System on screen, and on the Properties palette, click the Edit Type button.

    • Click the Duplicate button and name the new type: **MRAC Front Façade**.

By now you have likely noticed that we sometimes select the object and edit its properties, and other times we use the Families branch of the Project Browser. Ultimately, it is a matter of personal preference and both methods lead to the same result.

19. Beneath the "Construction" grouping, choose **System Panel: Glazed** for the Curtain Panel.

    • Change the "Join Condition" to **Grid 2 Continuous**.

20. Change the parameters of the "Grid 1 Pattern."

    • Verify that the "Layout" type is set to **Fixed Distance**.

    • Set the "Spacing" to **11'-0" [3300]**.

21. Change the parameters of the "Grid 2 Pattern."

    • Verify that the "Layout" type is set to **Fixed Distance**.

    • Set the "Spacing" to **12"-0" [3600]** (see Figure 9.39).

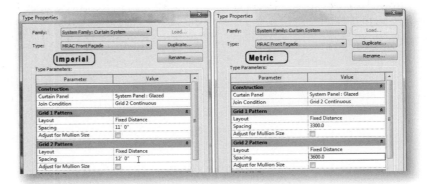

FIGURE 9.39 *Configure the Grid Pattern parameters for the new Curtain System type*

22. Click OK to accept and apply the changes.
23. On the Project Browser, double-click to open the *South* elevation view.

As we can see, the new spacing has been applied to the Curtain System in both directions. It may be easier to see the pattern clearly if we hide the column grid lines.

- Select one Column Grid line, on the Modify | Grids tab, on the View panel, click the **Hide in View** button and then choose **Hide Category**.

- Select one of the Section lines and repeat.

You should now be able to see the pattern clearly in the elevation view. Note that the spacing begins from one end of the Curtain Wall façade and is not centered. In some designs, this may be the desired result, but in this one, we want the pattern centered. This is controlled by the Justification parameter. Unlike the spacing parameters themselves (which were Type parameters), the Justification of the pattern is an instance parameter.

24. Select the Curtain System.

Be sure to select the Curtain System; press TAB as necessary.

- On the Properties palette, beneath the "Grid 1 Pattern" grouping, set the "Justification" to **Center**.

- Note the change to the justification (see Figure 9.40).

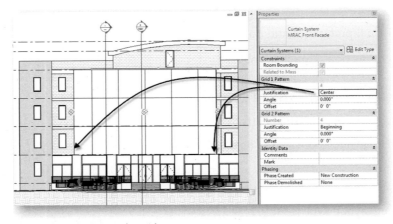

FIGURE 9.40 *Center the grid pattern*

NOTE

> The "Angle" and "Offset" parameters here are the same settings that we edited above using the on screen temporary dimensions in the "Adjust the Curtain Grid Spacing" topic.

25. Save the project.

Edit the Horizontal Mullions

We now have the overall horizontal and vertical spacing established. Next we'll address the shape and size of the mullions. In this sequence, we will edit the horizontal mullions to represent spandrels and the vertical Mullions to represent piers. We need to define a few new Mullion Types and then apply them to the horizontal and vertical orientations of the Curtain System. We can define these Types on the Project Browser.

26. On the Project Browser, locate the *Families* branch again. It should still be expanded from above—if not, click the plus (+) sign icon next to it to expand it.

    • Click to expand *Curtain Wall Mullions*, then *Rectangular Mullion*.

27. Right-click on **2.5" × 5" *rectangular* [50 × 150mm]** and choose **Duplicate**.

    • For the name, type **MRAC Spandrel** and then press ENTER.
    • Right-click on **MRAC Spandrel** and then choose **Type Properties**.

There are several parameters in the "Type Properties" dialog for Mullions. Let's take a look at several of them. Under the "Constraints" grouping are the following parameters:

• **Angle**—Use this constraint to rotate the Mullion relative to the Curtain Wall (see top left of Figure 9.41).

• **Offset**—Input a positive or negative value here to shift the position of the Mullion in or out relative to the Curtain Wall (see middle right of Figure 9.41).

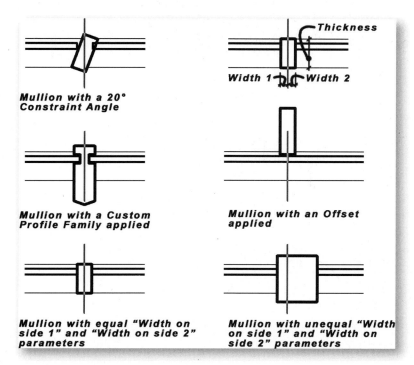

FIGURE 9.41 *Understanding Mullion type parameters*

Under the "Construction" grouping are the following parameters:

- **Profile**—The default shape for Mullions is rectangular. You can create custom profile Families that can be loaded into your project and then will appear in the list here (see middle left of Figure 9.41).
- **Position**—Two options exist for Position: Perpendicular to Face is the default condition. This orientation, as its name implies, sets the Mullion Profile relative to the Curtain Wall. Parallel to Ground rotates the Mullion Profile and is useful in conditions such as in a sloped glazing or a sloped curtain system (Not shown).
- **Corner Mullion**—This checkbox indicates if the Mullion defines a corner condition. This setting is read-only. You cannot make custom corner Mullions (Not shown).
- **Thickness**—This is the depth of the Mullion as measured along its axis perpendicular to the Curtain Wall (see top right of Figure 9.41).

Under the "Materials and Finishes" grouping, you can choose a Material for the Mullion from the list of Materials available in the project. Unlike the "Thickness" parameter listed above, width is divided into two separate width parameters under the "Dimensions" grouping (see top right of Figure 9.41):

- **Width on side 1** and **Width on side 2**—If you set both of these parameters to the same value, then your Mullion will be centered on the Curtain Grid line (see bottom left of Figure 9.41). If you specify different settings, you will essentially shift the Mullion relative to the Grid line (see bottom right of Figure 9.41). In either case, the overall width will be the total of both settings.
- In the "Type Properties" dialog, type: **1'-0" [300]** for the "Thickness," "Width on side 1," and "Width on side 2."

   Remember that the total width is the sum of both the "Width on side1" and "Width on side 2" parameters, so in this case, this makes the Mullion 2'-0" [600] wide.
- Click OK to dismiss the "Type Properties" dialog and return to the model.

The Mullion you just created will not appear in the model yet. All we did was create a new Mullion Type. We have not used it in the actual model yet.

28. On the QAT, click the Default 3D view icon (or open the *{3D}* view from the Project Browser).

   If you are not looking at the front façade, orbit the model to see it.

29. On the Project Browser, beneath the *Families > Curtain Systems* branch, right-click on the **MRAC Front Façade** Curtain System type and choose **Type Properties**.

As an alternative you can select the Curtain System element on screen (be sure to select the Curtain System and not the mullions or Grid lines—use the TAB if necessary). Choose Edit Types from the Properties palette.

- Beneath the "Grid 2 Mullions" grouping, choose **Rectangular Mullion : MRAC Spandrel** for each of the Mullion conditions: "Interior Type, "Border 1 Type," and "Border 2 Type."
- Click OK to see the result (see Figure 9.42).

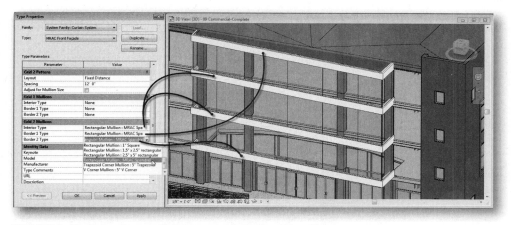

FIGURE 9.42 *Assign the new Mullion type to the horizontal Mullions*

Edit the Vertical Mullions

Repeating nearly the same process, we can edit the Vertical Mullions. We also need a new type here.

30. On the Project Browser, return to the *Families > Curtain Wall Mullions > Rectangular Mullion* node again.

31. Right-click on *2.5" × 5" rectangular [50 × 150mm]* and choose **Duplicate**.

   - For the name type **MRAC Pier Mullion** and then press ENTER.
   - Right-click on *MRAC Pier Mullion* and then choose **Type Properties**.
   - Set the "Thickness" to **1'-0" [300]**.
   - Set both "Width" parameters to **6" [150]** and then click OK.

This will make a 1'-0" [300] square pier Mullion.

32. Return to the "Type Properties" dialog for the *MRAC Front Façade* Type again.
   - Beneath the "Grid 1 Mullions" grouping, for "Interior Type," choose **Rectangular Mullion : Pier Mullion**.

     Leave "Border 1 Type" and "Border 2 Type" set to None.

   - Click OK to accept the changes and close the dialog.

Notice that the spandrels in the middle of the Curtain System are continuous horizontally. This is because we configured Grid 2 as continuous in the Type parameters. However, you can see that the top and bottom border Mullions are not continuous.

33. Return to the "Type Properties" dialog for the *MRAC Front Façade* Type.
   - Change the "Join Condition" to **Border and Grid 2 Continuous**.
   - Click OK to see the result (see Figure 9.43).

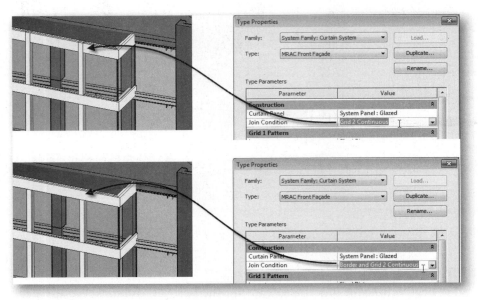

FIGURE 9.43 *Assign the new Mullion type to the horizontal Mullions*

You can also click Apply instead if you wish to stay in the dialog and continue to experiment.

34. Save the model.

Adjust Mullion Position

Take a close look at the Curtain Wall in the South elevation view and you will note that the size of the horizontal bays is not equal.

35. On the Project Browser, double-click to open the *South* elevation view.

Notice that the Level lines for Level 3 and Level 4 pass through the middle of the Spandrel Mullions, but at Level 2 and the Roof they do not. Since our façade is a Curtain System, and since the Curtain System is based on a Mass, we'll have to edit the Mass once more.

36. On the Massing & Site tab, click the **Show Mass** button.

- From the Project Browser, activate the *{3D}* View.
- Place your mouse near the edge of the Curtain System, press TAB.
- Continue pressing TAB until the **Mass: Front Façade: Front Façade** element pre-highlights and then click to select it.
- On the Modify | Mass tab, click the **Edit In-Place** button.
- Click on the top surface of the Mass.

In the center of the surface, a multi-colored shape handle will appear. You can drag the arrows to move the surface and constrain movement along the selected axis. We will see examples of this in Chapter 15. For this edit, we will simply rely on the temporary dimensions again.

- Click on the vertical temporary dimension (currently 36'-0" [10800]) and change it to: **37'-0" [11100]** and then press ENTER.
- Using the tab key to assist you, select the bottom surface and edit the temporary dimension to: **38'-0" [11400]** and then press ENTER.

37. On the ribbon, click the **Finish Mass** button.

The Mass is now 1'-0" [300] taller at both the top and the bottom than the Curtain System (see the left panel of Figure 9.44).

FIGURE 9.44 *Edit Mass, remake the Curtain System, and shift the start point of the bays*

- Select the Curtain System (TAB as necessary) and then on the Modify | Curtain System tab, click the **Update to Face** button.
- Update the Roof as well.
- Turn off Show Mass.

This adjusts the Curtain System to the correct height, but now we have an extra Mullion occurring at the top and the Mullions at the middle floors are not centered on the Level lines anymore (see the middle panel of Figure 9.44).

To correct this and get each of the Spandrel Mullions centered on a Level line, we need to shift the grid offset as we did above in the "Adjust the Curtain Grid Spacing" topic. We can do this with the temporary dimensions on screen, or in Properties palette. In the above referenced topic, we used the onscreen temporary dimensions. This time let's try the Properties palette.

38. Select the Curtain System (TAB as necessary).

- On the Properties palette, beneath the "Grid 2 Pattern" grouping, change the Offset to **1'-0" [300]** (see the right panel of Figure 9.44).

Looking at the right panel of Figure 9.44, it appears that this action solved both the centering of the middle levels and the extra Mullions at the top. However, if we were to investigate carefully (using the TAB key), we would discover that there are actually two Mullions at each location along the top and bottom edges. The reason for this is that the overall spacing of the horizontal bays (Grid 2) is 12'-0" [3600], yet the overall height of the Curtain System is 38'-0" [11400] after the last round of edits we made to the Mass element. In the last step, we shifted the grid so the difference of 2'-0" [600] is equally split at the top and bottom (see Figure 9.45).

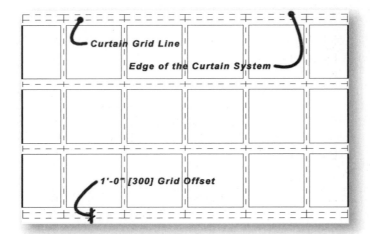

FIGURE 9.45 *With the spacing of 12'-0" [3600] between grids and the grid offset an equal space of 1'-0" [300] remains at top and bottom*

The preceding paragraph and figure adequately explain the presence of the extra Curtain Grid, but still do not explain why the two sets of Mullions are directly on top of one another. Revit treats interior Mullions differently than border types. Interior Mullions are centered on the Grid line. To change this default, you would need to edit the parameters of the Mullion Type, not the Curtain Wall or Curtain System. Border Mullions, on the other hand, are always shifted to fall completely within the Curtain System (or Curtain Wall).

While we could potentially leave the Mullions as they are, it would cause us difficulties later when working with the corners (and if we decided to do quantity takeoffs). Solutions include removing the extra Curtain Grid at the top and bottom or removing the extra set of Mullions on either of the two Curtain Grids at top and bottom. In this case, the easiest thing to do is to actually remove the border Mullions in the Type Properties dialog.

39. On the Families branch of the Project Browser, right-click the **MRAC Front Façade** Curtain System type and choose **Type Properties**.

   • Beneath Grid 2 Mullions, change Border 1 Type and Border 2 Type to None and then click OK.

A Revit warning will appear. The message indicates that some Mullions have become "non-type driven." This means that the Mullions in the model currently attached to the Curtain Grids at the top and bottom of the Curtain System are no longer being controlled by the Curtain System type. You have two choices. If you simply click OK, the Mullions will remain in the model but will no longer be pinned. The other option is to click the Delete Mullions button and remove them. This is what we will do here.

   • In the Revit Warning dialog, click the Delete Mullions button (see Figure 9.46).

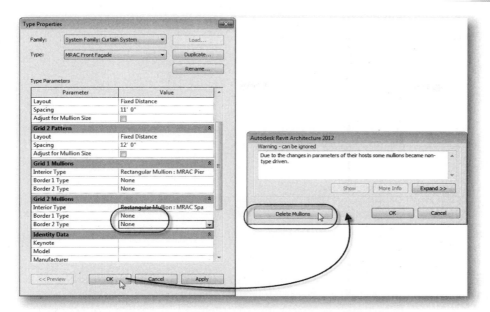

FIGURE 9.46 *Remove the border Mullion assignments and then delete the Mullions in the warning dialog*

If you would like to test that you no longer have double Mullions, select one of the Mullions and hide it with Temporary Hide/Isolate. When you are satisfied that there is only one Mullion in each location, you can reset the Temporary Hide/Isolate.

Sub-Divide Curtain Wall Bays

We can use virtually any of the techniques that we have covered so far to sub-divide the Curtain System that we have into smaller bays. Let's open the *South* elevation view and experiment with a few of these techniques.

40. On the Project Browser, double-click to open the *South* elevation view.

Our design currently has six bays—four equal ones in the center and two slightly smaller ones at the ends. Remember the fact that the centered pattern is a function of the element's instance properties, and not something that we can set in the Type. The spacing of the bays was determined by the Type and is at this point rather large. We can sub-divide these panels into smaller ones by simply adding Grid lines as we did with the Curtain Wall at the start of the chapter.

41. On the Home tab, on the Build panel, click the **Curtain Grid** tool.

- Move the pointer near the left edge of the Curtain System.

  A Grid line and temporary dimensions will appear.

- Using the temporary dimensions, click to create a horizontal grid line at the two-thirds mark (see Figure 9.47).

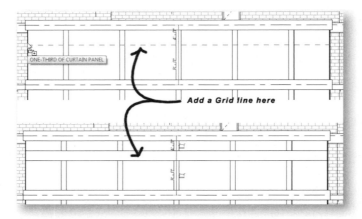

FIGURE 9.47 *Create a Grid line that divides the bay at the two-thirds mark*

To get the tooltips shown in the figure, you must have Tooltip Assistance set to at least Normal in the Options dialog. To open Options, choose it from the Application Menu.

Notice that the new Grid line automatically creates Mullions in the *Spandrel* Type. It would be better to use something a little less "heavy" for these intermediate Mullions. In fact, let's make a new Mullion type for these horizontal bands.

42. On the Project Browser, expand the *Families > Curtain Wall Mullions > Rectangular Mullion* node (as we did above).

43. Right-click on *2.5" × 5" rectangular [50 × 150 mm]* and choose **Duplicate**.

   - For the name, type **MRAC Horizontal Band Mullion** and then press ENTER.
   - Right-click on *MRAC Horizontal Band Mullion* and then choose **Type Properties**.
   - Set the "Thickness" to **1'-6" [450]**.
   - Set both the "Width" parameters to **2" [50]** and then click OK.

This will make a 4" × 18" [100 × 450] band Mullion.

44. Pre-highlight one of the new Mullions (added automatically above to the new Grid line), right-click, and choose **Select Mullions > On Gridline**.

This is a handy way to select an entire row of Mullions very quickly. Notice that when these Mullions highlight, that they are all pinned to the Curtain Wall. Before we can assign the new *MRAC Horizontal Band Mullion* to them, we must unpin them. You can click each of the pushpin icons individually, but this would be very tedious. Instead, use the tool on the ribbon.

45. On the Modify | Curtain Wall Mullions tab, click the *Unpin* tool.

The keyboard shortcut for Unpin Position is UP.

   - From the Type Selector, choose **MRAC Horizontal Band Mullion**.

46. Repeat this process to add horizontal Grid lines and Mullions to the other two floors as well.

47. Open the *{3D}* view and repeat the process to add MRAC Horizontal Band Mullions on the other two faces of the Curtain System.

   You can do the entire process in 3D. Orbit the model as required.

48. Save the project.

Swap Panel Types

Using a process nearly identical to the above, we can select groups of Panels and swap them out with a different type. For this example, we'll use a custom Panel included with the dataset for this chapter. Later in the next chapter, we'll learn how this custom Panel Family was built.

49. Continue in the *{3D}* view.

50. On the Insert tab, on the Load from Library, click the **Load Family** button.

- Browse to the *Chapter09* folder of the location where you installed the dataset files from the student companion.

- Select the *MRAC Curtain Panel.rfa* file and then click Open.

51. Pre-highlight one of the larger panels along the bottom row of the Curtain System.

- Right-click and choose **Select Panels > Along Grid 2** (see Figure 9.48).

FIGURE 9.48 *Select one Panel then right-click to select the entire row*

- On the Modify Curtain Wall Mullions tab, click the **Unpin** tool (or simply type UP).

- From the Type Selector, choose **MRAC Curtain Panel : 3 Bay**

You should now have three equal divisions in each of the selected bays. The loaded Family subdivides the bay equally. This is a simple illustration of what is possible. As stated above, we'll learn how to build this Family in the next chapter. There are many other possibilities. Pay a visit to the AUGI forums (*http://forums.augi.com/*), in the AEC forums section, click the Revit link, and then use the search feature to search for "Curtain Panel." If you wish, you can also search for "Spider" for several examples of Curtain Panels that integrate spider clamps. Try downloading some of these and loading them into your projects to test them out. After you have completed the next chapter, try your hand at making some on your own.

Returning our attention to the façade at hand, the end bays might look better if they were returned to the simple panel. You can easily do this by selecting them and changing the type back again.

52. Select the two end bays, and on the Type Selector, choose **System Panel: Glazed**.

TAB to pre-highlight one Panel, click to select it. TAB again to highlight the other, hold down the CTRL key, and then click to select it. You can also right-click to access additional selection options for panels and mullions.	**TIP**

To repeat the process on the other two floors, you could repeat the selection process and then unpin the Panels again. However, try this tip: select the Curtain System and edit its Type Properties. Change the Curtain Panel to None and click OK.

Notice that this does not remove any Panels that are already present in the model. It only affects how Panels are added automatically going forward (for example, when adding another Curtain System of this type). Now when you select Panels, they will no longer appear pined, therefore, unpinning will not be necessary.

- Select the same four bays on each of the other floors and swap in the **MRAC Curtain Panel: 3 Bay** type (see Figure 9.49).

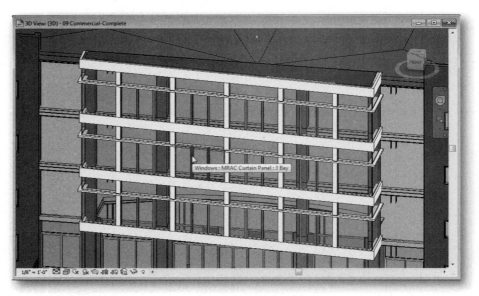

FIGURE 9.49 *Apply the custom panels to each of the other floors*

All of the techniques covered here for Curtain Systems work in nearly identical fashion for Curtain Walls. You are therefore encouraged to experiment on your own with a Curtain Wall type and try to duplicate some of the techniques covered here. Unfortunately, you cannot use the same type for both a Curtain System and a Curtain Wall. If you needed the same type for both a Curtain Wall and a System, you would have to build the type twice—one for the Curtain Wall and another for the Curtain System. Think of it as good practice.

Adding Corner Mullions

If you open one of the floor plans, you will notice that we need to address the corner condition of this Curtain System. Currently, the panels have a gap at the corner. Revit includes several corner Mullion Families in common shapes. We can experiment with each one here to find the one that suits our taste.

53. On the Project Browser, open the *{3D}* view.

- Zoom in on one of the corners of the Curtain System.
- On the Home tab, click the Mullion tool.
- From the Type Selector choose **L Corner Mullion: 5 × 5 Corner [L Mullion 1]**. On the Placement panel, be sure that the **Grid Line** button is selected.
- Click the Grid line at one of the corners.

A new corner Mullion will appear along the vertical grid. However, notice the way that the horizontal Mullions no longer join cleanly. Above in the "Edit the Vertical

Mullions" topic and specifically at Figure 9.43, we adjusted the behavior of the join conditions. At the time, the "Border and Grid 2 Continuous" option gave us the best result. As we refine the design, it appears that changing back to the Grid 2 Continuous option will give us a nicer result. This sort of progressive refinement is part of the process when working with Revit. As your design ideas progress, you can easily adjust settings and options and see the results in real-time.

54. Click the **Modify** tool and then edit the "Type Properties" for the **MRAC Front Façade** type again.

   • For the "Join Condition" choose Grid 2 Continuous and then click OK (see Figure 9.50).

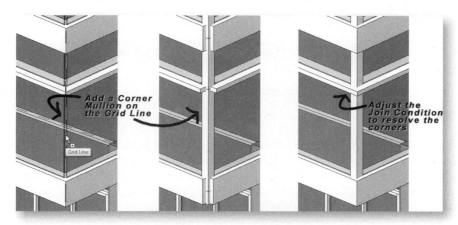

FIGURE 9.50 *Add a Corner Mullion and adjust the Join Condition*

   • Study the model in both the *{3D}* and *Level 2* plan views.

There are four Corner Mullion Families. To try an alternative, right-click one of the Corner Mullions and choose **Select Mullions > On Gridline**. Choose another type from the Type Selector. Try the others and choose the one you like best (see Figure 9.51).

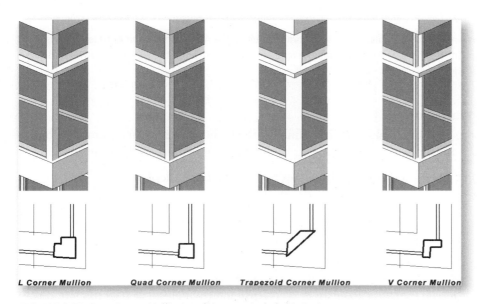

L Corner Mullion   Quad Corner Mullion   Trapezoid Corner Mullion   V Corner Mullion

FIGURE 9.51 *Four Corner Mullion conditions are included in Revit*

Further refinements are possible by creating new Corner Mullion types. Start by right-clicking the one you wish to start with on the Project Browser and then choose **Duplicate**. Give the new copy an appropriate name and then edit the properties. For example, if you want to vary the size of the L Corner Mullion, expand the Families > Curtain Wall Mullions item on the Project Browser. Right-click *5" × 5" Corner [L Mullion 1]* and choose **Duplicate**. Give the new type a good descriptive name. Right-click the new type and choose **Properties**. The size and shape of the Mullion is controlled by the Thickness, Leg 1 and Leg 2 parameters. You can also edit the material and identity data.

55. Add a Corner Mullion to the other side of the Curtain System.

- Add **Rectangular Mullion: 2.5" × 5" rectangular [Rectangular Mullion: 50 × 150mm]** Mullions at the two Curtain Grid Lines where the Curtain System meets the building.

  If you get an error, click Delete Elements. This refers to the very top and bottom again.

- Study the model in both the *{3D}* and *Level 2* plan views.
- Save the model.

Corner Mullions are System Families and cannot be edited or created in the Family Editor. Their profile is hardwired into Revit. You can, however, create new Types for each of the four built-in Families. (The concepts of Families and System Families will be discussed in detail in the next chapter.) When building custom Types for any of the built-in Corner Mullion System Families, the parameters available for edit do not include any way to customize the shape or graphics used to portray the mullion. Therefore, technically, there is no way to create completely custom Corner Mullions. However, some very clever work-around solutions to the problem have been devised by members of the Revit user community. One of the most popular online communities is found at the Autodesk Users Group International (AUGI) web site. There are dozens of forums hosted at www.augi.com supporting nearly all Autodesk products. If you are not a member of AUGI, you can join for free to gain access to this very dynamic community and its information-packed forums.

To create a custom corner mullion, you must actually create a custom Curtain Panel Family that has a Mullion "built-in." A very common mullion type overlooked by Revit is an angled mullion. An example of a Panel Family with integral custom angled corner mullion can be found on the AUGI forums. The thread can be found here: *http://forums.augi.com/showthread.php?t=42563&referrerid=90067*

This is simply one example. Spend some time searching the AUGI forums, the Autodesk discussion groups (*discussion.autodesk.com*), and the many other Revit sites on the web and you will find dozens of additional examples.

Edit Wall Profiles

A few more refinements remain to finalize the front façade of our commercial building. The most obvious one is closing in the building behind the Curtain System.

56. On the Project Browser, double-click to open the *Level 2* floor plan.
57. On the Home tab, click the **Wall** tool.

- On the Properties palette, change the Type to: **Exterior – EIFS on Mtl. Stud**.
- Set the Location Line to **Finish Face: Interior**.
- Make sure that Base Constraint is set to: **Level 2**.
- Change the Top Constraint to: **Up to Level: Roof**.
- Change the Top Offset to: **4'-0" [1200]** and then click OK (see the left side of Figure 9.52).

  Start on the right side of the plan (see the top right corner of Figure 9.52).

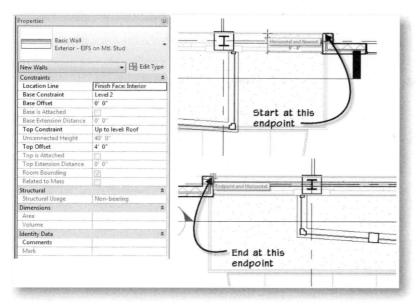

FIGURE 9.52 *Draw a new Wall on the upper floors*

58. Click the first point at the endpoint of the small vertical brick Wall.

    • Click the second point on the opposite side snapping to the endpoint of the other short brick Wall (see the bottom right corner of Figure 9.52).

You should now have a Wall filling in the space that was previously open on the upper floors. However, it fills in the space behind our Curtain System. To fix this, we'll use Edit Profile. This command was covered in Chapter 7 to modify the Wall under the Stair—we'll review it here.

59. On the Project Browser, double-click to open the *South* elevation view.

    • Select the new Wall and on the Modify | Walls tab, click the **Edit Profile** button (see Figure 9.53).

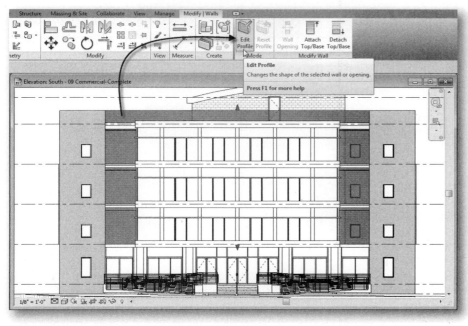

FIGURE 9.53 *Align the Wall to the face of the columns*

- Change the model graphics style to wireframe.

  This will allow you to see the Columns and Roof to help create your sketch.

- Using Figure 9.54 as a guide, draw sketch lines on the inside of the Columns and lower edge of the Roof. Split out the unnecessary segment.

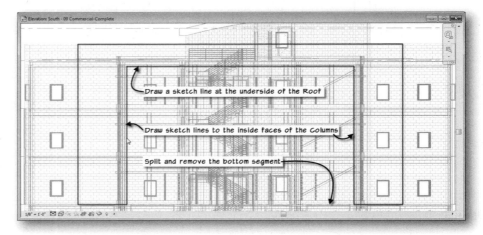

FIGURE 9.54 *Edit the sketch of the Wall profile*

Try using Pick Lines to get started and pick the edges you need from the Columns and Roof, then use Trim/Extend to corner to clean up.

- Click **Finish Edit Mode** and then change the model graphics style back to Hidden Line.

60. Add some punched Windows to the new Wall.

61. Edit the Floor sketches on each level to make them conform to the new front façade shape (see Figure 9.55).

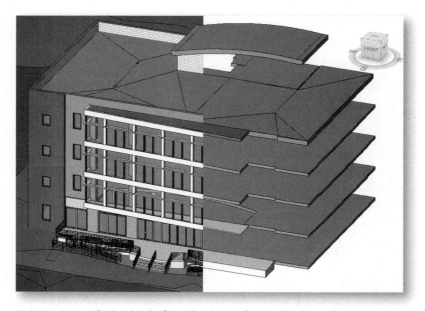

FIGURE 9.55 *Edit the sketch of the Floors to conform to the shape of the exterior shell*

If Revit asks you if you want to attach Walls to Floor bottoms, answer No. This would work well for interior partitions but not exterior shell walls. However, if you are prompted to join with the exterior Walls and cut them, you can answer Yes to this.

To add a floor under the Curtain System on the front façade, you can sketch it manually, or you can add one to the bottom face of the Mass. To do this, show Masses, select the Front Façade Mass, and then on the Modify | Mass tab, click the ***Mass Floors*** button. In the dialog that appears, choose Level 2 and then click OK. On the Home tab, click the drop-down button for Floor and choose the ***Floor by Face*** tool, select the floor plane created with the Mass Floors command, and then click the Create Floors button on the ribbon. You may need to also fine-tune the Mass shape and then update the Curtain System.

62. Save the model.

Whatever additional Curtain Wall explorations you wish to continue on to in this dataset, the specifics are left to you as an exercise. We will explore one more topic before ending this chapter—Stacked Walls.

WORKING WITH STACKED WALLS

At the start of the chapter, we built a custom Wall type and applied it to the exterior Walls. We can go even further and create a special kind of Wall called a "Stacked Wall." A Stacked Wall simply uses two or more Basic Wall types stacked on top of one another to form a more complex design.

CREATE WALL TYPES OR IMPORT THEM?

Import—To save you time, both of the custom Wall types that we need in addition to the one that we already created are provided in a separate project file included with the Chapter 9 files. The name of the file is: *MRAC Wall Types.rvt* [*MRAC Wall Types-Metric. rvt*]. You can open this project file, select the Walls on screen and copy them to the clipboard. Return to the *09 Commercial* project, open the *Level 1* floor plan view, and paste the Walls. Be sure to actually place them in the model off to the side and then click the Finish button on the Tools panel of the Modify Model Groups tab to complete the paste. Once complete, you can safely delete the two Walls. The result will be that two new Wall types will be added to your project: *MRAC Exterior Base* and *MRAC Exterior Parapet*.

Create—If you prefer to create the Wall types yourself, they are very similar to the one that we created in the "Creating a Wall Type" and the "Edit Wall Type Structure" topics. Refer to those topics and create two new Wall types based on *MRAC Exterior Shell*. To create an *MRAC Exterior Base*, change the Finish 1 layer from brick *to Masonry – Stone* and make its thickness 7 5/8" [190]. For *MRAC Exterior Parapet*, change the thickness of the Structure layer to 3 5/8" [90]. With this layer selected, click the Insert button to add a second Structure layer within the core. Make the thickness of this one 4" [100]. Change the material of both Structure layers to *Masonry – Brick* (see Figure 9.56).

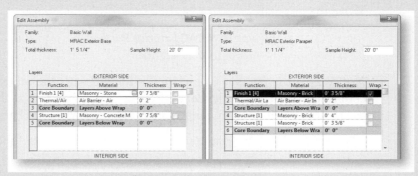

FIGURE 9.56 *Composition of the two custom Wall types*

Whether you choose to copy and paste the Wall types into the project from the provided file or create them yourself is not critical. In either case we will use these two Wall types next to create a custom Stacked Wall type.

In the simple example that follows, we will create a Stacked Wall that combines the Wall type we created at the start of the chapter with two others provided. This will give more articulation to the exterior shell Wall forming a base condition at the first floor and a parapet condition at the roof.

Create a Stacked Wall Type

1. On the Project Browser, expand the *Families > Walls > Stacked Wall* branch.

2. Right-click on ***Exterior - Brick Over CMU w Metal Stud* [*Exterior - Brick Over Block w Metal Stud*]** and choose **Duplicate**.

    *   Right-click on ***Exterior - Brick Over CMU w Metal Stud (2)* [*Exterior - Brick Over Block w Metal Stud (2)*]** and choose **Rename**.

    *   For the name type **MRAC Brick Exterior with Stone Base** and then press ENTER.

    *   Right-click on ***MRAC Brick Exterior with Stone Base*** and then choose **Type Properties**.

    *   Beneath "Type Parameters" click the Edit button next to "Structure."

The Structure of a Stacked Wall is very simple. You designate one or more Wall types that you want to "stack" and how tall each should be. You can insert additional types by clicking the Insert button. Use the Up and Down buttons to move the various Wall types relative to one another. Each type will be assigned an explicit height except for one (use the "Variable" button to choose which is variable). Only one can be variable and it will adjust to the actual height of the individual Wall instance in the model.

3. Click in the Name field next to Type 1.

    *   From the pop-up menu, choose ***MRAC Exterior Shell*** (this is the type we created at the start of the chapter).

    *   For Type 2, choose MRAC Exterior Base.

    *   Change the height of Exterior Base to **6'-0"** [**1800**].

    *   At the bottom of the dialog, click the Preview button.

    *   In the Offset field next to each type, input **4"** [**100**].

Be sure to offset both. This shifts the Wall type overall to compensate for the additional thickness that the base material adds. In this way, the Wall will remain in the same relative position when we swap it into the model.

4. Select item number 1, and then click the Insert button.

This shifts the other two types down and inserts a copy of MRAC ***Exterior Shell*** above the original.

    *   Change the type to ***MRAC Exterior Parapet***.

    *   Change the height to **4'-0"** [**1200**].

Notice how the parapet type does not line up with the others. This is because the core is not as thick and the default for the "Offset" setting at the top of the dialog is to align the Wall types by their Core centerlines. We can change this.

5. From the "Offset" list at the top, choose **Finish Face: Interior** (see Figure 9.57).

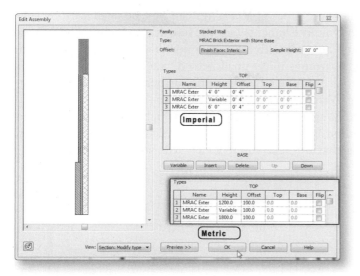

FIGURE 9.57 *Configure the Stacked Wall type*

- When you are done studying the settings, click OK twice to dismiss the dialogs.

6. Open the *Level 1* floor plan view and select one of the **MRAC Exterior Shell** Walls in the model.

    - Right-click and choose **Select All Instances ≫ Visible in View**.
    - Use the CTRL key and add the exterior masonry Wall at the stair core to the selection as well.
    - From the Type Selector, choose **Stacked Wall: Brick Exterior with Stone Base** (see Figure 9.58).

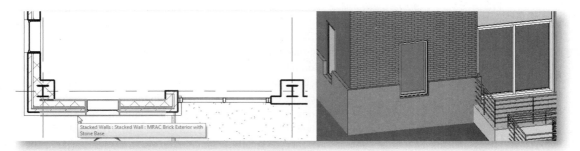

FIGURE 9.58 *Apply the Stacked Wall to the exterior shell walls*

Dealing with Revit Warning Messages

You might get warnings. The first is likely about the dimensions we have been adding. Some of the Witness lines are no longer valid. It is safe to click Remove References in this warning. The next refers to Walls that cannot remain joined—this is most likely due to a change in the join condition between the EIFS Wall placed above and the new Stacked Wall. These types of Warnings can best be dealt with by simply choosing the option to Unjoin Elements (see the left side of Figure 9.59). Do so now. Most Warnings should be dealt with as soon as possible. Some are more problematic than others, but all warnings if left unaddressed can have adverse effects on your model's performance. To examine either warning further, click the Expand button on the right side of the Warning (see Figure 9.59). Note that in the resulting dialog box, you can expand the Warnings to see specifically which elements are involved. Additionally you can see their Element ID's (handy for selecting them if necessary) and there is a Show button that will take you to a view that shows the condition.

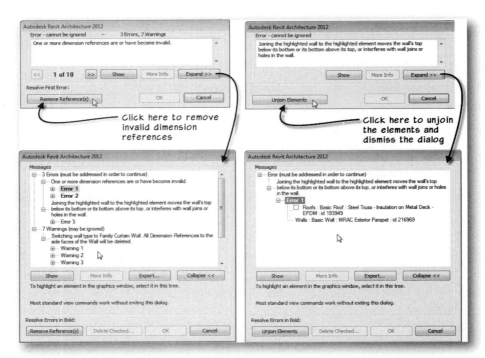

FIGURE 9.59 *Examine the Warnings*

 **MANAGER NOTE**

Some Warnings are easy to miss since they disappear as soon as another object or item is picked, or they can be ignored deferring action till later. In the press of a deadline, it is easy to forget to review such warnings. For this reason, it is good practice to periodically review Warnings that have built up in the Model over time. This can be done on the Manage tab in the Inquiry panel with the *Warnings* tool. Any unresolved Warnings will show up there. As these conditions are resolved they will drop out of the dialog box. While it is extremely rare to have a Model with no unresolved warnings whatsoever, it is desirable to keep them to an absolute minimum. As an example, after resolving the Warnings above, if you were to review any remaining Warnings you would find one that is currently unresolved, related to the number of risers in a monolithic stair. This is actually a common condition and since the design constraints of the Project require the stair remain as modeled, this would be one that would be allowed to remain.

7. Click the Remove References and Unjoin Elements buttons in their respective warning dialogs.

  • Open the *Level 2* floor plan view and zoom to one side of the **building**.

  • Notice that the Curtain Walls on the sides of the building are no longer cutting the Walls.

- On the Modify tab, click the *Cut* tool.
- Select the exterior Wall first and then click the Curtain Wall.
- Repeat on the other side.

8. Open the *Level 1* floor plan view.

- Select one of the dimensions that lost a reference and on the ribbon, click the *Edit Witness Lines* button.
- Click on the elements required to add back the missing witness lines and then click in an empty space to finish (see Figure 9.60).

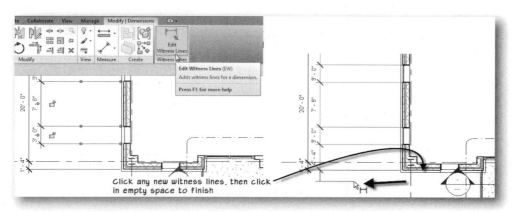

FIGURE 9.60 *Cleaning up the Wall Warning*

When the back Wall of the core is changed to the new Type, it has the parapet and the cap like the others. As a consequence, it no longer attaches to the underside of the high Roof. The best solution is to create another Stacked Wall type without the parapet Wall Type. Simply duplicate the Stacked Wall and name it something like: **MRAC Brick Exterior with Stone Base (No Parapet)**. Edit its Type Properties and remove the parapet Wall Type. Apply this to the core Wall. The specifics are left to the reader as an exercise.

Add a Wall Sweep

To complete the design of our shell Wall, we need to add a parapet cap at the top of the front Wall and it might be nice to add a soldier course. To do this, we'll use Wall Sweeps. You can add a Wall Sweep in two ways. They can be added to the Wall Type parameters so that they automatically show on all instances of the Wall type throughout the model or you can use the Host Sweep tool which allows you to add Sweeps to the model manually at whatever location you designate. In this example, we'll add them to the *Exterior – EIFS on Mtl. Stud* Type.

9. On the Project Browser, expand the *Families > Walls > Basic Wall* branch.

- Right-click the *Exterior – EIFS on Mtl. Stud* Type and choose **Type Properties**.

  Make sure that the preview is showing and choose **Section: Modify type** from the **Preview option.**

- Click the Edit button next to Structure.
- At the bottom of the dialog, click the Sweeps button.
- In the "Wall Sweeps" dialog, click the Add button.

- From the Profile list, choose **Parapet Cap-Precast: 16" Wide [M_Parapet Cap-Precast: 450mm Wide]**.
- Choose **Concrete - Cast-in-Place Concrete** for the Material.
- Choose **Top** from the "From" list (see Figure 9.61).

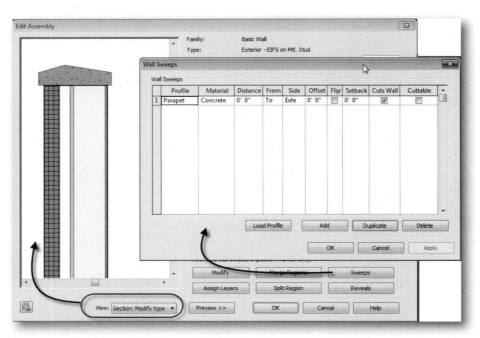

FIGURE 9.61 *Add a Parapet Cap to the Wall type*

10. Click OK three times to complete the Sweep.

If you built the *MRAC Exterior Parapet* Wall Type instead of importing it, then repeat the process to add the parapet cap to that Wall Type as well.

Add a Host Sweep

11. On the Project Browser, open the *{3D}* view.
12. On the Home tab click the drop-down button on the **Wall** tool and choose the **Wall Sweep** tool.

- On the Properties palette, click the Edit Type button.
- Click Duplicate. Name the new type **MRAC Soldier Course**.
- Place a checkmark in the "Cut by Inserts" checkbox.
- Choose Wall **Sweep-Brick Soldier Course: 3 Bricks [M_Wall Sweep-Brick Soldier Course: 3 Bricks]** from the Profile list.
- Choose **Masonry – Brick Soldier Course** for the Material.
- Click OK.

13. Following the prompt on the Status Bar, mouse over the brick Wall and click when the sweep is where you want it.

- Click one or more Walls to add the soldier course (see Figure 9.62).

FIGURE 9.62 *Add a Host Sweep soldier course to complete the exterior shell*

- On the ribbon, click the **Modify** tool or press the ESC key twice.

You can edit the Sweeps with the control handles that appear when you select them. You can add and remove Walls from the Host Sweep afterward using the tool on the Wall Sweep panel. You can add Reveals the same way. Sweeps integral to Wall Types cannot appear in schedules. Host Sweeps can appear in schedules. To learn more about Host Sweeps and Reveals, consult the online help.

14. In the *South* elevation view, click the Reveal Hidden element icon.

- Restore the Grid and Sections categories and then toggle the reveal mode.

15. Save the model.

Continue to make any other refinements or experimentations that you wish. When you are finished, close and save your project file.

SUMMARY

- Revit Architecture provides three kinds of Wall Family: Basic Wall, Curtain Wall, and Stacked Wall.
- You can create a custom Wall type and edit it to add and delete Layers (Components).
- Curtain Walls can be drawn like Walls or even embedded within other Walls.
- Use Cut Geometry to manually embed the Curtain Wall within the thickness of a Wall.
- Curtain Wall Grids can be sketched manually or defined within the Curtain Wall type.
- Mullions can be added to Curtain Grid lines.
- Mullion Joins can be edited to suit your preferences.
- Default joins can be assigned in the type or manually edited per Mullion.
- Walls can be used as panels for Curtain Walls.

- A Curtain Wall type can contain a pre-defined pattern of horizontal and vertical Grid lines.
- Curtain Systems have most of the same features as Curtain Walls but can conform to more free-form shapes such as the faces of Massing elements.
- Custom Curtain Panel Families can be used to create more complex Curtain Wall and Curtain System designs.
- Corner Mullions come in four pre-defined varieties.
- A Stacked Wall contains one or more Basic Walls stacked on top of each other.
- Wall Sweeps can be added to Wall types or added manually as Host Sweeps and used to represent bands, moldings, parapets, cornices, etc.

Working with the Family Editor

INTRODUCTION

All elements that you create and use in a Revit project belong to a Family. As noted in Chapter 1, a Family is an object that has a specific collection of parameters and behaviors. Within the limits established by the Family and its parameters, a potentially endless number of "Types" can be generated. Understanding Families and how to manipulate them is an important part of learning Revit. Families are the cornerstone of the Revit parametric change engine and the topic of this chapter.

OBJECTIVES

While there are several kinds of Families in Revit, the primary focus of this chapter will be "Component Model Families." Component Families (also referred to as "Loadable Families") are simply Families that describe some physical component in the model. Conceptually however, topics discussed in this chapter apply to any kind of Revit Family, including annotation Families such as tags and title blocks. To illustrate this point, and as noted in the "Loading Custom Elevation Tags" topic in Chapter 4, we will also explore a brief tutorial on creating a custom elevation tag. After completing this chapter, you will know how to:

- Explore the Families contained in the current project
- Insert instances of Families in your model
- Manipulate Family Types
- Create custom Families
- Create custom Parametric Families

KINDS OF FAMILIES

In the "Revit Architecture Elements" heading of Chapter 1, a detailed discussion of the various kinds of elements available in Revit is presented. To review, Revit elements include primarily model elements and view-specific elements. Model elements include both Hosts and Components. View-specific elements include Detail items and Annotations. Figure 1.1 presents a summary of these items graphically.

Both Model and Annotation Families include System Families and non-System Families. System Families are Families that are "built into" the software, i.e., "hard-wired," and cannot be manipulated by the user in the interface. This can include built-in-place construction like Walls and Floors, but also includes items like Views, Project Data, and Levels. System Families cannot be created or deleted. Their properties are pre-defined in the software. However, most System Families like Walls, Floors, and Roofs can have more than one Type. A Type is saved and named variation of a particular Family with pre-assigned values. A Family can contain one or more Types; each with its own unique user-editable settings. While we cannot, for example, create or delete Wall Families, we can add, delete, and edit the Types associated with each of the provided Wall Families. In the case of Walls, there are three Families: Basic Wall, Stacked Wall, and Curtain Wall. We worked with each of these in the previous chapter. Each of the Wall Families can (and often does) have more than one Type. For example, in the out-of-the-box template files, there are several pre-defined Basic Wall Types such as: Exterior – Brick on CMU, Generic 6 "and Interior – 5 ½" Partition (1hr). The Basic Wall definition simply means that the Wall has the same structure along its entire length and height. The actual make-up of this structure can vary widely as the sample names above imply. The Stacked Wall has the same structure along its length, but is allowed to vary in its height by "stacking" two or more Basic Walls on top of each other. Finally the Curtain Wall can vary in both length and height by defining detailed grid patterns directly into the structure of the Wall and/or including other Wall Types within regions of the Wall's mass.

Other System Families vary considerably in their specific composition and features, but at the conceptual level, they share the same basic characteristics—the overall behavior of the object is defined by the system and cannot be changed; however, the specific object-level parameters can be manipulated via the creation and application of Type and/or Instance variations.

As already noted, System Family is a broad term that refers to any Family that is built into the software and not editable at the Family level. Any System Family that is also a model element is also referred to as a "Host." A Host is an element that can receive, support, or provide structure for other model elements. A "rule of thumb" to help you identify which model Families are System or Host Families is to think of those parts of the building that are typically assembled on site from a collection of raw or basic materials —the so called "built in place" items. Examples include Walls, Floors, Roofs, Stairs, etc. Of course, there are exceptions to this "rule of thumb," such as columns. In Revit, columns are Component Families, not System Families. Columns in the real world either can be made off-site and then erected (like steel columns), or they can be made on-site (like cast-in-place concrete).

Component Families include all model elements that are not System Families. Other "loadable" Families include Families that are not model elements like annotation, detail components, and title block Families. Component Families can be "Host Based" (require a Host), or they can be freestanding (not requiring a Host). Unlike System Families, we *can* create, delete, and modify Component Families. This is accomplished in the Family Editor interface, and each Family thus created can also be saved to its own unique file (with an RFA extension). System Families cannot be stored as RFA files. Like System Families, Component Families can contain multiple Types. Unlike System Families, Component Families are completely customizable. You can create nearly unlimited parameters and behaviors for such a Family. Since they are created outside of the project editor context, Component Families can be loaded into any project. This is why they are also referred to as loadable. The inverse of our System/Host Family rule of thumb described above is the Component Family rule of thumb. To identify typical Component Families, the guideline is "the parts of a building that are typically built in a shop or factory and installed in the building." Examples include Furniture, Windows, Doors, Curtain Wall Panels, Electrical and Plumbing fixtures, etc. There are exceptions to this rule as well. The other parts of the Curtain Wall, including the wall itself, the Curtain Grids, and parts of the Mullions (corner Mullions), are System Families.

There is another type of Family in Revit called the "In-Place Family." In-Place Families are very similar to the Component Families in terms of creation, editing, and strategy. However, an In-Place Family is created directly within a project (not in a separate Family file as Component Families are), and it *cannot* be directly exported to other projects. In-Place Families can even be created for System Family categories like Walls and Roofs. The only time you should consider creating an In-Place Family is for an element that is unique to a particular project with no possibility that you will ever want to reuse it in future projects. This could be effective for unique "one-off" existing conditions in projects or very specialized design scenarios. In-Place Families are not intended for cases where they need to be used multiple times, copied, rotated, or mirrored, etc., even within the same project. If your needs require this, you should create the item as a Component Family in the Family Editor and then load it into your project(s) instead. If there is any doubt about whether a Family should be a Component Family or an In-Place Family, it is probably best to create it as a Component Family, as duplicating In-Place Families has a detrimental effect on project performance and size.

As has already been noted in the introduction, many of the concepts covered in this chapter might apply equally to System Families, Component Families, Annotation Families, Detail Component Families, and even In-Place Families. However, for the purposes of the following discussions and tutorials, we will limit our discussion mostly to the use and manipulation of Component (Model) Families. For techniques on manipulating editable variables of System Families, refer to the previous chapter where several examples of manipulating and/or creating Wall types were presented. The basic procedure for creating or editing a Type (whether belonging to a System or Component Family) is nearly identical in all cases. A Titleblock Family was created in the "Create a Custom Titleblock Family" topic of Chapter 4. In this chapter, we will explore one Annotation Family example in the "Creating Custom Elevation Tags" topic below. For an example of an In-Place Family, look back to the Fireplace creation example in Chapter 3. Finally, Revit also includes a special kind of Family called a Conceptual Mass Family. Chapter 16 is devoted this and other topics related to working in the conceptual massing environment.

Since the rest of this chapter will focus mainly on Component/loadable Families, for simplicity's sake, we will refer to Component/loadable Families as simply "Families" for the remainder of this chapter.	**NOTE**

FAMILY LIBRARIES

A well-conceived template project is a critical component in successful Revit implementation (refer to Chapter 4 for more information). A well-stocked library of commonly used Families is equally important. A "Library" is nothing more than a series of folders on your hard drive or network server that contain Revit content. Revit ships with a large library of ready-to-use items. These include Component Families,

Annotation symbols, and Detail Component items. Some of these items are included in sample template files (see Chapter 4) and therefore automatically become part of each newly created project, and others are provided in Family files (RFA) in the library folders.

The default United States installation creates the "Imperial Library" that is located by default in the *C:\ProgramData\Autodesk\RAC 2012\Libraries\US Imperial* folder (see Figure 10.1). Please note that the *"Program Data"* folder is hidden in Windows™ by default. To make it visible, click on the Start button and type: **Folder Options** in the search field. On the View tab, show hidden files.

FIGURE 10.1 *An example of one of the folders in the Revit Imperial Library*

There are many items in the library. For example, the Imperial library contains over 1,200 Family files. As we have seen in previous chapters and as we will discuss in more detail below, a Family can contain one or several "Types." Therefore, the 1,200 Imperial Family files represent potentially several thousand readily available component elements that can be added to our projects. It is a good idea to become as familiar as you can with the provided content. The reason is simple; it is always easier to use or modify something that exists than it is to create it from scratch. You may also find it useful to explore the libraries for other countries' components and unit systems since the items in Imperial and Metric libraries are not identical. When you install Revit, several language packs are available. Try installing some of the others and explore the content they contain. You can browse through the library in Windows Explorer or directly from Revit.

There are a few ways to access Family libraries from within Revit. If you are in a project, simply click the Insert tab and, on the Load from Library panel, click the ***Load Family*** button. As an alternative, for Component Families, a ***Load Family*** button is presented on the ribbon when you add the item. A Load button is also available within the "Type Properties" dialog for any Component Family. Click either of these buttons to open a dialog and browse to an appropriate library folder and Family file. This should take you directly to your default library. In other open dialogs, you can click the shortcut icon on the left (i.e., *Imperial Library* or *Metric Library*) to open your library. (These shortcuts are referred to as "Places.") Double-click a subfolder to view its contents. You can select a file (single-click) to see a preview on the right, or if you wish, you can use the Views icon at the top of the dialog to browse by Thumbnails (see Figure 10.2).

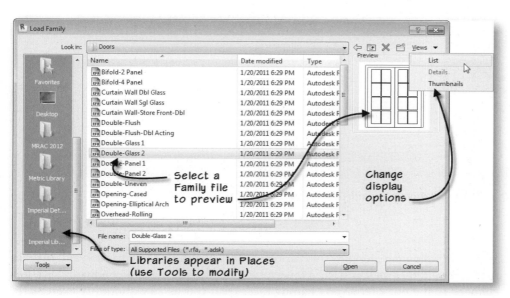

FIGURE 10.2 *Access a Library from within Revit from any load or open dialog*

No matter what method you use to browse your library, you can always open a Family file directly into the Revit interface. In Windows Explorer, simply double-click on the file you wish to edit. From Revit, you can choose **Open > Family** from the Application menu. The Family file will open into the Revit interface. The interface when a Family file is loaded is slightly different than when a project file is loaded. In this state, the interface is referred to as the "Family Editor." While in the Family Editor, the ribbon will show Family-specific tabs and tools. Some of the other functions will appear differently as well. You will still have a Project Browser, but it typically includes only a few Views that have been developed for editing the current Family. These views are only seen while in the Family Editor and do not appear in projects to which the Family is loaded (see Figure 10.3).

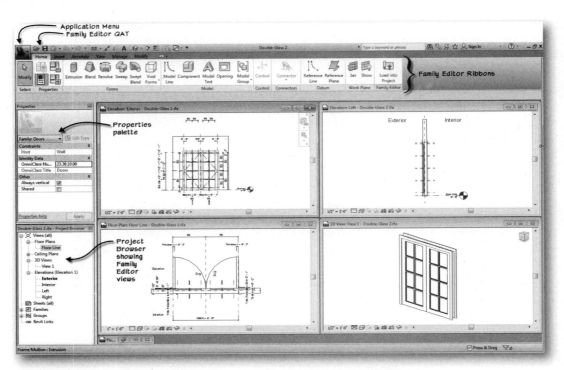

FIGURE 10.3 *The Family Editor (shown with Double-Glass 2.rfa Door Family loaded)*

If you are following along in this passage within Revit, feel free to view and open as many Family files as you wish. However, please do not make any edits or save any changes to the out-of-the-box Family files. Treat this as an exploratory exercise. In the lessons that follow later in this chapter, we will have the opportunity to load, edit, and build our own Families. The purpose of the current discussion is to give you familiarity with what has been provided with your software.

If you want to edit your library locations, you can do so by choosing Options from the Application menu. In the "Options" dialog, click the File Locations tab. There you can change your default template file and the location of the Family template files, and access your "Places" which are the icons that appear on the left side of the open dialogs. In the "Places" dialog, you can add a path, edit a path, delete a path, and change the order in which they appear. You can also edit your places directly from any open dialog using the Tools pop-up in the lower left corner, or you can use the "Options" dialog. From the Application menu, choose **Options** and then click the "File Locations" tab. Click the Places button to add or edit a Library shortcut, delete a shortcut, and move items up and down in the list.

In addition to any libraries that have been installed with your product on your local system, you may also have access to other libraries maintained by your firm's IT personnel. Libraries of this type are typically stored on the company network and accessible via your local or wide area network. Furthermore, Autodesk maintains a portal

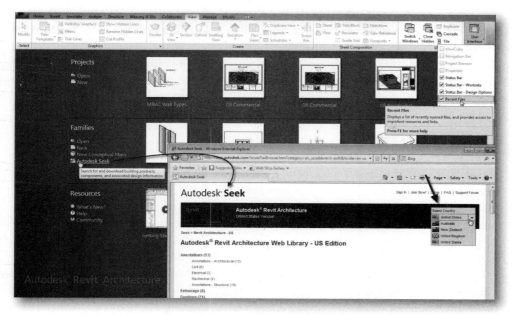

FIGURE 10.4 *Libraries show as shortcuts on the left of "Open" dialogs, add more in Options*

for accessing Revit content on the Internet called Autodesk Seek. Autodesk Seek is a repository of design content provided both by Autodesk and many third party vendors and product manufacturers. Many content providers offer their content in RFA format as well as DWG and other popular design formats. You can access Seek directly from the Insert ribbon tab in the Revit project environment. Simply type a search in this toolbar and press ENTER. In addition, you can always type **http://seek. autodesk.com/** into your default web browser to access Seek outside of Revit. You can also access Seek via the Recent Files page in Revit. Recent Files appears when you have no projects or Families open in Revit. This is the page that greets you when you first launch the application. In the Families area you will find a link for Autodesk Seek. This link opens your web browser to the Seek page showing the library for the region where you live. Other libraries are available from a drop-down list. If you have a Revit file open, you can access Recent Files from the ribbon. Click the View tab, and then on the Windows panel, click the ***User Interface*** drop-down button and choose **Recent Files** (see Figure 10.4). Autodesk also maintains archived libraries from previous versions of Revit, or other content not published to Seek. You can find the legacy Autodesk Web Library at **http://revit.autodesk.com/library/html/**. Feel free to visit these sites now and explore.

Several other online libraries can be found through a quick Web search. Many third-party websites maintain large libraries of Revit content that you can explore and download. Some are free and others charge a fee. Some of the more popular sites are listed in Appendix B of this book. You are encouraged to run a Google search for Revit, Revit Families, and Revit Content. The sites noted here and in the Appendix were accurate at the time of publication, but due to the ever-changing nature of the Web, it is always best to verify sites with a current search.

If you become serious about creating Revit content, you will also want to check out the *Revit Model Content Style Guide* available for download on Autodesk Seek. Visit

Seek at the address indicated above and click the Manufacturer FAQ link at the bottom of the page. Several documents are available detailing best practices for the creation of Revit content. If you begin making content seriously, particularly content that you intend to share inside or outside your office, it is highly recommended that you read and follow these guidelines. The document includes recommendations for naming, level of detail, materials, and a variety of other important topics.

FAMILY STRATEGIES

Do take the time to explore your Family library. As you analyze each of the library items, many of the items will seem useful. You will also find those that you will deem not useful. Still others will prove useful after they have been modified in some fashion. If you take the time to perform this analysis of the included library content before committing any time and resources to customizing existing items and/ or building new items from scratch, you can save yourself a great deal of effort and time. Perform the same analysis on any content you download from content websites. The other benefit to this process is that you will undoubtedly discover ways of performing certain tasks or representing certain items that you had not considered on your own. In other words, reverse engineering existing Families can prove to be a tremendous learning experience. Just remember not to save changes to existing files. Always make a copy or "Save As" first; then save your changes to the copied file.

Nearly everything you do in Revit involves the use or manipulation of a Family. We have already seen several examples of this in the previous chapters. We "used" Families whenever we added something to our model. We manipulated Families whenever we edited the Type and/or clicked the "Duplicate" button in the "Type Properties" dialog. Whenever you wish to add an element to your model, the element you add will be part of a Family. Always try to locate and use a Family that already exists and is available to you before you modify or create custom Families. The basic strategy of Family usage is to maintain the following priorities:

- **1st—Use Existing**—See if what you want already exists somewhere (in the current project or in a library or on the web) and simply use it.
- **2nd—Modify Existing**—If the precise Family (or Type) you want does not exist, find a Family (or Type) that is close to what you need, duplicate, and modify it.
- **3rd—Create New**—If necessary, create a new Family to represent what you need.

This basic approach is somewhat obvious and logical. However, it is surprising how often users will either resort immediately to building custom components without first checking the libraries, or do the opposite and settle for some less than ideal component in their models. Follow this simple three-step guideline, and you will always have the right Family for the job. If you follow the recommendations here and become familiar with the libraries available to you, it is likely that you will often only need to "Use" or "Modify" a Family rather than "Create" a new one. The remainder of this chapter is devoted to tutorials that will illustrate this basic three-step approach.

ACCESSING FAMILIES IN A PROJECT

If you wish to use an existing Family in your project, the first step is to explore what is available. The first place you should look is the current project. You can quickly see all of the Families available in the current project in the Project Browser.

Install the Dataset Files and Open a Project

The lessons that follow require the dataset included on the Aubin Academy Master Series student companion. If you have already installed all of the files from this site, skip to step 3 to begin. If you need to install the files, start at step 1.

1. If you have not already done so, download the dataset files located on the CengageBrain website.

    Refer to "Accessing the Student Companion site from CengageBrain" in the Preface for information on installing the dataset files included in the Student Companion.

2. Launch Autodesk Revit Architecture from the icon on your desktop or from the *Autodesk > Revit Architecture 2012* group in *All Programs* on the Windows Start menu.

You can click the Start button, and then begin typing Revit in the Search field. After a couple letters, Revit Architecture should appear near the top of the list. Click it to launch to program.	**TIP**

3. On the QAT, click the Open icon.

The keyboard shortcut for Open is CTRL + 0. **Open** is also located on the Application menu.	**TIP**

*   In the "Open" dialog box, browse to the location where you installed the *MasterRAC 2012* folder, and then the *Chapter10* folder.

4. Double-click *10 Commercial.rvt* if you wish to work in Imperial units. Double-click *10 Commercial Metric.rvt* if you wish to work in Metric units.

    You can also select it and then click the Open button.

Many interior Walls have been added on several of the floors of this project since the last chapter. If you would like to try your hand at adding those Walls yourself, please visit Appendix A for instructions on how to do so. Feel free to do this before continuing.

Accessing Families from Project Browser

There are a few ways to see which Families are already loaded into a project. Whenever you choose a tool from the ribbon, the list of Family/Type combinations will appear in the Type Selector. The Family will be listed first with a small preview and gray header and the Types will be listed beneath this. We have seen this already in several earlier chapters. While this method is effective for a particular category of element such as a Door, or a Window, the easiest way to see a list of loaded Families is via the Project Browser. The Project Browser has several major branches. The first branch is the Views branch. We have spent nearly all of our time in previous chapters on this branch. Beneath this one are specialized view types such as Legends, Schedules, and Sheets. The Families, Groups, and Revit Links branches are at the bottom. (In the previous chapter, we used the Families branch to provide quick access to the Curtain Wall and Curtain System items we were editing). If you expand the Families branch, you will see each element category currently loaded in the project. Expand any category, such as Doors, and you will see each Family of that category. Finally, expanding one step further reveals all of the Types for each Family.

5. On the Project Browser, double-click to open the *Level 3* floor plan view.

To make the contents of the Project Browser easier to read, drag the edge of the project Browser window to widen it. You can also tear it off and leave it floating on screen and even move it to a second monitor if you have one attached. The Project Browser will remember such customizations between sessions!

6. Select a Door element on screen.

- Open the Type Selector and study its contents (see the left side of Figure 10.5).

7. On the Project Browser, collapse the *Views (all)* and *Sheets* branches (click the minus (–) sign icon).

- Expand the *Families* branch, and then expand *Doors*.

- Expand **Single-Flush [M_Single-Flush]** (see the right side of Figure 10.5).

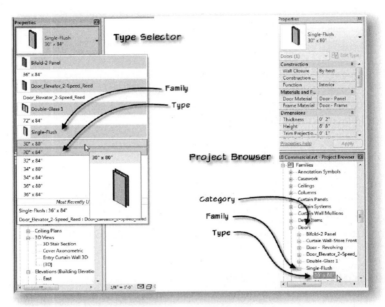

FIGURE 10.5 *Accessing Families in a Project via the Type Selector and the Project Browser*

- On the ribbon, click the **Modify** tool or press the ESC key twice.

As you can see, all of the Door Family/Type combinations will be listed in the Type Selector and the Project Browser. In addition to simply taking inventory of Families and Types in the project, you can also use the list on Project Browser to manipulate and interact with Families. Use the right-click menu to do this. You can right-click on the *Family* and *Type* branches of the Project Browser tree. (Right-clicking the *Category* branch will not produce a menu.) Again recall that we used this technique to edit Mullion Families in the previous chapter.

8. Right-click on the Type **36" × 84" [0915 × 2134mm]** and take note of the menu that appears (see Figure 10.6).

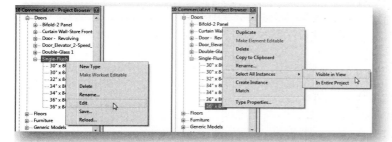

FIGURE 10.6 *Right-click a Family or Type in Project Browser to access menu options*

When you right-click on the Family name, you will only get a menu on Component (Loadable) Families. If you right-click a System Family, no menu will appear. A menu will always appear when you right-click types—either Component Family Types or System Family Types. This is one way you can tell if an item is a Component Family or a System Family. If a menu appears, it is a Component Family. If no menu appears, then it is a System Family.

New in this release, you can save all Families in a project to separate Family files (RFA) by right-clicking on the *Families* branch and choosing the Save command. This command will create RFA files for all component Families in the current project. You can also find the command on the Application Menu: **Save As > Library > Family**.

- From the right-click menu, choose **Select All Instances > Visible in View**.

All of the Doors in the current view that use this Type will be selected (mainly in the core area). There are two options of this command, all instances visible in view and all instances in entire project. The new "Visible in View" option is often a little safer in most cases as you will only be selecting elements that you can see in the current view. When using the "In Entire Project" option, it selects *all* instances in the model, not just those visible in the current view. Depending on what you do with the selection, this could be a little risky.

There are many other commands available on the right-click menu as well. You can Duplicate the current Type, Rename it, or edit its Type Properties. You can also create an Instance directly from this menu, rather than first clicking the tool on the ribbon and choosing the Type from the Type Selector. In some cases this can be quicker; the end result is the same, it is simply a matter of personal preference.

- On the ribbon, click the ***Modify*** tool or press the ESC key to deselect the Doors.

Let's add some furniture to this plan using the right-click option on the Families branch.

9. On the Project Browser, beneath Families, expand the *Furniture* branch and then the *Desk* [*M_Desk*] Family.

- Right-click on ***72" × 36" [1830 × 915mm]*** and choose **Create Instance**.
- Place the new Desk in one of the offices (see Figure 10.7).

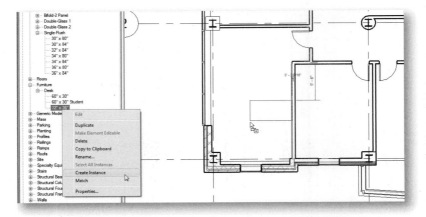

FIGURE 10.7 *Use the Family Tree right-click menu to add a Desk to the plan*

- Repeat in other offices. Press the SPACEBAR to rotate the Desk before placement as required (see Figure 10.8).

| TIP | You can also use Copy, Mirror, or Rotate commands on the Modify tab as needed. |

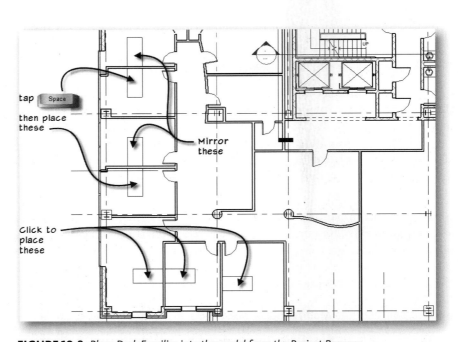

FIGURE 10.8 *Place Desk Families into the model from the Project Browser*

- On the ribbon, click the **Modify** tool or press the ESC key twice.

Match Type Properties

One of the useful functions on the Type's right-click menu is "Match." With this command, you can apply the Type parameters of the item highlighted in Project Browser to an element already in the model. This is a quick way to "paint" a Type's properties onto existing element. We can try this out on the entrance to the suite on Level 3. Currently there is a single Door. Let's make it a double glass Door.

10. On the Project Browser, beneath *Families*, expand *Doors* then *Double-Glass 1* [*M_Double-Glass 1*].

    • Right-click on **72″ × 84″ [1830 × 2134mm]** and choose **Match**.

       When the cursor is moved into the plan view window, it will change to a paintbrush shape.

    • Click on the Door at the entrance to the suite (see Figure 10.9).

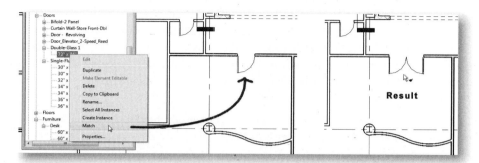

FIGURE 10.9 *Match a Type from Project Browser to an element in the model*

The Door will change to a double glass Door.

    • On the ribbon, click the **Modify** tool or press the ESC key twice.

11. Save the model.

ACCESSING LIBRARIES

So far we have limited our exploration to Families that are already part of our current project. As was mentioned, however, you can find potentially limitless Families in external libraries. You have both the out-of-the-box library installed with your software and also remote libraries accessible via the World Wide Web. The process of using such resources is nearly identical to that already covered with the additional step of first locating and loading the Family in the appropriate library.

Remember, a library is nothing more than a collection of files and folders stored on your local system or a remote server. To access and place a Family in a project, you must first load it into your project. We have already discussed several ways to do this in the "Family Libraries" topic above.

Load Families from "Preferred" Libraries (Places)

Continue in the same file and view as the previous topic.

    • On the Insert tab, on the Load from Library panel, click the **Load Family** button (see Figure 10.10).

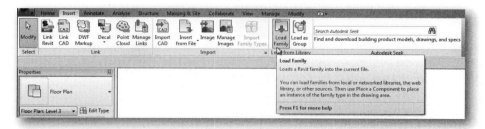

FIGURE 10.10 *Load a Family from a Library*

The "Load Family" dialog will appear starting in the default library folder. Depending on the options you choose during installation, there may be additional libraries available to you. Each library that you have installed or added will appear as a shortcut on the left of the dialog. Think of these as your "preferred" libraries. To learn how to edit your preferred libraries, called "Places," see the "Family Libraries" topic above.

- On the left side, click the *Imperial Library* [*Metric* Library] shortcut icon if necessary.

If your preferred library did not open automatically, use the shortcut icons on the left or the drop-down menu at the top to locate your desired library folder. The library folder will typically contain several subfolders. Each folder can contain additional folders or Revit Family (RFA) files. In some office environments, your CAD or BIM Manager may have installed the default libraries in alternate locations than those mentioned here. Please check with them to find the correct location.

TIP	If you access a folder that is not among your shortcuts, you can add it to Places quickly using the Tools button in the open and load dialogs (see Figure 10.4 above).

- Double-click the *Furniture* folder, select the *Credenza.rfa* [*M_Credenza.rfa*] file, and then click Open.

NOTE	If your version of Revit Architecture does not include either of the libraries mentioned herein, both Family files have been provided with the dataset files from the student companion download. Look for them in the *Library\Imperial Library\Furniture* [*Library\Metric Library\Furniture*] folder wherever you installed the dataset files on your local system.

12. On the Project Browser, beneath Families, expand *Furniture* to see the newly loaded Credenza Family in the list (see Figure 10.11).

FIGURE 10.11 *The newly loaded Family will appear among the others in Project Browser*

- Click on **72" × 24"** *[1830 × 0610mm]* and drag it to the model window.

This is an alternative to the right-click approach above that achieves the same end. Use whichever method you prefer.

- Place a Credenza in each office.

  Remember to tap the SPACEBAR to rotate them as you place them.

TIP	If you want to rotate to align with an angled edge such as the Curtain Wall at the front of the building, place the mouse over the Curtain Wall (or other angled item) and then tap the SPACEBAR. The Credenza will align to the angle of the Curtain Wall. Each time you press the SPACEBAR, it will rotate again. You may need to repeat a few times to get the correct angle.

Using Autodesk Seek

In addition to the libraries installed with the product, several websites are available containing Revit content and Families. Accessible from directly within the product is Autodesk Seek. We last used the built-in Seek functionality in Chapter 7 when inserting elevators. You can access Seek on the Insert tab of the ribbon.

13. On the Insert tab, on the Autodesk Seek panel, click in the search field, type: **chair** and then press ENTER (see Figure 10.12).

FIGURE 10.12 *Execute a search of Seek directly from the Revit ribbon*

You default web browser will open to the Autodesk Seek page and the search for files with the keyword "chair" will execute.

The search will yield several results. Some of the files are from Autodesk, others are provided by product manufacturers. Icons appear to the right of the descriptions indicating which file formats are available. Items might include DWG files, PDFs, Word documents, DXF files, 3ds max files, and of course Revit RFA files. Feel free to select any chair that you like that has an RFA file. For the purposes of this tutorial, we will select one provided by Autodesk for this example.

14. Scroll down and click on Chair-Executive.

This will open the Chair-Executive page. On the right side, the quantity of available files is listed. In this case, there are 4 files for download. They are all basically the same chair, but each a different format.

15. Place a checkmark in the box for the latest version (2011 was the latest at the time of publication) and then click the Download selected button (see Figure 10.13).

FIGURE 10.13 *Download a Family from Seek and Open it in Revit*

You may be asked to agree to the usage and download terms if you have never used Seek before. Please read the agreement before agreeing. Depending on your browser, you will be prompted to download the file or open it directly in Revit. The choice is up to you.

> If you open the file, it will open in the Family Editor:
>
> • Click the **Load into Project** tool on the ribbon.
>
> If you save the file to disk:
>
> • On the Insert tab, click the **Load Family** button, browse to the saved location and load the Family.

16. Using any placement method already covered, place a chair at each desk.

TIP	To make sure that the desks and credenzas are oriented correctly, you can cut some sections or create a third floor 3D view. To do this, copy the {3D} view, rename and orient it as instructed in the "Create a Working View" topic in Chapter 9. Mirror or rotate items as necessary.

17. Save the model.

There are many websites devoted to the distribution of Revit Families and general tech support and discussions. See Chapter 17 for a further list of websites offering such resources.

EDIT AND CREATE FAMILY TYPES

Until now, we have only worked with existing Families and Types. This was the first strategy outlined in the "Family Strategies" topic above. In many cases, you will need to edit existing Types and create your own Types within existing Families. This is our second strategy from the "Family Strategies" topic above. You should try both of these approaches before resorting to creating a completely new Family. There will certainly be situations where it is appropriate to create new Families. (We will look at examples in detail below.) For now, let's look at what we can do with Types first.

View or Edit Type Properties

In some cases, you will want to understand more about the particular Family or Type you are selecting before you place it in the model. You can view the Type Properties or edit the Family directly from Project Browser. This is also accomplished via the right-click menu.

1. On the Project Browser, beneath *Families*, expand *Furniture* then *Desk [M_Desk]*.

   - Right-click on **72" × 36" *[1830 × 915mm]*** and choose **Type Properties**.

We have seen enough examples of this dialog in previous chapters to know that if we make any edits here, they would apply at the Type level and therefore affect *all* Desks currently in the model (since they all currently share this Type).

   - In the "Type Properties" dialog box, click on the "<< Preview" button at the bottom left corner of the dialog.

This will expand the width of the dialog to include an interactive viewing pane on the left side.

   - If necessary, at the bottom of the viewing pane, click the "View" pop-up menu and choose **3D View:View 1** (see Figure 10.14).

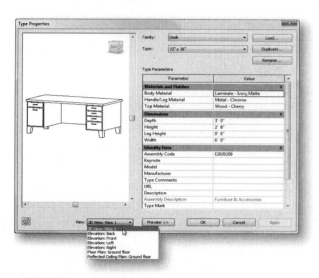

FIGURE 10.14 *Open the view pane and choose a 3D viewpoint*

There are several other view options, such as plan, elevation, and ceiling views. Try them all out if you wish and then return to the 3D view when finished. If you need to get a better look in any view, you can use standard navigation techniques. Right-click in the viewer to get a menu of standard zoom and scroll commands. There is also a "Steering Wheels" icon and the ViewCube.

2. At the lower left corner of the dialog, click the "Steering Wheels" icon.

   - Place your pointer over the Orbit area and then drag the mouse in the viewer to spin the Desk model.

   - Try other modes if you wish and then close the Steering Wheels when finished.

Take a close look at the parameters available at the right of the dialog. First, you can switch the Family you are viewing by choosing another from the Family list at the top. Likewise, you can choose any Type from the Type list beneath it. Changing either the Family or the Type will be reflected immediately in the viewer. Leave the Family set to Desk, but try choosing each of the other two Types one at a time to see the viewer react. Return to the Type: *72" × 36" [1830 × 915mm]* when finished. Notice that all three Types look essentially the same. This is critically important to understand. Study the various parameters listed in the tables below the Family and Type lists.

These are the "Type Parameters." The shape of this desk with its two pedestals, four legs, closed back and top are all characteristics of the Family called: "Desk [M_Desk]." The overall width, height, and even the height of the legs on the other hand are parameters of the each Type. As you choose a different Type from the list, one or more of these values will change accordingly. It is like choosing between several models of a certain make car. A Honda Civic comes in several trim models, some have higher quality wheels, or better sound systems, but they are all *Civics*. In this analogy, "Civic" is the Family and "LX" or "EX" is the Type. Returning to our Desk on screen, go slowly through each of the three types again and study the viewer and especially the change in Dimensions as you do. What this means is that if you simply want a different width or height Desk, or perhaps wanted brass hardware rather than the default chrome, you would simply choose or create a new Type. If, however, you wanted a different "make" of desk, with only a right or left pedestal for instance, you would need to choose or create a new Family.

- Click Cancel in the "Type Properties" dialog to dismiss it without making any changes.

Create a New Type from Project Browser

For the group of offices running vertically along the left, let's create a new Type for this Desk [M_Desk] Family that is a slightly smaller size, yet a bit larger than the other Types currently available.

3. Right-click on **72" × 36" [1830 × 915mm]** and choose **Duplicate**.

This will create a new Type named **72" × 36" (2) [1830 × 915mm (2)]** with the name highlighted and ready to be renamed. (Alternately, if you prefer, you can right-click the Family name and choose New Type, rename it, and the result will be the same.)

- Name the new Type: **66" × 30" [1650 × 762mm]** and then press ENTER.

You can give the Type any name you want, but usually it is helpful to include in the name what makes this Type different from others, in this case the size. This is standard procedure in all of the provided Revit content and a good practice to follow.

4. Right-click on **66" × 30" [1650 × 762mm]** and choose **Type Properties**.

- In the "Type Properties" dialog, change the Depth to **2'-6" [762]**.

  Notice that when you click in this field, a temporary dimension appears in the viewer to indicate how this parameter will apply (see Figure 10.15).

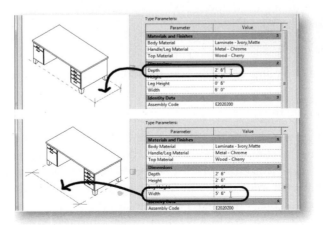

FIGURE 10.15 *Edit the Width and Depth of the new Type*

- In the "Type Properties" dialog, change the Width to **5'-6" [1650]** and then click OK.

5. Right-click on the new Type, and choose **Match**.

  - Click on each of the Desks in the offices along the left side of the plan to change them to the new Type.

  - Move the Desks if necessary to align with the Walls.

You can use the Align tool here and then apply a lock constraint to keep the Desk aligned to a particular Wall. Be careful not to "over" constrain the model however.

TIP

6. Save the model.

Because we can see that there is more than one Type in this Family, with different names, we have a good indication that this family is parametric, and new Types with different parameters, for example sizes or materials, can be created.

If there is only one Type, and it is the same name as the Family, that typically means that particular Family is not parametric. While you could make additional Types from such a Family, you would be limited to editing identity data and materials. For an example, edit the properties of the Chair-Executive. The only editable parameters are the Materials of the chair's components and the standard identity data information. Changing the width, height, and depth is not possible at the Type level for this Family. Another way to say this is that to edit the dimensions of the executive chair, you would be required to open and edit the geometry of the Family file itself.

CUSTOMIZING FAMILIES

While a great deal of content is available to us in the installed and online libraries, we often need components in our projects for which a suitable Family is not readily available. In these situations, we can create our own custom Family. Custom Families can be simple or complex. In this topic, we will look at various examples of creating and using custom Families in our projects.

Duplicate an Existing Family

When you decide to build your own Family, it can be useful to start with an existing Family that is close to the Family you wish to create. Doing so will only require you to save as and edit the existing Family which can be more expedient than starting from scratch. For example, flanking the corridor in the middle of our tenant suite is a secretarial space with room for two workstations. Perhaps the client would like desks in this area that have a CPU cubby rather than the two drawer pedestals occurring in the Family we currently have loaded. We can duplicate the existing Family, and then modify it to make this change. This is much more efficient than modeling the entire desk over again.

7. On the Project Browser, beneath *Families*, expand *Furniture* and then right-click on *Desk* [*M_Desk*].

  - Choose **Edit** from the right-click menu (see Figure 10.16).

As an alternative, select one of the desks in the model, right-click and choose **Edit Family** (or click the Edit Family button on the ribbon with the element selected).

TIP

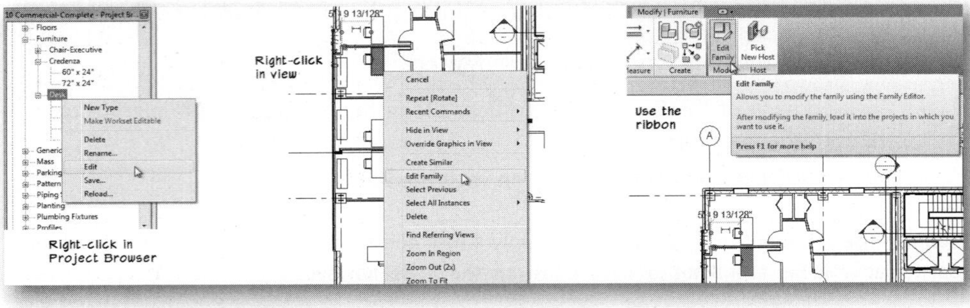

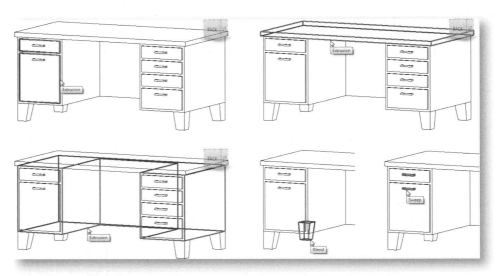

FIGURE 10.16 *Edit a Family from Project Browser or the model*

- In the message dialog that appears, click Yes to confirm opening the Family file.

The *Desk* [*M_Desk*] Family will open into the Family editor with a 3D view active. (You are no longer in the commercial building project file. You are now in the Desk Family file where you can edit it directly).

8. Move your Modify tool over each part of the desk and pause for the tool tip to appear (see Figure 10.17).

FIGURE 10.17 *The Desk is made from various Solid Forms*

Kinds of Forms in the Family Editor

The desk is made from a collection of Solid Extrusions, Sweeps, and Blends. An Extrusion is a 3D form created by a 2D closed shape that is "pushed" along a perpendicular path. A Sweep is an extrusion that follows a path, which can be any shape. A Blend is form that starts with one closed shape and then "morphs" into a second closed shape. The transformation from the bottom shape to the top shape occurs along a perpendicular path. In addition to the three solid forms used in this Family, we also have Revolve and Swept Blend. Revolve is a shape derived from rotating a 2D closed shape about an axis. Swept Blend is basically a Blend that can have a non-perpendicular or curved path. The path is limited to one segment, but it can be any shape you wish.

Each of these five basic shapes can be made as solids or voids. Solids create physical material in the model, while voids carve away from the solid form in order to achieve more complex resultant forms. In addition to this "raw material," a Family can also contain other Families. These "nested" Families are added as Components to the Family within the Family Editor. (We will see an example of a nested Family below.)

Edit an Extrusion

9. On the Project Browser, expand *Views (All)* and then expand *Elevations*.

   • Double-click to open the *Front* elevation view.

As you can see, navigating the Views in the Family editor is nearly the same as within a project, except the names of the views vary and the quantity of views is typically fewer.

10. Select the Extrusion on the left (that represents the drawer fronts) and then hold down the CTRL key and select the handles as well (see Figure 10.18).

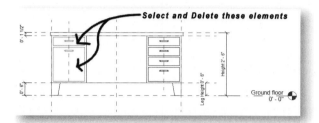

FIGURE 10.18 *Select and Delete the Drawer Fronts and Handles on the left*

   • Press the DELETE key to erase both elements.

The author of this Family chose to create a single extrusion that contained two shapes—one for the top drawer and another for the bottom. While this is a perfectly valid approach, it forces us to edit both drawers together. In your own Families think about such issues carefully. This approach may ultimately be the best, but in some cases, you may decide to instead model each drawer separately.

The pedestal on the left will become a cubby for a CPU. This is why we have deleted the drawer fronts. We now need to make the cubby a bit narrower and then cut a void from it to complete the cubby.

11. Click on the Extrusion that comprises the major form of the desk.

Notice all of the shape handles that appear. We can simply drag one of these to make the cubby narrower (see Figure 10.19).

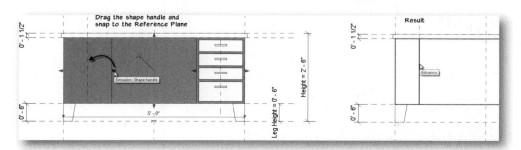

FIGURE 10.19 *Use the Shape Handles to resize the Cubby Pedestal*

12. Drag the shape handle (indicated in Figure 10.19) to the left and snap it to the Reference Plane (dashed vertical line in the middle of the pedestal).

- Click the small open padlock icon that appears to apply a constraint to this Reference Plane.

This action "locks" this side of the extrusion to this Reference Plane. If the Reference Plane later moves, the edge of the extrusion will move accordingly. Let's try it out.

13. Deselect the Extrusion and then select the Reference Plane.

- Edit the temporary dimension that appears to **1'-8" [500]** (see Figure 10.20).

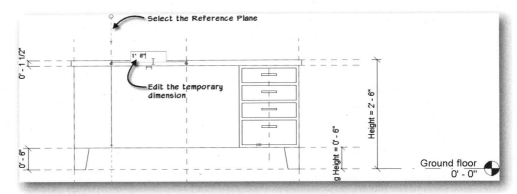

FIGURE 10.20 *Move the Reference Plane to move the constrained Extrusion edge*

Notice that the Reference Plane moved and pulled the shape of the geometry with it.

- On the ribbon, click the **Modify** tool or press the ESC key twice.

Adding a Void Form

Let's add the void next.

14. On the Home tab of the ribbon, on the Forms panel, click the **Void Forms** tool and then choose **Void Extrusion** from the drop-down.

Since we are working with solids and voids that are 3D forms, we need to indicate to Revit our preferred working plane. A Working Plane establishes a 2D surface in the model in which we can sketch our 2D shapes. In this case, since we are working on the Front elevation, a Work Plane parallel to Front makes the most sense (see Figure 10.21).

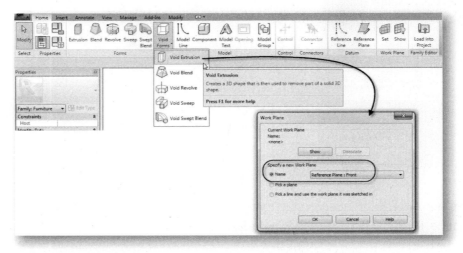

FIGURE 10.21 *Choose a Reference Plane in which to sketch the Void shape*

- In the "Work Plane" dialog, choose **Reference Plane : Front** from the "Name" list and then click OK.

15. On the Create Void Extrusion tab, on the Draw panel, click the Rectangle icon.

   - On the Options Bar, in the "Offset" field, type **3/4" [19]**.
   - Using Figure 10.22 as a guide, snap to one corner of the CPU pedestal.
   - Press the SPACEBAR to flip the sketch to the inside (if necessary) and then snap to the opposite corner as shown.

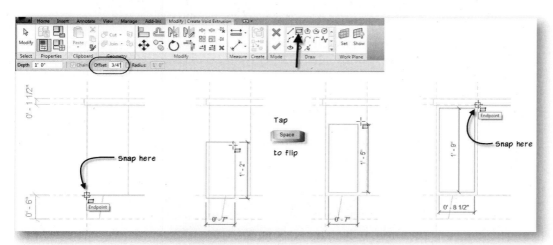

FIGURE 10.22 *Sketch the Cut Extrusion shape*

   - On the ribbon, click the ***Finish Edit Mode*** button (green checkmark).

16. On the Project Browser, beneath 3D views, double-click to open the *View 1*.

   - Deselect the Void.

From this view you can see that the Void is cutting away from the overall form of the desk. However, you can also see that it does not project back very far into the desk's depth. We can easily adjust this. (To see the Cut Extrusion more clearly, pass the Modify tool over the cubby to pre-highlight it.) The floor plan view might be a good location to work for this edit.

17. On the Project Browser, double-click to open the *Ground Floor* plan view.

The Cut Extrusion will appear among the other shapes.

   - Click the Cut Extrusion shape in plan. (It will be in the upper right in plan view.)
   - Use the Shape Handle to stretch the Void back to the rear plane of the desk (it should snap automatically) (see Figure 10.23).

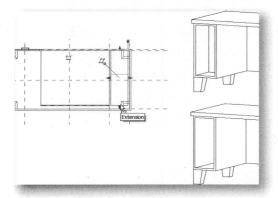

FIGURE 10.23 *Stretch the depth of the Void using Shape Handles*

18. Return to the 3D view to see the results.

With the Cut Extrusion still selected note the Depth field on the Options bar shows the current dimension. The Depth can also be edited here as well. Also note that when the Void is selected, it will not appear to cut the solid. When you deselect the Void, it will resume cutting.

Save a New Family

We are ready to save the results and load our new Family into the project.

19. From the Application menu, choose **Save As > Family**.

Revit imposes no limitations on where you can save Family files, however, the location where you save Family files is a very important consideration particularly in team environments. Check with your IT support personnel regarding the preferred location for saving Family files. Sometimes firms have a "check-in" process for newly created content. It is also common to have prescribed naming procedures. Follow whatever guidelines or practices are in place in your company. For this exercise, we will simply save the Family file (RFA) to our *Chapter 10* folder. If you decide later that you wish to use this Family in real projects, you can copy it from this location to a suitable location on your company network.

- In the "Save As" dialog, browse to the *Chapter10* folder in the location where you installed the dataset files from the student companion.
- In the "File name" field, type: **MRAC Desk-Secretary.rfa** for the name and then click Save.

CAUTION

It is very important to use Save As rather than Save. If you simply save the Family, it will overwrite the existing Desk [M_Desk] Family from which we started.

Before we load this Family back into the project, let's edit its Types. Remember, the Family controls the available parameters and physical form of the element. However, the Type(s) can have specific values for the established Family parameters. Following the lead of the Family from which we created this one, our Types will have predefined values for the sizes.

20. On the Home tab, on the Properties panel, click the **Family Types** button.

The "Family Types" dialog will appear. At the top is a drop-down list showing each of the existing Types inherited from the *Desk [M_Desk]* Family. Some of these are no longer necessary.

21. From the "Name" list, choose **60" × 30" Student [1525 × 762mm Student]** and then click the Delete button (see Figure 10.24).

- Repeat for the **72" × 36" [1830 × 915mm]** Type.

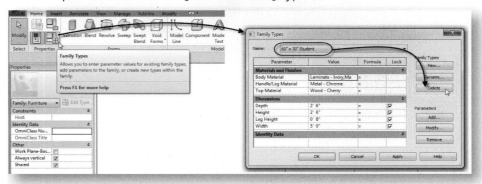

FIGURE 10.24 *Delete unneeded Types*

The remaining Types can stay for this Family.

22. Click OK to dismiss the dialog.

    • On the QAT click the Save icon (or choose **Save** from the Application menu).

> Since we have already saved this file as *MRAC Desk-Secretary.rfa*, we can simply choose Save this time to update the file. Save As would create a second copy of the file, which would be unnecessary.

NOTE

Load the Family into the Project

On the ribbon, notice that there is a gray tinted panel named: "Family Editor." This panel appears on all tabs of the ribbon. Go ahead and click through some of the tabs to see for yourself. Therefore, you will always have access to the *Load into Project* button (on this panel) regardless of the ribbon tab you have active.

23. On the ribbon, click the *Load into Project* button.

If prompted, check the commercial project and click OK. This will bring it to the front and add this Family to the Project Browser ready to be placed in the model.

24. Add two of the new desks to the model as shown in Figure 10.25.

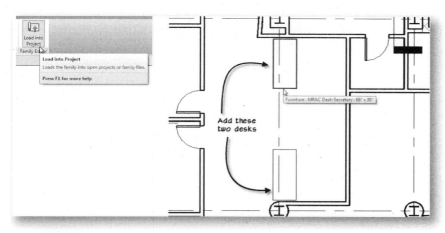

FIGURE 10.25 *Add two new desks to the model*

From the floor plan view, we cannot really see any of the edits we made. Let's add a Camera view to the project so we can get a look at our new Component Family.

25. On the View tab, click the drop-down on the 3D View button and then choose the *Camera* tool.

    • Click a point behind the desks and then drag toward one of the desks.
    • Click again beyond the center of one of the desks (see Figure 10.26).

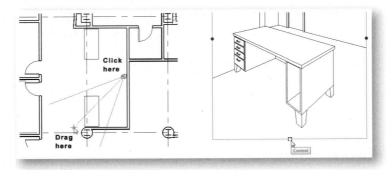

FIGURE 10.26 *Add a Camera view from the plan*

- Use the Control handles to adjust the size of the 3D perspective view so you can see the whole desk. You can also orbit the view to get a better vantage point.

TIP	If your desk is facing the wrong way, return to the *Level 3* plan view, select the desk, and then tap the SPACEBAR a couple times, or use the **Rotate** tool.

Edit a Family

Since this desk is now a computer-type desk, we should probably add a keyboard shelf in the middle. To do this, we simply repeat the process above and return to the Family Editor. We can model the shelf with a simple Solid Extrusion.

26. On the Project Browser, double-click to open the *Level 3* floor plan view.

- Select one of the Secretary Desks; on the Modify | Furniture tab, click the **Edit Family** button (or right-click and choose **Edit Family**).

27. On the Project Browser, double-click to open the *Ground floor* plan view.

28. On the Home tab, on the Forms panel, click the **Extrusion** button.

- Using the rectangle option on the Draw panel, sketch a rectangular shape for the keyboard shelf in plan (see Figure 10.27).

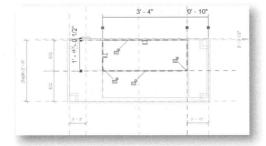

FIGURE 10.27 *Sketch the keyboard shelf*

You do not need to close the lock icons.

- On the Properties palette, beneath Constraints, set the "Extrusion Start" to **2'-1"** **[625]** and the "Extrusion End" to **2'-1 3/4" [644]**.
- On the ribbon, on the Mode panel, click the **Finish Edit Mode** button.

29. Return to the *View 1* 3D view (see Figure 10.28).

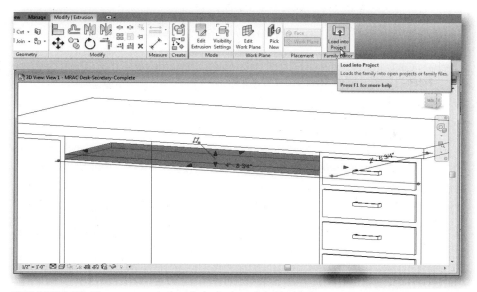

FIGURE 10.28 *The secretary desk with the keyboard shelf added*

30. Save the Family, click Yes to overwrite the existing family if prompted, and then click the ***Load into Project*** button.

- In the "Family Already Exists" dialog, choose the "Overwrite the existing version and its parameter values" option.
- On the Project Browser, right-click the *3D View 1* and choose Rename.
- Type: **Camera at Secretarial** for the new name.
- Double-click to open the *Camera at Secretarial* 3D view.

Notice how the update to the desk now appears in the project.

- Save the project.

Congratulations. You have completed your first custom Family. In the coming topics, we'll create a new Family from scratch and get into more advanced Family creation techniques.

BUILDING CUSTOM FAMILIES

In most cases, the preceding process will enable you to produce the Family you need by leveraging your existing library content. However, sometimes it will be necessary or easier to start from scratch. In this case, you will simply create a new Family file and begin modeling the item you require. All new Families are created from pre-defined Family templates. Revit ships with a large collection of pre-made Family templates from which to choose. It is important to select the Family template which best corresponds to the kind of Family you wish to create. This is because the template you choose determines the category of the Family and whether or not it requires a host. While possible to modify the category later, you *cannot* change the hosting behavior of a Family once it has been created. There are also other less obvious behaviors that Families inherit from their templates as well, so choose your template carefully.

Creating Custom Elevation Tags

If you completed Chapter 4, one of the steps we conducted in the process of setting up the commercial project was to import a custom elevation tag. This was accomplished in the "Loading Custom Elevation Tags" topic. In this passage, we will revisit that topic and go through the process of creating the elevation tags from scratch. Creating annotation Families like the elevation tag is straightforward. These are simple 2D Families, but many of the broader concepts also apply to more complex 3D Families, so this is a good "warm-up" exercise.

1. On the QAT, click the Switch Windows drop-down and then choose **Desk-Secretary.rfa - 3D View: View 1**.

   - From the Application menu, choose **Close**.

     If prompted to save, choose Yes.

2. On the QAT, click the Switch Windows drop-down and then choose **10 Commercial.rvt - Floor Plan: Level 3 [10 Commercial-Metric.rvt - Floor Plan: Level 3]**.

   - From the File menu, choose **Close**.

     If prompted to save, choose Yes.

   - If any other files are open, like the Executive Chair loaded from Seek, switch to them and close them too.

Following this process, you will basically have closed all open files. The Recent Files screen should display in response.

3. From the Application menu, choose **New > Annotation Symbol**.

This will open the "New Annotation Symbol – Select Template File" dialog box to the folder that contains all the available Annotation Family templates.

To create a custom elevation tag, we actually need to build two separate Elevation Mark Families: Elevation Mark Pointer and Elevation Mark Body. We will start with the pointer.

4. From the list of available templates, choose the *Elevation Mark Pointer.rft* template file and then click Open (see Figure 10.29).

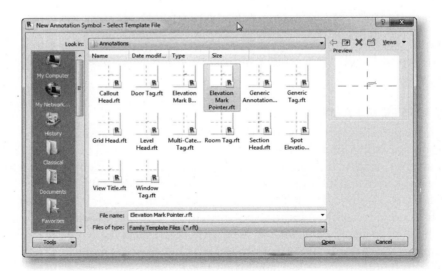

FIGURE 10.29 *Create a new Annotation Family and choose an appropriate template*

Most annotation Family templates contain two Reference Planes marking the insertion point of the Family and a descriptive note with some instructions. In this case, the note reads:

"Place elements/labels to represent the pointer element of an elevation mark.

> The direction of the pointer is vertical from the intersection of the ref planes.
>
> Insertion point is at intersection of ref planes.
>
> Delete this note before using."

Pay close attention to these instructions as they will help us build our elevation tag successfully. We will build our tag to match the guidelines in the US National CAD Standard. The final result can be seen in the tag we used in commercial project. We'll start with a filled region. A Filled Region is simply a two-dimensional shape with an outline and filled in with a pattern. The NCS elevation tag calls for a triangular shaped solid filled pointer surrounding a round tag.

5. On the Home tab, click the **Filled Region** tool.

   - On the Draw panel, click the Center-ends Arc tool.
   - Click to place the center of the arc at the intersection of the two Reference Planes.
   - For the radius, type **5/16"** [**8**] and then press ENTER.
   - Use the mouse to indicate a 180° arc pointing up (see Figure 10.30).

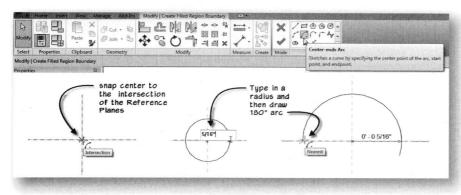

FIGURE 10.30 *Begin the outline of the Filled Region with an arc*

6. On the Draw panel, click the Circumscribed Polygon tool.

   - On the Options Bar, change the number of sides to **4**.
   - Snap the center point to the intersection of the Reference Planes again.
   - Draw out the radius at a 45° angle and snap it to the arc (see the left side of Figure 10.31).

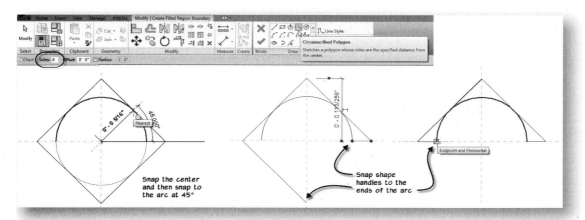

FIGURE 10.31 *Draw the triangular shape with a 4-sided polygon*

7. On the ribbon, click the **Modify** tool or press the ESC key twice.

- Select one of the lower diagonal lines.
- Click and drag the small blue shape handle up and snap it to the end of the arc.
- Repeat on the other side (see the right side of Figure 10.31).

This gives us the basic shape required, but we have to fine-tune it just a little. Revit will not allow us to complete the shape as is, because the diagonal lines intersect the arc. We have two options to deal with this. We can either create three separate shapes by breaking the sketch at the point of intersection (along the 45°) or we can slightly nudge the sketch lines to form a small gap. Let's do that method here.

8. Select the two diagonal lines (select one, then hold down the CTRL key and select the other).

- Zoom in close to the shape and then on the keyboard; press the up arrow key one time.

This will "nudge" the lines up slightly. The amount of the nudge depends on the zoom level. So if you are unhappy with the result, undo, zoom in or out, and try again.

9. Select all of the lines and the arc (the easiest way to do this is to pre-highlight one, press TAB, and then click to select the chain).

- On the Line Style panel of the ribbon, choose <Invisible Lines>.
- On the ribbon, click the **Finish Edit Mode** button.
- Verify that Filled Region: Solid Black is chosen on the Type Selector and then deselect the element (see Figure 10.32).

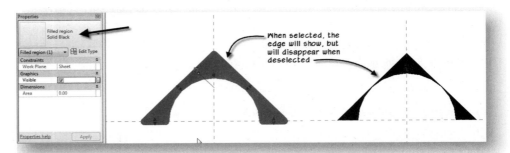

FIGURE 10.32 *The completed filled region shape for the elevation tag*

10. Delete the red text note and then save the file.

- Save it to the *Chapter 10* folder and name it **NCS Elevation Tag Arrow**.

11. From the Application menu, choose **New > Annotation Symbol**.

- From the list of available templates, choose the *Elevation Mark Body.rft* template file and then click Open.

This template appears nearly the same as the other. However, this time the text note reads:

> "Place elements/labels to represent the body of the elevation mark.
>
> Load a pointer family and place instances where you wish arrows to be available in the project.
>
> Insertion point is at intersection of ref planes.
>
> Delete this note before using."

12. On the Home tab, click the Line tool and then choose the circle tool on the Draw panel.

    - Click to place the center of the arc at the intersection of the two Reference Planes.
    - For the radius, type **5/16″ [8]** and then press ENTER.
    - On the Draw panel, click the Line tool and draw a horizontal line across the diameter of the circle (snap to the quadrant on either side).

13. On the Home tab, click the **Label** tool.

    - Click a point on the vertical Reference Plane in the upper part of the circle to place it.
    - In the "Edit Label" dialog, click Detail Number and then click the Add parameter(s) to label icon in the middle (see Figure 10.33).

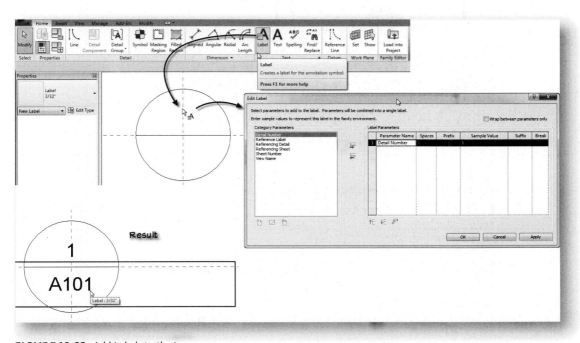

FIGURE 10.33 *Add Labels to the tag*

    - Click OK to complete the label.
    - Create a second Label in the lower portion of the circle for the Sheet Number parameter.
    - On the ribbon, click the **Modify** tool or press the ESC key twice, and then fine-tune placement of both parameters as necessary.

A Label is special text that will report one or more parameters in the tag. We saw an example of this in the "Create a Custom Titleblock Family" topic in Chapter 4.

14. On the QAT, click the Save icon.

    You will be prompted to name the file.

    - Save it to the *Chapter 10* folder and name it: **NCS Elevation Tag**.

15. On the QAT, click the Switch Windows tool and choose NCS Elevation Tag Arrow.
16. On the ribbon, click the **Load into Project** button.

You should only have one other file open; the new NCS Elevation Tag Family file. However, if you are prompted to select a project, choose NCS Elevation Tag.

17. The Place Symbol command should run automatically; if it does not, on the Home tab, click the **Symbol** button.

   - Place an instance of NCS Elevation Tag Arrow onscreen and the click the **Modify** tool or press the ESC key twice.

   - Move it so that it is positioned properly with the center of the circle. (The **Align** tool works well for this.)

   - Copy and rotate three more copies keeping each centered properly on the circle but ending up with one pointing in each direction (see Figure 10.34).

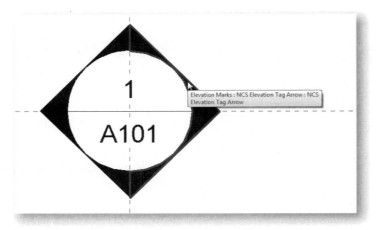

FIGURE 10.34 *The completed NCS elevation tag*

18. Delete the red text note and then save the file.

That completes our custom elevation tag. Since the custom tag is already a part of our *10 Commercial* project, to test it out you can create a new project file. Simply create a new project from the default template and then follow the steps in the "Loading Custom Elevation Tags" topic of Chapter 4 to load your tag and apply it to the default elevation tags.

19. Close all files before continuing to the next topic. (The Recent Files screen should reappear.)

Create a New Family File

Let's turn our attention back to model Families. Keeping with the furniture layout on the third floor of the commercial project a little longer, let's create a custom reception desk for the lobby to the office suite.

> **NOTE** It may be tempting to make your first Family a Door or Window or some other more common element. However, this is not recommended. Furniture is chosen here because such elements tend to be free-standing (not hosted), many have simple straightforward geometry (like desks, shelves and storage units), and their parametric requirements are often limited as well to overall dimensions like width and height. Doors and Windows are much more complex than they first appear, with many complex relationships and parameters, making them a challenging place from which to begin your Family editing explorations. Start simple with items like those showcased here and then work your way up to more complex objects.

1. From the Application menu, choose **New > Family**.

   Browse to your *Templates* folder if necessary.

This will open the "New Family" dialog box to the folder that contains all the available Family templates. Familiarize yourself with this list—there are many from which to choose. Remember the choice of template is important, so take the time to make a good choice (see Figure 10.35). Also note the *Annotations* folder in this dialog. This contains the annotation templates that we saw in the previous exercise.

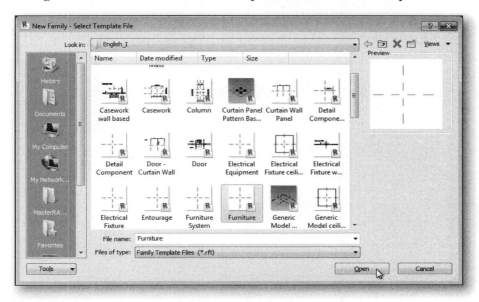

FIGURE 10.35 *Create a new model Family and choose an appropriate template*

Notice that some categories have more than one template. For example, in the list you can see that there is *Electrical Fixture.rft*, *Electrical Fixture ceiling based.rft*, and *Electrical Fixture Wall based.rft*. All three create Families of the electrical fixture category, but the ceiling based one requires a Ceiling host object and the wall based requires a Wall host. *Electrical Fixture.rft* requires no host and can be placed freestanding in a project.

As you can see, your choice of template determines two important characteristics of the new Family: its category and if it requires a host. Any of the templates whose name ends in "based" require a host. The most logical category for our reception desk is furniture. Therefore, we will select the *Furniture.rft* template. Unlike electrical fixtures, there is only one furniture template and it does not require a host.

- Select *Furniture.rft* [*Metric Furniture.rft*] and then click Open.

> **NOTE**
> If your version of Revit Architecture does not contain either of these templates, they have been provided in the *Templates* folder with the dataset files installed from the student companion files.

As with the previous exercise, we are now in the Family Editor. The difference here is that the current view contains only the two Reference Planes and no note. It is often easier when building Families to work with tiled windows. Most Family templates open several Views at once—a plan, a 3D, and two elevations. If you open the Switch Windows drop-down button on the QAT, you will notice that this Family loads with four open views. This is common practice for working in model Families, and it is often more convenient to work if we tile the windows. (If you see more than the four windows from the new Family on the Switch Windows drop-down, please close them first.)

2. On the View tab, on the Windows panel, click the **Tile** Tool.

TIP	The keyboard shortcut for Window Tile is **WT**.

- From the Navigation Bar, choose **Zoom All to Fit** from the Zoom tool.

TIP	The keyboard shortcut for Zoom All to Fit is **ZA** (see Figure 10.36).

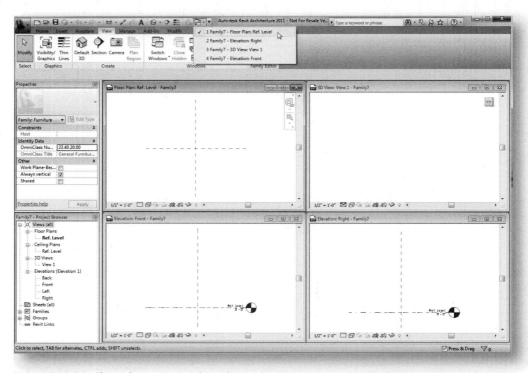

FIGURE 10.36 *Close other projects and Families so that only the current Family file's views show in the Window menu, then tile and zoom all windows*

While it is not necessary to close the other projects and tile the Views, it will be easier to work on the Family in this environment. As we make changes in any view, we will see the results immediately in the others. Since we will be making many three-dimensional edits, this will be very helpful.

3. On the QAT, click the **Save** icon.

- In the "Save as" dialog, be sure you are in the *Chapter 10* folder, type **MRAC Reception Desk** for the Name and then click Save.

Build Family Geometry

As we have already seen, there are two methods to create objects in our Family: adding Solid forms and adding Void forms. Solid forms represent the actual physical materials from which our Family object is created. Void forms carve away from Solid forms to help us create more complex shapes. Both Solid and Void forms can be sculpted in five ways—Extrusion, Blend, Revolve, Sweep, and Swept Blend. (Refer back to the "Kinds of Forms in the Family Editor" topic above for a more detailed description of each one.) We will see examples of each of these in this exercise.

Typically, the first step in building a custom Family is to lay out a series of Reference Planes. The Reference Planes provide structure and form to the Family's geometry. All geometry added to the Family is typically constrained to one or more Reference Planes in some way. In this way, the Reference Planes can be moved (flexed) by a change in parameters, and they will in turn affect the shape of the geometry. This is how parametric Families are created. You may recall seeing some examples of this in the desk Family that we modified earlier. While we will build a parametric Family below that fully utilizes Reference Planes in the way outlined in this note, the following Family example will not make use of Reference Planes to drive geometry. The focus of the next exercise is to acquaint you with the various solid geometry forms available. Reference Planes and parameters could easily be added to this Family. If you wish to explore further at the completion of this exercise, you are encouraged to do so.

NOTE

Create a Solid Extrusion

The first form we will build is a simple extrusion for the work surface.

Make sure that the *Floor Plan : Ref. Level* view is active. (The title bar will appear bold—click the title bar to make it active.)

4. On the Home tab, on the Forms panel, click the **Extrusion** button (see Figure 10.37).

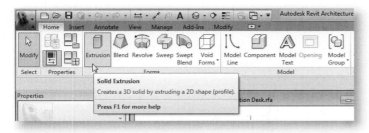

FIGURE 10.37 *Create a Solid Extrusion*

In the center of the plan view are two Reference Planes that were part of the original Furniture template. Like the annotation Family, these mark the insertion point and we will use them to center our geometry.

5. On the Modify | Create Extrusion tab, on the Draw panel, click on the Rectangle icon.

- On the Options Bar, type **-0' 3/4" [-19]** in the "Depth" field.

Be sure to use a negative value for the "Depth." We will discuss the reason below.

- Draw a rectangle that is roughly centered on the two Reference Planes. (The exact size is not important yet.)
- Edit the temporary dimensions to make the rectangle **6'-0" [1800] × 2'-6" [750]** (see Figure 10.38).

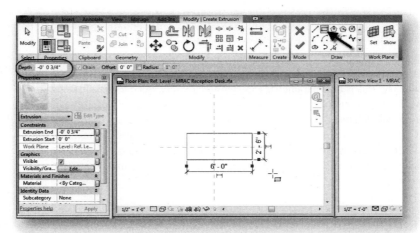

FIGURE 10.38 *Sketch a rectangle near the center of the plan and edit the temporary dimensions*

6. On the ribbon, click the **Modify** tool.

- Select all four of the sketch lines.
- On the Modify panel, click the **Move** tool.
- For the Move Start Point, click the Midpoint of one of the edges.
- For the Move End Point, click the intersection with Reference Plane (see Figure 10.39).

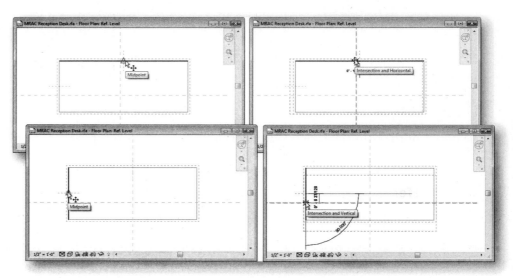

FIGURE 10.39 *Move the sketch to center it on the Reference Planes*

7. Repeat the Move operation in the other direction.

The rectangular sketch should now be centered on the Reference Planes. You can use the **Measure** tool on the QAT to verify if necessary.

- On the Mode panel, click the **Finish Edit Mode** button.

Adjust the View Windows

Now that we have some geometry built, we can adjust our view windows to show it better.

8. From the Navigation Bar, **Zoom All to Fit** (or type **ZA**).

> The Zoom tool on the Navigation Bar should "remember" the previous mode used, so Zoom All to Fit should already be active, and simply clicking the Zoom tool should be enough. However, if you change the mode, you may need to click it a second time to actually execute the command.

Notice that the 3D view shows the model from the side.

- Click in the *3D View: View 1* window and then drag the ViewCube to orbit the model (or hold down the SHIFT key and drag the middle (wheel) button on the mouse).
- If you prefer, you can also use the Steering Wheels to rotate the 3D view.
- Zoom in closer or otherwise fine-tune the view in each window to your liking.
- On the View Control Bar (for the 3D View), change the View Graphics to Shaded (see Figure 10.40).

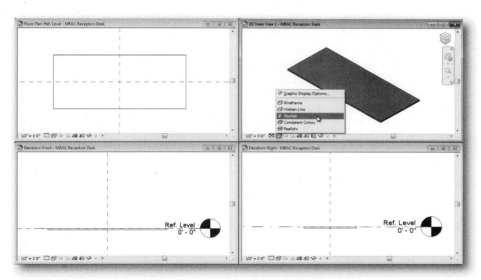

FIGURE 10.40 *Adjust all of the view windows to show the new geometry clearly*

- Save the Family.

Create a Reference Plane

If you study the model, you will notice that the desktop surface is sitting on the floor. While it is possible to edit the parameters of the extrusion to set it at the correct height, we will get more control if we create a new Reference Plane and designate it as the Work Plane for the Solid Extrusion.

9. Click on the title bar of the *Elevation: Right* view window to make it active.
10. On the Home tab, on the Datum panel, click the **Reference Plane** tool.

- Click two points horizontally above the Ref. Level (floor) plane.
- Edit the temporary dimension to **2'-6" [750]** (see Figure 10.41).

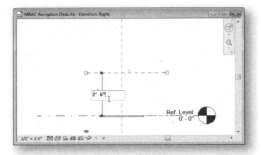

FIGURE 10.41 *Add a Reference Plane in elevation and set it to a appropriate height*

- On the ribbon, click the **Modify** tool or press the ESC key.

This will give you a Reference Plane that is 2'-6" [750] off of the floor. This is the height of the top of our work surface. This is the reason we used a negative Depth for the work surface extrusion above. It will make more sense to the users of this Family should height adjustments be required, that they are adjusting the top of the work surface. We now need to name this Reference Plane in order to make it eligible to be the Work Plane for the extrusion.

11. Select the new Reference Plane and then on the Properties palette, for the "Name" parameter, type **Worksurface Height** and then click Apply.

You will see the name near the end of the Reference Plane while it is selected.

12. Select the work surface Extrusion element.

- On the Modify | Extrusion tab, on the Work Plane panel, click the **Edit Work Plane** button.
- In the "Work Plane" dialog, from the "Name" list, choose **Reference Plane: Worksurface Height** and then click OK (see Figure 10.42).

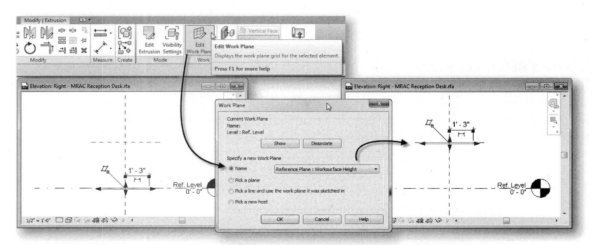

FIGURE 10.42 *Change the Work Plane of the work surface Extrusion to the new Reference Plane*

Notice the work surface extrusion will shift up to the new height in the elevation and 3D Views. If you move the Reference Plane, the work surface will move with it. Try it out, but be sure to return it to the 2'-6" [750] height when done experimenting.

13. Adjust the zoom in all Viewports.

  • Save the Family.

Create a Solid Blend

Now that we have our desktop surface and it is sitting at the desired height, let's add some legs. For the main part of the leg, we will use a Solid Blend to give them a bit of a taper. We will square them off at the top later using a simple extrusion.

Make the *plan* view window active.

14. Using Zoom In Region (from the Navigation Bar or right-click), zoom in on the top left corner of the work surface.

15. On the Home tab, on the Forms panel, click the **Blend** button.

  • On the Options Bar, type **1'-10"** **[550]** in the "Depth" field and then click the Rectangle icon.

  • Sketch a simple square near the corner of the desk.

  • Using the temporary dimensions and the left side of Figure 10.43 as a guide, make the shape 2" [50] square set 2" [50] away from the corner as shown.

If you have trouble with the temporary dimensions, add some permanent dimensions. They will become editable when the sketch lines are selected. You can delete them after you complete the shape, or simply leave them. They will get absorbed into the completed form.	**TIP**

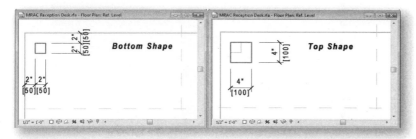

FIGURE 10.43 *Sketch the Base (Left) and Top (Right) of the Solid Blend*

A Solid Blend is basically an extrusion that transforms from one shape to another along the length of the extrusion. In this case, we will use two squares of varying sizes that share a common offset from the desktop corner. This will give a tapered shape to the final leg form. When you build a Blend, you first sketch the shape of the bottom, then the shape of the top.

16. On the ribbon, click the **Edit Top** button.

  • Using the Rectangle option again, sketch a **4"** **[100]** square with the top left corner aligned to the square at the base (see the right side of Figure 10.38).

You should be able to snap to the upper left corner and use the temporary dimensions to draw the other corner accurately. Or you can select the sketch before editing the top, and copy it to the clipboard. Edit the top, paste aligned to get a copy of the sketch, and then select an edge and use the temporary dimensions to edit the size of the square.	**TIP**

17. On the ribbon, click the **Finish Edit Mode** button (see Figure 10.44).

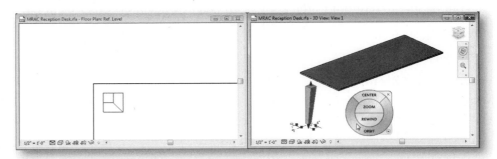

FIGURE 10.44 *Finish the Blend and view the results*

There is a noticeable gap between the top of the leg and the work surface (orbit the 3D view as necessary to see this). We will fill this in with an extrusion.

18. Click on the title bar of the *3D View : View 1* window to make it active.

- Deselect the blend.
- On the Home tab, on the Work Plane panel, click the **Set** button (see the right side of Figure 10.45).

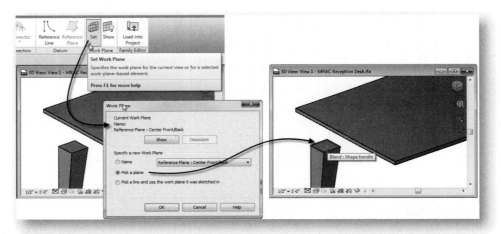

FIGURE 10.45 *The Work Plane icon allows you to change the Work Plane for the current solid form*

While it is most common to use Reference Planes and Levels as Work Planes, you can also use the surfaces in the geometry of your model as a Work Plane. In this case, we will use the top of the Blend.

- In the "Work Plane" dialog, choose the "Pick a Plane" radio button and then click OK.
- In the 3D view, click on the top of the Solid Blend leg (see the left side of Figure 10.45).

After you click the plane, you can click the Show button on the Work Plane panel. This will highlight the current Work Plane by tinting it blue. This gives you a good visual cue that the Work Plane has been set satisfactorily.

19. On the Home tab, on the Forms panel, click the ***Extrusion*** button.

- On the Modify | Create Extrusion tab, on the Draw panel, click on the Rectangle icon.
- Sketch a rectangle on the top of the leg (blend) snapping endpoint to endpoint (see Figure 10.46).

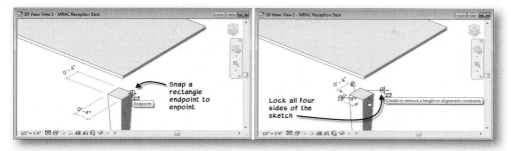

FIGURE 10.46 *Draw the Extrusion sketch on the Work Plane*

After sketching this shape, a lock icon will appear on each edge. You can close these locks to constrain the shape of the Solid Extrusion to the top shape of the Blend. This can be handy if you anticipate making edits to the Blend. This would keep the extrusion at the top of the leg coordinated with these changes. However, in some cases, such a constraint can have adverse affects. This could occur for example if you changed the shape of the Blend top to something other than rectangular. In this case, Revit might have trouble maintaining the constraints. If this were to occur, a warning dialog would appear at the time of edit. For the purposes of this tutorial, the decision is not critical. Regardless of your choice for this exercise, do keep this option in mind for future reference in your own Families.

- On the Properties palette, set the "Extrusion End" to **4" [100]** and then click Apply.

> There are two ways to set the height of the Extrusion; we can use the Depth field as we did earlier, or you can set the Extrusion Start and the Extrusion End on the Properties palette. The end result is the same.

NOTE

- On the Mode panel, click the ***Finish Edit Mode*** button.

The extrusion will appear at the top of the leg but will be too short to reach the work surface. In the next step, we will stretch the height of this element and constrain it to the work surface.

20. Click on the title bar of the *Elevation: Front* window to make it active.

The extrusion should still be selected; if it isn't, select the extrusion.

- Using the Shape Handle at the top, stretch it up to the bottom edge of the work surface.
- Click the lock icon that appears to apply the constraint (see Figure 10.47).

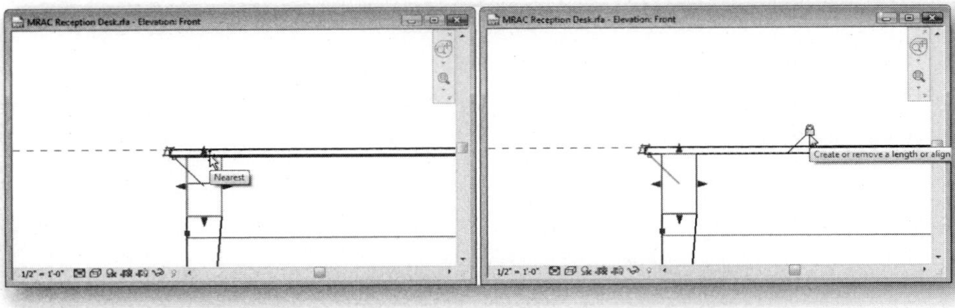

FIGURE 10.47 *Stretch the top edge of the leg and constrain it to the work surface*

Mirror the Leg

Now that we have one completed leg, we can mirror it to create the one on the other side.

21. Click on the title bar of the *Elevation: Right* window to make it active.

22. Select both pieces of the leg (the Blend and the Extrusion).

    • On the Modify | Multi-Select tab, on the Modify panel, click the **Mirror - Pick Axis** tool.

    • Click the vertical Reference Plane (Center Front/Back) as the Axis of Reflection (see Figure 10.48).

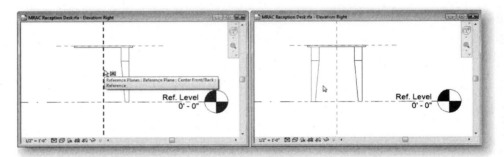

FIGURE 10.48 *Mirror the leg to create one on the other side*

Using Swept Blend to Make a Fancier Leg

We could simply mirror both of the legs again in the *Front* elevation. Before we mirror them, let's vary the Blend of one of the legs to make a slightly "fancier" design.

23. Deselect the leg, make the *Plan* view window active, and zoom in on the top left corner of the work surface.

24. On the Home tab, on the Datum panel, click the **Reference Plane** tool.

    • Draw a 45° Reference Plane through the middle of the leg as shown in Figure 10.49.

    • Select the new Reference Plane, and on the Properties palette, name it: **Leg Path**.

FIGURE 10.49 *Draw a 45° Reference Plane through the middle of the leg and name it*

25. Click on the title bar of the *3D View: View 1* window to make it active.

   •   On the Navigation Bar, click the Steering Wheel icon.

   •   Click the small menu pop-up icon on the lower-right of the Steering Wheel and choose **Orient to a Plane** from the menu (or right-click on the ViewCube).

The "Select Orientation Plane" dialog will appear. This dialog looks nearly identical to the "Work Plane" dialog pictured in Figure 10.42 and Figure 10.45 above. From the list of named planes, you can select the Reference Plane we just created.

   •   From the Name list, choose **Reference Plane: Leg Path** and then click OK (see Figure 10.50).

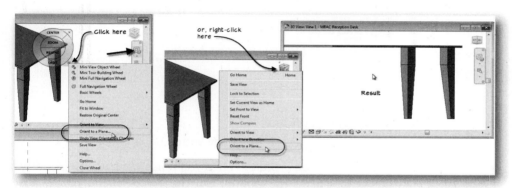

FIGURE 10.50 *Use the Steering Wheel to orient to the Reference Plane*

We now have the 3D view looking directly at the plane that the Reference Plane was drawn in above. Since we named the Reference Plane, we can make it a Work Plane to assist us in building the next object.

26. On the Work Plane panel, click the *Set* button.

   •   In the "Work Plane" dialog, choose the **Reference Plane: Leg Path** from the Name list and then click OK.

27. On the Home tab, on the Forms panel, click the *Swept Blend* button.

A Swept Blend is nearly identical to a Blend. It also has two sketched shapes, one at the top and the other at the bottom. However, instead of connecting these two shapes along a perpendicular path whose height we specify, we instead draw the path using either a straight line of any angle or a curve. In this case, we'll draw a gentle curve. The only limitation of the Swept Blend is that the Path can only be a single segment. To work around this, we'll use a spline curve in this example.

   •   On the Modify | Swept Blend tab, on the Mode panel, click the *Sketch Path* tool.

- On the Modify | Swept Blend > Sketch Path tab, on the Draw panel, choose the Spline icon.
- Click the start point at the bottom of the existing leg (labeled 1 in Figure 10.51).
- Click the next point straight up along the leg slightly above the first (labeled 2).
- Click the third point to left of the existing leg about half way up (labeled 3).
- Click the last two points on the existing leg near the top of the existing blend (labeled 4 and 5).

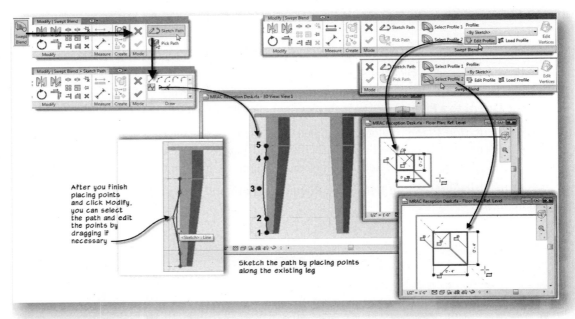

FIGURE 10.51 *Sketch the path and profiles of the Swept Blend*

- On the ribbon, click the Modify tool.
- On the Mode panel, click the **Finish Edit Mode** button.

A swept blend, like a blend, has two profiles. When you finish the path, the Modify | Swept Blend tab will reappear on the ribbon. On the Swept Blend panel, <By Sketch> will appear for the Profile choice. Two options are available for the profiles. You can sketch them or you can use a Profile Family. A Profile Family is simply a Family that contains a predefined profile shape. This can be useful if you use the same profile shape frequently or if you are building a form that represents a common element like moldings. The Profile list will show any Profile Families loaded in the current Family. You can use the Load Profile button to load external Profile Families into the current Family. Refer to the "Build or load a Mullion Profile Family" topic below for an example.

In this case, we will simply sketch the profile like we did for the blend above. Therefore leave the Profile set to <By Sketch>.

28. Click on the title bar of the *Plan* view window to make it active.

- On the Swept Blend panel, click the **Edit Profile** button (see the top right of Figure 10.51).

This allows us to edit the sketch of Profile 1.

- With the rectangle tool, trace the small square (of the existing blend) and then click the **Finish Edit Mode** button. Do not close the lock icons.

Before you can complete the swept blend, you must define both profiles 1 and 2.

29. On the Swept Blend panel, click the **Select Profile 2** button.

   - Click the **Edit Profile** tool, and using the rectangle shape, trace the large square. Do not lock it.
   - Click the Finish Edit Mode button (see the right side of Figure 10.51).

30. Click the **Finish Edit Mode** button again to complete the Swept Blend.

   - Delete the original Blend (see Figure 10.52).

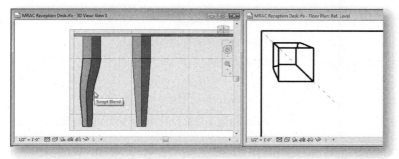

FIGURE 10.52 *Complete the Swept Blend*

31. Click on the title bar of the *Elevation: Front* window

   - Using a window selection box, select both legs (top and bottom portions) and mirror about the (Center Left/Right) Reference Plane.
   - Save the Family file.

Create a Pencil Drawer

Using another extrusion, we can add some geometry to represent the structural support of the work surface and a pencil drawer.

32. Click on the title bar of the *Floor Plan : Ref. Level* window to make it active.

   Zoom as required to see the whole desktop.

33. On the Home tab, on the Work Plane panel, use the **Set** tool to set the Work Plane to **Reference Plane : Worksurface Height**.

34. On the Home tab, click the **Extrusion** Tool.

   - On the Draw panel, click the Rectangle icon.

35. On the Properties palette, type **-6 3/4"** [**-169**] for the "Extrusion End" and type: **-3/4"** [**-19**] *for the* "Extrusion Start."

   Both negative.

   - On the Options Bar, type **3"** [**75**] in the "Offset" field.
   - Trace the desktop snapping from one corner to the one diagonally opposite (see Figure 10.53).

   Tap the SPACEBAR after the first corner if necessary to make the sketch lines offset inside the desktop shape rather than outside.

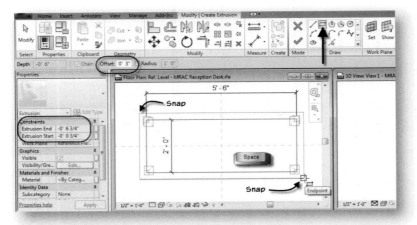

FIGURE 10.53 *Sketch a rectangle offset from the desktop and adjust its constraints*

- On the Mode panel, click the **Finish Edit Mode** button.

36. Working in the *Elevation: Front* window create a 1/2" [12] deep extrusion for a pencil drawer using the face of the extrusion just drawn as a Work Plane (see Figure 10.54). The exact size and position on the drawer are not important.

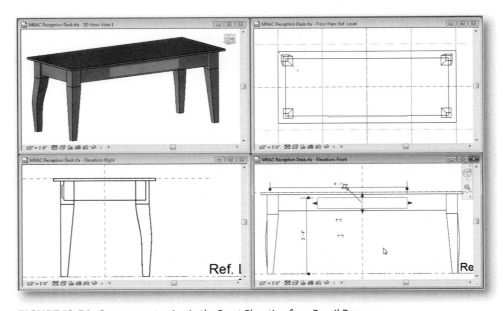

FIGURE 10.54 *Create an extrusion in the Front Elevation for a Pencil Drawer*

Create a Solid Revolve

To add a drawer pull to the pencil drawer, we will use a Solid Revolve. A Solid Revolve is a form in which you sketch the profile of the form and then spin the profile around an axis. To create a simple drawer pull, you draw half of the cross section of the pull, and then place the revolution axis at the center of the pull.

37. Click on the title bar of the *Floor Plan : Ref. Level* window to make it active.

- Zoom in on the Pencil Drawer in the plan view.

38. On the Home tab, click the **Revolve** tool.

   On the Draw panel, verify that the **Boundary Lines** tool is active and on the Options Bar verify that "Chain" is selected.

   • Starting at the midpoint of the pencil drawer, sketch the form shown in Figure 10.55. The depth should be about 3/4" [19] but the exact dimensions and shape are not critical.

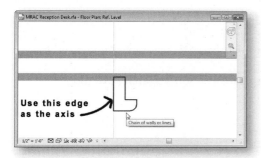

FIGURE 10.55 *Sketch the shape of the drawer pull*

39. On the Modify | Create Revolve tab, on the Draw panel, click the **Axis Line** button.

   • Trace the vertical edge of the shape drawn above.
   • On the Mode panel, click the **Finish Edit Mode** button.

If you study the result in all Views, you will notice that the drawer pull is at the height of the work surface. This is because Revit used the "Worksurface Height" Reference Plane as the Work Plane for this element. As a result, you cannot move the drawer pull to the correct height—try it in the Front view to see for yourself. If we had lots of hardware elements for this desk, we could create a new Reference Plane and associate the elements to it instead. In this case, we will simple disassociate the Work Plane from the Revolve which will allow us to move it freely.

40. Click on the title bar of the *Elevation: Front* window to make it active.

   • Select the drawer pull.
   A small Work Plane icon appears attached to the element.

   • Click the Work Plane icon to disassociate the Work Plane.
   • Move the drawer pull to center it on the height of the drawer (see Figure 10.56).

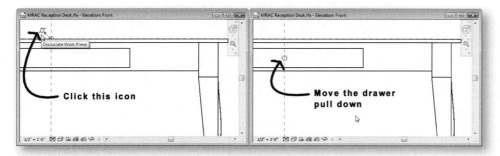

FIGURE 10.56 *Disassociate the Work Plane and move the drawer pull*

Create a Solid Sweep

Now let's create a privacy screen using the Solid Sweep command.

41. Make the *Floor Plan: Ref. Level* window active.

- Zoom out enough to see the whole desk.
- On the Home tab, click the **Sweep** tool.

Creating this form requires a few steps similar to the swept blend. For a sweep, we need a path and a profile. When you begin the command, the Modify | Sweep tab will appear on the ribbon. The first step is to designate the path. You can do this by sketching or you can use the pick option to set the path from existing geometry.

42. On the Modify | Sweep tab, on the Sweep panel, click the **Sketch Path** tool.

The Modify | Sweep > Sketch Path tab will appear.

- On the Draw panel, click the "Pick Lines" icon.
- On the Options Bar, in the "Offset" field, type **2" [50]**.

As with the "Offset" option in other commands, as you hover over a line in the model, a green dashed line will appear temporarily indicating the side to which the sketch line will offset. As you create sketch lines here, be sure to offset to the inside of the desk surface.

43. Offset the three outside edges of the desk (not the one with the pencil drawer) to create a "U" shaped sketch (see Figure 10.57).

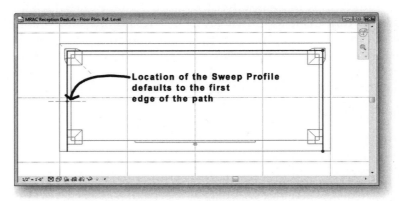

FIGURE 10.57 *Create Sketch lines by picking the edges of the desktop*

- On the Mode panel, click the **Finish Edit Mode** button.

These three lines will be the "path" of the Sweep. We have seen Sweeps in other chapters. A Sweep "pushes" a profile along a path. It is similar to an extrusion except that the path does not have to be a simple straight line as it does in an extrusion. As we saw above with the swept blend, when you finish the path, the Modify | Sweep tab will appear automatically. The profile of a sweep can be either a sketch or a loaded profile.

44. On the Sweep panel, click the **Edit Profile** tool.

The profile travels perpendicular to and along the path. This means that since we sketched the path in plan, we must sketch the profile in elevation. When you click the Edit button, Revit will prompt us to select an appropriate view in which to sketch the profile. Please note that it is also possible to load an existing Profile when creating a Sweep instead of sketching it. This allows you to create "Profile" Families and store them in your library so that you can quickly create Sweeps from your most common profiles. In this example, we will sketch the profile as a custom shape just for this desk.

- In the "Go To View" dialog, choose *Elevation: Front* and then click Open View.
- Zoom the *Front* elevation view as necessary to center the red dot on screen.

The red dot indicates the start point of the profile. It appears automatically on the first segment of the Path. If you are not satisfied with the default location of the profile, you can move it, but be sure to move it before you start sketching the profile.

45. Using the Line and Arc tools and the "Chain" option, create a profile similar to the one shown in Figure 10.58.

| The exact shape of the profile is not critical. Just make sure to make a closed shape. | NOTE |

| Remember that you can zoom in to make sketch lines automatically snap to a smaller increment. The Profile in the figure has a thickness of 1/2" [12]. | TIP |

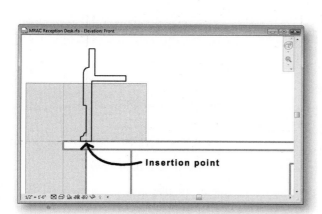

FIGURE 10.58 *Sketch the Profile of the Sweep*

- On the Mode panel, click the **Finish Edit Mode** button.
  This returns you to the Modify | Sweep tab.
- On the Mode panel, click the **Finish Edit Mode** button.

46. Zoom All Viewports to Fit (see Figure 10.59).

| Try setting the Viewports to hidden line display to see the final product more clearly in plan and elevation. You can also use the Join Geometry tool to merge solid forms together. | TIP |

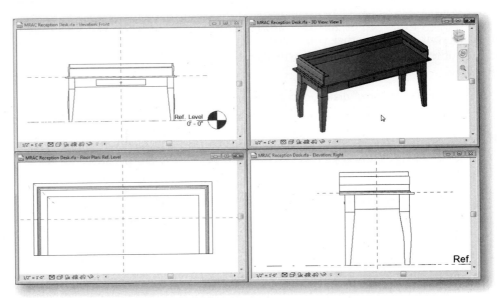

FIGURE 10.59 *Completed reception desk with privacy screen*

47. Save the Family file.

LOAD THE CUSTOM FAMILY INTO THE PROJECT

At this point, we have completed the geometry of our Custom Family and are ready to load it into our commercial project. The process has been covered before. Let's review it now.

Load the Commercial Project

We closed the project above to make it easier to tile the windows in the Family Editor. We now need to reopen the project.

1. Reopen the Commercial project.
2. On the Project Browser, double-click to open the *Level 3* floor plan view.

   • Zoom in on the reception space (in the middle of the plan).

Load a Family

The Family file is still open. We can switch over to it and then load the Family back into the project. We could also use the ***Load Family*** tool on the Insert tab, but since the Family file is still open, it is easier to load from there.

3. Hold down the CTRL key and then press TAB.

This will cycle through open windows with each separate press of the TAB key.

Repeat the CTRL + TAB if necessary until the Reception Desk Family file comes into view.

4. On the Family Editor tab, click the ***Load into Project*** button.

The screen will switch back to the commercial project. The Modify | Place Component tab should become active as Revit puts you directly in placement mode.

5. Add an Instance of the Family to the reception space (see Figure 10.60).

Place the mouse over the curved Wall and then press the SPACEBAR. This will allow you to rotate it as you place it.	**TIP**

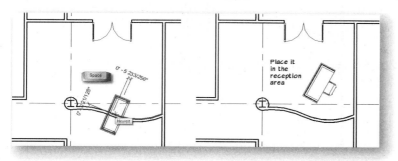

FIGURE 10.60 *Load the Family and then add an Instance to the commercial project*

6. From the Type Selector, choose the Executive Chair.

- Highlight the edge of the reception desk, tap the SPACEBAR to match the rotation, and then place one with the reception desk.
- On the ribbon, click the **Modify** tool or press the ESC key twice.
- Save the project.

If you look at the Furniture branch beneath Families on the Project Browser, you will see that the Reception Desk Family is now loaded. Notice also that its only Type shares the same name as the Family. Even though we did not create any Types for this Family, all Families must have at least one Type. Revit simply makes one for us using the same name if we do not create one.

The reception desk tutorial was designed to introduce you to the basics of the modeling forms available in the Family Editor. All five forms were explored, and even though we did not create any Void forms for the reception desk, we saw an example above in the "Adding a Void Form" topic. Voids can be made in any of the five forms just like Solids. The reception desk Family is a simple non-parametric Family that cannot be edited in the project at all. To make even simple edits like changing the width, height, or depth, we would need to return to the Family Editor. The kind of Family we have created here is sometimes referred to as a "stand-alone" Family or a "Family of one."

If we chose to, we could certainly make the reception desk a "parametric" Family. We would need to follow a slightly different procedure when building it, and doing so would allow us to expose certain dimensions and other parameters to the users of Family directly in the project editor. For example, if our reception desk could be ordered in multiple sizes, we could make the dimensions of the desk parametric in the Family Editor and then adjust our geometry to be controlled by these parameters. In this way, a user of the Family could manipulate the exposed parameters without needing to open the Family Editor, nor understand the complexities of creating or editing Families. Our next topic is devoted to the subject of parametric Families.

BUILDING PARAMETRIC FAMILIES

Throughout the course of this chapter you have been exposed to a variety of Family editing and creation scenarios. Families can be very simple or very complex. When you are first learning their scope and potential, it is helpful to start with simple examples and work your way up to more complex ones. This is exactly the approach this chapter has been following. As such, we have yet to cover one of the most powerful and useful aspects of Families. That is their ability to be "parametric." Like much terminology that is associated with Building Information Modeling, the term *parametric* is often misused or misunderstood. Since it is nearly impossible to avoid hearing reference to the term in nearly any document, seminar, or training session on Revit, it will be helpful to take a moment to properly define the term.

Parametric is the adjective form of the noun "Parameter." Browsing "Merriam-Webster Online Dictionary" we find the following definition for the term:

> **Parameter**—any of a set of physical properties whose values determine the characteristics or behavior of something.

This particular definition of "parameter" was selected from a few available variations because it is the most appropriate in the context of its use in describing the behavior of elements in Revit. When we describe something in software such as Revit as being "parametric" we are therefore simply saying that the thing in question is characterized by its associated set of "physical properties," each of which hold "values determining characteristics of the element's behavior." In the specific case of Revit, each element has one to several available parameters. We input values into these parameters to determine the specific characteristics of the element (Wall, Door, Roof, Furniture, etc.) which we are editing. Not specifically mentioned in the Webster's definition but implied by the use of the term *parametric* in software is the ability to modify parameters at any time. Therefore, the ability for a particular parameter's value to determine the characteristics of behavior is not limited to the point of creation, but rather is a "living" parameter with ongoing influence over the element. This behavior is what makes the notion of "parametric" so significant in software like Revit. This dynamic interaction across the whole system (not just with Model and Annotation Elements, but Views and Schedules as well) is often referred to in Revit as its "Parametric Change Engine."

Having outlined our definition of the term "parametric," a Parametric Family is simply a Family that has editable parameters. Typically, such a Family will also have Types—though it is not required. Therefore, as we will see, Family Parameters can be Type- or Instance-based. We could certainly continue in our Reception Desk Family file and add parameters to the desk, but for simplicity, we will explore these topics, in a new Family file.

Create a New Family

As we did above, we will create a new Family file based on one of the provided Revit Family templates. We are going to create a simple binder bin to hang on the wall behind the reception desk. This object will be geometrically simpler than the reception desk, so we can focus on adding and working with parameters.

1. If you still have the Reception Desk Family open, from the QAT, switch to the Reception Desk Family, and then from the Application menu, choose **Close**.

NOTE Closing this way closes the Family file while remembering which views were open. This means that the next time you open the Family, all of the previously open views (plan, elevation, and 3D) will open again. If you click the close box in the corner of the view window, you are closing only that view. You would need to close each view this way and then when closing the last open view, you would also actually be closing the Family file as well. However, the next time you open the Family, only one view (the last one you closed) would open automatically.

In the Commercial project, make sure that the *Level 3* floor plan view is open; if it is not, please open it now.

2. Maximize the *Level 3* floor plan view.

- On the QAT, click the **Close Hidden Windows** icon.

  If you open the Switch Windows drop-down, the only open document listed should be *10 Commercial – Floor Plan: Level 3*.

- Minimize the *Floor Plan: Level 3* view.

3. From the Application menu, choose **New > Family**.

  Browse to your Templates folder if necessary.

- Select *Specialty Equipment wall based.rft* [*Metric Specialty Equipment wall based.rft*] and then click Open.

If your version of Revit Architecture does not contain either of these templates, they have been provided in the *Templates* folder with the dataset files installed from the student companion files.

NOTE

---

4. On the View tab, click the **Tile** button, and on the Navigation Bar, choose **Zoom All to Fit**.

The keyboard shortcut for Window Tile is **WT** and for Zoom All To Fit it is **ZA**.

TIP

- On the QAT, click the Save icon.
- In the "Save as" dialog, browse to the *Chapter10* folder, type **MRAC Binder Bin** for the Name, and then click Save.

Like the Furniture Family template used above, this Family opens with four view windows and some existing Reference Planes on screen. In addition, a temporary Wall element is included in this template. This Wall is used for reference while working in the Family Editor and will not be included with the binder bin when it is loaded into a project. The Wall is provided since this is a "wall-based" (hosted) Family. We have chosen the "Specialty Equipment" template in this exercise for its simplicity. Because there are not many Reference Planes, parameters, or settings in this template, we can learn to build them ourselves. We could make the argument that the wall-based Casework template would be more appropriate for a binder bin. It would be difficult to refute this argument. This is particularly true since Casework Families display in plan even when they are above the cut plane and Specialty Equipment Families do not. However, we will discuss an acceptable workaround to this situation below in the "Making Families above the Cut Plane Display in Plan" topic. In some cases the choice of template is simply a judgment call, and a clear understanding of what the design/usage intentions for the Family are that will help you make the final decision. In this case, a wall-based Furniture category might have been ideal, but Revit does not provide such a template. Remember: when you build your own Families, study the list of provided templates carefully before making your choice. Below in the "Enhance Families with Advanced Parameters" topic, we will revisit this issue and take a slightly different approach.

For the educational value of learning to create/use Reference Planes and parameters, we will stick with Specialty Equipment template here. If you open the Casework templates, you will note that these items are already provided. When you build your own Families, choose the template that gives you the most overall benefit. Having many of the required Reference Planes and parameters pre-defined can be a big time-saver. Also, it is important to note that you can start your Family using the Specialty Equipment template, and later choose the *Family Category and Parameters* button on the ribbon and select a different category. In some cases, it will make sense to do this (as we will see in the "Enhance Families with Advanced Parameters" topic below). However, use caution with this approach since changing the category also resets the subcategories including any custom ones you may have added (see "Creating Subcategories" below). Therefore, where possible, it is best to make this decision early in the process of editing your Family.

One final consideration on the choice of template is whether or not you wish to have the element you create "cutable." Model elements can interact with the cut plane in plan and section views. Cutable elements will use a different lineweight (often bolder) when such an element intersects the cut plane. If an element is not cutable, only its projection lineweight setting will apply. Projection lineweight is the setting used when the element (or a portion of the element) is seen beyond the cut plane. The easiest way to tell if an element is cutable is to look in Object Styles. On the Manage tab, click the Object Styles button. Cutable elements have a lineweight setting in the Cut column, and non-cutable elements do not have this field available (see Figure 10.61).

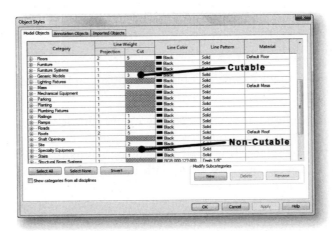

FIGURE 10.61 *Determining cutable elements in the Object Styles dialog*

So as you can see, the choice of template for your Family controls many aspects of your Family's behavior. Here is a quick summary:

- Assigns the Family Category.
- Determines if the Family requires a host or is free-standing (not hosted).
- Determines (via Category) if the Family is cutable or not.
- Often includes starting Reference Planes and possibly parameters.
- Establishes default parameters and subcategories.

If you are serious about becoming a proficient Family author, take the time required to become familiar with as many of the Family templates and their built-in behaviors as possible.

Create Reference Planes

We have used Reference Planes in several places throughout this book. Their inclusion in Families serves a vital purpose. If we wish to create a parametric Family, the procedure is to associate the parameters with various Reference Planes (or Reference Lines—which are similar but finite in length) and then lock the geometry to these Reference Planes. In this way, we can vary the dimensions that control the locations of various Reference Planes, which in turn manipulate the geometry constrained to them. This makes the Family's geometry parametric! Our first task will be to demonstrate this here; therefore, we must add some Reference Planes to our Family.

5. Click on the title bar of the *Elevation: Placement Side* window to make it active.

NOTE

> View names in the various Family templates are deliberately generic. For example, the plan view is often called "Ref. Level" because the Level included in most Family templates is there for reference only. In the Furniture template, we had elevation names like "Left" and "Front." Here we have "Placement Side" and "Back Side" instead of "Front" and "Back." This is typical of a wall-based template and helps you orient yourself relative to how the Family will later insert within projects.

6. On the Home tab, on the Datum panel, click the **Reference Plane** tool.

   • Create four Reference Planes, two vertically on either side of the existing one, and two horizontally as shown in Figure 10.62.

Do not be concerned with the precise locations. Simply sketch the Reference Planes in the approximate locations indicated in the figure. We will adjust the temporary dimensions later.

7. Click the **Modify** tool to complete the command.

   • Select the top Reference Plane, on the Properties palette, in the "Name" field type: **Top**.

   • Click Apply to complete the change, and then repeat for each of the other three Reference Planes, using the names indicated in the figure.

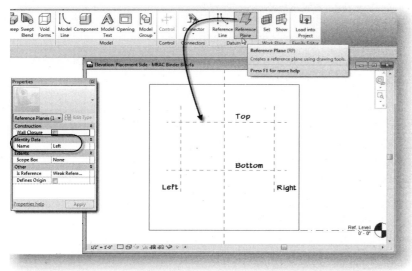

FIGURE 10.62 *Create four Reference Planes in elevation and then name them*

While you are on the Properties palette, you can optionally change the "Is Reference" parameter. Choices include the six cardinal directions (Left, Right, Front, Back, Top, and Bottom). When you use one of these options you establish the Reference Plane as that absolute edge of the Family. For example, if you set the Reference Plane to Right, then that becomes the right edge of the Family. If you later swap in a different Family, it can automatically align properly if the new Family also has a "right" Reference Plane. There are many interesting and powerful features of Reference Planes that time and space will not permit us to explore in this chapter. For a terrific explanation of Reference Planes, visit: **http://revitoped.blogspot.com/2006/03/once-upon-reference-plane.html** at the Revit OpEd blog by Steve Stafford. You can find a vast collection of useful articles, tips, and musings on all things Revit at Steve's blog.

8. On the Home tab, click the ***Extrusion*** tool.

   • On the Options Bar, set the "Depth" to -**1'–2"** [**–350**] and then click the Rectangle icon.

   • Snap to the intersection of the Top and Left Reference Planes for the first corner.

   • Snap to the intersection of the Bottom and Right Reference Planes for the other corner (see Figure 10.63).

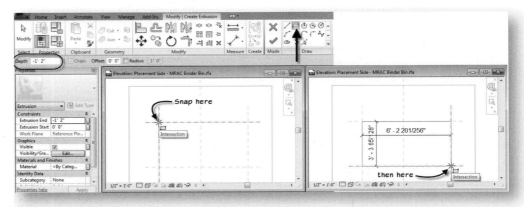

FIGURE 10.63 *Sketch a Rectangle snapping to the Reference Planes*

Four open padlock icons will appear—one on each edge of the rectangle.

   • Close each of the padlock icons to constrain the rectangle shape to the Reference Planes.

   • On the Mode panel, click the ***Finish Edit Mode*** button.

Take note of the extrusion in each view window.

9. In the 3D view window, change the display to Shaded.

10. In the *Elevation: Placement Side* view window, move one of the Reference Planes (see Figure 10.64).

The exact amount of the move is not important. What is important is that the shape of the Solid Extrusion will adjust when you move the Reference Planes. This is because we constrained the edges of the sketch to the Reference Planes.

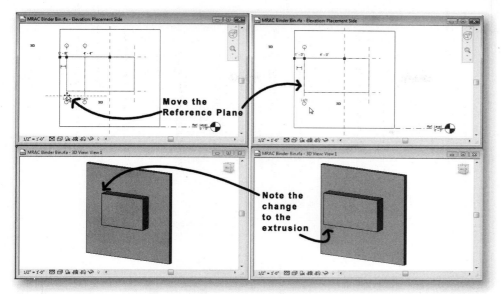

FIGURE 10.64 *Sketch a Rectangle snapping to the Reference Planes*

- Undo the Move to return the Reference Plane to its previous position.
- Save the Family file.

Create Dimension Parameters

The first step in creating our parametric Family is complete—we created some geometry that is constrained to our Reference Planes. The next task is to create dimension label elements and associate them to parameters.

11. Make the *Elevation: Placement Side* window active.

- Maximize the view window (you can do this with the icon in the top right corner of the window or just double-click the title bar).
- Zoom the window to fit (type **ZF** on the keyboard or choose the command from the Navigation Bar).

12. On the Modify tab, on the Measure panel, click the ***Aligned Dimension*** tool.

- Highlight and then click the Ref. Level line.
- Highlight and then click the Bottom Reference Plane (see Figure 10.65).

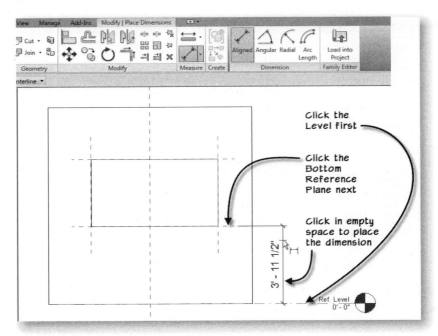

FIGURE 10.65 *Place the first Dimension element*

- Click in a blank space between the two references to place the Dimension.

It is important to make sure you do not highlight any objects on your last click, or the dimension will assume you want to add that item to the dimension chain. Click in empty space to complete a dimension. Also, be sure to dimension the Reference Plane and *not* the bottom of the extrusion.

13. Repeat the process to add another Dimension between the Top and the Bottom Reference Planes (see Figure 10.66).

We want two separate dimension elements here, not one continuous one.

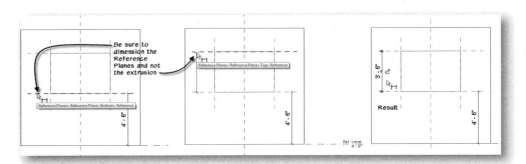

FIGURE 10.66 *Place another vertical Dimension between the Top and Bottom Reference Planes*

14. Cancel the Dimension command (click Modify or press ESC) and then select the Left Reference Plane.

- Edit the temporary dimension between the Left Reference Plane and the Center (Left/Right) Reference Plane to **2'–6"** [**750**].
- Select the Right Reference Plane and edit the temporary dimension to the same value.

Again notice that the extrusion geometry adjusts as well.

15. On the Annotate tab, on the Dimension panel, click the ***Aligned*** tool.

- Highlight and then click the Left Reference Plane.
- Highlight and then click the Center Reference Plane.
- Highlight and then click the Right Reference Plane.

    This will give you three witness lines and two dimensions in a continuous string.

- Click in a blank space above the binder bin to place the Dimension.
- Click the "Toggle Dimension Equality" control (see Figure 10.67).

FIGURE 10.67 *Toggle on Dimension Equality for the horizontal string*

The reason that we first edited the values to be equal before toggling the dimension equality is because this ensures that the center Reference Plane will not move. By default, when you toggle on the dimension equality control, spacing is calculated from the outermost elements requiring the internal elements to shift. Since our center Reference Plane marks the insertion point of our Family, it is undesirable to have it move.

16. Add one final Dimension horizontally between the Left and Right Reference Planes only—do *not* include the Center one in this string (see Figure 10.68).

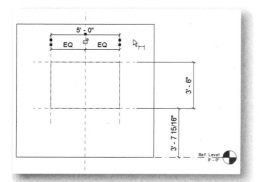

FIGURE 10.68 *Add the final dimension between the left and right Reference Planes*

- On the ribbon, click the **Modify** tool or press the ESC key twice.

Now that we have created several Dimensions, we will create parameters that will control them. Recall that you can manipulate model geometry in Revit by editing the values of permanent dimensions and temporary dimensions. When you apply parameters to Dimensions in a Family file, you are doing the same thing. Let's take a look.

17. Select the overall horizontal Dimension element (the one between left and right, not the EQ one).

Take note of the "Label" item on the Options Bar (and the Properties palette). We use this drop-down list to assign parameters to the selected Dimension. You can also right-click and choose **Label** if you prefer.

18. On the Options Bar, click the drop-down list next to "Label" (currently <none>) and choose <**Add parameter**>.

The "Parameter Properties" dialog will appear; verify that the "Family Parameter" radio button is selected at the top.

- In the "Parameter Data" area, type **Width** in the "Name" field and accept the remaining defaults (see the right side of Figure 10.69).

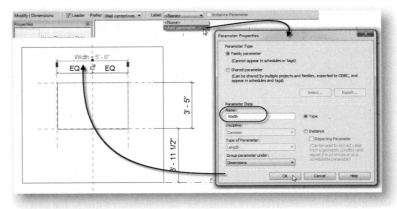

FIGURE 10.69 *Create a custom "Width" parameter*

The "Group parameter under" value is assigned automatically based on the type of parameter. So in this example, since we added a Length parameter (by choosing a linear dimension) Revit automatically assigned this parameter to the Dimensions grouping. You can always modify this if you wish, but in most cases the default will be suitable.

We also made this a Type parameter by accepting that default. This means that a parameter named "Width" that will appear beneath the "Dimensions" grouping of the "Type Properties" dialog box when this Family is used in a project. This means that if the value is edited, it will affect all elements in the model that use that Type. If we chose "Instance" instead, the parameter would apply to the individual elements in the project and could be unique for each one—we will create an instance parameter below.

- Click OK to complete the parameter.

Notice that the Label "Width" now appears in front of the Dimension text. This tells us that this Dimension is controlled by this parameter (see the left side of Figure 10.69).

19. Repeat the process on the vertical Dimension between the Top and Bottom Reference Planes.

  • Name the parameter **Height**, and accept the other defaults as well.

20. Select the vertical Dimension between the Ref. Level and the Bottom Reference Plane and repeat the process once more.

  • Name the parameter **Mounting Height**.

    Let's make this one an Instance parameter this time.

  • Click the Instance radio button and then click OK (see Figure 10.70).

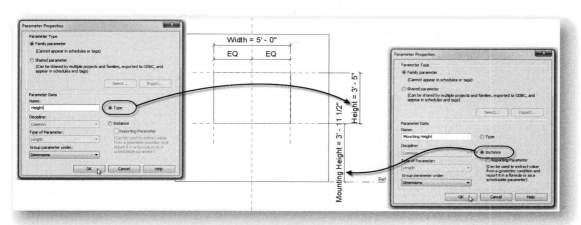

FIGURE 10.70 *Create another Type parameter and one Instance parameter*

21. Save the Family.

Add Family Types and Flex the Family

All that remains now is for us to test our Family parameters. In Revit, this is typically referred to as "flexing" the Family. When you flex the values of the parameters, you can see if the Reference Planes and geometry are moving as expected. We do this with the Family Types command. This command allows us to not only test to see if our parameters are behaving properly, but to also create some actual Types that will load into projects when we load the Family; thus the name of the dialog. Remember, just like the out-of-the-box Families that we have used so far, users will be able to add new Types to our custom Family later. But it is still a good idea to stock it with a few preferred variations ahead of time.

22. On the Home tab, on the Properties panel, click the ***Family Types*** button.

Notice that the three parameters are listed at the top of the dialog beneath a "Dimensions" grouping exactly as we specified. To flex the parameters, move the dialog to the side so you can see the view window in the background. Type numbers into the "Value" field next to each parameter and then click the Apply button. The Dimensions in the view window will adjust to the new sizes.

  • Change the Width to **4'–0" [1200]**.

- Change the Mounting Height to **4'-6" [1350]**.
- Change the Height to **2'-0" [600]** (see Figure 10.71).

In this exercise, we have waited until after adding several parameters to "flex" them. It is a very good practice, however, to perform this procedure immediately following the creation of each parameter. In this way, you can catch mistakes or issues before they become potentially compounded by other parameters and/or geometry. Some Families can get very complex. Making sure that you test it thoroughly as you build will help you avoid potentially hours of frustration.

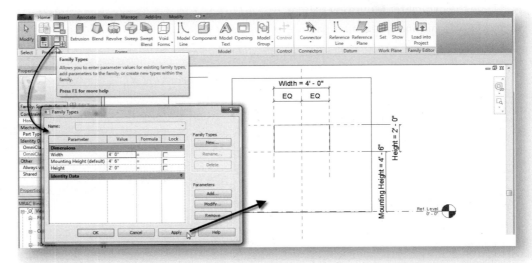

FIGURE 10.71 *View the custom parameters in the Family Types dialog and edit the values to flex the model*

If something did not work correctly, close the dialog and go back through the previous steps to find your error and then try again.

In the "Family Types" dialog, you can add Types and also add or edit Parameters. Let's add some Types.

23. On the right side of the dialog, click the "New" button beneath "Family Types."

- In the "Name" dialog, type **48 wide 24 high [1200 wide 600 high]** and then click OK.

The new name will appear in a list at the top.

24. Click New again to create **48 wide 18 high [1200 wide 450 high]**.

- Set the Height to **1'-6" [450]**; leave the other dimensions the same and then click Apply.

   Create additional Types if you wish. Click Apply after each one to test them.

- After you have created all of your Types, choose each one off the list, and then click Apply one at a time as a final test.

Notice that creating Types serves two important functions. First, you will likely wish to have some typical Types available to your project team when this Family is loaded, so adding them in the Family Editor allows you to pre-stock the Family with Types. Second, having a few Types with different values makes your subsequent testing go more quickly. To flex the Family, you just choose a different Type from the list at the top and then click Apply.

> If your Family requires fewer than half a dozen Types, follow the procedures outlined here to add them to your Family. However, if you need more than this, you should consider the use of a Type Catalog instead. A Type Catalog is a TXT file saved in the same folder as the Family RFA file. When you load a Family with an associated Type Catalog, a dialog will appear presenting a list of all available Types. From this list, you can choose just the Types you wish to load into your project. This can be very helpful in cases where the Family contains dozens or even hundreds of variations. Loading such a quantity of Types can inhibit performance and workflow. So a Type Catalog makes working with such Families much more manageable. Furthermore, as the Family author, Type Catalogs make managing and creating Types much easier. Since they are just text files, they can be edited outside of Revit in a program like Excel where creating multiple Types with similar values can be achieved much more efficiently. We will not build a Type Catalog in this book, however, in the "Detail Components" topic in the next chapter; you will have an opportunity to load a Family that contains an associated Type Catalog. So you will get to see the user experience of using them at that time. If you wish to learn more about creating a Type Catalog, please search the online help.

- Click OK when finished.

Add Hardware

We are almost ready to load our Family into our project. Before we do, our bin could use some finishing touches. Let's start with some hardware.

25. Tile the view windows again.
26. In the *Elevation: Placement Side* view, create a Reference Plane 2" [50] above the Bottom Reference Plane.

- On the Properties palette, name it **Hardware Height**.
- Add a Dimension element between the Bottom Reference Plane and the new one.
- Click to close the padlock control on the dimension (see Figure 10.72).

When you toggle the dimension equality control or close the padlock control, you are applying a constraint to your Family rather than a parameter.

> **Remember It This Way:** Both Parameters and Constraints are rules in your Family. A parameter is a value that a user can manipulate in the project without editing the Family. A constraint "built-in" to the Family can only be changed by editing the Family.

TIP

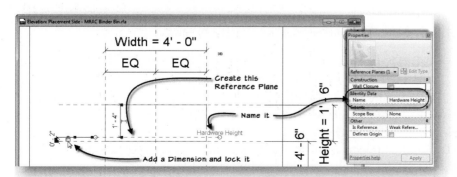

FIGURE 10.72 *Add a new Reference Plane, dimension it, and lock the dimension to constrain it*

The specifics of creating the hardware will be left to you as a practice exercise. You can make a knob using a Solid Revolve like the one we created for the reception desk above, or you can make a wire pull from a Solid Sweep. Just remember to constrain the position of the hardware to the appropriate Reference Plane(s). For example, if you create a Sweep, start in the plan view, edit the Work Plane, and choose the Hardware Height Reference Plane from the list. Sketch the path centered on the bin in plan; then switch to the elevation and create the profile. A simple square profile is sufficient (see Figure 10.73).

TIP	For a quick way to create the profile, sketch a 4-sided circumscribed polygon. This will make it easy to center and draw quickly.

If you opt for the revolve, you can build it nearly identically to the way we built the one for the reception desk above, except that you will want to set the Work Place to the Hardware Height Reference Plane first.

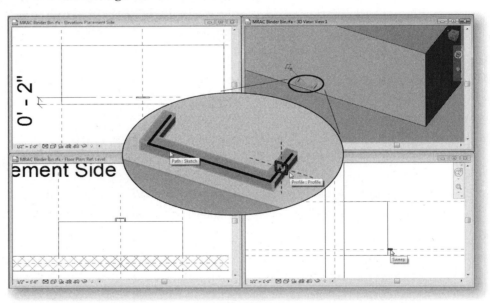

FIGURE 10.73 *Add a Solid Sweep wire pull for a handle*

When you have finished the Solid, open the "Family Types" dialog and choose each of your pre-defined Types in succession from the list and apply to flex the model. The wire pull should remain 2" [50] from the bottom edge and centered on the bin as you flex the model.

27. Zoom all Views to Fit when finished and save the Family.

Creating Subcategories

Revit classifies elements in a fixed hierarchy starting with Categories, then Families, then Types, and finally Instances. We have been working with Families, Types, and Instances throughout this chapter. Categories however are the broadest classification of elements in Revit. They are often sub-divided into Model, Annotation, and Imported object groupings for convenience. Each Family belongs to a Category. When you built each of the Families in this chapter so far, your choice of template (first Furniture and then Specialty Equipment) determined the Family's category. The list of Categories available is predetermined by Revit—we cannot rename, add, or delete Categories. However, we can add and delete "Subcategories." A subcategory gives us more detailed control over the visibility of Model and Annotation elements by allowing parts of an element to appear differently from the whole. For example, the "Doors" Category includes subcategories for "Elevation Swing," "Frame/Mullion," and "Panel" (among others). Each of these subcategories gives us visibility control over these common sub-components globally for all Door elements. When you build or modify a Family, you can assign the various components within the Family to any of the available subcategories. You can see a list of available categories and subcategories in the "Object Styles" dialog.

28. On the Manage tab, on the Settings panel, click the **Object Styles** button (see Figure 10.74).

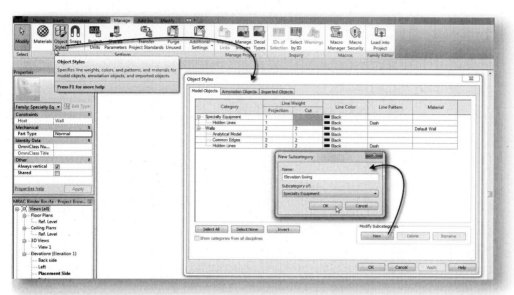

FIGURE 10.74 *Object Styles shows the available categories and subcategories; new subcategories can be added*

Since we are currently in a Family file, the list of available categories includes only those relating to the Family we are editing—in this case Specialty Equipment. (If you wish to see a more complete list, you will need to repeat this command later in a project file.) Specialty Equipment contains one subcategory called, "Hidden Lines" by default. You should consider adding subcategories for overall conditions that would apply to most elements of this category. For example, if you were to edit a Door Family, you would notice that there is a subcategory for "Plan Swing" and another for "Elevation Swing." Since most drawings need to represent the swing of the doors regardless of the kind of door, the swing subcategories provide a useful way to control all door swings universally throughout the project without needing to edit several Families individually. If you study most of the subcategories provided with the software, you would find they share this kind "global edit" strategy. While most of the subcategories you will likely require are already included in the provided Family templates, there will be situations when you decide to create new ones. In the case of our binder bin Family, it might be useful to have a subcategory to indicate the door swing much like the Door Families have. This is a somewhat gray area as not all Families that belong to Specialty Equipment need this subcategory. For example, the elevator Family we added to the model back in Chapter 7 was Specialty Equipment, but it is unlikely that it would need an Elevation Swing. You will have to exercise some judgment here. The best advice that can be given is to add subcategories sparingly and after some careful consideration. Having subcategories is not bad, but having too many and ones that are too specific to a particular Family can be undesirable to the larger project team.

29. At the bottom of the dialog, in the "Modify Subcategories" grouping, click the "New" button.

  - Name this new subcategory **Elevation Swing** and then click OK.
  - Change the Line Pattern to **Dash**.

30. Create another new subcategory of Specialty Equipment named **Overhead Items**.

  - Make its Line Pattern **Overhead** and then click OK.

To use the new subcategory, you edit the element on the Properties palette and choose the subcategory you wish. If you do not assign a subcategory to an element, it simply associates with the parent category. In this case, everything we have added to the current Family belongs to the Specialty Equipment category and no subcategory. In the next topic, we'll discuss Symbolic Lines and assign them to our new subcategories.

Add Symbolic Lines

Symbolic Lines are special Revit elements that can be added to Families to embellish the 2D views. In projects, we have model elements and annotation elements. Model elements appear in all views, and annotation (including detail) elements appear only in the view to which they were added. In Families, model elements behave exactly as they do in projects. However, the views that we see in the Family Editor are not transferred over to the project when we load the Family. Therefore, drafting lines would not be transferred either. Instead, we use Symbolic Lines in the Family Editor. Symbolic Lines will appear in all views parallel to the view in which they are created. So if you add a Symbolic Line to a front elevation, it will appear in all sections or elevations that face the same direction, but not in any other views. The use of Symbolic Lines allows us to add detailing and embellishment that is view-specific like Detail Lines, but also somewhat parametric like model geometry. In this example, we will use Symbolic Lines to indicate the swing direction of the binder bin when it opens in the front facing elevation.

Don't confuse these with Model Lines. These are available too, but Model Lines are treated like model geometry and show in all views, not just the one they are drawn in. In other words, the Symbolic Lines that we will draw will show only when viewing the binder bin head on. If we used Model Lines instead, they would also show in 3D views or skewed elevations.

31. Make the *Elevation: Placement Side* window active.

  - Zoom in on the bin.
  - On the Home tab, on the Work Plane panel, click the *Set* tool.

This opens the "Work Plane" dialog (shown in Figure 10.45 above) and allows you to set the Work Plane for the Symbolic Lines.

  - In the "Work Plane" dialog, choose the "Pick a plane" option and then click OK.
  - Click on the front face of the bin. To do so it is often easiest to select the edge of the face you are trying to select. All four edges of the front face will pre-highlight when the correct face is selected (use the tab key as necessary to aid in selection).

    If you wish, on the Work Plane panel, click the *Show* button (to display the Work Plane).

32. On the Annotate tab, click on the ***Symbolic Line*** tool.

    • On the Subcategory panel, from the Subcategory list, choose ***Elevation Swing***.

Notice how the new subcategory added in the previous topic now appears here. If you forget to choose it while you are adding the element, you can edit the properties of the lines later as noted above. On the Options Bar, the "Placement Plane" drop-down should read: **Extrusion**. This is because we just set the face of the extrusion as the Work Plane.

On the Options Bar, verify that "Chain" is selected.

    • Sketch two lines as indicated (see Figure 10.75).

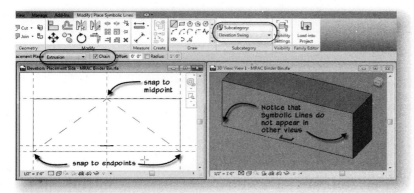

FIGURE 10.75 *Add Symbolic Lines to the front elevation using the new Elevation Swing subcategory*

Notice that the Symbolic Lines do not display in the 3D view. This is by design. Symbolic lines by definition only display in views parallel to the view in which they are created.

    • Stay in the Symbolic Line command (don't cancel yet).

If you recall the default Mounting Heights that we assigned above were 4'-6" [1350] for each Type. While we did make these Instance Parameters (which means that each bin we add to our model can have its own Mounting Height), initially when we place them, they will fail to show in most plan Views. This is because the default cut height in plan is 4'-0" [1200], which is below the lowest point of our bin's geometry. Assuming that you want to see the bins displayed in plan views even though they are mounted above the cut plane, you can add additional Symbolic Lines in the plan view that show an outline of our bin in a dashed line style to indicate that it is mounted above.

33. Make the *Floor Plan: Ref. Level* window active.

    • Check the Placement Plane on the Options Bar.

It probably remembers the Hardware Height plane since we have switched back to a floor plan. If you switch to a view that is impractical for the current Work Plane, Revit will default to another logical choice, like the previously used plane in this case.

    • From the Placement Plane list, choose: **Level: Ref. Level**.
    • On the Modify | Place Symbolic Lines tab, choose ***Overhead Items*** from the Subcategory drop-down.
    • On the Draw panel, click the Rectangle icon and then trace the bin geometry.
    • Close all four padlock icons (see Figure 10.76).

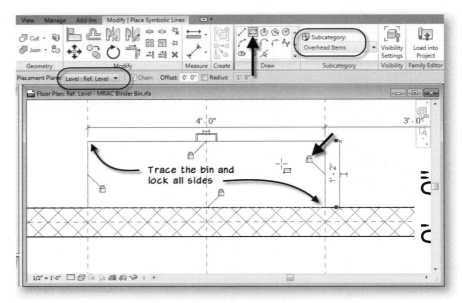

FIGURE 10.76 *Trace the plan with Symbolic Lines using the new Overhead Lines subcategory*

- On the ribbon, click the **Modify** tool or press the ESC key twice.

Making Families above the Cut Plane Display in Plan

We now have the Symbolic Lines that we need, but there is one other important criterion determining whether they will display in projects. Since none of the geometry of the bins is being cut by the cut plane, Revit simply will not show any geometry in plan views. This is despite the presence of our Symbolic Lines. In other words, the interaction with the cut plane is the first criterion that Revit considers. Once it is determined which objects are cut (and therefore should be displayed), Revit then considers *how* to display them. In our current Family, all geometry is above the cut plane. To trigger the Family to display our Symbolic Lines, we need some part of the Family to intersect the cut plane. To achieve this, we can draw an invisible line element passing through the Cut Plane. In this way, Revit will recognize that we wish to display this Family in plan. It is an effective work-around, conveying our intent to the system. The "invisible line" will be visible in the Family Editor, but not in the Project Editor.

The BIMManager Note on p. 538, at the start of the binder bin tutorial, discussed our choice of template and Family Category. It indicated that, in some cases, you may wish to choose another Category for your Family, such as Casework. It should be noted that there are three Revit Categories that will display above the cut plane without need for the workaround covered here. Those include: Windows, Casework, and Generic Models. If cutting behavior is your primary concern, Casework might be a viable option for this Family. If you wish to explore this, you can change the Category of the Family. However, realize that this would delete the subcategories that we have already added. So once again, it is best to make such a decision early in the process of building your Family. The longer you wait to make such a change, the more likely it is that you will need to redo some aspects of your Family. Please note that our binder bin could really be thought of as Specialty Equipment, Furniture, or Casework. So there is no right or wrong Category to choose here. Discuss the issues with your team and try to build some standard guidelines for all Family authors to follow. The most important issue is consistency. If everyone on the team is building things the same way, it goes a long way toward making your overall workflow much more predictable. Regardless of the ultimate choice you make regarding Category, the following workaround for trigging overhead display will be useful in many of your Families. It has applications like the one illustrated here, to display light fixtures like wall sconces, light switches in reflected ceiling plans, etc.

34. Click on the title bar of the *Elevation: Placement Side* window to make it active.

   - Change the Work Plane to **Reference Plane: Back**.
   - On the Home tab, on the Model panel, click the ***Model Line*** tool.
   - From the Subcategory drop-down, choose *<**Invisible lines**>*.

35. Draw a Model Line from the Ref. Level to the Reference Plane at the bottom of the bin (see Figure 10.77).

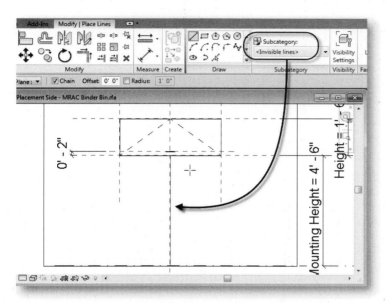

FIGURE 10.77 *Draw an <Invisbile Lines> Model Line in elevation view*

Make sure you snap to the Reference Plane and not the extrusion. Use TAB if necessary.	**TIP**

   - On the ribbon, click the ***Modify*** tool or press the ESC key twice.

With the placement of this Model Line, Revit will "see" this object when the plan is cut at 4'-0" [1200]. Once the display of the object is triggered by interaction with the cut plane, the Family will display in plan. One final step remains. Recall the discussion above in the "Create a New Family" topic regarding cutable Families. The Specialty Equipment category is not cutable. This means that interaction with the cut plane for a specialty equipment element simply determines if the element should display or not. Now that we have added the invisible line, we have established that the Family should display in plans. However, since it is not a cutable Family, it will display all of the Family's geometry indiscriminately—not just the dashed Symbolic Lines as we would prefer. To make sure that only the Symbolic Lines display in plan and not the bin geometry we can use the visibility settings in the Family to instruct the bin and hardware geometry not to display in plan views. The Symbolic Lines, on the other hand, require no further intervention since they will display only in plan views automatically (having been created in plan).

36. Select the bin Extrusion.

    • On the Modify | Extrusion tab, click the **Visibility Settings** button.

    • In the "Family Element Visibility Settings" dialog, clear the checkmark from the "Plan/RCP" box and then click OK (see Figure 10.78).

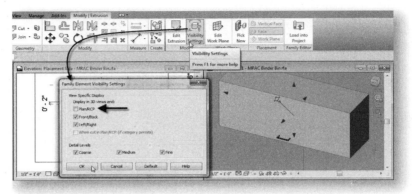

FIGURE 10.78 *Turn off the visibility of the extrusion in plan views*

37. Repeat the process for any hardware geometry.

    • Zoom all Views to fit.

38. Save the Family file.

Congratulations, you have just created a parametric Family from scratch! For the final test, let's load it into the project.

Load the New Family into the project

Make sure you are satisfied with the geometry and have tested the various Types. Be sure to save the Family before you continue.

39. On the ribbon, click the **Load into Project** button.

This will restore the *Level 3* view of the Commercial Project. The Modify | Place Component tab will be active allowing you to place the new Family immediately.

40. Zoom in on the reception space.

    • Since this is a wall-based component Family, you will need to click on a Wall to place it (see Figure 10.79).

At this point, we can test several of the behaviors of this new Family. The first test is verifying that it displays correctly in plan. It should appear as a dashed rectangle. Add a camera view to the reception area to look at the binder bin in 3D. (If you need a refresher on creating a camera view, refer to the "Load the Family into the Project" topic and Figure 10.26 above.) The 3D view will give you an overall look at the space.

Check the Families branch of the Project Browser under *Specialty Equipment* to see the newly loaded Family. Return to the plan and cut a section through the reception space looking at the binder bin. Open the new Section view. You should see the Symbolic Elevation Swing Lines in that view. Notice that, by default, both binder bins share the same mounting height. The mounting height parameter is an instance parameter, so try adding more than one bin to the project, and then edit the properties of one of them (on the Properties palette) to set the mounting height differently. Open Object Styles and beneath Specialty Equipment, experiment with changing the settings—turn off the Elevation Swing and/or change Line Styles. You can also edit these settings per view by clicking the Visibility/Graphics tool on the View tab (or just type **VG**). Finally, you can add, edit, or delete Family Types on the Project Browser as we did above.

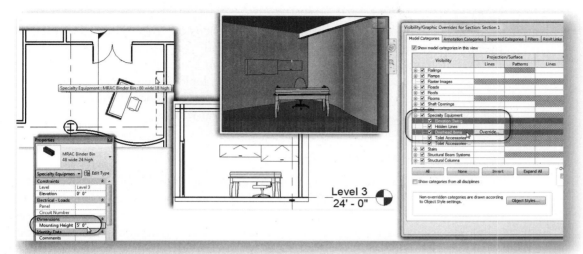

FIGURE 10.79 *Add a binder bin to the commercial project third floor reception space*

41. When you have finished experimenting with the binder bin, close the Family file.
42. Save the commercial project and close the Binder Bin Family.

CREATE A CURTAIN PANEL FAMILY

In the previous chapter, in the "Swap Panel Types" topic, we loaded a custom Curtain Panel Family into the commercial project and used it in the Curtain System design at the front of the building. In the short exercise that follows, we will walk through the process used to create that Family.

Understanding Curtain Panel Family Templates

The process begins the same as any other new Family. The only difference is the template we'll choose.

1. From the Application menu, choose **New > Family**.

There are four Curtain Panel templates: *Curtain Wall Panel.rft* [*Metric Curtain Wall Panel.rft*], *Door - Curtain Wall.rft* [*Metric Door - Curtain Wall.rft*], *Window - Curtain Wall.rft* [*Metric Window - Curtain Wall.rft*], and *Curtain Panel Pattern Based.rft* [*Metric Curtain Panel Pattern Based.rft*]. The first three will create a Curtain Panel that will appear on the Type Selector when working with Curtain Walls and Curtain Systems. The fourth is a special kind of Curtain Panel used in the conceptual modeling environment. You can learn more about the conceptual modeling environment in

Chapter 16. The Door and Window templates are a little more specific and share characteristics of Curtain Panels and Doors or Windows, respectively.

- Choose the *Window - Curtain Wall.rft* [*Metric Window - Curtain Wall.rft*] template and then click Open.

  This one only opens one view by default.
- Open the plan and *Exterior* elevation views; Tile and Zoom them to Fit.

Look at both the plan and exterior elevation views. Notice that several Reference Planes are already present as well as an EQ dimension. There is not, however, an overall width dimension. This is because the width of a Curtain Panel is determined automatically by the Curtain Grids when it is assigned to a Curtain Wall in a Project. This is true for the height as well. If you look at the exterior elevation, the panel size is the space defined by the left and right Reference Planes horizontally, the top Reference Plane, and the Reference Level vertically. Therefore, we simply need to draw our panel geometry within this space.

2. On the Manage tab, click the **Object Styles** button.

In the topic above, we learned about "Creating Subcategories." Notice that this Family actually belongs to the Windows category and as such has all of the Window subcategories. The Door Curtain Panel template would likewise be categorized as a Door sharing its subcategories, and the Curtain Panel template is assigned to the Curtain Panels category.

- Click Cancel to exit the "Object Styles" dialog.

Establish Reference Planes, Constraints, and Parameters

3. In the *Floor Plan: Ref. Level* view, draw two vertical Reference Planes in the middle of the panel.

   The spacing is not important yet. We'll correct this next.
4. Add a dimension including all Reference Planes *except* the center one.
   - Toggle the equality on (see Figure 10.80).

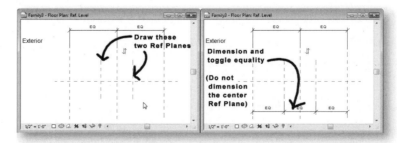

FIGURE 10.80 *Add two Reference Planes and center them*

The next several steps are illustrated in Figure 10.81.

5. Click the **Reference Plane** tool, and then on the Draw panel, click the **Pick Lines** tool.
   - On the Options Bar set the Offset to **1"** [**25**].
   - Offset a Reference Plane on each side of the two just created (four total).
   - Dimension the vertical reference planes and lock them as shown.

6. Change the Offset to **1 1/4" [31]** and Offset the Center (Front/Back) Reference Plane (running horizontally) to create a new horizontal one above it.

   • Edit the properties and name this new horizontal Reference Plane **Panel Offset**.

7. Create a dimension between the Center (Front/Back) Reference Plane and the Panel Offset Reference Plane.

   • Label this dimension with a new Parameter named **Panel Offset**.

   Leave it a Type parameter and grouped under Dimensions.

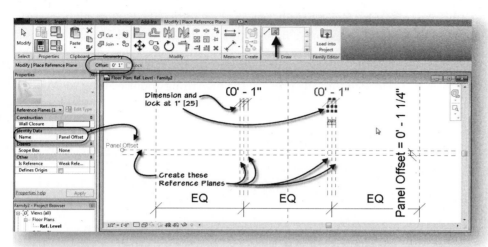

FIGURE 10.81 *Add Reference Planes, constraints, and parameters*

8. Save the Family as **MRAC Window Panel**.

Build Panel Geometry

9. In the Elevation: Exterior view, set the Work Plane to **Reference Plane: Panel Offset**.

10. From the Home tab, add an Extrusion.

    • Sketch three rectangles that snap to the Reference Planes and lock the sketch lines (see Figure 10.82).

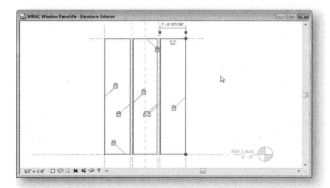

FIGURE 10.82 *Snap the sketch lines to the Reference Planes and lock them*

11. On the Properties palette, set the Extrusion End to **1/2" [12]**.

    • Set the Extrusion Start to **-1/2" [-12]**.

Notice that the Depth field on the Options Bar will now read 1" [24]. Revit simply calculates the resulting Depth from the two offsets.

- Next to Material, click on <By Category> and then click the small browse icon.
- In the "Materials" dialog, choose Glass and then click OK twice.
- Beneath the identity Data grouping, change the Subcategory to **Glass**.
- On the Mode panel, click the ***Finish Edit Mode*** button.

You can best see the results back in the plan view. The extrusion appears like three panes of glass. If you open the 3D view and turn on shading, the glass will be transparent. To complete the Panel Family, we need to add some mullions between the panes of glass. You may also want to lighten the weight of the glass in plan. We could repeat the Symbolic Line procedure covered above. However, since we have assigned the Extrusion to the Glass subcategory, we will have the ability to lighten all glass in plan view once it is loaded into the project. The nice thing about waiting to do it this way is that the change will apply to the glass of all Windows, not just this Family.

Build or Load a Mullion Profile Family

Profile Families are simple 2D Families that contain a closed profile shape. The shape can be drawn using any of the standard line and arc tools. Profile Families give us a convenient way to save commonly used shapes. Profile Families can be loaded into other Families and used to create sweeps and swept blends as noted in the "Using Swept Blend to Make a Fancier Leg" topic above.

12. On the Insert tab, on the Load from Library panel, click the ***Load Family*** button.

- Browse to the *Imperial Library\Profiles\Curtain Wall* [*Metric Library\Profiles\Curtain Wall*] folder.
- Choose *Curtain Wall Mullion-Rectangular-Center.rfa* [*M_Curtain Wall Mullion-Rectangular-Center.rfa*] and then click Open.

If you prefer, you can build your own mullion profile. To do so, start a new Family from the *Profile-Mullion.rft* [*Metric Profile-Mullion.rft*] template. In the Family, sketch a closed 2D shape and save the Family. Then use Load into Projects to load it into the Panel Family. You can also find another pre-made example in the *Chapter09* folder.

Continue in the Elevation: Exterior view with the Work Plane set to: **Reference Plane: Panel Offset**.

13. Create a Solid Sweep.

- Sketch the Path in elevation along the Reference Plane centered in the gap between glass panels.
- Finish the Path.
- On the Sweep panel, from the Profile list (currently set to: <By Sketch>), choose one of the three sizes listed (see Figure 10.83).

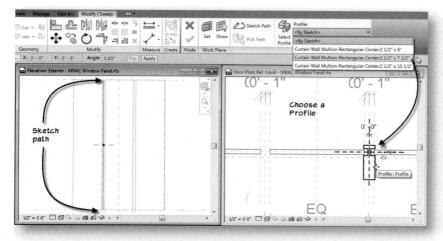

FIGURE 10.83 *Create a Sweep using a Profile for the mullion*

- Finish the Sweep.

There are other options. Notice that if you select the Profile, you can move it using both the X and Y offset fields on the Options Bar, or interactively in the view. It can also be flipped or rotated if necessary.

- Repeat for the other mullion.

14. Edit the Properties of the sweeps and set the subcategory to Frame/Mullion.
15. Save the Family (see Figure 10.84).

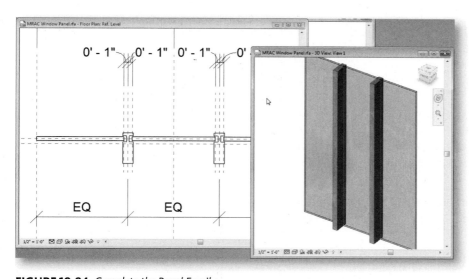

FIGURE 10.84 *Complete the Panel Family*

- Load the Family into the commercial project and swap it in for some of the Panels in the Curtain Walls or Systems.

Perform any further experiments you wish in the commercial project before continuing. Try editing the Curtain Grid spacing and see how the mullions in the Panel Family stay equally spaced. If you want to reduce the lineweight of the glass in plan views, open Object Styles (Manage tab) and expand the Windows Category. For the Glass subcategory, change the lineweight assigned to Cut to something smaller such as pen 1 or 2.

16. Close the Panel Family.
17. Save and close the commercial project.

ENHANCE FAMILIES WITH ADVANCED PARAMETERS

In the binder bin example above, we made a basic parametric Family by adding labels to dimension strings that in turn controlled the size of the object. This is the most basic way to add parameters to a Family. Adding such parameters allows you to create multiple Types from a single Family. Each Type simply saves alternate values of each parameter.

Dimensional parameters are the most common parameters used in Families and are the most straightforward to define. However, the potential of parameters goes well beyond the controlling of dimensions. In this topic, we will explore several "advanced" parameters. We will explore nested Families, visibility parameters, array parameters, and the use of formulas in parameters.

Create a Coat Hook Family

For use in this exercise, a Family file has been provided with the dataset files from the student companion. The file contains a wall-based coat rack. It was built from scratch from the same template used above for the binder bin. Similar geometry, Reference Planes, and Dimension Parameters have already been set up in that file. In this exercise, we will create a new Family for an individual coat hook. We will then open the provided wall coat rack Family file and load our coat hook Family into it. From there we will explore the various parameters mentioned above.

You should have closed all files at the completion of the previous exercise, and the Recent Files window should now be displayed in Revit.

1. From the Recent Files screen, in the Families area, click the "New" link to create a New Family.

    • Browse to your Templates folder if necessary.

    • Select *Generic Model.rft* [*Metric Generic Model.rft*] and then click Open.

Since the coat hook does not need to be inserted on its own in a project and will only be nested into another Family, Generic Model is a good choice for it. This will make it easier to isolate it in the project environment. There are several Generic Model templates. For this example, be sure to choose the non-hosted one.

NOTE If your version of Revit Architecture does not contain either of these templates, they have been provided in the *Templates* folder with the dataset files installed from the student companion files.

2. Save the Family and name it **MRAC Coat Hook**.
3. Make the *Ref. Level* floor plan window active.

    • Zoom in a bit toward the center.

4. On the Home tab, click the **Revolve** tool.

    • Accept all the defaults and draw a vertical line starting at the insertion point (the intersection of the vertical Reference Planes and the Level line).

    • Move straight up, type: **4" [100]**, and then press ENTER.

    • Zoom in more if necessary and draw a horizontal line to the right **1/2" [12]**.

5. On the Draw panel, click the Start-End-Radius Arc tool and Draw a shallow arc vertically back down to the Reference Plane and then close the shape with a small **1/2"** **[12]** long horizontal line (see Figure 10.85).

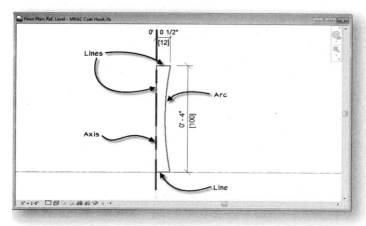

FIGURE 10.85 *Sketch the profile of the coat hook*

6. On the Draw panel, click the ***Axis Line*** tool.

   • Trace over the vertical Reference Plane.

   • On the ribbon, click the ***Finish Edit Mode*** button (see Figure 10.86).

It will be necessary to zoom the various view windows to study the results.

NOTE

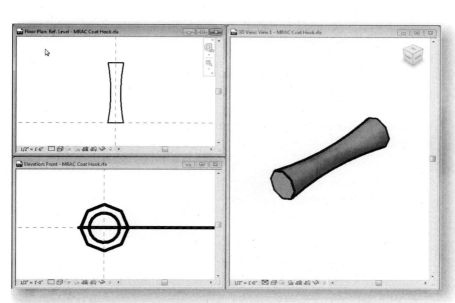

FIGURE 10.86 *Complete the Solid Revolve*

7. Save and Close the Family file.

Changing the Family Category

We now have the provided Wall Coat Rack Family and the Coat Hook Family that we just built. Let's open the provided Wall Coat Rack Family to continue our work.

8. In the *Chapter10* folder, open the *MRAC Wall Coat Rack.rfa* [*MRAC Wall Coat Rack-Metric.rfa*] Family file.

We have discussed the Category of Families in several instances in this Chapter. While it is maintained that the best practice approach is to choose a proper Category at the start of the Family creation process and not change it wherever possible, there are those times when it will become necessary or desirable to change the Category of a Family. In this case, the Family we have opened was created following a procedure very similar to the other Families created in this chapter and also used the *Specialty Equipment wall based.rft* [*Metric Specialty Equipment wall based.rft*] template file. This template was actually chosen more for the wall based behavior than the Specialty Equipment Category. It was desired to have this Family a wall based Family, but it would be preferable for it to use the Furniture Category. Revit does not include a wall based furniture template, so the solution employed here is to build the Family using another wall based template and then change the Category to Furniture. Doing so will maintain the wall based behavior but change the Category as desired.

Changing the Category is a simple process but does come with some warnings. Remember, any custom subcategories already added to the Family will not be preserved. Further, some Categories cannot (or should not) be changed. We cannot change Mass Families to other Categories and vice-versa. Doors and Windows and Curtain Panels should not be changed and typically will not behave properly if you do. Changing from Specialty Equipment to Furniture can be done without detriment, but in general it is best to exercise caution; save a backup copy of your Family and pin down the Category as early in the Family authoring process as possible.

9. On the Home tab, on the Properties panel, click the Family Category and Parameters button.

   • In the "Family Category and Parameters" dialog, choose Furniture from the Family Category list (see Figure 10.87).

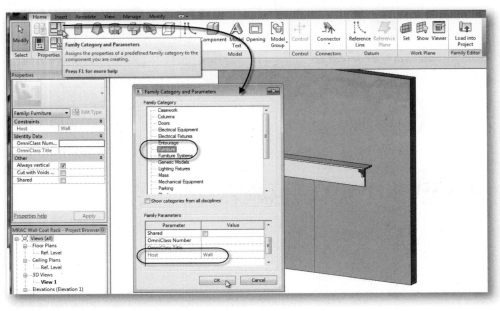

FIGURE 10.87 *Change the Category to Furniture*

If you look down at the bottom of the dialog, you will notice that the Host parameter is listed as Wall. This is grayed out, which means that we cannot change it. The hosting behavior is built in and assigned by the template. Furniture would typically not need to be hosted. This is why Revit does not provide a hosted furniture template. But the method shown here can effectively overcome this slight limitation.

- Click OK to complete the change.

10. Follow the procedure covered in the "Creating Subcategories" topic above to create a new Subcategory called: Overhead Items using the Overhead line pattern.

- In the floor plan view, select the Symbolic Lines, and on the Properties palette, change them to the Overhead Items subcategory.
- Save the Family.

Working with Nested Families

The parameters in this Family are very similar to the binder bin Family completed in the previous exercise. We have a Width parameter centered on the Center (Left/Right) Reference Plane with an EQ dimension. We also have a Height parameter and a Mounting Height. In addition to these familiar parameters, at the top of the shelf are two Dimensions, one on each side reading 5" [125]. We need to create a new Dimension Parameter for these. We will create one parameter named "Hook Inset" and apply it to both dimensions.

11. Open the *Elevation: Placement Side* view.
12. Select one of the 5" [125] dimensions at the top.

- Using the procedure in the previous lesson, create a new parameter named **Hook Inset**. Include it in the Dimensions grouping and make it a Type parameter.
- Select the other 5" [125] dimension and label it with the same parameter.
- Open the **Family Types** dialog and flex the Hook Inset parameter—set the value to **6" [150]** and then click OK.

Both Reference Planes should move a little closer to the center.

13. Open the *Floor plan: Ref. Level* view window.
14. On the Home tab, on the Model panel, click the **Component** tool.

An alert will appear indicating that no Families are loaded and requesting that you load one.

- In the dialog, click Yes to load a Family.
- Browse to the location where you saved *MRAC Coat Hook.rfa* and load it.

15. Place the hook anywhere on the Placement side of the Wall

Place it randomly; do not snap it to anything at this time.

- Click the Modify tool.
  Using the left side of Figure 10.88 as a guide:

16. On the Modify tab, on the Modify panel, click the **Align** button.

- Align the hook to the back plate of the Coat Rack in plan. Lock the padlock to constrain it.

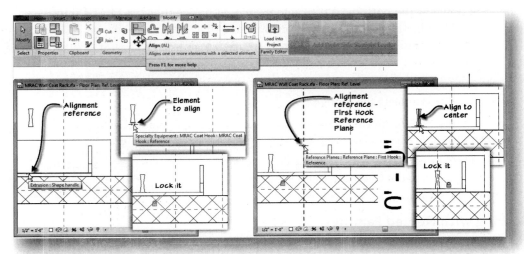

FIGURE 10.88 *Align the Hook to the back board of the Coat Rack and to the First Hook Reference Plane*

Using the right side of Figure 10.88 as a guide:

- Align the center of the hook to the "First Hook" Reference Plane (on the right side of the rack) and lock it.

<table>
<tr><td>**TIP**</td><td>Identify the correct Reference Plane by pre-highlighting it with the Modify tool (it will say "First Hook" on the tool tip and on the Status Bar).</td></tr>
</table>

When you align, be sure to click the Reference Plane first and then the middle of the hook. You may need to zoom in to the hook. When you zoom in on the hook, you will be able to highlight its centerline. Do *not* use align for the height of the hook. This we will achieve the correct vertical location with a new formula parameter next.

- On the ribbon, click the **Modify** tool or press the ESC key twice.

Create a Formula Parameter

If you were to study the elevation and open the "Family Types" dialog, you would notice that the height of the back plate of the coat rack is controlled by a "Height" parameter. We can create a formula based upon this variable height that keeps our hooks centered vertically on the back plate.

17. Open the *Elevation: Placement Side* view.
18. On the Home tab, on the Properties panel, click the **Family Types** button (see Figure 10.89).

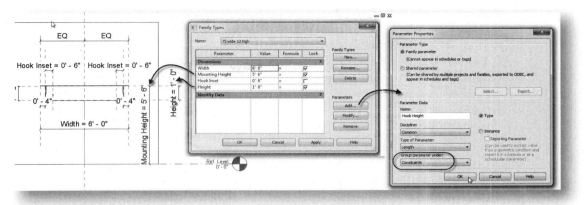

FIGURE 10.89 *Compare the Dimension Parameters to the elevation to see how they are applied and create a new parameter*

Move the "Family Types" dialog on screen so that you can see the *Placement Side* elevation in the background.

- In the "Parameters" area of the "Family Types" dialog, click the Add button.

Until now, we have added all of our parameters directly in screen in the Family Editor. Using the "Family Types" dialog is an alternative approach. Both methods yield the same result.

- Name the new parameter **Hook Height** and change the "Group parameter under" to: **Constraints**.
- Verify that the "Type of Parameter" is: **Length** and leave it a Type parameter.

In the previous method, when labeling a dimension, the type of parameter was automatically set to a Length parameter; but when adding it this way, we can make the type of parameter anything we like.

- Click OK to complete the parameter.

19. Back in the "Family Types" dialog, click in the formula field to the right of the Hook Height value field (see Figure 10.90).

- For the formula, type: **Mounting Height - (Height / 2)**.

These label names *are* case-sensitive.

NOTE

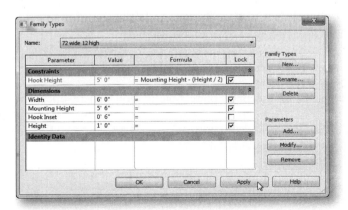

FIGURE 10.90 *Type a formula that subtracts half of the Height from the Mounting Height*

You will notice that once you click Apply, the result of the formula will be input as "read only" in the Value field (it grays out). Study the formula we just typed. We are asking Revit to take the Mounting Height, which you can see in Figure 10.89 goes from the floor to the shelf of the coat rack and subtract from it half of the coat rack height. This means that no matter what values the Mounting Height and Height parameters assume, the value of this new Hook Height parameter when measured from the floor will place the hooks centered on the back board of the coat rack.

20. At the top of the "Family Types" dialog, choose **72 wide 8 high [1800 wide 200 high]** from the Name list.

Two types are pre-defined in this Family, each with different Heights. Note the change in the Hook Height's automatically calculated Value.

- Click OK to dismiss the dialog.

21. Select the hook element onscreen (it is probably down at the floor).
22. On the Properties palette, in the Constraints grouping, locate the small button next to the Offset parameter and click it.

This far right column of the Properties palette only appears in the Family Editor. In this very narrow column are small buttons that allow us to associate parameters to the values shown (see Figure 10.91).

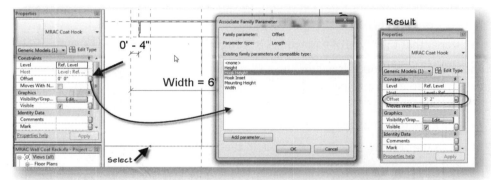

FIGURE 10.91 *A parameter button appears in the formula column next to the Offset field*

- Click this parameter button, and in the "Associate Family Parameter" dialog, select the "Hook Height" parameter and then click OK.

Again, notice that the read only value of the resolved formula fills in automatically. Notice also the small equal (=) sign that now appears on the button.

- Click Apply or shift focus away from the Properties palette to complete the operation.

The hook will move to the center of the back plate height. You can re-open the "Family Types" dialog and test the formula by changing Types and then clicking apply. If you wish, you can create additional types with varying Mounting Heights and back plate Heights to fully flex the new parameter.

23. Save the Family file.

Add a Visibility Parameter

Sometimes you add details to a Family that you don't want to display in all views. We can create a Visibility parameter that will allow us to control the display of components in each Type within the Family. In this case, we will make it possible to turn the hooks on and off in each Family Type.

24. Select the Hook component and look to the Properties palette again.

  - Beneath the "Graphics" grouping click the parameter button (in the same far right column) next to the "Visible" (checkmark) parameter (*not* the Edit button next to "Visibility").
  - In the "Associate Family Parameter" dialog, click the "Add parameter" button.

Here is another alternative way to create parameters. Again, the method of creation is a matter of preference with the results being the same in each case.

  - Name the parameter: **Show Hook**, group it under **Graphics**, leave it a "Type" parameter.

Notice that this time the "Type of Parameter" is assigned for us. Since we started from a Boolean value (a value that has two options only, like yes and no, true and false, or up and down) the only kind of parameter that this can be is a Yes/No type. In other words, Visible can only be checked or not checked.

  - Click OK twice.

Notice that an equals (=) sign appears on the button next to Visible and that the checkmark is now grayed out.

25. Return to Family Types, and create a new Type. Name it: **60 wide 8 high (Shelf only-no hooks)**.

  - Edit the Width and Height. Remove the checkmark from the Show Hook checkbox for this Type and then click Apply (see Figure 10.92).

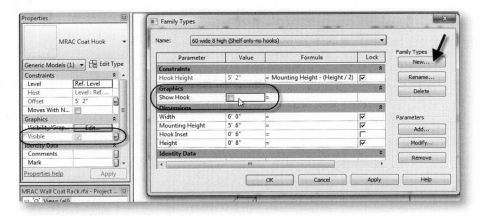

FIGURE 10.92 *Create a type that hides the hooks by unchecking the Show Hook parameter*

  - Click OK to exit the dialog.

> **NOTE** While you are working on the family in the Family Editor, invisible items will turn gray to indicate that it is invisible. To see the true behavior and have the items become invisible, the Family must be loaded into a project.

Create an Array of Hooks

By default, when you use the Array command in Revit, it groups the arrayed elements and maintains some of the parameters used to create the array. We can utilize this feature to make a parametric array of hooks in our Family.

Continue in the *Elevation: Placement Side* view.

26. Select the hook and then on the Modify | Generic Models tab, click the ***Array*** button.

    • On the Options Bar, be certain that a checkmark appears in the "Group And Associate" checkbox.

    • Leave the Number set to 2 and for the "Move To" option, choose the "Last" radio button.

Following the prompts at the Status Bar (bottom left corner of the Revit screen); we must indicate the start point and then the end point of the Array. We will use the Reference Planes for these.

    • For the First Point, click the "First Hook" (left side) Reference Plane.

    • For the End Point, click the "Last Hook" (right side) Reference Plane (see Figure 10.93).

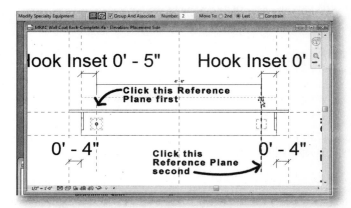

FIGURE 10.93 *Click each of the Reference Planes as the start and end of the Array*

A temporary dimension will appear with the Array prompting you for the quantity of items.

    • Type **3** for the "Array Count" and then press ENTER.

    • On the ribbon, click the ***Modify*** tool or press the ESC key twice.

There are now 3 hooks on the back plate of the coat rack. Click on any hook to edit the Array Count or to gain additional Group options on the ribbon. To test this out, try changing the Array Count with the temporary dimension if you wish. Earlier, we constrained the hook on the left to the "First Hook" Reference Plane using the Align command. We need to also constrain the last hook (the one on the right) to the "Last Hook" Reference Plane. Otherwise, if you choose a different Family Type changes the width of the coat rack, the last hook will no longer line up properly.

27. On the Modify tab, click the Align tool.

    • Following the process (illustrated in Figure 10.81 above) Align the hook on the right to the "Last Hook" Reference Plane and lock it.

    Zoom in on the hook as required.

28. On the Properties panel, click the **Family Types** button.

   - Create a new Type named **84 wide 12 high [2100 wide 300 high]**.
   - Edit the Width and Height parameters accordingly, turn on Show Hook, move the dialog out of the way, and then click the Apply button to see the change.

If everything has flexed properly, the coat rack will have gotten larger but the hooks should still be positioned properly with the fixed offset at the ends and the height being in the middle of the back board.

29. Click OK to close the dialog and then Save the Family file.

Making the Array Count Parametric

At this point, we have a coat rack that keeps our three hooks equally spaced even as we change its dimension parameters. As a finishing touch, let's change the Array Count into a parameter.

30. Select one of the hooks.

Like before, a temporary dimension indicating the current Array Count will appear. This dimension includes a horizontal line and a text field in which we can edit the count. If you select the horizontal dimension line, rather than the text field, you expose the dimension options on the Options Bar.

   - Click on the Array dimension line to select it.
   - On the Options Bar, next to "Label" choose **<Add parameter>** from the drop-down list.
   - In the "Parameter Properties" dialog, name the parameter **Number of Hooks** and group under Graphics.
   - Leave "Type" selected and then click OK.

Notice the label that now appears on the Array dimension. You will only see this with a hook selected.

31. Re-open the "Family Types," dialog and change the "Number of Hooks" value a few times.

   - Click Apply after each change and watch the hooks appear and re-space automatically.

This parameter can be different for each type.

32. For the **84 wide 12 high [2100 wide 300 high]** Type, set the Number of Hooks to **5** and then Apply.

   - From the "Name" list at the top, choose **72 wide 12 high [1800 wide 300 high]** Type, set the Number of Hooks to **4**, and then Apply.
   - Edit the other Types to appropriate values.

Don't worry about changing the quantity on the "No Hooks" Type. Even though they show here in the Family Editor, they will not appear when used in a project.

Use a Formula to Set the Array Count

Finally we can make the Number of Hooks a formula based on the Width of the coat rack.

33. If you closed "Family Types," re-open it now.

- In the formula field next to the "Number of Hooks" parameter, type: **(Width - (Hook Inset * 2)) / 10" [(Width - (Hook Inset * 2)) / 250 mm]** (see Figure 10.94).

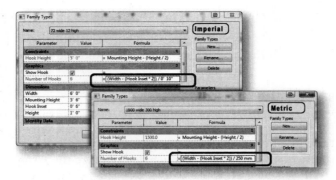

FIGURE 10.94 *Use a formula to calculate the Number of Hooks based on the Width*

You can test the formula in the "Family Types" dialog by changing the Width parameter and clicking Apply. Test out several values until you are satisfied. Be sure that the values of Width and Height match the Type name before you close the dialog.

Load the Residential Project

The Family file is complete and ready to load into a project. Let's load it into our residential project.

34. Open the Residential project for your choice of units:

- Open *10 Residential.rvt* if you wish to work in Imperial units.
- Open *10 Residential Metric.rvt* if you wish to work in Metric units

Like the Commercial project, many interior Walls have been added on several of the floors of this project since the last chapter. If you would like to try your hand at adding those Walls yourself, please visit Appendix A for instructions on how to do so.

By now we have learned a few ways to load Families into a project. You can start from the residential project, click the ***Component*** tool on the Home tab, and then click the ***Load Family*** button on the Place Component tab to browse to and load the *Wall Coat Rack.rfa* Family file. You can choose the ***Load Family*** button on the Load from Library panel of the Insert tab. Since the Family is still open in Revit, you can also return to one of the Family file's view windows from the Application menu (or CTRL + TAB) and then click the ***Load into Project*** button on the ribbon. The exact method is up to you.

35. On the Project Browser, double-click to open the *First Floor* plan view.

- Load the *Wall Coat Rack* Family (using your method of choice).
- Place an Instance in the room on the left on the interior vertical Wall.
- On the Project Browser, double-click to open the *Entertainment Room Camera*3D view (see Figure 10.95).

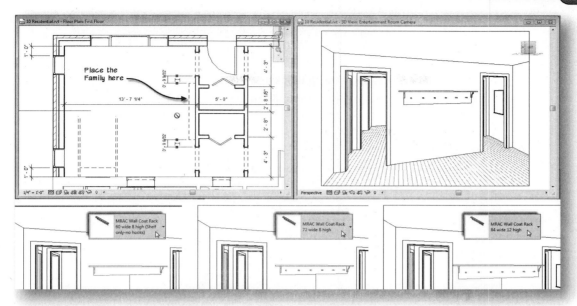

FIGURE 10.95 *View the Family in a 3D Perspective view*

From the Type Selector on the Properties palette, try the different Types to see them in the project. Note that the "No Hooks" Type now displays without hooks as we indicated. Remember to double-check each parameter and type. The final test of any Family is to flex it in a project. Some parameters, such as the hook Visibility parameter, can only be tested in a project. Thorough testing will prevent frustration later in your project cycle.

36. Save the project.
37. Switch to and close the *MRAC Wall Coat Rack.rfa* Family.

FAMILIES FROM MANUFACTURER'S CONTENT

As the final exercise for this chapter, let's remain in the residential project and create a custom Family for a new whirlpool tub on the second floor. Rather than build this Family from scratch, we will use drawing files (DWG) downloaded from a manufacturer's website. (In this exercise, the hypothetical manufacturer's files have been provided with the other dataset files, but the procedure would be the same if you visit and download from actual manufacturer's websites).

	NOTE
If any changes are required to the DWG files, you will need a copy of AutoCAD to open and edit them. Contact your Autodesk reseller for more information.	

---

38. Create a new Family file.

   - Choose the *Plumbing Fixture.rft* [*Metric Plumbing Fixture.rft*] Family template.
   - Minimize the residential project windows and then tile the Family windows and make the plan active.

39. On the Insert tab, click the **Import CAD** button.

- Browse to the *Chapter10* folder and choose *Whirlpool_Tub_P.dwg* [*Whirlpool_Tub_P-Metric.dwg*].

- From the Colors list, choose **Black and White**.

- From the Positioning list, choose **Auto - Origin to Origin** (see Figure 10.96).

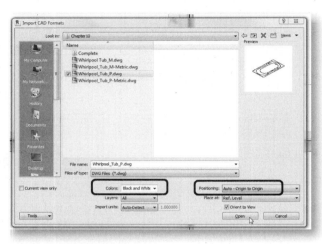

FIGURE 10.96 *Import the first CAD file*

- Click Open.

A 2D drawing of the plan symbol will appear. You can import any AutoCAD or Microstation drawing or even a SketchUp file this way. When you do, characteristics of the imported drawing, such as layers will be maintained. This particular file has only one layer, but if you link other drawing files, you can access their layers when selected on the ribbon and in the "Visibility/Graphics" dialog. We only want this symbol to appear in plan Views. To do this, we will edit the visibility of the imported element.

40. Select the imported drawing.

Note the pushpin icon. This appears since we chose the Origin to Origin option. When you use this positioning option, Revit pins the import so that the position is maintained. Think of it as "locking" the position relative to the origin of the host file. You can, of course, unpin it if you decide to move it. In this case we will leave it pinned.

- On the Modify | Imports in Families tab, on the Import Instance panel, click the **Visibility Settings** button.

- In the "Family Element Visibility Settings" dialog, clear the checkmarks from the "Front/Back" and "Left/Right" boxes and then click OK (see Figure 10.97).

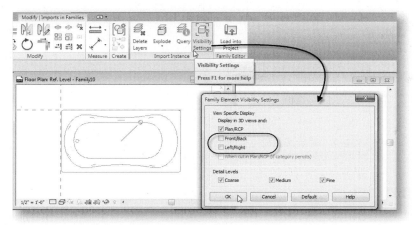

FIGURE 10.97 *Turn off visibility in elevation Views for the plan symbol*

41. Switch to the 3D view.

    Repeat the process to import another DWG file.

    • In the *Chapter10* folder, choose *Whirlpool_Tub_M.dwg* [*Whirlpool_Tub_M-Metric. dwg*] this time.

    Use the same color and positioning options.

This time the imported drawing contains a 3D model. We want this to show only in 3D and elevation Views.

42. Zoom all windows to fit. Select the imported 3D drawing.

    • On the Modify | Imports in Families tab, on the Import Instance panel, click the **Visibility Settings** button.

    • In the "Family Element Visibility Settings" dialog, clear the checkmark from the "Plan/RCP" box and then click OK.

43. Save the Family with the name **MRAC Whirlpool Tub**.

    • Click the **Load into Project** button.

44. Back in the residential project, when prompted if you would like to load a tag, choose No.

    • On the Project Browser, double-click to open the *Second Floor* plan view.

    • Demolish the bathtub at the top left of the plan. (On the Manage tab, click the **Demolish** tool, and then click the tub.)

    • On the Home tab, click the **Component** tool and place the **MRAC Whirlpool Tub** Family in the same room (Use the SPACEBAR if necessary to rotate it as you place it).

There is a demolished Wall in the bathroom space that is covering the new tub. You can apply element-level overrides to any object.

45. Select the demolished Wall; on the Modify | Walls tab, on the View panel, click the Override Graphics in View drop-down and choose Override By Element.

    • In the "View-Specific Element Graphics" dialog, select the "Transparent" checkbox and then click OK (see Figure 10.98).

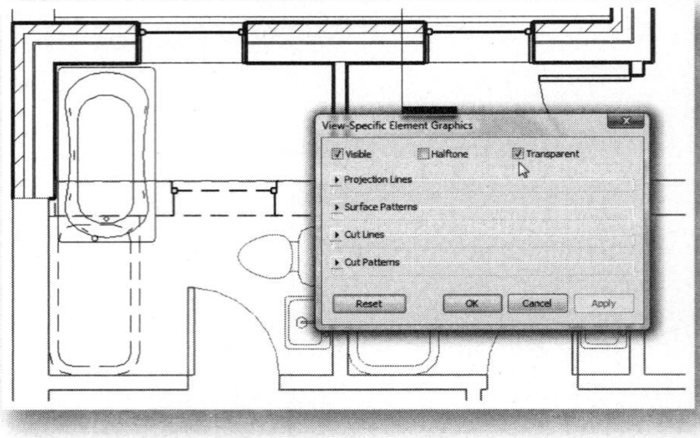

FIGURE 10.98 *Add the new Family to the project and make the demolished Wall transparent*

One final Family exercise dealing with angular parameters (in a Door swing) is presented in Appendix A. If you wish you can skip to that exercise now or save it for later. Also, Chapter 16 is devoted to the conceptual massing environment. While this is not exactly the same as the standard Family Editor, the conceptual massing environment is a modified and enhanced Family Editor environment. Feel free to explore that chapter now if you wish.

46. Save and Close the project and any open Family files.

SUMMARY

We have covered quite a bit of ground in this chapter. By now, you should have a good grasp of the power and flexibility of Families in Revit. Despite the lengthiness of this chapter, we have only scratched the surface of the potential inherent in Families. Continue to explore and customize your own Families. If you have not already done so, read and work through the exercises in the tutorials provided with the software (access them from the Help menu). Each of the exercises provided explores different kinds of Families from the ones explored here. While the process is similar, the more examples you work through, the more comfort and confidence you will gain with this critical and powerful part of the Revit software package. Please visit **www.paulaubin.com** for information on other Family Editor training resources. Also, remember that if your intention is to begin building Revit content seriously, download the *Revit Model Content Style Guide* from seek.autodesk.com for detailed guidelines and best practices.

- An extensive Library of Family Content has been included with your Revit software.
- Familiarize yourself with the provided Library before embarking upon any customization.
- In addition to the included Libraries, extensive Libraries and resources are available on the Internet.
- Autodesk Seek is a repository of Autodesk-provided, manufacturer-provided, and even user-provided content accessible directly from within the software.
- The simplest way to customize Families is to add or edit their types.
- Before building a custom Family from scratch, determine if you can save a copy and modify an existing one first.
- Editing an existing Family is accomplished by opening it in the Family Editor and making modifications and then saving a copy of the Family file.

- You can build a completely custom Family from one of the many provided Family template files.
- Family templates establish the basic framework, category, and behaviors of the Families you create—choose your template carefully.
- You can create a "singular" Family—a non-parametric Family with only one type, or a "parametric" Family—which has parameters allowing for interaction and multiple types.
- Add dimension parameters to your Families to allow for various "sizes" of the same basic Family geometry.
- Advanced parameters such as visibility controls, formulas, and parametric Arrays enable you to make very complex and robust Families.
- You can use manufacturer's drawing files directly in Families to create symbols and other items in your projects quickly.
- We have only scratched the surface—play, research, and explore!

Construction Documents

Most common words in Section III

While the benefits of Building Information Modeling describe an architectural design and delivery process that may one day put all members of the extended team in direct contact with a single digital model, we still have a way to go before we achieve this. This is not to say that current strides in BIM technology are not worthwhile; on the contrary, there is much value to be gained in implementing BIM strategies even on a small scale and within the internal design team. As you have hopefully seen in the previous chapters, BIM strategies can help the internal members of the design team create better coordinated and higher quality design decisions. This leads to fewer errors, higher quality projects and often time and cost savings. These benefits can be achieved internally even if the Building Information Model is not shared outside the four walls of your architectural firm.

If your extended design team is also using Revit and the BIM paradigm, so much the better for everyone involved. However, you will often find yourself working with professionals at different stages of BIM implementation. Some will be using Revit and BIM to its fullest potential, while others will continue to use traditional methods without use of Revit or BIM at all. This means that even though we may one day achieve "paperless" delivery and sharing of information, most firms and projects still rely heavily on traditional deliverables and methods even as they look toward the future. Further, the creation of Construction Documents (typically in the form of printed drawings) remains an important part of the accurate conveyance of design intent and further is almost always still contractually required. Therefore, Contract Documents (CDs) will remain relevant and necessary for some time to come whether they are delivered digitally or in paper form. With these issues in mind, this section explores the tools that help us produce these deliverables in Revit and ends with an exploration of the Worksharing features.

Section III is organized as follows:

Detailing and Annotation

INTRODUCTION

In this chapter, we will explore the detailing process and tools available in Revit. As the design development phase gives way to construction documentation, details are created to clarify basic plan, section and elevation views of a project and assist in conveying overall design and construction intent. Before Revit, such details had been drafted independently of the overall drawings with perhaps some tracing to help minimize redundant effort. In Revit, much of the detailing can begin within a fully coordinated model view. You will then add detail information directly to this live Revit view.

The process is simple—first create a callout or section view of the model at a large scale and then add additional drafted components, text, dimensions and other embellishments necessary to craft the detail and convey design intent. In most cases, such embellishments are at minimum drawn relative to an underlying building model view and in many cases remain automatically constrained or linked to model geometry in the view. It is important to understand, however, that unlike the other Revit views, all drafting components appear *only* in the view to which they are added.

OBJECTIVES

In this chapter, we will create detail drawings using several techniques. Working first from the Revit model, we add additional information to create a wall-floor-foundation section detail. A variety of tools will be explored to assist in this process. We will also create a detail in our project using a detail originally created in AutoCAD. This process allows you to utilize detail libraries that you may already have directly within Revit. Our exploration will include coverage of Detail Lines, Detail Components, Repeating Details, Filled and Masking Regions, and various annotations. After completing this chapter you will know how to:

- Modify Crop Regions and add View Breaks
- Add and modify Detail Lines
- Add and modify Detail Components and Repeating Details
- Add and modify Text and Leaders

- Add and modify Filled Regions, Masking Regions, and Break Lines
- Work with Drafting views
- Import legacy details into a Revit Architecture Drafting view

MODIFY WALL TYPES

To prepare us for the detailing tutorial that follows, we will modify the Wall Types currently in use in the residential project for the exterior walls. We will unlock the outer two components (called Layers) of the brick Wall and add a brick ledge to the foundation Wall. Doing so will allow us to create a brick shelf on the concrete Wall and extend the brick down to sit on it. These steps will make the model more accurately represent the construction, as well as making the detailing process a bit easier.

Install the Dataset Files and Open a Project

The lessons that follow require the dataset included on the Aubin Academy Master Series student companion. If you have already installed all of the files from this site, skip to step 3 to begin. If you need to install the files, start at step 1.

1. If you have not already done so, download the dataset files located on the CengageBrain website.

    Refer to "Accessing the Student Companion site from CengageBrain" in the Preface for information on installing the dataset files included in the Student Companion.

2. Launch Autodesk Revit Architecture from the icon on your desktop or from the *Autodesk > Revit Architecture 2012* group in *All Programs* on the Windows Start menu.

TIP	You can click the Start button, and then begin typing Revit in the Search field. After a couple letters, Revit Architecture should appear near the top of the list. Click it to launch to program.

3. On the QAT, click the Open icon.

TIP	The keyboard shortcut for Open is CTRL+O. **Open** is also located on the Application menu.

- In the "Open" dialog box, broswe to the location where you installed the *MasterRAC 2012* folder, and then the *Chapter11* folder.

4. Double-click *11 Residential.rvt* if you wish to work in Imperial units. Double-click *11 Residential Metric.rvt* if you wish to work in Metric units.

    You can also select it and then click the Open button.

Add a Brick Shelf

Let's start with the foundation Wall. By modifying the Wall Type, we can create a brick shelf to receive the bricks from the exterior Wall above. This will be achieved by adding a Reveal directly to the Wall Type.

5. On the Project Browser, double-click to open the *Basement* floor plan view.

There are four foundation Walls (in the new addition), three bounding the outside perimeter and another framing the right side of the passageway to the existing basement. We do not want to apply a brick shelf to the Wall in the passageway. Since we are going to edit the Wall type, the easiest way to prevent this is to create a new type for the Walls needing the brick shelf.

6. Using the CTRL key, select the three exterior foundation Walls (two vertical and one horizontal) (see Figure 11.1).

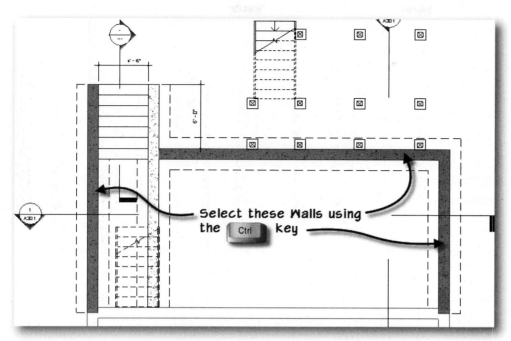

FIGURE 11.1 *Select the three exterior foundation Walls*

- On the Properties palette, click the Edit Type button.

  The "Type Properties" dialog will appear.

- Next to the Type list, click the Duplicate button.

A shortcut to this is to press ALT+D.

TIP

A new Name dialog will appear. By default "2" has been appended to the existing name.

- Change the name to: ***MRAC - Foundation - 12" Concrete (w Brick Shelf) [MRAC - Foundation - 300mm Concrete (w Brick Shelf)]*** and then click OK.

- At the bottom of the dialog, click the << Preview button.

A viewer window will appear to the left attached to the "Type Properties" dialog.

- From the "View" list (bottom left), choose **Section: Modify type attributes**.
- In the viewer, Zoom in to the top of the wall. (You can use the mouse wheel, right-click or the steering wheel icon in the lower-left corner.)
- On the right side of the dialog, at the top, click the Edit button next to Structure (see Figure 11.2).

This will open the "Edit Assembly" dialog and show the Wall Layers included in this Type. We can edit these Layers here as well as add other parameters such as Sweeps and Reveals (below we will use the same process to "unlock" some of the Wall Layers of another Type).

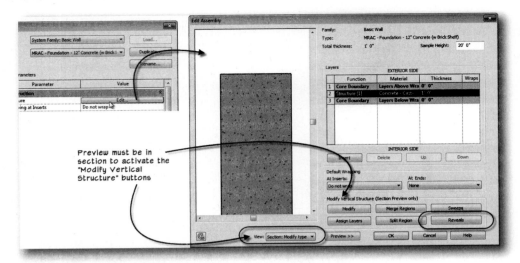

FIGURE 11.2 *Access the "Edit Assembly" dialog to edit the Wall Structure*

7. In the bottom right corner of the dialog, within the "Modify Vertical Structure" area, click the Reveals button.

NOTE The "Reveals" and other buttons in the "Modify Vertical Structure" area will not be available if you have not enabled the Section Preview as noted in the previous steps.

A Reveal is a profile-based extrusion that cuts away from the mass of the Wall. Profile Families were discussed in a few locations in the previous chapter, including the "Build or load a Mullion Profile Family" topic. You can create and load Reveal Profile Families in the same manner as discussed in those topics. In this case, we will use a Profile that has been provided with the dataset files from the *Aubin Academy Master Series: Revit Architecture 2012* student companion.

8. In the "Reveals" dialog, click the Add button to add a Reveal.

Item "1" will appear using the default profile. None of the Profile Families currently loaded in this project meet our needs for the brick shelf. Fortunately, we can load one from an external file directly from this dialog. One has been provided in the *Chapter11* folder for this purpose.

- From the Reveals Dialog click the "Load Profile" button.
- Browse to the *Chapter11* folder where you installed the datset files from the student companion.
- Select the file named: *MRAC Brick Shelf Reveal.rfa* [*MRAC Brick Shelf Reveal-Metric. rfa*] and then click the Open button.

This is a simple Reveal Family built using the *Profile-Reveal.rft* [*Metric Profile-Reveal.rft*] template. Feel free to open the file directly and study it or try your hand at building it yourself.

The Load Profile button is just a shortcut to loading the profile. You still need to assign it to the Reveal.

- Click in the Profile field and then click again on the down arrow to display the Profile list.
- Choose **MRAC Brick Shelf Reveal : 12" d × 6" w[MRAC Brick Shelf Reveal-Metric : 300 d × 140 w]**.
- For the "From" setting, choose **Top** and then click the Apply button (see Figure 11.3).

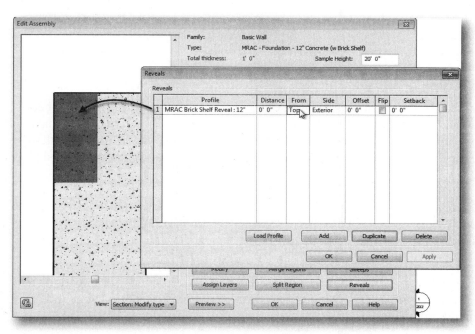

FIGURE 11.3 *Add a new Profile, set it to Top, and then click Apply to see the result*

In the "Edit Assembly" dialog in the background, you should see the Reveal Profile appear at the top left edge of the Wall in the viewer. Move the Reveals dialog out of the way if necessary.

- Click OK to return to the "Edit Assembly" dialog.
- Click OK two more times to return to the model view window.

9. On the Project Browser, double-click to open the *Longitudinal* section view.
   - Zoom in on the left side to study the results (see Figure 11.4).

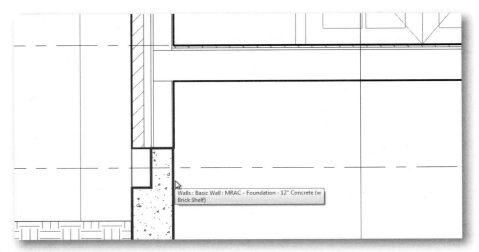

FIGURE 11.4 *Open the Longitudinal Section view to see the results*

Unlock Wall Layers

Now that we have a brick shelf in the foundation Wall Type, we should put some brick there. To do this, let's edit the brick Wall Type to unlock the outer material Layers (to allow their top and/or bottom offsets to move freely from the Wall's top and base offsets) and then project them down to sit on the foundation brick shelf.

10. In the section view, select the Left exterior brick Wall.
    - On the Properties palette, click the Edit Type button.
      The Preview window should already be open. If it is not, open it again.
    - Click the Edit button next to Structure.

The preview window should be showing the section view. If it is not, choose **Section: Modify type attributes** from the "View" list. You can use standard navigation techniques such as the wheel of your mouse or the right-click menu to zoom and scroll the model in the viewer.

11. Right-click in the viewer window and choose **Zoom In Region**. Zoom in on the lower portion of the Wall.
    - In the "Modify Vertical Structure" area, click the Modify button (see Figure 11.5).

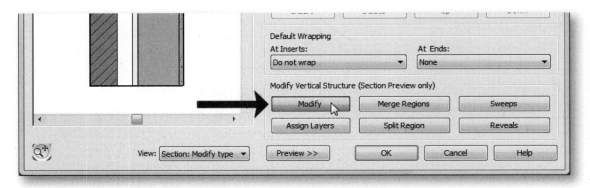

FIGURE 11.5 *Zoom in on the lower portion of the Wall and then click the Modify button*

12. Click the bottom edge of the Brick Layer.

The edge will highlight light blue to indicate that it is selected. A small padlock icon will appear on the edge. We can use this padlock icon to unlock the bottom edge of the Layer, which will allow it to be moved independently from the Wall itself in the model.

- Click the padlock (to open it) and unlock the bottom edge of the Layer.
- Repeat the process to unlock the bottom edge of the Thermal/Air Layer (next to brick) (see Figure 11.6).

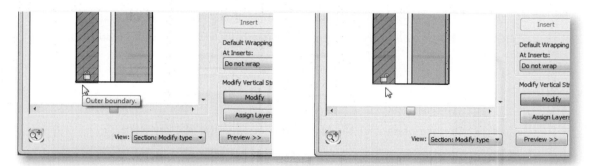

FIGURE 11.6 *Unlock the Brick and Thermal/Air Layers*

- Click OK twice to return to the model.

Upon returning to the model view window, you will note that the Wall now has two Shape Handles at the bottom edge (see Figure 11.7). If you don't see the shape handles, de-select the Wall, then select it again. You can use the second handle to modify the bottom edge of just the unlocked Layers. The other handle will continue to modify the Base Constraint of the entire Wall.

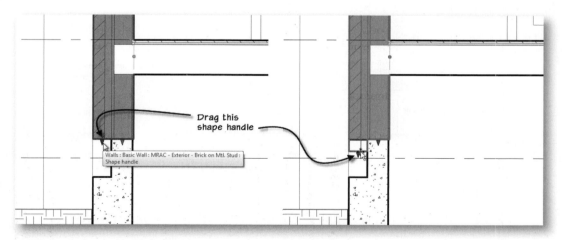

FIGURE 11.7 *Unlocking the Layers makes a second Shape Handle appear*

While the Shape Handle provides an easy way to edit the brick and air Layers, the Align tool provides a nice alternative too.

13. On the Modify | Walls tab, on the Modify panel, click the **Align** tool.
    - For the Reference line, click the bottom edge of the brick shelf on the foundation Wall (use the TAB key as necessary to make the proper selection).
    - For the Entity to Align, click the bottom edge of the brick Layer in the Wall above (see Figure 11.8).

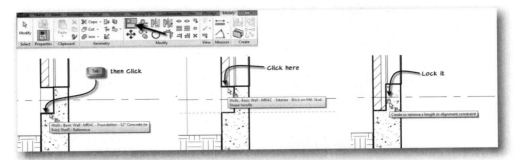

FIGURE 11.8 *Align the brick and air Layers to the bottom of the brick shelf on the foundation Wall*

- Click the small padlock icon that appears to lock the constraint and keep these elements aligned.
- On the ribbon, click the **Modify** tool or press the ESC key twice.

So now the brick and concrete have the proper relationship, but there is a bold line between them. This is because there are still two separate Walls here. If you want the graphics to merge together showing a thin line between all internal layers, use the **Join Geometry** tool.

14. On the Modify tab, on the Edit Geometry panel, click the **Join** tool.

- Select the Brick exterior Wall first and then the foundation Wall.
- On the ribbon, click the **Modify** tool or press the ESC key twice (see Figure 11.9).

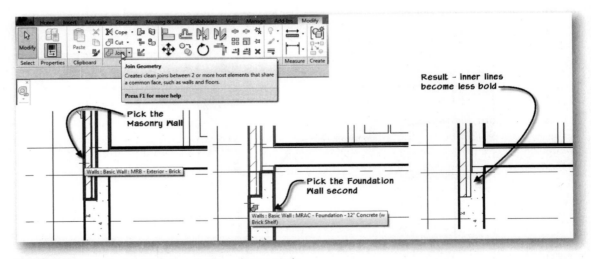

FIGURE 11.9 *Join the two Walls to make them cleanup nicely*

TIP	Be sure to click the Brick Wall first, and then join it to the Foundation Wall. Clicking in the opposite order will remove the customized bottom alignment.

You can repeat the procedure on the other two exterior Walls if you like. However, it is not necessary at this time.

- Save the project.

DETAILING IN REVIT ARCHITECTURE

Detailing in Revit Architecture is in many ways similar to detailing in traditional drafting. This is true regardless of whether you compare it to drafting created by hand on a drafting board or created in Computer Aided Design (CAD) software on a computer. The major difference is that in Revit Architecture you rarely start from scratch because you can base your detail on views that are automatically generated from your Revit model.

The typical Revit model includes enough data to generate a majority of the drawings that will be required in an architectural document set at an appropriate level of detail and accuracy. This is true for most plans, sections, and elevations. In the case of details, however, while it is theoretically possible to model all of the bricks, fasteners, joints, hooks, and other items that will actually occur in the building, the amount of effort (in man-hours) and the sheer size of the resultant model (in computer memory and hard drive requirements) would typically not yield a sufficient return on investment.

To keep the size of our models reasonable and to avoid spending additional and often unnecessary time modeling every bolt, screw, and piece of flashing, the strategy to detailing in Revit Architecture is instead a "hybrid" approach. In nearly all details you may create in Revit, you will be able to start the process with a cut (callout) from the model. This live view of the model portrayed at the scale of the detail will give you a starting point upon which to add detail components and other view-specific two-dimensional elements and annotations. By separating a detail into both live model elements and view-specific embellishments, we achieve the best of both worlds: we have an underlay that remains live and changes automatically with the overall building model and we have all of the additional data required to convey design intent occurring only on the specific detail view, thus saving on overhead and unnecessary modeling effort.

It is this process that will be discussed in detail in the following tutorial. In this exercise we will discuss the available tools and techniques using the wall and floor intersection from our residential project that we edited above.

Adding a Callout View

Continue from the previous exercise in the *Longitudinal* section view. If you closed the project or this view, reopen them now. Details are typically presented at larger scales than the drawings from which they are referenced. The Callout tools in Revit Architecture will allow us to create a detailed view at a larger scale of any portion of the building model.

> In the *Longitudinal* section view, be sure you can see the left exterior Wall from the first floor down to the footing. Zoom and Scroll as necessary.

1. On the View tab, on the Create panel, click the **Callout** tool.

   - From the Type Selector on the Properties palette, choose **Section: Wall Section**.
   - Click a point outside the exterior Wall on the left above the first floor and then drag a callout around the Wall to beneath the foundation (see Figure 11.10).

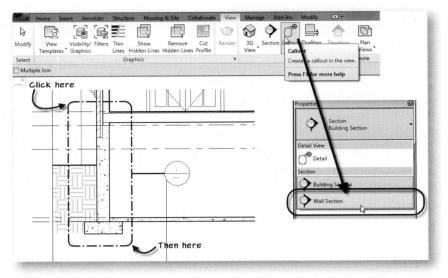

FIGURE 11.10 *Using the Callout tool, click two opposite corners to create the Callout view*

Revit Architecture will create a new branch on the Project Browser called Sections (Wall Section) and a new view called *Callout of Longitudinal*.

Like section and elevation markers, a callout marker will appear. When you deselect all elements, this callout will remain blue. As with the others, this indicates that you can double-click it to jump to the referenced view. You can also open the view from the Project Browser.

2. On the Project Browser, beneath *Sections (Wall Section)* right-click on *Callout of Longitudinal* and choose **Rename**.

   • Type: **Typical Wall Section** and then click OK.

3. Open up the callout view (you can double-click its name on Project Browser or its Callout symbol) (see Figure 11.11).

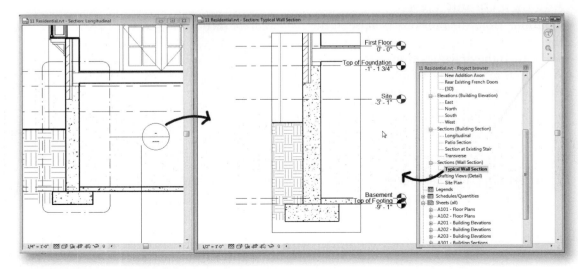

FIGURE 11.11 *Open the Callout view*

Notice that the boundaries of the crop region in the *Typical Wall Section* view match the extents of the callout boundary that we sketched in the *Longitudinal* section view. If this boundary is adjusted in either view, the boundaries in the other view automatically adjust. If you wish to see this, try tiling the *Longitudinal* section view and the *Typical Wall Section* view side by side and test it out. Remember to close hidden views or minimize other views first, so that when you tile, only the two sections will appear (see Figure 11.12).

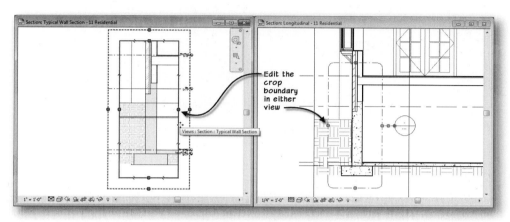

FIGURE 11.12 *Drag the Control Handles in either view to edit the extent of the Crop Region*

4. When you are finished experimenting, return the shape of the crop region to match approximately as shown in Figure 11.11.

   • Maximize the *Typical Wall Section* view and zoom to a comfortable size.

Adjust Scale and Annotation Visibility

Annotation is separate from the model geometry shown in a view. While the level of detail and graphical display characteristics of the model may vary from view to view, the model will display in *all* views unless you specifically override the display settings to hide it. Annotation, on the other hand, is applied on top of the model and occurs only in the specific view in which it is created. Model and annotation elements also differ from one another with regard to scale. Annotation appears at a consistent height relative to its desired plot size, while the model geometry adjusts its size relative to the assigned scale. All of this behavior occurs automatically.

5. On the View Control Bar (bottom of the window) change the scale to ¾" = 1' –0" [1:20] (see Figure 11.13).

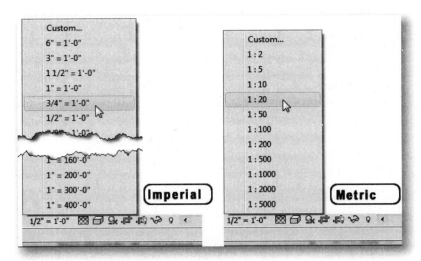

FIGURE 11.13 *Choose a larger scale for the detail view*

6. Select the Site Level Line.

   • On the Modify | Levels tab, on the View panel, click the Hide in View icon and choose **Hide Elements**. (Or you can right-click and choose **Hide in View > Elements** from the menu that appears.)

This operation does not have any effect on any other view. We have hidden this Level line in only the current view. The Site Level is not really relevant in the current Callout view, so by hiding the Level line, we eliminate potential clutter and confusion. In a similar fashion, we can adjust the location of the Level Heads and the length of the Level lines and again, the edits will be confined to *only* this view.

7. Click on any Level line.

   • Using the Control Handle at the end, adjust the end points so the Level text is completely outside the Crop Region (see Figure 11.14).

   • Click the Top of Footing Level line and then click the blue "squiggle" symbol near the Level Head.

   • Edit the blue grips to modify the Level line with an offset (see Figure 11.14).

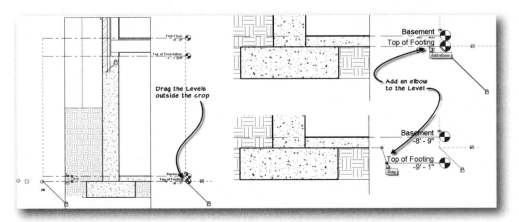

FIGURE 11.14 *Adjust the end points of the Level lines to move the text outside the Crop Region*

Notice that all Level lines move together when you drag one.

8. Save the project.

Detail Lines

Now that we have cut our detail callout and configured the Level lines and scale to our liking, we are ready to begin adding embellishments. We can draw a variety of view-specific elements directly on top of the section view of our model. We will start with Detail Lines. These are simple drafted elements much like the sketch lines with which you are already familiar. When you add a Detail Line, it appears only in the view to which you add it. If you wish to draft a line that appears in multiple views, use a Model Line instead. We will use Detail Lines to sketch in some flashing at the bottom of the wall cavity.

9. On the Annotate tab, on the Detail panel, click the **Detail Line** tool.

Notice the choices on the Modify | Place Detail Lines tab that appears are very familiar and match those that we have seen in many sketch-based objects so far.

- From the Line Style list on the ribbon, choose **Wide Lines**.
- On the Options Bar, verify that there is a checkmark in the "Chain" checkbox.

10. Zoom into the bottom of the wall cavity (at the brick shelf).
11. Sketch the line segments shown in Figure 11.15.

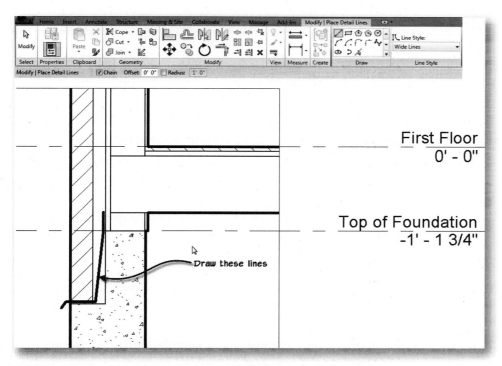

FIGURE 11.15 *Sketch Detail Lines to represent the flashing in the wall cavity*

To make it appear like flashing, it is best to avoid snapping these lines to the Wall. Unlike the sketch lines that we drew in previous chapters, these lines are complete "as is." They do not describe the shape of a more complex element like a Floor or a Stair. These are simply drafted lines placed on top of a model view, much like drafting directly on top of a Mylar background in traditional hand drafting.

12. On the Project Browser, double-click to open the *Longitudinal* section view.

- Zoom in to the same portion of the Wall and notice that this linework does not appear in this (or any other) view.

13. On the Project Browser, double-click to return to the *Typical Wall Section* view.

TIP	You can hold down the CTRL key and then press TAB to cycle through the open views.

All of the parts of the detail that we are going to create next could be created with Detail Lines following the same process. However, several other detailing tools are available to us. Let's look at them now.

Detail Components

Detail Components are simply two-dimensional view-specific elements that (like Detail Lines) appear only within the view in which they are placed. They are more useful and more powerful than simple Detail Lines in that they are Families and can be parametric. Like other Families, a Detail Component Family can have many Types built into it. The parameters can be as simple as Length and/or Depth, or include dozens of parametric dimensions. For Example, a "Wide Flange" Family file included in the out-of-the-box *Detail Component* folder contains hundreds of Types representing all of the commonly-available steel shape sizes. Another example that is a bit more pertinent to the detail that we are creating here is dimension lumber. Rather than attempt to edit the Wall type and begin adding three-dimensional framing elements, which would add a level of complexity to the model that is typically only needed in details, we can add predefined detail components to our detail view to represent this information more efficiently only in the views that require it.

14. On the Annotate tab, on the Detail panel, click the **Component** drop-down button.

- Choose **Detail Component** from the list.

Currently, there are no "Dimension Lumber" Families loaded in our project. Like other components in Revit, we can simply load them from the library.

- On the Modify | Place Detail Component tab, click the **Load Family** button.
- In the "Load Family" dialog, from default library folder (either the *Imperial Library* or the *Metric Library*) browse to:

  Imperial: *Detail Components\Div 06-Wood and Plastic\061100-Wood Framing*.

  Metric: *Detail Components\Div 06-Wood and Plastic\06100-Rough Carpentry\06110-Wood Framing*.

NOTE	If you do not have access to either of these libraries, the Family files mentioned in this tutorial have also been provided in the *Library* folder with the dataset files installed from the student companion.

- Double-click the file named *Nominal Cut Lumber-Section.rfa* [*M_Nominal Cut Lumber-Section.rfa*].

  The "Specify Types" dialog will appear.

Unlike other Families we have seen so far, this Family uses a "Type Catalog." A Type Catalog is used when Families have dozens or even hundreds of types. It is an external text file saved in the same folder as the Family that lists the parameter values of all of the possible types. This method allows you to load only the selected types from this Family file into your project, and not the potentially hundreds of types associated with this Family file.

- From the matrix of types listed at the right, hold down the CTRL key; click **2×6 [50×150mm]** and then **2×10 [50×250mm]** to highlight them.
- Click OK to load just these types into the project (see Figure 11.16).

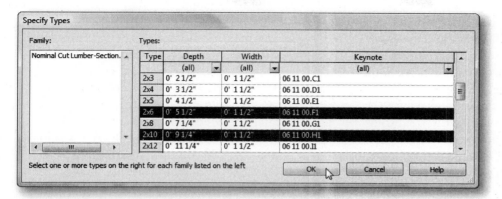

FIGURE 11.16 *Choose the Detail Component Family and specific types that you wish to load*

15. From the Type Selector, choose **Nominal Cut Lumber-Section: 2 × 6 [M_Nominal Cut Lumber-Section: 50×150mm]**.

   - Press the SPACEBAR three times.

   This will rotate it so the placement point is at the top left corner of the 2×6 [50×150mm].

16. Place two plates in the space between the floor joist and the foundation Wall (see Figure 11.17).

Use the Move or Align tools to assist in accurate placement.	**TIP**

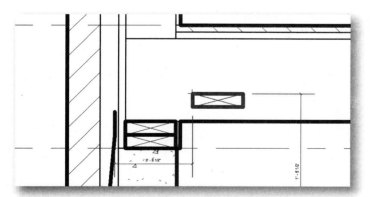

FIGURE 11.17 *Place a double top plate on the foundation Wall*

17. Repeat the process (or copy) to add a sill plate above the joist at the first floor.
18. On the Annotate tab, choose the **Detail Component** tool from the **Component** drop-down button again.

ADJUSTING THE LINE WEIGHT OF DETAIL COMPONENTS

Perhaps you have noticed how heavy the outline is around this particular component. The line weight in this case is a result of the subcategory assigned to the outline geometry in the Detail Component Family. In this particular Family the subcomponent is "Heavy Lines," which is set to a line weight of 5. If you wish to make this line weight less bold in your detail, you have three options: you can edit Object Styles in the current project and reduce the line weight assignment of the Heavy Lines subcomponent, use VG to edit the subcategory in this view only, or edit the Family and modify the outline to use a different subcomponent. Editing the Object Styles in the current project is quicker and easier, but not considered "best practice." While the desired line weight will be achieved, the effect will apply to all Families that use the Heavy Lines subcategory and its name "Heavy Lines" will no longer be applicable.

The best practice approach is to edit the Family and reassign the outline to a lighter subcomponent. To do this, select one of the *Nominal Cut Lumber-Section: 2 × 6 [M_Nominal Cut Lumber-Section: 503150]* elements on screen. On the Modify Detail Items tab, click the *Edit Family* button (you can also right-click to find this command). This will open the Family in the Family Editor. (The Family Editor was covered in detail in the previous chapter). Select the outline. The out-

line element is a Masking Region, which is a polygon object with an outline and solid opaque fill. On the Modify Detail Items tab, click the *Edit Boundary* button to edit the sketch of the Masking Polygon. If you are working in Imperial units, you will need to create the Medium Lines subcategory. To do this, click the Manage tab and then click **Object Styles.** Click the New button in the lower-right corner to add a new subcategory. Name the new subcategory **Medium Lines** and set its line weight to 3. Click OK to finish. On the Modify Detail Items > Edit Boundary tab, chain select the entire outline (four lines) on screen and then choose Medium Lines from the Type Selector. On the ribbon, click *Finish Region* and then save the Family. Finally, click the *Load into Project* button on the ribbon. Overwrite the existing Family when prompted.

A modified version of the Family named *MRAC Nominal Cut Lumber-Section.rfa [MRAC M_Nominal Cut Lumber-Section.rfa]* has been provided in the *Chapter11* folder. You can make the edits listed here or load the provided Family using the *Load Family* tool on the Insert tab.

- Change the Type to *Nominal Cut Lumber-Section: 2 × 10 [M_Nominal Cut Lumber-Section : 50×250mm]*.
- Use the spacebar to rotate if necessary and place a rim joist as shown in Figure 11.18.

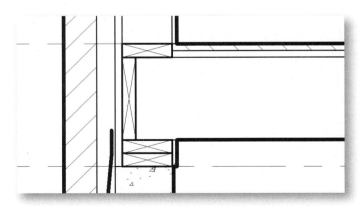

FIGURE 11.18 *Place a rim joist using a 2×10 [50×250mm]*

A frequent concern when creating these "hybrid" details is change management. What happens if, after placing Detail Components in several detail views, components in the model have to move, even if only slightly? To help alleviate the problem of having to revisit all of those details to make sure that all of the affected Detail Components get updated, you can lock the Detail Components to the model geometry that they are aligned with. After using the Align tool to position a component, click the lock symbol that appears. This will cause the Detail Component to move with the associated model component. Conversely, you cannot accidentally move a Detail Component once it has been locked to model geometry, so you don't have to worry about accidentally modifying the model by inadvertently editing a Detail Component. Keep in mind that you can "over constrain" your models as well. Even though there is a potential benefit to locking the detail components to the underlying geometry, in some cases you may experience errors later in the design process when moving model components if such a move causing the locked relationship to become invalid. In such a case, the user performing the edit may not understand the error, nor know the impact of clicking the Remove Constraints button. As always in BIM, you must strike a balance between potential benefits of a practice with the potential disadvantages.

MANAGER NOTE **BIM**

You use the same process to load and place any Detail Component. Revit ships with a very large collection of pre-made Detail Component Families. As we discussed in the previous chapter, set aside some time to get acquainted with what is provided. You can use the components in the library, modify them, or build your own. It is usually best to start with those provided before endeavoring to create your own. Let's continue to add to our detail by repeating the load and add process to add an anchor bolt.

19. On the Annotate tab, on the Detail panel, click the **Component** drop-down button and choose the **Detail Component** tool from the list.

   - Click the **Load Family** button and browse to:

     Imperial: *Detail Components\Div 05-Metals\050500-Common Work Results for Metals\050523-Metal Fastenings.*

     Metric: *Detail Components\Div 05-Metals\05090-Metal Fastenings.*

   - Open the *Anchor Bolts Hook-Side.rfa* [*M_Anchor Bolts Hook-Side.rfa*] file.

   - From the Type Selector, choose the **1/2"** [**M16**] type.

   - Place the anchor at the midpoint of the lower plate.

   - On the ribbon, click the **Modify** tool or press the ESC key twice.

20. Select the bolt that you just placed.

   - On the Properties palette, change the "Length" parameter to **1'-9"** [**525**].

   - Change the "Hook Length" to **3"** [**76**] and then click OK.

   - With the bolt still selected, on the Modify| Detail Items tab, click the **Mirror – Pick Axis** tool, clear the Copy checkbox, and mirror the bolt about its center.

   - On the Modify | Detail Items tab, click the **Align** tool; use the top of the plate as reference and align the bottom of the bolt to it (see Figure 11.19).

Fine-tune your placement as necessary to match the figure.

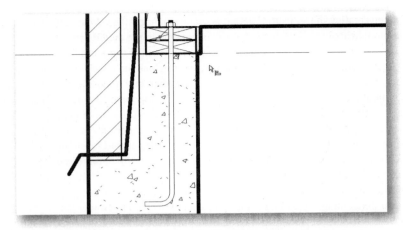

FIGURE 11.19 *Place an anchor bolt and adjust its location and parameters*

Repeating Detail Elements

Repeating Detail elements are Detail Components that automatically repeat about an invisible sketch line. This allows more rapid placement of Detail Components like studs, CMU, Brick, etc. In the detail that we are constructing, we can see the brick Layer of our Wall with the heavy cut line on the exterior and the diagonal fill pattern. This rendition is fine for general scales and overall plans and sections. However, at the scale of this construction detail, adding mortar joints will better delineate the brick veneer and suggest the individual bricks. While we could place one mortar joint and then array or copy it, a Repeating Detail Component is more expedient and the spacing can be edited later if necessary.

21. On the Insert tab, on the Load from Library panel, click the **Load Family** button.

    - Browse to the location where you installed the dataset files and open the *Chapter11* folder.
    - Select the *MRAC Mortar Joint with concave joint.rfa* [*MRAC Mortar Joint with concave joint-Metric.rfa*] file and then click Open.

Now that we have loaded a mortar joint Family, we will create a new Repeating Detail Type using this Detail Component. If you wish, feel free to open this Family in the Family Editor and study its composition.

22. On the Annotate tab, on the Detail panel, click the **Component** drop-down button and choose the **Repeating Detail Component** tool from the list.

Only one Type is available on the Type Selector—***Repeating Detail: Brick***. We are going to use this as the basis for a new Type.

    - On the Properties palette, click the Edit Type button.
    - In the "Type Properties" dialog, click the Duplicate button to create a new type.
    - Name the new Type **MRAC Mortar** and then click OK.
    - In the "Type Parameters" area, choose **MRAC Mortar Joint with concave joint: Brick Joint** from the "Detail" list.
    - Leave all of the other settings unchanged and then click OK to return to the view window.

23. Click at the bottom-left corner of the brick veneer and drag up past the top of the Crop Boundary and click again (see Figure 11.20).

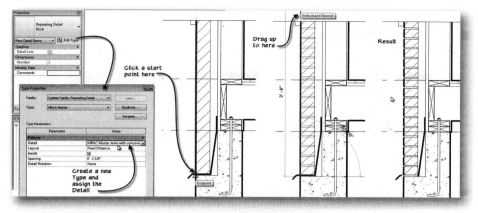

FIGURE 11.20 *Place a Repeating Detail for the Mortar Joints*

- On the ribbon, click the **Modify** tool or press the ESC key twice.
- Save the project.

Filled Regions

Filled Regions are two-dimensional shapes comprised of boundary lines and fill patterns. You can draw them any shape you like and use them to create, hatch, or cover up parts of the detail or other drawing. We will use a Filled Region here to illustrate the filled trench on the exterior side of the foundation wall.

24. On the Annotate tab, on the Detail panel, click the top part of the **Region** tool.

The Modify | Create Filled Region Boundary tab will appear with familiar sketch tools.

- From the Line Style panel on the ribbon, choose **Wide Lines**.

On the Draw panel, be sure that the **Line** icon is selected, and on the Options Bar that the "Chain" checkbox is selected.

25. Zoom into the bottom of the foundation Wall near the footing.
26. Sketch the shape shown in Figure 11.21. The exact dimensions are not critical.

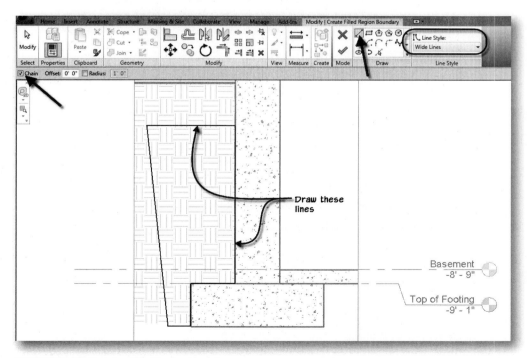

FIGURE 11.21 *Sketch a Filled Region boundary*

27. On the Draw panel, change the shape to Circle.
    • Add a sketched circle with a **2" [50]** radius as shown in Figure 11.22.

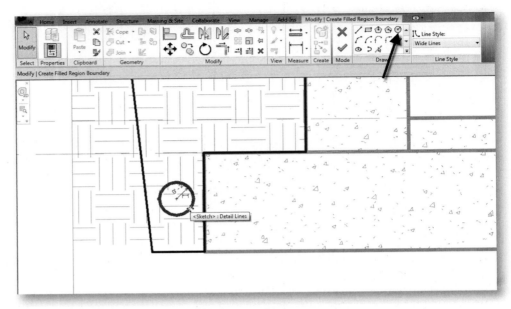

FIGURE 11.22 *Sketch a circle in the Filled Region shape*

28. On the Properties palette, click the Edit Type button.

    • From the Type list, choose **MRAC River Rock** and then click OK.
    • On the Mode panel, click the **Finish Edit Mode** button (see Figure 11.23).

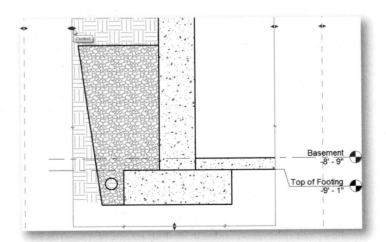

FIGURE 11.23 *Change the Filled Region to River Rock and then Finish the Sketch*

NOTE If necessary, you can widen the crop region to allow more room to draw the Filled Region.

29. Repeat the same process for adding the finished grade with the **MRAC Earth Disturbed** Filled Region (see Figure 11.24).

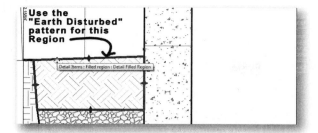

FIGURE 11.24 *Add another Filled Region*

Both of the Filled Region Types have been provided in the dataset for your use. They make use of custom Fill Patterns. You can edit and create Fill Patterns on the Manage tab on the ***Additional Settings*** drop-down. Simple fills can be created directly in the "Fill Patterns" dialog by designating the spacing of parallel lines. For more complex patterns, PAT files can be imported.

Adding Break Lines

Next let's drop in some Break Line components to hide part of the model. Break Lines have Instance parameters so we can individually adjust their size to fit the Detail.

30. On the Annotate tab, on the Detail panel, click the ***Component*** drop-down button and choose the ***Detail Component*** tool from the list.

   - Click the ***Load Family*** button and browse to:
     Both Imperial and Metric: *Detail Components\Div 01-General*.

   - Open the *Break Line.rfa* [*M_Break Line.rfa*] file.

   - Place a Break Line at the top of the detail to cover the top edge.

   - Press the SPACEBAR three times, and then place another Break Line covering part of the floor joist to the right (see Figure 11.25).

Break line components contain invisible Masking Regions that mask (cover up) the model objects beneath them. The concept of a mask is common in graphic design software and can be helpful in creating details.

   - Use the Shape Handles to make adjustments as necessary.

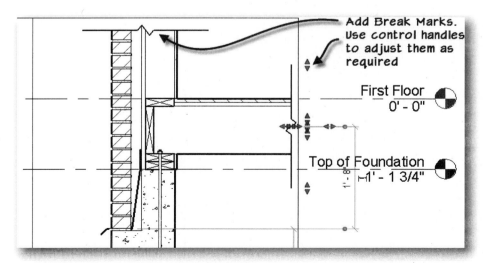

FIGURE 11.25 *Add Break lines with integral Masking Regions*

Batt Insulation

Next we'll place some batt insulation in the wall and floor.

31. On the Annotate tab, on the Detail panel, click the **Insulation** tool.
    - On the Options Bar, set the Width to **5" [130]**.
    - Click the first point at the bottom midpoint of the stud space, move up vertically and then pick the second point above the Crop Region (see Figure 11.26).

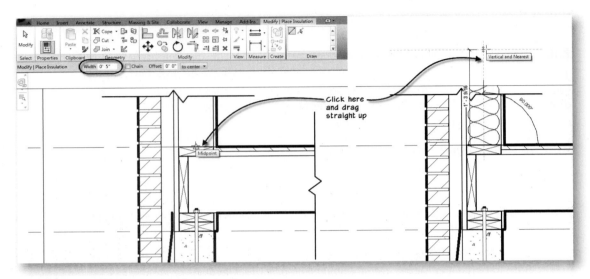

FIGURE 11.26 *Draw Insulation in the Stud cavity*

32. Press the ESC key one time (this deselects the previously drawn insulation, but remains in the command).

    - On the Options Bar, choose "to far side" from the drop-down list.
    - Click a point on the inside of the rim joist and drag to the right past the Crop Region (see Figure 11.27).

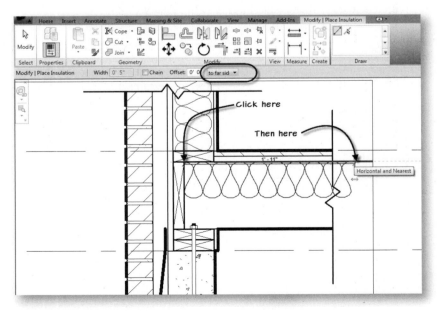

FIGURE 11.27 *Draw Insulation in the Floor cavity*

- On the ribbon, click the ***Modify*** tool or press the ESC key twice.

Notice that the insulation is not masked by the Breaklines. This is because there is an explicit display order for the view-specific Detail Components. The insulation is currently on top because it was added to the view after the break lines. We can shuffle the display order now.

33. Select the Break Line component on the right.

   - On the Modify | Detail Items tab, click the Bring to Front button (see Figure 11.28).

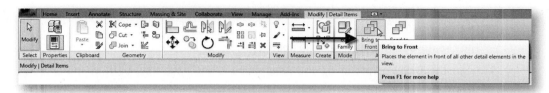

FIGURE 11.28 *Use the Display Order icons to shuffle the order of Components in the view*

   - Repeat this process on the other Break Line elements.
   - Use the **Send to Back** button on the Mortar Joints repeating detail item to bring it behind the flashing Detail Lines.

Edit Cut Profile

Sometimes you encounter a situation where the automatically created graphics do not suit your specific needs in a particular view. One such example is the keyway locking the foundation Wall to the Footing. It is possible to modify the model geometry to rectify this situation, but as an alternative, Revit Architecture provides us with the ***Edit Cut Profile*** tool. This tool gives us the ability to edit the path of the cut lines that Revit automatically generates. This type of edit is view-specific and two-dimensional. While it does not change the 3D shape of the model, it gives us a quick way to make the detail look the way we need without forcing us to model something that would have little or no benefit in other views. Since a key between the bottom of a foundation wall and the top of a footing would never be seen in any view other than a section or detail view, it would be difficult to justify the additional time or effort required to model it in 3D. Using the ***Edit Cut Profile*** tool we can make the section or detail appear as required more quickly and without the extra modeling overhead.

34. On the View tab, on the Graphics panel, click the ***Cut Profile*** tool.

   - On the Options Bar, select the "Boundary between Faces" option.

This option allows us to edit two boundaries—in this case the Footing's boundary and the foundation Wall's boundary—with one sketch. If we used the other option, Face, we would have to first edit the bottom face of the foundation and then go back, repeat the process and edit the top face of the footings.

   - Select the boundary line between the foundation Wall and the footing (see Figure 11.29).

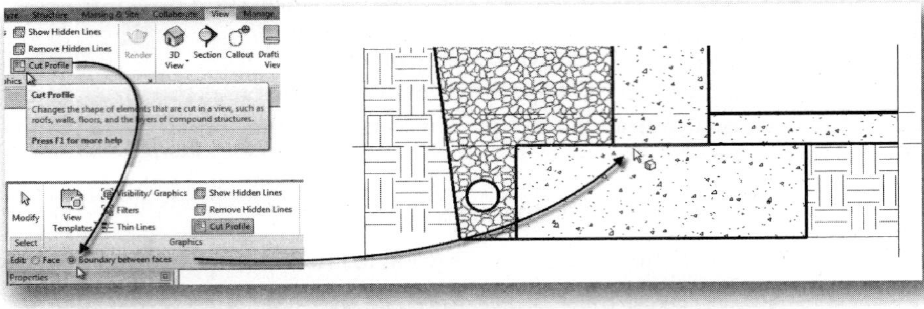

FIGURE 11.29 *Using the "Boundary between faces" option, select the face to edit*

The Create Cut Profile Sketch tab appears.

35. From the Draw panel, using the **Line** shape, sketch the new path as indicated in Figure 11.30.

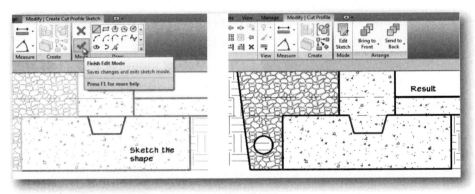

FIGURE 11.30 *Sketch the new shape using a Chain of Lines*

- On the Mode panel, click the **Finish Edit Mode** button.

In this case the fill pattern is the same on both sides of the Cut line, but if they were different you would notice that the fill pattern for the footing receded and the fill pattern for the foundation wall extended to fill in the key shape.

View Breaks

It is common that a detailed wall section will be too tall to fit on a Sheet. So it is typically broken into separate parts that crop away the repetitive areas. The crop boundary for any view includes "View Break" Controls and can be clipped to achieve this effect.

36. Select the Crop Boundary surrounding the section callout (it appears as a rectangle surrounding the drawing).

On each of the four edges of this Crop Region, a blue dot control handle appears at the midpoint and a "zig zag" break Control appears on either side of it. The "zig zag" controls allow us to truncate the view into smaller parts facilitating placement on a Sheet (see the left side of Figure 11.31).

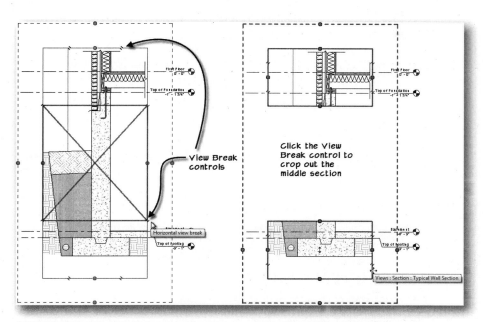

FIGURE 11.31 *View Break Controls allow you to crop out the middle portion of the section*

- Click one of the View Break Controls on a vertical edge (there are four total; you can pick any one) of the Crop Boundary (see the right side of Figure 11.31).

The view splits into two separate Crop Regions with a large gap in the middle. A blue arrow Control Handle appears in the middle of each View Break region. We can use these to move the two portions closer together. Notice that along the vertical edges of each of the two new Crop Boundaries the same types of Control Handles appear. You can continue to break them into additional sub-views as necessary. But all breaks must be along the same direction as the first one—vertical in this case. In this example two is enough so we will not break it any further. However, we need to adjust the top View Break so we can see the entire Anchor bolt.

- Using the blue dot Control Handle at the bottom of the upper View Break, drag the edge down a bit to show all of the anchor bolt.

37. Click on the View Break Control arrow (in the middle) of the top View Break and drag it down so the Crop Region is a little above the upper crop boundary of the lower View Break (see Figure 11.32).

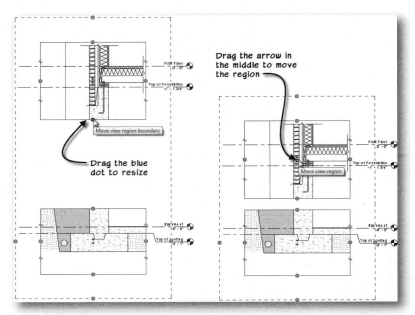

FIGURE 11.32 *Drag the upper view closer to the lower one using the Move View Region Control*

If you continue to drag so that you overlap the two View Breaks they will join back into one. This is how you "remove" the break.

CAUTION	Be sure to move the portions of the View Breaks with the Control arrow in the middle. Do not drag the edge of the Crop Region. Doing so will actually move the area of the callout in both this view and the referring Longitudinal section view.

Although the sub-views are truncated and closer together, distances are dimensionally correct. Look at the Level lines to the right of the views. The First Floor is at elevation zero (0) and the Top of Footing is at elevation -9'-1" [-2700]. Therefore, if we were to add a dimension from the top of the finished floor on the first floor to the top of the footing, it should read a distance of 9'-1" [2700]. Let's try it out.

38. On the Annotate tab, on the Dimension panel, click the ***Aligned*** tool.

   • Move the Dimension tool over the top edge of the Floor object at the First Floor.

Most likely the Level line will prehighlight. While we could dimension this point and still receive the correct value, we want to associate the dimension with the Floor element instead. We can use the TAB key here (like so many other places in Revit Architecture) to cycle to the element that we want.

   • Press the TAB until the top edge of the Floor prehighlights and then click.

The selected edge will remain highlighted while you complete the dimension operation.

   • Move down with the Dimension tool and over the top cut line of the Footing.

If the top cut line of the Footing does not automatically pre-highlight, use the TAB key again.

   • With the top edge of the Footing pre-highlighted, click to select it.
   • Move to the left Crop Region Boundary and click next to it (in the white space) to place the Dimension string (see Figure 11.33).

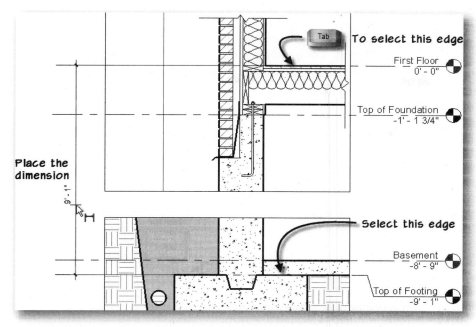

FIGURE 11.33 *Dimension the distance between the Floor and the Footing*

Notice that the Dimension displays the correct 9'-1" [2700] value from the top of the Footing to top edge of the finish Floor. As you can see, applying a View Break is a graphical convention only and has no impact on the dimensional accuracy of the model being displayed in each portion of the Crop Boundary.

- On the ribbon, click the **Modify** tool or press the ESC key twice.
- Add Break Line detail components on the foundation Wall at the break between the two halves of the detail.

39. Save the project.

ANNOTATION

Annotating a drawing with notes, dimensions, symbols, and tags is essential to communicating architectural design intent. Such annotations in Revit are view-specific elements. This means that these elements appear *only* in the view to which they are added. The exception to this is view indicators and Datum elements like section markers, elevation makers, Level Lines, Grids, and Callouts. These items are purpose-built to appear in all appropriate views and enhance the fully coordinated nature of a Revit project.

> It is possible to create a view where annotation appears simultaneously in it and another "dependent" view. This feature is used to facilitate large drawings that require matchlines to fit on a standard sized sheet. If you would like to learn more, please look up Dependent Views in the online help.

NOTE

Each view in Revit Architecture has a "View Scale" parameter and all annotations added to a particular view will scale and adjust accordingly. View indicators and Datum elements are included in this behavior. This means that no matter what the scale of the drawing, the annotation, view indicators and Datum symbols (level heads and grid bubbles) will be the correct size required for printed output. This behavior also applies to line weights and drafting patterns. Each graphical view you open will have its own scale. In addition, if the scale parameter of a view is changed, the text, line weights and drafting patterns will automatically adjust. The relative thickness of a particular line, or line weight, is controlled in a matrix of common plot scales. If desired, you can edit this matrix with the *Line Weights* command from the *Additonal Settings* drop-down button on the Manage tab. Drafting patterns will maintain their line spacing so the spacing always looks correct on printed output no matter what the scale is.

NOTE Model patterns do not change with the scale of the view; they are a fixed size relative to the model.

BIM MANAGER NOTE Try using the out-of-the-box settings as-is for a while before making any changes. You will likely find the out-of-the-box settings for scale and line weight to be adequate for most situations. If you do make changes, save these modified settings in a modified version of the standard Revit template file and make it your office standard. This is much more efficient than repeating your desired edits with each new project. A common practice is to keep a record of changes made in the Project that should become standard settings. Periodically, you can use the *Transfer Project Standards* button on the Manage tab to migrate these settings into your Project Template.

Create a Custom Text Type

A text element in Revit, like other elements, can have one or more Types. A text Type in this case is simply a grouping of parameters that control the look and formatting of the text. There are several parameters, many of which are similar to text in other computer software and are likely familiar to you.

Like other Families and Types, Text Types can be preconfigured and added to a Project Template. Additional Types can be added to the template or as a project progresses. The process for creating a new text Type is nearly identical to the one used for duplicating other element Types; you simply duplicate an existing one, rename it and modify its parameters. Let's create a new text Type for our project. In this example, we will create a "general note" text Type that is 1/8" [3] high and uses a different font.

NOTE In Revit, the height is its final plotted height—you are not required to calculate text size relative to the model.

40. On the Annotate tab, on the Text Panel, click the Text Types icon in the corner of the panel title bar (see the top half of Figure 11.34).

   • Next to the Type list, click the Duplicate button.

TIP A shortcut to this is to press ALT + D.

A new Name dialog will appear. By default "2" has been appended to the existing name.

   • Change the name to **MRAC Standard Notes** and then click OK.

You can use any font that is installed on your system. Since the choice of fonts can vary widely from one computer to the next, your system may not have the same fonts as those indicated here. Feel free to choose a different font if you prefer.

41. Beneath the "Text" grouping, from the "Text Font" list, choose the font of your choice.

   • In the "Text Size" field, type **1/8"** **[3]** (see Figure 11.34).

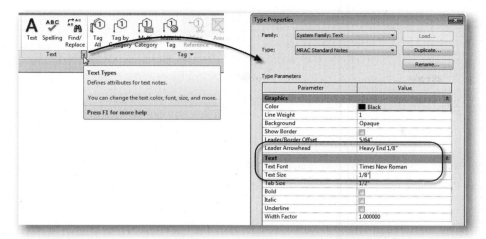

FIGURE 11.34 *Create a new Text Type*

This is the size the text will be when printed out. Beneath this, you can choose to make the text bold, italic, or underline if desired. The "Width Factor" setting is used to compress or stretch the text. This is a multiplier. When set to 1, the text draws in the way it was designed in the font. A value less than 1 will compress the text and a value greater than 1 will stretch it out.

In the "Graphics" area, you can change the color of the text as well as assign an arrowhead to be used when you create text with a leader line attached. Attaching a leader is done on the Options Bar. The "Leader Arrowhead" parameter is used to assign an arrowhead Type to the Text Type.

- From the "Leader Arrowhead" list, choose **Heavy End 1/8**" [**Heavy End 3mm**].

Arrowheads are System Families. You can add additional Types using the **Arrowheads** command from the **Additional Settings** drop-down button on the Manage tab.

- Click OK to complete the new type.

Placing Text

42. On the Annotate tab, on the Text panel, click the **Text** tool.

- On the Properties palette, choose **MRAC Standard Notes** from the Type Selector.

To place text in a view, simply click a point on screen or drag a rectangle at the location where you want the text to appear. If you choose one of the leader options on the ribbon, the first (and possibly second) click will be to place the arrow and elbow of the leader, then you click to place the text. If you click a single point, the text will flow in one continuous line without wrapping. If you click and drag two points, it will wrap to the width between the points. Regardless of your choice, you can always edit the wrapping of a text element later using the control handles on the text element. Pressing the ENTER key within a text element will insert a "hard" Return. This will move the cursor to the next line regardless of the automatic wrapping. It is similar to using a word processor.

43. On the Format panel of the ribbon, choose the No Leader option and then on the left side of the detail, click and drag a text region near the top, close to the crop region edge.

- Type **Standard Face Brick Veneer- see specification for color** (see Figure 11.35).

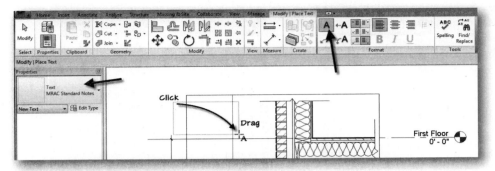

FIGURE 11.35 *Add a Text element and type in the desired note*

- Click next to the note (in the white space) to finish typing.
- Click the **Modify** button on the ribbon or press ESC.

> **NOTE** If a warning message appears and a text element disappears, you have created the text outside the annotation crop region (see Figure 11.36). The annotation crop appears as a dashed boundary outside of the view crop region and hides any annotation that falls outside its boundaries. Review the next topic for a more thorough explanation of the annotation crop region.

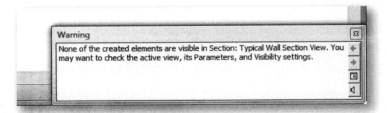

FIGURE 11.36 *Revit warns you when a newly created element is invisible*

Annotation Crop Region

In the exercises above, we made adjustments to the crop region of our detail call-out views to fine-tune how much of the model's geometry was included in the view. In addition to the crop region, we also have the annotation crop region. This region falls outside the normal crop region and affects only the annotation elements added to the view. When any portion of an annotation element intersects the annotation crop, the entire element disappears. While the annotation crop can be enabled in any plan, section, or elevation view, the most effective place to utilize it is in drawings that contain matchlines. In this way, if you have text or other annotation that occurs near the matchline, Revit can show a limited amount of the duplicate annotations on each matchline sheet. To see this feature in action, explore the dependent views feature in Revit. An example is shown in Figure 14.10 in Chapter 14. You can also look up dependent views and annotation crop regions in the online help.

In our current detail, we have no need to crop the annotation. Therefore we will simply turn off the feature in the current view.

- With nothing selected in the View window, on the Properties palette, beneath the "Extents" grouping, remove the checkmark from the Annotation Crop setting and then click OK (see Figure 11.37).

FIGURE 11.37 *Turn off the Annotation Crop*

If the notes you typed above were not showing, they should have now appeared.

> Another effective way to deal with the annotation crop in a detail view, such as the one we have here, would be to simply enlarge the annotation crop region using the control handles.

NOTE

Some blue control handles will appear attached to the text element while selected. You can use the one on the left to move the element (while leaving any arrow heads in place), the one on the right to rotate it, and the two small round ones on either side to resize and reshape the element and its word wrapping (see Figure 11.38).

44. Select the Text Note added above.

   • Use any of the control handles to fine-tune its placement.

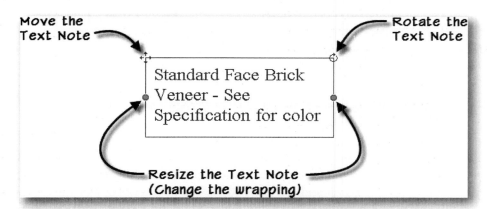

FIGURE 11.38 *Move, Rotate, or Resize a Text Element with its Control Handles*

Including Leaders with Text

To place a leader and arrowhead with a note, you can choose the appropriate option on the ribbon.

45. On the Annotate tab, click the **Text** tool again.

   • On the Modify | Place Text tab, on the Format panel, click the **One Segment** button.
   • Click the Leader at Bottom Right button (if it is not already active).
   • In the view window, click near the middle of the double top plate on the foundation Wall.

This is the location of the arrowhead for the Leader.

   • Drag to the left and click beneath the first note (a temporary guideline will appear to assist you).

This is the end of the Leader. A text object will appear. (If you selected the "Two Segments" icon instead, you would place two segments of the Leader line before typing would begin.)

- Type the next note, **Double Top Plate – (2) 2x6** and then click next to the note (in the white space) to finish typing (see Figure 11.39).

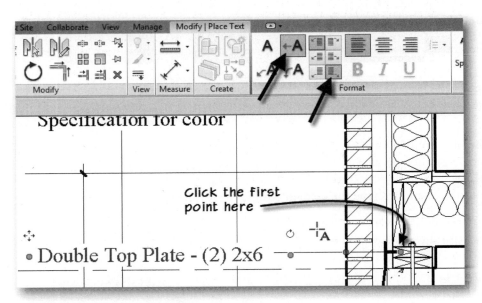

FIGURE 11.39 *Add a Text Element with a Leader*

- Using the Move handle, drag the text element to align it with the first one. A temporary guideline will appear to assist you.

When you drag a text element with a leader, be careful not to drag up or down as this will bend the leader line. This is because the leader's arrowhead stays attached to the element to which it points. If you want to move the entire thing (text and leader together), use the Move command (on the Modify tab of the ribbon) or the arrow keys on the keyboard (to nudge). Make any fine-tuning adjustments that you wish to the position of either text element.

- Use the grip on the right side of the Text Box to shorten it until the Text wraps to make two lines.
- On the ribbon, click the **Modify** tool or press the ESC key twice.

The Leader should remain attached to the bottom line of Text. You can determine the Leader location for either the left or right side of the text on the Format panel of the ribbon. In the process of adjusting the Text box, the Leader may now have shifted so that it is no longer horizontal, so you may want to move the Text up or down until it returns to its original orientation as shown in Figure 11.39.

The first text element we created does not have a leader attached to it. You can add leaders to existing text anytime.

46. Select the first text element (the brick veneer note).

- On the Format panel, click the **Add Right Straight Leader** tool (see Figure 11.40).

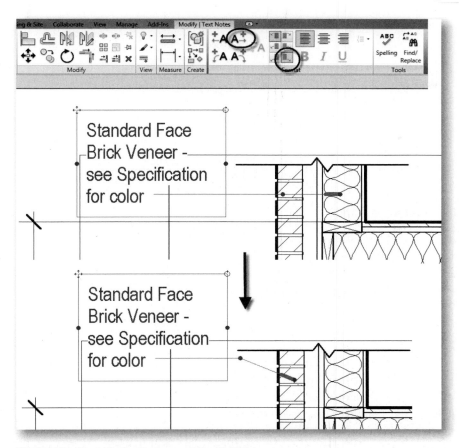

FIGURE 11.40 *Add a Leader to an existing Text Element*

A leader will appear attached to the text. You can then use the **drag** handles to modify its shape and adjust the location of the arrowhead.

- Make adjustments with the drag controls as necessary to move the arrowhead to point at the brick.

47. Using the **Text** tool with a leader option, add a note pointing to the batt insulation that reads: **Batt Insulation**.

- Adjust the position of the note and leader as required.

Sometimes you want to have the same note point to more than one location in the detail. You can add additional leaders to an existing text element. To do this, you simply select the text element and then click the appropriate icon on the Options Bar. To remove a leader you no longer need, click the "Remove Leader" button.

- With the Text selected, on the Modify Text Notes tab, on the Leader panel, click the **Add Right Straight Leader** button.
- Position the leader and its arrowheads as necessary to point at the insulation in the floor (see Figure 11.41).

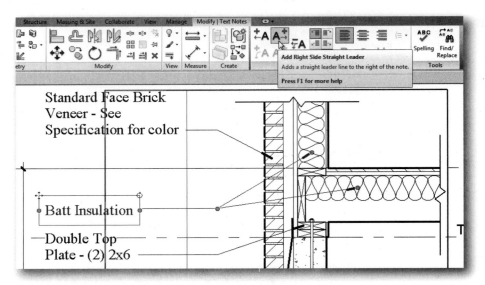

FIGURE 11.41 *Add a second Leader line to the Text Element*

Adding Keynotes

Adding text is not the only way to add notes to a detail (and other views). We can also use keynotes. Keynoting allows you to annotate your details using a pre-defined list of notes. The notes are organized in a keyed list, which is why they are referred to as keynotes. If your firm uses a keynoting system such as the AIA ConDoc system, keynotes provide a means to simplify the application of keyed notes and the compilation of all keys into a keynote legend for inclusion on titleblocks.

However, it is not required that you actually utilize the keys in order to use the keynote functionality. Even if you do not currently use a keynoting system, you may still find the keynoting tools useful. This is because rather than being required to type out each note you add to a project, with keynoting you choose the note from a pre-defined list of standard notes. Furthermore, you can even pre-assign keynotes directly to your materials, templates, and Family types in your office standard templates and library files.

Revit Architecture includes a sample keynote file organized in CSI format. You can use this list as is, edit it, or create your own. Creating or editing your own file is easy. The Keynote list is stored in a simple tab-delimited text file. If you wish to create your own file, look up the "Adding Additional Categories" topic in the online help for instructions and an example of the proper format. Of course using keynotes is optional, and to benefit fully from them, a certain amount of setup is required. You will have to decide if the benefits of doing so prove valuable enough to justify the initial configuration effort.

Before we begin adding keynotes to a project, you must choose a keynote file. You can use the same file for all projects in the office, or have different files for each project.

48. On the Annotate tab, click on the Tag panel titlebar.

    The Tag panel expands to reveal more tools.

49. Click the **Keynoting Settings** button (see Figure 11.42).

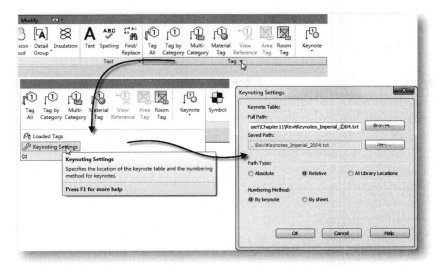

FIGURE 11.42 *Load a Keynote file and configure other settings*

Use the Browse button to load an existing file. Revit ships with files based on the CSI specification format. The *RevitKeynotes_Imperial.txt* [*RevitKeynotes_Metric. txt*] file is based on the traditional 16 section CSI format and the *RevitKeynotes_ Imperial_2004.txt* file (imperial only) is based on the newer CSI 2004 format and its 48 specification sections. These out-of-the-box keynote files are located in the *C:\ProgramData\Autodesk\RAC 2012\Libraries\US Imperial* folder for imperial and in the *C:\ProgramData\Autodesk\RAC 2012\Libraries\US Metric* folder for metric. You can use one of the provided files or create your own. Copies of these files are included with the dataset files from the student companion in the *Chapter11* folder.

When assigning your chosen file to a project, the path to the file can be set to absolute, relative, or set by library location. An absolute path writes the complete path back to the drive letter. A relative path assumes that the keynote file is located in the same location as the project file and therefore only writes the path relative to the location in which the project is saved. Using the "At Library Locations" option writes the path relative to the locations defined on the "File Locations" tab of the Options dialog. The Options command is on the Application menu (big "R").

Keynotes can be numbered using the keynote defined in the file or by sequential number relative to each Sheet in your document set. In other words, the "By keynote" method will use a fixed and predefined key. The "By sheet" method will compile the numbering uniquely for each Sheet of the set based on the notes actually used on that sheet.

Copies of the out-of-the-box keynote files have been included with the book dataset files installed from the student companion. For this reason, the "Relative" path type is configured for our project (as shown in the figure).

• Verify the settings, make any required changes, and then click OK.

50. On the Annotate tab, click the **Keynote** drop-down button and then choose the **Element Keynote** tool.

Move the cursor around on screen. Items that have a keynote assigned will appear as the mouse passes over them.

- Move the mouse over the Anchor Bolt element on screen and then click it.
- Click a point for the leader and then a point to place the keynote tag.

  It requires two clicks. If you want a straight line leader, click twice along the same line.

  A Keynote symbol will appear with the key for the associated note displayed.

Most of the out-of-the-box detail components (like the anchor bolt and studs we used here) already have keynotes assigned from Autodesk. If you click an item that does not already have a keynote assigned (like the Walls and Floors) then the keynotes dialog will appear. At the top of the dialog whatever keynote file you assigned above will appear in the title bar. A list of major categories will appear. Each contains additional sub-categories and notes. You can choose any appropriate note from the list for the item you are noting.

The note will appear within the keynote tag with a leader pointing to the anchor bolt detail item. Continue to keynote other items if you wish. You will only be prompted to select a note the first time you keynote an item. After assigning the note the first time, Revit will simply display that note on each subsequent instance you keynote.

Understanding Keynote Tags

The default keynote tag has four Types. Three Types display the key and the fourth displays the text of the note. Using the text display option, you can use keynote tags to speed up data entry without being required to actually use "keyed" notes (see Figure 11.43).

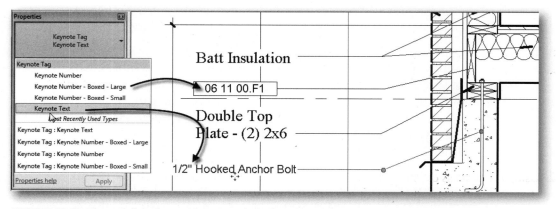

FIGURE 11.43 *The out-of-the-box Keynote Tag includes four variations*

51. Select one of the keynotes you have added.

- From the Type Selector, choose **Keynote Number** or **Keynote Text**.

A couple variations are shown in the figure. If you prefer a variation not shown, you can edit the Keynote Tag Family.

When you choose the Keynote Text option, you will notice that the text is center justified. To use right or left justified, edit the keynote Family, select the appropriate label element and change its properties to your preferred justification.

Types of Keynotes—Keynotes have three modes: Element, Material, and User. The Element option reads the keynote assigned to the element in the model such as the keynote assigned to a Wall or Door, not the individual layers or sub-components of the Wall or Door. To keynote the layers of a Wall or components of a Door, you would use the Material keynote option. This will read the keynote assigned to the Material of the selected sub-component. When you wish to override the pre-defined keynote setting, choose the User option. This option will always display the "Keynotes" dialog and prompt you to choose a note. Since this option is an override, it will not update if you edit the type or Material of the selected element.

Keynotes offer some compelling features, but they are not as mature as other features in the software. For example, certain items cannot be keynoted, like Drafting Lines, Repeating Details, and Batt Insulation. Furthermore, keynotes have not been pre-assigned to all of the out-of-the-box content. They have been assigned to the out-of-the-box detail component Families as we have seen, but not to the out-of-the-box model Families or Materials. This means that to fully benefit from the power of keynotes, a good deal of effort will be required to go through the library and assign keynotes to both Families and Materials. While you might be tempted to abandon the keynote functionality altogether based on these limitations, remember that the alternative to keynotes is to *manually* type every note. Once set up, having keynotes assigned to elements will save a great deal of time in production and will help to standardize the verbiage and phrasing used on notes throughout the office. So they remain worthy of your consideration.

Keynote Legend—If you want to compile a list of all the keynotes used on a particular sheet or throughout the entire project, you can create a keynote legend. A Keynote Legend lists all the keys and their corresponding notes. This can be a real time saver versus manually compiling such a list. You create a keynote Legend from the View tab. On the Create panel, click the ***Legends*** drop-down button and choose ***Keynote Legend***.

Finalizing the Detail

Our detail is nearly complete. With a few final edits, it will be ready to place on a sheet.

52. Using the process covered here, add additional notes or keynotes to the detail.
53. Using the ***Dimension*** tool, add dimensions to the footing and foundation Walls.

Remember to use your tab key as needed to select the required edges to dimension.	**TIP**

The Crop Regions around the detail are becoming a bit distracting. We can turn off their display.

54. On the View Control Bar (at the bottom of the view window) click the Hide Crop Region icon (see Figure 11.44).

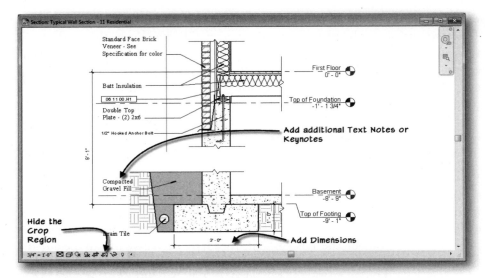

FIGURE 11.44 *Add additional Notes, Dimensions and hide the Crop*

Hiding an Element in the View

On the left side of this detail we see a gray vertical line. This is the edge of the chimney beyond. In cases like this, where some piece of the model that we would rather not see displays, we can hide it in this view. We have two ways to approach this. Both are view-specific overrides leaving the chimney unchanged in all other views.

Method 1

55. Select the Fireplace element.

   • On the ribbon, click the **Hide in View** drop-down and choose **By Element** (see the left side of Figure 11.45).

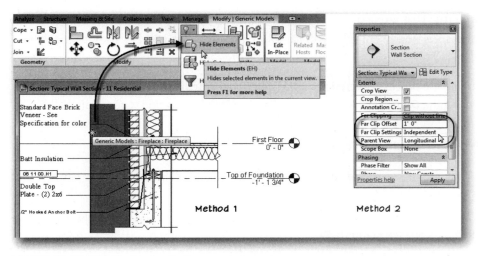

FIGURE 11.45 *Hide the fireplace in the current view only*

The fireplace will disappear. Should you need to make it reappear, click the small light bulb icon on the View Control Bar. This will make all invisible elements reappear tinted maroon. You can then select the chimney (or any maroon element) and choose the appropriate unhide command on the ribbon.

Method 2

56. Make sure you have no objects selected in the View Window.

- On the Properties palette, in the Extents grouping, for the "Far Clip Settings," change from Same as parent view to **Independent**.

  This makes the Far Clip Offset field editable.

- Change the Far Clip Offset to a small value like **1'-0"** [**300**] and then click OK (see the right side of Figure 11.45).

This method will crop out everything in the view beyond 1'-0" [300] from the cut plane.

Detail the Remainder of the Wall

To detail the rest of the Wall section, you can follow the same procedures as outlined here. Start by returning to the *Longitudinal* section view and create a new callout of the top portion. Use the View Break controls to crop the detail and remove the repetitive portions. Add masking break line Detail Components to each of the breaks. Hide the Crop Region of the view when finished. Begin adding Detail Components on top of the section cut as we did above, add drafting lines and edit the linework as required. Complete the detail with dimensions, notes and/or keynotes. Focus on the Wall connection at the second floor and the overall studs, rafters, joists, and insulation. When you are finished, the detail should look something like Figure 11.46.

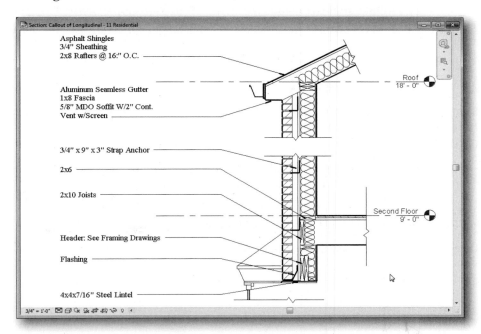

FIGURE 11.46 *Create additional Details using the same process*

Most of the components that you will need are already loaded into this project; however, for items like the steel angle at the Window lintel, you can simply load them in from the appropriate library. At the roof eave, you will need to rely more on Filled Regions, Drafting Lines, and Edit Cut Profile. Let's take a look.

57. From the View tab, click the **Callout** tool.

- On the Properties palette, choose **Detail View: Detail** from the Type Selector.
- Create a Callout bubble similar to the one shown in Figure 11.47.

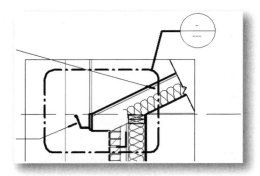

FIGURE 11.47 *Create a Callout View for the eave condition.*

58. In the Project Browser, expand the Detail Views branch.

   • Select *Detail 0* and right-click to rename it to: **Typical Eave Condition**.
   • With the *Typical Eave Condition* view still selected, on the Properties palette change the scale to **1 ½" = 1'-0"** and the Detail Level to **Fine**.
   • Double click on the *Typical Eave Condition* View to open it.

Using Edit Cut Profile to Modify Wall Layers

As you have seen, you can use the technique of adding Filled Regions and Masking Regions to cover unwanted geometry and then sketching Detail Lines on top for almost any situation. There is nothing inherently wrong with the procedure, but if the underlying model should change, the Filled and Masking Regions may no longer cover the intended geometry leading to errors in coordination and intent. Another approach is to modify the underlying geometry as it is displayed in this view. To do this, we use the **Edit Cut Profile** tool as we did above for the footing.

Work in the *Typical Eave Condition* view.

59. Select the Roof Level Marker; on the ribbon, click the **Hide in View tool** and choose **Hide Category** from the pop-up.
60. On the View tab, on the Graphics panel, click the **Cut Profile** button.

   • Pass the cursor over the Wall and when the stud layer pre-highlights, click the mouse.

   The Create Cut Profile Sketch tab appears and shows some now familiar sketch tools. The existing boundary of the stud layer will show as an orange outline.

61. Using the Lines icon on the Draw panel, draw the cut profile (see Figure 11.48).

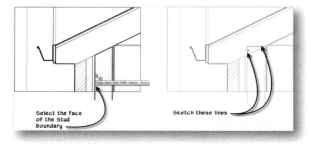

FIGURE 11.48 *Sketch the new edge of the Cut Boundary*

A small arrow handle will appear on the sketch line. It should be pointing to the inside of the stud to indicate that you wish to keep everything below the sketch line. If it points outside the stud, click it to reverse it. Be sure that the line touches the edges of the stud component on both sides.

- On the Modify | Cut Profile panel, click the **Finish Edit Mode** button.

The result should look like the stud shown in Figure 11.49

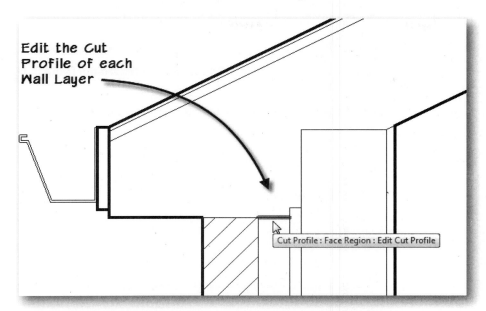

FIGURE 11.49 *After the Cut Profile Edit, the shape of the Wall reflects the change*

62. Use the Cut Profile tool again to edit the face of the Brick, Air Gap and Sheathing layers so that your detail looks like Figure 11.49.

    The profile line for the Sheathing layer should be 5/8" [16] above the profile lines for the Air Gap and the Brick.

63. Use Detail Lines, Masking Regions, Detail Components and Repeating Details to add embellishment to the Detail (see Figure 11.50).

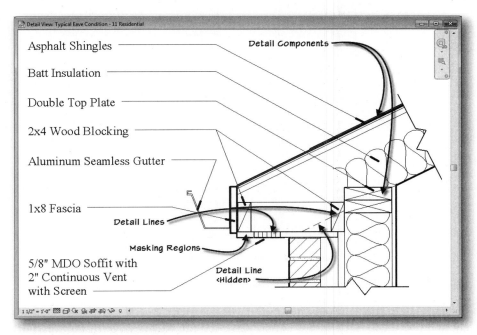

FIGURE 11.50 *Finishing the Eave Detail.*

64. Make any additional edits and then save the project.

Add a Detail Sheet

Once we have created one or more Detail views, we can add them to Sheets in the same fashion as other views. We explored this process back in Chapter 4. Let's review the steps here to create a new Detail Sheet that contains our Typical Wall Section detail.

65. On the Project Browser, right-click the *Sheets (all)* branch and choose **New Sheet**.

- In the "Select a Titleblock" dialog, choose **MRAC D 22 × 34 Horizontal [MRAC A1 metric]** and then click OK.

This will create "G101 - Unnamed." This is because the last Sheet we created was Sheet G100.

66. On the Project Browser, right-click on *G101 – Unnamed* and choose **Rename**.

- In the Number field, type: **A601.**
- In the Name field, type: **Details** and then click OK.

| **TIP** | You can also click directly on the (blue text) values in the titleblock and edit them directly on screen without right-clicking the sheet on Project Browser. |

67. From the Project Browser, drag the *Typical Wall Section* detail view and drop it on the Sheet.

- Click a point to place the detail. Move it around as desired to fine-tune placement.

68. On the Project Browser, double-click to open the *Longitudinal* section view (if you prefer, you can also open *A301 – Sections* Sheet instead).

Notice that the callout annotation has automatically filled in to indicate that the detail is number one on Sheet A601. This will also remain coordinated automatically (see Figure 11.51).

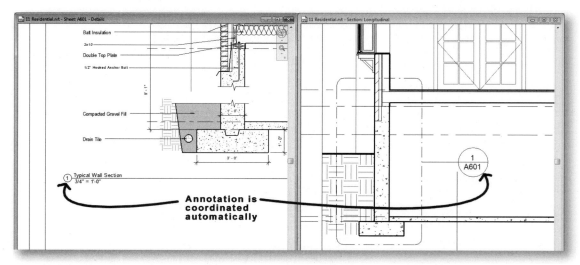

FIGURE 11.51 *Annotation will coordinate automatically after adding the detail view to a Sheet*

69. Repeat the process to add the other details done in this chapter to this Sheet.

If you want to align the views to one another on the sheet, it can be tricky sometimes to select and move the right thing. When you click the viewport, both the viewport and the title will highlight and move together. However, if you click just the title, you can move it independently. Revit will try to give you alignment guidelines as you drag items on screen.

In some cases, you will add details to the Sheet and then later wish to reorganize or renumber them. To do this, you edit the View Properties of the view in question. Edit the value of the "Detail Number" parameter in the "Instance Properties" dialog for the view. Be sure to type a number not yet in use—Revit will not allow you to duplicate an existing number. To swap the numbers of two details, first edit one to a unique value, edit the other to the value originally used by the first, and then edit the first to the number originally used by the second. If you make such a change, open the *Longitudinal* section view and note that the new numbers are reflected there as well. A change in one location is a change everywhere in Revit!

You can edit the view's Properties directly from the Sheet if you wish. Expand the Sheet entry on the Project Browser to see a listing of all views already placed on a particular Sheet. Click the name listed and the properties for the View will be listed on the Properties palette. You can also double-click the view from there to open it.

Create a Custom View Title

Some firms like to see the sheet where a detail is referenced. You can customize the View Title Family to include this information automatically.

70. Expand the *Families* branch on the Project Browser.

   • Expand *Annotation Symbols*, right-click *View Title*, and choose **Edit**.

Annotation Families are much simpler than the component Families we worked with in the previous chapter. Here you can add linework, text, and labels. Labels report the values of parameters. In the case of View Title Families, we can report the name and number of the detail, the scale and the referring detail, and the sheet.

71. On the Home tab, on the Text panel, click the **Label** tool (see Figure 11.52).

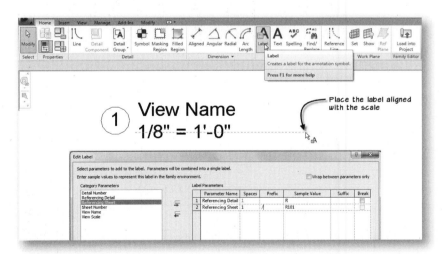

FIGURE 11.52 *Add a Label to the View Title Family*

The line under the view title is actually part of the viewport object back in the project. Therefore, as you can see, there is no line here. You will have to approximate the correct location of the new label relative to the line. We'll place it next to the scale Label in this case.

   • In the "Edit Label" dialog that appears, choose Referencing Detail and then click the Add Parameter(s) to Label icon in the middle of the dialog.

   • Select Referencing Sheet and add it to the label as well.

- In the Prefix field for Referencing Sheet, type: / (a forward slash) and then click OK.

This will make the Label display the Referencing Detail parameter, then a slash, then the Referencing Sheet like this: R/R101.

72. Edit the Element Properties of the new Label and change the Horizontal Align to Left.

- Use the control handles to reduce the width of the field and move it as necessary (see Figure 11.53).

FIGURE 11.53 *Add Labels for the Referencing Detail and Sheet*

If you wish, add additional graphics, text, or labels.

73. From the Application menu, choose **Save As > Family**.

- Browse to the *Chapter11* folder and save the Family as: **MRAC View Title.rfa**.
- On the ribbon, click **Load into Project**.

74. On the *A601 – Details* sheet, select all three viewports, on the Properties palette, click the Edit Type button.

- Click the Duplicate button.
- Name the new Type: **MRAC Viewport with Referencing Title**.
- Change the Title entry to **MRAC View Title** and then click OK (see Figure 11.54).

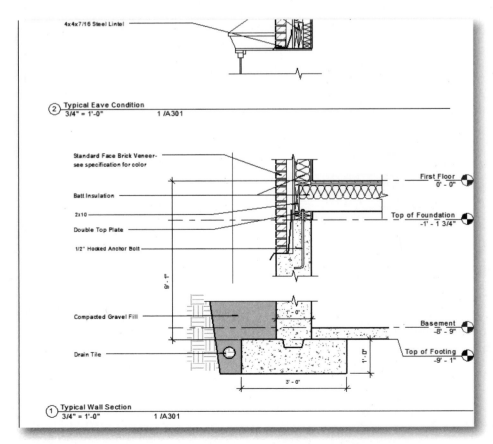

FIGURE 11.54 *Load the Custom View Title into the project and apply to the viewports*

75. Save the project.

Notice that the two wall details reference back to the A301 sheet, but the Typical Eave Condition references A601.

DRAFTED DETAILS (NOT LINKED TO THE MODEL)

In some cases, you will want to add a detail to a project that does not require a call-out underlay from the model. There might be several situations where this is appropriate. Examples include typical details that are generic in nature such as a typical head, jamb, or sill detail. Other examples might include a carpet transition, typical blocking condition, Wall type details or just a simple diagram of something related to the project but not specific to a particular area in the model. To create these kinds of details in Revit, we use a drafting view. A drafting view is like a simple blank sheet of paper. You can draw your detail on this blank page using any of the tools covered so far like detail components, filled regions, masking regions, drafting lines, and text. You can even add view references from other views if appropriate.

Creating a Drafted Detail

In this example, we will create a drafting view and a simple carpet transition detail. This can be created either with or without a view reference callout in our floor plans. You can create it as a typical, unreferenced detail by creating a new **Drafting View** on the View tab. If you want to reference the drafting view from a particular area of the plan, you can create a drafting view from the section and callout tools. To do this, you choose the "Reference other View" setting on the Options Bar before drawing the section or callout. For this example, we will create an unreferenced detail. In the next sequence, we will create a referenced one using the section.

1. On the View tab, on the Create panel, click the **Drafting View** tool.

   - In the dialog that appears, type **Floor Transition Detail** for the name.
   - Choose **3"=1'-0" [1:5]** and then click OK (see Figure 11.55).

FIGURE 11.55 *Create a new Drafting View*

A new drafting view will be created and opened. When Revit opens the new drafting view the most obvious characteristic is that the view is empty, showing no model geometry. A drafting view is like a blank sheet of paper. There are no automatically generated graphics from the model.

2. On the Annotate tab, on the Detail panel, click the **Component** drop-down button and choose the **Detail Component** tool from the list.

   - Click the **Load Family** button and browse to:
     Imperial: *Detail Components\Div 06-Wood and Plastic\061600-Sheathing.*

Metric: Detail Components\Div 06-Wood and Plastic\ 06100-Rough Carpentry\ 06160-Sheathing.

- Open the *Plywood-Section.rfa [M_Plywood-Section.rfa]* file.
- From the Type Selector, choose the ¾" [**19mm**] type.

If you have trouble finding this file, or if you did not install the default library files, all of the Families noted in this section are provided with the dataset files from the student companion. You will find them located in the same folder structure as noted here in the *MRAC\Imperial Library [MRAC\Metric Library]* folder. Feel free to load the required detail components from there instead.

- Following the prompts, create a horizontal length of plywood approximately 10" [250] long across the middle of the screen (see Figure 11.56).

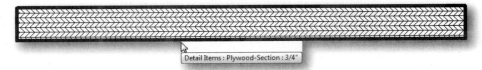

Detail Items : Plywood-Section : 3/4"

FIGURE 11.56 *Draw the plywood subfloor*

- Zoom in on the component after you draw it.

3. Click the ***Load Family*** button again and browse to:
   Imperial: *Detail Components\Div 09-Finishes\096000-Flooring\096400-Wood Flooring.*
   Metric: *Detail Components\Div 09-Finishes\09600-Flooring\09640-Wood Flooring.*

   - Open the *Wood Strip Flooring-Section.rfa [M_Wood Strip Flooring-Section.rfa]* file.
   - From the Type Selector, choose the **1×3 [19 × 76mm]** type.

4. Place the item on the top edge of the plywood.

   - Repeat the process to load four more Families from:
     Imperial: *Detail Components\Div 09-Finishes\096000-Flooring\096800-Carpeting.*

     Metric: *Detail Components\Div 09-Finishes\09600-Flooring\09680-Carpeting.*

     Carpeting-Section.rfa [M_Carpeting-Section.rfa]

     Carpet Reducer at Flooring-Section.rfa [M_Carpet Reducer at Flooring-Section.rfa]

     Carpeting Tack Strip-Section.rfa [M_Carpeting Tack Strip-Section.rfa]

     Carpet Pad-Section.rfa [M_Carpet Pad-Section.rfa]

With the first three, simply place them on screen in approximate locations for now. The *Carpet Pad–Section.rfa [M_Carpet Pad–Section.rfa]* Family behaves like the *Plywood–Section.rfa [M_Plywood–Section.rfa]* Family above did. You must click two points to place it. You can click two points along the top edge of the plywood for this component.

5. Move and copy the carpet and wood flooring components on screen to match Figure 11.57.

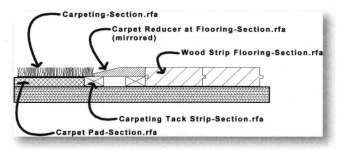

FIGURE 11.57 *Lay out the detail components to create the basic detail*

6. Add a Break Line Detail Component to the end of the detail.

   • Edit the Element Properties of the Break Line and change the Dimensions parameters as follows:

TABLE 11.1

Jag Width	1/2" [12] 3/4" [18]
Right	1" [25]
Left	1" [25]
Masking Depth	2" [50]

Set the Jag Width and Jag Depth first and the Right and Left last. This will avoid Revit's displaying error messages.	**TIP**

   • Copy the Break Line to the other side, and then press the SPACEBAR twice to flip it.

7. Add notes or keynotes to complete the detail (see Figure 11.58).

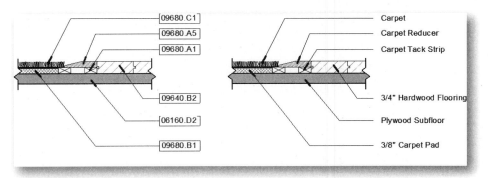

FIGURE 11.58 *The completed detail showing keynotes on the left and text notes on the right*

8. Add the detail to the *A601 – Details* sheet.

If you want this detail to be a typical detail, edit its View Properties and change the "Title on Sheet" parameter to "Typical Floor Transition Detail." Otherwise, if you prefer to call it out from the plan, you can open the *First Floor* plan view, zoom in on an appropriate area and then click the **Section** tool. From the Type Selector, change the view type to Detail View: Detail. Before you draw the section, check the "Reference other View" box on the Options Bar and choose *Drafting View: Floor Transition Detail* from the list of views. Draw the section line. The callout will read detail 3 on sheet A601 (see Figure 11.59).

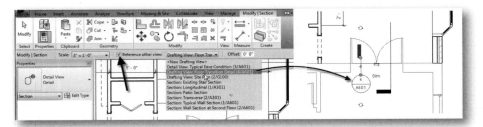

FIGURE 11.59 *You can optionally add a section callout that references the existing Drafting View*

9. Save the file.

WORKING WITH LEGACY DETAILS

Often details are reused from one project to the next. These "standard" details are typically kept in libraries for easy reuse and retrieval. In the days before computer design and drafting software, such a library would be a three-ringed binder from which photocopies were made. With computers, these standard details are stored digitally. If your firm has been using CAD software for a while, you likely already have such a digital library of standard details. You can use these legacy files directly in your Revit Architecture projects. You simply import the DWG or DGN files and place them on Sheets like other details.

Create a Referenced Section View

In this tutorial we will assume that the handrail of the existing Stair will be replaced with a new one. To show this, we will create a "Referenced Section View" to create a Section marker callout of a handrail detail within a stair section view. However, instead of creating the actual section view in Revit Architecture or drawing an unreferenced drafting view as we did above, the Referenced Section will link to a drafting view containing an AutoCAD file.

1. On the Project Browser, double-click to open the *Section at Existing Stair* section view.

- Zoom in on the area of the Stair between the First Floor Level and the Second Floor Level.

2. On the View tab, click the **Section** tool.

   - From the Type Selector, choose **Detail View: Detail**.
   - On the Options Bar, set the "Scale" to **6"=1'-0" [1:2]**.
   - Place a checkmark in the "Reference other View" checkbox, and verify that the menu is set to **<New Drafting View>** (see Figure 11.60).

These settings instruct Revit to create a new Drafting view instead of the typical live section view of the model. The detail marker will point to this new drafting view.

   - Drag the section line through the Railing as shown in Figure 11.60.

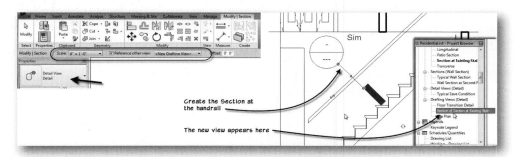

FIGURE 11.60 *Create a new Section View set to reference a New Drafting View*

Notice that a new Drafting view was created on Project Browser beneath *Drafting* views.

3. On the Project Browser, right-click the new Drafting view and choose **Rename**.

   - Name the view **New Railing Detail** and then click OK.

4. In the section view window, double-click on the detail head to open this Drafting view.

We again have a blank page upon which to work. Drafting an image that makes sense relative to the detail cut location is up to you. The only reference back to the model is the callout. We have already seen how we can draft something from scratch. Now let's look at importing a legacy CAD file.

Import a Detail Drawing

5. On the Insert tab, on the Import panel, click the **Import CAD** button.

   - In the "Import CAD Formats" dialog, browse to the *Chapter11* folder and choose *Typical Handrail Detail.dwg* [*Typical Handrail Detail-Metric.dwg*].
   - In the "Layer/Level Colors" area, choose **Black and white**.
   - In the "Positioning" area, choose the **Manual-Center** option (see Figure 11.61).

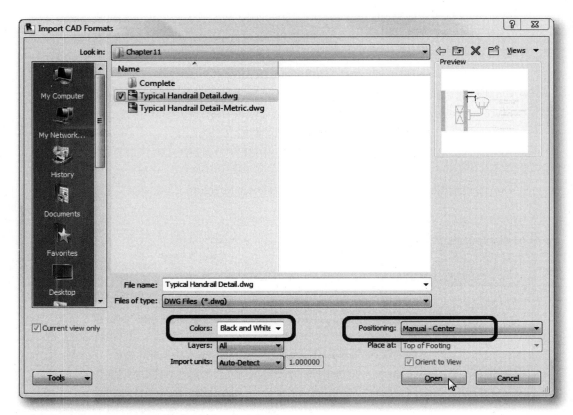

FIGURE 11.61 *Import a DWG file for the handrail detail*

Import embeds the file into the Revit project and does not maintain a link to the file. If the file were changed outside of Revit, you need to re-import the modified file. If you wish to link to the file instead, use the Link CAD tool on the Link panel instead. Linking makes it possible to reload the file later via the "Manage Links" dialog if the original file changes.

6. Click Open to import the detail, and then click a point on screen to place the detail in the view.

   • Verify that the scale of the current view is **6"=1'-0" [1:2]** as indicated above. If it is not, please change it.
   • Zoom to fit.

Notice that if you change the scale, it has an impact on how the line weights of the imported view display. If you wish to experiment with the way that the line weights import, click the small dialog launcher icon on the Import panel titlebar to open the "Import Line Weights" dialog. Revit uses the line weights built into the CAD file as is. If there are no line weights assigned to the CAD file's layers, then it looks to the colors of the layers and assigns lineweights as listed in the "Import Line Weights" dialog for each color.

This is a typical detail and there is no need for any changes. If we needed to make edits, we could select the detail, and then on the ribbon, click the ***Explode*** button to convert it to individual Revit Detail Lines and Text so we could edit it. However, it is best to avoid this and make such edits in the original file using its native application instead of using Explode. Exploding an imported file can increase file size adversely and create inaccuracies in the file. Most CAD/BIM Managers would look unfavorably on exploded CAD files in live Revit projects, so please consider carefully before you proceed.

If you choose to explode imported CAD files, you will discover that many element types are added to your file beyond what you see on screen or what you would otherwise expect. For example, you will likely end up with many line styles, text styles, and other elements bearing names reminiscent of the original CAD file's layers. In addition, regardless of whether you choose to explode the file, you will get Materials bearing names like Render Material 63-0-255 in your Material list. In general, these items will not cause you difficulty but they can increase the size of your files and cause confusion among team members. If you have decided to explode a CAD file, consider the following procedure. First, if you have access to the CAD program that created the file, open the file there first and clean up the geometry as much as possible. This includes deleting unneeded geometry and layers, purging the file, and resaving it. Next, import the CAD file into a new Revit project. Explode the CAD file in this temporary project and perform additional cleanup. This will include reassigning linework to appropriate Revit Line Styles, changing text to Revit text Types, etc. Please note that CAD dimensions and text leaders will not become Revit dimensions or leaders, so if you want actual leaders and dimensions, you will need to recreate these items. Once you have cleaned up the file to your satisfaction, you can select all of the elements and copy and paste them back to a drafting view in your original project. In general, importing CAD files into Revit should not become a long-term practice. CAD files in a Revit project can unnecessarily bloat the file and cause performance problems. Over time, you will find it beneficial to convert your standard CAD details to Revit format if you wish to continue using them in your Revit projects. For more information on this and other critical model management tips, you can download the *Model Performance Technical Note* at http://usa.autodesk.com/adsk/servlet/pc/index?siteID=123112&id=8480751. While this document was written for Revit 2010 at the time of this writing, it is still relevant and the topics discussed in it apply to Revit 2012 as well.

7. On the Project Browser, double-click to open the *A601 - Details* Sheet view.

   • Drag the Drafting view and drop it on the Sheet (see Figure 11.62).

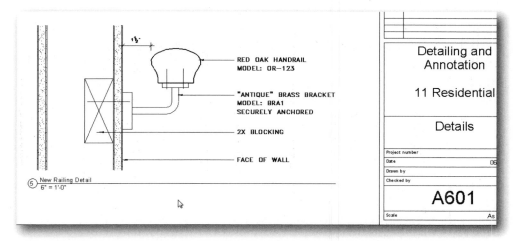

FIGURE 11.62 *Add the Detail view to the Details Sheet*

This will become detail 5 on the sheet. If you return to the *Section at Existing Stair* view, you will see that this number and sheet reference have appeared automatically in the callout.

8. Save the project.

ADDITIONAL DETAILING TECHNIQUES

Except for Drafting views, all views in the Revit Architecture project are generated directly from the building model. While Revit does a very good job of interpreting this model geometry into abstracted two-dimensional representations such as plans and elevations, there are often items that we wish to manipulate in order to create the Architectural drawings we are accustomed to producing. We have already seen all of the techniques that are used to perform such edits. Until now we have used these techniques and tools only on detail views. However, you can add drafting embellishment on any Revit view including plans, sections, and elevations. All such edits, including those made with the Linework tool, Filled Regions, Masking Regions, Detail Components, Edit Cut Profile, and Drafting Lines can be done on any view. More importantly, such edits apply *only* to the view in which they are applied.

Embellishing Model Views

Let's make a few enhancements to one of our elevation views.

9. On the Project Browser, double-click to open the *East* elevation view.

One common architectural drafting convention is to show the new foundation in an elevation as dashed below grade. We can achieve this using a combination of the Linework tool and adding Drafting Lines. Let's start with the footing.

10. Select the Terrain element, and then on the View Control Bar, choose **Hide Element** from the Temporary Hide/Isolate menu (sun glasses icon).

A cyan colored boundary will appear around the viewport. New in 2012, a label reading "Temporary Hide/Isolate" will appear in the upper corner. Remember that this is the temporary hide/isolate command. The cyan boundary appears as long as some elements are temporarily hidden. Temporary hide/isolate does not affect printing and is reset when the model is closed.

11. Select each of the Level Heads that do not have associated views (the ones that are black), right-click and choose **Hide in View > Elements**. (You can also use the ribbon tool.)

This will hide levels for "Top of Footing" or "Bottom of Stair" etc. This is the permanent hide command. These elements will stay hidden even after closing and re-opening the model. Permanently hidden elements also do not print. To reveal hidden elements and unhide them, click the light bulb icon on the View Control Bar. If you try this now, a maroon colored border will surround the screen and the label in the corner will change to reflect the new mode. The three hidden Level Heads will appear maroon in color as well, and the temporarily hidden terrain, which is still hidden, will appear cyan in color. Click the light bulb again to exit the mode.

12. On the Modify tab, on the View panel, click the **_Linework_** tool.

   • Choose **_<Hidden>_** from the Line Style drop-down list.
   • Click on each of the edges of the footings to change them to <Hidden> lines.
   • Do not change any of the vertical foundation lines yet (see Figure 11.63).

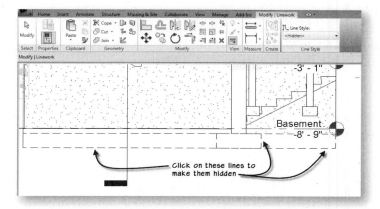

FIGURE 11.63 *Change the display of the footing lines to <Hidden> with the Linework tool*

You may need to pick more than once in the same general spot or a little to either side since there is more than one footing in the same spot in the elevation. If you are unhappy with the result, you can instead use the <Invisible lines> Type and then draw a continuous Drafting Line in on top. Be sure to lock the constraint padlock icon to keep the Drafting Line associated with the position of the footing

13. On the View Control Bar, choose **Reset Temporary Hide/Isolate**.

The terrain model will reappear. Notice that the footing still shows dashed through the terrain and is no longer hidden.

14. With the ***Linework*** tool still active, choose **<*Hidden*>** from the Type Selector again.

   • Click on one of the vertical lines of the foundation Walls.
      With it still highlighted, a drag handle will appear at either end.

   • Drag the top handle down to the point where it intersects the terrain (see Figure 11.64).

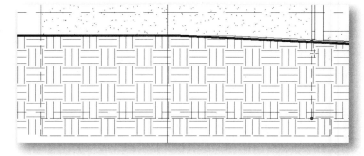

FIGURE 11.64 *Change the Linework of the foundation walls and edit the extent of the change with the drag handles*

   • Repeat for other vertical foundation Wall edges.

Later if you wish to change the linework to a different linetype or return it to its default setting, you can use the Linework tool again. To restore the default, use the <By Category> option from the Type Selector.

If you wish to modify the way that the terrain displays, you can use a Filled Region to trace over it. Draft additional linework as desired to complete the elevation. You can add notes, dimensions and tags as required. If the Linework tool is not working for a particular edge, you can try Masking Regions and Drafting Lines.

15. On the Annotate tab, on the Tag panel, click the **Tag by Category** tool.

    • On the Options Bar, clear the "Leader" checkbox.

    • Click on each of the Windows in the new addition. (Do not tag the Windows of the existing house.)

16. Add some text or keynotes to the patio on the right or to indicate materials of the elevation such as brick veneer and roof shingles (see Figure 11.65).

TIP	Try the **Keynote > Material** option to keynote the materials in the Wall instead of the entire Wall.

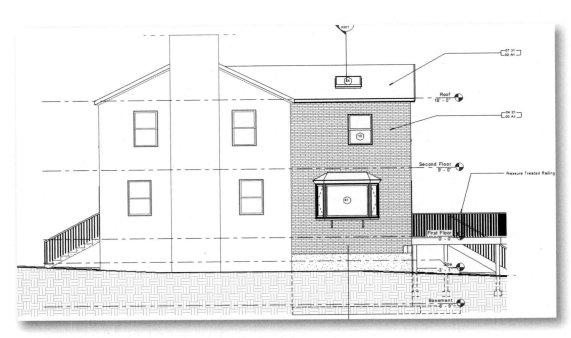

FIGURE 11.65 *Add tags and notes to complete the elevation*

17. Perform similar edits in other elevations if you wish.
18. Save the project.

Controlling Display of Items Beyond

To show depth in elevations, it is a common architectural convention to lighten the line weights of objects as they recede from view. Unfortunately, Revit does not offer an automated way to do this in elevations and sections. However, we can use the override graphics feature to manipulate the graphical display manually. Like all graphical overrides, edits you make are view-specific. So they will apply only to the view in which you make them. It may take you a little time and effort to fine-tune the elevations to display as desired, but you should be able to achieve acceptable results. We'll do a quick example here to illustrate the concept and process.

19. On the Project Browser, double-click to open the *North* elevation view.

    • Feel free to repeat any of the previous edits (foundation display, notes, etc) on this elevation before you proceed.

    • On the right side, at the upper patio, select the gable Roof, its fascia boards and gutter, and the two upper Windows.

    • On the ribbon, on the View panel, click the **Override Graphics in View** button and choose **Override By Element from the pop-up**.

    • In the "View-Specific Element Graphics" dialog, expand the Projection Lines item and change the Weight to **1** and then click OK (see Figure 11.66).

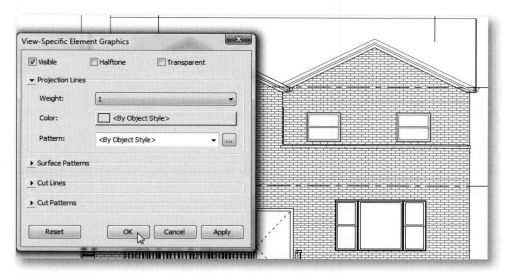

FIGURE 11.66 *Override the graphics of selected elements in the view*

This change may be hard to see without zooming in. We also want to lighten the brick hatch on the Wall beyond at the patio in the same area.

20. Select the Wall beyond at the patio in the same area.

    • Choose **Override Graphics in View > Override By Element** again.

    • Check the Halftone checkbox and then click OK.

Notice that because of the way that the Wall joined, we are seeing the end of the perpendicular Wall on the right side. We can override this Wall the same way or we can edit the join of this Wall on the Second Floor to fix this.

21. On the Project Browser, double-click to open the *Second Floor* plan view.

    • On the Modify tab, click the ***Wall Joins*** tool.

    • Click on the intersection of Walls at the lower-left side of the patio.

    • On the Options Bar, click the Miter radio button and then click the ***Modify*** tool to accept the change (see Figure 11.67).

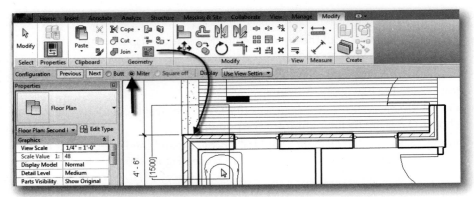

FIGURE 11.67 *Override the graphics of selected elements in the view*

When you return to the *North* elevation, you will notice that selecting the Wall now gives better results.

22. Repeat any of these procedures on the remaining elevations and then save the project.

There is a feature called Silhouettes which will override the profile edges around your model in the current view. To access it, click the Visual Style pop-up on the View Control Bar (the tool that sets hidden line or shading) and choose **Graphic Display Options**. In the "Graphic Display Options" dialog, you can choose a Line Style for Silhouettes. However, this feature lacks the capacity to control which edges receive the effect. Revit will determine which edges are silhouette edges and which are not. Give it a try and see if the results are satisfactory. If not, you can use the Linework tool to override the outline of elements in elevation the same way we dashed the footings.

Legend Views

As our final exploration in this chapter, we will look at another type of drafting view: the Legend View. This kind of view, as its name implies, is used to create symbol legends in your project. When working in a legend view, you can add Legend Components. Legend Components are symbolic versions of all Families and Types in your project, and as such are only graphical representations, not actual model elements.

23. On the View tab, on the Create panel, click the **Legends** drop-down button and then choose the **Legend** tool.

    • In the "New Legend View" dialog, type Door Types for the name, choose **1/4"=1'-0" [1:50]**, and then click OK.

Like the other drafting views we have created, a blank page will appear. The unique feature of a legend view is the availability of the **Legend Component** tool on the Annotate tab. We can use this tool to place a symbolic representation of any Family in the project. In this case, we are building a Door Types Legend, so we want to add elevation views of each kind of door, but do not want to add actual Doors, which would throw off the count in the Door Schedule later. This is where the Legend Component comes into play.

24. On the Annotate tab, on the Detail panel, click the **Component** drop-down button and choose the **Legend Component** tool from the list.

- From the Family list on the Options Bar choose **Doors: Single Flush: 36" × 80" [Doors: M_Single Flush: 0915 × 2032mm]**.
- From the View list on the Options Bar choose **Elevation: Front**.
- Click a point on screen to place the symbol.
- Repeat the process to place each of the following:

  Doors: Bifold-2 Panel: 36" × 80" [Doors: M_ Bifold-2 Panel: 0915 × 2032mm]

  Doors: Bifold-4 Panel: 72" × 80" [Doors: M_ Bifold-4 Panel: 1830 × 2032mm]

  Doors: Double-Glass 2: 68" × 80" [Doors: M_ Double-Glass 2: 1730 × 2032mm]

Line them up next to one another. You cannot tag the symbols because they are not real doors, but you can add text and dimensions where appropriate.

25. Add labels, notes, and dimensions as appropriate (see Figure 11.68).

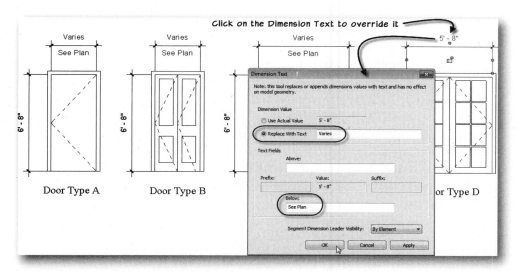

FIGURE 11.68 *Add notes and labels to complete the legend*

To edit the dimension values as shown in the figure, simply click on the dimension text and then edit the appropriate fields in the dialog that appears.

You can drag this legend view onto any sheet like the other views. Perhaps the most appropriate sheet for this legend would be the door schedule sheet. Since we have not created a door schedule yet, we will wait until the scheduling chapter for that task.

26. Save the project and close the file.

SUMMARY

Understanding the relationship between modeled elements and drafted elements is an important concept in Revit Architecture. Creating the basic model geometry can be accomplished in nearly any convenient view and as we have seen throughout this book and will remain coordinated as changes occur in all views. Drafting and annotation, on the other hand, occur in only the currently active view. This means that we can apply additional embellishments on top of an automatically generated model to explain and clarify design intent. We can modify the display of underlying model geometry using the *Linework* tool, element level graphic overrides, or view-specific display settings. Finally, we can create drafting views, which contain only drafting elements and no model geometry. Using a combination of these techniques, we can fine-tune any Revit view for inclusion in our complete set of architectural construction documents.

- Detailing occurs at many levels in Revit: as part of the model, as view-specific embellishment on top of the model, and as completely independent drafting views.
- You can use Wall Type edits such as unlocking layers, adding Sweeps, and adding Reveals to add details to the Walls that show throughout the model.
- Create callout views of any overall view to create the starting point for a construction detail.
- Add view-specific detail component (2D) Families, drafting lines, repeating details, filled and masking regions and batt insulation to embellish and add content to the underlying model callout view.
- Each view has its own scale and visibility settings.
- Detail Components and Detail Lines are view-specific embellishments that are used to convey design intent.
- Repeating Detail Components save time by adding an array of Detail Components at a predefined spacing.
- Use Masking Regions in any view to mask unwanted portions of the model.
- Use Filled Regions to apply patterns to areas and draw view-specific embellishments.
- Add Break Lines and adjust Crop Regions to isolate "typical" portions of detail views.
- Annotate the detail with dimensions, text notes and keynotes.
- Keynotes reference an external keynote file and help to maintain consistency in noting and reduce repetitive typing.
- Keynotes can display either the key or the note. You can add a keynote legend to list all notes used in the project or sheet by sheet.
- Use Cut Profile to modify the automatically created profile of building model elements within a particular view.
- Adding details to a sheet automatically numbers them and keeps the annotation coordinated.
- You can draw isolated two-dimensional details that do not link to the model. Use drafting views for this purpose.
- Import legacy CAD details and add them to drafting views to leverage existing detail libraries.
- Edit any section or elevation view using similar techniques to those used to create and modify details.
- Use the Override Graphics in View command to indicate elements that appear "beyond" in an elevation or section.
- A legend view is a special kind of drafting view that allows symbolic representations of any project Family to be added and annotated.

Working with Schedules and Tags

INTRODUCTION

Schedules are an important part of any architectural document set. Generating schedules is often a laborious process of manually tabulating the hundreds—or sometimes thousands—of items in a project that require presentation in a schedule. For example, in firms that do not use BIM software, a door schedule involves the painstaking process of manually listing each door specified in a project and then typing in detailed information about each entry—even those bits of information (such as size) that should be easily queried from the plans. In Revit, Schedules are generated automatically from the building model data already in the project in real-time. A Schedule in Revit is simply another view of the project that differs from plans and elevations only in its presentation as tabular information rather than graphical information. You can create a Schedule from nearly any meaningful information in the model and like all Revit views, you can edit in this view and see the change instantly in all views.

OBJECTIVES

In Chapter 4, we added some typical Schedules to our commercial project. In this chapter, we will explore the workings of these Schedules as well as create additional Schedules not yet in our project. After completing this chapter, you will know how to

- Add and modify a Schedule view
- Edit model data from a Schedule view
- Place a Schedule on a Sheet
- Work with Tags
- Work with Rooms and Color Schemes

CREATE AND MODIFY SCHEDULE VIEWS

In this chapter, we will return to our commercial project and look deeper into the Scheduling tools in Revit. As was mentioned above, we added Schedules to this project back in Chapter 4. However, we simply imported those Schedules into the commercial project from an existing template project. We'll start with a quick review of those Schedules and then take a detailed look at how to create a new Schedule view from scratch. Even though we are exploring these tools in the commercial project, they would work equally well in the residential project. At the completion of this chapter, feel free to open the residential project and add some Schedule views there as well.

Install the Dataset Files and Open a Project

The lessons that follow require the dataset included on the Aubin Academy Master Series student companion. If you have already installed all of the files from this site, skip to step 3 to begin. If you need to install the files, start at step 1.

1. If you have not already done so, download the dataset files located on the CengageBrain website.

   Refer to "Accessing the Student Companion site from CengageBrain" in the Preface for information on installing the dataset files included in the Student Companion.

2. Launch Autodesk Revit Architecture from the icon on your desktop or from the *Autodesk > Revit Architecture 2012* group in *All Programs* on the Windows Start menu.

TIP	You can click the Start button and then begin typing **Revit** in the Search field. After a couple letters, Revit Architecture should appear near the top of the list. Click it to launch to program.

3. On the QAT, click the Open icon.

TIP	The keyboard shortcut for Open is CTRL+O. **Open** is also located on the File menu.

• In the "Open" dialog box, browse to the location where you installed the *MasterRAC 2011* folder, and then the *Chapter12* folder.

4. Double-click *12 Commercial.rvt* if you wish to work in Imperial units. Double-click *12 Commercial Metric.rvt* if you wish to work in Metric units.

   You can also select it and then click the Open button.

View Existing Schedules

You will recall that the commercial project already has a few Schedules in it that were added back in Chapter 4. One of the most significant benefits of Revit Schedules is that once you have added them to the project, they maintain themselves. In other words, even though we have not looked at the Schedules since we added them back in Chapter 4, they have changed quite a bit. Let's have a look at how they have been shaping up.

5. On the Project Browser, beneath *Schedules/Quantities*, double-click to open the *Door Schedule* view.

Wow! Every Door added to the project since Chapter 4 has been automatically added to the Schedule. This is a live view of our project that is filtered to show only Doors and present them in a list rather than a drawing. That's all there is to a Schedule in Revit.

There is still plenty of work to be done to this Schedule, but a good deal of it is already done. Let's focus on the Door Number column for the time being. Notice that the numbers are not sequential and some are missing. Since a Schedule is a live view, you can simply click in the Door Number field and edit the value. The results are the same as if you were editing the item from a plan or another view, but editing in a Schedule is typically faster and more efficient. You can work with a Schedule and plan view tiled next to each other to assist in editing. You can also select a line item in the Schedule and then click the ***Highlight in Model*** button on the Modify Schedule/Quantities tab. This highlights and zooms in on the selected item. If you do not have a suitable view window open, Revit will prompt you to allow it to search for and open alternate views. You can cycle through several views to find the best one.

6. Close or minimize all other windows, Open the *Level 3* floor plan view, and then tile just the Schedule and the third floor windows.

   The easiest way to do this is to click the Close Hidden Windows icon on the QAT, open the *Level 3* floor plan next, and then tile the windows (WT).

   • Select Door 118 in the Schedule (see Figure 12.1).

      Notice how it highlights in the plan.

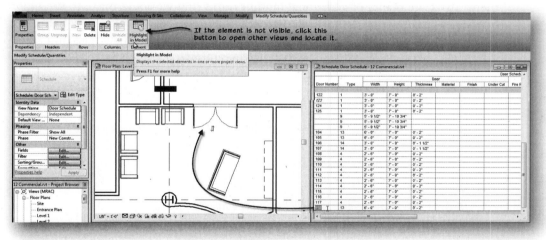

FIGURE 12.1 *Selecting an item in the Schedule highlights it in the plan*

Spend some time going back and forth between the plan and Schedule views. Get comfortable with the interaction. If you wish, you can renumber the Doors so that they correspond better to the project. For example, you can renumber all the third floor Doors starting with the number 301 and incrementing from there. Open the other plans to edit their numbers as well. In some cases, you might find it easier to edit the numbers in the plan using Door Tags. We will discuss tags in detail below in the "Adding Door Tags" topic. If you wish, you may skip there now for more information and assistance on numbering the doors. You can also skip renumbering for now and come back to it later. The exercise will not be adversely affected if you do not renumber the doors at this time.

A common door numbering scheme is for the numbers of the door to match the room numbers. If a room has more than one number, then the doors use the room number plus a lettered suffix. This is the approach advocated by the United States National CAD Standard (NCS) and the preferred method for many an architectural firm. Unfortunately, there is no automated way to use this approach in Revit. If you wish to number doors this way, you are forced to number them manually to match this standard. However, Avatech Solutions, a national Autodesk Reseller and solution provider, has developed a series of utilities for Revit that provide this functionality. Please visit: www. avatech.com and search for "Door Mark Update" or "Avatech Utilities for Revit." They have several other useful utilities, including: "Room Renumber." This tool makes updating room numbers easier and works well with the Door Mark Update utility. Autodesk Subscription Customers also have access to the Room Book Extension which offers many useful tools for working with and quantifying Rooms. If you are on subscription, check out this utility as well.

7. After editing the numbers, click on the titlebar of the Schedule view to make it active.

   - On the Properties palette, click the Edit button next to Sorting/Grouping.
   - On the Sorting/Grouping tab, choose **Mark** from the Sort by list and then click OK.

If you renumbered the Doors, they should now be sorted in the order in which you numbered them. If you decided to skip that task, do not worry; when you renumber the doors later, they will automatically resort in the Schedule as you renumber them. There are plenty of additional modifications that we could make to the Door Schedule. For now however, we will leave them for later and create a new Schedule. Before continuing, feel free to open any of the other existing Schedules and study how they have come along since we set up the project.

Add a Schedule View

8. Click the titlebar of the *Level 3* floor plan view to make it active.

In Chapter 10, we added some furniture to this plan. As you can see, the layout has been refined a bit. In this topic, we will create a furniture Schedule. To add a new Schedule, look for the *Schedules* tool on the View tab of the ribbon, or new in this release, we can also right-click on the Schedule/Quantities item on Project Browser.

9. On the Project Browser, right-click the *Schedule/Quantities* branch and then choose **New Schedule/Quantities**.

The "New Schedule" dialog will appear where we can select a Category. The "Category" list is a fixed list of categories built into Revit. In the "Name" field you can type any suitable name to describe the contents of the Schedule—usually the default is acceptable. When you choose the "Schedule building components" radio button, you get a list of building elements from the model with each one occupying a row of the Schedule. The "Schedule keys" option creates a user-defined list of common elements useful in filling in other Schedules. For example, you could define a certain kind of Door or chair, including its manufacturer, model number, finish, etc. When you begin the task of filling in your door or furniture schedule, you can choose a key and it will input all the other fields automatically saving you much of the redundant data entry. We will look at this option in the "Using Schedule Keys to Speed Input" topic below. In a multi-phase project, you can indicate which Phase the Schedule should report. The default is the New Construction Phase.

   - In the "New Schedule" dialog, select "Furniture" from the Category list.
   - In the "Name" field, accept the default name of "Furniture Schedule."

- Accept the remaining defaults of "Schedule building components" and "New Construction" for the Phase and then click OK (see Figure 12.2).

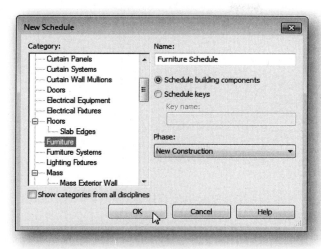

FIGURE 12.2 *Create a Furniture Schedule*

The "Schedule Properties" dialog appears next. The first tab (Fields) presents a list of "Available Fields" on the left side and the fields actually appearing in the Schedule on the right. This list includes all of the parameters available for the items you are scheduling (furniture in this case). To add a field to a Schedule, simply select it from the list and then click the Add button to move it to the "Scheduled fields" list on the right. Once you add fields, you can adjust the display order on the right. Each field will become a column in the resultant Schedule view.

- In the "Available fields" list, choose Type Mark and then click the Add button.

> **"Mark"** is the term used for the "Number" or other unique designator of the items in the Schedule. Most categories of elements include both a Mark, which is unique per instance, and a "Type Mark" which like other type parameters is common to all elements of the type.

NOTE

- Repeat for "Manufacturer," "Model," "Cost," "Count," "Family and Type," and "Comments."
- Be sure the order of fields is as listed in Figure 12.3.

  If you need to adjust the order, use the "Move Up" or "Move Down" buttons.

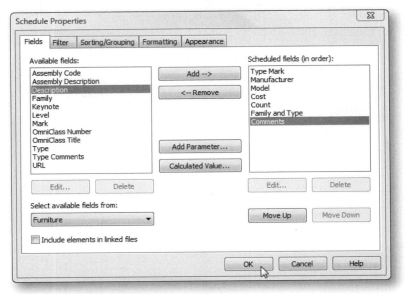

FIGURE 12.3 *Add Fields to the Furniture Schedule*

10. Click OK in the "Schedule Properties" dialog to create the Schedule.

A Furniture Schedule view will appear beneath the *Schedules/Quantities* branch of the Project Browser and it will open on screen. The name of the Schedule will appear at the top and directly beneath it; each field appears as a column header. If you scroll the Schedule up or down, the title and header will remain at the top as hidden rows come in and out of view. (This is like "Freeze Panes" in Microsoft Excel.) In fact, the overall look of the Schedule view closely mimics Excel.

11. Close all project views except the *Furniture Schedule*.

| HINT | Maximize the Furniture Schedule, and then use the Close Hidden Windows icon on the QAT (or View tab). |

- Open the *Level 3* floor plan view.
- On the View tab, click the **Tile** button (or type WT).

Now that we can see the plan and the Schedule together on screen, we can explore some of the basic behaviors of Schedules in Revit.

Edit Model Elements from the Schedule

This has been stated before, but it is worth repeating: elements in a Schedule are the same elements that we see graphically in the model. A Schedule is just another view of the model—a tabular view rather than graphical view.

12. Zoom in on the plan view to just the tenant suite on the left side (as close as you can while still seeing all furniture).

- Click on any field in the Schedule view.
- Repeat on as many elements as you wish.

Notice how the corresponding furniture element highlights in the plan view. This is simply review of what we did with the Door Schedule above.

13. Click in the "Type Mark" column for any element.

Notice that a standard text cursor appears in the field and you could begin typing a value.

- Click in the Type Mark field and drag down through several fields.

(Do not click first and then drag; click and drag in a single motion.) Notice how the corresponding elements in the plan highlight simultaneously. The schedule can be an effective way to select elements.

- Click in the "Manufacturer" column.
- Click in each field in succession.

"Cost" is like "Mark" and is a simple text field. Notice that "Manufacturer" and "Model," however, appear with a drop-down list icon. If you try to open the list, it will appear empty. The way these fields work is that you can type in any value, as in the plain text fields, but these values you type will begin populating a list that you can choose from in subsequent edits. Let's try it.

14. Widen the Family and Type column.

- To do this, click and drag the edge between the Family and Type and Comments headers. (Double-click to maximize the width of the column.)

15. Scroll as necessary and locate the first instance of **MRAC Desk-Secretary** and then click in the "Comments" field next to it.

- Type: **Include Keyboard Tray and Footrest Option** and then press ENTER (see Figure 12.4).

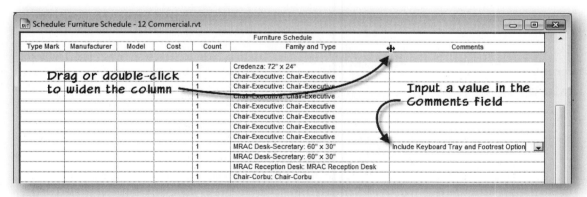

FIGURE 12.4 *Input a value in a list field*

There is a second secretarial desk beneath the one we just edited.

- Click in the "Comments" field of the second secretarial desk and then click the drop-down list icon.

Notice that the note typed above now appears in the list. Each new item you type will be added automatically to this list making it easier to input existing values—simply choose them from the list. The parameter you are editing in the Schedule can be either an Instance parameter like the one edited here, or it can be a Type parameter. Instance parameters are applied independently to each instance of the item in the project. In this case, it would be possible to order a keyboard tray and footrest for one of the secretarial desks but not the other. With a Type parameter, the value applies to all instances of the Type in question. This is the case with "Manufacturer," "Model," and "Cost." The assumption here being that the Family and Type in question represents a particular item that can be ordered from a catalog, purchased, and installed in the project. Therefore, the Manufacturer, Model, and Cost would be the same for all instances of that item. For our purposes, these behaviors are perfectly logical. However, in your own projects, if you wish to modify these behaviors, you must edit the Families of the elements in question (refer to Chapter 10 for more information on editing Families and for examples of both Type and Instance parameters). Let's edit a Type parameter next.

16. Click in the "Manufacturer" field for one of the desks.

   - Type: **Acme Furniture** and then press ENTER.

     A dialog will appear indicating that this change will apply to all instances of this Type (see Figure 12.5).

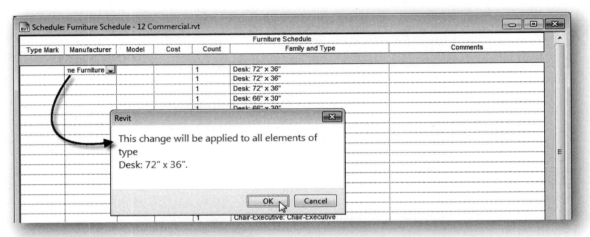

FIGURE 12.5 *Applying a change to a Type parameter*

   - Click OK to accept this change.
   - Scroll down in the Schedule and notice that the change has applied to several desks.
   - Repeat the process to add a Model designation and Cost to the same Desk.

17. Choose the "Acme Furniture" value from the "Manufacturer" column for the ***Chair-Executive [M_Chair-Executive]*** Family.

   - Again, when you make this edit, it will apply at the Type level. Click OK to confirm the change.
   - Add a Model and Cost to the Chair as well (see Figure 12.6).

					Furniture Schedule	
Type Mark	Manufacture	Model	Cost	Count	Family and Type	Comments
	Acme Fur	ExecuCha	1495.00	1	Chair-Executive: Chair-Executive	
	Acme Fur	ExecuCha	1495.00	1	Chair-Executive: Chair-Executive	
	Acme Fur	ExecuCha	1495.00	1	Chair-Executive: Chair-Executive	
	Acme Fur	Manager'	1200.00	1	Desk: 72" x 36"	
	Acme Fur	Manager'	1200.00	1	Desk: 72" x 36"	
				1	Credenza: 72" x 24"	
				1	Credenza: 72" x 24"	
	Acme Fur	Manager'	1200.00	1	Desk: 72" x 36"	
				1	Credenza: 72" x 24"	
				1	MRAC Desk-Secretary-Complete: 66" × 30"	Include Keyboard Tray and Footrest Option
				1	MRAC Desk-Secretary-Complete: 66" × 30"	Include Keyboard Tray and Footrest Option
				1	MRAC Reception Desk: MRAC Reception Desk	
	Acme Fur	ExecuCha	1495.00	1	Chair-Executive: Chair-Executive	
				1	Chair-Corbu: Chair-Corbu	
				1	Chair-Corbu: Chair-Corbu	

FIGURE 12.6 *Changes to Type parameters apply to all instances*

18. Using the same process, assign a Manufacturer, Model, and Cost to the Corbu and Breuer chairs.

Be sure to click on the in-cell drop-down and note how the list includes each new item you type in that field.

19. Save the project.

Sorting Schedule Items

Since many of the values in Schedule are identical for all instances of a particular item, it might be nice to sort and group some of these values in the Schedule instead of listing each element separately and showing so much repetition.

On the Properties palette, beneath the "Other" grouping in the "Instance Properties" dialog, are five items each with an "Edit" button next to them. These five items correspond to the five tabs of the "Schedule Properties" dialog (shown in Figure 12.3). We only looked at the "Fields" tab above. Each of these tabs can be accessed here anytime. Using the settings for Sorting/Grouping, we can modify the Schedule to group elements with common parameters into a single entry. We can also sort the Schedule in a variety of ways.

- On the Properties palette, next to "Sorting/Grouping," click the Edit button.

This dialog is shown below in Figure 12.8.	**NOTE**

20. At the top of the dialog, from the "Sort by" list, choose **Manufacturer** and then click OK.

- Scroll to the bottom of the Schedule to see the results.

All of the blank fields (which are listed first in an alphabetic sort) are now at the top of the list, and the "Acme Furniture" items are next. The Desks and Chairs are still interspersed. If you wish, you can sort by more than one criterion.

21. Return to the Sorting/Grouping tab of "Schedule Properties" dialog (click the Edit button next to Sorting/Grouping on the Properties palette again).

- At the top of the dialog, beneath "Sort by," from the "Then by" list choose: **Model** and then click OK (see Figure 12.7).

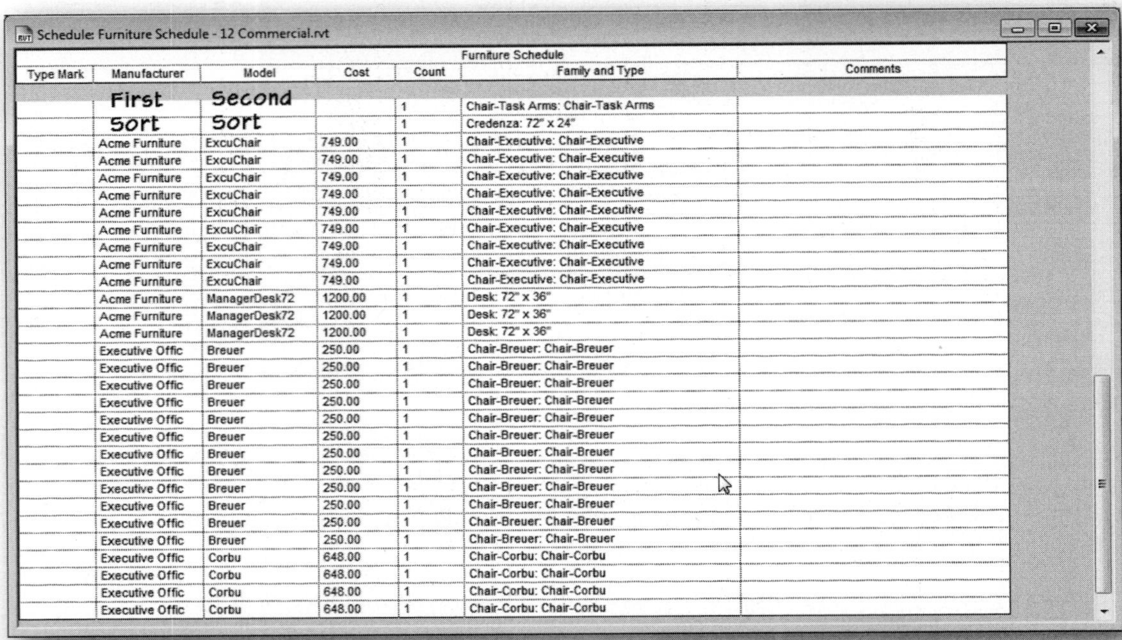

FIGURE 12.7 *Sorting the Schedule first by Manufacturer, then by Model*

Several items in the Schedule still do not have parameters assigned. Let's edit another item and see how the Schedule responds with our current sort settings.

22. Locate an instance of the "Chair-Task" in the Schedule and edit the "Manufacturer" to **Furniture Concepts**.

   - Click OK in the dialog that appears.

Notice that the Schedule immediately re-sorts to accommodate the new value.

   - Input a Model and Cost as well.

Grouping Schedule Items

In addition to sorting the items in the Schedule, we can also group them when all of the values are the same. It would be easier to work with the Schedule if we simply had one listing for each type of chair that reported how many there were instead of showing each as a separate line item. We need to make a simple change to display the Schedule this way.

23. Return to the Sorting/Grouping tab of "Schedule Properties" dialog.

   - At the bottom of the dialog, clear the "Itemize every instance" checkbox.
   - Click the Fields tab.
   - Move the Count field to the bottom of the list and then click OK (see Figure 12.8).

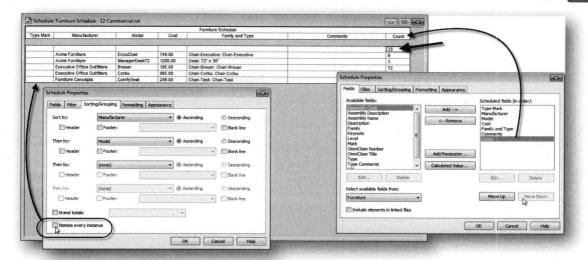

FIGURE 12.8 *Group items and show the count*

There are nearly limitless ways that we can format and display the same data. Notice the way the blank fields at the top (Count 33 in the figure) show none of the values even though we have data (particularly Family and Type) for some of the values. This is because we are sorting on the Manufacturer and Model fields and those fields are blank for the 33 items. If you edit the Schedule Properties and sort first by Family and Type instead, the result will change dramatically (see Figure 12.9).

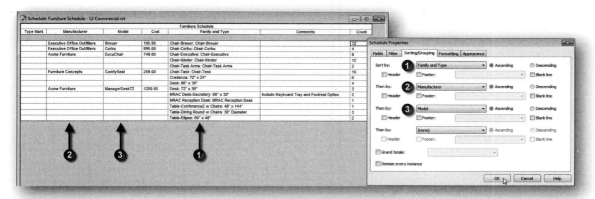

FIGURE 12.9 *Sort by Family and Type instead of Manufacturer*

Look at the Acme Furniture items. There are three separate entries because there are three separate Families assigned to this manufacturer. Add up the items with blank Manufacturer, Model, and Cost and you will see that the total is still 33. However, only now can we continue to edit them because each Family and type is listed separately.

Duplicating a Schedule View

The current format of our Schedule view is useful for editing, but is perhaps not ideal for reporting furniture and making budgetary and purchasing decisions. Sometimes you will find it useful to have two versions of the same view—one for working purposes, the other for presentation and printing. This is true of any kind of view, Schedule, plan, or elevation, etc.

24. On the Project Browser, right-click *Furniture Schedule* and choose **Duplicate View > Duplicate**.

   - Right-click *Copy of Furniture Schedule* and choose **Rename**.
   - In the "Rename View" dialog, type **Working - Furniture Schedule** and then click OK.

Now when we wish to modify an item, we can open the "Working" version of the Schedule to see each Family and type listed separately, which will make it easier to select and edit what we need. Regardless of where we make the edit, it will show in all views—both furniture Schedules and the model. Below we'll edit the original Schedule to format it more appropriately for printing, but first let's complete the modifications to our furniture.

25. Edit the remaining items to add Manufacturers, Models, and Costs.

   - Add a code in the Type Mark column for each item. They can be any alpha-numeric codes you like (see Figure 12.10).

		Furniture Schedule (Working)				
Type Mark	Manufacturer	Model	Cost	Count	Family and Type	Comments
CH-1	Exclusive Office Outfitters	Breuer	195.00	12	Chair-Breuer: Chair-Breuer	
CH-2	Exclusive Office Outfitters	Corbu	895.00	4	Chair-Corbu: Chair-Corbu	
CH-3	Acme Furniture	ExecuChair	1495.00	4	Chair-Executive: Chair-Executive	
CH-4	Exclusive Office Outfitters	GuestChair	225.00	12	Chair-Kinder: Chair-Kinder	
CH-5	Acme Furniture	WorkChair	850.00	8	Chair-Task Arms: Chair-Task Arms	
CH-6	Furniture Concepts	ComfySeat	249.00	10	Chair-Task: Chair-Task	
CRD-1	Office Systems	Low Credenza	899.00	7	Credenza: 72" x 24"	
DSK-1	Acme Furniture	WorkDesk	749.00	4	Desk: 66" x 30"	
DSK-2	Office Systems	Manager's Desk	1200.00	3	Desk: 72" x 36"	
DSK-3	Acme Furniture	CompuDesk	549.00	2	MRAC Desk-Secretary-Complete: 66" × 30"	Include Keyboard Tray and Footrest Option
DSK-4	Quality Millwork	Custom	10500.00	1	MRAC Reception Desk: MRAC Reception Desk	
TBL-1	Furniture Concepts	Board Room Table	9999.00	1	Table-Conference2 w Chairs: 48" x 144"	
TBL-2	Acme Furniture	Break Table	225.00	3	Table-Dining Round w Chairs: 36" Diameter	
TBL-3	Office Systems	Huddle Table	949.00	2	Table-Ellipse: 60" x 48"	

FIGURE 12.10 *Finish inputting values for all empty fields*

Notice that now when you click in a field and make an edit, you no longer get a warning about editing multiple items. This is because you are actually selecting all of these items when you click in the grouped field in the Schedule. You can see this visually if you like by tiling the plan next to the Schedule and then selecting a line in the Schedule. You should see all the corresponding items in the plan highlight.

26. Save the project.

Headers, Footers, and Grand Totals

The Count Field is not the only way to have Revit tabulate the quantity of items. If you wish to show totals above or beneath the groupings, you can add Headers and Footers and Grand Totals to the Schedule as well.

27. Close the *Working - Furniture Schedule* and return to the *Furniture Schedule*.

   - On the Properties palette, click the Edit button for Sorting/Grouping.
   - Change the first criterion back to **Manufacturer**.
   - Beneath Manufacturer, place a checkmark in the "Header" checkbox.

- Change the next criterion to **Model** and then place a checkmark in the "Footer" checkbox and accept the default of: **Title, count, and totals**.

- Change the third criterion to **(none)**.

- At the bottom of the dialog, place a checkmark in the "Grand totals" checkbox and accept the default of: **Title, count, and totals** (see Figure 12.11).

FIGURE 12.11 *Enable Headers, Footers, and Grand totals*

28. Click OK to return to the Schedule view window and study the results (see Figure 12.12).

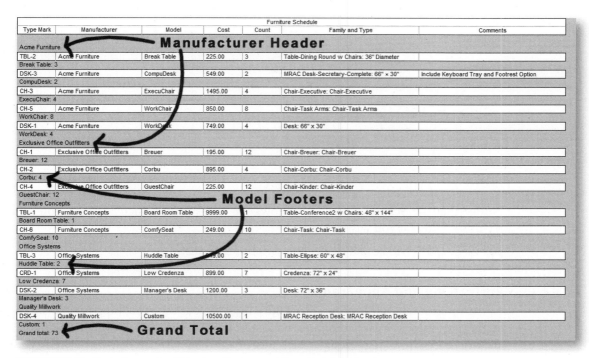

FIGURE 12.12 *Study the results of Headers, Footers, and Grand totals in the view*

Several other combinations are possible. You can change either Manufacturer or Model to display Headers or Footers or both. You also have control over whether it shows totals, counts, and/or titles. The Grand total can be turned on or off. Notice, with these items enabled, that the information in the Count field is now somewhat redundant. You can return to the Fields tab and remove the Count field from the Schedule if you wish. Finally, if you wish to put a bit of space between each sort criterion, you can use the "Blank line" checkbox. In some cases, this will make your subtotals easier to read, particularly when there are many items in the Schedule. If you would like to see for yourself, edit the properties once more and place a checkmark in the "Blank line" checkbox for the Manufacturer sort criterion only.

The Filter Tab

So far we have explored the high level of flexibility in our two Schedules from just the Fields and Sorting/Grouping tabs. There are other settings that we can apply in the remaining three tabs of the "Schedule Properties" dialog. For example, sometimes you only need part of the information available. Suppose you were on the phone with your sales representative from Office Systems. You could create a version of your furniture Schedule that listed only the items that you are specifying from this manufacturer.

29. Right-click the *Furniture Schedule* on Project Browser and choose **Duplicate View >
Duplicate**.

   - Rename it **Furniture - Office Systems**.
   - On the Properties palette, click the Edit button next to "Filter."
   - From the "Filter by" list, choose **Manufacturer**.

About a dozen operations are possible: Equal to, begins with, etc.

   - Leave the next list set to "equals."
   - From the list below Manufacturer, choose: **Office Systems** and then click OK (see Figure 12.13).

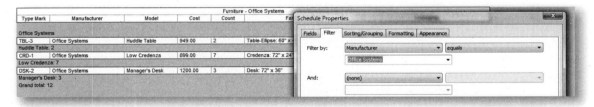

FIGURE 12.13 *Set the Schedule to Filter by a particular Manufacturer*

Notice that the Schedule now only shows the items from Office Systems. You can try other Filters if you wish.

30. Close this Schedule and reopen the *Furniture Schedule*.

Formatting and Appearance

In the interest of completeness, let's take a brief look at the remaining two tabs. The settings on these tabs are straightforward. On the Formatting tab you can configure the orientation and alignment of each individual field. On the Appearance tab, you configure the look of the overall Schedule.

31. On the Properties palette, click the Edit button next to Formatting.

32. Click on the Cost field in the list at the left.

    - Change the "Alignment" to **Right**.
    - Place a checkmark in the "Calculate totals" checkbox.

33. Click on the Family and Type field in the list at the left.

    - Place a checkmark in the "Hidden field" checkbox.

34. Click on the Count field in the list at the left.

    - Change the "Header orientation" to **Vertical** and "Alignment" to **Center** (see Figure 12.14).

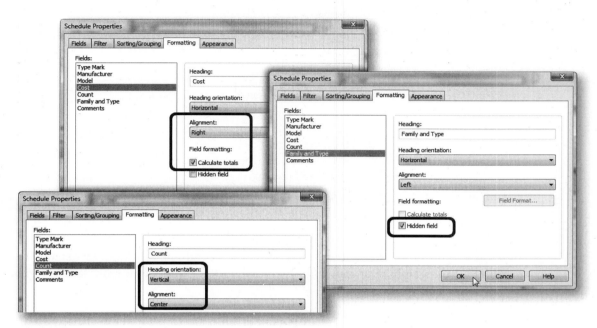

FIGURE 12.14 *Set the Formatting options of several fields*

35. Click OK to return to the Schedule view window.

Notice that most of the formatting shows immediately in the view window. However, the changing of the "Header orientation" to Vertical for Count does not. It will only appear when we add this Schedule to a Sheet.

Add a Schedule to a Sheet

To see the vertical orientation of the Count field and to see the effects of most of the settings on the Appearance tab, we need to add the Schedule to a Sheet. You add a Schedule view to a Sheet the same way as any other view.

36. On the Project Browser, right-click the Sheets branch and choose **New Sheet**.

    - Choose the E1 30 x 42 [A1 Metric] title block and then click OK.
      The new Sheet will appear as G101 – Unnamed and should open automatically.
    - Rename the Sheet: **A603 – Schedules**.

We have two Schedule Sheets (A601 and A602) but they have several Schedules on them already. Creating a new Sheet will give us plenty of room to work.

37. From Project Browser, drag *Furniture Schedule* and drop it on the Sheet.

   - Click a point near the top left corner of the Sheet to place it.

Notice that with the Schedule still selected, there are control handles at the top of each column. You can use these to interactively change the width of each field (see Figure 12.15). Notice also that you can see that the Count field header is in fact oriented vertically.

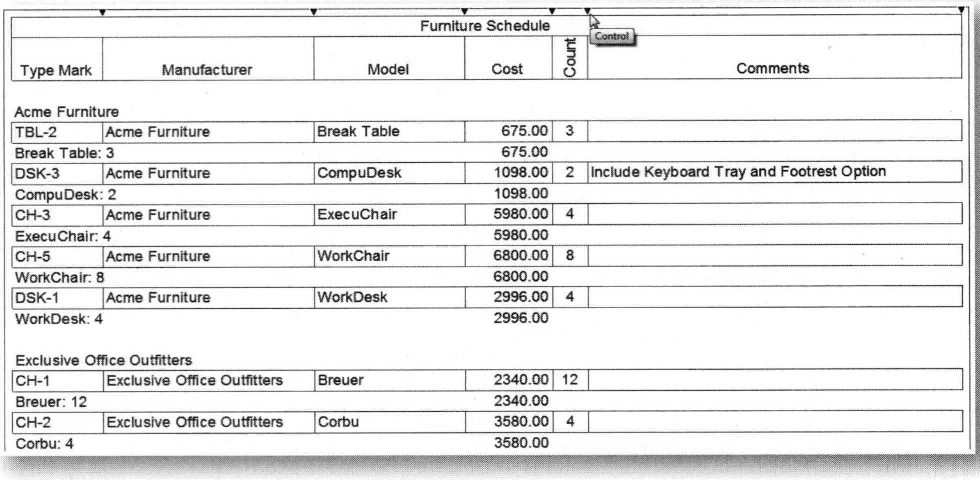

| Furniture Schedule | | | | | | |
Type Mark	Manufacturer	Model	Cost	Count		Comments
Acme Furniture						
TBL-2	Acme Furniture	Break Table	675.00	3		
Break Table: 3			675.00			
DSK-3	Acme Furniture	CompuDesk	1098.00	2		Include Keyboard Tray and Footrest Option
CompuDesk: 2			1098.00			
CH-3	Acme Furniture	ExecuChair	5980.00	4		
ExecuChair: 4			5980.00			
CH-5	Acme Furniture	WorkChair	6800.00	8		
WorkChair: 8			6800.00			
DSK-1	Acme Furniture	WorkDesk	2996.00	4		
WorkDesk: 4			2996.00			
Exclusive Office Outfitters						
CH-1	Exclusive Office Outfitters	Breuer	2340.00	12		
Breuer: 12			2340.00			
CH-2	Exclusive Office Outfitters	Corbu	3580.00	4		
Corbu: 4			3580.00			

FIGURE 12.15 *Edit the width of the fields interactively on the Sheet*

Since we created a new Sheet and have plenty of room, let's resize each of the field widths to prevent the data from wrapping to a second line as it does by default. To do this, you simply drag the control handles to the right. Work from left to right across the Schedule.

   - Resize each column as required to prevent wrapping.

38. On the Project Browser, select the *Furniture Schedule* in the list.

   You do not need to double-click or open it.

   - On the Properties palette, click the Edit button next to "Appearance."

If you wish to have the grid lines show continuously across the header and footer areas and in the other "white" spaces, check the "Grid in headers/footers/spacers" checkbox.

   - Place a checkmark in the "Outline" checkbox and then next to it, choose **Wide Lines** from the list.
   - Click OK to see the result (see Figure 12.16).

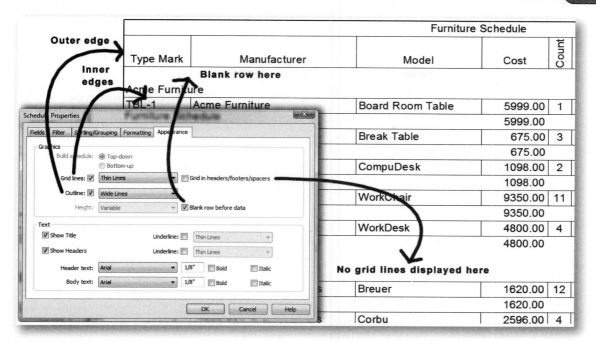

FIGURE 12.16 *Edits on the "Appearance" tab affect the look on the Sheet*

Return to the Appearance tab and make other edits if you wish to fine-tune the look of the Schedule on the Sheet. The figure shows that we can control the inner grid lines independently from the outline. Also, grid lines for headers and footers have a separate checkbox. The text settings should be self-explanatory. Notice that there are settings for the title, headers, and body text. You can change the fonts, heights, and linework drawn for each component.

> **Unlike graphical views, if you want to print a Schedule, you must drag it to a Sheet first and then print the Sheet.**

NOTE

Split a Long Schedule

In some cases, the data in the Schedule grows beyond the edge of the Sheet. You can break the Schedule into pieces to make it fit the Sheet better. (An example occurs on the existing *A601 – Schedules* Sheet).

39. On the Project Browser, double-click to open the *A601 - Schedules* Sheet view.

The Wall Schedule and possibly the Window Schedule overrun the bottom of the Sheet. We can split it to make the Sheet legible.

40. Select the Wall Schedule on the Sheet.

    • On the right side is a small control handle (looks like a "Z"); click this control to split the Schedule.

Using the other control handles that appear, you can make additional adjustments as necessary. The Schedule can be split multiple times if required. To remove a split and put the two pieces back together, simply drag and drop one piece on top of another (see Figure 12.17).

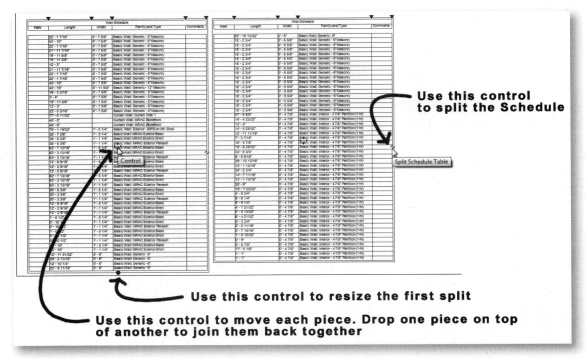

FIGURE 12.17 *Split the Schedule and make other adjustments if needed*

If after splitting a Schedule it is still too long for the Sheet, you can use the Filter tab to edit the Schedule view. Filter by a number range or by floor level to shorten the overall Schedule. Then duplicate this Schedule view and Filter the new copy to include the rest. Drag the two (or more Schedules) to different Sheets. For example, in a 20-story building, you could create a Schedule that included levels 1–10 and then duplicate it and change the filter to include levels 11–20.

  41. Save the project.

EDITING THE MODEL

Returning to our Furniture Schedule, let's see how it keeps up to date as we make modifications to the model.

Modify Elements in the Model

Locate the conference room in the third floor plan (bottom middle). The current table has 12 chairs around it. However, the room looks like it could accommodate a larger table. Let's take a minute to understand how the conference table Family is constructed.

  1. Close all views except *Level 3* and the *Furniture Schedule,* and then Tile them.

- In the *Level 3* view, zoom in on the conference table.
- Place your mouse over the conference table, note the way it pre-highlights, and note the name that appears in the Status Bar.

    The conference table is a Family: Type named: ***Table-Conference2 w Chairs: 48" × 144" [Table - Conference w Chairs: 1200 × 4500mm]***.

- Place your mouse over one of the Chairs.

The table will continue to pre-highlight.

- Press TAB.

Notice how the chair now pre-highlights independently.

In Chapter 10, we learned about nested Families when building the coat rack with nested coat hooks. This conference table Family and its chairs are another similar example of a nested Family. The difference here is that the nested chairs are set to be "Shared" Families. When you make a nested Family shared, it is part of the host Family but can also be selected and scheduled independently. Another way to see this is to select the chairs in the Schedule.

2. Locate the conference chairs (the quantity is 10) in the Schedule and click that row.

   The 10 conference chairs should highlight in the plan (see Figure 12.18).

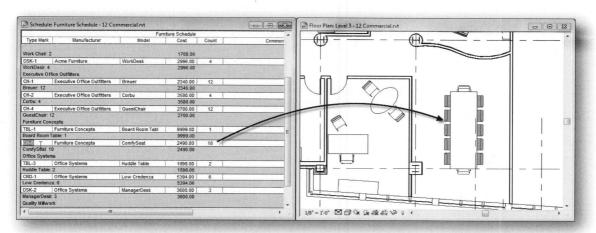

FIGURE 12.18 *Nested "Shared" Families can be selected and scheduled independently*

3. Select the conference table in the plan view.

   - From the Type Selector, choose the next larger size - ***Table-Conference2 w Chairs: 54" × 168" [Table - Conference w Chairs: 1200 × 5100mm]***.
   - Move it if necessary to fit the room better.

     Notice the change in quantity of chairs both in plan and in schedule (see Figure 12.19).

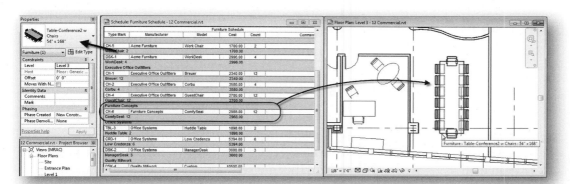

FIGURE 12.19 *When choosing a different conference table, the chair count immediately updates*

While the chairs took care of themselves, the conference table needs attention. The new type we selected does not have values for Manufacturer, Cost, and the other type-based fields. You can see this in Schedule because an empty field appears at the top. We can certainly edit the values in the Schedule as we did previously. However, we can also edit from the model.

Editing Schedule Parameters from the Model

We edited all of the Manufacturer and other identity data from the Schedule. While this is typically the most efficient way to make such edits, it is possible to edit these data fields using the same methods that we have employed when editing other types. Namely we can select an object and edit its Type Properties, or we can right-click the item from the Families branch of the Project Browser.

4. On the Project Browser, expand the *Families* branch and then the *Furniture* branch.

Beneath the Furniture branch, all of the Furniture Families used in the project are listed. ***Chair-Task*** [***M_Chair-Task***] is the chair that repeats along the sides of the conference table. ***Chair-Task Arms*** [***M_Chair-Task (Arms)***] is used at the ends. The conference table is ***Table-Conference2 w Chairs*** [***Table - Conference w Chairs***].

5. Beneath the ***Table-Conference2 w Chairs*** [***Table - Conference w Chairs***] Family, right-click on the ***54" × 168"*** [***1200 × 3050mm***] type and choose Type Properties.

   • Input values for the Manufacturer, Model, and Cost (see Figure 12.20).

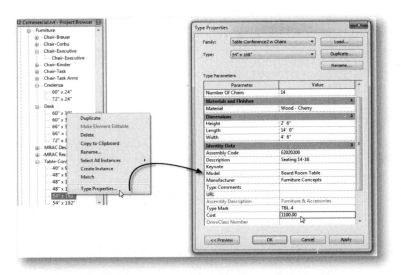

FIGURE 12.20 *Edit the type parameters of the new conference table*

6. Click OK to apply the change and update the model and Schedule.

Using Schedule Keys to Speed Input

In the "Add a Schedule View" topic, Schedule Keys were briefly discussed. Schedule Keys allow you to define a named collection of parameter values that you can add to an item in a Schedule in one step. They are very useful in Door, Window, and Room Schedules where there are many instance-based parameters. Let's return to our predefined Door Schedule for a brief demonstration of this powerful feature.

7. On the View tab, on the Create panel, click the ***Schedules*** drop-down button and choose the ***Schedule/Quantities*** tool.

   • In the "New Schedule" dialog, choose Doors for the category, change the Name to **Door Frame Style Schedule**, and then select the "Schedule Keys" radio button.

   • Change the name to **Frame Style** and then click OK (see Figure 12.21).

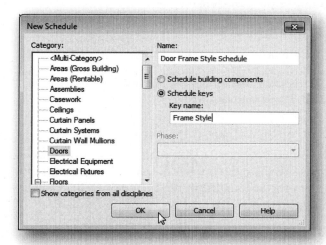

FIGURE 12.21 *Create a new Door Frame Style Schedule Key*

Notice that the Available Fields list includes all of the instance parameters for Doors. Key Name is already added on the right. This field is required.

8. Add the five frame fields: "Frame Type," "Frame Material," "Frame Finish," "Jamb," and "Head."

   • Click OK to finish.

   • Adjust the widths of the columns in the Schedule window.

9. On the Modify Schedule/Quantities ribbon tab, click the ***New*** button.

   • Input appropriate values in the first three fields. Leave the Jamb and Head fields blank for now.

   • Add at least one additional Row. Add more if you like (see Figure 12.22).

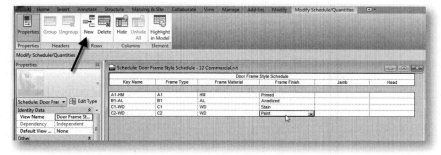

FIGURE 12.22 *Add Rows and input standard values*

The Key Name is any unique identifier you like. In other Schedules or when editing element properties, when you choose this key, the other five values will fill in automatically. We'll see this next.

10. On the Project Browser, right-click the Door Schedule and choose **Duplicate View >**
   Duplicate.

   • Rename the duplicate **Working - Door Schedule**.

11. On the Properties palette, *Working - Door Schedule* click Edit for Fields.

   • Add the new "Frame Style" field and move it up to just before Frame Type.

| TIP | Select Hardware on the right before you click Add. The new field will insert below the selected one. You can always use the Move Up and Move Down buttons if necessary. |

   • Click OK.

Your duplicated Door Schedule should now have a Frame Style column. The value currently reads "(none)" for all rows. All the setup is now complete. To use the Frame Styles, we simply choose them from the pop-up list in the Frame Style field and all the corresponding values will fill in automatically.

   Close, open, and tile views as desired to assist in editing the Schedule.

12. Select any Door and choose a Frame Style (see Figure 12.23).

12 Commercial.rvt - Schedule: Door Schedule (Working)

Door Schedule (Working)

Door Number	Type	Width	Height	Thickness	Material	Finish	Under Cut	Fire Rating	Hardware	Frame Style	Type	Material	Finish	Jamb	Head	Comments
		Door									**Frame**					
3002	1	3' - 0"	7' - 0"	0' - 2"						(none)						
3003	1	3' - 0"	7' - 0"	0' - 2"						(none)						
3004	1	3' - 0"	7' - 0"	0' - 2"						(none)						
3005	1	3' - 0"	7' - 0"	0' - 2"						(none)						
3006	8	3' - 6"	8' - 0"	0' - 1"						(none)						
3007	8	3' - 6"	8' - 0"	0' - 1"						(none)						
3101	11	6' - 0"	7' - 0"	0' - 2"						B1-AL	B1	AL	Anodized			
3102	4	2' - 6"	7' - 0"	0' - 2"						C2-WD	C2	WD	Paint			
3103	4	2' - 6"	7' - 0"	0' - 2"						C1-WD	C1	WD	Stain			
3104	4	2' - 6"	7' - 0"	0' - 2"						C1-WD	C1	WD	Stain			
3105	4	2' - 6"	7' - 0"	0' - 2"						A1-HM	A1	HM	Primed			
3106	4	2' - 6"	7' - 0"	0' - 2"						(none)						
3107	4	2' - 6"	7' - 0"	0' - 2"						A1-HM						
3108	4	2' - 6"	7' - 0"	0' - 2"						B1-AL						
3108A	14	3' - 0"	7' - 0"	0' - 1 1/2"						C1-WD						
3109	4	2' - 6"	7' - 0"	0' - 2"						C2-WD						
3109A	14	3' - 0"	7' - 0"	0' - 1 1/2"						(none)						
3110	4	2' - 6"	7' - 0"	0' - 2"						(none)						
3111	4	2' - 6"	7' - 0"	0' - 2"						(none)						
4001	1	3' - 0"	7' - 0"	0' - 2"						(none)						
4002	1	3' - 0"	7' - 0"	0' - 2"						(none)						
4003	1	3' - 0"	7' - 0"	0' - 2"						(none)						
4004	1	3' - 0"	7' - 0"	0' - 2"						(none)						
4005	1	3' - 0"	7' - 0"	0' - 2"						(none)						
4006	8	3' - 6"	7' - 0"	0' - 1"						(none)						
4007	8	3' - 6"	8' - 0"	0' - 1"						(none)						
5001	1	3' - 0"	7' - 0"	0' - 2"						(none)						
5002	1	3' - 0"	7' - 0"	0' - 2"						(none)						
5003	1	3' - 0"	7' - 0"	0' - 2"						(none)						
5004	1	3' - 0"	7' - 0"	0' - 2"						(none)						
5005	1	3' - 0"	7' - 0"	0' - 2"						(none)						

FIGURE 12.23 *Choosing Frame Styles automatically fills in the pre-assigned values*

   • Repeat for as many Doors as you wish.

Recall that our *Door Frame Style Schedule* is also controlling the Jamb and Head detail fields. However, at this time, we don't know which Sheet will have those details. A wonderful benefit of this process is that you can return to the *Door Frame Style Schedule* at any time and fill in the detail designation. The change will be reflected immediately by all Doors using that Frame Style.

13. On the Project Browser, open the Door Frame Style Schedule.

   • Edit one or more Jamb and/or Head detail entries. Input something like **1/A501** or **2/A501** (see Figure 12.24).

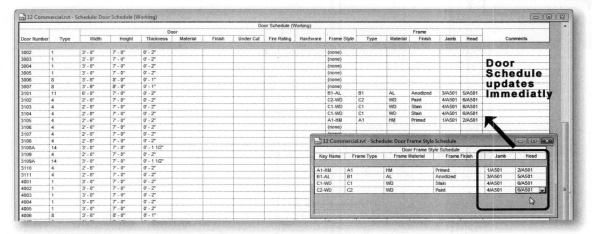

FIGURE 12.24 *Edit the Key Schedule later and the other Schedules referencing it update immediately*

TIP

To apply your Frame Styles even faster, drag through several items in the Schedule to select the corresponding Doors. Switch to the floor plan view, and then on the Properties palette you can choose a Frame Style that will in turn apply to all Doors in the selection.

You can make another Schedule Key for the remaining Door instance properties. While it is possible to make a single Schedule Key control all the properties, having two Keys gives more flexibility to mix and match door types and frame types. For example, a Door of type A-WD could in one instance use a frame type A1-HM and in another use C2-WD.

TIP

Make your Door Schedule easier to read. Using the procedure covered in the "Headers, Footers, and Grand Totals" topic, edit Fields, add the Level field. In Sorting/Grouping, sort first by Level with a Header, and then by Mark (no header or footer for Mark). Finally, on the **Formatting** tab, make the Level field a Hidden Field.

You can keep working in the Door Schedule if you wish. As noted, another Key Schedule will help you quickly fill in the "Door" columns. The Fire Rating column is a type-based parameter, so when you edit it, Revit will warn you that the value applies to all elements of the same type. Finally, if there are some Doors on the Schedule that should not be included, you can apply a filter to your Schedule as we did in the "The Filter Tab" topic above to exclude them (for example, if you did not want to include the Curtain Wall Doors at the front facade).

To add the filter, simply edit each of the Doors you want to exclude and add some text to the Comments field. It can be anything, even "Don't Schedule." After adding this text, edit the Properties of the Schedule; click the edit button for Filter, filter by Comments, change to **does not equal**, and then type "Don't Schedule." Click OK twice to view the results. The filter is case-sensitive, so type it exactly the same each time. The only trouble with the method is that it is manual. You need to add the filtering comment to each Door. However, you can select one such Door in a plan view, right-click and choose **Select All Instances > In Entire Project**. With them all selected, add the comment on the Properties palette. You could also filter by a Type property like Type Comments instead. This might make it easier to manage.

MANAGER NOTE

The need to occasionally apply a filter highlights one of the "dangers" of using online or manufacturer's content. You never know exactly what you will get and you cannot assume that all Families are created equally. Frankly, the same is true even with some of the out-of-the-box content. In several places in this book you have been encouraged to explore content from various sources, but also to open it, edit it and try to understand how it is built. Review the procedures covered in Chapter 10 for working with existing Families and just remember that you should always analyze any outside content and perform the required "clean-up" steps on it before incorporating it into a project or your office standard. Also, consider downloading, reading, and following the procedures outlined in the *Revit Model Content Style Guide* available at seek.autodesk.com. This document makes many best practice recommendations for building Revit content and is a prerequisite for any manufacturer who posts content to Seek.

14. Save the project.

WORKING WITH TAGS

Tags serve an important role in Architectural documentation. They provide a means of easily and uniquely locating and identifying elements in a project. All sorts of tags are needed in construction document sets: Room Tags, Door Tags, Window Tags, etc. Revit makes adding and managing Tags simple. Furthermore, the data displayed in the Tags comes directly from the objects in the same way as the data that feeds Schedules. Therefore, one need only add Tags to a view and let Revit handle the rest.

Adding Door Tags

We have spent some time on our Door Schedule, but so far, we have not added any Door Tags. Adding Tags to existing elements in the model is simple.

1. Continue in the *Level 3* floor plan.

- On the Annotate tab, on the Tag panel, click the ***Tag by Category*** tool.
- Slowly move the pointer around the model pausing over different kinds of elements. Do not click yet (see Figure 12.25).

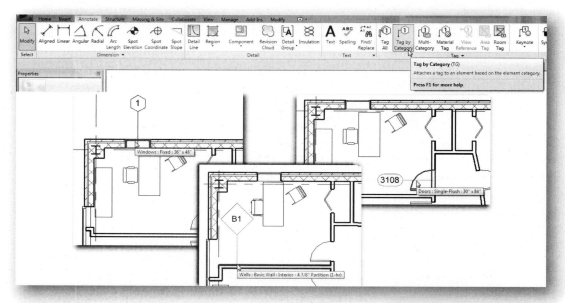

FIGURE 12.25 *Tag by Category automatically uses the correct Tag for the element you select*

Notice that the shape of the tag on your cursor changes with each element that you pre-highlight. Your project can have a different Tag loaded for nearly every type of element. Therefore, if you pre-highlight a Wall, it will automatically place a Wall Tag—pre-highlight a Door and you get a Door Tag and so on. You can place Tags with or without a leader attached; the default (on the Options Bar) is with a leader. If an element pre-highlights but no Tag appears, it means that this type of Tag has not been loaded into your project yet. If you click such an object, you will be prompted to load an appropriate Tag.

- On the Options Bar, clear the "Leader" checkbox.
- Pre-highlight and then click on one of the Doors to the offices at the left.

Notice that when the Tag appears, it already has a number in it. The Doors were numbered automatically when they were added to the model. If you followed the suggestion in the "View Existing Schedules" topic, then you have already renumbered the Doors more logically. The tag simply queries the element (the Door in this case) for the instance mark value and then displays the value in the tag. (You can create custom Tags that query and display nearly any parameter in an element.) You can continue placing Tags manually using this tool and just by clicking on them. However, there is a quicker method.

2. Cancel the command and deselect the object.

- On the Annotate tab, on the Tag panel, click the ***Tag All*** tool.
- In the "Tag All Not Tagged" dialog, click on Door Tags and then click OK (see Figure 12.26).

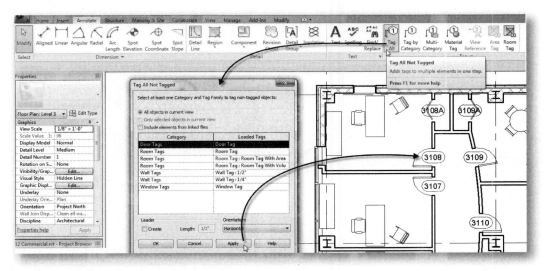

FIGURE 12.26 *Tag all Doors that are not currently Tagged*

Door Tags will appear on all Doors in the currently active view.

Loading Tags

Sometimes you want to Tag elements in your model, but an appropriate Tag is not loaded in your project. Like other Families, you can simply load a Tag from a library. Let's load a furniture Tag and then Tag some of our furniture. The Tags panel on the ribbon is an expandable panel. This is indicated by the small drop-down arrow on the panel title bar. Click here to access additional Tag-related tools.

3. Click on the Tags panel titlebar.

The panel will expand revealing additional tools.

- Click the **Loaded Tags** tool.

Scroll through the list. All Tag types are listed. Tags that are already loaded are listed next to their respective category. Element categories with no Tag loaded will be indicated.

4. Click the Load button.

- Browse to your library (Imperial or Metric) and open the *Annotations* folder.
- Open the *Architectural* folder, select *Furniture Tag.rfa* [*M_Furniture Tag.rfa*], and then click Open.

- Click OK to dismiss the "Tags" dialog.

5. On the Tags panel, click the **Tag by Category** tool.

- On the Options Bar, place a checkmark in the "Leader" checkbox.
- In the text box, type **3/8"** [**9**].
- Tag some of the chairs on the right in the reception area (see the left side of Figure 12.27).

If your Tag appears blank, you need to return to the "Duplicating a Schedule View" topic above and add values to the Type Mark fields for each piece of furniture. If you did complete this step, your Tag will appear with your Type Mark code already displayed.

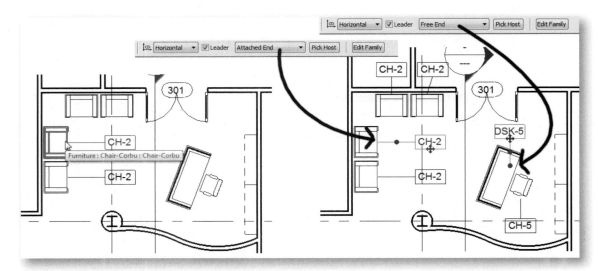

FIGURE 12.27 *Add Tags to the furniture in the reception space*

After you place a Tag, you can click on it and use the four-sided arrow control handle to move it and the round handles to add elbows to the leader. If you want complete control over the leader, select the Tag and on the Options Bar, change the Leader type to Free End. This gives you an additional control handle at the end of the leader allowing you to drag it freely (see the right side of Figure 12.27).

Create a Furniture Plan View

You do not have to Tag all pieces of furniture in this view. In fact, this might be a good time to create a separate Furniture Plan for the third floor and perhaps even create an enlarged "Typical Office Furniture Layout" view. Doing so at a larger scale would eliminate many redundant tags in this view and allow us more room to place the Tags. At this point in the text, you should have the skills required to complete this task on your own. Therefore, we will review the overall steps only.

1. On Project Browser, right-click the *Level 3* plan and choose **Duplicate View >** **Duplicate with Detailing**.

This command creates a copy of the view including all the Tags we have added so far. Duplicate would create a view of just the model geometry without any annotations.

2. Rename this plan **Level 3 Furniture Plan**.
3. In the original *Level 3* plan, delete the Furniture Tags added so far.

Make a window around all the tags added, click the Filter button, click Check None and then check only Furniture Tags and click OK. Delete the tags.	**T I P**

---

4. In the *Level 3 Furniture Plan*, delete all of the Door Tags.

The easiest way to do this is to make a selection box around the entire plan and then click the Filter selection button. Click Check None, and then check only the Door Tags box. Click OK to complete the filter, and then delete the Tags.

5. Tag all of the unique furniture in the view—reception, secretarial areas, conference room, etc.
6. Create a Callout plan at lager scale of one typical office—call it: **Typical Office Furniture Plan**.

   • On the View tab, on the Create panel, click the Callout button.
   • Drag two opposite corners around the area you want to callout.

7. Add Tags in the view to complete it (see Figure 12.28).

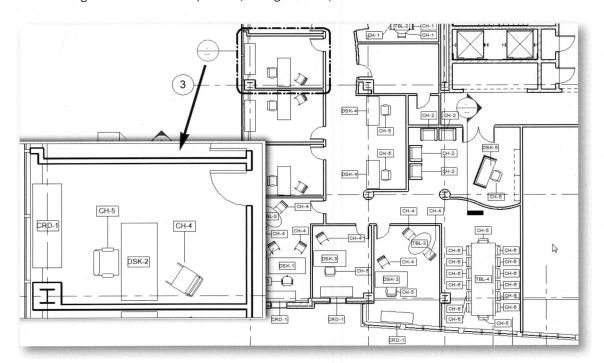

FIGURE 12.28 *Create a separate furniture plan and enlarged plan Callout view and finish adding Tags*

Editing View Visibility Graphics

Now that we have a separate furniture plan, we might not want to see the furniture in the *Level 3* plan view any more.

8. On the Project Browser, double-click to open the *Level 3* floor plan view.

   - On the keyboard, type **VG**. (You can also click the ***Visibility/Graphics*** tool on the Graphics panel of the View tab.)
   - Clear the checkbox next to "Furniture" (see Figure 12.29).

TIP

If you only wish to hide an element category as we are doing here, a shortcut is to select one of the elements, right-click, and choose **Hide in view > Category**. This will achieve the same result as opening the "Visibility/Graphics Overrides" dialog and un-checking the element. You can also find this tool on the ribbon when an element is selected.

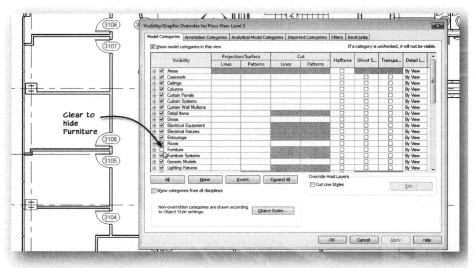

FIGURE 12.29 *Visibility/Graphic Overrides apply to the current view only*

Visibility/Graphic overrides apply to the current view only. This extensive dialog lists all Revit elements grouped into major categories shown on each of the tabs at the top of the dialog. Take note of the message at the bottom left corner of the dialog. This entire dialog and all of its settings is devoted to overriding the default graphical settings. Anything that is not overridden is drawn according to the settings in the Object Styles dialog. This dialog can be accessed directly from Visibility/Graphic Overrides by clicking the Object Styles button. Object Styles can also be accessed on the Manage tab. In most firms, you can typically expect that the settings in Object Styles have been thoroughly configured in the project template ahead of time.

If you study just the Model Categories tab, you will see that in addition to being able to turn object categories on and off (using the checkbox) you can also override the line patterns and fills that each category uses. Furthermore, most categories have subcategories giving you even more fine control over visibility and graphic settings.

   - Move the dialog out of the way so that you can see the plan in the background. Click the Apply button.

     Notice that the furniture is now hidden in this view.

Perhaps you would like to see the furniture after all, but make it less prominent.

- Check the box next to Furniture again to turn it back on and then place a check-mark in the Halftone column as well.
- Click Apply to see the result (see Figure 12.30).

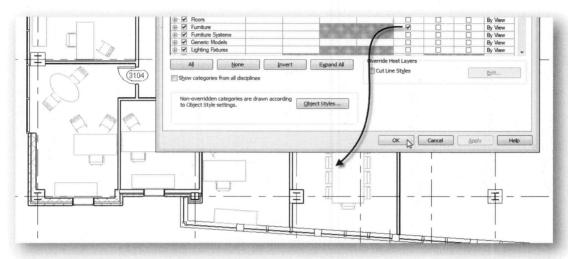

FIGURE 12.30 *Display the Furniture as Halftone*

While we will not go into detail on each item of the "Visibility/Graphic Overrides" dialog, a few additional tips are worth mention. If you wish to see all of the categories and their subcategories, click the Expand All button. This will expand the entire list in one step. Any item can be overridden. For example, if you wanted to change the line weight, color, or pattern of the furniture, you select the Furniture item and then click one of the Override buttons that appear. A dialog will appear with additional override options (see Figure 12.31).

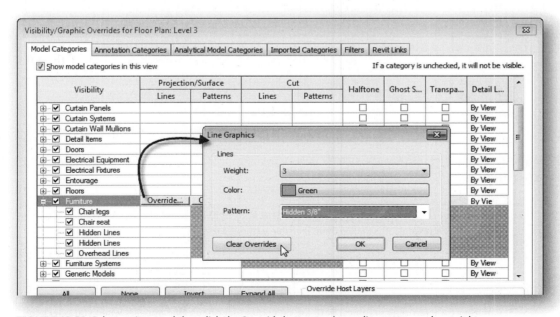

FIGURE 12.31 *Select an item and then click the Override button to change line pattern, color, weight, or fill patterns (not shown)*

You can make similar edits on the Annotation Categories tab. However, since annotation elements are simpler, there are fewer options to edit. If you have imported DWG files, you will be able to turn on and off the layers in those files on the Imported Categories tab. The Revit Links tab gives extensive control over the visibility of a linked Revit file. Unfortunately, we cannot explore this or the Filters tab in detail at this time. If you wish to learn more about any of the settings in this dialog and each of its tabs, please consult the online help.

9. Be sure that the settings for Furniture meet your preferences and then click OK.

The "Visibility/Graphic Overrides" dialog allows us to manipulate the display at the category level. In some cases, you will wish to modify the display of individual elements independently of their category. For example, in Chapter 10, we built a binder bin Family for the reception area and used the Specialty Equipment template to do so. As a result, the binder bins in the reception area did not respond to the display modifications that we just made because they are not furniture. We could edit the Family and change its category to Furniture, but this would be a bit of a drastic way to solve the problem. Instead, we can use element-level overrides.

10. In the Reception area, select both binder bins.

- On the ribbon, on the View tab, click the ***Override Graphics in View*** drop-down button and choose **Override by Element**.

  If you halftoned the furniture, do the following:

- Place a checkmark in the Halftone box and then click OK.

  If you turned off the furniture, do the following:

- Remove the checkmark from the Visible box and then click OK (see Figure 12.32).

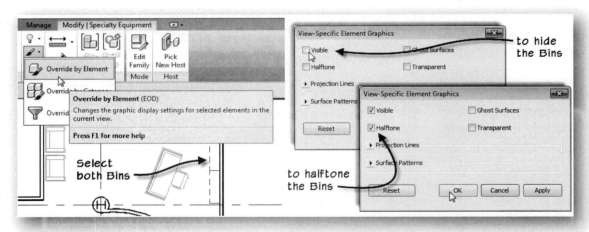

FIGURE 12.32 *Halftone or turn off individual elements*

TIP	If you want to turn the elements off, a shortcut is to right-click and choose the **Hide in view >** **Elements** command. This makes them invisible without a dialog.

TIP	When an element(s) is selected, you can also find a button on the ribbon for hiding elements and another for overriding graphics. Both are on contextual ribbon that appears when you select the object on the View panel. Finally, the shortcut to hide elements is **EH** and to hide categories is **VH**.

11. Save the project.

ROOMS AND ROOM TAGS

A Room element is a non-graphic selectable element in your model used to represent an actual room or other defined space in your project. Rooms are used for the obvious spaces that you would consider rooms like offices, kitchens, corridors, closets, and lobbies. However, you can define a Room for any program space you wish to label or quantify as such. Rooms are considered model elements even though they do not display graphically by default. Rooms are useful to generate room Schedules, reports, perform area calculations, volume calculations, finish Schedules, and of course for Room Tags. Rooms are used for any occupied space. In most instances, Rooms can be created automatically from the bounding Walls and other elements already in your model. However, not all geometry can "bound" a Room, and in open plan situations Revit will be unable to correctly determine the proper shape and boundary of the Room without a little manual intervention. There are also situations where the shape of the Room you wish to define has no Walls or other "hard" boundaries. This may be the case in open plan layouts or outdoor spaces. In these cases, you can manually sketch Room Separations. We'll explore both techniques here.

Continue in the *Level 3* floor plan view.

1. On the Home tab, on the Room & Area panel, click the **Room** tool.

Study the options on the ribbon, the Options Bar, and the Properties palette (see Figure 12.33). Many options should be familiar from working with other tools. The first option is Tag on placement. While we previously turned this off when placing Doors and Windows, here we will leave it on.

- **Type Selector**—Use this drop-down to choose from a list of loaded Room Tags.
- **Upper Limit** and **Offset**—You can configure these settings on either the Options Bar or the Properties palette. These settings are used to determine the height of a Room. This will come into play if you enable volume calculations.

NOTE

Volume calculations can be enabled by expanding the Rooms & Areas panel of the Home tab, then clicking the Area and Volume Computations button. However there is a performance penalty for doing this. Best practices are to leave the calculations set to Areas only until and if you need to perform volume calculations for energy analysis or to export a gbXML file to Green Building Studio, Ecotect, or some other analysis application, then set it back to Areas only when done.

- **Dimensions**—On the Properties palette, these values will appear "read only" (grayed out) and will report the actual area and perimeter of the Rooms as you place them or later select them.
- **Room**—Rooms behave differently than other model elements in a few ways. You can add Rooms to the model graphically in a plan view or you can add them in a Schedule view. Adding Rooms to the model's database is *not* the same as placing them physically in the model. In some cases, it may be useful to create several Rooms based on a building program before the plans have been laid out or the Rooms are actually placed. This is accomplished in a Schedule using the **Add** (Rows) tool (similar to adding rows to the Key Schedules above). If you have previously added Rooms to a Schedule but have not yet placed them in a plan, they will appear in this list. If there are no unplaced Rooms in your model, the only choice here will be "New," which will create and place a new Room in a single step.

- **Highlight Boundaries**—This button highlights all objects on screen that can bound Rooms. This gives you a quick way to determine if adding bounded Rooms will be successful. It is possible to create a Room without boundaries, but you will not be able to query area, perimeter, or volume calculations from it.
- **Tag on Placement**—Toggle this button on to add a Room Tag as Rooms are added to the project. Even if you toggle this off, you can still add Tags to Rooms later using the methods covered above.
- **Tag Orientation**—When "Tag on Placement" is selected, this option sets the orientation of the Tag as either horizontal or vertical.
- **Leader**—Check this to add a Leader to your Tags as they are placed.

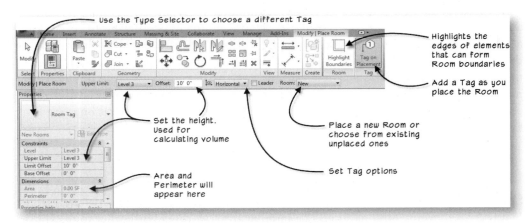

FIGURE 12.33 *The Place Room tab and Options Bar for adding Rooms*

- Zoom In Region around the two spaces to the left of the Stair tower in the upper left corner of the plan.
- Move the cursor into the room at the corner of the plan (see Figure 12.34).

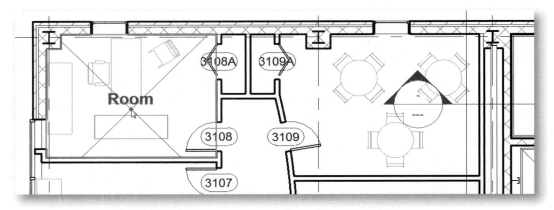

FIGURE 12.34 *A Reference graphic (the "X") appears as you are placing the Room*

- Click in the space to Create the Room and its Tag.
- Move into the space across the hall to the right and then click again.

Notice that Revit attempts to align the Tags as you place Rooms. While the Room tool remains active, the existing Rooms will remain highlighted with blue shading on screen. This will make it easier to place the remaining ones while keeping track of the ones already placed. Also, note that the first Room was automatically numbered as "1" and the next one sequentially numbered "2." The problem with this is that you would likely want the reception space to be the first number and then number sequentially from there. Not surprisingly, you can renumber the Rooms at any time and their Tags will automatically update. However, knowing that Revit will number sequentially and using a little strategy as we place Rooms can save us the effort of renumbering in many situations.

2. Zoom out and move the mouse into the reception space. Don't click yet. Notice how the corridor flows into the reception, secretarial spaces, and conference room. In this open plan configuration, we need to create some manual boundaries before we can add proper Rooms.

   • On the ribbon, click the ***Modify*** tool or press the ESC key twice.

3. Click on the drop-down button on the Room tool (lower half) and choose the ***Room Separation Line*** tool.

   • Sketch a horizontal line segment above column line 2 to enclose the reception space.

   • Press ESC once, and then sketch a vertical line segment next to column line C to enclose the conference room.

   • Snap both of these to the endpoints to the existing Walls and then to the intersection with the Column.

   • To enclose the secretarial space, use a three-point arc and sketch it as indicated in Figure 12.35.

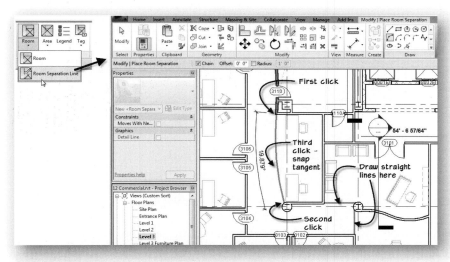

FIGURE 12.35 *Sketch Room Separation lines as indicated*

NOTE

Room Separation lines in the figure have been enhanced for clarity. You can make this edit if you wish. Choose **Line Styles** from the Additional Settings drop-down button on the Manage tab and edit the line weight and color of the <Room Separation> line style.

4. Test your Room Separation by returning to the ***Room*** tool and placing a Room in the reception area.

It should now give the proper shape Room. Revit automatically uses the next number in the sequence as you can see (3 in this case). If you want a different pattern to your numbering such as 3101, 3102, etc., edit the first number to 3101 and then return to the **Room** tool. It will pick up at 3102 and continue sequencing from there.

- On the ribbon, click the **Modify** tool or press the ESC key twice.

5. Click on the Room tag in the reception area.

The text label will turn blue to indicate that you can edit it.

- Click on the number 3 to edit it.
- Change the value to 3100 and then press ENTER.

6. On the Home tab, click the **Room** tool again.

- Add the remaining Rooms starting with the conference room and continuing clockwise around the tenant suite.
- When you get to Room 1 and 2, click Modify or press ESC.
- Renumber Room 1 and Room 2 to fit in sequence with the other Rooms.
- Continue adding the remaining Rooms, but skip the two closets at the top of the plan (see Figure 12.36).

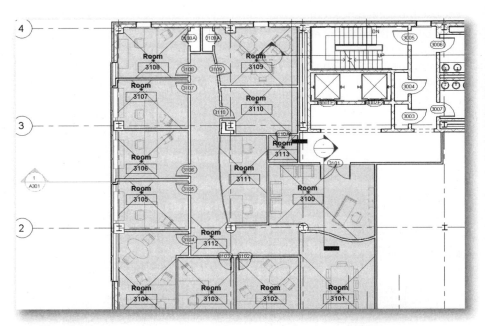

FIGURE 12.36 *After renumbering the reception space, add the remaining Rooms*

- On the ribbon, click the **Modify** tool or press the ESC key twice.

We could add Rooms to the two closets at the top of the plan if we wanted, but they do not really need to be treated as separate Rooms. Let's incorporate their area into the Rooms they are attached to.

7. Zoom in on Room 3109 (the break room to the left of the stair tower).
8. Select the small vertical Wall with the bifold Door between the room and the closet.

- On the Properties palette, clear the checkmark next to Room Bounding.
- Click Apply or simply shift focus back to the drawing window.
- Move your cursor around in the middle of Room 3109 until the Reference indicator (the "X") pre-highlights and then click (see Figure 12.37).

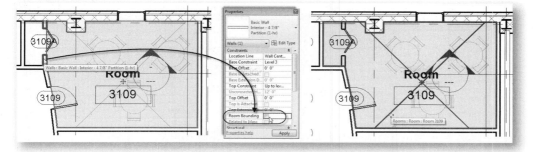

FIGURE 12.37 *After renumbering the reception space, add the remaining Rooms (continued)*

Notice that the Room now flows into the closet ignoring the closet Wall. You can repeat this process on the other office. To do this you would first have to split the long vertical Wall (use the Split tool on the Modify tab). Otherwise, the entire length of the Wall would become non-bounding, adversely affecting all of the offices below it. If you try toggle off Room Bounding for this Wall before splitting it, Revit will generate a warning about "Multiple Rooms in the same enclosed region." If you get such a warning, read it carefully and then click Cancel and check your Room Boundaries again carefully. Remember, you can highlight all eligible Room Bounding elements with the button on the ribbon when the Room tool is active. This will help you spot the problem areas.

If you are doing a lot of work with Rooms, you can turn on the Interior Fill and/or Reference (the "X") subcategories in the view to make them easier to see and select. These are both off by default and only display when pre-highlighting and selecting a Room as we have seen. If you decide to do this, type VG to open the Visibility/Graphic Overrides dialog, expand Rooms and check either or both of these items. When you are finished working with Rooms, be sure to turn them back off or they will print that way. As an alternative, you can duplicate your floor plan and create a "Working – Level 3" plan first; then display the subcategories there only.	**TIP**

Edit Room Names

To edit the Room names, simply click on the Tag, click on the blue name value, and then type in the new value. The process is the same as editing the numbers.

9. Click on the room label of the Tag in the reception area.

An editable text field should appear on screen.

- Type **Reception** and then press ENTER.
- Rename the Conference room next (it is directly below Reception).

You can continue to rename Rooms this way, but in the case where several Rooms have the same name, like the offices, there is a quicker way.

> Move your mouse around inside one of the offices and near the Tag.

As you move your mouse around inside a Room, you can make the reference indicator (the "X") pre-highlight. When you see this reference, you can click to select the Room. Once a Room is selected, you can edit it on the Properties palette the same as any other model element.

10. Pre-highlight and then select one of the offices.
11. Using the CTRL key, select all of the remaining offices (2 at the bottom, 1 in the bottom left corner, and 4 at the left).

- On the Properties palette, beneath for the Name type: **Office** and then click Apply (see Figure 12.38).

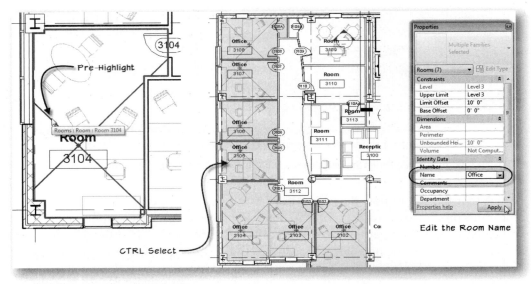

FIGURE 12.38 *Rename all of the Offices*

Viewing a Room Schedule

Back in Chapter 4, we added several Schedules to this project—among them a Room Schedule. At the start of this chapter, that Schedule was still empty. Let's take a look at that Schedule now to see how it reflects the data that we just added to the model.

12. On the Project Browser, double-click to open the *A601 - Schedules* Sheet view.

13. Zoom In Region on the Room Schedule (top right corner of the Sheet).

Earlier when we made adjustments to the Wall Schedule (also on this Sheet), the Room Schedule was empty. Now that we have added Rooms to our model, this Schedule has been filled in to list these Rooms. Notice that this is really a room finishes Schedule and that currently there is no finish information listed. We can add finish data directly in the Schedule as we did in the furniture Schedule, or we can select a Room and edit the Properties in the model. You cannot edit a view directly on a Sheet however. If it is a graphical view, right-click and choose **Activate View**. For a Schedule, right-click and choose **Edit Schedule**.

14. Select the Schedule on the Sheet, right-click, and choose **Edit Schedule**.
    This will open the Schedule view on screen.

> **NOTE** You can, of course, go directly to the Schedule on the Project Browser. It is not necessary to first open the Sheet; this is simply an alternative method.

15. Close or minimize all views except the *Room Schedule* view and the *Level 3* floor plan.

    • Tile the windows.

> **NOTE** For the next several steps, refer to Figure 12.39.

16. Select one of the Rooms whose name still reads "Room" in the Schedule.
    The Room will highlight in the plan.

    • Type an appropriate name in the Schedule.

    • Repeat until all Rooms are named.

For the utility space just above Reception, we will need to move the Tag outside the Room to make it more legible. However, if you try to move a Room Tag outside of its Room boundaries, a warning will appear. To do this, you must first enable the leader option on the Options Bar. Then you can safely move the Tag outside the Room.

17. Select the Utility Room Tag.

   • On the Options Bar, enable the Leader option, and then using the move control on the Tag, drag the Tag outside the space (see Figure 12.39).
   • With the Tag still selected, click the Edit Type button on the Properties palette.
   • For the Leader Arrowhead option, select an arrowhead of your choice.

18. On the Properties palette, edit the Room Schedule to sort by Number.

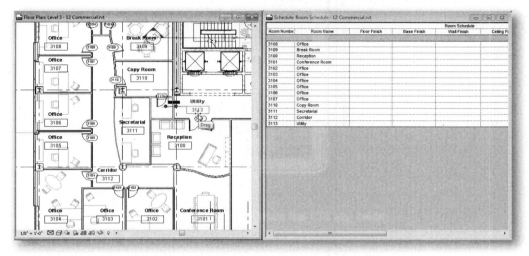

FIGURE 12.39 *Finish renaming, renumbering, and adjusting Tag locations*

Adding Room Finish Information

As noted previously, the Room Schedule is really a room finish schedule. Using the procedure covered in the "Using Schedule Keys to Speed Input" topic, create a Room Style Key Schedule that includes all of the finish fields: "Base Finish," "Ceiling Finish," "Ceiling Height," "Floor Finish," and "Wall Finish." Create a few key styles such as those shown in the top portion of Figure 12.40.

Once you have defined a few styles, you can add the Room Finish Style key field to your Room Schedule and use it to input the finishes quickly on the basis of room type. For example, the two large offices at the bottom of the plan could be assigned the Executive Office key, while all the others get the Standard Office key. Define and assign as many keys as you wish. Remember, if you change a value in the Schedule Key later, the Room Schedule will automatically reflect the change. You can use any of the methods discussed so far to make the selections and edits. Also, remember that you can select many objects in the plan and then use the Filter tool to remove everything but Rooms from the selection. This should make selection quicker for the Rooms with common settings. When you are finished, your Room Schedule should look something like Figure 12.40. Please notice in the figure that you can manually edit custom values in the fields as long as you leave the Room Style set to "none." If you prefer, you can define another Room Style, but in spaces that are unique like the reception or the conference room, it is just as easy to not assign a Room Style and instead assign the values of each finish directly.

Schedule: Room Finish Style Schedule - 12 Commercial.rvt

Room Finish Style Schedule					
Key Name	Base Finish	Floor Finish	Wall Finish	Ceiling Finish	Ceiling Height
Executive Office	WB-1	CPT-2	P-3	ACT-2	9' - 0"
Public Area	VB-2	CPT-3	P-2	ACT-1	9' - 0"
Standard Office	VB-1	CPT-1	P-1	ACT-1	9' - 0"

Schedule: Working - Room Schedule - 12 Commercial.rvt

Working - Room Schedule								
Room Numbe	Room Name	Room Finish	Floor Finish	Base Finish	Wall Finish	Ceiling Finish	Ceiling Height	Comments
3100	Reception	(none)	CPT-2	VB-1	P-3	ACT-2	10' - 0"	
3101	Conference Room	(none)	CPT-1	WB-1	P-2	Open	10' - 0"	
3102	Office	Executive	CPT-2	WB-1	P-3	ACT-2	9' - 0"	
3103	Office	Executive	CPT-2	WB-1	P-3	ACT-2	9' - 0"	
3104	Office	Executive	CPT-2	WB-1	P-3	ACT-2	9' - 0"	
3105	Office	Standard	CPT-1	VB-1	P-1	ACT-1	9' - 0"	
3106	Office	Standard	CPT-1	VB-1	P-1	ACT-1	9' - 0"	
3107	Office	Standard	CPT-1	VB-1	P-1	ACT-1	9' - 0"	
3108	Office	Standard	CPT-1	VB-1	P-1	ACT-1	9' - 0"	
3109	Break Room	Public Are	CPT-3	VB-2	P-2	ACT-1	9' - 0"	
3110	Copy Room	Public Are	CPT-3	VB-2	P-2	ACT-1	9' - 0"	
3111	Secretarial	Public Are	CPT-3	VB-2	P-2	ACT-1	9' - 0"	
3112	Corridor	Public Are	CPT-3	VB-2	P-2	ACT-1	9' - 0"	
3113	Utility	(none)	VCT-1	VB-1	P-1	ACT-1	9' - 0"	

FIGURE 12.40 *Create a Room Finish Key Schedule and then use it to input values for the finishes of all of the Office spaces*

Tagging Other Views

If you return to the furniture plan, you will note that it does not have Room Tags. At first this may seem strange, until we recall that annotation is always view-specific. This means that Tags and other annotation will appear *only* in the view to which it is added. However, it is very easy to add the Tags to any plan that needs them.

19. On the Annotate tab, click the ***Tag All*** tool.
    - Select the Room Tag item (first one).

Since there are already many furniture tags in this view, it might be useful to add leaders to the Room Tags so we can move them around.

- Place a checkmark in the Create checkbox beneath "Leader" at the bottom of the dialog and then click OK.

You should get a Tag in every Room. You will need to move them around to make the plan legible, but the Tag All tool does save considerable time over manually tagging. In some cases, you might find copy and paste quicker. If you have two plans that can accommodate tags in the same location, select all the Tags in the original plan (use Filter to select quickly), press Ctrl-C, and then use Paste Aligned (Modify tab) in the other view with the Current View option. This will copy all the Tags exactly as they appear in the first plan. You can then fine-tune placement by moving those that require it. Remember, if a Room Tag needs to move outside the boundary of its Room, enable the Leader option first.

20. Save the project.

QUERYING DATA

You should now be getting fairly comfortable with using Schedules. We have explored many examples and you are no doubt beginning to see just how powerful Schedules are. One of the biggest benefits to scheduling in Revit is all the robust data we can readily extract from our model. This data extraction is the "I" in BIM. While the potential data we can extract is virtually limitless, in this topic, we will focus on a couple very simple queries.

Reporting the Room in the Furniture Schedule

We have a detailed furniture Schedule and a detailed Room Schedule. While we have seen how easy it is to tile a Schedule and plan side by side and quickly find elements by selecting them in the Schedule, it might be useful to have Revit report directly in the Schedule in which Room each piece of furniture is located.

1. On the Project Browser, right-click the *Working - Furniture Schedule* view and choose **Duplicate View > Duplicate**.

    • Rename the view: **Furniture Location Schedule**.

2. On the Properties palette, edit the Fields.

    • At the bottom left corner of the Fields tab, click the pop-up menu beneath "Select available fields from" and choose **Room**.

Notice that all of the Room fields are now available to add to the Schedule. All we need is Room Name and Number, so we'll add just these.

    • Add Room: Name and Room: Number to the right side.

    • Use the Move Up and Move Down buttons to locate the fields where you want them and then click OK twice (see Figure 12.41).

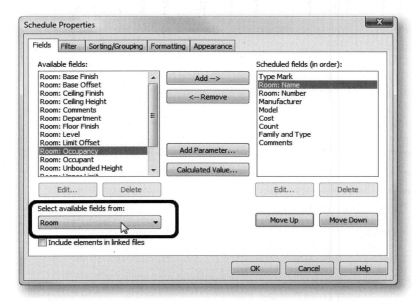

FIGURE 12.41 *Add fields from the Rooms category to the Furniture Schedule*

Room names and numbers now appear in the Schedule, but because of the way that we are currently sorting and grouping, the data appears incomplete. A quick modification on the Sorting/Grouping tab will take care of this.

3. Edit the Sorting/Grouping.

    • Change the first sort criterion to **Room: Name**. Enable the Header option and also the Blank Line option.

    • Change the second criterion to **Room: Number**. Enable the Header option.

    • Make sure that the third criterion is set to **Model**.

    • On the Formatting tab, make both the Room: Name and Room: Number fields hidden fields.

This removes the redundancy. When you use a field as a header, you can hide it as a column.

    • Click OK to see the result (see Figure 12.42).

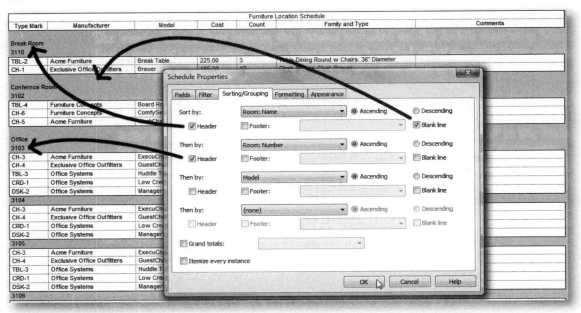

FIGURE 12.42 *Sort and group by Room fields*

There are many other ways that this Schedule can be sorted, grouped, or formatted. Please feel free to experiment further. Remember, if you are uncertain how a particular combination of settings will look, make a duplicate of the Schedule first.

Conditional Formatting

Another very useful way that you can highlight important data in a Schedule is through the use of conditional formatting. This allows you to shade cells in color that meet certain criteria that you designate.

4. On the Properties palette, click the Edit button next to Formatting.

- Select the Cost field and then on the right, click the Conditional Format button.
- For the Test, choose Between from the drop-down list.
- Type in a range of values such as **250** and **850**.
- Click the color swatch next to Background Color and choose a color.
- Click OK to dismiss the dialogs and see the change.

Notice how cost values that fall within this range highlight in the color you indicate. You can try other conditions for additional fields if you wish. This can be a powerful way to quickly locate certain data in a detailed Schedule.

Material Takeoffs

Most model elements have materials assigned to their components. These materials control how the components draw in various view types. For example, a brick material controls the cross hatching in plan and the brick pattern in elevation as well as the brick photo texture in rendering. In addition to the display characteristics, materials can also have identity data assigned to them. A Material Takeoff is a Schedule that can query the material properties assigned to model elements. For example, using a Material Takeoff, we can ask Revit for the volume of concrete used in our project. Like other schedules, we can limit the query to a single category like Walls, or we can do a multi-category takeoff from many element categories at once.

5. On the View tab, on the Create panel, click the ***Schedules*** drop-down button and choose the ***Material Takeoff*** tool.

   • In the "New Material Takeoff" dialog, accept the <Multi-Category> choice on the left, name the takeoff **Concrete Takeoff** and then click OK.

Scroll through the list of fields. Since we did a multi-category Schedule, the fields listed will be those that are common to *all* element types. Also, several Material fields will be available. These will all be prefixed with the word "Material:" in front of the field name.

   • Add the following fields: "Family and Type," "Material: Name," "Material: Description," and "Material: Volume."

   • Click OK to create the takeoff.

   • Double-click the divider between each header to widen all columns to the widest item in the list.

The resultant takeoff is very long and includes far more than concrete. Scroll through the entire list. Notice that our best chance for isolating just the concrete is the Material: Name column. Any of the Concrete materials use the word "Concrete" in the name. We do have to be careful however as we also have several materials named "Concrete Masonry Units," which we don't really want to include in the volume of concrete. Therefore, we need to be specific about the filter criterion we establish.

6. On the Properties palette, click the Edit button next to Filter.

   • For the first Filter by criterion, choose **Material: Name**.

Several operators are available. The default is "equals."

   • Change the operator to **begins with**.

   • In the text field, type **Concrete** (see Figure 12.43).

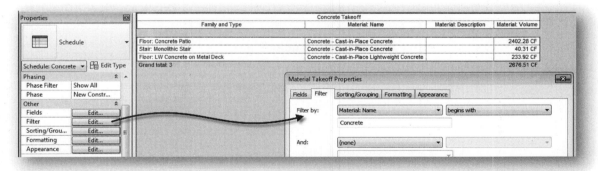

FIGURE 12.43 *Add filtering to limit the takeoff to just concrete materials*

Another approach that would yield the same result would be to filter by Material: Name containing "Cast-in-Place." Please note that, in either case, filters are case-sensitive. To fine-tune the takeoff further, edit the Formatting options. Set the alignment of the volume column to right and select the Calculate totals option. On the Sorting/Grouping tab, you have to turn on Grand totals for this to display (each of these enhancements is shown in the figure).

Filtering and totals work in any Schedule. The only difference between a Schedule and a Material Takeoff is the inclusion of the material fields in the Material Takeoff.

You can get pretty good raw data from your model directly in Revit Schedules. There is even the possibility of creating formula fields and custom parameters. Both of these are out of the scope of what we will cover here. However, even with this functionality, you might find that you need to process the raw data in a way that is not easy to accomplish directly in Revit. In such cases, you can export any Schedule to an Excel compatible file. This file is a static delimited text file (TXT) export, not linked back to the model. However, you can open such a file in Excel or any program that can import delimited text files and process the data in nearly limitless ways.

7. Open the Door Schedule (or any Schedule you wish to export).
8. From the Application menu, choose **Export ≥ Reports > Schedule**.

- If necessary, browse to the *Chapter12* folder.
- Accept the default file name and then click Save.
- In the "Export Schedule" dialog, accept all defaults and then click OK.

You can open the resultant file in Microsoft Excel or any other spreadsheet or database program. Please note that such exported reports do not remain linked to your Revit model. As your project progresses and changes are made, you will need to re-export the reports.

ADD A COLOR SCHEME

When working with Rooms (or Areas—see next topic) we noted how they only display on screen during creation and selection. In many cases, this is appropriate particularly in printing construction documents. There are times however when graphical indication of Rooms (or Areas) can be valuable, for example, when creating a programmatic color-coded diagram. Revit gives us this ability with Color Schemes.

Create a New View with Alternate Room Tags

Let's make a color-coded third floor plan.

1. On the Project Browser, right-click on *Level 3* and choose **Duplicate View > Duplicate with Detailing**.

- Rename the new view **Level 3 Color Plan**.

2. Make a window selection around the entire plan.

- On the ribbon, click the ***Filter*** button.
- In the "Filter" dialog, click the Check None button.
- Place a checkmark in only the Door Tags box and then click OK.
- Delete the selected Door Tags.

Remember, each view's annotation is unique to that view. We have only deleted the Door Tags in the *Level 3 Color Plan*, not the *Level 3* plan.

3. Make the same selection again and return to the filter dialog.

- Click the Check None button again.
- Place a checkmark in only the Room Tags box this time and then click OK.

4. From the Type Selector, choose ***Room Tag: Room Tag With Area [M_Room Tag: Room Tag With Area]***.

All of the Room Tags in this view will now display the area beneath the Tag (see Figure 12.44).

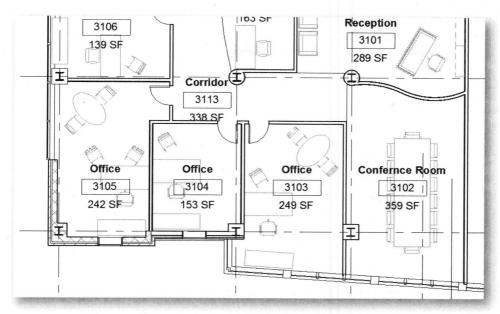

FIGURE 12.44 *Swap the Room Tags for a Type that displays the Area*

The Room Tag used here was part of the template from which the project was created. You can create a Room Tag (or any Tag) that displays any of the parameters you wish. While we will not create one here, building a Tag Family is nearly identical to the process we used to create the custom elevation tag in Chapter 10. In fact, creating a custom Tag Family is also like creating a custom title block Family like we did in the "Create a Custom Cover Sheet Titleblock Family" topic in Chapter 4. If you wish to experiment on your own, select one of the Tags in this project that is similar to the one you wish to create. On the ribbon, click the Edit Family button to launch the Family Editor and load the Tag for editing. Save the file as a new name and manipulate it to suit your needs. The text values are special elements called "Labels." You can add or edit Labels in the Family editor and have them report any of the parameters available. Look up "Labels" in the online Help for more information. If you wish to add a parameter that is not included on the list available, you must create a custom Parameter. This must be done with a "Shared Parameter" in a Shared Parameter File (the **Shared Parameters** tool is on the Manage tab). A Shared Parameter File is simply a text file saved on a hard drive or network server. The advantage of this file is that you can store your custom parameters in this file, save it to a common network server location, and then all users in the firm can access and use these parameters in their projects, Schedules, and Tags. For more information on creating and working with Shared Parameters, see the online Help.

There should only be one Shared Parameter file for the entire firm. The file should be stored on a network drive accessible to all users. These points are very important to guarantee that custom parameters work properly. Having more than one Shared Parameter file will almost certainly cause problems, annoyances, or even outright failures in the implementation of custom parameters in projects. The Shared Parameter file is merely the holding place for the parameter definition. It is *not* the parameter itself. A single Shared Parameter file can store hundreds of custom parameters for an unlimited number of projects and users. Therefore, please heed this recommendation and implement a single network-based Shared Parameter for your office.

Adding a Color Scheme

By adding a Color Scheme to a view, we can graphically represent any of the data fields in our Room elements.

5. Deselect all objects and then direct your attention to the Properties palette.

When there are no objects selected, the Properties palette displays the properties of the current view. In this case, we are seeing the properties of the *Level 3 Color Plan*. At the top of the Properties palette, just below the Type Selector, you should see Floor Plan: Level 3 Color Plan on the drop-down list.

*   Scroll down on the Properties palette and locate the Color Scheme item and click the button (labeled <none>) next to it.

You can make a color scheme based on any of the Room's data fields. There are two predefined schemes: Name and Department.

*   Click on the Name scheme on the left and then click OK (see Figure 12.45).

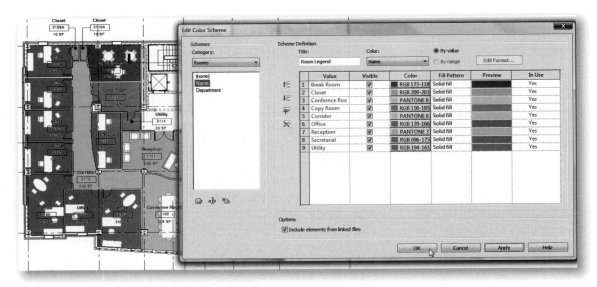

FIGURE 12.45 *Add a Color Scheme to the Level 3 Color Plan view*

As you can see, it is easy to add a color scheme. However, a scheme based on Room Names does not convey much useful information. Let's learn to create our own scheme.

Create a Custom Color Scheme

Using the Room Tag with Area as we did above is a useful way to display the area of each Room directly in the plan, but it might also be useful to show this data more graphically using color coded shading. Let's build a custom Color Scheme to convey this information.

*   On the Properties palette click the Color Scheme button again (it is now labeled "Name" instead of <None>).

6. Select the Name item in the list at the left.

*   At the bottom of the list area, click the Duplicate icon.
*   In the "New Color Scheme" dialog, type **MRAC Area by Range** and then click OK.
*   At the top of the dialog, change the Title to: **Room Area Legend**.
*   Beneath "Color," choose **Area** from the drop-down list.

Notice all of the Room parameters are included on this list. You can make a color scheme from any parameter on this list.

- In the warning that appears, click OK.

There are two ways to create an Area scheme: by the actual values in the model (By value) and by ranges that you define (By range). By value is the default; we'll build a By range scheme.

7. Click the By range radio button.

All existing values will be replaced by just two.

- Click in the "At Least" column for the last item (currently 20.00 ft² [20m²])
- Change the value to **50.00 ft² [5m²]**.
- On the left side of the table, click the Add Value icon.

A new 100.00 ft² value will appear.

8. Select the 100.00 ft² [10m²] value, and then click the Add Value icon again.

A new 150.00 ft² [15m²] value will appear.

9. Keep selecting the last item and clicking the Add Value icon until you reach 300.00 ft² [30m²] (see Figure 12.46).

Make sure you select the last item in the list each time before you click Add Value or the new value will be added in between the one you have selected and the one after it.

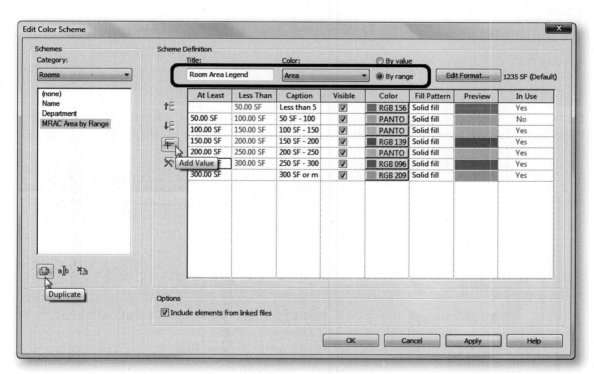

FIGURE 12.46 *Add several values to the Area color scheme*

By default, the colors are assigned automatically. This is the easiest way to see immediate results. If you prefer to designate your own colors, you can click the color buttons next to each item. We will not do that for this exercise.

10. Click OK to accept the default color scheme and display it in the view.

Notice that the colors in the view change to reflect the new scheme.

Add a Color Scheme Legend

It may not always be obvious what the colors in a plan represent. So to make it easier to understand the color scheme, we can add a legend.

11. On the Home tab, on the Room & Area panel, click the *Legend* tool.

   • Zoom out a bit allowing room to the left side of the plan.

   • Click a point to the left of the plan to place the Color Fill Legend (see Figure 12.47).

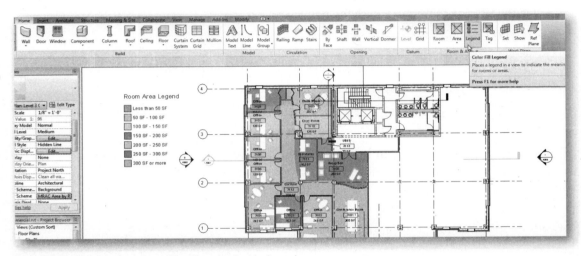

FIGURE 12.47 *Complete the Color Scheme and add a legend*

Create Another Color Scheme

You can continue to experiment by creating new color schemes. Simply return to the Properties palette and click the Color Scheme button. There you can make new duplicates and configure them. Let's do one more.

12. Return to the Properties palette and click the Color Scheme button.

   • Click the Duplicate icon and name the new Scheme **MRAC Floor Finish** and then click OK.

   • Change the Title to: **Floor Finish Legend** and then from the "Color" list choose: **Floor Finish**.

   • Click OK in the warning dialog.

   • Click OK again to return to the plan view and see the results.

In this case, the colors show where the floor finish varies based on the values we assigned earlier in the chapter. Feel free to create other schemes and vary the colors or other settings. If you want to make any of these permanent, duplicate the floor plan view first. After you complete the "Working with Area Plans" topic below, you can make a color scheme for your Area Plan as well.

Exporting Reports

In some jurisdictions detailed area analysis with triangulated area proofs are required. You can generate such a report from Revit in HTML format.

13. From the Application menu, choose **Export > Reports > Room/Area Report**.

   - In the dialog that appears browse to the *Chapter12* folder.
   - Click the Options button in the lower right corner. At the bottom of the "Area Report Settings" dialog, select the "Report window area as a percentage of room area" checkbox.
   - If you wish, edit the fonts and/or colors to be used in the report and then click OK.
   - Accept the default file name and then click Save.

14. Launch your web browser and open the HTML file created to view the report.

The default report is generated from the currently active view window. In the "Export Room Area Report" dialog, you can change the Range to a selection of views if you wish to generate a report from several floors.

WORKING WITH AREA PLANS

Using the alternative Room Tag and color scheme shown in the previous topic is a quick way to see the area of each Room. We can also see the area when we select a Room on the Properties palette and we can also add it as a field to our Room Schedule. This area is certainly useful information, but we have limited control over how the areas of Rooms are actually calculated. For Room area computation, you can use one of four methods: wall finish, wall centers, wall core, or core centers. However, this setting is global and all Rooms will be calculated the same way. To make your choice, expand the Room & Area panel on the Home tab and then click the Room and Volume Computations button.

If you need to calculate areas more precisely than is possible with Rooms, or if you need to follow code requirements or leasing area standards such as BOMA, you need to use Area Plans instead of Rooms. An Area Plan is a special type of Revit floor plan that allows for much more control in calculating the square footage of a plan. The process is as follows:

1. Create an Area Plan.
2. Generate Area Boundary Lines (similar to Room Separation Lines).
3. Add Area elements (similar to Rooms).
4. Assign properties such as Area Type.
5. Generate Schedules.

Let's run through the process. Area Plans are required before any of the other steps can be performed. Area Plans are tied to Area Schemes. There are two Area Schemes built into the software and you can copy them in the "Area and Volume Computations" dialog (Home tab, expanded Room & Area panel). In this example, we will use the built-in Rentable Area Scheme.

1. On the Home tab, on the Room & Area panel, click the Area drop-down button and choose Area Plan.
   - In the "New Area Plan" dialog that appears, make sure the Type is set to Rentable, select Level 3, and then click OK.

The Rentable Area Scheme has some rules built into it based on common rentable area requirements for commercial buildings. When you create the Area Plan, Revit will therefore offer to generate some of the Area Boundary Lines you need automatically based on the exterior Walls of your model. The lines thus generated will try to logically choose the most appropriate face of each Wall, be it the centerline, exterior face, or face of glazing. It is not a bad idea to answer Yes to this question and see what Revit determines. You can always delete the auto-generated lines later and draw them over again manually if necessary.

- In the dialog that appears, click Yes.

Fine-tune display as desired (turn off underlay and crop out unwanted site items). Some purple lines will appear on the exterior Walls. Some are at the face of the glazing on the Curtain Walls; others are at the centerlines of the Walls. Area elements behave almost identically to Rooms. They are added in similar fashion by clicking within an enclosed region bounded by Area Boundary Lines (the purple lines we have here). However, since the goal an Area Plan is often to determine the overall square footage of a plan and determine how much to charge to each tenant, you will not create boundaries at each room and office. Rather, on the third floor we have here, we only need one overall Area element to represent the entire tenant space. Another can be placed in the vacant space to the right of the plan and two in the core: one for the vertical penetrations and the other for the floor common area. So our next task is simply to draw the required boundary lines for these areas. Let's start with the front façade curtain Wall. Revit did not draw this boundary automatically.

2. Click the Area drop-down button and choose Area Boundary Line.

   - On the Modify | Place Area Boundary tab, click the Line tool.
   - Zoom in on the front façade and trace the inside face of the glazing on all three sides (see Figure 12.48).

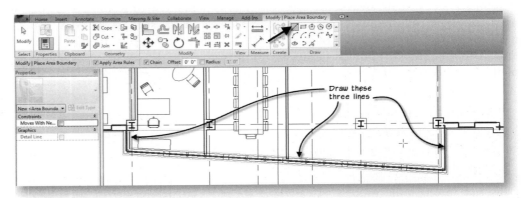

FIGURE 12.48 *Draw Area Boundary Lines along the inside of the glazing at the front façade*

3. Switch the Pick Lines tool and click the Walls that separate the tenant space from the rest of the plan (the demising walls).

   - Trim the lines as required.

On the Options Bar there is an "Apply Area Rules" checkbox. With this selected for the Pick Lines tool, Revit will draw the line along the center of the picked Wall. To get the face of a Wall for example in the core area, uncheck this box.

4. Using either Pick Lines or Line, draw the remaining lines as shown in Figure 12.49.

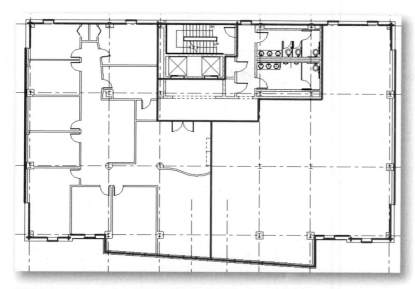

FIGURE 12.49 *Draw Area Boundary Lines to separate the tenant space and the core areas*

Using techniques discussed earlier, the Furniture, Area Separation Lines, and Sections have been hidden in this view to make it more legible. We are now ready to add Areas.

5. On the Home tab, on the Room & Area panel, click the Area drop-down button and choose Area.

   Move the cursor around the screen and notice that the Area behaves just like Rooms. They will find the region defined by the Area Separation Lines automatically.

   - Place an Area in each region (four total).
   - Click on the Tag for the Area in our office tenant space and edit the name to: **Tenant 3A**.
   - Name the remaining Areas:
   - The tenant space on the right: **Unoccupied**.
   - The toilet/lobby space: **Level 3 Common**.
   - The stair and elevator as: **Vertical Penetration**.

6. Select the Tenant 3A Area and on the Properties palette, change the Area Type to: **Office Area**.

   Repeat for the other Areas:

   - Set Unoccupied to: **Office Area**.
   - Set Level 3 Common to: **Floor Area**.
   - Set Vertical Penetration to: **Major Vertical Penetration** (see Figure 12.50).

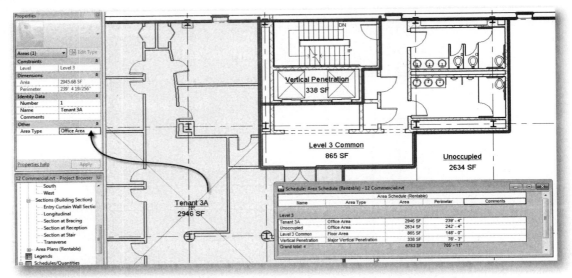

FIGURE 12.50 *Draw Area Boundary Lines to separate the tenant space and the core areas (continued)*

As a finishing touch, you can create a rentable area Schedule. Follow the procedures covered already in this chapter. You can see an example in the inset of the figure. If you decide to print your Area Plans, you can print them as-is or apply a color scheme like we did in the previous topic to make them display more graphically. Color schemes can apply to either Rooms or Areas. When you apply a color scheme to an Area Plan, you can color-code the Areas based on the Area Type or any other criteria of the Area objects. Give it a try!

SUMMARY

- A Schedule is simply a tabular view of your live building model data.
- You can edit elements referenced in a Schedule directly from the Schedule or in the model
- Changes to model elements are reflected immediately in both the graphical views and the Schedules.
- Schedules can contain any combination of fields and be grouped, sorted, and filtered.
- Add headers, footers, and totals to your Schedules to break them up and make the data more legible.
- Adding Schedules to Sheets is a simple drag and drop process.
- You can adjust the size and formatting of Schedule columns on the Sheet.
- Add breaks to the Schedule on the Sheet to wrap a long Schedule into two or more columns.
- New elements added to the model will immediately appear in Schedules as they are added.
- Schedule Keys allow you to efficiently manage repetitive groups of related instance property values shown in Schedules.

- Material Takeoffs report the parameters (such as name, area, and volume) of materials assigned to model elements in your project.
- You can filter Material Takeoffs and other Schedules to show just a sub-set of related data.
- Tags report data in similar fashion to Schedules.
- Add Tags manually or all at once using the **Tag All** tool.
- Create custom views to show different Tags and graphics.
- Create Room elements to generate Room Tags, Area takeoffs, and finish Schedules.
- Use Room Separations to manually define the shape of Rooms.
- Color Schemes can be used to color-code and display graphically nearly any data that appears in the Rooms.
- Room data can be exported in Room Reports that present Room information graphically in an HTML report.
- Any Schedule can be exported to Excel compatible delimited text files.
- To gain more control over precise square footage calculations, use Area Plans and Area objects.

chapter 13

Ceiling Plans and Interior Elevations

INTRODUCTION

The goal of this chapter is to round out our construction document set. Reflected ceiling plans will be the primary focus of the chapter with a brief exploration of interior elevations at the end. Reflected ceiling plan views are included in the default project template file used to start both projects in this book. Therefore, we simply need to open these views and add appropriate model data and annotation. The default templates include only exterior elevations. Therefore, we will need to indicate which rooms we wish to elevate and create the required interior elevation views and any embellishments they require.

OBJECTIVES

Ceiling plans are very similar to other plans. The View Properties are the only major difference between the two. We will add Ceiling elements to the model and ceiling view specific annotation. After completing this chapter, you will know how to:

- Add and modify Ceiling elements
- Understand Ceiling Types
- Add and modify Ceiling Component Families
- Manipulate the View Properties of the Ceiling Plan
- Create Interior Elevation views

CREATING CEILING ELEMENTS

Ceiling elements in Revit are used in ceiling plans to convey the material of the ceiling plane—such as acoustical tile ceiling, gypsum board, or other ceiling treatment. Ceiling elements are very similar to Floors and Roofs in composition. The default tool when creating Ceilings allows you to pick a point within a closed space in the model (much like creating Rooms, demonstrated in the previous chapter). However, they are sketch-based elements, and can utilize all the typical functions like the "Pick Walls" and other sketch methods on the ribbon during creation and editing.

Install the Dataset Files and Open a Project

The lessons that follow require the dataset included on the Aubin Academy Master Series online companion. If you have already installed all of the files from this site, skip to step 3 to begin. If you need to install the files, start at step 1.

1. If you have not already done so, download the dataset files located on the CengageBrain website.

   Refer to "Accessing the Student Companion site from CengageBrain" in the Preface for information on installing the dataset files included in the Student Companion.

2. Launch Autodesk Revit Architecture from the icon on your desktop or from the *Autodesk > Revit Architecture 2012* group in *All Programs* on the Windows Start menu.

You can click the Start button, and then begin typing **Revit** in the "Start Search field. After a couple letters, Revit Architecture should appear near the top of the list. Click it to launch to program.	**TIP**

3. On the Quick Access Toolbar (QAT), click the Open icon.

The keyboard shortcut for Open is CTRL+O. **Open** is also located on the File menu.	**TIP**

   • In the "Open" dialog box, browse to the location where you installed the *MasterRAC 2012* folder, and then the *Chapter13* folder.

4. Double-click *13 Commercial.rvt* if you wish to work in Imperial units. Double-click *13 Commercial Metric.rvt* if you wish to work in Metric units

   You can also select it and then click the Open button.

Creating an Acoustical Tile Ceiling

Suspended acoustical tile ceilings in commercial office buildings are constructed in one of two ways. In the first method, walls are built past the height of the ceiling tiles (to a fixed height or all the way to the deck) and each room contains its own ceiling. In the second method, the walls are built only up to the height of the underside of the ceiling (underpinned) with the ceiling plane being continuous across the tops of the Walls.

In Revit, when each room contains its own ceiling and the walls continue past the ceiling plane height, Ceiling elements can usually be created quickly with the "Auto Ceiling" function of the ***Ceiling*** tool. This is the default function of the Ceiling tool. When you have an underpinned Ceiling, you can instead use the Sketch Ceiling option to draw the shape of the overall Ceiling plane using any of the available sketching tools. We will explore both creation methods as we continue to refine the third floor of our commercial project.

Revit usually creates both a floor plan and a ceiling plan view whenever you create a new Level in a project. The template from which we originally created the commercial project includes such views. In this sequence, we will work in the ceiling plan views for the first time. Be sure that you work in a Ceiling plan view when adding Ceiling elements. If you do not, you will receive a warning like the one shown in Figure 13.1.

> **Warning** ☒
>
> None of the created elements are visible in Floor Plan: Level 3 View. You may want to check the active view, its Parameters, and Visibility settings, as well as any Plan Regions and their settings.

FIGURE 13.1 *Attempting to add Ceilings in floor plan views yields a warning*

5. On the Project Browser, expand the *Ceiling Plans* branch.

    • Double-click to open the *Level 3* ceiling plan view.

NOTE Again, please be sure that you are opening the *Level 3* ceiling plan view and *not* the *Level 3* floor plan view.

Don't let the heavy vertical and horizontal lines on screen confuse you. These are the beams in the linked Structural model. You can hide or unload the linked file if you find them distracting, but there is no need since the Ceiling elements will cover the joists as we add them.

6. On the Home tab, on the Build panel, click the **Ceiling** tool.

    On the Status Bar, a message will appear "Click inside an area bounded by walls to create ceiling." On the Modify | Place Ceiling tab, a single button labeled **Sketch Ceiling** appears. On the Properties palette the Type Selector appears (see Figure 13.2).

    • **Type Selector**—Use this drop-down to choose the kind of Ceiling you wish to add.

    • **Level** and **Height Offset from Level**—Ceiling height is measured from the current Level by default. The Height Offset is the actual height of the Ceiling plane in the space. The Ceiling's thickness will extend up from that point.

    • **Room Bounding**—If you have Room Volume computations enabled in your project, (see Figure 13.15) the Ceiling will bound the top extent of Room elements in the space.

    • **Automatic Ceiling**—This is the easiest way to add a Ceiling. As noted on the Status Bar, simply click in a region bounded by Walls. Revit will create the sketch of the Ceiling automatically from these Walls.

    • **Sketch Ceiling**—Use this option when you wish to draw a custom-shaped Ceiling or when you do not have Walls on all sides of the Ceiling or if the Ceiling needs to extend past Walls.

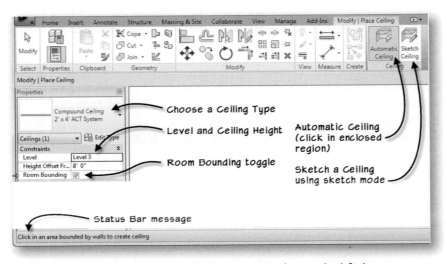

FIGURE 13.2 *Using the Ceiling Tool, "Sketch" is an option, but not the default*

7. From the Type Selector, choose **Compound Ceiling : 2' × 4' ACT System [Compound Ceiling : 600 × 1200mm grid]** if it is not already selected.

   • Move the cursor around the screen pausing within various Rooms—do not click yet.

Notice that the behavior is similar to the **Room** tool from the previous chapter. However, unlike the **Room** tool, this tool does not have the "Room Separation" option, nor can it recognize those Room Separation lines already in the model. Therefore, if the auto-detecting routine does not automatically recognize the Room you need, you can use the **Sketch Ceiling** tool noted above. We will try the sketch option a little later. For now, we will add a Ceiling in a Room that is recognized by the auto-detection routine.

   • Move the cursor into one of the two offices flanking Column Line 3 on the left side of the plan.

     The Room boundary should highlight automatically.

   • Click in the Room above Column Line 3 to add the Ceiling (see Figure 13.3).

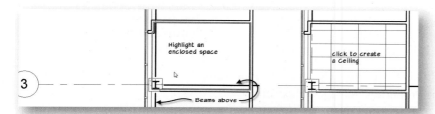

FIGURE 13.3 *Add a Ceiling to an office—notice how it hides the Beams above*

   • On the ribbon, click the **Modify** tool or press the ESC key twice.

Explore Ceiling Type Properties

8. Click on one of the Ceiling grid lines.

Notice that each line is individually selectable. However, they behave together as a unit. The auto-creation routine will attempt to center the grid in the room left to right and top to bottom. You can move any grid line if you like, and the entire grid pattern will move with it. Use this technique to apply custom centering (see below).

   • With a grid line selected, look to the Properties palette.

Like other elements, the Ceiling's properties appear on the palette. Notice that the "Height Offset From Level" is to 8'-0" [2600] above the associated Level. Notice also that Revit calculates the area, perimeter, and volume of the Ceiling. You cannot change these values directly, they are the result of the Ceiling's shape and composition. But like other Revit elements, if you move a bounding Wall or edit the Ceiling Type to change its thickness, these numbers would adjust accordingly. Try it if you like, just be sure to undo before continuing.

   • Change the Height Offset from Level value to **9'-0" [2700]** and then click Apply (see Figure 13.4).

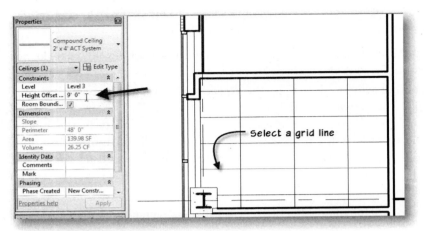

FIGURE 13.4 *Ceiling height above the current Level can be easily adjusted*

Notice the Room Bounding checkbox. This is selected by default. Like many other elements, Ceiling elements can be used as Room boundaries. Below, in the "Study a Ceiling in Section" topic, we will discuss this setting in more detail. Rooms and the Room Bounding setting were discussed in the "Rooms and Room Tags" topic in Chapter 12. Look there as well for additional information.

9. On the Properties palette, click the Edit Type button.

    The "Type Properties" dialog will appear.

Notice that a Ceiling element's Type has an Edit "Structure" button like many other "layered" Types. If you were to edit this Structure, you would notice that it is comprised of a simple Core element with a Finish Layer on the bottom. The grid pattern comes from the Material assigned to this Finish Layer (see Figure 13.5).

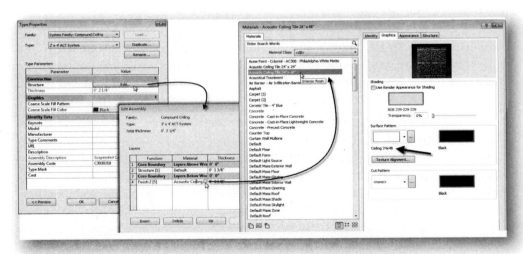

FIGURE 13.5 *The Type Properties of the ACT Ceiling uses a "Ceiling Tile" Material*

We do not need to make any changes to the material, structure, or type properties. However, such exploration is educational, particularly when you realize that editing a Ceiling type is nearly identical to editing other layered element types such as a Wall, Floor, or Roof type. Keep this in mind if you need to create custom Ceiling types in your own projects.

10. Cancel the Materials, Structure, and Type Properties dialogs when finished exploring.

11. On the Home tab, click the **Ceiling** tool again.

You can easily change the height of the Ceiling before you place it. Just make the edit on the Properties palette before you click to place the Ceiling.

- On the Properties palette, change its Height Offset to **9'-0" [2700]**.
- Add another Ceiling using the same technique in the office below Column Line 3.
- On the ribbon, click the **Modify** tool or press the ESC key twice.

You can verify that is was created at the desired height by selecting one of the grid lines again.

Move, Rotate, and Align Ceiling Grids

We could continue and add additional Ceilings in the other Rooms, but for now we will work with just these two.

12. Select one of the vertical grid lines of the Ceiling you just added.

- On the Modify Ceilings tab, click the **Move** tool and then move the grid line (see Figure 13.6).

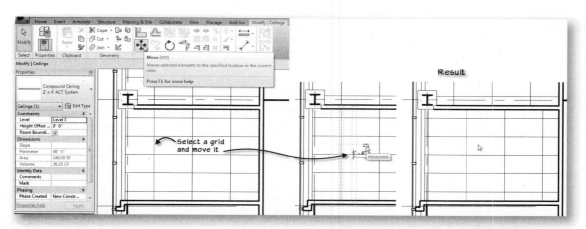

FIGURE 13.6 *Move a Ceiling grid line*

The exact amount of the move is not important.

As noted above, notice that the entire grid will reposition with this move, not just the selected grid line. Also, only the grid pattern is affected by this move, not the boundary of the Ceiling element, which still conforms to the shape of the Room. You can use other typical modification techniques as well, such as Rotate. If you need your grid pattern at a different angle, simply click the **Rotate** tool and rotate the grid line. The rest of the grid will follow (see Figure 13.7).

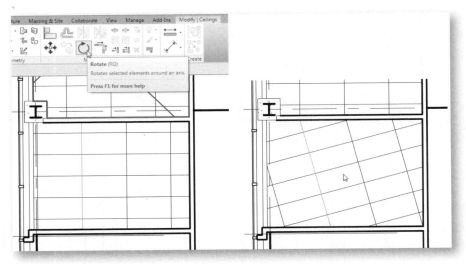

FIGURE 13.7 *Rotate a Grid using the Rotate tool*

The precise amount of the rotation is not important.

You can use the Align command to create alignment between the grids in two Rooms.

NOTE   It is not necessary to undo the rotation first. Align will rotate the grid back to horizontal for you. If you prefer, or find it easier to visualize, you can undo the rotation first.

13. On the Modify tab, click the ***Align*** tool.

- Select one of the vertical grid lines in upper office.
- Click on a horizontal grid line across the hall in the other Room to align it (see Figure 13.8).

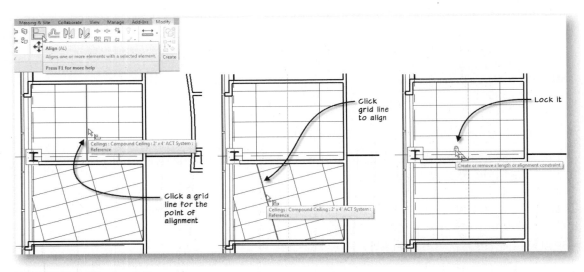

FIGURE 13.8 *Align two Ceiling grids*

14. Click the padlock icon that appears to constrain the alignment.
15. Select one of the vertical grid lines.

- Move it left or right.

Notice that the Ceiling grids in both Rooms move together. When you move this way, you are simply moving the Model Pattern that is applied via a Material to the Ceiling, *not* the Ceiling itself. This is why the pattern remains clipped to the shape of the Room as you move or rotate it.

| This behavoir applies to any patterned material. If you wish to shift or align a brick pattern on Wall for example, you can use the same techniques. | NOTE |

Create a Ceiling via Sketch

Some of the spaces that we have will prove problematic for the automatic Ceiling creation option. The routine does not recognize Columns very well. Also, the Curtain Walls can sometimes make finding closed boundaries difficult. To create the next several Ceilings, we'll use the Sketch option. In addition, we will sketch a single continuous underpinned Ceiling for the non-office spaces in the suite.

16. Zoom out to see the entire plan.
17. On the Home tab, click the **Ceiling** tool again.

   • Move your mouse around the screen and look for rooms where you can generate a ceiling boundary automatically.

Of the remaining offices, you will only be able to auto-generate two of them. The two rooms at the top of the plan (one office and the break room) will work as well, but notice that the Ceiling will flow into the closets. This is because of the modification we made to the Room Bounding property back in Chapter 12. If we want separate ceilings in the closets, we will have to sketch them instead. We'll start these two with an auto-ceiling and then modify them below. Figure 13.9 shows which rooms should work with Auto Ceiling and which will require a manual sketch.

   • Create Ceilings using the pick point method in those rooms that give a clean boundary (see Figure 13.9).

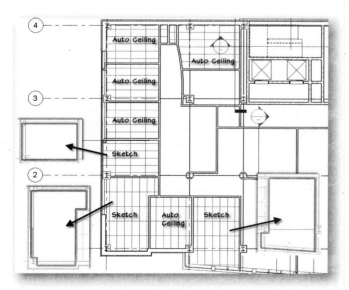

FIGURE 13.9 *Sketch the remaining office space Ceilings, align and rotate them as desired (insets show sketch lines)*

The two lower left offices at the corner and the lower one next to the conference room will not work well with the automated option. We will sketch these instead.

18. Continue with the Ceiling tool, which should still be active.

- On the Modify | Place Ceiling tab, click the **Sketch Ceiling** button.

  The Create Ceiling Boundary tab appears with the normal compliment of sketch tools.

By now, the sketch tools in Revit should be very familiar to you. The basic steps are summarized here, but please feel free to use whatever techniques you prefer to sketch the inside shape of the Room. On the Draw panel, there are two "mode" buttons: **Boundary Line** and **Slope Arrow**. Most of the time you will use the Boundary Line mode. This allows you simply to sketch the boundary of the Ceiling element using the typical shape icons. The Slope Arrow mode allows you to create a non-horizontal ceiling. Use this for sloping ceilings such as cathedral ceilings. While we are not covering this method at this time, feel free to open the residential project and try this mode out later. If you wish to return to the automatic boundary method at any time, click the **Auto Ceiling** button.

- On the Draw panel, click the **Pick Walls** icon.
- Zoom in on the corner office and then Pick the Walls surrounding the Room.
- Use the Flip controls to change the side of the Sketch lines as required.
- Use **Pick Lines** to sketch around the Column.
- Use **Trim** to cleanup the sketch (see Figure 13.10).

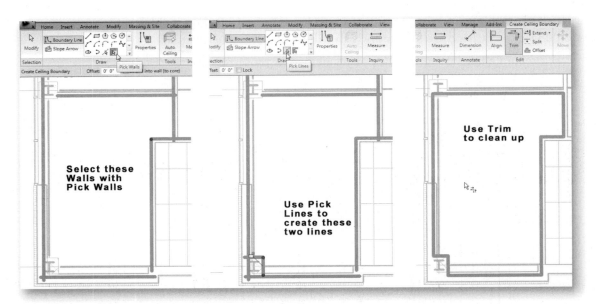

FIGURE 13.10 *Sketch the inside shape of the office space Ceiling*

19. On the ribbon, click the **Finish Edit Mode** button.

A new Ceiling will appear in the office.

20. Repeat this process to add Ceilings to each of the remaining offices (see Figure 13.9).

When sketching a complex shape like the office at the lower right, use Pick Lines to get one edge on each side and then use Trim to cleanup. Do not select several edges along the Curtain Wall for example. One will do.	**TIP**

Notice that Revit will typically choose what it considers to be the best orientation for the ceiling grid. If you wish, you can rotate the grid 90° to orient the offices all the same. To do this, remember, simply select one of the grid lines, and then click the ***Rotate*** tool and rotate the line 90°. The remainder of the grid will match the new orientation.

21. Using the process covered above, Rotate and Align the Ceiling grids to one another in a logical pattern (see Figure 13.9).

To align several Ceilings to one another, choose the "Multiple Alignment" option on the Options Bar when using the Align tool.	**TIP**

22. Make a window selection of all the Ceilings, use the ***Filter*** tool to filter everything except Ceilings.

   • Edit the Properties of the selected Ceilings; make sure that the Height Offset is set to: **9'-0" [2700]**.

23. Save the project.

Edit an Existing Ceiling

If you need to select the actual Ceiling element to edit the sketch, place the cursor near the edge of the Room and use the TAB key to pre-highlight and then select the Ceiling element. Once you select the edge, the Modify | Ceiling tab will appear and you can click the ***Edit Boundary*** tool to return to the sketch. New in this release, you can just select one of the pattern lines and then click the ***Edit Boundary*** tool.

If you have any trouble selecting a Ceiling to edit (like ones without patterns), another alternative is to use a crossing windows selection at the Wall near the edge of the Ceiling and then use the Filter tool to filter out everything but the Ceiling.	**TIP**

24. Select a Ceiling pattern line in the office beneath Column Line 4.

25. On the ribbon, click the Edit Boundary button.

   • Use Trim to exclude the closet from the sketch.
   • Delete the unnecessary sketch lines and then click the Finish Edit Mode (see Figure 13.11).

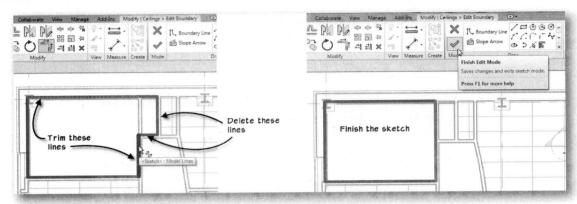

FIGURE 13.11 *Edit the sketch of the two Ceilings at the top of the plan to remove the closets*

26. Repeat the process on the break room across the hallway. Here you can move the sketch line at the back of the closet to enclose the shape or draw a new line.

Create an Underpinned Ceiling

The Ceilings that we just sketched were all contained wholly within single Rooms. You can use the same procedure, using sketch lines to encompass the overall perimeter of several spaces at once. When you do this, you will have a single continuous Ceiling grid that can represent an underpinned ceiling.

27. On the Home tab, click the **Ceiling** Tool.

- On the Modify | Place Ceiling tab, click the **Sketch Ceiling** button.
- Sketch the outline shown in Figure 13.12.

  Don't include the Conference Room. We will do this one later.

You can create most of the required edges with the **Pick Walls** tool. The rest you can build with the **Lines** tool or the **Pick Lines** tool. **Trim** to clean up the shape.

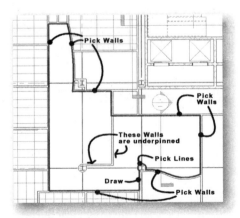

FIGURE 13.12 *Sketch an underpinned Ceiling across several Rooms*

28. On the Properties palette, change the "Height Offset From Level" to **9'-6" [2850]** and then click Apply.

- On the Mode panel, click the **Finish Edit Mode** button.

TIP	If you try to complete the sketch and warning appears, it usually means you missed a trim somewhere. Simply click the Continue button in the warning and look for an untrimmed corner.

Study a Ceiling in Section

At this point we have several Ceiling elements in our project. We have studied their Type parameters, moved, rotated, and aligned them with one another and discussed the difference in selecting and manipulating the Model Pattern vs. the Ceiling itself.

29. On the Project Browser, double-click to open the *Longitudinal* section view.

If you prefer, double-click the Section Head in the plan instead.

TIP

30. Zoom in on the third floor at the left side of the section.

Notice that the Ceiling in the offices at the left is a bit lower than the one in the public spaces. Also, notice that the Walls currently go all the way to the floor deck above. Since we have made a single continuous underpinned Ceiling in the public spaces, we might want to adjust these Walls to stop at the Ceiling height. The section view helps us spot the problem, but the edit is best accomplished in the ceiling plan view.

31. On the Project Browser, double-click to open the *Level 3* ceiling plan view.

-   On the Modify tab, click the **Split** tool and split the Wall at the back of the secretarial area as indicated in Figure 13.13.
-   Select the two Walls in the underpinned area (see the right side of Figure 13.13).

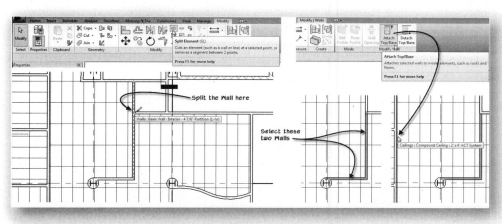

FIGURE 13.13 *Split Walls as shown and then select the underpinned Walls*

32. On the Modify | Walls tab, on the Modify Wall panel, click the **Attach Top/Base** button.

-   Move your mouse to the edge of the Ceiling.
    The edge of the Ceiling will pre-highlight.
-   Click when the Ceiling pre-highlights to attach the selected Walls to the Ceiling.

You could also edit the properties of the Walls and change their Top Constraint to Unconnected at a height of 9'-6" [2850]. However, if you anticipate that the Ceiling height will change at some point, the Attach option is better as it will stay "connected" to the Ceiling. Further, the graphics in the section will be nicer with Attach.

NOTE

Now that we have adjusted the height of the Walls, the model more accurately reflects our design intent. You can see this best by returning to the *Longitudinal* section view (see Figure 13.14). If you want the Walls on the left and right to cleanup nicely with the Floor above, you can repeat the Attach Top/Bottom step right here in the section view. Attach them to the Floor element.

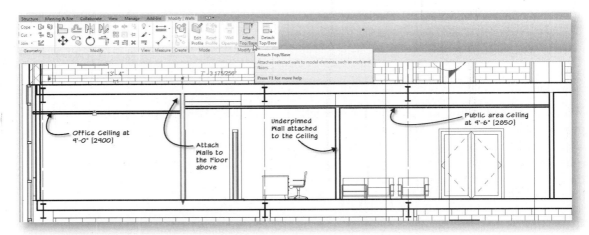

FIGURE 13.14 *Study the results in the Longitudinal Section view*

There is one further thing to notice while we are in the *Longitudinal* section view. Move your mouse near the middle of the office space. When your mouse is over the Room element, it will pre-highlight as it does in plans. Notice that its height ignores the Ceiling element; even though the element is set to Room Bounding as noted in the "Creating an Acoustical Tile Ceiling" topic above (see the left side of Figure 13.15). On the Home tab, expand the Room & Area panel and click the ***Area and Volume Computations*** button. This opens the "Area and Volume Computations" dialog. For Volume Computations, choose the Area and Volumes option and then click OK. Pre-highlight the Room again and notice that it now conforms to the height of the bounding Ceiling element (see the right side of Figure 13.15).

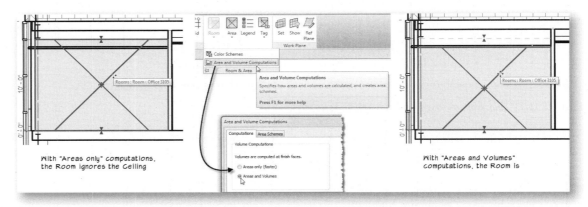

FIGURE 13.15 *Enabling Area and Volume computations is required to make the Ceiling's room bounding behavior kick in*

33. On the Project Browser, double-click to open the *Level 3* ceiling plan view.

Edit Element and View Visibility

If you study the two underpinned Walls in the secretarial area you will notice that even though the Walls now sit beneath the height of the Ceiling plane, they continue to hide the Ceiling as if they were going all the way through. This is simply the default behavior of the graphics and there is an easy fix.

34. Select the same two Walls again.
35. On the View panel, click the Override Graphics in View drop-down button and choose ***Override by Element***.

You can also right-click and choose **Override Graphics in View > By Element**.

TIP

- Place a checkmark in the "Transparent" checkbox and then click Apply (see Figure 13.16).
- New in the release, there is also Ghost Surfaces. Try this for a more subtle effect.

If you like, you can also override the Cut Lines Weight to make it slightly lighter. The transparent and Ghost Surfaces features can also be useful in 3D views when you wish to display elements that would otherwise be obscured by elements in front of them. Another application would be to make items like Furniture display without

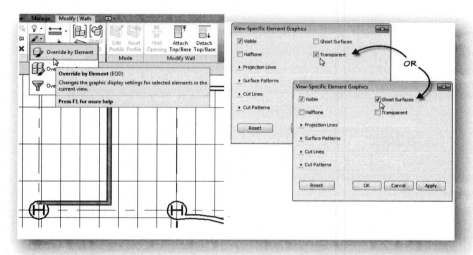

FIGURE 13.16 *We can make the Walls display correctly in ceiling plan by making them transparent or using Ghost Surfaces*

their white mask on shaded plans. For example, recall the "Add a Color Scheme" topic in Chapter 12. If you return to the Level 3 Color Plan, you will notice that the Furniture is displaying with white masking in front of the color fills. This is despite the fact that it is set to halftone. To remedy this, select any piece of furniture, click the Override Graphics in View drop-down button again, but this time choose ***Override by Category***. In the "Visibility/Graphic Overrides" dialog, place a checkmark in the Transparent column next to Furniture (or Ghost Surfaces) and then click OK.

If you want to designate some of your Walls as being a few inches above the ceiling plane, there is no need to modify the ceiling plan graphics, but you may want to consider selecting the Walls in question and modifying their properties to represent this. To do this, select a Wall and on the Properties palette, change the Top Constraint to **Unconnected**. Set the Unconnected height to 6" [150] above the Ceiling height.

Be aware that if you previously attached the top of the Wall to the Floor or Ceiling, you will need to Detach it before this change will become apparent. Attach takes precedence over the height settings of the Wall.

CAUTION

If you go to all this trouble to make the Wall heights correct, you probably would like to indicate the wall types in the drawing with a Wall Tag. Tags were covered in Chapter 12. You can find an additional exercise for adding the Wall Tags in follow-up to our exercises in this chapter in Appendix A.

Adding a Drywall Ceiling

For the Conference Room, we will use the same basic process as the other Ceilings, but simply choose a different Ceiling type.

36. On the Home tab, on the Build panel, click the **Ceiling** tool.

   - From the Type Selector choose **Compound Ceiling : GWB on Mtl. Stud [Compound Ceiling : Plain]**.
   - On the Modify | Place Ceiling tab, click the **Sketch Ceiling** button.
   - Follow the process outlined above to create a Ceiling sketch in the Conference Room.
   - Click **Finish Ceiling** to create the Ceiling.

37. Add drywall Ceilings to the two closets.
38. Add 2 × 4 Ceilings to the Copy Room and the Utility Room.

As you can see, except for choosing a different Type, the process is the same as with Ceiling grids.

Switch to a Different Grid Size

Like other elements in Revit, you can change the Type used on an existing element at any time. Suppose you preferred a 2' × 2' [600 × 600] ceiling grid layout instead of the 2' × 4' [600 × 1200] one we used. With an existing Ceiling selected, you can simply choose a different Type from the Type Selector. While we will not cover the steps here, you can also duplicate and modify an existing Type, assign a different pattern, and achieve other Ceiling designs not included in this project file. Feel free to experiment with this on your own later if you wish. The process is nearly the same as creating a new Wall or Floor Type.

39. Select one of the grid lines on the large underpinned Ceiling in the middle of the plan.

   - From the Type Selector, choose **Compound Ceiling : 2' × 2' ACT System [Compound Ceiling : 600 × 600mm grid]**.
   - Align the public area ceiling grid to the office grids if you like.

40. Save the project.

ADDING CEILING FIXTURES

Now that we have completed adding Ceiling elements to our project, we can add some lights and other fixtures to the ceiling plan. To achieve the optimal lighting layout, you may need to adjust the location of your grid lines. You can use your Move and Rotate commands as noted above to this. If you rotate a grid and an error appears indicating that constraints are no longer satisfied, simply click the "Remove Constraints" button and then, if desired, use the **Align** tool again to reapply constraints. This may occur if you previously used Align and locked the grids in one room to another.

Adding Light Fixtures

Let's start with a simple fluorescent lighting fixture in the offices.

1. Make adjustments to the position of grid lines in the offices at the left if necessary to accommodate a suitable lighting layout.

2. On the Home tab, on the Create panel, click the **Component** tool.

   - On the Place Component tab, click the **Load Family** button.
   - Browse to your library folder and open the *Lighting Fixtures* folder.
   - Select the *Troffer – 2 × 4 Parabolic.rfa* [*M_Troffer - Parabolic Rectangular.rfa*] Family file and then click Open.

> As mentioned in previous chapters, if you do not have these Families in your installed libraries, you can find them in the *Library* folder with the dataset files installed from the online companion.

NOTE

3. Place a light fixture randomly in the office.

   - Use the Align command to align it to the grid lines.

You do not need to lock them when you Align. Later if you move the grid, the lights will move too automatically. This is because the lights are hosted by the Ceiling.

   - Select and then Copy the light as appropriate for the space (see Figure 13.17).

FIGURE 13.17 *Align the light to the grid and then copy it to create a lighting pattern*

If you are not satisfied with the position of your lights relative to the grid lines, you can move the grid line one-half a tile or one-quarter of a tile in either direction. The lights will move with them.

The light fixtures are Families that use the Light Fixture Family template. They contain two-dimensional graphics for the ceiling plan, three-dimensional graphics show in other views. Also, they actually contain illumination information that is used when you generate a rendering of your Revit model (refer to Chapter 17 for information on rendering in Revit). If you cut a section of the room and its light fixtures, you will see that the fixture actually cuts the Ceiling element (shown on the right side of Figure 13.17).

> In a large project with hundreds or even thousands of light fixtures, you may want to consider creating a 2D only symbol for the majority of your light fixtures and reserve the use of the 3D one only for spaces that will show in 3D and/or be rendered. The cut that the light fixture performs on the Ceiling, when repeated extensively, can prove detrimental to computer performance.

NOTE

4. Select all of the lights in the office and copy them to the other offices.

- Either copy or use the Component tool to place additional lights (see Figure 13.18).

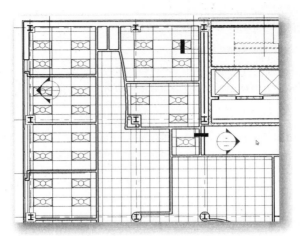

FIGURE 13.18 *Copy the lights to other offices*

Display Doors in Reflected Ceiling

Some firms prefer for Doors to show in RCP to make it easier to locate light switches, exit signs, and other RCP equipment relative to door swing. To do this, you need to adjust the Cut Plane of the ceiling plan view.

5. Deselect all elements so that the Properties palette shows the properties of the current view.

- Beneath the Extents grouping, click the Edit button next to View Range.
- For the Cut Plane Offset, change the value to **6'-6"** [**1950**] and then click OK (see the top of Figure 13.19).

This shows Doors, but they display the same as they would for floor plans. You might also consider editing the Door graphic defaults in the Visibility/Graphics override dialog.

- On the View tab, click the ***Visibility/Graphics*** tool (or type **VG**).

6. In the "Visibility/Graphic Overrides" dialog, expand Doors.

- In the Cut column next to the Panel subcomponent, click the Override button. Choose a fine dashed line pattern such as Dash 1/16" [Dash] and then click OK.
- Repeat for the Plan Swing component. Override the Projection Lines this time (see the bottom of Figure 13.19).

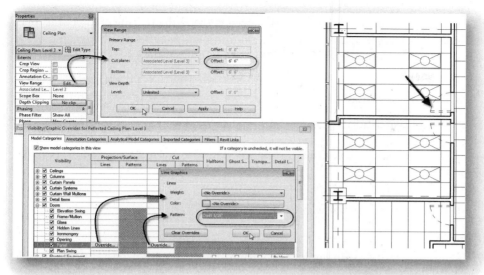

FIGURE 13.19 *Optionally edit the Cut Plane height in the "View Range" and the graphic display of Doors*

The results appear to the right of Figure 13.19. You can choose different line patterns if you prefer, and/or edit the color or line weight as well. You can also use the Halftone checkbox for Doors instead. The exact graphical settings are left to you and your company standards.

> If you make such an edit and wish to use it on other floors of the building, you can save a View Template. On the View tab, on the Graphics panel, click the View Templates drop-down button and choose **Create Template from Current View**. Give it a name and click OK twice. To use it on other ceiling plans, use the **Apply Template to Current View** command. To apply to several views at once, select them with the CTRL key in the Project Browser, then right-click and choose **Apply View Template**.

TIP

Adding Light Switches

Sometimes it is nice to see light switches on the ceiling plans instead of floor plans. The out-of-the-box light switch Families provided with Revit work better in floor plans. However, it is simple to modify them. A modified version is provided with the dataset for Chapter 13. You add switches the same as any Component Family.

7. On the Home tab, click the **Component** tool.

   • Click the **Load Family** button again, browse to the *Chapter13* folder and then open the *Switch-Single (rcp).rfa* [*MRAC_Switch-Single (rcp)-Metric.rfa*] Family file.

This Family file is a modification of the standard Revit Architecture library symbol with modifications to allow it to display in reflected ceiling plan. We discussed Symbolic Lines and forcing overhead items to display in the "Making Families above the Cut Plane Display in Plan" topic in Chapter 10. In summary, in order for any elements to display in a plan view (ceiling plan in this case), there must be some element that passes through the Cut Plane height that triggers Revit to display its 2D plan representation. The two ways to deal with the situation are to lower the Cut Plane of the ceiling plan view (currently set at 6'-6" [1950] after the modification in the previous topic) or to add an element to the Family file that passes through this default Cut Plane and provide dedicated plan graphics in the Family. The Family file provided in the *Chapter13* folder has a simple Model Line drawn vertically and set to the <Invisible Lines> Line Style that passes through the reflected ceiling plan Cut Plane. This triggers Revit to display the graphics for this symbol in the RCP. Detailed steps were provided for a similar Family in Chapter 10.

If you prefer, you can edit the View Properties of the Level 3 ceiling plan view again and lower the View Range further and then use the out-of-the-box Family.

8. Place a light switch adjacent to the latch side of the door in one of the offices.

- Place or copy additional switches in the other offices.

Adding Wiring

Revit Architecture does not include a wiring element. (Revit MEP does include wiring.) To show the wiring connection to the switches, simply draft Detail Lines on the view. Remember, Detail Lines are view-specific and will show only in this reflected ceiling plan.

9. On the Annotate tab, on the Detail panel, click the **Detail Line** tool.

- On the Line Style panel, from the Line Style list, choose <Hidden>.
- On the Draw panel, click the **Start-End-Radius Arc** icon.

10. For the first point, click near the switch.

- For the end point, click near the middle of a light fixture, and then click a third point somewhere in between (see Figure 13.20).

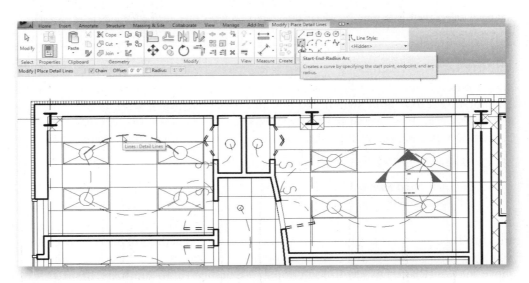

FIGURE 13.20 *Sketch three-point arc Drafting Lines for the wiring*

11. Continue drawing arcs to connect all the lights in the room to the switch.
12. Repeat in the other Rooms.

- On the ribbon, click the **Modify** tool or press the ESC key twice.

Complete the RCP

At this point, you can continue adding elements to the reflected ceiling plan as needed. Following procedures covered in this and previous chapters, you can add text, dimensions, and tags as necessary. If you wish to add Room Tags, they will "see" the Rooms that we have already added to this model in the previous chapter including the Room Separation lines. If the Room Tag does not fit comfortably in the Room, add it with the "Leader" option. Then you can move it outside the boundaries of the Room. The fastest way to add the Tags is with *Tag All* tool (see the "Tag All Not Tagged" topic in Chapter 12). You can even copy and paste the tags from one view to another. Use Paste Aligned > Current View to put them in the same spot. If you add dimensions, you can click on the value and add manual edits, prefixes, and suffixes. For example, you may want to add a note indicating the point of beginning (P.O.B.) for the grid. Add a dimension, click its text, and then add the text above, below, or as a prefix or suffix (see Figure 13.21).

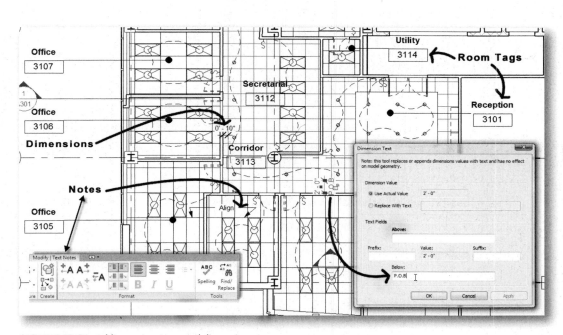

FIGURE 13.21 *Adding notes, tags, and dimensions*

You can add Text Notes with the tool on the Annotate tab. Simply click on screen and begin typing. Types on the Type Selector control the font and height of the text. You can use the tools on the ribbon to add leaders to the left or right side of the text. You can add multiple leaders on each side of the text and they can even have various attachment points like top, middle, and bottom.

13. When you are finished add elements to your reflected ceiling plan, save the project.

CREATING INTERIOR ELEVATIONS

No construction documents set would be complete without interior elevations. Adding such views in Revit is easy to do. Our reception area, the corridor, and the toilet rooms in the core are all good candidates.

Add an Interior Elevation

1. On the Project Browser, double-click to open the *Level 3* floor plan view.
2. On the View tab, on the Create panel, click the **Elevation** tool.

   - From the Type Selector, choose **Elevation : Interior Elevation**.
   - On the Options Bar, from the "Scale" list, choose **1/4" = 1'-0" [1:50]**.

3. Move the cursor into the secretarial area—do not click yet.

   - Move the cursor around the space and watch the orientation of the elevation head change dynamically (see Figure 13.22).

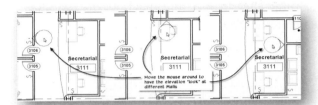

FIGURE 13.22 *Elevation Heads orient automatically to the nearby Walls*

The orientation of the elevation symbol automatically orients to a nearby Wall. This makes creating single elevations easy, especially to angled or curving walls. Move slowly along the curved Wall to see this clearly. In this case, we want four elevations—one in each direction.

4. When the symbol is pointing right, click to place the elevation tag in the room.

   - On the ribbon, click the **Modify** tool or press the ESC key.

You will notice that on the *Interior Elevations* branch of the Project Browser a new view appears named *Elevation 1 – a* (see Figure 13.23).

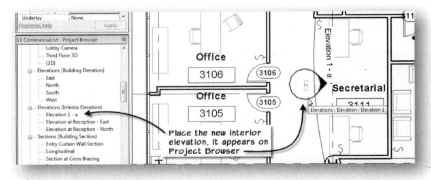

FIGURE 13.23 *Add a new Elevation view to the Secretarial space*

5. Double-click on triangle part of *Elevation 1 – a* to open it.

Notice that the view is cropped nicely to the size of the Room. You can adjust the cropping if you need to fine-tune it. If you don't want to see the section markers, right-click them and choose **Hide in View > Category** (or use the tool on the ribbon). Depending on where you clicked the elevation marker, you may or may not be seeing the desks. You can move the symbol in plan and adjust the location of the elevation cut line. It will appear as a long blue line when the triangluar portion is selected.

Add Additional Interior Elevations

We can add additional elevations of the same elevation symbol very easily.

6. On the Project Browser, double-click to open the *Level 3* floor plan view again. Zoom to the Reception space.

In a previous chapter, we added a few interior elevations to this space. They are currently hidden in this view.

- Add a leader to the Room Tag and move it outside the space.

- On the View Control bar, click the Reveal Hidden Elements icon (small light bulb).

- Select the elevation marker and the two elevations, and on the ribbon, click the Unhide Elements button.

- Click the Toggle Reveal Hidden Elements Mode button to turn off the mode.

7. Click on the Elevation Marker (the circle part of the marker) in the middle of the reception space.

A series of checkboxes will appear surrounding the symbol and ghosted arrows pointing to up to four possible elevation directions. To add another elevation view, simply check one of the boxes.

- Place a checkmark in each of the two other checkboxes surrounding the elevation symbol (see Figure 13.24).

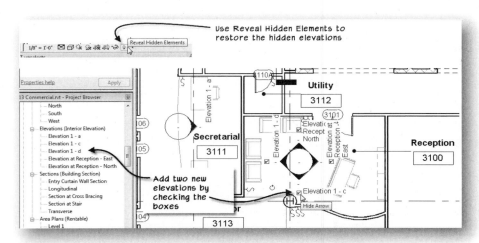

FIGURE 13.24 *Add new Elevation views by checking the direction checkboxes*

The new views will appear on the Project Browser.

8. On the Project Browser right-click on the *Elevation 1 – d* view and choose **Rename**.

  - Name it **Elevation at Reception – West**.
  - Repeat for each of the other two elevations.

If you want to remove the labels from the elevation tag in the plan view, it requires a custom Interior Elevation Tag. The process to build one was covered in the "Creating Custom Elevation Tags" topic in Chapter 10. Feel free to review that topic now and try your hand at it. However, a custom Tag has been preloaded into the dataset. All we have to do is assign it to the Interior Elevations. To do so, select one of the elevation view markers (the triangle part), and then on the Properties palette, click the Edit Type button. For the Elevation Tag item, click on the current selection, then the small browse icon. From the Elevation Mark list, choose **Elevation Mark Circle - MRAC:1/2" Circle** and then click OK twice.

You can complete these views in any way you wish. Add notes, dimensions or tags. Drop them onto a new Sheet to round out the set (see Figure 13.25). When you drag them to a Sheet, the numbers and sheet references will automatically fill in.

TIP	If you want the name of the elevations on the Project Browser to vary from the name shown on the titlebar on the Sheet, you can select the view on the Project Browser and then in the Properties palette, change the Title on Sheet parameter.

Continue to work in the project, making further enhancements if you wish.

9. When you are finished, save and close the project.

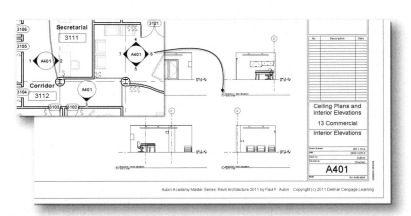

FIGURE 13.25 *Create Interior Elevations and add them to a new Sheet*

SUMMARY

Reflected ceiling plans and interior elevations are two important parts of a complete construction documentation set. In this chapter, we have taken a brief look at the steps to create both of these important document types.

- Ceiling elements are easily created from existing Walls by picking points within a bounded space.
- If bounding Walls are not available, or if you wish to create a custom Ceiling shape, you can Sketch the Ceiling using standard sketch tools.
- Ceiling types control the structure of a Ceiling element like other "layered" types.
- Acoustical tile ceilings are simply Ceiling types that have a finish layer assigned to an appropriate Material and surface pattern.
- The grid lines in the Material use a Model Pattern. Moving or rotating the pattern does not change the Ceiling element shape.
- To create an underpinned Ceiling, use the **Sketch Ceiling** tool.
- Adjust the height of Walls to coordinate with the underpinned Ceiling after creation.
- Drywall Ceilings are created with the same process and simply use a different type.
- You can change grid size by swapping the type.
- Light fixtures and switches are component Families that you load and place in the Ceiling plan.
- Add wiring using Detail Lines.
- Add Interior Elevations using the Elevation View tool.
- Check more than one box on the interior elevation tag to add additional elevations.

Printing, Publishing and Exporting

INTRODUCTION

After all the hard work you put into building and annotating your models, you may be ready to generate some output. Revit can output data in a variety of forms. The most common method of output is, of course, Printing (or plotting). We also can output our design digitally in a variety of formats. In this chapter, we will look at the various ways we can generate output from Revit.

OBJECTIVES

Creating output from Revit is simple. Several options are available on the Application menu. After reading this chapter, you will know how to:

- Configure Print Setup options
- Print to a hard copy printer
- Create a multi-sheet DWF file
- Export your model in various formats

DATASET

For the purposes of the topics covered in this chapter, you can open any Revit project. Versions of the commercial and residential projects in Imperial units have been provided in a folder called *Chapter14*. You can practice printing and exporting from these projects if you wish, or open the "Complete" version of either project in metric or imperial units from any chapter folder instead. Feel free to open and print your own Revit projects as well.

THE APPLICATION MENU

Output from Revit can be grouped into three overall categories: Printing, Publishing and Exporting. Each of these items has a menu with sub-options on the Application menu (see Figure 14.1).

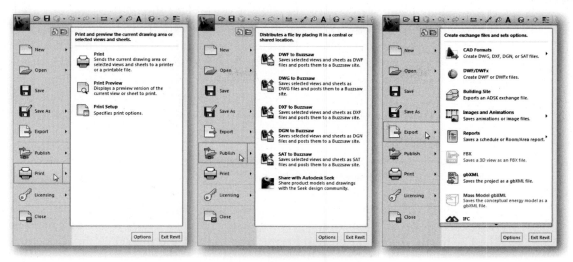

FIGURE 14.1 *The Application menu include three output sub-menus*

The Print sub-menu has commands for outputting to hard copy devices such as plotters and printers. The Publish sub-menu has specialized functions to send your project to Autodesk Buzzsaw or Autodesk Seek. Under Export, you find many commands to export some or all of your project data in many popular design file formats such as AutoCAD (DWG), Design Web Format (DWF), and a variety of others.

PRINT SETUP

Printing is perhaps still the most popular output. To prepare for your print, you should typically start with Print Setup. The "Print Setup" dialog box has many settings that can be configured to enhance the quality of printed output. Choose **Print Setup** from the Application menu to access this dialog (see Figure 14.2).

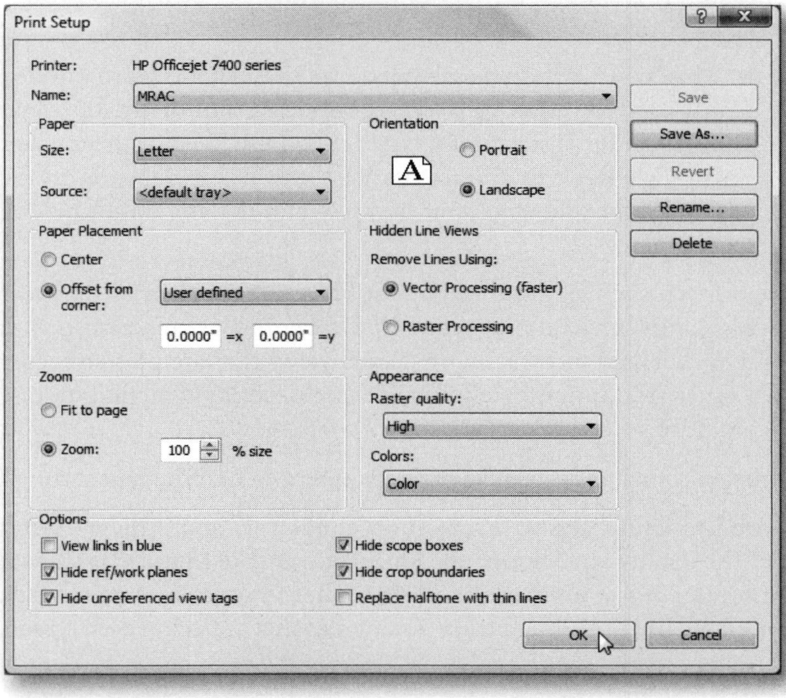

FIGURE 14.2 *The Print Setup dialog*

If you make changes to any of the default settings in this dialog, you can click the SaveAs button on the right and give the configuration a new name. The Print Setups are stored in the project. This custom configuration will then be available to you when you print the project in the future (MRAC is shown in the drop-down in Figure 14.2 as an example). If you want to read a description on any element in the dialog, click the question mark (?) icon at the top right and then click it on the item in question. A help window will appear with a detailed description of the item.

TIP	If you create custom Page Setups, they can be shared with other projects using the Transfer Project Standards tool on the Manage tab of the ribbon.

Be certain that the printer you wish to use is listed at the top of the dialog before you configure the options in the dialog. If it is not listed, click Cancel, choose Print from the File menu, choose the desired Printer from the list and then click the Close button. Reopen the Print Setup dialog and continue.

Paper: Set the correct paper size. Only sizes that the printer driver supports will be available.

Orientation: Set to Portrait (vertical) or Landscape (horizontal).

Paper Placement: Center will work for most printers. Otherwise, use Offset from corner and its associated options. You may need to experiment with your printer and driver to find the right combination for correct placement. If you make such adjustments, be sure to save the configuration with a descriptive name to preserve your efforts. (It is also not a bad idea to write the offsets down somewhere, just in case.)

Zoom: To print to the view's current scale it must be set to 100% of size. To print a half-sized set change the value to 50%. Keep in mind that everything will be smaller including all annotations.

Fit to page will also potentially affect the actual size and scale of the printed output. Fit to page might be fine for check plots, but if you want your print to be "to scale," choose 100%.

Hidden Line Views: Most construction document views in Revit are set to Hidden line. For example, plan views, elevation views, sections views, and many 3D views default to Hidden Line. Printing is frequently faster if you use "Vector processing" as indicated. The time it takes Revit to process the Vector print job depends on the size of the project, the type of model objects in the view, and the amount of hidden line removal that must be processed.

The Autodesk support web site has this to say about Raster Processing: "Using vector processing to plot complex views may result in display inconsistencies, such as lines that are missing or merged incorrectly, or shaded views that are plotted incorrectly, among other issues. You can often eliminate these issues by switching to raster processing when plotting from Revit."

Run some test prints on actual projects in your environment to find the best settings.

Appearance: If you are using Raster Processing, choose an appropriate Raster Quality. The higher the quality you choose, the longer it will take to print. If you are printing to a color printer and want black lines, remember to set the colors to black lines. This will result in all lines printing black—even ones that are colored on screen.

Setting the colors to grayscale will print any lines in the view that are black in solid black ink, and every other color will be a value of gray. If the gray printed output does not look exactly the way you want, try setting the colors to Color and then set the printer driver to black and white. This often gives slightly different grayscale results that might be useful.

Options: This tells Revit to print or hide certain elements in the view. Hide reference/ work planes is very useful. You would not typically want these items to print; likewise for scope boxes and crop boundaries. The "Hide unreferenced view tags" option will hide any elevation, section, or callout tag view that has not yet been placed on a Sheet. This is a very useful feature (see Figure 14.3).

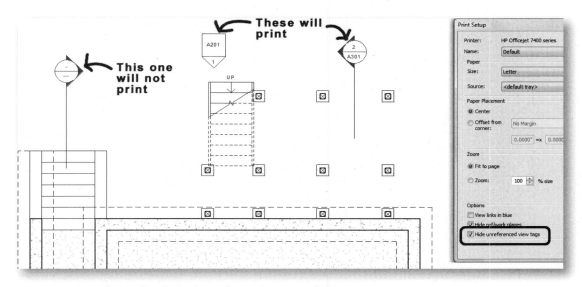

FIGURE 14.3 *Understanding the "Hide unreferenced view tags" feature*

Name: (See the top of the "Print Setup" dialog.) This is the name of the Print Setup settings. Click Save or Save As to create a new Print Setup. This is retained in the project for future use. You can also rename and delete existing named setups with the buttons on the right.

PRINT

When you are ready to print, choose **Print > Print** from the Application menu. You can print from any Revit view. Most often, you will want to print Sheets because they are designed for this purpose and include title blocks and borders. However, if you wish to print from another view you can do that as well.

Always choose your printer from the list first. Click the Properties button if necessary and make any edits to the printer's unique properties (these properties vary by printer device, which is why you want to choose the printer first). For the "Print Range" you have the option of the current view, the "Visible Portion" of the current view (for example, the zoomed-in area on screen), or a range of Views or Sheets. If you choose the "Select views/sheets" option, the "Select" button will become available. Click on this button to choose a range of Sheets or views to print (this is sometimes referred to as batch printing). Under Options choose "Reverse print order" if your printer prints pages face up. If a physical printer is used (as opposed to a PDF or other "digital" printer), you may also set the number of copies to be printed and whether they are collated. Click the Setup button to open the "Print Setup" dialog (seen above) and make additional edits. Click OK to print (see Figure 14.4).

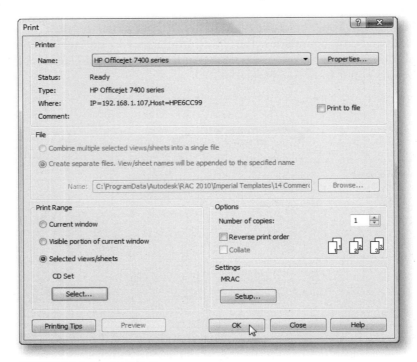

FIGURE 14.4 *The Print dialog*

Print to File: If your output device requires it, you can select this option and then use the fields in the "File" area to create a plot file.

Current window: prints the entire extent of the currently active view or sheet on screen.

Visible portion of current window: prints only the part of the view that is displayed within the current window on screen. To set what you want to print, close the Print dialog, re-size, and zoom the view's window accordingly. The proportion of the window is also important and should be similar to the paper size and layout (portrait vs. landscape). If the window proportions do not match the paper proportions, the printed output may exclude part of the view or might include extra white space.

Selected views/sheets: is a mechanism for batch printing. It allows you to pick multiple views and or Sheet views and print them at the same time. You can save the selection of views and or Sheet views in named sets that can be re-used later (see Figure 14.5).

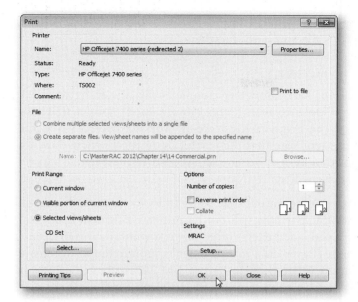

FIGURE 14.5 *Create lists of views and/or sheets to plot*

At the bottom of the dialog, you can choose to show only Sheet, only Views or both. Buttons on the right help you make selections. Select sheets and views individually, or use the SHIFT and CTRL keys to assist you. If you create a set that you would like to be able to print again, click the SaveAs button at the right to save it for future use.

The final command on the Print sub-menu is Print Preview. This works exactly as it does in other programs, and generates a preview on screen of what the print will look like with the current settings. If you wish, feel free to configure your choices in both the Page Setup and Print dialogs and go ahead and print. Depending on how many views or sheets you print, it should take a few minutes to generate your prints. If you wish to conserve paper, try printing a PDF or a DWF. You will need a PDF print driver such as Adobe Acrobat or one of the many alternatives to create a PDF. Revit does not supply such a driver.

PRINTER DRIVER CONFIGURATION

Since Revit prints using the Windows printing mechanism like most Windows software it is important that the correct printer driver be used. The correct driver is the driver specifically made for the physical printer and model used. To obtain the correct drivers for your printers contact the printer manufacturer and ask for the Windows printer driver for the model printer you have and follow the manufacturer's instructions for installing the driver. Most drivers are available for download from the printer manufacturer's web site.

Frequently Printing/Plotting Services or Bureaus will ask for a plot file in the HPGL format. This will not work with Revit. Inform the plotting service that you are printing from Windows-based Revit and must use the Windows printer driver for the printer model that they intend to use. Alternatively, you can ask if they can print a

PDF or DWF file. Most service bureaus can print PDFs and DWFs. In this case, simply print the PDF or DWF file and send it to them.

TROUBLE SHOOTING PRINTING

If your printed output does not look correct or complete, the first thing to check is whether the Print Preview looks correct. If the Preview looks correct that tells us that Revit, Windows, and the Printer Driver are all working correctly and the problem is at the physical printer. Note that the resolution of the Preview may not be correct because the high resolution is being displayed on your screen at a lower resolution and that is to be expected.

A common problem includes printed output that does not look complete. Either all the text will be missing or whole portions will not be printed. This is indicative of the printer's memory being overloaded. Revit often sends a combination of vector, raster, and other data formats. Raster data in particular is more memory intensive than vector data. As a result, it is common to have a 100 MB or larger print job. Some printers/plotters do not have enough onboard RAM or a built-in hard disk to handle this.

Some printer drivers have an option that says something like "Process print job in the computer's memory". To resolve the issue, try using this setting. If your printer driver does not have this setting you can either install more memory in the physical printer, use a printer that has a built-in hard disk and more memory or add a separate RIP (Raster Image Processor) device to aid the printer. Many currently available new large format printers have built-in RIPs.

PUBLISH TO BUZZSAW

You can publish to Autodesk Buzzsaw and Autodesk Seek. Autodesk Buzzsaw is a hosted document management system. It allows you to host project information, manage information flow and all kinds of documents including CAD and BIM files, RFIs, contracts and more. Visit www.autodesk.com/buzzsaw for more information. If you have a Buzzsaw account, you can publish your Revit project directly to your Buzzsaw site in either DWF or DWG format. Choose either DWF to Buzzsaw or DWG to Buzzsaw from the Publish sub-menu on the Application menu. Make your choices in the dialogs that appear and then publish. Open your Buzzsaw site to see the published file. You or your recipients can download the published files for offline viewing, or they can view directly in Buzzsaw with its embedded viewers (see Figure 14.6).

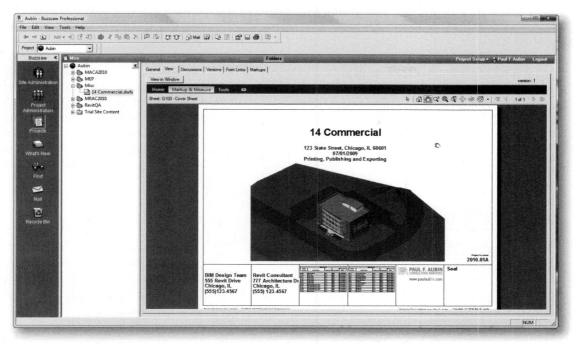

FIGURE 14.6 *After publishing a DWF to Buzzsaw, it can be viewed online directly in the Buzzsaw window*

PUBLISH TO AUTODESK SEEK

Have you created a Family that you just need to share? You can also publish directly to Autodesk Seek. Choose this option from the Application menu and Revit will prompt you to open a Family file or use one you already have open. Follow the prompts on screen to categorize the Family and then upload it to Seek. You will next be presented with a web page instructing you on how to complete your submission. You must register to post files to Seek. Log in to your account, fill in the required data and then complete the publication. Your Family will now be available for anyone to download from the Autodesk Seek portal.

EXPORT TO CAD

Export is the most extensive menu of the output choices. You can export your Revit model to popular CAD file formats like DWG and DGN, to DWF files, gbXML, IFC, ODBC and more. Time and space will not permit us to go into all formats in detail. Do take the time to look at the sub-menu on the Application menu and look up any formats that interest you in the online help. For this chapter, we will focus on two common formats: DWG and DWF. DWG is the industry standard for CAD data storage and transmission. It is likely that some if not most of your consultants and partners on any given project work directly in this format using AutoCAD. Most existing CAD data is saved in this format as well. When it comes time to share your design data with team members outside your firm, DWG will often be a desired, or even required, output. Fortunately, exporting Revit projects to DWG is simple. The basic steps are as follows:

1. **Configure your output settings**—From the Application menu, choose **Export** > **Options** > **Export Setups DWG/DXF**. This opens the "Modify DWG/DXF Export Setup" dialog. This dialog is completely redesinged and enhanced in this release. You can configure the way Revit should treat Layers, Lines, Patterns, Text, and more. Several layer

standards are included on the Layers tab such as the industry standard US National CAD Standard (NCS) guidelines (www.buildingsmartalliance.org/ncs/). You can edit any of the items in the dialog to match your company's or your recipient's company standard (see Figure 14.7). At the bottom left corner of the dialog, you can click the icons to create or duplicate setups and create new ones. Use the tabs across the top to change the various settings. You can map Revit line styles to AutoCAD linetypes on the Lines tab. On the Patterns tab, map Revit fill patterns to AutoCAD hatch patterns. Establish font mappnig on the Text & Fonts tab. On Colors, you can use either the AutoCAD color index or true RGB colors. If you export 3D views, the Solids tab controls if the export will create meshes or ACIS solids. Configure unit conversion options on the Units & Coordinates tab, and finally on General, you can edit a variety of settings such as the DWG file format used.

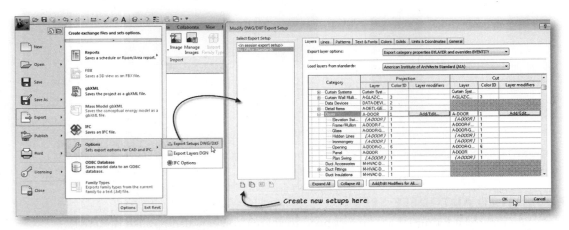

FIGURE 14.7 *If necessary, configure the layer mapping used to map Revit categories to DWG Layers*

2. **Export to DWG**—From the Application menu, choose **Export > CAD Formats > DWG**. If you created an Export Setup in the "Modify DWG/DXF Export Setup" dialog, it will be listed at the top left corner. You can also click the browse button to open the dialog and configure one. Any view lists you created in Print Setup will appear in the drop-down at the top right. Several options appear next to "Show in list" to help you filter the list of views and/or sheets to be exported. Select the checkboxes for the views you want to export in the Include column. If you want to save the list for later, use the icons at the top of the list (see Figure 14.8).

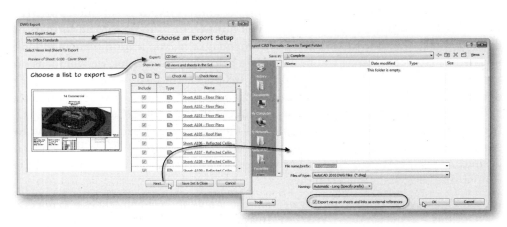

FIGURE 14.8 *Export one or more CAD files and configure their properties*

3. **Choose name and location to save the file(s)**—After you select your list of views to export, click Next and a save dialog will appear where you can browse to a folder to save the files, choose your file naming options, and optionally have Revit XREF views onto sheets. This is a very handy feature as most AutoCAD users set up sheet files that include a titleblock in Paper Space with each drawing externally referenced onto the sheet. When Revit exports sheets, it can emulate this setup by creating DWG files of each Revit view and then a separate DWG for the sheet view with XREFs to the separate DWGs. This setting is recommended.

If you have AutoCAD or Microstation, go ahead and try exporting one of the book projects to DWG or DGN. Configure any settings you wish to experiment or match your office standards. Open the resulting files in CAD and study the results. Revit does a nice job of creating CAD files following industry standard best practices.

EXPORT A DWF FILE

While many projects require output to actual paper sheets, digital submissions such as DWFx files are becoming more popular. A DWF/DWFx (Design Web Format) file is a highly compressed, two- and three-dimensional, vector-based file format designed for viewing and distributing design files over the Internet and by email. What makes the DWF file so powerful is that it is a vector-based, high-quality graphics file that is read only. This means it can be distributed to consultants and clients without fear of unauthorized editing. DWF preserves access to critical design data and graphics from the original model. DWF files can be referenced as backgrounds in AutoCAD. DWF files can also be embedded in web pages for viewing in a browser with the plug-in provided free from Autodesk. Anyone with a copy of the Autodesk Design Review software, available as a free download from www.autodesk.com/designreview, can view, zoom, pan, query, digitally mark up and print the DWF file. In Windows 7, DWFx files can be viewed native without any additional software required. Creating a DWF is simple, because it is the same as printing to a hard copy device.

In addition to being a good way to distribute a digital plot of your project, most Autodesk products also import DWF/DWFx files and take advantage of the data within them. For example, a DWF file can be XREFed into an AutoCAD file to scale and snapped to. Autodesk Quantity Takeoff (QTO) opens DWF files and can perform a one-click takeoff from the data in the DWF file. This provides a nice way to perform quick takeoffs of your Revit projects. Finally, Navis Works can import a DWF file directly as well. The other alternative for Navis Works is an NWC export. This option gets installed to your Revit Add-Ins tab if you have Navis Works installed on your system.

In order to create a DWF file from Revit follow these steps:

1. **Export to DWF**—From the Application menu, choose **Export > DWF**. Listing options are the same as they were for DWG. Build your desired list of views.

2. **Configure options**—Click the DWF Properties tab. There you can set additional options. For example, you can select the "Rooms and Areas in a separate boundary layer" option to make these elements selectable in the DWF file. You can also click the Project Properties tab and edit the project information. This is the same information you will find on the Manage tab of the ribbon.

3. **Choose name and location to save the file(s)**—After you select your list of views to export, a save dialog will appear where you can browse to a folder to save the files,

and choose your file naming options. If you selected multiple views, you can have them all included in a single multi-sheet DWF by selecting the "Combine selected views and sheets into a single dwf file" option (see Figure 14.9).

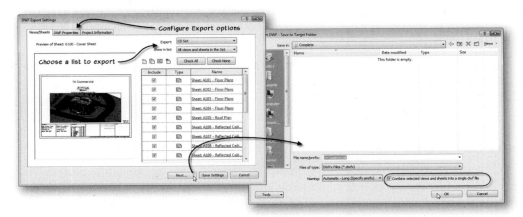

FIGURE 14.9 *Create a Multi-sheet DWF file*

When export is complete, open the DWF file with Autodesk Design Review software and view the results. Sheets and 2D views will appear as digital "plots." However, you will be able to select elements and query them directly in Design Review. The 3D views will be interactive and you can use the familiar steering wheel and ViewCube to navigate them. Model elements can be selected and their properties can be viewed, but not edited. Again, if you are using Vista or Windows 7, and you create a DWFx, you can just double-click it in Windows Explorer.

PUBLISHING TOOLS

Revit includes several tools to make Sheet composition and printing easier. While space here does not permit detailed tutorial coverage of the following tools, a quick mention is appropriate.

Dependent Views and Matchlines: When a drawing is too large to fit on a Sheet, it is customary for the drawing to be split into sections and a matchline to be employed. To accommodate this need, Revit provides the Dependent view feature. With this feature, you can duplicate a large view into several dependent views. Unlike the standard duplicate options, dependent views share their annotation. Annotation added or editing in either the parent or child view is reflected in both. To set up dependent views, right-click the overall view and choose **Duplicate View > Duplicate as a Dependent**. Repeat for as many sections as you need. For example, to matchline a plan into east and west sections, you will have one overall view (referred to as the Host view) and two dependent views. The dependent views are shown indented beneath their parent on the Project Browser. Adjust the crop regions as appropriate in each of the dependents (see Figure 14.10).

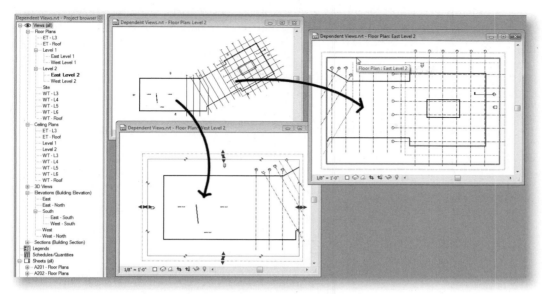

FIGURE 14.10 *Simple example of a large plan split into two dependent views*

The figure also shows the dependent view rotated to match the Sheet orientation. To do this, select the crop region boundary and use the ***Rotate*** tool. The crop region will remain horizontal when you are finished rotating and the model within it will rotate.

Scope Boxes: In complex buildings, you may have multiple Levels or Grids in different sections of the building. By default, all datum elements show across the full extent of the building. While it is possible to stretch the handles and adjust them manually, this can become tedious on large projects. A Scope box is a three-dimensional box that you sketch in plan. Use the handles to adjust its size. Give the scope box a unique name. All datum elements have a Scope Box property. Views also can use Scope Boxes to control the extent of their crop region. If you have no Scope boxes, the value of this property remains None. If you add Scope Boxes, you can edit the properties of the datum (or view) and assign it. The Grid or Level will now only display in views that intersect the Scope Box (see Figure 14.11).

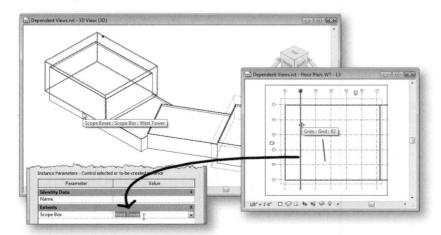

FIGURE 14.11 *Simple example of a Scope Box*

Revision Tracking: Despite every Architect's best efforts, revisions are a part of any document set. Once you have issued your drawings for bid and the inevitable revision packages must go out, use the ***Revision Cloud*** tool on the Annotate tab and the coordinated revision table available on the View tab; click the small icon on the Sheet Composition panel titlebar to manage them (see Figure 14.12).

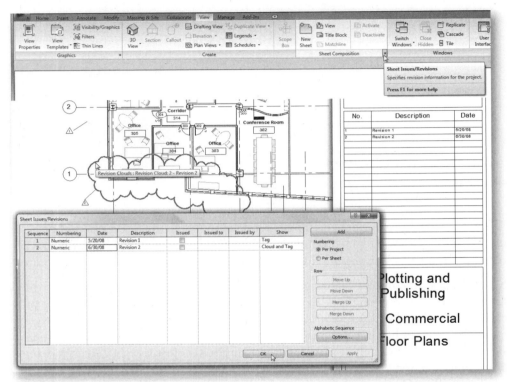

FIGURE 14.12 *Revision tracking coordinates, revision clouds, and tags with the table in the title block*

You add each new revision in the "Sheet Issues/Revisions" dialog (access it from the Sheet Composition panel as shown in the figure). When you add revision clouds (Annotate tab), they will appear automatically in the table on the Sheet.

There are many other export options. Feel free to explore some on your own. We exported some of the reports back in Chapter 12.

SUMMARY

Output from Revit includes many useful options. The overall process is simple and straightforward, regardless of the specific output desired.

- Like other Windows software, choose Print Setup to configure the Print options.
- Save your choices for future use in the project.
- Choose Print from the Application menu to print to paper.
- Use Print to create PDF output as well. A PDF print driver is required.

- Your project files can be published directly to Buzzsaw or Seek.
- Choose **Export** > **DWF/DWFx** from the Application menu to create a DWF or DWFx file.
- You can create a multi-sheet DWF file of your entire document set in one step. This can be an excellent way to distribute document sets to project stakeholders and reduce expensive paper, energy, and ink/toner at the same time!
- Revit provides tools for matchlines and tools for revision tracking.

INTRODUCTION

The term "Worksharing" applies collectively to the various techniques used in the Revit platform to work in teams of multiple individuals and firms. Worksharing typically refers to the internal segmentation of a project using Worksets, Central and Local Files but can also include processes we have already seen in this book such as linked Revit (RVT) files and linked AutoCAD (DWG) files. Worksets is the Revit toolset that allows a single project file to be divided into smaller pieces with which team members can work independently and simultaneously without impeding the work of others. Care must be taken when enabling Worksets to ensure that a strategy appropriate to the team dynamics is established. In this chapter, we will introduce the concepts and key terminology used in Worksharing as well as briefly discuss general Worksharing issues and strategies.

OBJECTIVES

Many resources are available to the reader for learning and understanding the concept of Worksharing. This chapter will provide an overview of the salient concepts and suggestions for further reading in the Revit Help system and online resources. After completing this chapter you will understand:

- Key Worksharing tools available
- Workset Terminology
- Where to find additional Worksharing Resources

WAYS TO SHARE WORK

Architectural projects usually involve teams of professionals either within the same firm (under the same roof) or dispersed among several companies and/or physical locations. Whether you simply need to load a CAD file as a background for your own design work or you need to manage a fully coordinated Revit model among several members of your firm, Revit has tools and capabilities suited to the task.

LINKING AND IMPORTING

In earlier chapters, we explored two forms of linking: linked RVT files and linked CAD files. If you need to simply keep track of work being done in another application such as AutoCAD or Microstation, then file linking is the appropriate solution. DWG (AutoCAD) and DGN (Microstation) files can be either linked or imported into your Revit model. If you wish to maintain the ability to update the file periodically as the original author of the file makes changes, choose to link the file. In this scenario, you simply reload the linked file when you receive an updated version from your consultant or teammate. If you instead need to use the geometry in the CAD file to assist you in creating your Revit model and have no need to reload changes in the future, you can Import. This places a static copy of the file within your Revit model. You can leave this imported file intact as a single element in the Revit model or even choose to explode it. If you explode it, Revit will convert the imported geometry into simple Revit Drafting and Model Lines. You should not explode these files unless absolutely necessary, as it will increase overhead and memory demands on your system depending on how large and extensive the imported files are. A better approach when exploding is deemed necessary to create a temporary Revit file first. Import the CAD data to this file, explode it, and then clean it up. This might involve reassigning linework to office standard Line Styles, deleting unnecessary items and other measures. Once the geometry is suitably cleaned up, you can copy and paste it to your project file. Examples of linking and importing DWG files can be found in Chapters 6, 8, and 11.

While linking to CAD files does help bridge the gap between your firm and those not using Revit software, it is always better if you can get the entire project team working in the same software and file format. Therefore, wherever possible, having all team members using Revit is preferable. They can be using any flavor of Revit including: Revit Architecture, Revit Structure, or Revit MEP. However, it is critical that all team members use the same version (2012 for example). You can still work in separate models and utilize linking in this scenario as we did in Chapter 6. In fact, this is often the preferred workflow for managing various disciplines. In the case of RVT files, you will typically link the file. Should you decide that it is desirable to "merge" two RVT files into a single file, it is possible to bind a link. This converts it to a Group but breaks the link to the original file. You can also use Copy and Paste or Group Save and Group Load instead. Open one of the files, copy all of the elements that you wish to merge, and then paste them into the other project. Group all of the model, annotation, and datum elements you wish to merge, save the group file out to a RVT file and then load that RVT file into the other project. Remember, model and annotation elements will actually be stored in separate model and detail Groups. Chapter 6 covers the process in more detail.

Revit also has the ability to import and export IFC files. The Industry Foundation Class (IFC) standard attempts to define a universal standard file format for the storage of building model data. IFC is currently the only bridge format that allows Walls, Doors, Windows, Roofs, Floors, and other building components to be preserved when imported and exported to and from Revit and other BIM software packages. While the technology is promising, it still cannot preserve all of the nuances that each software package introduces into its building models. If you absolutely must work in a team that uses Revit and other BIM packages together, then spend some time testing out the IFC import and export commands and settings found on the Application menu.

COORDINATION MONITOR AND INTERFERENCE CHECK

When you link two Revit files together (such as a Revit Architecture and Revit Structure file), you can use the Coordination tools to keep track of duplicate elements or elements that rely upon one another in each file. For example, if the Structural Columns are in the Revit Structure file and the partitions and the Architectural Columns are in the Revit file, these elements can get out of synch as users in each model make changes. Using the **Copy/Monitor, Coordination Review,** and **Interference Check** items on the Collaborate ribbon tab, you can copy elements between files, watch them for changes, and make updates to keep both linked projects synchronized. Look for more information about Copy/Monitor and Coordination Review in Chapter 6.

We have not looked at Interference Check yet, so let's take a quick look. In the *Chapter 15* folder of the dataset files, a version of our commercial project has been provided. You can use this for a quick example, or any other suitable project. The file is named: *B Commercial.rvt.*

> **NOTE** *B Commercial.rvt* is in Imperial units. If you wish to use Metric units, you can open a file from one of the earlier chapters.

An Interference Check searches the model for conditions where the 3D geometry of one object is interfering with (in the same space as) the geometry of another. An interactive report is generated when you run the command that allows you to select the elements in question to highlight them in the model. This allows you to assess the problem and work on solutions. To run an Interference Check, click the Collaborate tab, and then the Interference Check drop-down button. Choose Run Interference Check from the menu that appears. In the "Interference Check" dialog, two columns appear. You can run interferences between any combination of selected Categories in the same project or between the current project and a linked Revit file. Running the check between the current project and a linked project is the more common scenario, but in complex projects, running it between different elements in the current project can be very valuable as well.

Figure 15.1 shows an example comparing the Stairs in the current project with all of the structural elements in the linked *Commercial-Structure.rvt* file. You simply choose the desired project files from the drop-downs at the top (labeled no. 1 in Figure 15.1), and then check the Category or Categories that you wish to check against in the list beneath each file (item 2 in Figure 15.1). When you click OK, Revit will analyze the elements in your model and look for clashes between the Categories you selected (item 3 in Figure 15.1).

When the "Interference Report" dialog appears, you can expand any condition listed to see the elements that interfere. If you have an appropriate view open on screen, you can click on one of the items listed in the clash and see it highlight onscreen (item 4 in Figure 15.1). If you do not have an appropriate view open, you can click the Show button to have Revit search for an appropriate view. If you click the Export button, you can generate an HTML report of the clashes found. An example of the report is provided in the *Chapter 15* folder. You can use the element IDs in the report to select elements later if necessary. To select by element ID, use the Select by ID button on the Manage tab.

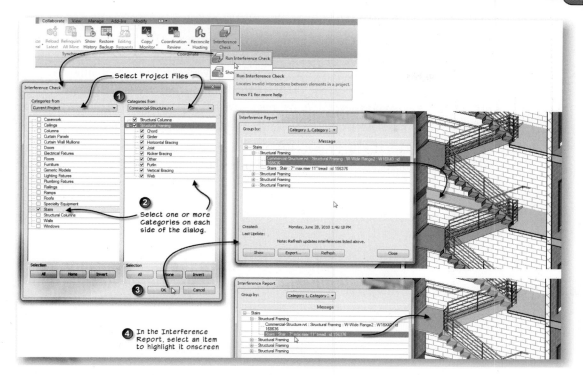

FIGURE 15.1 *Display the Worksets toolbar*

As you fix the problems listed, you can click the Refresh button to run the report again. The Interference Check tool offers a nice way to spot potential problems before they go too far in the design process. For much more complete tools and functionality, consider evaluating Autodesk Navisworks. Revit models can be exported directly to Navisworks for robust interference checking, timeline simulation, and rendering functionality.

WORKSETS

Even if all work on a project is carried out within the physical walls of your own office, it is typical that more than one member of your firm will need to access and work in the project at the same time. In this situation, enable "Worksets" to allow different members of the project team to work simultaneously on the same project file. Each user saves their own "Local file," which remains connected back to a master "Central file" that keeps all changes and team interactions coordinated. While Workset use is most common on a team project, it can be used in any Revit project and some Sole Practitioner firms find it useful as a tool for managing project data and visibility of elements. Whatever your specific situation, achieving success with Worksets requires a clear understanding of each tool, careful pre-planning and ongoing management.

> **NOTE**
>
> The following is a brief introduction to the concept of Worksets. It is intended to introduce the concepts, define the terminology, and suggest some common scenarios for their usage. A comprehensive tutorial-based exploration of the topic falls out of the scope of this text. Explore the online help and various online resources on the topic and consider attending a formal training session at a local training provider or Autodesk reseller. To learn more about hands-on training opportunities, visit the author's web site at: www.paulaubin.com.

Getting Started with Worksets

Before using Worksets they must be enabled in a project. After it is turned on, some configuration will be required. Typically this task should be performed by a single member of the project team knowledgeable in Worksets and their nuances. This person is often referred to as the "Project Coordinator" or some similar title. They might be a CAD Manager, BIM Manager, or simply a Project Architect on the project who has good knowledge of Revit.

To enable Worksets the first time, click the ***Worksets*** button on the Collaborate tab, or use the control located on the Status Bar at the bottom of the Revit screen (see Figure 15.2).

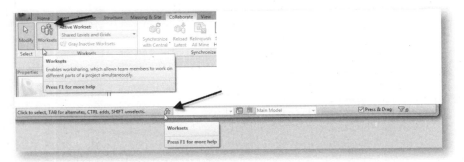

FIGURE 15.2 *Access the Worksets dialog*

Every member working on the team should have an understanding of the basic concepts of a Workset-enabled project. In addition, it is highly recommended that each member of the team reviews the help resources provided with the software and multiple team members practice together on a sample Workset project. The easiest way to find the topic in the Help file is to type: Working in a Team in the help field at the top right corner of the application frame and then press ENTER. The "Working in a Team" topic will appear on the list. Read the complete topic to become familiar with all terminology and features.

UNDERSTANDING WORKSETS

The concept behind Worksets is simple. Normally, in a computer environment, only one user at a time can access a particular file. Since by default, Revit places all project elements within a single file, this would naturally impede teamwork. Using Worksets, there is still a single "Central" file that houses all project data, but from this Central file, each team member must save a local copy that maintains its association with the Central file. In this way, when any team member saves work from their local copy back to the Central, those changes are synchronized with the Central file and become available to all members of the team. Simply having the Central and Local files is not always enough, however; the vast collection of potentially editable elements within the project files must be managed. Worksets are used for this purpose. A Workset is basically a named group of elements that can be "checked out" by an individual user. Once checked out, a Workset becomes "read-only" to other members of the team. This allows those members to view any part of the project that they wish, but only edit those parts that they have checked out. When saves back to Central are made, users can choose to "relinquish" their Worksets and/or check out other ones. The concept is similar to a public library or video rental store. If you wish to check out a particular book or video, it must be in the store at the time of your visit. If someone else has already checked out the item, you must wait for him or her to return it before you can check it out.

The biggest challenge involved in a Revit Workset-enabled project is deciding how to organize the Worksets and which specific elements they will contain. You want to have enough Worksets for the quantity of team members, but don't want to have so many that management of them becomes problematic. Any Worksets you or your team members create are referred to as "User-Created" Worksets. Certain other Worksets are created and maintained automatically by Revit. These include "Views," "Families," and "Project Standards" Worksets.

- **View Worksets:** For each view, a dedicated View Workset is created. It automatically contains the view's parameters (scale, visibility, graphics style, etc.) and any view-specific elements such as text notes, dimensions, detail elements, etc. View-specific elements *cannot* be moved to another Workset. To move them, you would have to cut the elements in question from their current view and paste them into a new view.

- **Family Worksets:** One Workset is created for the definition of each loaded Family in the project. (This is not for the individual instances in the project, but rather their presence on the Project Browser tree.)

- **Project Standards Worksets:** One Workset for each type of project setting, such as Materials, Line Styles, etc.

These Worksets are created and maintained automatically by Revit—no user interaction is required.

- **User-Created Worksets:** Elements that are part of your model will occupy one or more "User Defined" Worksets. These contain all of the building model elements in a project. The quantity and composition of User Defined Worksets is completely user defined; thus the name. By default, Revit will create two such Worksets when Worksets is first enabled: "Shared Levels and Grids" and "Workset1." All of the Levels and Grids are moved automatically to the "Shared Levels and Grids" Workset and all of the rest of the model geometry is moved to "Workset1." You can keep these two, rename them, add more or even delete them. However, at least one User Defined Workset must remain.

An element can only belong to one Workset at a time. You can move a model element from one Workset to another, but you cannot assign it to more than one at the same time. This is why careful planning is important. The Project Coordinator must therefore attempt to anticipate the needs of the project team and create Worksets to house model elements in a way that supports these needs. While there are no set "rules" regarding this, there are common best practices.

The factors to consider include:

- Project size
- Team size
- Team member roles
- Default Workset visibility
- Project performance issues

Each of these factors may have an impact on the use and composition of each Workset. Larger projects and larger project teams will typically have more Worksets.

This stands to reason. However, even small projects can benefit from Worksets, so this is not the only factor. When you open a project that has Worksets enabled, you can optionally choose which Worksets to load before opening. Choosing to open only those Worksets needed for a particular task can help files load more quickly and preserve valuable computer resources. You can also take advantage of Workset visibility to hide entire Worksets in one or more views as appropriate. Therefore, any or all of these factors can play an important role in determining the ultimate composition of Worksets in a particular project. Also, remember that each project is unique and while you may follow many common strategies from one project to the next, you must always be flexible enough to allow for specific circumstances that may arise in a particular project.

Enabling Worksets

As was mentioned above, it is typically advisable for a single experienced member of the project team to take responsibility for Workset management and setup. Worksets must be enabled before they can be used. This is a one-time task per project. The next step is to create a Central file and save it in a location that all team members can readily access. This location is typically on a network server on the company LAN (Local Area Network).

NOTE While it is possible to host a shared project over a WAN (Wide Area Network) it is only advisable if considerable bandwidth is available to all team members. In most cases, bandwidth limitations make hosting a project on a WAN impractical. Some firms are seeing success setting up remote stations in the physical office that hosts the central file, and using remote desktop for users in other offices to access the project instead. Other tools available to enable working across a WAN are known as WAN accelerators, such as Riverbed Technology's Steelhead accelerators. See www. riverbed.com for more information.

Another solution is the new Revit Server technology available in Revit 2012. Revit Server will be discussed briefly below. With it you can replicate and coordinate Central files across a wide area network to leverage extended project teams in separate physical locations.

To Enable Worksets, click the Collaborate tab and then click the **Worksets** button. As an alternative, you can click the Worksets button on the Status Bar at the bottom of the screen. A dialog like the one in Figure 15.3 will appear.

FIGURE 15.3 *Worksharing must be enabled. Revit will suggest two User Defined Worksets to start*

By default, two User-Created Worksets will be suggested by Revit: one for Shared Levels and Grids containing those elements, and another called "Workset1" that will contain all of the other elements. You can accept these names or choose alternate names. You can also rename a Workset later (provided that no users are accessing the project at the time you choose to rename). It will take a moment for Worksharing to be enabled; once complete, the "Worksets" dialog will appear and list each Workset in the project. Additional User-Created Worksets can be added at this time, or later. You can also use this dialog to list all of the "Views," "Families," and "Project Standards" Worksets (see Figure 15.4).

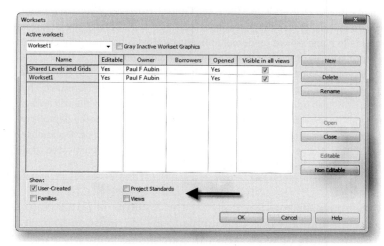

FIGURE 15.4 *Use the "Worksets' dialog to view and edit Worksets*

Create a Central file

Before users can begin working in a Workset-enabled project, there must be a Central file. This file is the main "hub" of the project. It should be located on a network server accessible to all team members. After Worksharing has been enabled, you use the **Save As . Project** command on the Application menu to save a Central file. The first time that Save as is chosen after enabling Worksets, the file saved automatically becomes a Central file. After that, if you wish to create a new Central file for any reason, you must choose **Save As . Project** from the Application menu, click the Options button, and then place a checkmark in the "Make this the Central location after save" checkbox (see Figure 15.5).

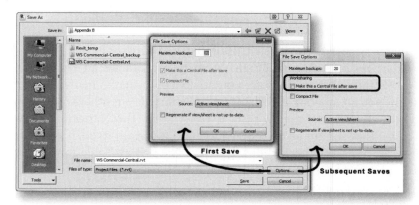

FIGURE 15.5 *Making a file the Central file*

There is some debate as to what the "proper" name for a Central file ought to be. It is common practice in many firms for the word "Central" to be added to the name. This has been the defacto standard for many years. Recent releases offer an alternative. Regardless of how you choose to name files, the critical issue is that it be clear to all team members which file is the Central file so that they do not open it and work directly in it. The traditional reason for appending the "Central" suffix has always been to help make it clear to users which file is the Central. However, if you create your Local file using the built-in functionality (see the "Creating a Local File" topic below), the inclusion of the word "Central" in the file name will be rendered unnecessary. This is because when you let Revit create a Local file automatically, it will automatically use the same file name as the Central file and append a suffix with your user name, for example, if the "Central" file is named **ABC Office Towers.rvt**. When you create a local file, it will become **ABC Office Towers_User Name.rvt**. This should be clear enough to distinguish one from the other. Further, since the Central file will always be on the network server and the Local file is typically in the **My Documents** folder (by default) or some other local folder like: **C:\Revit Projects**, this should also clarify the difference. Remember, whatever naming convention and procedure you adopt, make sure it is clearly communicated to all project team members and that all are required to follow it.

While it is possible to move or make a copy of the Central file, it is highly recommended that you maintain only one Central file to avoid confusion to project team members and potential loss of work. You can and should regularly back up the Central file using whatever method is currently in place in your firm, but avoid simply making a separate copy to another location on the network as users may mistakenly open this file and create Local files from it. Also, avoid "Detach from Central" except in situations where you are positive you will not need to save changes back to the Central file.

It is really very important that users understand these issues. If user A creates work synchronized to Central file A, and user B works in a detached file or synchronized to Central file B, there is no way for these two to synchronize their work with one another. Such a situation would largely defeat the whole purpose of using Worksharing. Please make sure that all team members understand this lest they decide to do a save as and work on a quick study off to the side. This may be acceptable practice in a CAD environment, but it will cause grief and rework in a Revit environment.

Editable and Open Worksets

The "Worksets" dialog has a column labeled "Editable" and another labeled "Open." Both will read either Yes or No next to each Workset (you can see an example in Figure 15.4). Both of these designations are potentially confusing. When you make a Workset "editable" you are making it editable *only* to yourself—in other words, you are "checking it out." This "locks" the Workset to other users. They can still borrow from your Workset if they submit an editing request and you grant it. So, if Editable says Yes, you have the Workset checked out. If it says No, it is available for anyone to check out or borrow from.

When a Workset is "Open" (Open reads Yes) it will be visible on screen. If Open reads No, then the Workset will not be visible. Choosing not to open certain Worksets is a strategy frequently employed by firms working on large projects to help speed load times.

CREATING ADDITIONAL WORKSETS

After the Central file has been saved, the next step in setting up your Central file is to decide if you need any additional Worksets, and if so, how many. It may not be necessary to create any additional Worksets at all. This is because Revit allows element-level locking. In Revit this is called "Element Borrowing" (see the "Borrowing Elements" topic below). What this means is that when you select an element to edit in the model, Revit checks to see if the element itself or the Workset to which it belongs is locked by another user. If it is not, the element is borrowed in your name, and then you are allowed to edit. Dedicated Worksets give you broad control over a collection of elements belonging to that Workset. If you are personally responsible for all of the furniture and wish to prevent anyone else from making edits to it, you can create a dedicated Furniture Workset and "check it out." This makes you the only user able to edit items in the Workset even if you are not actively editing them.

If you prefer a less absolute approach, use Element Borrowing. Any element not checked out by another user is available to edit—even if you don't have the Workset checked out or current. Simply select an element and begin editing it. Revit will allow you to borrow the element, leaving the rest of the elements in its Workset available to other users. See the "Borrowing Elements" topic below for more information.

It is recommended that you only create the Worksets that you are certain your team will require. Rely on Element Borrowing for the majority of editing and save dedicated Worksets for those items that need the extra level of "locking" control or as a tool to manage visibility and/or performance. If you are not sure which Worksets you need, try not creating any for the first project and leave all model elements on Workset 1. Doing so will have the team using Element Borrowing exclusively. When and if you run into limitations with this approach, consider adding additional Worksets. This approach is rapidly becoming "best-practice" at many large architectural firms who find that additional Worksets add complexity to the project but not always a commensurate level of value or benefit.

In the case where you have determined additional Worksets are required by the project team, you return to the "Worksets" dialog to add them. On the Collaborate tab, click the **Worksets** button (or the button on the Status Bar) and then click the New button to add a Workset. You can name each descriptively. An important consideration when creating new Worksets is their default visibility. If you wish to have the Workset automatically visible in all project views by default, then check the "Visible by default in all views" box when creating it. However, to help increase performance on large projects, you can opt to leave this setting disabled and allow users to control the visibility of each Workset manually (see Figure 15.6).

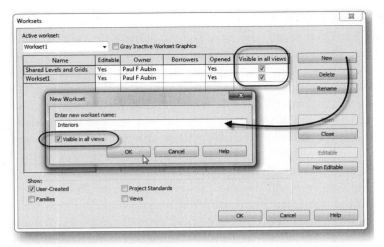

FIGURE 15.6 *You can disable visibility by default of a new Workset if desired*

When setting up the initial Worksets, try to divide building elements into logical groupings. These will often be "task-based" to support the work of the team member who will author its contents. For example, in a typical commercial office building like the project constructed in this book, you might create a Workset for the exterior shell of the building, the core elements, the lobby, and one for interior elements. Furniture might also be separated as might other specialty equipment (see Figure 15.7).

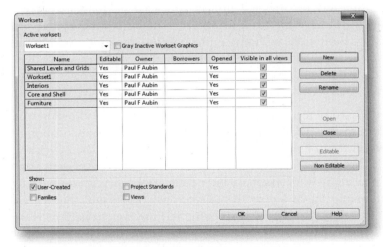

FIGURE 15.7 *Some Worksets added to the commercial project*

You can add new Worksets at any time in a project. It is however a good idea to try to establish the basic Workset organization as early as possible. This will make it easier for project team members to become comfortable with the project and its organization. As new Worksets are added later in the project, be sure that these and their functions are clearly communicated to the project team. Also remember, that if you are uncertain on whether a particular Workset is needed, leave it out initially and see how the project goes. If it becomes evident that the Workset is required, you can add it at a later time.

Moving Items to Worksets

When you start a new project with a collection of typical Worksets, there will be no model geometry, so creating the Worksets will be the final step of setting up Worksharing. If an existing project has Worksets enabled after geometry has already been added (say after a schematic design phase), existing elements in the model must be moved to an appropriate Workset(s). For instance, if a Workset named "Interiors" has been created, then any interior elements such as partitions, doors, etc., should be moved to this Workset.

NOTE

> **Important:** Following your initial setup, it will be up to the team members to create new elements on the appropriate Workset as they work. This includes Shared Levels and Grids. Revit only places Levels and Grids on this Workset automatically the first time it is enabled.

To move existing elements to a different Workset, select one or more elements, on the Properties palette choose the desired Workset from the "Workset" list (see Figure 15.8). When Worksets are enabled in a Revit project a Workset parameter will be automatically added to every model element and appear on the Properties palette.

FIGURE 15.8 *Move a selection of elements to a different Workset*

NOTE

> Your selection must contain only model elements for the Workset field to be available on the Properties palette. You can select a multiple selection of Walls, Doors, Windows, etc., but be sure that you have not inadvertently selected any tags, view tags, notes, or dimensions. These items are always view specific and are automatically kept in the View Workset for the current view and cannot be moved to a different Workset. Use the Filter tool to remove such items from the selection.

TIP

> If you make a selection of one or more elements, and attempt to change their Workset and the Workset option is grayed out, try the following: First, make sure that you only have model elements selected. You should be able to edit it now. New in 2012, you no longer need to make the elements editable first. The option still exists to right-click in the drawing area and choose Make Elements Editable. This will borrow the selection of elements and allow you to change the Workset. Starting with 2012, this should no longer be required. See the "Editing Workset Elements" and the "Borrowing Elements" topics below for more information on element borrowing.

Workset Tooltips

When you work in a Workset-enabled project, the tool tips that appear when elements are pre-highlighted will also report the Workset in front of the usual **Element Category : Family : Type** designation (see Figure 15.9).

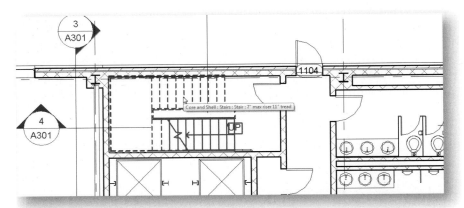

FIGURE 15.9 *Tool tips in Workset projects will report Workset : Element Category : Family : Type*

Workset Visibility

Worksets can be set to display or be hidden in each view of the project. When Worksets are created, the default visibility setting is assigned to the Workset. In the topic above, this issue was noted as one of the considerations to factor into Workset creation. It was also noted that new to Revit 2011, we can now change the default visibility anytime by reopening the "Worksets" dialog. To see which Worksets are visible in a particular view, open the "Visibility/Graphics Overrides" dialog. In a Workset-enabled project, a Worksets tab will appear in the dialog. On this tab, you can choose which Worksets you wish to see in the current view (see Figure 15.10). Next to each Workset listed will be three options: Show, Hide, and Use Global Settings. Use Global Settings is the default and will use whatever setting is enabled in the "Worksets" dialog. The Show and Hide settings override the global setting accordingly.

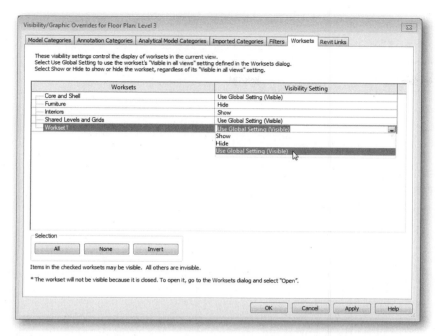

FIGURE 15.10 *Choose which Worksets you wish to display in a particular view in the Visibility/Graphics dialog*

This functionality offers a powerful way to manage the specific visibility of a collection of elements assigned to the same workset in a project. To fully realize the potential of this functionality, take care in the planning stages to determine the default visibility of each Workset in each view of the project. Users can change the settings later, but it is a good idea to establish effective defaults ahead of time.

Remember, anything you do in the "Visibility/Graphic Overrides" dialog is an override for the current view only. If you want to temporarily turn off a Workset throughout the project, close it instead. See the next topic for more details.

Close Worksets and the Central File

Following your initial setup, you should relinquish all Worksets. If you don't, team members will try to edit elements and be told that they are locked by you! As a final step in your Central file creation, relinquish (check back in) all the Worksets you created. Fortunately this is easy to do as a consequence of saving the Central file. Close the "Worksets" dialog. On the Collaborate tab of the ribbon, on the Synchronize panel, click the ***Synchronize with Central*** button. This button is also available on the QAT. The "Synchronize with Central" dialog will appear. In the "After synchronizing, relinquish the following worksets and elements" area, place a checkmark in the "User-Created Worksets" checkbox and then click OK (see the left side of Figure 15.11).

FIGURE 15.11 *Synchronize with Central to complete the setup*

Relinquishing the Worksets will make them non-editable to you and available for editing to other team members. If you return to the "Worksets" dialog, you can confirm that "No" now appears in the Editable column for all Worksets (see the right side of Figure 15.11).

The final step is to close the Central file. In normal use and project modeling, you and all the team members should avoid opening and working directly in the Central file at all costs. Only under special circumstances (usually related to data recovery) would you want to open the Central file directly. The integrity of the Central file is critical to your project's success. Please close the Central and do not reopen it! All project work will be performed in a Local file (see the next topic).

BIM *MANAGER NOTE*

A notable exception to the "never work in the Central file" rule is regularly scheduled model maintenance. Many recommendations are made in a technical document published by Autodesk. You can find it online at: http://images.autodesk.com/adsk/files/revit_tech_note.pdf. In general, it recommends that regular maintenance be performed on the Central file including: Audit, Compact, and recreating all Local files. These tasks are best performed by the project data coordinator or other person knowledgeable in Revit and Revit worksharing.

REVIT SERVER

Revit Server was introduced mid-release last year and is included with the 2012 product line. Many firms have multiple physical offices and need to share Revit projects across the wider extended project team. Achieving this has always proved challenging. The Revit Server offers a solution to this problem. Without major capital expenditures, distributed teams now have the option to do server-based worksharing while achieving similar speeds and experience to file-based worksharing on a LAN. This solution requires some setup and investment from an IT perspective, but should cost considerably less and perform better than WAN acceleration alone. Companies that already own such technology might see further improvement when coupling this with a Revit Server setup, but due to the protocol used in moving data to and between Revit Servers, it is thought to be of minimal benefit.

A Revit Server setup follows a hub-and-spoke network design, where one single Central Server communicates with any number of Local Servers. In turn, each Local Server communicates with only one Central Server. Users create Local files as normal. In its simplest form, one Central Server could be located in the office housing the largest number of team members, with any quantity of Local Servers distributed in other locations. From the perspective of how this technology works, it is of little difference to user experience where the Central Server is based since team members on any LAN segment all save to their Local Servers. These differ in their behavior from a Central Server: Once data is saved to a Local Server, it is immediately relayed to the Central Server. However if data is first saved to a Central Server, it will not be synchronized until such time when remote users open a copy through their Local Servers. So by strategically locating the Central Server, network traffic could be reduced, assuming most projects would be started on the Central Server. Once a project is opened from a Local Server, any changes to the project on the Central Server are immediately synchronized with that Local Server and any others that have previously been used to access that data. It is very important to note that only byte-level data moves across the WAN. This process is very efficient since only the required amount of digital information necessary to describe the changes since the last save needs to be replicated. By tuning the frequency of file saving, user experience could be improved to the point where all this underlying technology and infrastructure becomes almost completely transparent.

The Autodesk Online Help and Wiki provide very good detailed information on hardware requirements and proper setup. It is imperative to have the expertise of IT readily available to ensure a successful installation. On the software side, Revit Server has a very small footprint and the same files are used to set up all the servers. First, you will establish and name a Central Server, after which you will set up any number of Local Servers that point to it.

BIM Managers and assigned team members in each location will be permitted access to the Revit Server Administration utility, which is a web browser-based tool that requires Microsoft Silverlight. From this utility, Revit Central files can be moved around, renamed, and locked to prevent access during maintenance periods. Similarly, folders can be created, renamed, deleted, and locked through this administration utility, whose access can be limited to authorized users at the server level. Note that when this utility is accessed at any location across the WAN, the folder structure on the Central Server will be displayed. Local Servers are mostly transparent to the everyday user, but they can also be promoted to Central Servers in case of an extensive network/server failure or prolonged connectivity issues.

To save and open server-based projects, users have to connect to the Server on their LAN. This is a one-time operation, after which a Revit Server icon becomes available in the Open dialog. From then on, opening projects is very similar to file-based worksharing projects, with the only difference being the path to follow in finding and opening a project. Local files will still need to be created at the typical frequency. If you work in an organization with multiple physical locations, then you should definitely take a look at Revit Server.

Special Thanks to David Baldacchino for writing this passage on the Revit Server. You can find David's blog at: http://do-u-revit.blogspot.com.

Creating a Local File

Each member of the project team must create and work in a Local copy of the Central file. From this Local copy, users can check out the Worksets in which they need to work, borrow elements, and work alongside other team members. When a user checks out a Workset or borrows an element, the element is locked in the Central file and becomes read-only to other team members until that user relinquishes it.

Creating a Local file is easy. On the QAT, click the Open icon. Browse to the Central file and select it. At the bottom of the "Open" dialog is a checkbox labeled: "Create New Local." This checkbox is selected by default. Verify that it is selected and then click Open (see the left side of Figure 15.12). When this box is selected, Revit will create a Local file from the selected Central file. The Local will be created in the default location indicated in the "Default path for user files" setting on the File Locations tab of the "Options" dialog. If you want to view or change this setting, choose the Options button from the Application menu (see the right side of Figure 15.12).

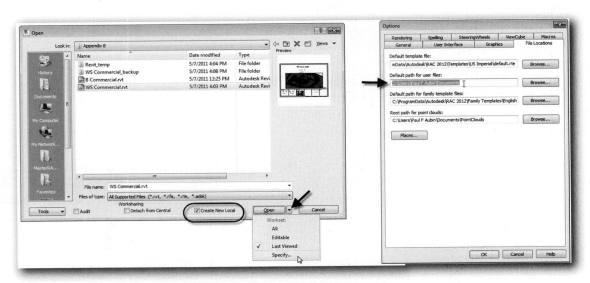

FIGURE 15.12 *Create a Local copy of the Central file in the default location*

It is very important that you do not deselect the "Create New Local" checkbox. If you do so, you are opening the Central file! Remember, rule number 1 is that opening the Central file should be avoided at all costs. Rule number 2 is that refer to rule number 1.

Feel free to open "Options" and change the path for local files if you wish. The exact location is a matter of personal preference. While you are in the "Options" dialog, take note of your user name on the General tab. This is the user name that Revit will use for all Worksharing functions. Every team member needs a unique user name. The default is your Windows login.

When you open your Local file, you can choose which Worksets you wish to open. In larger projects, this can make load times quicker. To do this, click the small drop-down next to the Open button (see the bottom of Figure 15.12). Editable loads the Worksets you have checked out. Last Viewed opens the same ones that were open the last time you opened the file, and Specify allows you to select the ones you want to load. After choosing "Specify," click the Open button and an "Opening Worksets" dialog will appear. Choose the Worksets that you wish to open by using the SHIFT and CTRL keys to select multiple items. With various Worksets highlighted, click the Open or Close buttons as desired (see Figure 15.13).

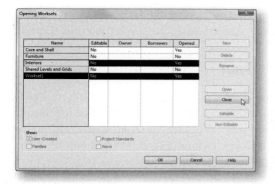

FIGURE 15.13 *Open and Close Worksets in the "Opening Worksets" dialog*

> When you first create the Central file, you can indicate an option for Open workset default to make Specify or another choice the default option. This default will apply to all users and their Local files.

When you click OK on this dialog, the model will load with only those Worksets that you specified. You can see this by opening views that require a Workset that you did not load. The required elements will not appear in that view. Should you realize that you need to open a Workset that you did not choose to open initially; you can simply launch the "Worksets" dialog, and then select the Workset you need and click the Open button. You can Close Worksets that you no longer need in the same fashion. Opening and closing this way is project-wide (in all views). Closing unneeded Worksets also reduces the load on the computer's resources, potentially increasing performance. The alternative was illustrated in Figure 15.10, where you can actually turn the display of Worksets on and off in the "Visibility/Graphic Overrides" dialog for individual views.

The first time you have Revit automatically create a Local file using the procedure covered here, it will save the file by appending your user name (from the "Options" dialog) to the original file name. Subsequently, when you follow the same method again (select the Central and let Revit create a new Local), you will be prompted either to overwrite the existing file or create a new one with a date and time stamp (see Figure 15.14). You should re-create your Local file on a regular basis to keep the

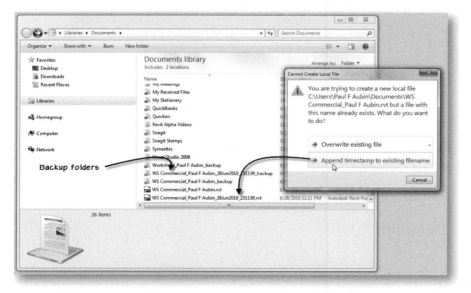

FIGURE 15.14 *Creating another new Local can overwrite the existing or append a date and time stamp*

file healthy. Using the date and time stamp option essentially gives you an additional backup copy. However, always work in the latest Local file. Save the local file and Synchronize with Central frequently.

Editing Workset Elements

When you first open your Local file and choose the Worksets you wish to Open, none will be editable. Opening a Workset does *not* automatically make it editable (check it out). This action is achieved in the "Worksets" dialog. There are a few approaches you can take to editing the model. In general, if an element that you wish to edit is not being edited by another user, Revit will allow you to edit it, even if you do not have the associated Workset checked out. As noted above, this is called "borrowing." It happens automatically and is completely managed by Revit, except for Synchronizing with Central and relinquishing the borrowed elements. That happens when you select the Synchronizing with Central button. If another user is actively editing this element, or has the Workset checked out (set to Editable), you will need to issue an "Editing Request" to that user. The other user can then choose to allow your edit or refuse your request. In any case, a good line of communication between you and your team is crucial.

Editing Requests

When you need an element that is checked out by someone else, you can contact the user via phone, email, instant messenger, text message, or face-to-face to coordinate the edits. New in 2012, you can use the improved "Editing Request" feature; the owner of the element will be alerted by pop-up message on their screen. This will instruct them to check the "Editing Requests" dialog and respond to pending requests.

TIP	If you an Autodesk Subscription customer, you have access to the Worksharing monitor. Visit the Subscription portal for more information. The Worksharing Monitor is a separate application that provides tools to manage Worksharing.

The flow is as follows. If you try to edit an element that is already owned by someone else or is part of a Workset that is checked out by someone else, you can click the "Place Request" button in the error that appears. An alert will appear on the other user's screen. They can then click the Editing Request icon on the Status Bar to open the "Editing Requests" dialog. There they can grant or deny the request. The decision will appear on the requesting user's screen as a pop-up message (see Figure 15.15).

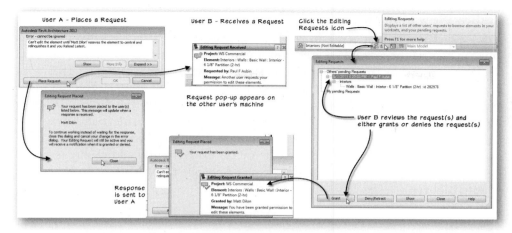

FIGURE 15.15 *Place and grant requests to borrow elements*

Borrowing Elements

You can "check out" a Workset by making it Editable in the "Worksets" dialog. Launch the "Worksets" dialog, select one or more Worksets, and click the Make Editable button (see Figure 15.16). However, as noted previously, it is better to rely on borrowing. Notice in the figure that the user "Matt Dillon" is listed as a "Borrower" to the Interiors Workset. Borrowing occurs in real-time as you edit elements in the model. If no one else is editing the element already or if the Workset to which it belongs is not locked for editing, Revit will "borrow" the element from the Workset and allow you to make changes. From that point on, no one else on the team will be able to edit that element until you relinquish it. When you Synchronize with Central, the element will be relinquished automatically (unless you specify otherwise) and the changes will be updated to the Central file.

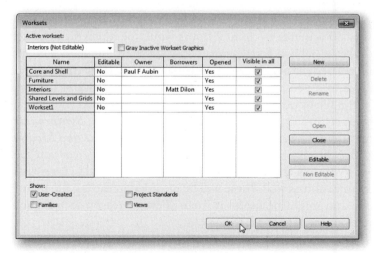

FIGURE 15.16 *Make a Workset editable*

Adding New Elements

Any new elements that you add to a model will be added to the active Workset. You can choose the active Workset from the drop-down list on the Worksets panel (Collaborate tab) or from the drop-down on the Status Bar (see Figure 15.17). In Figure 15.16, Interiors is the active Workset (see the top left corner). You can make a Workset active even if it is not editable (as you can see in the figure). In other words, you do *not* need to check out a Workset to add new elements to it. Always pay attention to the active Workset! This is very important since you don't want to add several objects to the wrong Workset. If you do inadvertently add elements to the wrong Workset, you can always select them and change their Workset on Properties as noted above.

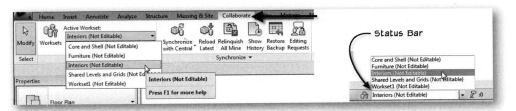

FIGURE 15.17 *Change the active Workset on the Collaborate tab, the Status Bar, or in the "Worksets" dialog (not shown)*

Synchronize with Central

You should save your work at regular intervals regardless of whether you are working in a Workset-enabled project or not. When you work in a Local file, there are two types of save: Save and Synchronize with Central. When you choose **Save** from the Application menu, the QAT or press CTRL + S, you are simply saving your Local copy of the project file. This is very important to do and it is a good idea to do this regularly, for example, every 15–30 minutes. Also at regular intervals (but perhaps not quite as frequently), you should Synchronize with Central. You will find this tool on the QAT and the Synchronize panel of the Collaborate tab. It is a split-button with two options: the ***Synchronize and Modify Settings*** and the ***Synchronize Now*** tools. Synchronizing every 30 minutes to up to 2 hours would be a practical choice, or whenever you would like to publish your changes so your team members can get access to your additions and edits. Synchronize with Central updates the Central file with all of the changes that you have made in your Local file and also retrieve changes made to the Central file by other team members since you opened your Local copy, last synchronized, or performed a Reload Latest. If you choose the ***Synchronize Now*** tool, default options are used without prompting. The ***Synchronize and Modify Settings*** tool opens the "Synchronize with Central" dialog. This dialog gives you the option to relinquish your Worksets and borrowed elements. You can add comments to document your synchronization as well. These will be recorded in the log file maintained by the Central file and might be useful if there is ever a reason to roll back changes to a previous version (see Figure 15.18).

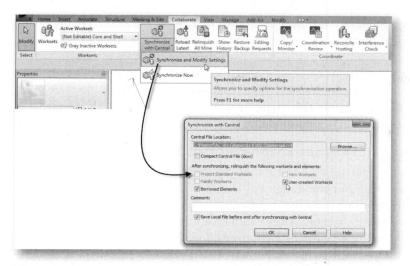

FIGURE 15.18 *The "Synchronize with Central" dialog offers options to relinquish borrowed elements and Worksets and to save your Local copy after save*

As a general rule of thumb, you should always check the "Save the Local File before and after synchronizing with central" check box. This will save your Local copy before and after synchronizing with the Central file. The Local save before is a Fail Safe procedure that ensures your work is saved in case the Synchronize fails and crashes Revit. The Local save after helps keep your file on disk up-to-date once all the additions and edits by your team members are merged into your Local file during the Synchronize with Central process. You should also relinquish your borrowed elements. You may wish to keep your Worksets checked out (User-Created Worksets) if you are still actively editing them. Otherwise, you should relinquish those as well. Comments are also a good idea as noted above.

The Synchronize with Central process involves three sub-processes: reloading the latest data in the current Central file; merging the additions and edits in your Local file with the latest data in the Central file; and finally writing the combined data back to the Central file. During this process the Central file is locked so no other team members can access it. The duration of the Synchronize with Central depends on the quantity of elements added and edited being merged in both the Local and Central files, their complexity, and the time it takes to read and write the data to the Central file over your network.

Reload Latest

Sometimes you may leave your Revit session inactive for a while with a Workset-enabled project loaded. You may also open a Local file that has not been synchronized in a while. If you have not made any changes, but know that your team has continued to work, you can use the Reload Latest tool. This tool downloads the latest changes from the Central file and updates your local file but does not push any changes from your local file to the server. If you have made changes, then it is better to simply Synchronize with Central which both uploads your changes and downloads your teammates changes.

Relinquish All Mine

Sometimes you make changes that you decide you don't want to keep. In such a case, your edits have still resulted in elements being borrowed from the Central file. Therefore, unlike most other software where you could simply close the local file and choose not to save, in Revit, you must relinquish all of your borrowed elements and discard your changes by not saving. If you simply close your file without saving, you will still be listed as a borrower and/or owner of one or more Worksets and elements. Naturally, it is very important to be sure that you do not want to save the changes. Once you choose to close a file and not save, there is no undo. The changes are gone. So make certain first. If you do this, make sure you relinquish all your Worksets so that other team members can gain access to those elements.

Default 3D Views in Workset-Enabled Projects

When you click the Default 3D View tool, Workset-enabled projects will create a new 3D view as usual, but the name will also include your user name. This happens automatically when you click the tool (see Figure 15.19). Notice in the Project Browser the new 3D view named: {3D Paul F Aubin} in the figure. Each user will get their own 3D view like this one when they click the tool. When you next synchronize, View Workset will be checked in the relinquish area. This will occur anytime you create a new view in a Workset-enabled project.

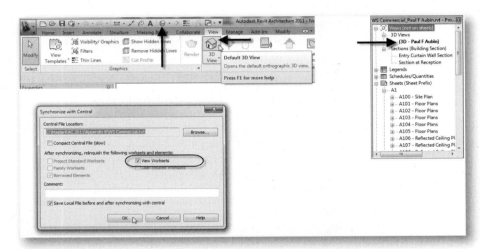

FIGURE 15.19 *Default 3D view names are appended with your user name in Workshare projects*

Worksharing Display

One of the most useful enhancements to 2012 is the Worksharing Display modes. With this feature, you can color-code the onscreen display of elements to reflect checkout status, owners, model updates, and Worksets. In Workshare-enabled projects, the Worksharing Display Options icon will appear on the View Control Bar. Click it to configure the settings and choose a display mode (see Figure 15.20). There are four modes and each can be customized in the "Worksharing Display Options" dialog.

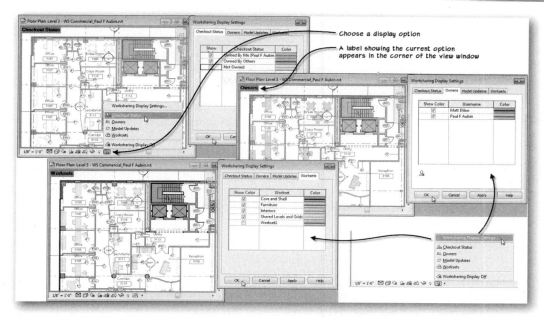

FIGURE 15.20 *Worksharing Display Options display useful Worksharing information in color codes onscreen*

Checkout Status: This mode shades elements that are checked out (either as part of a Workset or borrowed). Three colors are used: a color for your elements, one for other users, and a third for elements that are not owned by anyone.

Owners: It shades elements based on the user who owns them. Each user who has edited the project is listed with their own color.

Model Updates: This mode displays updated and deleted elements using different colors. This can be helpful to determine which elements are available for edit.

Worksets: It shades each dedicated Workset in a unique color. This can be a useful way to identify if elements are on the wrong Workset.

For each display mode, you can choose which conditions to show and change the color(s) used in the "Worksharing Display Options" dialog.

WORKSET TIPS

Keep the following tips in mind when working in Workset projects.

- Never work in the Central file. If you ever have to open the Central file for any reason, perform your required tasks quickly and get out quickly. Inform your team before opening the Central file and have them Synchronize with Central and close their files first. Any maintenance or other work performed directly in the Central file ought to be performed by the Project Data Coordinator exclusively.

- If you have recently opened the Central file (please review the previous tip), realize that it will appear on the Recent Files screen like other projects. However, if you click the icon for the Central file on Recent Files, it will *not* create a Local file, but will open the Central again! Therefore, click the Open icon instead, and browse to the Central to trigger Revit to create a new Local.

- Save often! Save both your Local copy and Synchronize with Central often.

- Revit projects can get large. Central files can get even larger. While it is important to save often, Synchronize with Central times can be very long on big projects. Plan your synchronizations to take advantage of down times.

- Always relinquish all borrowed elements and Worksets before leaving work for any extended period of time: going to lunch, attending a meeting, leaving at the end of the day, going on vacation, etc. Nothing makes a coworker crankier than trying to edit an element locked by someone who is out of the office.

- If you have been away from your computer for a while, and left a project file open (which is not recommended) reread the previous tip! After that, perform a ***Reload Latest*** (Collaborate tab, Synchronize panel). This command synchronizes your local copy with the latest saved changes of the Central file without saving your additions and changes back to the Central file.

- Check the Review Warnings dialog (Manage tab, Inquiry panel) on a regular basis. Perhaps once a week. Sometimes excessive warnings can have a negative impact on performance and even in extreme cases corrupt your Central file.

- Recreate your Local file on a regular schedule. You can recreate it as often as you like, (even every day) but in general you ought to create a new Local file at least once a week. Remember, Revit will automatically offer to date and time stamp your previous Local file for you (see the "Creating a Local File" topic).

- You can never have too many backups of your Central file or Local files. Hard drives are much cheaper than hours of recreating lost work.

- Placing linked files on a dedicated Workset can be an effective way to manage performance. While you can certainly unload a linked file, this change would affect all users the next time you Synchronize with Central. Alternatively, using a Workset to manage Links, each user can control which Worksets are active and displayed in their own work session. Therefore you can turn off the "Links" Workset on your local file without affecting your colleagues.

- Be careful with using the Activate View feature on Sheets. While this is a handy command, sometimes annotation added in this mode can loose its association to the proper View Workset and not display properly or disappear altogether. It is recommended that you expand the Sheet on the Project Browser and then double-click the view you wish to edit to open it directly instead.

- In most cases, trying to load and Synchronize with Central files over a Wide Area Network (WAN) is prohibitively slow due to the significantly higher bandwidth requirements. Consider setting up a remote desktop connection instead.

- When archiving a Central file, do not just copy it to an archive folder. Copying the file makes it a Local file which remains associated with the Central file. To make an archive copy, open the Central file (yes, just this one time) and choose **Save As** from the Application menu. Click the Options button and choose the "Make this a Central file after save" option as shown in Figure 15.5. You can also use the "Detach from Central" option to achieve the same goal.

- If you or a team member wishes to open a project "read only" to take measurements, print, or perform other work that they do not wish to save, open the file with the "Detach from Central" box selected in the "Open" dialog. Use this option with caution however. There is *no* way to change your mind later and "re-connect" a detached file with the Central. Once you have detached it, it cannot be reattached!

GOING FURTHER

This appendix explains the basic concepts of Worksharing and Worksets. The best way to learn and implement this multiuser mechanism is to practice with at least one other individual. This will give you a better sense of the nuances of working in a Workset-enabled project. Two copies of the commercial project have been provided with the dataset files in the *Chapter 15* folder. *WS Commercial.rvt* is a Central file.

If you wish to experiment with this file, make a Local copy and then explore. If you want to try your hand at making a Central file, open *B Commercial.rvt* and walk through the procedures outlined at the start of this chapter to makes a Central file.

Feel free to experiment in these files. If you have a colleague who can help you, each of you can create a Local copy and work simultaneously. Each of you should make changes, Synchronize with Central, and then reload from the Central file. Try borrowing elements, try to edit the same element as your colleague, and see what happens. Be sure to use the new Worksharing Display Modes as well.

SUMMARY

Working in teams is an important part of Architectural production. Using linked files, Element Borrowing, and Worksets, Revit provides the means to accomplish sharing of data and managing coordinated team projects. When implemented with care, Worksets provide an invaluable toolset to the extended Revit project team.

- AutoCAD and Microstation files can be linked into Revit models.
- Import CAD files when you do not need them to update. Link them when you wish to be able to update them regularly in their native software.
- Worksets provide the means to sub-divide a project into parts so that multiple team members can work simultaneously.
- Enable Worksets in the project and create User-Created Worksets.
- Try using Element Borrowing instead of creating extensive user Worksets.
- Move existing elements to appropriate Worksets.
- Save the project file as a Central File on a network server accessible to all users.
- Each user creates a Local file from the Central file in which to work.
- Decide which Worksets to Open when opening the Local file.
- Decide which Worksets to make Editable (which is like "checking them out").
- You can edit elements that are not in an editable Workset if they are not being edited by other users.
- Editing an element in a non-editable Workset is called "borrowing."
- Borrowed elements and Editable Worksets can be relinquished during Synchronize with Central operations.
- Save your Local file and Central files often.

Conceptual Massing
and Rendering

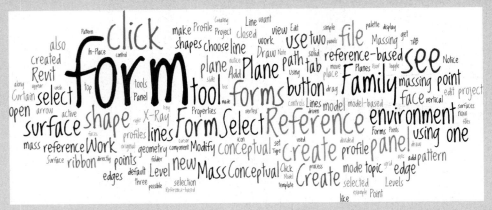

Most common words in Section IV

Conceptual Modeling tools in Revit offer exciting new potential for the design phases of a project. Introduced in 2010 and packed with new enhancements in the releases since then, this collection of tools allows you to explore design studies in a freeform three-dimensional environment. Explorations can be explored and presented independently in the Conceptual Modeling environment, or such studies can be loaded into project to form the basis for a project's design. The Conceptual Modeling environment is presented here in its own complete chapter.

Section IV is organized as follows:

Chapter 16 Conceptual Massing
Chapter 17 Rendering

Conceptual Massing

INTRODUCTION

Revit's massing environment offers tools to explore conceptual design concepts in a free-form 3D environment. The massing environment is a variation of the Family Editor and, therefore, shares many of its traits with all other Revit Families. The environment allows you to easily create and edit forms, create complex parametric shapes and patterned surfaces, and apply adaptable sub-components to patterned surfaces. The conceptual massing environment is meant to facilitate working/designing/creating/experimenting all while remaining in a 3D (axonometric) view. This is why, when you open up the conceptual massing environment, the default three-dimensional view *{3D}* is active. It is possible to make all of the shapes and forms that available in the standard Family Editor, but the tools and workflows are different and there is an enhanced ability to edit and manipulate the forms in real-time without sketch modes. Complex interactive and **parametric** form modeling and design rationalization is also possible.

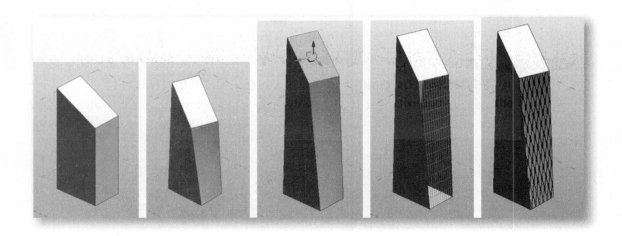

OBJECTIVES

Discussion of the mass modeling capabilities in Revit could fill an entire book on their own. Time and space here will therefore not permit complete coverage. The goal of this chapter is to introduce you to the basic workflow and features to get you started using this exciting toolset. In this chapter you will:

- Learn how to access the Conceptual Massing Environment
- Learn how to Create Forms
- Understand the difference between Model and Reference-based forms
- Create a Conceptual Mass Loadable Family
- Work with Divided Surfaces, Patterns, and Points
- Understand the difference between Driving and Driven points

IN-PLACE VS. LOADABLE FAMILIES

The conceptual massing environment is available both as an external environment in which you work in a specialized Family Editor (creating a "loadable" Massing Family) and in the In-Place Family mode where you work directly within the context of a project. Use the In-Place mode if your massing depends on the context of other elements in your project file, or if you prefer to design your massing within the context of the project elements. As with any In-Place modeling, the forms you make remain within the current project and cannot be saved out for use in other projects. There are a few limitations with the conceptual massing environment when working in In-Place mode; you will not be able to view Reference Planes in 3D views (this is typical with Revit) and you won't be able to see Levels in 3D views. However, you will be able to reference the levels and reference planes through the ***Set Work Plane*** tool, and you can also reference geometry that exists in the project (such as Walls, Floors, Topography, etc.) by selecting the geometry and Creating Forms from the project geometry. A very simple example of the in-place Massing Environment is found in the "Add an In-Place Mass" topic in Chapter 4 and the "Editing Massing Elements" topic in Chapter 9.

Use the external Family Editor environment when you want maximum flexibility in form manipulation, when you want to be able to load the Conceptual Mass into multiple projects, or when you want to iterate a design without necessarily loading it into a project environment.

ACCESSING THE CONCEPTUAL MASSING ENVIRONMENT

The massing environment is accessed in one of four ways (two for loadable mass Families and two for In-Place massing Families). To create a "loadable" massing Family, you must use a Conceptual Mass template such as the: *Mass.rfa* [*Metric Mass.rft*] Family Template. Family Templates were discussed in Chapter 10. You can access this template directly using the "New Conceptual Mass" link on the Recent Files page or the **New > Conceptual Mass** command on the Application menu. In either case, the "Select Template" dialog will open to the *Conceptual Mass* folder of your Revit *Templates* folder (see Figure 16.1).

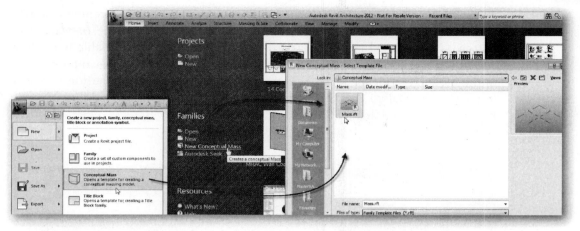

FIGURE 16.1 *Create a "Loadable" Massing Family from the Conceptual Massing Family Template*

The *Mass.rfa* [*Metric Mass.rft*] Family Template file will open in the new conceptual mass environment.

The *Adaptive Component.rfa* [*Metric Adaptive Component.rft*] Family Template file creates a Family that can adjust to the unique conditions in its context. The potential of the Adaptive Component is vast and could fill an entire chapter on its own. We will explore a very simple kind of adaptive Family below in the "Stitching Borders of the Divided Surface" topic. To learn more, explore the online help and be sure to visit the BuildZ blog at: http://buildz.blogspot.com/. BuildZ is devoted almost entirely to the conceptual modeling tools in Revit, and its author Zachary Kron provided invaluable assistance in updating this chapter.

If you are already working in a project and wish to create an In-Place Massing Family instead, you can do so from the tool on the ribbon. On the Home tab, from the **Component** tool drop-down button, choose the **Model In-Place** tool. In the "Family Category and Parameters" dialog box, choose Mass and then click OK (see Figure 16.2).

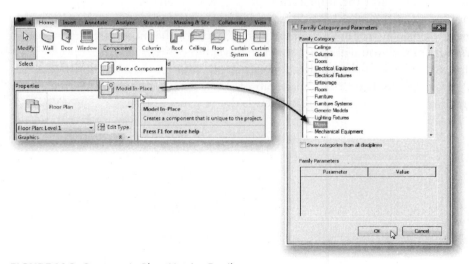

FIGURE 16.2 *Create an In-Place Massing Family*

The second method for accessing the In-Place conceptual mass environment is from the Massing & Site tab on the Ribbon. From this tab, click on the ***In-place Mass*** button. Creating an In-Place Mass this way will automatically enable the "Show Mass" mode. A dialog will appear indicating that masses will be shown temporarily in the model (see Figure 16.3). You can also click the Show Mass button to enable it without the dialog. The Show Mass tool was discussed in the "Editing Massing Elements" topic in Chapter 9.

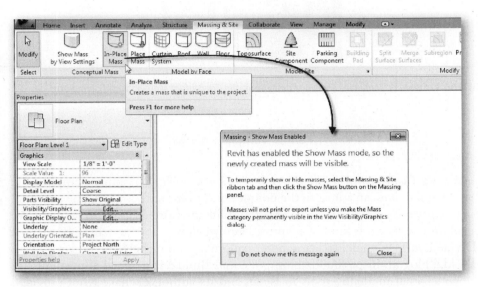

FIGURE 16.3 *Creating an In-Place Mass enables the temporary Show Mass display mode*

THE MASSING ENVIRONMENT INTERFACE

When you create a conceptual mass in the conceptual mass environment, you will default to an axonometric view. The environment includes one Level and two vertical Reference Planes by default. Similar to other Family templates, the two Reference Planes are pinned and the intersection of the two Reference Planes indicates the insertion point of the mass when it is loaded and placed in a project. Levels in this environment display in 3D and appear in the "back" of the model area; therefore, if you spin the model, you will notice that the Level lines always make an upside down V-shape at the back of the model. You can create additional Levels and Reference Planes using the tools on the Create tab of the ribbon. Unique to the conceptual massing environment, you can create Levels directly in a 3D view.

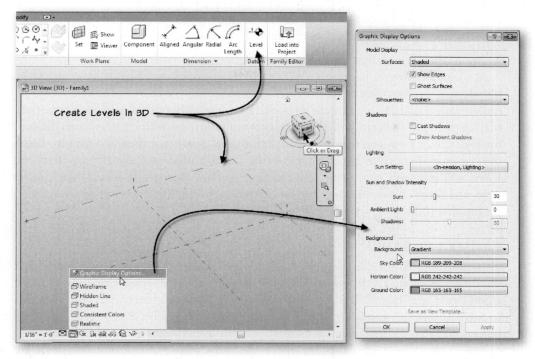

FIGURE 16.4 *The Conceptual Massing enviroment uses a shaded display with gradient background and shows the edges of datum elements*

Shaded With Edges is the default shading mode in the Conceptual Massing environment. A gradient background is also enabled. You can adjust the colors of the gradient background in the "Graphic Display Options" dialog box accessed from the Visual Styles pop-up icon on the View Control Bar (see Figure 16.4). Other than these obvious visual differences, the Conceptual Massing environment user interface is largely similar to the Family Editor discussed in Chapter 10.

The Conceptual Mass Family is workplane-based by default, which means that when it is loaded into and placed in a project environment (or another mass Family environment) it will look for host faces, rather than Levels or Reference Planes as many Families require. (Family hosting behaviors are discussed in Chapter 10.) This makes the Family very flexible with regard to placement. Another unique feature to the conceptual mass Family environment is that Project Location is exposed; this allows you to perform still-frame Sun and Shadow studies from directly within the Conceptual Mass environment. Other kinds of Families must first be loaded into a project environment before you can perform a Sun and Shadow study. The Sun Path feature is also available directly in the massing environment.

CREATING FORMS IN THE CONCEPTUAL MASS ENVIRONMENT

There are two form creation methods: Model-based forms are drawn using model lines and reference-based forms are drawn using reference lines or other referenced geometry. You make the choice between creating a model-based or reference-based form when you create a new form. Each has advantages and limitations. You can find complete details below in the "Model-based and Reference-based Forms" topic. For now, we will first look at building the basic kinds of forms available before delving into the differences between model line and reference line creation methods.

Many forms are possible in the conceptual modeling environment. Like the standard Family editor (Chapter 10), you can create extrusions, blends, sweeps, swept blends and revolves. You can also create both solid and void forms. In the conceptual environment, you can also create lofts, surface (or mesh) models, and even patterned surfaces. Unlike the standard Family editor, you don't need to choose a specific tool for the type of form you want to create before you create it; rather, you simply sketch the required shapes and use a single "Create Form" button to make any type of form. Revit will interpret your geometry and give you an appropriate form. When more than one form is possible, you will be offered a choice. We'll look at several examples in this topic.

PROJECT VASARI

This past year, Autodesk released Project Vasari. This is a technology preview that incorporates much of the functionality in the Revit Conceptual Environment in a stand alone application. The project Vasari website (www.projectvasari.com) has this to say: "Autodesk® Project Vasari is an easy-to-use, expressive design tool for creating building concepts. Vasari goes further, with integrated analysis for energy and carbon, providing design insight where the most important design decisions are made." You can download it and try; Project Vasari for free! Much of the content of this chapter will apply as written to Vasari and Revit. Feel free to perform the tutorails that follow in Vasari if you wish.

Install the Dataset Files and Open a Project

The lessons that follow require the dataset included on the Aubin Academy Master Series online companion. If you have already installed all of the files from this site, skip to step 3 to begin. If you need to install the files, start at step 1.

1. If you have not already done so, download the dataset files located on the CengageBrain website.

   Refer to "Accessing the Student Companion site from CengageBrain" in the Preface for information on installing the dataset files included in the Student Companion.

2. Launch Autodesk Revit Architecture from the icon on your desktop or from the *Autodesk > Revit Architecture 2012* group in *All Programs* on the Windows Start menu.

TIP	You can click the Start button, and then begin typing Revit in the Search field. After a couple letters, Revit Architecture should appear near the top of the list. Click it to launch to program.

3. On the QAT, click the Open icon.

TIP	The keyboard shortcut for Open is CTRL+O. **Open** is also located on the Application menu.

   • In the "Open" dialog box, browse to the location where you installed the *MasterRAC 2012* folder, and then the *Chapter 16* folder.

4. Double-click *Form Examples-01.rfa*.

   You can also select it and then click the Open button.

NOTE	Units are not a critical aspect of understanding the conceptual modeling environment. For this reason, only an Imperial units dataset is provided for this chapter.

Please assume that forms created are model-based forms unless noted otherwise. (Please refer to the "Model-based and Reference-based Forms" topic below for more information.)

NOTE

The Intent Stack

In the conceptual massing environment you can make many types of forms. Three-dimensional forms can be both solid and surface models. Forms created from solids behave as if they are carved from a single piece of solid material like a block of wood or poured from concrete. If you were to section a solid model, it would show material all the way through. A surface model behaves as though it were constructed from thin material like an eggshell or cardboard box. Surface models can be either closed or open shapes. A closed shape would be like a cardboard box. If you sectioned it, it would be hollow inside. An open shape would appear more as if it were made from sheet metal not defining an enclosed form.

Whether you get a solid or surface form depends on the geometry from which you create it. We will see examples of both below. Many forms are possible, such as Extrusions, Lofts, Sweeps, Revolves, and Swept Blends. (Each of these will be covered in the next topic.) No matter what type of form you are making, the process is the same; select some lines and click on the Create Form button. Depending on what lines and/or elements are selected, Revit will determine which type of form can be made from the selection set. If more than one form is possible from the selection set, Revit will provide a set of thumbnail sketches directly onscreen referred to as the "intent stack." Each will show one possible solution from the selection. You can choose from the forms presented by clicking the desired thumbnail. Let's start our exploration of the **Create Form** process with a look at the Intent Stack.

You should have the *Form Examples-01.rfa* open. If not, please open it now.

The *Form Examples-01.rfa* file contains the two standard Reference Planes included in the *Mass.rfa* template file and a total of three Levels (the standard Level 1, plus Levels 2 and 3). The only geometry in this file is two Model Lines. Creating a form from these two simple Lines can actually produce three separate results.

5. Select both the lines on screen. (Select only the Lines, not the Reference Planes or Levels).

   • On the Modify | Lines tab of the ribbon, on the Form panel, click on the **Create Form** button.

Three sketches appear on the screen; each sketch represents a possible form-intent that can be created based on the selection set (in this case, two parallel lines).

If you place your cursor over one of the sketches, the form in the main view will preview that form in wireframe.

6. Click on one of the sketches to create that form (see Figure 16.5).

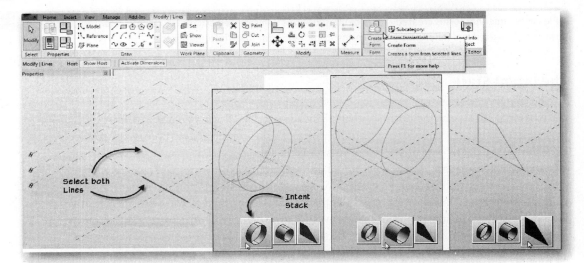

FIGURE 16.5 *When more than one form is possible from the selection set, the Intent Stack will display allowing you to choose the form you intended*

As noted above, Revit will determine if your selection will yield a solid or surface form. In this case, all three possible forms will yield surface models: a round hollow cylinder in the first two choices and a flat wedge-shaped plane in the third instance. If you wish to try more than one form, undo after creating and then repeat the process to choose a different intent.

Creating Extrusions

If you select only one Line, and then click the **Create Form** button, your result will be a simple plane.

> If you created a form from both Lines, undo it now to return to the two original Lines.

1. Select the long Line (lower one), and then on the Create tab, click the **Create Form** button.

The Line will "extrude" to a flat plane running parallel to Level 1. When Revit determines the type of form that can be created this time, there is only one option: a single straight line will yield a flat plane. No Intent Stack will display since there is only one option.

2. Select the other Line (the short one) and then click the **Create Form** button.

This time, the plane will be created perpendicular to the Levels. The reason that one Line extrudes horizontally and the other vertically has to do with how they were drawn and, more specifically, which work plane was active at the time of creation (see Figure 16.6). You can see the Work Plane on the Properties palette with the Line selected.

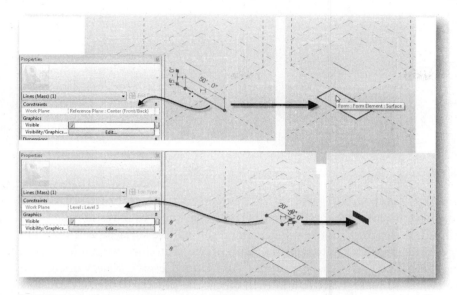

FIGURE 16.6 *The work plane in which a Line is drawn determines the direction that it will extrude*

3. Undo the creation of both planes (return to the original Lines).
   - Select the long line.

On the Properties palette, notice that the Work Plane for this Line is: Reference Plane: Center (Front/Back).

   - Select the other Line.

Notice that the Work Plane this time is Level: Level 3. When Create Form yields an extrusion, the direction of the extrusion will be perpendicular to the Work Plane.

   Deselect all elements.

Unlike the standard Family Editor and Project Editor environments where you must use the Set (Work Plane) tool to designate the active Work Plane, in the Conceptual Massing environment, simply selecting a Level or Reference Plane makes it the current Work Plane.

4. Select Level 1.
5. On the ribbon, on the Draw panel, click the rectangle or polygon icon.
   - Draw the shape any size.
   - Click the Modify tool and then select the shape you just drew.

Notice that the default selection behavior in the massing environment is chain selection. So when you pre-highlight the rectangle or polygon, it automatically highlights the whole shape. (You can still select just one edge if you like; use the TAB key. This is discussed below).

   - Click the **Create Form** tool (see Figure 16.7).

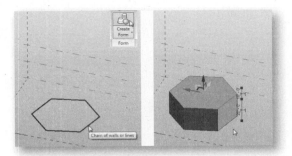

FIGURE 16.7 *Draw any closed shape and use Create Form to extrude it to a solid*

When you have a closed shape defined by multiple lines, your result will be an extruded solid. If you draw an open shape, the result will be surfaces as if each line had been extruded separately to a plane. Also, notice that since we made Level 1 the active Work Plane, the shape extruded "up" from Level 1. Select one of the Reference Planes to make it active and then repeat the process to create extrusions perpendicular to the one created here. Feel free to try a combination of open and closed shapes (see Figure 16.8).

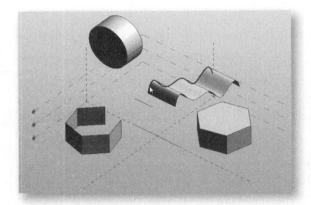

FIGURE 16.8 *Experiment with a series of open and closed shapes on different Work Planes*

6. Close the *Form Examples-01.rfa* file.

    Saving this file is optional. If you created a form that you wish to use again, please save the file.

Creating a Loft Form

A Loft is made by selecting two or more profiles and clicking on the Create Form button. The form will blend between the shapes selected. The profiles can be either open or closed or a combination of open and closed loops.

1. From the *Chapter 16* folder, open *Form Examples-02.rfa*.

Two sets of profiles are included in this file. On the left side, there are three closed shapes. If you select any of these three shapes and edit their properties, you will note that each has its Work Plane associated to a Level. The three open shapes on the right side are associated to the three corresponding Reference Planes.

2. Pre-highlight one of the closed shapes on the left (see Figure 16.9).

Chain select is the default behavior in the Conceptual Massing environment. To select just one segment of the chain, you would use the TAB key. This is opposite of the order in the project environment.

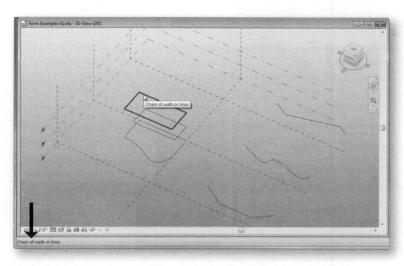

FIGURE 16.9 *A collection of shapes provided to experiment with lofting*

• Press the TAB key.

Just one segment of the shape should now be pre-highlighted. Press TAB again to pre-highlight the chain.

3. Select all three closed shapes on the left side (you can use a window selection box for this).

4. On the Modify | Lines tab, click the **Create Form** button (see Figure 16.10).

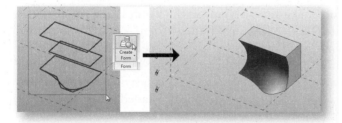

FIGURE 16.10 *Create a loft form from the three closed shapes*

You will get a very different result if you select only two of the three shapes. Try undoing and then selecting only the top and bottom shapes. The result will have less curvature on the front face.

5. Select the three open shapes on the right and then click the **Create Form** button (see Figure 16.11).

You can use any selection method to select the shapes; remember your SHIFT, CTRL and TAB keys as well as your window and crossing selection boxes.

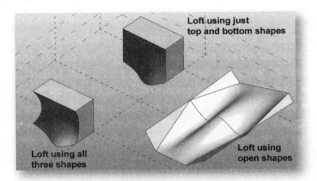

FIGURE 16.11 *Various loft examples*

Feel free to experiment further before continuing.

6. Close the *Form Examples-02.rfa* file.

Saving this file is optional. If you created a form that you wish to use again, please save the file.

Creating a Revolve Form

A revolve form is made by selecting a linear axis and another shape that is co-planar to the axis. The shape can be either a closed loop of lines (which will create a solid revolve) or an open loop of lines (which will create a surface revolve). The axis can be either a reference-based line or a model-based line. The revolve will default to a full 360 degrees, but this can be changed on the Properties palette.

NOTE You cannot use the Add Profile tool on revolve forms. See the "Using Add Profile" topic below.

1. From the *Chapter 16* folder, open *Form Examples-03.rfa*.

This file has a straight vertical line at the intersection of the two Reference Planes (which is also the insertion point of the Family as noted above) and a simple closed shape next to it. You will also notice that the Center (Front/Back) Reference Plane is shaded in this file. On the Home tab of the ribbon, on the Work Plane panel, notice that the ***Show Work Plane*** tool is active. This makes the current Work Plane shaded and therefore visible in the view window. This tool is a simple toggle. You can try it now if you like. Toggling it off will make the Reference Plane invisible again (unless selected). The ***Show Work Plane*** tool is view specific. So if you switch to other views on the Project Browser, the active Work Plane will not be shaded unless you toggle the ***Show Work Plane*** tool on.

2. Select both the closed shape and the vertical line (use any selection method).

   • Click the ***Create Form*** button (see Figure 16.12).

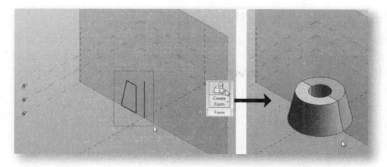

FIGURE 16.12 *A Revolve requires a shape and a straight line axis*

This geometry produced a hollow form since the axis did not touch the shape. If you want to experiment, undo the form creation and try moving the closed shape toward the axis line. Revit gives you an ignorable warning about overlapping lines. This is fine. Try the Revolve again and note the new result (see the top half of Figure 16.13).

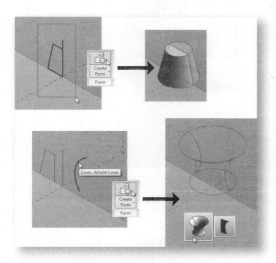

FIGURE 16.13 *Try some Revolve variations*

If you want to try another variation, undo again and this time draw an open shape on the same plane as the axis line. (The Work Plane is already shaded, but if you want to be sure, simply select the Center, or Front/Back, Reference Plane.) Try drawing a simple arc. When you create a form this time, you should get an intent stack. Choose the Revolve option (see the bottom half of Figure 16.13).

Feel free to experiment further before continuing.

3. Close the *Form Examples-03.rfa* file.

   Saving this file is optional. If you created a form that you wish to use again, please save the file.

Creating a Sweep Form

A sweep form consists of an explicit user-drawn path and a profile (open or closed loop of lines) drawn perpendicular to the path. The sample file provides some geometry to get us started.

1. From the *Chapter 16* folder, open *Form Examples-04.rfa.*

This file has two paths (the arcs drawn on Level 1) and two shapes: one closed and one open. As you can see, the current Work Plane is at the end of one of the arcs. You can use the **Set Work Plane** tool on the Home tab to set the Work Plane. If you pre-highlight the endpoint of one of the arcs, that point will become the active work plane (see below). Let's start by creating a solid sweep and then a surface sweep from the shapes already in this file.

2. Select the closed shape and the arc in the middle of the view window.

   • Click the **Create Form** button.

   • Repeat for the open shape and its arc (see Figure 16.14).

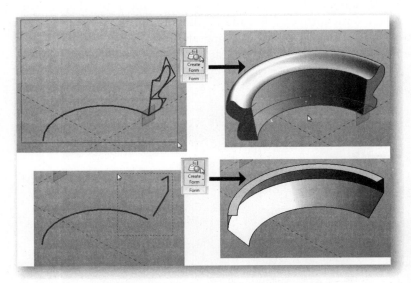

FIGURE 16.14 *Create sweeps from the provided shapes*

Since there were already shapes provided in the file, we didn't need to worry about the Work Plane. However, if you undo the two forms, we can start again with the paths and draw new shapes to sweep. This will give us an opportunity to learn about the two methods to set the Work Plane perpendicular to the path: Method 1: Use the Set Work Plane tool to set the endpoint of the path to be the current work plane, and then draw your profile on that work plane. Method 2: Place a Reference Point anywhere along the path, and then set the active work plane to the plane of that point.

Undo the creation of the previous sweeps or reopen the file without saving.

3. On the Create tab, on the Work Plane panel, click the **Set Work Plane** tool.

   • Click on one of the endpoints of the arc to set the plane of the endpoint to be the current work plane (see Figure 16.15).

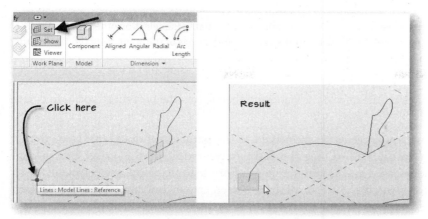

FIGURE 16.15 *Set the endpoint of the arc as the current Work Plane*

4. Draw a shape on this Work Plane.

The exact shape is up to you. It can be closed or open.

- Select the path and the new shape and then Create Form (see Figure 16.16).

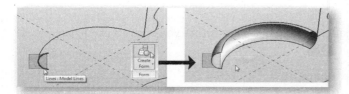

FIGURE 16.16 *Draw a new shape at the endpoint of the path and create the swept form*

Let's now try method 2. In method 2, you can place a point anywhere along the sweep path and use that point as a Work Plane.

For this exercise, work on the second arc path or undo the previous form and work on that path, whichever you prefer.

5. On the Home tab, on the Draw panel, click on the **Point Element** tool.

- Click anywhere along the arc path to place the reference point.

> Reference points are described in more detail in the "Divided surfaces, Patterns, Components, Points" topic.

NOTE

- On the ribbon, click the Modify tool or press ESC.
- Select the new Reference Point.

Like selecting Levels or Reference Planes, selecting the new Reference Point will make it the active Work Plane. You can also use the **Set Work Plane** tool on the Create panel as we did for the other path as well (see Figure 16.17).

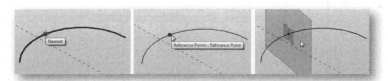

FIGURE 16.17 *Add a Reference Point and then select it to make it the current Work Plane*

The plane that is perpendicular to the host path should be first in the selection order automatically. If it is not, you may have to cycle through the various work planes of the point using the TAB key to get the desired plane.

6. Draw a shape on the new Work Plane and then follow the procedure above to create a new sweep.

The profile does not have to be connected to the path, but it does have to be drawn perpendicular to the path. Once the form is created, if it is a model-based form (see the "Model-based and Reference-based Forms" topic below), selecting any face, edge, or vertex of the form will display on-screen arrow controls, which can be used to modify the shape of the form. See the "Global and Local Controls" topic below for more information.

If the profile is large and the path has tight curves, you may see an error message about self-intersecting forms and the form will fail to be made. If this happens, try making your profile smaller.

Feel free to experiment further before continuing.

7. Close the *Form Examples-04.rfa* file.

Saving this file is optional. If you created a form that you wish to use again, please save the file.

Creating a Swept Blend Form

A swept blend form is similar to the sweep form. However, instead of having just one profile, the swept blend has two or more profiles which blend together while following along the explicitly drawn path.

1. From the *Chapter 16* folder, open *Form Examples-05.rfa*.

   • Select all the elements on screen and then click the **Create Form** button (see Figure 16.18).

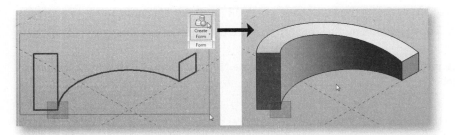

FIGURE 16.18 *A Swept Blend is made from two profiles and a path*

Both sweep forms and swept blend forms can have paths consisting of multiple line-segments. These are referred to as "multi-segment sweeps" or "multi-segment swept-blends." However, you cannot add a profile (see below) to a form made from a multi-segment path.

You can combine many of the techniques that we have explored so far. For example, undo the previous form and you can add a Reference Point to the path like we did for the sweep above. Once you have the Reference Point, we can make it active and draw another profile on that Work Plane.

Undo the previous form.

2. On the Home tab, click the **Point** tool.

   • Add a Reference Point on the path.

3. Select the new Reference Point to make it the active Work Plane.

   • Draw a new close shape on this Work Plane. You can make it a simple rectangle or something fancier if you prefer.

4. Select the three profiles (two original rectangles at each end and the one you drew) and the arc path.

5. Click the **Create Form** button (see Figure 16.19).

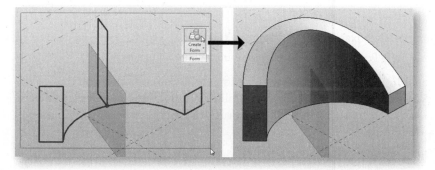

FIGURE 16.19 *Add an additional profile to the path and create a new form*

Technically there is little difference between the loft and the swept blend. In the examples that we explored here, the only difference was the path. In the case of the loft, we used an "implicit" path meaning that Revit interpolated the path based on the profiles we selected. In the case of the swept blend, we had an "explicit" path meaning that we actually drew the path using a Model Line (arc) element.

Editing Forms in the Conceptual Design Environment

When you select the whole or any part of a form, the Modify Form tab will appear on the ribbon with some additional editing tools. Selection of existing forms defaults to the sub-element level, meaning that as you move your mouse over a form, the edges, surfaces, and vertex points will pre-highlight under your cursor. If you wish to select the entire form, press TAB to cycle to Form Element selection. Selecting either a sub-element or the complete form gives you access to the same Modify | Form ribbon tab. If you select a sub-element, you will also get an onscreen drag handle for further direct manipulation. In this topic, we will look at the tools on the Form Element panel of the Modify | Form tab and explore the onscreen direct manipulation techniques.

Using X-Ray

Select any portion of a form, and the X-Ray button becomes available on the ribbon (it is also available on the right-click menu). X-Ray changes the visibility of the selected form. When X-Ray is toggled on for an element, the form becomes transparent and you see the underlying structure, or bones, of the form. The profile(s) of the form are shown in purple, and the path appears in black. Each vertex is also accentuated with a round filled vertex handle. All of the faces, edges, and vertices of the form continue to be available for editing with the use of the arrow controls and temporary dimensions. Even if you deselect the element, X-Ray remains active although the contextual Modify | Form tab disappears. Reselect any part of the form to re-display the contextual tab and gain access to the X-Ray toggle on either the ribbon or right-click.

1. From the *Chapter 16* folder, open *Form Examples-06.rfa*.

This file has a collection of the forms that we created in each of the other files above.

2. Select the revolve form in the upper left corner.
   - On the Modify Form tab, click the **X-Ray** button.

Take notice of the black line in the center of the form. This is the original path (axis) for this revolve. Paths display in black in X-Ray mode. Notice that this line is a continuous black line. Explicit paths like the one used for this revolve display as a continuous (solid) line.

3. Select the extrude (hexagon shaped) form on the right.
   - On the Modify Form tab, click the **X-Ray** button (see Figure 16.20).

Notice that enabling X-Ray for a new selection disables it for the previous selection. The **X-Ray** tool is an object-specific visibility mode; only one Element can be in X-Ray mode at a time. When an object is in X-Ray mode, every view will display the object in this mode.

Notice also that the path element (black line) this time is dashed. This indicates that the path is "implicit."

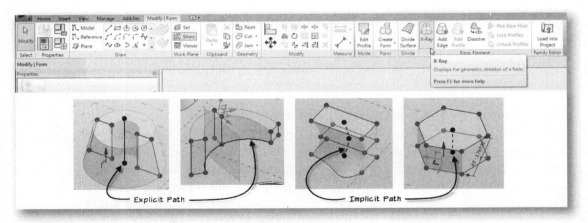

FIGURE 16.20 *Study the various forms in the dataset in X-Ray mode*

X-Ray mode is a useful mode in which to edit forms because you can see all of the faces of the form, manipulate its "skeleton," and still get a sense of the form's overall shape through the shaded faces.

Global and Local Controls

By now you have noticed the drag controls that appear when you select a sub-element of a form. These controls have two or three arrows for direct manipulation and sometimes some other specialized arrows as well. Red, green, and blue arrows allow you to drag the selected element while constraining it along a particular axis. Standard 3D modeling conventions use the labels X, Y, and Z for length, width, and height, respectively, with respect to the overall coordinate system (global) of the file. The labels U, V, and W are also used when the coordinates are localized to a particular element (local). Revit allows you toggle between the global (X,Y,Z) coordinates and the local (U,V,W) coordinates using the SPACEBAR. When you toggle to the local coordinates, the arrows change color to orange. (Try this on the hexagon shape, for example.) Working in X-Ray mode and using the global and local controls, let's manipulate some of the forms.

4. Try selecting an edge of the form or a corner vertex of the form.

Notice that when you select a face, edge, or vertex, the same arrow controls appear indicating that you can drag the sub-element in any of the three directions indicated by the arrows.

You can always select the whole form by placing your cursor over the form, press the TAB key once, and then click on the form. In other words, the whole form is always first in the tab-selection order. Once the whole form is selected, you have access to common editing tools like Move, Copy, Rotate, and Mirror on the Modify Form contextual tab of the ribbon. You have access to these tools with a sub-element selected as well, but the transformation would apply to the selected sub-element only.

5. Select the revolve form in the upper left corner.
   - On the Modify Form tab, click the **X-Ray** button.
   - Using the TAB key, select the outside vertex at the bottom.

6. Try some of the manipulations indicated in Figure 16.21.

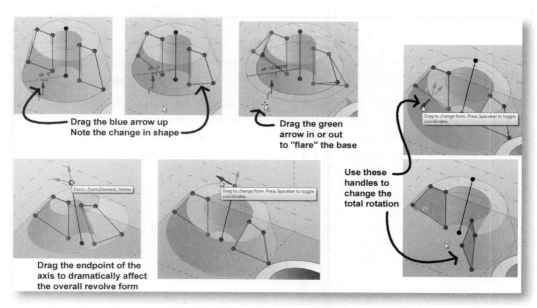

FIGURE 16.21 *To directly manipulate a form, click an edge, surface, or vertex point and then use the arrow controls to manipulate the shape*

The figure shows just a few possibilities. Don't forget to toggle the axis from global to local using the SPACEBAR and see the effect on your manipulations.

Using Add Edge

The ***Add Edge*** tool allows you to add a vertical or angled edge to the face of a form. It does *not* work on the top or bottom faces of forms. You will find the ***Add Edge*** tool on the Modify | Form tab of the ribbon when a form is selected.

7. Select the extruded hexagon form.
   - On the Modify | Form tab of the ribbon, click the ***Add Edge*** tool.
   - Hover the cursor over a vertical face of the form.

A preview of a single, vertical edge will appear. Move the mouse around to see the preview edge move around. Notice that if you move your mouse over the top surface, the preview edge will disappear.

8. When the preview edge is where you like it, click once to place the edge (see Figure 16.22).

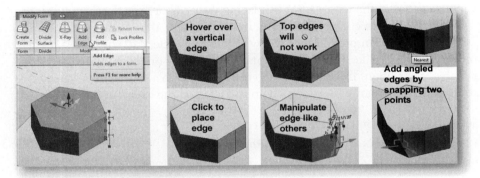

FIGURE 16.22 *Place vertical edges by clicking. Place angled edges by snapping two points at the top and bottom of a vertical edge*

If you want an angled edge, you can click two points instead. Start the ***Add Edge*** tool the same way. Instead of clicking once on the vertical face of the form, click on the top edge of the vertical face to place the first point. Click again on the bottom edge of the same face to place the second point. Use object snaps to gain more accuracy.

Edges can be applied to Extrusions, Lofts, Sweeps, Revolves, Swept Blends, and multi-segment path sweeps. You can also delete existing edges and user-created edges. To delete an edge, simply select the edge and then press the DELETE key. For example, if you delete on edge of a box form, the resulting shape is a wedge (see Figure 16.23).

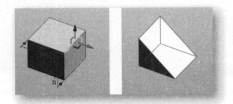

FIGURE 16.23 *You can delete an edge of a form*

Using Add Profile

You can add a new profile to an existing form using the ***Add Profile*** tool. Add profile will work for all form types except revolve and multi-segment path sweeps.

9. Select a form or any sub-component of the form (a face, edge, or vertex).
10. On the Modify | Form tab, click the ***Add Profile*** tool.

   • Move your cursor over the form and click once to place the profile.

   • Once the Profile is placed, select an edge or vertex of the profile and drag it using the arrow controls (see Figure 16.24).

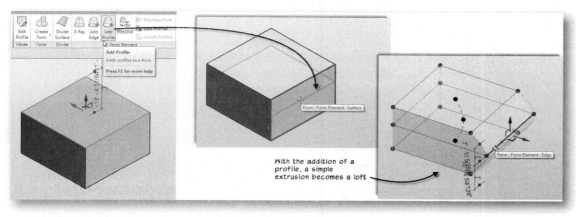

FIGURE 16.24 *Add a Profile to a form and manipulate the resultant edges*

Profiles can be added to Extrusions, Lofts, sweeps, and swept blends. They cannot be added to revolves or multi-segment path sweeps.

Try additional manipulations. For example, notice the small right-angle symbol at the intersection of each pair of arrows on the arrow controls. Dragging a single arrow constrains movement along a single axis. Dragging the right-angle control instead will allow free movement in the plane defined by the two adjacent arrows. For example, to move within the X,Y plane, drag the right-angle symbol between the red and green arrows.

Edit Profile

The Edit Profile method allows you to edit the profiles of a form using familiar sketch-based tools and techniques. To use the tool, select the edge of a form, and then on the Modify | Form tab, click the ***Edit Profile*** button. If you select the entire form before clicking this tool, Revit will display a select profile cursor. Simply click on an edge to begin editing. The form will gray out, and the selected profile will appear in sketch mode. Use any of the normal editing tools on the ribbon to edit the shape of the profile. When you are finished, click the ***Finish Edit Mode*** button (see the left side of Figure 16.25).

Dissolve Forms

When a form is selected, you will find the *Dissolve* tool on the Form Element panel. Simply select a form, and then click the *Dissolve* tool. The form will be replaced with a collection of shapes and points that would be required to create the same form. This tool gives a fall back when X-Ray or Edit Profile proves insufficient to make the desired manipulations to a 3D form. You can use Dissolve to start over, while preserving your original shapes (see the right side of Figure 16.25). Note that each profile will be accompanied by a point. This point will define a work plane for each profile, making it easier to perform further manipulations.

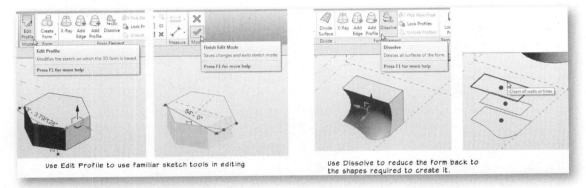

FIGURE 16.25 *Add a Profile to a form and manipulate the resultant edges (continued)*

Other Manipulations

Several other tools are available on the Modify Form tab. You can lock a profile so that it cannot be manipulated using the *Lock Profiles* tool. On the Modify panel, you can use the standard move, copy, rotate mirror, and scale tools on most form and sub-element selections. Continue experimenting in the current file before continuing on to the next topic.

Feel free to experiment further before continuing.

11. Close the *Form Examples-06.rfa* file.

Saving this file is optional. If you created a form that you wish to use again, please save the file.

MODEL-BASED AND REFERENCE-BASED FORMS

The conceptual massing environment is capable of creating two related, but unique, types of forms: **model-based forms** and **reference-based forms**. Model-based forms are drawn using model lines, and reference-based forms are drawn using reference lines or other referenced geometry. You make the choice between creating a model-based or reference-based form when you create a new form. On the Create Tab, on the Draw panel, buttons are provided for each mode. Select the *Model* button to create a model-based form. Select the *Reference* button to create a reference-based form. Revit defaults to model-based forms, so if you want a reference-based form, be sure to select the *Reference* tool first (see Figure 16.26).

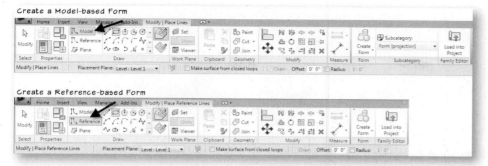

Create a Model-based Form

Create a Reference-based Form

FIGURE 16.26 *Use the Line and Reference buttons to indicate the kind of shape to draw*

Model-based forms and Reference-based forms each have unique attributes. When you create a form using model lines (refer to the next topic: "Creating Forms in the Conceptual Mass Environment" for more information), the model lines used to sketch the basic shape of the form will be absorbed into the form. From a conceptual point of view, the model lines are replaced with the solid or void form created from them. However, it is possible to access the profiles and shapes later. You can use the X-Ray mode (see the "Using X-Ray" topic), Edit Profile tool (see the "Edit Profile" topic), or Dissolve the form (see the "Dissolve Forms" topic). Each of these techniques is one step removed from the form itself.

Model-based forms have the most real-time flexibility because you can select any vertex, edge, or face of the form and move it using the control arrows (refer to the "Editing Forms in the Conceptual Design Environment" topic). Model-based forms are "unlocked" by default, which means there are no constraints maintaining relationships between faces of the form (see Figure 16.27). If you delete a model-based form, the initial sketch(es) you made to create the form will also be deleted; if you want the original shapes back, be sure to dissolve the form first.

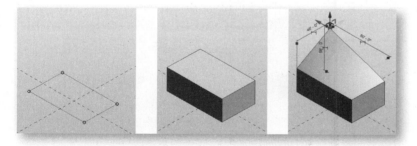

FIGURE 16.27 *Example of a model-based form*

When you create a form from reference lines or other reference geometry (refer to the "Create a Pattern-Based Curtain Panel Family" topic below), the reference lines used to sketch the basic form will remain available and selectable after the form is created. The form continues to remain dependent on the reference-based lines (edges or faces) and, in order to edit the shape of the form, you must edit the reference lines. Reference-based forms are "locked" by default, which means that all profiles that are parallel to the original reference lines will always remain parallel to—and exactly match the size of—the reference lines (see Figure 16.28). When you delete a reference-based form, the underlying reference lines will remain in the canvas and can then be edited and reused to create another form.

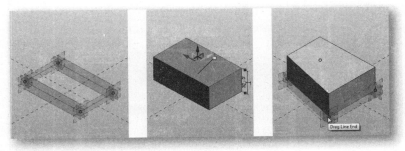

FIGURE 16.28 *Example of a reference-based form*

If you select a vertex of this form, notice that you do not get arrow controls. This is because if you were able to drag the arrow controls of the vertex, the top face of the form, which was initially parallel to the original reference lines, would no longer be parallel to the original reference-line profile. Also, notice the on-screen Lock icon which indicates that the profiles of the form are locked. You can unlock the form by selecting the form and then either click on the on-screen Lock icon or click the ***Unlock Profiles*** button on the Modify | Form tab of the Ribbon. If you unlock the profiles, the base of the form still follows changes made to the reference lines but the rest of the form becomes free-form like model-based forms (see Figure 16.29). It should be noted that you can use the Edit Profile tool to edit the unlocked profiles of reference-based forms. As a result, unlocked reference-based forms still do not give you as much freedom of manipulation as you get with a model-based form.

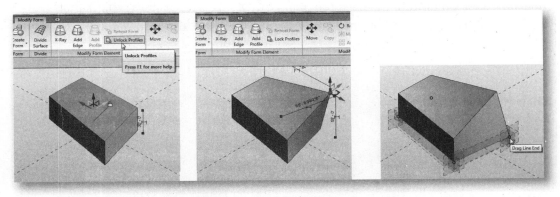

FIGURE 16.29 *Example of a reference-based form with profiles unlocked*

The difference between a model-based and reference-based forms is most apparent when used to create Curtain Panels for a patterned surface (refer to the "Divided Surfaces, Patterns, Components, Points" topic below), for example, if you are creating a panel that has a fixed thickness and you want to load this panel into a pattern across a non-planar face. The panel will conform to the shape of the non-planar face, but will maintain its thickness if it is reference-based. If it is model-based, the thickness and shape of the panel can vary from one side of the panel to the other (see Figure 16.30).

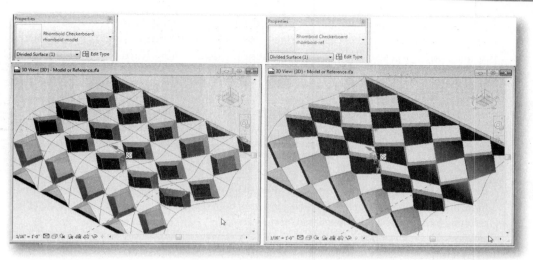

FIGURE 16.30 *Comparing a surface with model-based panel and a reference-based panel*

The model shown in Figure 16.30 is provided with the dataset files from the student companion. Look in the *Chapter 16* folder for the Revit file named: *Model or Reference.rfa*. If you open this file, you can select the surface, and then on the Type Selector (on the Properties palette) you can switch between the **rhomboid-model** and **rhomboid-ref** panels provided in the file.

On the left side of the figure a model-based panel is loaded. Notice that the panels appear to taper from one side of the face to the other. On the right side of the figure, a reference-based panel is used. Notice that the top face and bottom face of the panel are the same size—this is a result of the reference-based form's having locked profiles. As the top face of the component adapts to the changing size of the contoured, patterned surface, the bottom face of the component also conforms because it is locked to the top face.

In addition to forms created from reference lines, reference-based forms can also be created from the edges and faces of existing geometry. For example, if you select the top edge of one form and the top edge of another form and then click on the ***Create Form*** button, the result will be a surface spanning between the two edges that is a reference-based form. If you select the surface and delete it, the edges that were selected to create the form will not be deleted (see Figure 16.31).

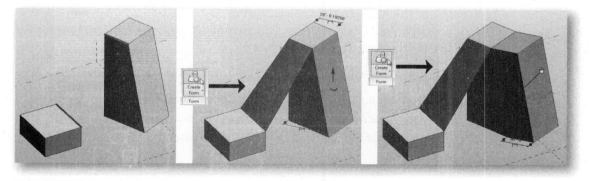

FIGURE 16.31 *Example of a reference-based form with profiles unlocked*

Reference-based forms can also be created from imported geometry. You can import two-dimensional or three-dimensional geometry; select lines, edges, or faces of the imports; and then use the selected elements to create new forms in the conceptual mass environment.

> **NOTE** For this to work, the imported geometry must be in ACIS solid format and not polymesh.

> **NOTE** If any line in your selection set is a reference line, then your resulting form will be a reference-based form. In other words, if your selection set consists of both model lines and reference lines, then the resulting form will be a reference-based form.

Model and Reference Form Rules of Thumb

Here are some tips and considerations for when to use a model-based form and when to use a reference-based form:

- Create reference-based forms when you create a Curtain Panel using the *Curtain Panel Pattern Based.rft* [*Metric Curtain Panel Pattern Based.rft*] template file.
- If you expect to make frequent edits to profiles of a form, consider reference-based forms. You can edit the profiles of model-based forms in X-Ray mode or using the Edit Profile tool. (Remember, to use X-Ray mode, you must enable it first and it only works on the selected element, one at a time).
- You can also use the Dissolve form functionality to restore the original components of a form to modify it. This offers more flexibility than X-Ray mode as you are essentially starting over with the original forms.
- If you have a solid idea of what you want your final form to look like, create a reference-based form.
- If you are brainstorming design ideas for the massing of a new building, and desire the most flexibility in free-form editing in real time on the screen, create a model-based form.

> **TIP** Before creating a form from model lines, you always have the option of changing model lines into reference lines by selecting the model line(s), editing the Instance Properties, and, under Identity Data, checking the box next to "Is Reference Line." However, once you have created a form from the model lines, this option will no longer be available.

CREATE A NEW CONCEPTUAL MASS LOADABLE FAMILY

At the start of the book, in the Quick Start chapter, we worked on a simple pavilion house model. Using that model as our concept, let's explore some options for the pavilion in our conceptual massing environment.

Create a New File and Add Levels

1. From the Application Menu, choose **New > Conceptual Mass**.
   - Choose the *Mass.rfa* [*Metric Mass.rft*] Family Template file and then click Open.
2. On the Home Tab of the ribbon, click on the **Level** button.

Creating a new level in a 3D view in the conceptual Mass editor is a one-click process. After clicking on the *Level* tool, move your cursor over the canvas area and you will notice a temporary dimension from the existing Level 1 to your cursor.

• Use the temporary dimension to locate 10'-0" above Level 1 and then click once to place the new Level.

If you prefer, you can simply type **10 [3000]** and press the ENTER. The new level will automatically be placed at 10'-0" [3000] above Level 1.

• Create another Level **10'-0"** [**3000**] above the Level you just created.
• On the ribbon, click the *Modify* tool or press the ESC key twice.

The two new levels will be named "Level 2" and "Level 3." Just as we saw in the "Working with Levels" topic of Chapter 4, floor plan views for each new Level will be added to the Project Browser. You can see the name of any Level in the view window when you select it. You can also edit its properties to see and change the name.

Create a Solid Form

Now that we have a few Levels, we will create the basic mass form of the pavilion.

3. Select the lowest Level on screen.

Since we just drew Levels, the last Level that we drew is the active Work Plane. In the Conceptual Massing environment, as we saw above, you need only select a Datum (Level or Reference Plane) to make it the active Work Plane.

4. On the ribbon, on the Draw panel, click the *Model* tool, and then click on the *Rectangle* icon (see Figure 16.32).

Recall from the "Model-based and Reference-based Forms" topic below that the *Model* tool creates model lines, making forms created from them model-based forms. If you prefer reference-based forms, you would click the *Reference* tool instead.

FIGURE 16.32 *Using the Model drawing mode, choose the Rectangle draw tool*

5. Draw a rectangle starting at the intersection point of the two reference planes.
   • Make the rectangle 20'-0" by 20'-0" [6000 by 6000].
   • Select the rectangle and then click on the *Create Form* button.

This will create an extrusion of arbitrary height. We want to modify the height of the extrusion to be at Level 3 (or 20'-0" in height).

   The top face of the solid extruded form should be selected; if it is not, select it now.

6. Click and drag the blue arrow control up.

This drags the height of the top face up. Notice that when you get close to one of the Levels, the level line highlights, allowing you to snap to the height of the level (see Figure 16.33).

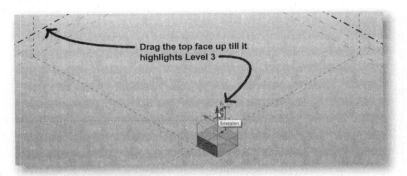

FIGURE 16.33 *Drag the top face up and snap it to Level 3*

You may also use the temporary dimensions to type in the desired height of 20'-0"
[6000].

Sculpt the Top Surface

Let's edit the top face of our basic rectangular form to create a curved, asymmetrical barrel vault. There are several approaches we could take for such a form. For this example, we'll take a look at using a Void form to carve the shape we want from the existing cube form.

7. On the Home tab, on the Draw panel, click the Start-End-Radius Arc shape.

Next to the draw tools are two buttons: Face and Work Plane. These are two drawing modes for all of the Line drawing tools. Face is active by default. This means we can draw directly on the faces of our existing geometry.

Place your cursor over the vertical side face of the solid extrusion and notice that the outline of face highlights.

This indicates that we will be drawing on this face.

8. Draw an arc on the side face (see Figure 16.34).

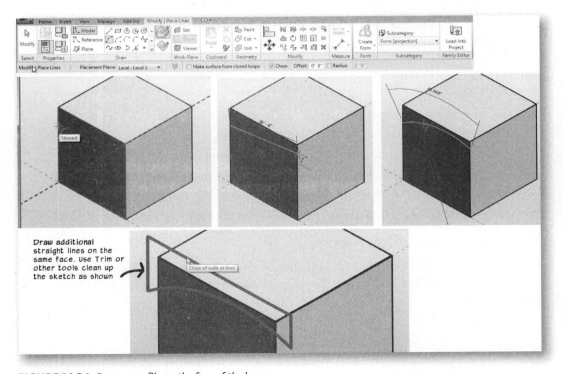

FIGURE 16.34 *Draw a profile on the face of the box*

9. Switch to the Line draw tool and create a closed shape as shown in the figure.

Be sure you are drawing on the vertical face. If you need to, draw the straight lines close to the 3D form and then move them out to match the figure. Next, we want to create a similar profile on the opposite side of the box extrusion.

10. Orbit the model around to see the opposite face.
    - On the ribbon, on the Work Plane panel, click the **Set Work Plane** button.
    - Click the back face of the box (now facing us).

This is an alternative method to using the existing face as a work plane.

11. Draw a similar profile on the current work plane.

Snaps will help you create the exact same profile. To make it more interesting, vary the slope of the curve slightly and use the **Trim** tool to clean up. If you look at the *North* or *South* elevation views, you should have the two arcs slightly off axis from one another (see Figure 16.35).

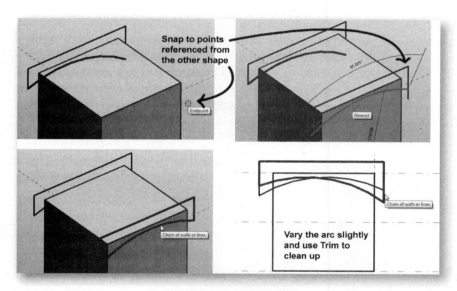

FIGURE 16.35 *Draw a profile on the opposite face of the box*

12. Select both profiles.
    - Click the drop-down button on the Create Form tool and choose **Void** (see Figure 16.36).

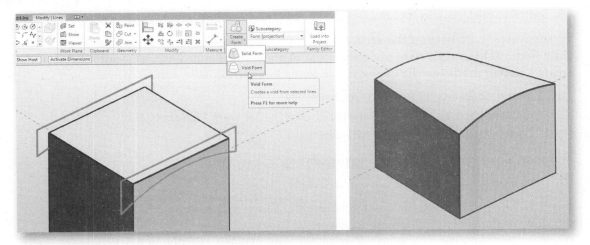

FIGURE 16.36 *Create a Void Form from the two profiles*

A lofted void form is made and automatically cuts the solid extrusion. If you need to modify the void form, you can still access it by putting your cursor over the area of the void form and pressing TAB to cycle through the selection options. When the void form pre-highlights, click to select it. Once the void form is selected, you can select edges, faces, and vertices of the void form and drag to edit the shape of the void by using the arrow controls just like solids (see Figure 16.37). You can even enable X-Ray mode or use Edit Profile for the selected void form. If necessary, you can even Dissolve it.

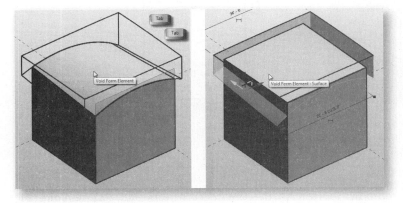

FIGURE 16.37 *TAB to select the void form. Once selected, you can select the sub-elements*

If you had clicked on the Create Form button and not specified a Void form, then a solid loft form would have been created by default. Whether a form is a Solid or a Void is an instance property of the form and can easily be toggled on the Properties palette. Simply change the Solid/Void parameter under Identity data (see Figure 16.38). The exception to the ability to swap between Solid and Void is if a form is already participating in a geometry combination (meaning it is joined or cut by another form). If this is the case, then the Solid/Void parameter will not exist for that form on the Properties palette.

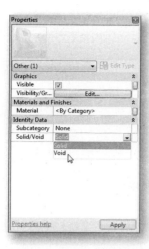

FIGURE 16.38 *You can edit the Solid/Void parameter for solids and voids not yet participating in a geometry combination*

13. Save the Conceptual Mass Family and name it: **Pavilion Concept.rfa.**

Add a Bay Window

Continuing on the same side of the model, let's add a contemporary bay window to the pavilion. Make sure that the active Work Plane is still the back face of the box. If it is not, use the ***Set Work Plane*** tool to set it now.

14. On the Draw panel, click the Pick Lines tool.

 • On the Options Bar, type **2'-0"** [**600**] in the Offset field.
 • Offset the three straight edges inward (see the left side of Figure 16.39).
 • Change the Offset to **4'-0"** [**1200**] and offset the curve downward.
 • On the Modify tab, use the ***Trim*** tool to clean up (see the right side of Figure 16.39).

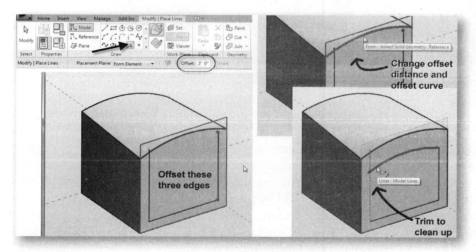

FIGURE 16.39 *Using Pick Lines with an Offset, create a new shape within the Work Plane face*

15. Select the new profile shape and on the ribbon click ***Create Form***.
    Temporary dimensions will remain after the form is created.

 • Edit the depth of the extrusion to **4'-0"** [**1200**] (see Figure 16.40).

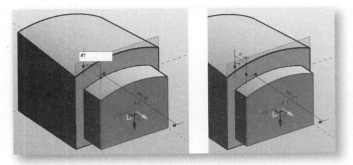

FIGURE 16.40 *Create an extruded form and edit the depth temporary dimension*

16. Select the curved edge at the top of the bay extrusion. Use the TAB key as necessary.

You will get a green and orange arrow control and a second orange control with a round dot at the end rather than an arrowhead. This control allows you to interactively change the radius of the selected arc edge. The orange arrow will move the curve within its plane and the green arrow will move the curve closer or further away from the overall building (or perpendicular to its plane).

- Using the green arrow control, move the curve in toward the building slightly about 1'-0" [300].
- Using the Change Radius control, drag slightly up (reducing the radius slightly).
- Select the vertex on the left, and using the green arrow control drag it away from the building slightly (see Figure 16.41).

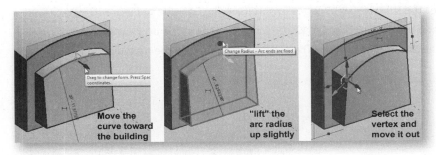

FIGURE 16.41 *Make several free-form manipulations to the bay form*

You can continue to fine-tune the forms in this model if you like. Feel free to use any of the techniques covered so far. You can work in X-Ray mode, add edges, add profiles, and create additional forms. You can use Edit Profile or even Dissolve. If you use Edit Profile, note that profiles will be projected to the work plane. Sketching is always performed on a 2D plane. Freeform modifications will be reapplied when you finish editing the profile. When you are satisfied with your explorations, we'll load this Family into a project and see how it fits on the site.

17. Save your Conceptual Mass Family.

Load a Conceptual Mass Family into a Project

To test out our design, let's open the original Pavilion project from the Quick Start chapter and then load our conceptual massing Family and test it out on the site.

18. From the *Chapter 16* folder, open the *15 Pavilion.rvt* project file.

19. From the QAT, click the Open Documents icon and then choose *Pavilion-Concept - 3D View: {3D}*.

You can cycle through open view windows by holding down the CTRL key and pressing TAB.

- In the conceptual mass environment, on the ribbon, click on the **Load into Project** button.

Revit will load the family into the project environment and automatically turn on mass visibility in the project. A message will appear alerting us of this.

- Read and then dismiss the message to continue.

The Mass model will be attached to the cursor ready to place in the project. It will probably be difficult to place it in the {3D} view however. The orientation will probably be angled in a strange way. If you tap the SPACEBAR, it will rotate, but in 90° increments. You can click the checkbox on the Options Bar to Rotate after placement. This will allow you to place it and then immediately rotate it. Let's switch to the *Site* plan view instead.

If you have trouble placing the mass in the 3D view, try opening the Site plan view and place it there.

20. On the Project Browser, double-click to open the *Site* plan view.

21. On the Massing & Site tab, on the Conceptual Mass panel, click the **Place Mass** tool.

The bay window should be oriented to the right.

- Press the SPACEBAR as necessary to rotate the model correctly (with the bay to the right).

- On the ribbon, click the Modify tool or press ESC.

  Fine-tune the placement using move and/or rotate as required (see Figure 16.42).

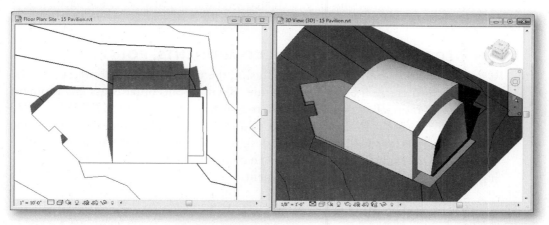

FIGURE 16.42 *Place the Mass in the project and position it on the site*

You are now ready to begin applying Walls, Roofs, and Curtain Systems to the conceptual mass. The process to do so was covered in the "Add an In-Place Mass" topic of Chapter 4, and the process to update such geometry when the underlying mass has changed was covered in the "Editing Massing Elements" topic of Chapter 9. Using techniques covered in those exercises, apply Walls, Roofs, and Curtain Systems to the model. You can also use the Mass Floors tool on the Modify Mass tab to add floor planes at each Level and then add Floors to the mass model similarly to the process for adding Roofs (see Figure 16.43).

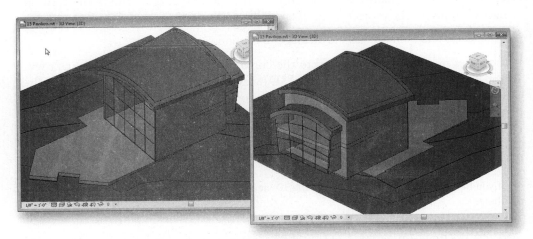

FIGURE 16.43 *Optionally add Walls, Floors, Curtain Systems, and Roofs to the mass model*

Some cleanup will likely be required after you apply the elements to the mass. You may want to adjust the overhangs of the Roof, attach the Walls to the Roof, and create a custom Curtain System. When you have finished adding the elements, turn off the *Show Mass* tool. These are just suggestions. The specific steps are left to you as an additional exercise.

Feel free to experiment further before continuing.

22. Close all files.
    Saving this file is optional. If you created a form that you wish to use again, please save the file.

DIVIDED SURFACES, PATTERNS, COMPONENTS, POINTS

Once a complex form has been created using the tools described above, it must be rationalized to create a buildable structure. Revit's Conceptual Mass environment allows us to rationalize complex geometric surfaces through the Divided Surface and Patterning tools. It also provides a tool to apply adaptable components to the surface divisions for further parametric design studies and ease of transition to fabrication. We will study several iterations of a complex roof structure, all while remaining in the conceptual massing environment.

Build a Complex Roof Surface

In this tutorial, we will create a complex roof surface, divide the surface, apply a pattern, and then create and load a component onto the patterned surface.

1. In the *Chapter 16* folder, open the file named *Complex Roof.rfa*.

The file contains an additional (not in the template), vertical Reference Plane named "End 1" parallel to the default "Center (Left/Right)" Reference Plane. The first profile has also been begun with a few Model Lines already in the file. To begin, we will finish the profile on one of the Reference Planes and then copy it to the parallel reference plane.

2. On the Home tab, on the Draw panel, click the **Spline Through Points** tool.

   • Draw a curving spline to close the loop on Reference Plane "End1."

   • Place the first point at the end of the short vertical line and the last point at the end of the tall vertical line (see Figure 16.44).

DRIVING POINTS

The **Spline Through Points** tool allows you to place points in 2D or 3D space. The points then drive the shape of the resulting spline. The points you place are considered "Driving Points" because when you adjust their location using the arrow controls, the spline will conform to the modified point location. The points drive the shape of the spline. After placing the initial points, if necessary, adjust the points (or delete some of the points) to create a curved spline similar to the one in the figure.

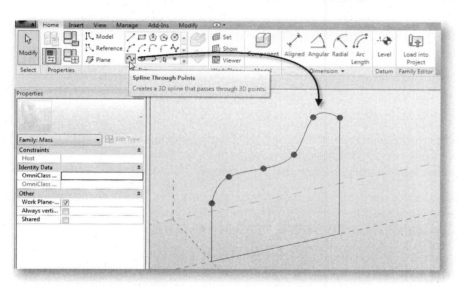

FIGURE 16.44 *Draw a Spline Through Points to close the shape*

   • On the ribbon, click on the Modify tool or press ESC.

3. Click to select the "Center (Left/Right)" Reference Plane to make it the active Work Plane.

4. On the Draw panel, click the **Pick Lines** tool (see Figure 16.45).

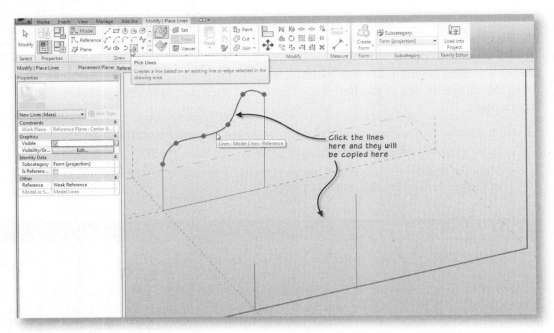

FIGURE 16.45 *Using the Pick Lines tool make copies of each of the lines*

5. Click on each of the three straight lines making up the remainder of the closed loop on reference plane "End1."

As you click the lines, a copy of each line will appear on Reference Plane "Center (Right/Left)."

- On the ribbon, click on the Modify tool or press ESC.

6. Select both closed loops and then click the **Create Form** button to create a solid loft form.

7. Select the form and then click on the X-Ray button.

As we saw above, this will allow you to see the structure of the form, including the splines, points, and implicit path. At this point, you can select any of the sub-components and modify its location by dragging one of the control arrows.

8. Select the tall vertical line on the "Center (Right/Left)" Reference Plane and drag it away from the form, so that the end face of the form is slanted (see Figure 16.46).

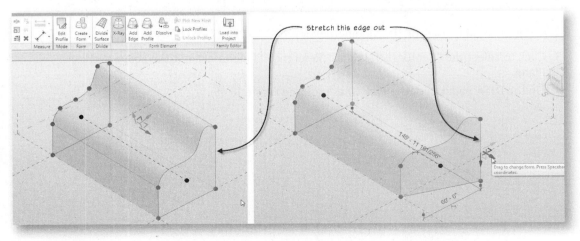

FIGURE 16.46 *Select a vertical edge and stretch it away from the form to make a slanted surface*

For more information on modifying forms, see the "Editing Forms in the Conceptual Design Environment" topic above.

9. Make sure that the form or a sub-element is selected and then toggle off the X-Ray mode on the Modify Form tab of the ribbon.

Applying a Divided Surface to a Form

Next, we will divide the top surface of the form, pattern it, create a component, and add the component to the panels.

10. Select the top (wavy) face of the form.

   • On the Modify | Form tab, on the Divide panel, click the **Divide Surface** tool.

This creates a default divided surface. The surface displays as a grid subdivided in the "U" and "V" directions. Recall from the "Global and Local Controls" topic that U and V are used to describe the "X" and "Y" (or width and depth) of a local coordinate system, in this case, the local coordinate system of the top wavy face.

The spacing of the U and V coordinates can be adjusted in a few ways: on the Options Bar, on the Properties palette, or by clicking on the coordinate icon located at the center of the divided surface. Additional modifications such as rotation angle, justification, layout orientation, and offset are accessible on the Properties palette as well (see Figure 16.47).

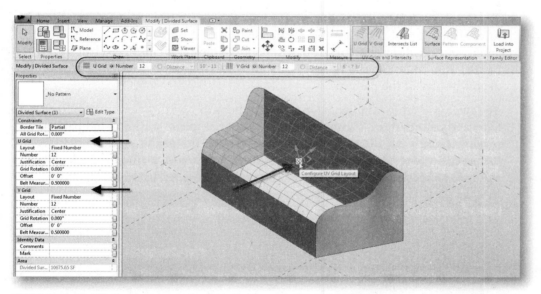

FIGURE 16.47 *Edit the parameters of a divided surface directly, on the Options Bar or on the Properties palette*

11. On the Options Bar or Properties palette, change the U-spacing to: **21** and the V-spacing to: **10**.

For a divided surface, the Type Selector on the Properties palette is used to choose a pattern. By default, no pattern is applied and the Type Selector simply reads: _No Pattern. If you open the Type Selector, you will see several patterns to choose from.

12. From the Type Selector, choose the **Rectangle Checkerboard** pattern (see Figure 16.48).

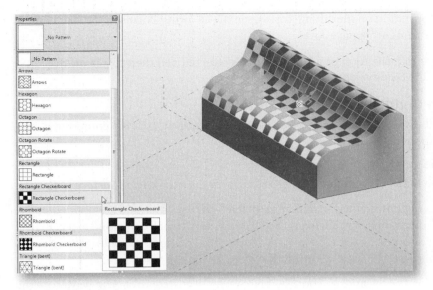

FIGURE 16.48 *Apply a pattern to the surface*

Patterns are System Families and it is not possible to make a new user-defined pattern. However, it is possible to edit an existing pattern. We will do this below. When you apply a pattern to a surface, the divided surface grid will automatically turn off. On the Surface Representation panel of the Modify Divided Surface tab are tools to control what is displayed for the divided surface. By default, when a pattern is applied, the grid is turned off. You can choose to have both displayed by clicking the **Surface** button on the ribbon. Similarly, the **Pattern** button on the ribbon will toggle off display of the pattern.

You will see little difference in turning both on with our current Rectangle Checkerboard pattern. Figure 16.49 shows an example of a Rhomboid pattern displaying along with the divided surface gridlines. The gridlines of the surface are displayed in blue because they are considered a reference element. The pattern (in this case a rhomboid pattern) is displayed in black.

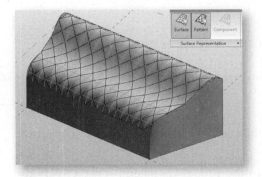

FIGURE 16.49 *Displaying both the surface grid and a pattern*

13. Select the surface and from the Type Selector, choose the Rhomboid Checkerboard pattern.

Create a Pattern-Based Curtain Panel Family

In this sequence, we will create a parameterized component using a predefined Family template.

1. From the Application menu, choose **New > Family**.

Your default family templates location will be displayed.

- Select the *Curtain Panel Pattern Based.rft* [*Metric Curtain Panel Pattern Based.rft*] file and open it.

This family template is a special curtain panel template that has all of the tools and capabilities of the Conceptual Mass environment; it is designed to be used with a patterned surface. We will use this environment to draw a single component that will then load into our surface pattern and conform to each "cell" of the pattern. The template file consists of a grid (blue), several points, and Reference Lines connecting the points.

2. Select the blue grid onscreen and notice that on Properties palette, you have access to all of the same patterns as you did in the conceptual mass file (see Figure 16.50).

The points and the Reference Lines (green) define how the component will lay out when it is applied to the patterned surface in the Mass family file. When you build a pattern-based Curtain Panel, it is important to create reference-based forms (refer to the "Model-based and Reference-based Forms" topic for more information). The reason for this is that as the cells of the pattern change shape in the massing Family, the panels must conform to the new shapes. This can only happen if the panels are reference-based forms. The Family template starts with these Reference Lines to get us started.

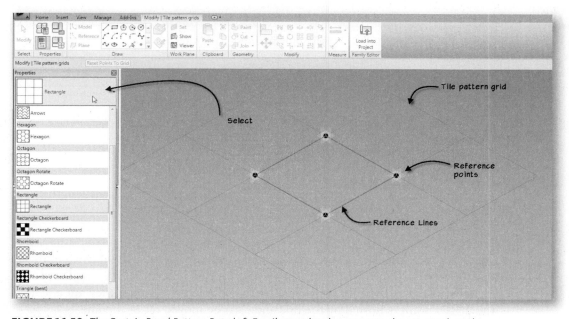

FIGURE 16.50 *The Curtain Panel Pattern Based.rft Family template has access to the same tools as the conceptual mass environment*

When building a pattern-based Curtain Panel, you will always want to choose the same pattern that you will be using in the conceptual mass family.

3. Select the blue grid.

- From the Type Selector, choose the Rhomboid Checkerboard (see Figure 16.51).

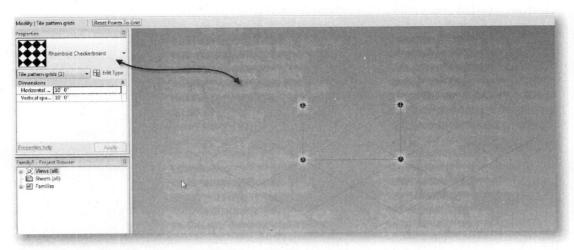

FIGURE 16.51 *Change the pattern to match the mass family*

Notice that the rhomboid shape is defined using four cells of the grid. This means that when the component is loaded into the conceptual mass model and placed on a patterned surface, a single component will span four cells of the divided surface. This will help you plan your U and V spacing back in your main massing file.

DRIVEN POINTS

A driven point is a point whose location is controlled (driven) by the line (or surface) upon which it is placed. If the line or surface moves, then the point will move along with the line. By default, placing a reference point along an existing line creates a Driven Point.

4. On the ribbon, on the Draw panel, click the **Point** icon.

- Pre-highlight one of the green Reference Lines on screen.
- Move the mouse along the Reference Line.

Notice that the point will attach to the line and stay constrained to its length.

- Click to place the point on the Reference Line.

Notice the relative size of the Reference Point. It is smaller than the others that were already in the file. Above we discussed "driving" points. The larger points are driving points. They "drive" the shape of attached geometry. The smaller points are driven points. They are controlled by the geometry to which they are attached. The driven point is hosted by the line and can't be moved off of the line.

5. Select the new Reference Point (the small one).

- Try to drag it off of the Reference Line.

Notice that it will move freely along the length of the Reference Line but will not move off of it.

6. With the Reference Point selected look at the Properties palette (see Figure 16.52).

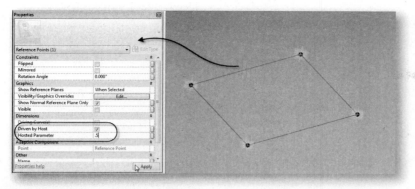

FIGURE 16.52 *Change the Hosted Parameter to control the location of the driven point along the Reference Line*

Beneath the Dimensions grouping, notice the "Driven by Host" parameter. This is a simple checkbox. You can easily convert a driven point to a non-driven (free to move) point by deselecting this checkbox. The "Normalized Curve Parameter" determines where (as a percentage of its total length), along the line, a driven point will remain. For example, if we change the Normalized Curve Parameter to be ".5" then the driven reference point will be located at the midpoint of the line, even if the length of the line changes.

7. Change the "Normalized Curve Parameter" to **.5** and then click Apply.

   • Select one of the points at the corners of the green square and drag it.

Notice how, as the length of the Reference Line changes, the driven Reference Point now remains at its midpoint.

   • Undo the change to return to the default square shape.

> Remember that we are actually working in the Family Editor. As such, it is also possible to place a labeled dimension between the driven point and the end of the line. This sets up a parametric relationship like the many we explored in Chapter 10. We can also click the small buttons in the right column of the Properties palette to assign parameters to any of the values shown. When forms are hosted on one of the work planes of the driven point, the form will react when the parameter is flexed. *We have the same potential for constraints and parameters in the conceptual massing environment as we do in the traditional Family Editor: more, in fact, since Reference Points give us even more opportunities to establish key relationships.*

For now, we will create a simple non-parametric example. If you want to try making a parametric panel Family, you can apply the concepts discussed in Chapter 10 to any Family you create in the conceptual massing environment.

8. Select the driven Reference Point on screen to make it the active Work Plane (see Figure 16.53).

If you wish, you can click the Show button on the ribbon to verify that the Work Plane is set to the Reference Point.

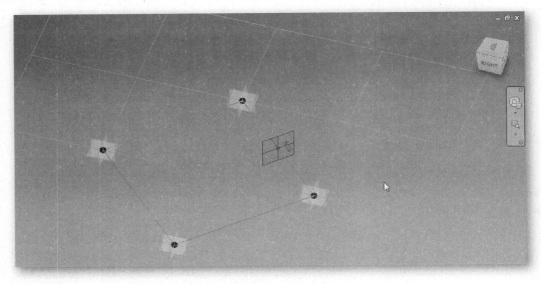

FIGURE 16.53 *Set the Work Plane to the driven Reference Point*

9. On the Draw panel, click the **Reference** tool.

- Click the Rectangle shape icon.
- Draw a rectangle **1'-0"** [**300**] tall by **4"** [**100**] wide.
- Use the Move tool to center it on the Reference Point (see Figure 16.54).

TIP	You can select individual lines (use TAB) of the rectangle to use temporary dimensions to get them located properly. Use the ViewCube to adjust the view as necessary.

Another option to help you view and position elements is the Viewer window. When you click this button on the ribbon, a small window will open oriented looking directly at the current work plane automatically. Sometimes this can be just the tool you need to draw a complex shape. All of the normal editing works in the viewer. Pay attention to the orientation however. As you can see in the figure, it opened at a 90° rotation. When you are finished drawing the shape, simply toggle the Viewer back off.

Remember, we are using reference lines in the panel Family so that it will conform to the shape of the pattern in the massing Family to which we later load it.

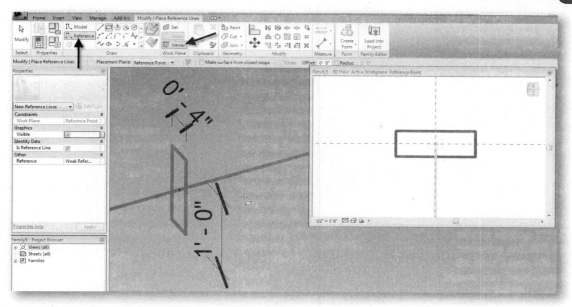

FIGURE 16.54 *Draw a retangular profile of Reference Lines*

10. Select all of the elements in the view and then click the **Create Form** button (see Figure 16.55).

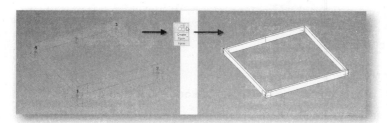

FIGURE 16.55 *Create a form from all of the Reference Lines*

Once again, in this example we have created a reference-based form. When working in the *Curtain Panel Pattern Based.rfa* file, you should always create reference-based forms. Reference-based forms will give you the most predictable and controllable outcome when the curtain panel is loaded into the Mass family. Reference-based forms ensure that the panel will conform to (and remained locked to) the reference lines which in turn conform to the shape of the pattern along the UV pattern grid.

11. From the Application menu, choose **Save As > Family**.

   • If necessary, browse to the *Chapter 16* folder.
   • Name the file *Roof Panel.rfa* and then click Save.

12. On the ribbon, click the **Load Into Project** button.

The *Complex Roof.rfa* conceptual mass Family file will come to the front, and the new panel Family will now be loaded into the Complex Roof Family file.

Apply the Custom Family Panel

13. Select the patterned face.
14. From the Type Selector, beneath **Rhomboid Checkerboard**, choose the **Roof Panel** type (see Figure 16.56).

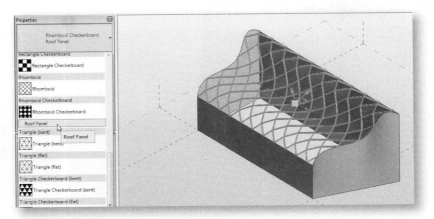

FIGURE 16.56 *Apply the new panel type to the surface pattern*

The component begins to define a structure for the glazed roof. You could create several panel variations and load them all into the conceptual mass and quickly evaluate several design iterations. As a final exercise in this brief exploration of conceptual massing, let's add some glazing to the panel Family and then load it into the Complex Roof conceptual mass.

15. On the QAT, click the Open Documents icon.
    • From the list of documents on the right, choose **Roof Panel.rfa – 3D View: {3D}**.

16. Select the sweep on screen (the outer frame).
    • On the View Control Bar, click the Temporary Hide/Isolate icon and choose **Hide Element** from the pop-up.
    • Click to select the chain of Reference Lines.
      When you select one Reference Line, all four in the chain should select.

17. Click the **Create Form** button.
    • From the Intent Stack, choose the simple plane.

This will give you a simple plane that we can assign to a glass material. If you prefer your glazing to have a thickness, choose the extrusion option from the Intent Stack and set it to a very thin thickness. If you decide to change from a plane to an extrusion later, you can click the plane and use the arrow controls to extrude it.

18. Select the plane, and assign it the Glass Material on the Properties palette.
    • Back in the view window, Reset the Temporary Hide/Isolate.

19. Save the Family and then **Load Into Project** again.

    • In the "Family Already Exists" dialog, choose the "Overwrite the existing version and its parameter values" option (see Figure 16.57).

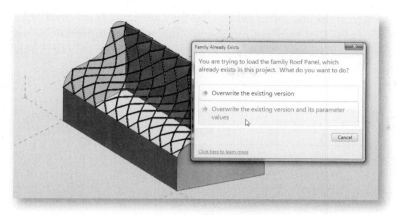

FIGURE 16.57 *Reload the modifed panel Family*

You can still modify the form using any of the form modification tools (Add Edge, Add Profile, drag arrow controls of faces, edges, or vertices) and the pattern and component will update to conform to the new surface. You can also change the properties of the divided surface including the U and V spacing, the rotation angle, and mirror, or flip the component from the Instance Properties of the surface.

Stitching Borders of the Divided Surface

The edges of divided surfaces can sometimes be problematic. Look closely at the edges, you still may not be pleased with the results at the edges. To help remedy this, we can use custom Curtain Panels to "stitch up" the borders of a divided surface. We will explore a very simple application of the feature as a way to create a more acceptable edge condition.

1. From the Application menu, choose **New > Family**.

   • Select the *Curtain Panel Pattern Based.rft* [*Metric Curtain Panel Pattern Based.rft*] file and open it.

The process of stitching up the edges involves a few tricks. First, you have to create a custom panel for use at the corners. In our case, a triangular shape should do nicely. Next we must eliminate all of the partial panels that Revit added automatically at the edges. Revit does an admirable job of trying to adapt the panels to the irregular shaped portions of the pattern like the edges, but we'll still need to eliminate them and add a custom triangular panel manually. Finally, each of the points in the Family template is numbered. If we drag the panel from the Project Browser into the canvas area, we are able to manually place the panel and "stitch" the points in numerical order to place them. In this way, we can apply a panel to each edge condition that requires it in the precise way that the design calls for.

Before we can explore all of this, we need to build a new panel.

2. Select the blue grid.

   • From the Type Selector, choose the Triangle (Flat) pattern.

The grid, reference lines, and points will adjust accordingly.

3. Select one of the reference points at the corners.

Notice, when you pre-highlight and select them, that they are actually referred to as: Adaptive Points. Notice also that they are numbered. This will be important below when we perform the stitching process. Keep this in mind and make a mental note of the sequence of the numbering (see Figure 16.58).

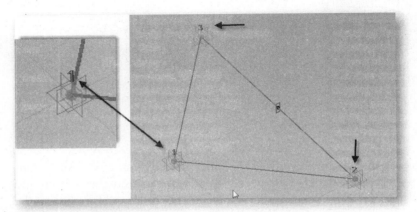

FIGURE 16.58 *Adaptive Points are numbered to indicate their insertion sequence*

The geometry for this panel will be nearly identical to the one we built above.

4. Add a point at the middle of one of the Reference Lines like we did in the previous curtain panel exercise above.

5. On the Draw panel, click the **Reference** tool.

    • Click the Rectangle shape icon and draw a rectangle like we did above: **1'-0"** [**300**] tall by **4"** [**100**] wide.

    • Be sure to center it on the Reference Point (similar to Figure 16.54).

    • Repeating the procedure above, create a simple plane from the three original Reference Lines and assign it to a Glass Material.

6. From the Application menu, choose **Save As > Family**.

    • If necessary, browse to the *Chapter 16* folder.

    • Name the file *Roof Panel Edge.rfa* and then click Save.

7. On the ribbon, click the **Load Into Project** button.

    You should now be back in the *Complex Roof* file.

8. On the Project Browser, expand the Families branch and then expand the Curtain Panels branch.

    • Expand Roof Panel Edge.

Like other Families, if you do not create any Types in the Family, Revit will automatically create a single Type using the same name. Having both the Family and Type the same name might be confusing.

    • Right-click the Type (indented beneath the Roof Panel Edge Family), and choose **Rename**.

    • Name it: **Triangular Edge**.

9. Drag the newly renamed Triangular Edge Type and drop it in the canvas area.

The panel will be attached to your cursor at point number 1. Recall that we took note of the numbers above. Using object snaps, click where you want point 1, then point 2, and finally point 3 (see Figure 16.59).

FIGURE 16.59 *When you drag a panel from Project Browser, you can place the points in order*

If you zoom in and look carefully, you will notice that it is difficult to get the points to snap to the precise locations required. To make it easier to place the edge panels, we can temporarily hide the other panels.

10. Delete the triangular panel.

11. Select the freeform surface.

- On the ribbon, on the Surface Representation panel, click the Component toggle button.

- On the Surface Representation panel itself, click the Display Properties dialog launcher icon.

- On the Surface tab, check the Nodes checkbox and then click OK (see Figure 16.60).

12. Repeat the drag and drop process from the Project Browser and place a triangular panel at the edge.

Revit will easily snap to the nodes now displayed. Repeat the process until all edge panels are added.

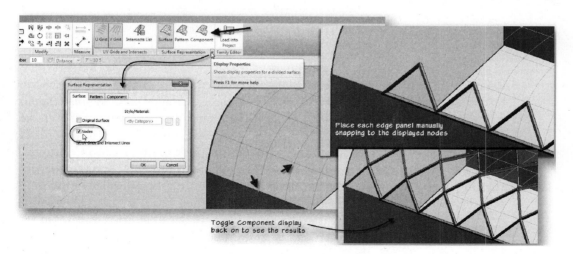

FIGURE 16.60 *With the Component display off and Nodes on, placement becomes much easier*

- When you have added all the panels, toggle the Component display back on to see the final result.

Using Intersects for Divided Surfaces

When you divide a surface, Revit calculates the direction of the U and V grids. The automatic orientation is not always ideal. With the Intersects tool, we can have the grid lines coincide with Levels and Reference Planes.

13. Select the end wall facing us in the 3D view.

14. On the ribbon, click the *Divide Surface* tool.

    Notice that the irregular shape of this surface has generated a grid that is not ideal.

15. Following the process covered in the "Create a New File and Add Levels" topic, add five Levels each **12'-0"** [**3600**] apart.

16. Select the divided surface.

    • On the UV Grids and Intersects panel, click the *Intersects List* button.

    • Check all of the Levels and then click OK.

    • On the ribbon, turn off the U Grid (see Figure 16.61).

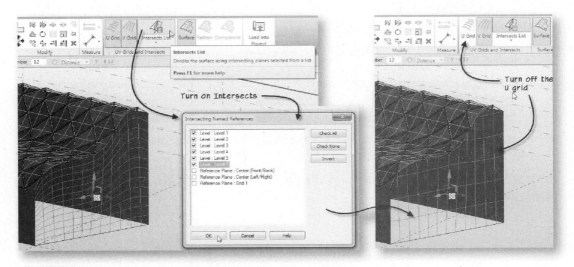

FIGURE 16.61 *Turn on Intersects with Levels and turn off the U Grid*

There is much more to explore in the conceptual massing environment. Feel free to continue experimenting in this or any of the other files used in this chapter. Try making more Curtain Panels and applying them to the new surface if you like. When you get a design to a point where you are ready to incorporate it into a project, you can simply load it into a project file like we did above in the "Load a Conceptual Mass Family into a Project" topic. When you toggle off the Show Mass mode, any curtain panels you have in your massing study will continue to display in the project. Therefore, unlike Walls, Floors, or Roofs, it is not necessary to apply geometry over the divided surface/curtain panel studies. They can be used directly in the project. However, if you wish to edit them, you will have to return to the Massing Environment and edit them there.

17. Close all files.

    Saving these files is optional. If you created forms that you wish to use again, please save the file(s).

SUMMARY

The lessons in this chapter merely scratch the surface of the possibilities in the Conceptual Massing Environment. The aim of this chapter has been to introduce you to the key concepts and get you acquainted with the possibilities. Please continue to explore and experiment in your own files. Consult the online help and the tutorial files that ship with the software and those posted at Autodesk.com for more information. Also consult Appendix B for additional online resources that you will find useful to the Revit learning experience.

- All features in this Massing Environment are also available in Project Vasari, which is available as a free download from www.projectvasari.com.

- In the conceptual Mass environment you can create and edit forms freely while maintaining the level of accuracy needed ultimately to build the forms.

- The conceptual massing environment allows you to design in a 3D axonometric view, while making it easy to view and edit Levels, Reference Planes, and Work Planes directly on screen.

- Create extrusions, lofts, revolves, sweeps, blends, and swept blends, and interactively edit the forms in real time using the on-screen arrow controls.

- Forms can be modified using the ***Add Edge*** and the ***Add Profile*** tools.

- Forms can be viewed and edited in a new element-specific visual style called "X-Ray mode," which reveals the skeleton profiles and paths of the form, making them readily available for editing.

- Edit forms using Edit Profile which mimics sketch mode in many other Revit tools, or use Dissolve to return to the original shapes, make modifications, and recreate the 3D forms.

- Surfaces (whether planar or non-planar) can easily be rationalized using the Divide Surface tool, which creates a modifiable UV grid on the surface.

- Divided surfaces can be further rationalized by applying patterns and loading components onto the surface.

- The *Curtain Panel Pattern Based.rft* Family template provides a guided environment for creating components that can then be loaded into the Mass family and placed on the patterned surfaces.

- Curtain Panels can be "stitched" to the edges of divided surfaces by dragging them from the Project Browser into the canvas and placing the points sequentially onscreen.

- Grids can be applied to divided surfaces using the Levels and Reference Planes.

- Conceptual Mass elements can be loaded into a project and have Walls, Floors, Roofs, and Curtain Systems applied to them using the "By Face" tools.

- Conceptual Masses can be further analyzed in the project environment with the use of the ***Mass Floors*** tool and Schedules.

Rendering

INTRODUCTION

The rendering tools in Revit Architecture allow you to take one of your 3D views and easily create a photorealistic rendering of your model. Since rendering is an integral part of Revit, you need only maintain a single model for both documentation and visualization purposes. This speeds the process of rendering because no exporting is required and you do not have to build and maintain two models. Quick renderings can be created at any point in the design process to help make design decisions and verify that your design intent is being maintained.

Revit uses the Mental Ray (visit **www.autodesk.com/mentalray** for more information) rendering engine to create an accurate representation of the lighting conditions (both daylight and artificial) and materials applied to the objects in the 3D view you are rendering. Multiple renderings or even animated rendered walkthroughs can be created as presentation tools to show a client or to aid the design team in understanding the design.

OBJECTIVES

The aim of this chapter is to familiarize you with the overall rendering process and capabilities native to Revit. We will explore working in 3D views, lighting, materials, and creating the rendering. At the completion of this chapter you will know how to:

- Set up Camera views
- ConFiguredaylight and artificial lighting
- Work with materials' rendering properties
- Launch the render dialog and create a rendering

REVIT RENDERING WORKFLOW

Regardless of the type of final rendering you wish to generate, the process follows a common workflow. We will begin by outlining at a high level each of the overall steps required to achieve a rendering in Revit.

- Define a 3D view

You can only render three-dimensional views in Revit. Creating a 3D view is the first step in the rendering process. Isometric and perspective views can both be used. Elevations and sections may also be rendered if a 3D view is created and the orientation of the view is set to view the model in an elevation or section. You can use the ViewCube for this. A well-composed 3D view is often the most critical step in creating a good rendering. A nicely composed image is often the difference between good renderings and great renderings.

> Recall that you can orient any 3D view to match an existing plan, section, or elevation view in your project. Therefore, if you want a plan or section rendering, you need only create a 3D view, right-click the ViewCube, and then orient it to one of your existing plan views. Refer to either the "Edit Railings" topic in Chapter 7 or the "Create a Working View" topic in Chapter 9 for more details.

NOTE

- Assign Materials

Materials are typically assigned to elements directly in the model as part of the element type. Many Families and their types have default materials already applied, but you may need to make modifications to these default materials or create new materials for use in your project. Materials give both color and texture to your renderings. A large library of predefined materials is included with Revit.

- Define Lighting

You can define both natural (day lighting) and artificial lighting for your rendering. To use natural light, you indicate the longitude/latitude position of your model and the time of day and time of year. Artificial lights will include information such as their intensity, color, etc. Artificial lighting Families can even include photometric IES files provided by the lighting manufacturer to add more realism to the effects created by the light. A photometric IES file accurately represents the shape and pattern of the light, as well as its intensity and color. A background for the rendering can be included and defined when lighting is defined. The background settings do have impact on the natural lighting calculations done during the rendering process.

- Define the Render Settings

The rendering settings include both the quality and size of the output. Both of these have a direct impact on the time it takes to render your final image. Understanding how the rendering will be used can help you make informed choices about both size and quality of renderings. Revit includes default quality presets to get you started and you can customize them as required. Use the "region" option to define a smaller portion of the view for Revit to render. This will allow you to quickly see a portion of the rendering to determine if your material, lighting and rendering selections are producing the desired results. Once you are satisfied, the "region" option can be turned off.

- Render the Image

Grab a cup of coffee and render the image. Rendering can be a lengthily process depending on many factors. So plan accordingly and make sure you are satisfied with all your settings before you start the final render. Render a file at night and, if all goes well, when you return to the office in the morning a high-quality rendering will be waiting for you!

- Adjust Image

The Mental Ray rendering engine calculates lighting levels through the entire scene. This means you can make minor adjustments to the rendering after the rendering process has completed! You can adjust basic exposure of the image as well as shadow intensities and color values. Making adjustments can save time because the entire view may not need to be re-rendered if the exposure controls can help you get the results you want.

- Save/Export Rendering

Once complete, renderings can either be saved to the project or exported to an external file. Saving a rendering to a project will make it part of the project file. This makes them easy to find, but such renderings can't be modified further in Revit or exported to a file. Renderings exported to a file can be manipulated further in an image editor such as Photoshop. Further, you can still use a saved image in Revit by inserting it into Revit using the *Image* tool on the Insert tab.

MODEL PREPARATION

When you intend to generate a rendering from a 3D model, certain considerations must come into play. For example, the more geometry your model contains the longer the render process will take to complete. Geometry is not just the number of objects in the rendered model, but the number of "faces" a model contains. For example, a model of a simple cube form would contain 6 "faces." An organic free-form element may contain hundreds of faces. Unlike some rendering applications, Revit does not have the ability to drop faces of objects that are not directly visible. This means that careful consideration must be made when creating or using Families with detailed 3D geometry. The geometry in question may serve a necessary purpose for the documentation, coordination, cost estimating, or any number of other functions that Revit BIM models are often asked to serve. This makes it a challenge to create a single model suitable for design, documentation, and rendering.

Fortunately, creating a model that will perform well when rendered typically involves the same considerations you want to make to optimize overall Revit performance. For example, it is important to remember that not everything needs to be modeled in 3D. Consider if the 3D geometry is serving a vital purpose. There may be items that do not need to be displayed in 3D to serve their intended function. Even if you plan to show, count, schedule, and dimension an element in multiple views, it does not necessarily need to contain detailed 3D geometry. A model Family can be just as successful relying on symbolic lines in plan and elevation views as it would be using 3D geometry. Sometimes such an approach can offer a nice compromise between modeling and performance. Again, making such decisions when building your model will have a positive impact on the overall performance of your model, not just your renderings, so it is good to get in the habit of thinking this way. Just because something *can* be modeled in 3D, does not mean it *should* be modeled in 3D.

To help decide if something should be modeled in 3D you can ask the following questions:

- Will this item show in a rendering?
- Do you need to check interference or clashes (see Chapter 15) for the item in question?
- Will this item need to be shown in more than one view? (floor plan and section at the same scale for example)

If you answer yes to any of these questions then you can probably make a case to model at least the overall form of the object. If you answer no to these questions then you may want to consider using 2D drafting items to represent the object. Using 2D drafting items for objects when it makes sense can reduce the object and face count of your Revit models, enhancing both rendering performance and overall Revit model performance. In some cases, it can also cut down on the amount of view-specific graphical overrides you may need to apply.

When you decide to model an element in 3D, another option that can help increase performance is the use of Worksets. As we discussed in Chapter 15, Worksets are specifically designed to allow multiple users to work within a Revit project. With some planning, however, Worksets can also be utilized to improve performance while rendering. Worksets allow you to break the model up into smaller pieces by assigning groups of elements to a particular Workset. As we discussed in Chapter 15, you can choose which Worksets you wish to load when you open a project. This allows you to load the model partially in order to increase performance. For example, if you consider rendering as a criterion when devising your Worksets, you might decide to have all of the exterior model elements in one Workset separate from the interior elements. When you are rendering the exterior of the building you would only need to load the exterior Workset, which would take less time than loading the entire model.

Another model setup consideration to enhance rendering performance is to set up dedicated views specifically for rendering. Having dedicated rendering views offers many advantages. You can control the detail level of the view (Coarse, Medium, or Fine). If the Families in the project have been created with multiple levels of detail, then setting to a lower detail level can reduce the geometry in the view even while higher levels of detail are used in other views not being rendered.

You can also turn off the visibility of model categories not needed in the rendering. View templates can be used to quickly apply changes to other rendering views. Refer to the "Editing View Visibility Graphics" topic in Chapter 12 for more information on editing the Visibility/Graphics of a view, and refer to the "Renaming Levels" topic in Chapter 4 for information on using View Templates (see Figure 17.1).

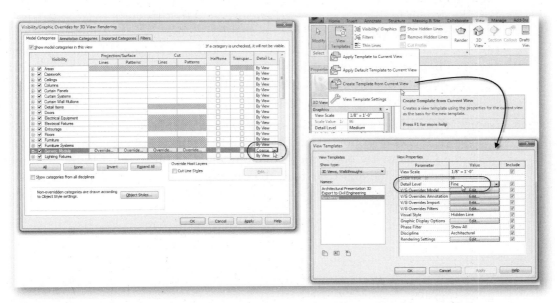

FIGURE 17.1 *Edit the View Visibility/Graphic Overrides and use View Templates to speed setup of other views*

Another tool you can use to limit geometry in a view is a section box. A section box will limit the extents of the 3D view cropping off portions of the model not required for rendering. This can be especially useful when rendering interior views. The section box can be used to crop out the entire model except the interior space you are rendering. If you create a camera view, use the far clipping to crop out portions of the model.

3D VIEWS AND CAMERAS

A 3D view can either be an isometric view or a camera view which generates a perspective view based on camera placement position and camera target position. We have created several 3D views elsewhere in this book. Refer to the "Viewing the Model in 3D" topic in Chapter 3 for an example of creating an isometric view with the Default 3D tool and refer to the "Load the Family into the Project" topic in Chapter 10 for an example of creating a Camera view.

Once you have a 3D view, you can adjust the viewing angle using any of the techniques already covered. Some topics to review include the "View the Model in 3D" in the Quick Start chapter and the "Navigating in Views" topic in Chapter 2. Once you have positioned a view the way you like, you can save the view to the Project Browser by right-clicking the Home icon on the ViewCube and choosing **Save View**. Apply the View Template you made above to the new views. This will allow you to have multiple views of your model that can quickly be accessed and specifically set up and optimized for rendering. Once the view is saved you will also want to set the home view on the ViewCube. Right click the home icon on the ViewCube and choose **Save View**. This allows you to quickly reset the view to the pre-set position in case you accidently make changes to the view and would like to reposition it where it started.

You can create Camera views from any view, but it is easiest to place a camera while in a plan view. When placing a camera, the first click is where the camera will be positioned, or where you would be standing if you imagine yourself taking a picture. You can control the height of the camera position on the Options Bar before you click the point. The height of the target point can also be modified from the Options Bar. The default for both is at eye level (about 5'-6" [1650]), which makes a 2-point perspective. If you vary the height of the points, you can create a 3-point perspective. The positioning of a camera can be changed after it is placed. To do this, open any view but typically a plan works best. On the Project Browser, right-click the Camera view and choose **Show Camera**. Edit the Camera using the settings on the Properties palette, or the controls on screen.

Isometric views have a scale parameter which determines how big they will be when dragged to a sheet or printed. Camera views use a Crop Size to determine their size. With the Camera view open, click the crop boundary on screen. On the Modify Cameras tab, click the *Size Crop* tool. You can either edit the field of view or maintain the current proportions. The size you indicate will be the size that the view will print (see Figure 17.2).

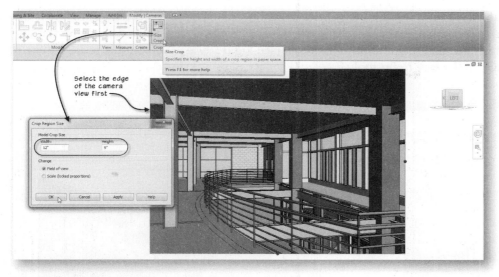

FIGURE 17.2 *Edit the size of the Camera Crop*

MATERIALS

One of the key components to creating a good rendering is the application of convincing materials. Revit ships with the Mental Ray Materials library. This gives a good basic set of architectural materials you can use right away in your projects. The material library can also be modified to create new materials to match your project's specific needs. The Autodesk Material Library will be installed when you install Revit. This is a shared library of materials that is used by many Autodesk products and includes many high quality rendering materials. Revit uses these for the appearance settings of Revit materials.

Materials are applied to elements in your model as either type or instance parameters. For layered elements such as Walls, Floors, Ceilings, and Floors, you apply materials to each individual layer in the assembly at the type level. Loadable component Families such as Doors, Windows, Furniture, and Equipment can use material parameters either at the type or instance level (see Figure 17.3).

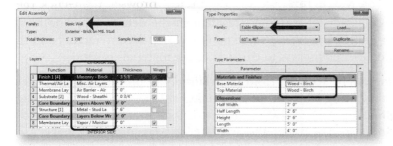

FIGURE 17.3 *Materials can be assigned to element types and/or instances*

Another way materials can be applied to objects is with the ***Paint*** tool. Found on the Modify tab, Geometry panel, the ***Paint*** tool will allow you to assign any material to a single face of an element. This simply applies an override to the assigned material for just the selected face. This can be useful if one side of a wall has one color of paint and the other side has a different color of paint. One wall type can be created and the ***Paint*** tool used to get the render appearance correct. If the entire face is not the same material, you can first use the ***Split Face*** tool to sketch a shape on the surface and then paint another material (see Figure 17.4). New in this release, there is also a ***Remove Paint*** tool.

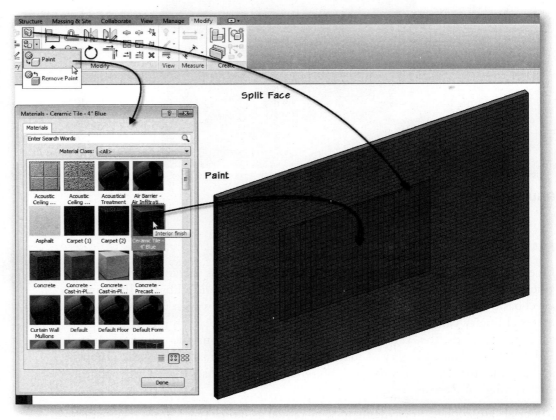

FIGURE 17.4 *Use Split Face and Paint to apply materials to selected areas of a face*

The final way in which materials can be defined is at the category level. When the material assignment is: <By Category> the category level material will be used. Use Object Styles on the Manage tab to view and assign category level materials.

Creating and Editing Materials

The out-of-the-box collection of materials (resident in the default template files) offers a good starting point. However, at some point you will need to modify and/or create your own materials. The process to edit and create materials is similar.

The "Materials" dialog can be accessed from the Manage tab, on the Settings panel. You will also access this dialog when assigning a material to an element. The dialog lists all materials defined in your project. If the list is long, you can use the search field at the top to find a specific material. The Material Class drop-down also can be used to further refine a search or simply list only materials from a specific class.

All of the material appearances in Revit are based on the Mental Ray materials and the shader that material uses. A shader is a set of rules and settings used by the rendering process to recreate the look required by a material/appearance. The values on this tab will determine what controls and settings are available for a particular material. For example the settings for a wood material will be quite different from the controls you will see on a material based on metal (see Figure 17.5).

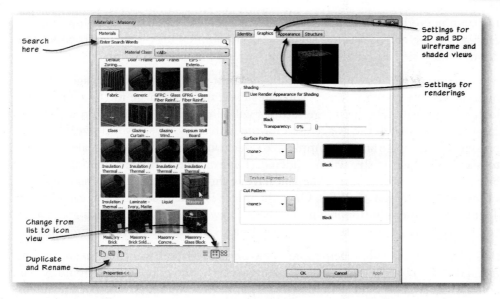

FIGURE 17.5 *The Render Appearance tab of the Materials dialog contains settings used for rendering*

The first step to editing a material is to check if a material already exists with the look you want. Do not confuse the list of materials in the "Materials" dialog on the left (which are the materials available in your project) with those available in the Autodesk Materials library. The Autodesk Materials library is available to all projects (and all Autodesk products installed on your system). The safest way to experiment is to select a material on the left side, click the Duplicate icon at the bottom and rename it. Then on the right side, click the Appearance tab. This will open the Appearance Property Sets tab. Here you can see all of the predefined Autodesk Materials that ship with the product. This library can be searched and grouped by material class to help you limit the choices and find the material you would like to use.

You will want to find the render appearance that most closely matches the material you would like to use. This will get you very close to your final render appearance, minimizing the changes you need to do in order to get the look you would like. You can change the scene used in the preview window to a number of preset scenes; this will give you a quick idea of how the material will look in a particular condition. The preview window is a live rendering of the material so it may take some time to update (see Figure 17.6).

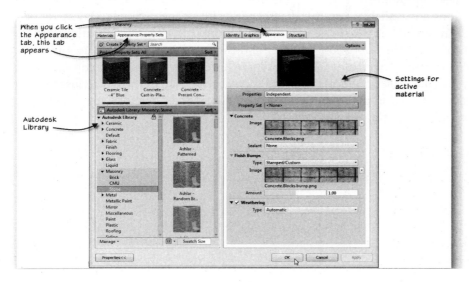

FIGURE 17.6 *Click Replace to access the Render Appearance Library*

Depending on the render appearance selected, you will have different properties you can alter. These will be settings such as (but not limited to) the images used for rendering, image size and rotation information, and bump information.

Most materials will use some kind of image file. If you can obtain an image file of the material you are trying to create you can substitute your image file to get a custom look for the material. The image settings will also include sizing information. You can use this to match the image used for the material to a real world size. Using the preview of the image and the sample size value you can get the image scaled properly to match a real world size when rendered. (For example, if you are creating a custom brick material, and the image file you have is three bricks wide by nine bricks tall, you would set the size to 2'-0"). Once the image is scaled properly you can click the Texture Alignment button just under the preview. This will open a dialog and allow you to align image to the surface pattern. Images may also be rotated if required.

Be sure to read the online help topics on materials for more information and recommendations on building materials. Starting with the provided library should get you going in most cases. Add custom materials only after you have ruled out the out-of-the-box offerings.

LIGHTING

Playing just as important role in rendering as materials and modeling is the lighting of the model. Without good lighting all of the hard work you have put in on modeling and defining materials is at risk of being wasted. Depending on the scene, you might have natural light/daylight, artificial lights, or both. The Mental Ray rendering engine uses global illumination. This is a fancy term for figuring out how light from all sources bounces around in a scene. The lighting conditions must be defined accurately in order to achieve the best results in this kind of rendering.

The daylight for a scene is easy to set up. In your rendering 3D view choose Graphic Display Options from the View Control Bar. The "Graphic Display Options" dialog has been completely revamped in Revit 2012. At the top, in the Model Display area, you can choose the visual style you want, whether to show edges and the optional silhouette setting. Shadows can be enabled and if your video card supports it, you can enable ambient shadows. In the Lighting area, the Sun Setting has a button labeled <In-session, Lighting> by default. You can click this button to set an actual global location, time, and date (see Figure 17.7). In the "Sun Settings" dialog several options are available. You can do a generic still image, a Single Day study or a Multi-Day study. For the single and multi-day options, click the Location Browse button to access the "Location Weather and Site" dialog. This dialog uses Internet Mapping Service powered by Google. All you have to do is type the address of your project into the Project Address field and then click search. In the figure, just the city was input, but you can put in the actual street address and Google Maps will find the address and assign it to your project.

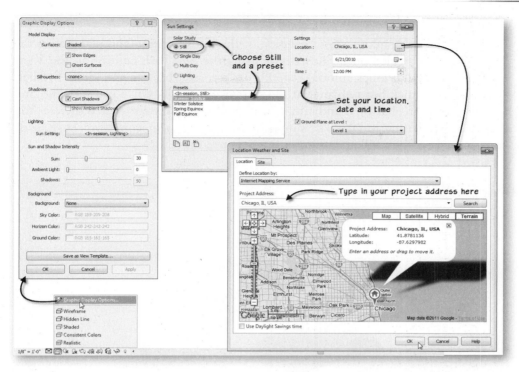

FIGURE 17.7 *Set the position of your project by city, date, and time and optionally enable shadows in the view*

Back in the "Graphic Display Options" dialog you can enable the "Cast Shadows" option. This will apply shadows directly to the 3D view. This is not a fully rendered view, but by turning shadows on in a view before rendering you can get an idea of how the shadows will be rendered in the scene. You can also choose Realistic shading mode from the Surfaces drop-down list at the top to preview the render materials directly in the viewport. This option with shadows on is the closest you can get to previewing the way the scene will look rendered before performing the actual rendering. Any adjustments you make in this dialog can be saved directly to a View Template using the button at the bottom of the dialog. The View Template can then be applied to other 3D views.

Realistic shading is only supported if you have set your graphics mode to Direct3D, which must be supported by your video card. For a list of supported graphics cards, visit the Autodesk Web site: http://usa.autodesk.com/adsk/servlet/index?siteID=123112&id=16470651.

NOTE

Leaving shadows turned on in a view can reduce performance, so it is suggested to leave shadows turned off in view until you need them for printing or previewing the lighting conditions for a rendering. You can quickly turn them on and off in the view using the toggle icon on the View Control Bar.

If you are working on a daytime exterior rendering, it is likely that this is all of the lighting setup you will need to do. If working on an interior or an evening image, you will also need to define artificial lighting.

There are many options in the "Sun Settings" dialog to explore. You can create animated solar studies directly in Revit. To do this, choose the Single Day or Multi-Day option. Choose your dates and time ranges and click OK. From the Application menu, you can choose **Export** > **Images and Animations** > **Solar Study** to create an AVI file of the sun moving across your project over time.

NOTE

Sun Path

Revit includes a graphical representation of the Sun's position and path across the sky. It can be toggled on in most views using the Sun Path control on the View Control Bar. For a quick tutorial covering its use, refer back to the "Solar Studies" topic in the Quick Start chapter at the start of this book. Once you turn the Sun Path on, you can click on it and interact with the many controls located therein. You can change the time and the date simply by dragging the sun icon. If you have shadows turned on, they will update accordingly in real time.

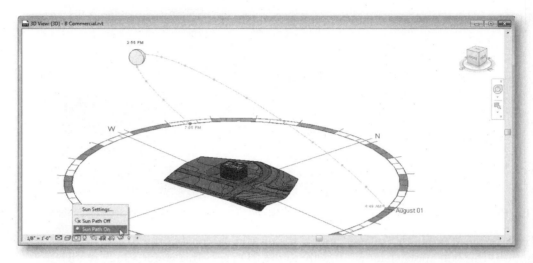

FIGURE 17.8 *Use the View Control Bar to turn on a graphical representation of the Sun Path directly in the view window*

Artificial Lighting

Artificial lighting for a scene is established by lighting Families placed into the project. Lighting Families include both the actual geometry of the fixture itself, plus the lighting information required for it to cast light into the scene. There are many lighting Families shipped with Revit. We added several from the out-of-the-box content to the third floor of our commercial project in Chapter 13. You can create or open a Camera view from that project and render one of the third floor spaces. If it is an exterior space, it will use both the artificial light and any sunlight settings you define. Interior spaces without windows will use the artificial light only.

The out-of-the-box light fixtures have photometric information defined as type parameters of the fixture. This photometric information affects the intensity, color, and other properties of the artificial light source in the rendering. New types can be created or existing types altered in order to create different lighting conditions or fixtures to match your needs. Remember that a light fixture Family in Revit gives you two things: geometry that represents the actual fixture and the lighting information that casts light in the space during rendering. Lighting manufacturers sometimes provide Revit Families of their fixtures. If you are specifying a particular fixture, contact the manufacturer to see if they have Revit files available. In many cases, they will at least have IES files available. An IES file (sometimes referred to as a photometric web file) describes the actual physical and geometric distribution of light from a fixture. Using lights that have these files assigned to them are capable of describing lighting conditions much more accurately. Any light fixture Family can have an IES file embedded within it. When you render a space with such a Family as the light source, the accurate geometric distribution of lighting will be generated automatically and used in the rendering.

RENDER

Render is where the rubber meets the road. The model is ready to go, you have created your view, materials have been applied, and lights have been defined in the model; it is now time to render. The render dialog is available from the View Control Bar. Click the "tea kettle" icon to open the "Rendering" dialog and create your first rendering. You can change the quality, size, and lighting conditions used to perform the rendering (see Figure 17.9).

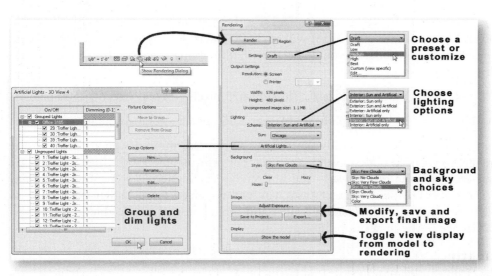

FIGURE 17.9 *Set the position of your project by city, date, and time and optionally enable shadows in the view*

Refer back to Figure 17.9 for the next several items.

Quality

The Quality drop-down lists several presets. These presets allow you to quickly choose a quality for the render without having to configure the specific rendering settings. This makes achieving results very quick and easy. In general, the lower the quality, the faster the rendering will complete. If you want to run a quick test to see if you have your lights and materials right, choose Draft. If you are ready for the high quality render and have time to wait, choose Best. You can also choose Edit to make changes to a wide variety of settings. For your first few renderings, it is recommended that you stick with one of the presets. They offer enough variation in the quality vs. time to render tradeoff to give you satisfactory results. If you wish to edit the settings, consult the online help for more information.

Output Settings

Output Settings determine the size of the final rendering. Computer images like digital photos and renderings are measured in pixels. In general, the more the pixels in an image, the higher the quality of that image. However, the tradeoff is increased file size of the rendering file and increased time to render. When running test renders at any quality setting, choose Screen resolution to start. Screen resolution is much smaller than printer resolution and will therefore take less time to process. However, if you export the finished screen resolution rendering and try to print it, the results even at Best quality will likely not be satisfactory. Understanding your intended final use of the rendering is important to make the right choice in output settings.

If the image will only be seen on a screen such as in a Power Point presentation, then Screen resolution can be safely used. If you intend to print the image, then select the Printer option and your desired DPI setting. Remember from the "3D Views and Cameras" topic above that the size (in inches) of the Camera view is determined with the *Size Crop* tool. The DPI setting will determine how many pixels are rendered for each inch in the final image. For example, in Figure 17.2, the view was set to 12" wide by 9" tall. (This size will appear in the Render dialog when you choose the Printer option.) If you choose 75 DPI, the final width of the image will be 900 pixels. If you choose 300 DPI it will be 3,600 pixels! Furthermore, when the height and width are considered, you will realize that doubling the resolution actually quadruples the total pixels.

At 75 DPI   Width 900 pixels   Height 675 pixels   Total size: 607,500 pixels

At 150 DPI Width 1,800 pixels Height 1,350 pixels Total size: 2,430,000 pixels

In general, 75 DPI is basically the same as screen resolution. If you are printing the image, you should not choose this setting. For printing, try 150 DPI first. If you are satisfied with the results, you need not render any higher. In some cases, you may opt for 300 DPI, but in most cases, you will not need to go higher. Rendering to 600 DPI will take considerably longer and often you will not be able to notice any difference in the final printed image.

Lighting

The Lighting section of the "Rendering" dialog includes three settings. First choose your lighting Scheme. There are six options here. If you are rendering in exterior scene, choose one of the Exterior options, and if you are rendering an interior scene choose one of those options. This setting sets a default exposure for the scene based on the lighting condition you indicate. The exposure can be adjusted both before and after the rendering is complete if you are not satisfied with the results.

If you select an option which includes sunlight, you will be able to select the sun position. This is the same list we saw earlier. If artificial lights are part of the lighting selection, the Artificial Lights button will be available. Click this button to see all of the lights in your model. The "Artificial Lights" dialog is shown on the left side of Figure 17.9. You can turn lights on or off in this dialog by checking or clearing the checkbox. You can also dim lights here by typing in a value between 0 and 1. A value of 1 indicates the light is on and not dimmed; a value of 0 means the light is fully dimmed (off). A value of .5 would shine the light at half intensity.

If you have added many lights, the list could be quite long. To help you manage the list, lights can be grouped. Click the New button to create a new group. Name it descriptively. Next, using the SHIFT and/or CTRL keys, select one or more lights in the list and then click the Move to Group button. The advantage of grouping the lights is that you more readily decide which lights should be on or off for a given scene. For example, if you are rendering a single office space, you would not need to use the lights in the neighboring office spaces. Every light you leave on in the scene takes processing power. It makes no sense to calculate lights in another room. Therefore you will want to turn off lights that are not in your current rendering, as it will save much time in the rendering process.

Background

The background setting allows you to specify the amount of clouds in the sky from a list of preset choices. A solid color can also be specified for the background. You can also load a photograph as a background. In the Background area of the Rendering dialog, choose **Image** from the Style drop-down. In the "Background Image" dialog, click the Image button and locate a suitable image on your hard drive. (see Figure 17.10).

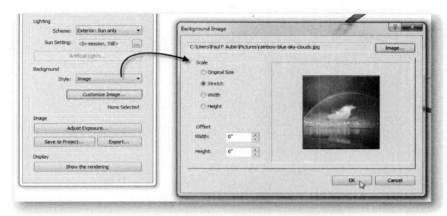

FIGURE 17.10 *Use an image on your hard drive as a background for renderings*

Keep in mind the orientation of your 3D view, the vantage point, time of day, and lighting. In order for the image you select to be believable in your final rendering, the perspective and shadows need to match. For the Scale settings, you can make the image stretch to fit your rendering size, or leave it set to its original size.

Image

Click the Adjust Exposure button to edit the overall exposure, highlights, midtones, and shadows of the rendered image. Settings can be changed prior to rendering or after. If you render first, you can edit the exposure and using the Apply button get immediate feedback on screen.

Generate the Rendering

Once you have configured all the setting to your liking, click the Render button at the top of the dialog. This will begin processing the results, which, depending on your settings, can take minutes or hours. If you are not certain about the settings you have chosen, select the "Region" checkbox before clicking the Render button. This will display a rectangle on screen. Click and use the handles to adjust its size. Surround a small portion of the view and then click the Render button. This will allow you to check shadow/lighting, material, and quality of the rendering without taking the time to render the entire scene.

During the rendering process Revit will display a rendering progress bar. You will see how long the render process has been running and the number of artificial lights and number of daylight portals included in the rendering. Once the rendering has started, you cannot do other work in Revit. Try to plan your render times to occur during off hours, such as in the evening after business hours.

Post-Rendering Exposure Adjustment

When the rendered image is complete, you can still adjust the exposure. If the results are not as expected you can still use the Adjust Exposure button to make adjustments. Making adjustments here can often get the results you are looking for without the long process of altering lighting and re-rendering the scene. The exposure adjustments can be immediately applied to the scene by pressing the Apply button in the dialog (see Figure 17.11).

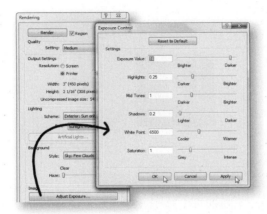

FIGURE 17.11 *Adjust Exposure after the rendering is complete*

Exposure Value is like the F-Stop on a camera. You can make the overall scene darker or lighter by making changes to this setting. The pixels of an image fall into three broad categories: Highlights, Midtones, and Shadows. Highlights are the brighter sections of the scene. Midtones are the middle range and Shadows are the darkest areas of the image. Adjusting any one of these does not affect the others. This can give you finer control over the image than the Exposure Value alone would. The White Point setting will shift the colors in the scene to a warmer or cooler end of the color spectrum. If images are orange or red in color you can shift the white point values cooler to get a more balanced appearance. Saturation controls the color intensity of the image. A value of 0 will result in a grayscale image, while 1 is the default. Higher than 1 will make the colors much more intense.

Any changes made to the exposure values will be remembered by the camera for this view unless changes are made in the lighting section. When you are done with the rendered image you can click the Show Model button at the bottom of the dialog to normal model display. The button will toggle between Show Model and Show the Rendering. If changes occur in the model they will not be reflected in the rendering. You will have to render again to include them.

OUTPUT

The onscreen rendering is not permanent. If you quit Revit, the rendering will be lost. You have two options for saving your rendering images (shown at the bottom right of Figure 17.9). Use the Save to Project button to create a raster image of the rendered output and add it to a branch on the Project Browser. Such views cannot be further modified, but they can be added to a sheet and printed with the project. To create an external image file, use the Export button. This will save the image to an image file on your hard drive. You can choose between popular formats like BMP, TIF, JPG, and PNG.

Exporting renderings to a file is a more flexible alternative to saving the rendering to the project. This allows you to modify the image outside of Revit in any image editing program. If you need the rendering as part of the Revit document set, you can import the image file back into the project (Insert tab, Image button) and place on a sheet.

The rendering tools inside of Revit are very powerful and can produce compelling images with minimal extra effort. The strong connection to the model which is being used for design and documentation is the strongest argument for using the rendering tools in Revit. You can always be sure your renderings will reflect the building at its current stage of development (see Figure 17.12).

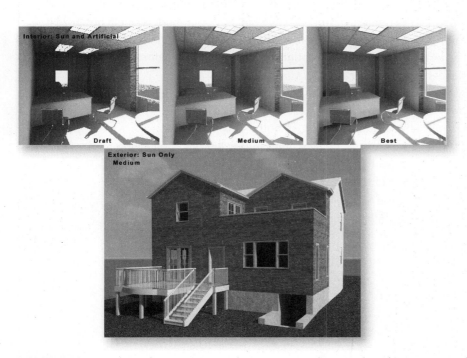

FIGURE 17.12 *Sample Renderings: Commercial project interior above, residential project exterior below*

While Revit Architecture can generate compelling images, its focus is on modeling and documenting a building project. If you decide at some point that you need a more robust rendering solution to incorporate into your BIM process, you may wish to consider Autodesk 3ds Max Design, the primary focus of which is photorealistic rendering and animations. Additionally, Max Design provides the ability to perform accurate lighting analysis studies. Autodesk FBX files can be exported from Revit Architecture and directly imported into Max Design, maintaining all of your lighting and material assignments and settings established in Revit Architecture, eliminating potentially hours or days of rework.

WALKTHROUGHS

Revit also has the ability to create simple walkthrough animations. Start in a plan view, click the View tab and choose the *Walkthrough* tool from the *3D View* drop-down button. Click a point to start the walk and then continue clicking points to draw a path for the walkthrough. Click the Finish button to complete the path. A Walkthrough branch will appear on the project browser. Walkthrough views are basically animated camera views. They share many characteristics with other cameras and have additional controls to make them move along the walkthrough path. To play the walkthrough on screen, open the Walkthrough view from the Project Browser. Select the crop border on screen and then click the *Edit Walkthrough* button on the ribbon. On the Modify Cameras tab, you will have several VCR-style buttons to play, fast forward, and reverse the walkthrough. Finally, you can export the walkthrough to an AVI file using the **Export > Images and Animations > Walkthrough** command on the Application menu.

SUMMARY

- The process to generate renderings in Revit is simple.
- Work with a single model that you are already using for design and documentation.
- Consider the rendering process when building your model to help optimize it.
- Course level of detail, Visibility/Graphic overrides, Section boxes, and Worksets can all be effective ways to optimize model geometry before rendering.
- Rendering can be performed from any 3D view.
- Assign Materials from the vast library provided with the software or create your own.
- Use daylight, artificial lighting, or both.
- Most lighting Families include photometric lighting data giving renderings realistic lighting qualities.
- Render to one of several preset quality settings.
- Adjust exposure to fine-tune results.
- Save an image file of your rendering for the greatest flexibility.

SECTION

V

Appendices

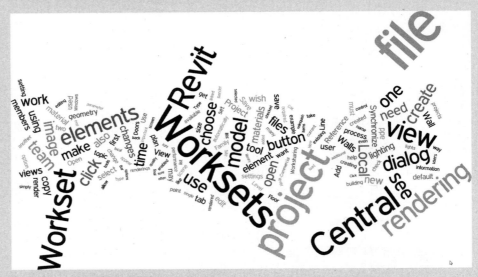

Most common words in Section V

The Appendices include additional important exercises and resources. Appendix A offers some additional practice exercises that space did not permit in the main text of the book. Appendix B provides additional resources that you might find useful as you begin using Revit Architecture in your daily work. Also be sure to explore the contents of the dataset files installed from the online companion for other useful files and resources.

Section V is organized as follows:

Appendix A   Additional Exercises

Appendix B   Online Resources

Additional Exercises

INTRODUCTION

Contained in this appendix are some additional exercises that you can perform to gain more practice with tools and techniques covered in the book. With the exception of the existing conditions of the residential project, most of the attention in Chapters 3 through 9 was paid to the exterior of the buildings. The residential dataset provided in Chapter 5 and all subsequent chapters includes the existing conditions for the Second Floor. The commercial dataset for Chapter 10 includes interior partitions and other refinements not detailed in the tutorials. This appendix addresses these gaps by providing the minimal guidance required to assist you in creating the elements provided at the start of these chapters' datasets. Some other suggested exercises are also included to help round out the projects with some finishing touches.

OBJECTIVES

The exercises presented in this appendix are intended for additional practice and are meant to be self-directed. As such, you're given the overall goal of the exercise, the critical dimensions and some guidelines and tips to assist you. The exercises are optional and if you prefer you can skip them. The exercises are intended to be completed following each chatper listed in the headings below. After completing this appendix you will have:

- Gained additional experience with Wall layout

- Added Windows, Doors, and other embellishments to the residential dataset

- Modified the toposurface in the commerical project to add roads

- Completed the third floor tenant build-out on the commercial project

- Added plumbing fixtures to both projects

- Practiced several other skills covered throughout the book

RESIDENTIAL PROJECT

The residential project requires refinements on all three levels (Basement, First, and Second). You can begin completing these exercises following Chapter 3.

Chapter 3—Second Floor Existing Conditions

Following the same procedures and techniques outlined in Chapter 3, create the existing conditions for the second floor of the residential project. You can use any of the techniques covered, including drawing Walls, offset, copy, trim, split, and the use of temporary dimensions. Use Figure A.1 as a guide.

1. From the *Chapter03\Complete* folder, open *03 Residential-Complete.rvt* [*03 Residential Metric-Complete.rvt*].

   • Save a copy of the file to the *AppendixA* folder.
   • Work on Level 2 of the project.
   • Use the same Wall, Door, and Window types as used on the first floor.
   • Use the procedures covered in Chapter 3 to import appropriate plumbing fixtures.

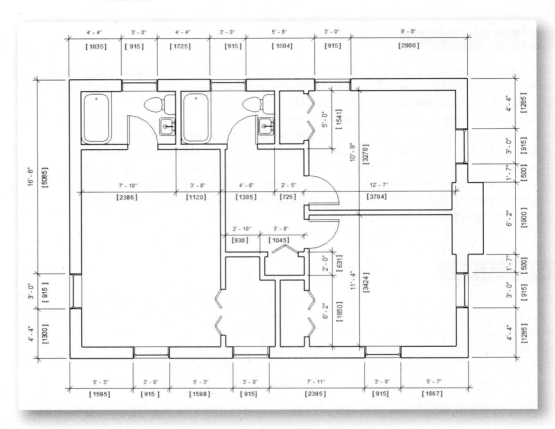

FIGURE A.1 *Layout the Second Floor existing conditions of the residential project*

RESIDENTIAL PROJECT—CHAPTER 10

Chapter 10 starts with all of the new construction interior Walls, Doors, and Windows complete for the residential project. If you wish to build these items yourself, complete these exercises.

1. From the *Chapter08\Complete* folder, open *08 Residential-Complete.rvt* [*08 Residential Metric-Complete.rvt*].

   • Save a copy of the file to the *Appendix A* folder.

Basement Plan—New Construction

1. Start in the *First Floor* plan view.

   • Select the three Walls of the addition, edit the properties and change the Base Constraint to **Site** with a **6" [150]** Base Offset.
   • Delete the Footings.
   • In the West and North elevation views, edit the Profile of the Walls and remove the step from the bottom.

- Copy and paste aligned the same three Walls to the *Basement* plan. Ignore any warnings.

- Select the Walls on the west and the north one at a time and click the Reset Profile button on the ribbon.

- Select all three Walls, edit their properties and change the type to ***Foundation - 12"** **Concrete [Retaining - 300mm Concrete]***. Set the Base Offset to **0**, the Top Constraint to **Up to level: Site** with a **6" [150]** Top Offset.

2. Offset a copy of the west Wall inside the plan. Trim and or Split the Walls as required and use the handles to stretch the two west Walls to match the dimensions in the figure (see Figure A.2).

- Add a monolithic Stair from the Basement level to the height of the terrain at the rear of the house. (Refer to Chapter 7 for information on creating Stairs.) Cut a section to assist in placement and sizing.

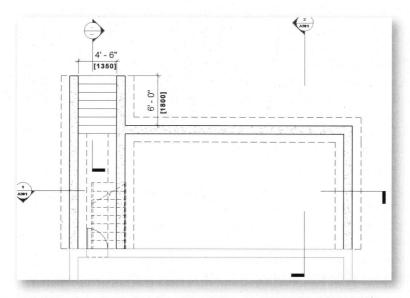

FIGURE A.2 *Layout the Basement plan new construction of the residential project*

- Add new footings to the Walls as shown. The "Add Continuous Footings" topic of Chapter 5 discusses adding wall foundations and making them display in dashed line.

If you want to hide the demolished Stair, you can edit the floor plan view Properties and choose **Show Complete** for the Phase Filter.

First Floor Plan—New Construction

1. Open the *First Floor* plan view.
2. Create a new Wall type.

- Make the Core use a Structure layer of Wood Stud **3 1/2" [90]**.

- Add a Finish layer of Finishes - Interior - Gypsum Wall Board on each side **1/2" [12]** thick.

- Draw the Walls in the center of the plan using the new type.

- Add Windows and Doors as shown. Assume that all inserts are centered unless dimensioned otherwise.

- Load the Bay Window from Autodesk Seek or from the dataset files.

- Use the Demolish tool on the two Windows between the addition and the existing house (see Figure A.3).

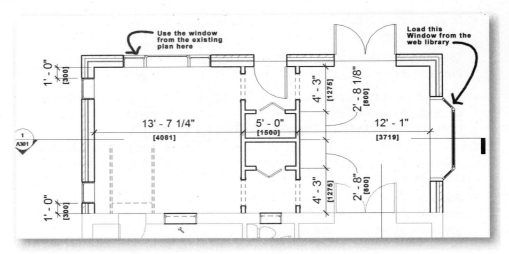

FIGURE A.3 *Add Walls, Windows, and Doors to the First Floor*

Second Floor Plan—New Construction

Demolition needs to take place in the bathrooms between the existing house and the new addition. However, the challenge is that currently, this is one continuous Wall from ground to roof. To demolish just the portion on the second floor, we need to copy and paste the Wall like we did for the basement, adjust the height parameters, and then demolish the portion we don't need.

1. Stay on the *First Floor* plan view.

   - Copy and paste aligned (to the same place) the Wall between new and existing. Ignore the warning.
   - With the Wall still selected, edit the properties and change the Base Constraint to **Second Floor**.
   - When the Warning appears, click the Delete Instances button.

The warning is about the Doors and Windows on the lower portion of the Wall that no longer have a host. It is OK to delete them because pasting gave us copies, but ultimately need only one in each location.

   - Remaining in the *First Floor* plan, select the Wall again.
   - Edit its properties and change the Top Constraint to **Up to level: Second Floor**.
   - You will also need to open the *Transverse* section, select the Wall, and on the ribbon, click the Detach button. Select the Roof to detach the Wall from the bottom of the Roof.

When the process is done, you should have two Walls stacked on top of each other and the Doors and Windows on both first and second floors should still be intact (see Figure A.4).

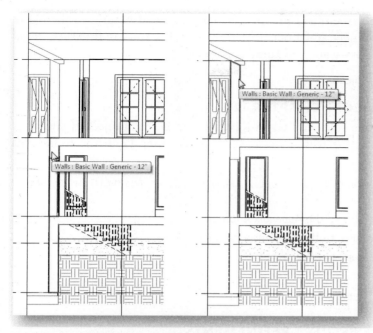

FIGURE A.4 *Add Walls, Windows, and Doors to the First Floor*

1. Open the *Second Floor* plan view.
   - Split the Walls between the new addition and existing house.
   - Demolish the left side of the Wall and demolish the remaining Window in the Wall as well.
   - Draw a new Wall using the custom type build for the first floor between the two bathrooms.
   - Add a Wall between the bathroom and new bedroom.
   - Add a closet to the new bedroom.
   - Add Doors and Windows as shown. Choose shorter height Windows for the bathrooms and set their sill heights at **3'-8" [1100]**.

Assume that all Windows and Doors are centered unless noted otherwise (see Figure A.5).

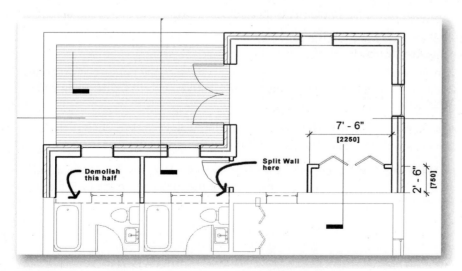

FIGURE A.5 *Add Walls, Windows, and Doors to the Second Floor*

Feel free to add additional embellishments, such as upgraded fixtures in the bathrooms and counters, cabinets, and appliances in the kitchen. You can find such items in the out-of-the-box library and on Autodesk Seek.

1. Save and close the project.

RESIDENTIAL PROJECT—CHAPTER 12

Chapter 12 explores Schedules and tagging in the commercial project. If you wish you can follow many of the techniques covered in Chapter 12 in the residential project.

1. From the *Chapter11\Complete* folder, open 11 *Residential-Complete.rvt* [*11 Residential Metric-Complete.rvt*].
   - Save a copy of the file to the *Appendix A* folder.
   - Add various Schedules to the residential project.
   - Add Rooms to the residential project.
   - Add Room Tags.
   - Add a Schedule Sheet.

COMMERCIAL SITE PROJECT REFINEMENTS

The Commercial project requires some refinements to the site file created in Chapter 6; interior wall layout on the first and third floors. The core walls will be added for all floors, a double-volume space for the lobby of the building and we also have the opportunity to add wall tags to the Walls on the third floor following the exercises in Chapter 13 where the Wall heights and their relationships to the ceilings were adjusted.

Commercial Project—Chapter 6

The Commercial Site project built in Chapter 6 made a single Toposurface. The underlying CAD file however defines roads parking lots and grass areas. The material applied to a Toposurface by default is earth. To make the Toposurface appear more realistic in shaded and rendered views, we can split the surface and apply materials to it.

1. Continue in the commercial site project file you created in Chapter 6. Open *06 Commercial Site.rvt* [*06 Commercial Site Metric.rvt*] file.
   - Save a copy of the file to the *Appendix A* folder.
2. Start in the *Site* floor plan view.
   - Change the display to wireframe.
   - Open the Visibility/Graphic Overrides dialog (VG), click the Imported Categories tab and turn off all layers in the *Commercial-Site* DWG except **C-Site-Pkng** and **C-Site-Road**.

While not required, you may wish to increase the lineweight and/or change the color of these layers as well to make them stand out better.

3. On the Massing & Site tab, click the **Split Surface** tool. Select the Toposurface and then use the Pick Lines tool to chain-select one of the road lines in the CAD file (see Figure A.6).

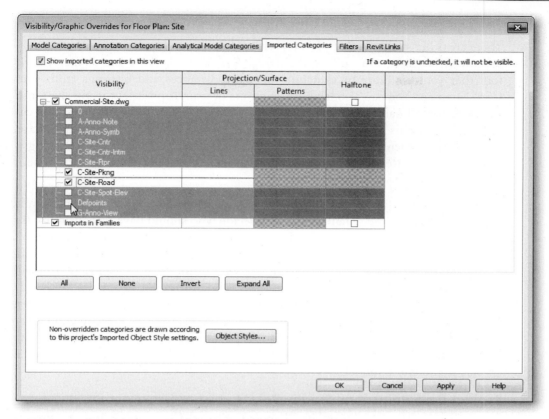

FIGURE A.6 *Chain-Select one side of the road in the CAD file linework*

- Click Finish Edit Mode.
- Repeat for the other three corners of the surface on the sides of the road intersection.

You will end up with five total Toposurfaces after this operation—four grass areas in the four corners and the roads which will also include the parking lot and building lot area. You cannot use Split surface for the building lot area. For that you need Subregion. Use Split Surface to cut a single surface into two pieces. Use Subregion to create one (or more) surface region within another.

1. Click the **Subregion** tool, use the **Pick Lines** tool and chain select the curb line surrounding the building. Finish the surface.
2. Using the CTRL key, select the Subregion surface and the four split regions.
   - On the Properties palette, change the Material to **Site – Grass** file (see Figure A.7).

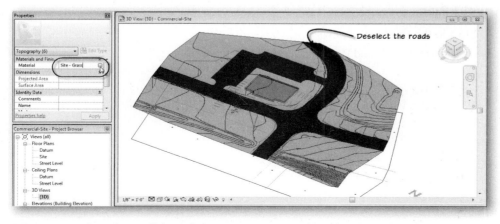

FIGURE A.7 *After splitting all the surfaces, assign grass to everything but the roads*

3. Repeat the process to assign **Site – Asphalt** to the roads surface.

There are other site tools that you may wish to explore. If you decide to join two surfaces together later, you can use the *Merge Surface* tool. There is a *Property Line* tool that you can draw by sketching or using distances and bearings from your Civil Engineer. You can also label the contours of your Toposurface using the *Label Contours* tool. Finally, the *Graded Region* tool will allow you to calculate basic cut and fill. To use this tool, you have to move your Toposurface to the Existing phase. The grading is then applied in a later phase such as New Construction. Phasing is discussed in the Residential project in Chapters 3 and 7. Feel free to experiment with any of these tools now.

For the parking lot, you can add parking components. A Parking Component Family is already loaded in the default template. Decide who should have responsibility for laying out parking. If it is the Civil Engineer, you can add parking to this project, if it is the Architect, you may want to add it directly to the Commercial project instead. If you add it to the Site project, you should work in the Street Level view. You will need to turn on the Parking category in the Visibility/Graphic Overrides (VG) dialog first. You can also apply the Site Plan View Template to the view if you prefer. The Parking component places a single parking space. Move, rotate, copy, and array it as appropriate. The *Site Component* tool can be used to add trees and other site accoutrements. Again, it is up to you to decide if these should be in the Site model or the Architectural model.

Project Base Points and Survey Points

In the "Linked Projects" topic in Chapter 6, Shared Coordinates and the Project Base Point and Survey Point icons were discussed. Provided in the *Appendix A* folder are some additional files that you can use to explore this topic. Three files are provided: *Site.rvt, Building.rvt,* and *Grid.dwg.* Grid is linked into the Site file and can be moved around to get a sense of how the various points behave. Feel free to open these files and explore.

Commercial Project—Chapter 7

The commercial project in Chapter 7 starts with Walls in the building core and a concrete entrance terrace in front of the building. You can do this exercise before you do the tutorials in Chapter 7 for the commercial project if you wish to gain additional practice with Wall layout.

1. From the *Chapter06\Complete* folder, open *06 Commercial-Complete.rvt [06 Commercial Metric-Complete.rvt]* file.
   - Save a copy of the file to the *Appendix A* folder.

Level 1 Plan

1. Start in the *Level 1* plan view.
   - Draw the Walls and Doors shown in (see Figure A.8).

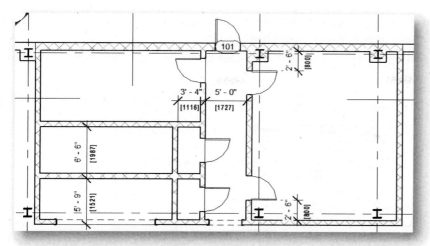

FIGURE A.8 *Add Walls and Doors to create a lobby on Level 1*

All Walls shown in the figure go from Street Level to the Roof level. Use the ***Generic – 8"*** ***Masonry [Generic – 225mm Masonry]*** Wall Type. The Doors are ***Single-Flush:*** ***36" × 84" [0915 × 2134mm]***. Center the ones in the closets, place the one in the Stair Core 6" [150] from the corner and place the others at the dimensions indicated. The two openings are the Opening-Cased Family. You need to create two types. The large one should be named: 144" × 84" [3600 × 2134mm] and the other named: 48" × 84" [1200 × 2134mm]. Set the width and height parameters to match.

COMMERCIAL PROJECT—CHAPTER 8

The provided dataset for Chapter 8 includes a double-height space at the front of the building. This simple task involves editing the Floor on Level 2 and adding a guardrail.

1. From the *Chapter07\Complete* folder, open *07 Commercial-Complete.rvt* [*07 Commercial Metric-Complete.rvt*] file.

   - Save a copy of the file to the *Appendix A* folder.
   - Edit Floor element on Level 2 to create double-height space.
   - Add guardrail.

Use (see Figure A.9 as a guide to create the Floor sketch.

COMMERCIAL PROJECT—CHAPTER 9

The commercial project requires a lobby plan, restroom layouts, and some tenant build-out on the third floor. The exercises that follow will progress through these tasks.

1. From the *Chapter08\Complete* folder, open *08 Commercial-Complete.rvt* [*08 Commercial Metric-Complete.rvt*] file.

   - Save a copy of the file to the *Appendix A* folder.

Level 1 Plan

1. Start in the *Level 1* plan view.

The lobby will follow the shape of the floor slab already in place on Level 2. The easiest way to do this is to add Level 2 as an underlay to Level 1.

- Edit the View Properties of Level 1 and add Level 2 as an underlay.

- Draw Walls using the pick option with an offset of approximately **1'-6 [450]** and click each of the gray lines from the Floor object on Level 2.
- Draw the additional Walls, Doors, and Openings indicated (see Figure A.9).

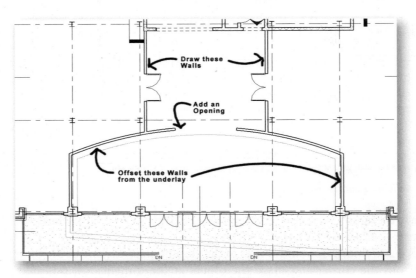

FIGURE A.9 *Add Walls and Doors to create a lobby on Level 1*

Toilet Rooms

1. Add Walls and fixtures to the toilet rooms in the core area.

- Select all the components added, Group them, and then copy and paste aligned to the other floors (see Figure A.10).

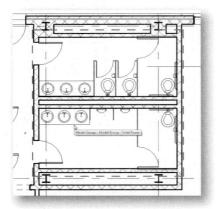

FIGURE A.10 *Create a toilet room layout, Group it, and copy it to the other floors*

Level 3 Plan

1. Open the *Level 3* plan view.
   - Using the **Interior - 4 7/8" Partition (1-hr) [Interior - 123mm Partition (1-hr)]** Wall type, add the Walls for the interior build-out as shown in the figure.
   - Add Doors. Assume **6" [150]** offset from corners or centered in the room as appropriate. Load any Door types not currently resident in the project (see Figure A.11).

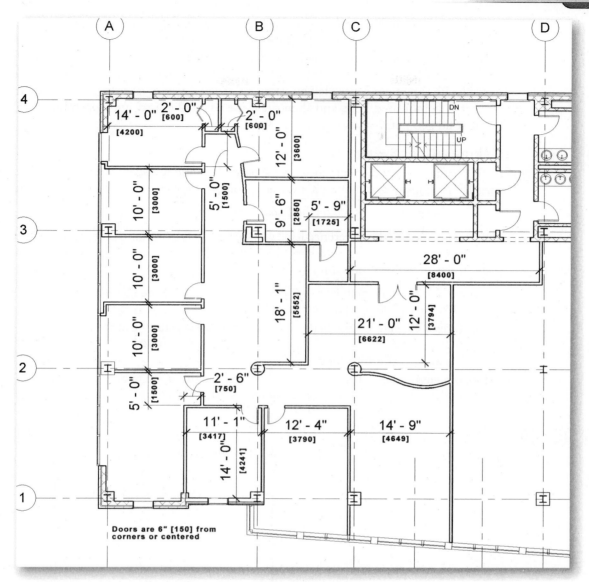

FIGURE A.11 *Add Walls and Doors for interior build-out on the third floor*

Dimensions shown in the figure are to the face of the Wall. Wall thickness dimensions are not shown to maintain clarity.

NOTE

CHAPTER 10—BUILD A FAMILY WITH ANGULAR PARAMETERS

To add an angular parameter to a Family, you need to use a Reference Line instead of a Reference Plane. Even though Reference Planes appear on screen to have a start and end point, they are really infinite. This makes it difficult to use them to control rotation in a Family. A Reference Line, on the other hand, has a finite length and both a start and end point. You can assign angle parameters to them and use them to control the rotation of elements within a Family such as the swing of a Door.

In addition to a start and end point, a Reference Line has four integral work planes associated with it. In order to use a Reference Line to control a rotation parameter effectively, you need to establish two things: first, you must constrain the end of the Reference Line that you wish to be the point of rotation. If you do not, the rotation parameter will behave unpredictably. Second, you must set one of the work planes of the Reference Line current and then add your solid geometry to this plane. To add symbolic lines that are controlled by a Reference Line, you must first make the solid geometry as indicated here, and then draw the Symbolic lines and constrain them to the solid geometry.

Making a Door Swing

In this tutorial, we will create an angular parameter and associate it to a Reference Line for the purposes of defining a variable door swing parameter in a Door Family.

1. Create a new Family based on the *Door.rft* template.

The trick to making the Reference Line work properly is being sure that you apply the constraints and parameters to it very carefully. In the Family Editor, it is sometimes easier to do this if you hide some of the geometry.

1. Select the Wall, and on the View Control Bar, choose the **Hide Element** item from the Temporary Hide/Isolate control.

   • Draw a 45° Reference Line (Home tab, Datum panel) starting at the intersection of two of the Ref Planes where you want the hinge point to be.
   Make it 3'-0" [900] long.

   • Dimension the end points of the Reference Line (from end to end).
   Use the TAB key to select the endpoint at each end (see Figure A.12).

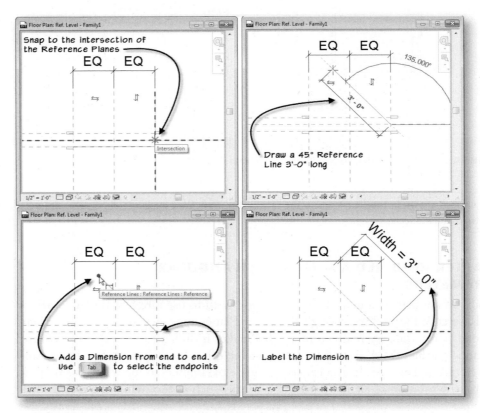

FIGURE A.12 *Add a Reference Line and label it with a parameter*

- Select the dimension and label it with the Width parameter.

2. Use the Align tool and the TAB key technique to align and lock the end of the Reference Line to each of the Ref Planes (see Figure A.13).

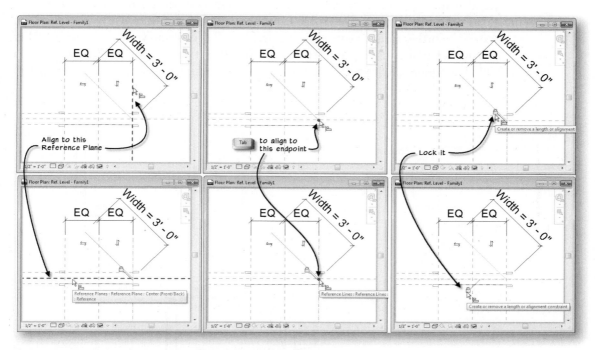

FIGURE A.13 *Align Reference Lines to lock them to the Ref Planes*

This step is important. By locking the end point (hinge point) of the Reference Line in both the horizontal and vertical directions, we ensure that the hinge point of the Door will move with the Door as expected. Having applied the Width parameter above further ensures that the Reference Line will flex as expected with the rest of the Door.

1. Flex the model. Modify the Width and then set it back to 3'-0" [900] before closing the Family Types dialog.

ADD A SWING PARAMETER

1. On the Annotate tab, on the Dimension panel click the Angular button.
   - Click the horizontal Ref Plane first and then click the Reference Line. Place the Dimension.

2. Select the angular Dimension and then from the Label option, choose **‹Add parameter›**.
   - Name the parameter **Swing**, group it under **Graphics**, make it an **Instance** parameter, and then click OK.

3. Flex the Model. Try different Widths and different Swing angles.
   When satisfied that everything works properly, reset the Width to 3'-0" [900] and the Swing to 45°.

It is simple to draw door geometry and constrain it to the Reference Lines. The Reference Line has four integral Work Planes. There is one horizontal, one vertical, and one at each end point. You simply click the Work Plane tool, choose the "Pick a Plane" option, and then select the plane upon which you wish to draw. It is recommended that you leave the Reference Line oriented at 45° for this. Cut a section at 45° as well, parallel to the Reference Line. Then open this view to work. If you work in one of the orthographic views, you can accidentally constrain your geometry to other nearby Ref Planes and geometry making it difficult to flex the model later. If you build your door panel extrusion on the 45°, you can easily avoid this (see Figure A.14).

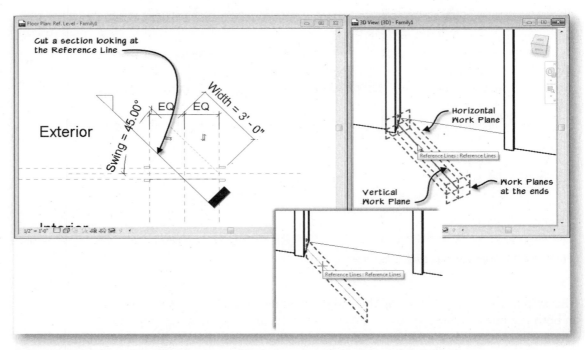

FIGURE A.14 *Reference Lines have four integral Work Planes. Set the vertical one as the Work Plane*

Create the section, open it, and then set the vertical work plane of the Reference Line current. Draw a solid extrusion on this plane. Snap it to the ends of the Reference Line using the Align tool with the TAB key to get the endpoints. Use the Thickness parameter for the height of the extrusion. To create a 2D plan version, draw Symbolic Lines using the pick lines option and constrain them to the edges of the solid extrusion.

An example of a door panel using the techniques in this tutorial is provided in the *Appendix A* folder. The Family is named: *Door Panel w Swing.rfa*. This file was created using the *Generic Model.rft* template. The Family contains only the Reference Line, Reference Planes, and the door panel geometry. You can insert the entire Family as a nested Family in any other Door Family. Open the file to explore the techniques covered here and feel free to incorporate it into your own Families.

Chapter 13—Edit Wall Type Parameters and Add Wall Tags

In Chapter 13, we modified the heights of some of the Walls in the commercial project third floor plan. We made some go all the way to the deck above, some stopped at the ceiling creating an underpinned condition and we discussed a third possible condition where the Walls could be adjusted to an unconstrained height to make them stop a few inches above the ceiling plane. Each of these conditions would typically be assigned a different Wall Type designation in a construction document set. To do this in Revit, we need to simply duplicate the existing Wall Type for each condition, edit the Type Mark and then add Wall Tags.

1. From the *Chapter13\Complete* folder, open *13 Commercial-Complete.rvt* [*13 Commercial Metric-Complete.rvt*] file.

   - Save a copy of the file to the *Appendix A* folder.
   - Select each of the underpinned Walls (refer to Chapter 13), and on the Properties palette click the Edit Type button.
   - Duplicate the Type and give it a unique name.
   - Input a value for the Type Mark such as: A1.
   - Select one of the other Walls in the suite. In section, detach its top from the Floor above. With it still selected, duplicate its Type and change the Type Mark to A2.
   - Edit the Type Properties of a different Wall (one that still goes to deck). Edit the Type Mark to A3.
   - Using Tag by Category, add Wall Tags to each Wall using a 3/8" [9] leader (see Figure A.15).

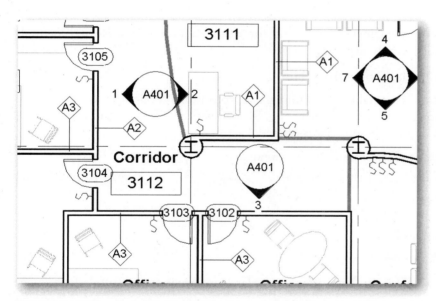

FIGURE A.15 *After duplicating Wall Types and inputting Type Marks, the Wall Tags wil report the Type designations*

Online Resources

In this appendix are listed several online Web sites and other resources that you can visit for information on Revit and related topics. All URLs listed are also provided with the dataset files as Internet shortcuts that you can simply double-click. Look in the *Appendix B* folder.

WEB SITES RELATED TO THE CONTENT OF THIS BOOK

http://www.paulaubin.com

The Web site and blog of the author. Includes information on ordering this book and Aubin's other books like *The Aubin Academy Master Series: Revit MEP*. Check there for ordering information and addenda.

If you are interested in video training for Revit Architecture, a Revit Essentials course is available at lynda.com.

http://www.lynda.com/trial/paubin

Paul F. Aubin is available for training and consulting services at your office or virtually through the use GoToMeeting. Please visit the Web site and use the contact form to inquire about services offered.

http://www.autodeskpress.delmar.cengage.com

Web site for Autodesk Press. Visit for information on other CAD titles, online resources, student software, and more.

AUTODESK SITES

http://www.autodesk.com

Autodesk main Web site. Visit often for the latest information on Autodesk products.

http://discussion.autodesk.com

Autodesk Discussion Groups main page: Online community of Autodesk users sharing comments, questions, and solutions about all Autodesk products.

http://au.autodesk.com/?nd=home

Autodesk University (AU) is Autodesk's annual convention for users. Held in Las Vegas at the end of November, attendance frequently tops 10,000! Classes are offered by industry experts from around the world on all Autodesk products including over 300 sessions on Revit alone. In addition to the live conference, AU Online is a thriving user community for information exchange, networking, and a resource to download course content and materials. Consider attending AU this year if you can, and if you do, drop by one of my classes and say hello!

http://seek.autodesk.com

Referenced elsewhere in this book, the Autodesk Seek portal is an online repository of content from manufacturers and users alike. Content can be downloaded in many popular formats including RVT, DWG, PDF, 3DS, and more.

USER COMMUNITY

http://forums.augi.com/forumdisplay.php?f=93

Autodesk Users Group International Revit focused forums and information.

http://www.revitcity.com

Web site devoted to all things Revit.

http://forums.cadalyst.com/index.php

Online user forum hosted by Cadalyst Magazine and moderated by Paul F. Aubin.

http://www.cadalyst.com

Main home page for Cadalyst Magazine. View magazines online or subscribe to print edition.

http://modocrmadt.blogspot.com/2005/01/bim-what-is-it-why-do-i-care-and-how.html

Excerpts from Chapter 1 come from an essay on this Blog (Web Log) maintained online by Matt Dillon. Matt also offered editorial assistance on several chapters in this edition. Matt is an expert in AutoCAD Architectural, Revit Architecture, and other BIM technologies. Matt was gracious enough to provide permission to quote his essay in this book.

http://revitoped.blogspot.com

Steve Stafford's Blog ~ Revit OpEd = OPinion EDitorial ~ My view of things Revit, both real and imagined. This is one of the most popular and frequently visited Revit Blogs. Steve's insight and knowledge are tremendous. Visit often for tips, issues, opinions, explanations, and other Revit resources.

In addition to his own content, Steve links to dozens of other blogs from his. Be sure to check them out.

http://do-u-revit.blogspot.com

Dave Baldacchino's Blog. Another excellent Revit resource. David's classes at Autodesk University on advanced formula editing in Revit Families were terrific. Visit his blog for good tips and Revit resources. Thanks again to David for providing the information on Revit Server for Chapter 15.

MISC ONLINE RESOURCES

http://www.GreenBuildingStudio.com

Green Building Studio is an online web service that provides architects and engineers using Autodesk Revit, Autodesk Architectural Desktop, or Autodesk Building Systems with early design stage whole building energy analysis and product information appropriate for their building designs.

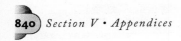

http://www.e-specs.com

Built around our e-SPECS technology, which links the project drawings to the specification documents, InterSpec has a variety of products and services to help you manage your construction specifications in the most accurate, efficient, and cost-effective manner possible.

http://www.buildingsmartalliance.org

The BuildingSmart Alliance is responsible for the US National CAD Standard and new National BIM Standard. Visit for the latest information and to purchase and download copies.

http://www.aia.org

Web site of the American Institute of Architects.

http://www.aecbytes.com

Offers analysis, research, and reviews of AEC technology.

http://buildz.blogspot.com/

Zach Kron's amazing blog on the 3D massing possibilities of Revit and Vasari. In Zach's words: "These posts are mainly about the Revit platform, with recommendations on how to enjoy it."

INDEX